Galectins in Cancer and Translational Medicine

Galectins in Cancer and Translational Medicine

Special Issue Editor

Armando Bartolazzi

MDPI • Basel • Beijing • Wuhan • Barcelona • Belgrade

MDPI

Special Issue Editor
Armando Bartolazzi
St. Andrea University Hospital
Italy

Editorial Office
MDPI
St. Alban-Anlage 66
4052 Basel, Switzerland

This is a reprint of articles from the Special Issue published online in the open access journal *International Journal of Molecular Sciences* (ISSN 1422-0067) from 2017 to 2018 (available at: https://www.mdpi.com/journal/ijms/special_issues/galectins_cancer)

For citation purposes, cite each article independently as indicated on the article page online and as indicated below:

LastName, A.A.; LastName, B.B.; LastName, C.C. Article Title. *Journal Name* **Year**, *Article Number, Page Range.*

ISBN 978-3-03897-408-6 (Pbk)
ISBN 978-3-03897-409-3 (PDF)

Contents

About the Special Issue Editor

Armando Bartolazzi, M.D., Ph.D., has worked since 1991 as a specialist in oncology, particularly in the area of diagnosis and follow-up of cancer patients (outpatient care). He has more than 30 years of experience in cancer diagnosis, mostly in the field of solid tumors. From November 2001 to June 2018, he was Professor of Pathology with a highly specialized profile in thyroid pathology at St. Andrea University Hospital, Rome, Italy and Research Associate at the Department of Oncology-Pathology, Cancer Center Karolinska, Karolinska Hospital, Stockholm, Sweden. Since June 2018, he has held the position of Undersecretary of State, Italian Ministry of Health. Prof. Bartolazzi has 30 years of experience in the production, characterization, and clinical applications of monoclonal antibodies directed to tumor-associated antigens. Some of these reagents are routinely used in several national and international institutions for immuno-diagnosis, differential diagnosis of tumors, characterization of cancers of unknown origin, and intra- and post-operative evaluation of minimal residual disease. From 1998 to 2001, he was a member of the Scientific and Technical Committee (STC) at the National Cancer Institute Regina Elena (IRCCS), Rome, Italy. Since February 2015, Prof. Bartolazzi has been listed in the TIS (Top Italian Scientists), Via-Academy, http://www.via-academy.org. He was President of the 68th session of the WHO regional Committee for Europe, Rome, Italy, 17–20 September 2018.

International Journal of
Molecular Sciences

MDPI

Editorial

Galectins in Cancer and Translational Medicine: From Bench to Bedside

Armando Bartolazzi [1,2]

[1] Pathology Research Laboratory, St. Andrea University Hospital, via di Grottarossa 1035, 00189 Rome, Italy; armando.bartolazzi@ki.se or a.bartolazzi@sanita.it; Tel.: +39-065-994-5778; Fax: +39-065-994-5331
[2] Ministry of Health—Lungotevere Ripa 1, 00168 Rome, Italy

Received: 3 September 2018; Accepted: 25 September 2018; Published: 27 September 2018

Galectins (also worded S-type lectins) are an evolutionarily conserved family of carbohydrate-binding proteins characterized by the presence of β-galactoside-binding sites, generally represented by one or two 130 amino acid carbohydrate recognition domain (CRD). No other types of folded protein domains are structurally present [1,2]. These molecules are functionally involved in several physiological processes, among which inflammation, immune-response, RNA splicing, gene transcription, apoptosis, signaling, cell migration, and differentiation [3]. Moreover, these pleiotropic molecules are also in the position to play relevant biological roles in different diseases, among which fibrosis, cancer, and heart diseases deserve a special mention [3]. Galectins can be found both intracellularly and in the extracellular milieu and are functionally active in converting glycan-related information into cell biological programs. Many biological activities of galectins, in fact, are mediated by carbohydrate-dependent interactions with glycoconjugates, mostly occurring extracellularly. Galectins-mediated extracellular functions require the CRD binding to saccharides associated with cell surface glycoproteins, for example, on T-lymphocytes, stromal cells, and endothelial cells (i.e., TCR, integrins, CD44, CD36, CD13, ganglioside GM1) and/or to extracellular glycoproteins (i.e., laminin, fibronectin, vitronectin, elastin, tenascin, lymphokines). Considering the fact that different cell types (both normal and malignant cell types) generally express a specific "galectin signature", it appears that the overall function of different galectins can vary considerably according to the biological context. Specific galectins in fact can modulate the T-cell immune-response, and galectin-1 has been reported to induce T-cell apoptosis via interaction with TCR, whereas galectin-3 suppresses apoptosis and increases T-cell proliferations. The interaction of galectins with specific extracellular matrix components (glycoproteins) may affect, in turn, cell adhesion, migration, and homing. This is particularly important for tumor cells, which use specific galectin-ligand interactions for promoting tumor growth, progression, and immune-escape [2–4].

However, several intracellular functions mediated by galectins seem to be carbohydrate-independent (i.e., interactions with cytosolic or nuclear targets). This has been demonstrated in particular for galectin-3, which was found in the nucleus and cytoplasm of different cell-types. Galectin-3 interacts with Ras and Bcl-2 in the cytosol and these molecular interactions play a key role in regulating cell growth and apoptosis. Interestingly, such galectin-3-mediated functions can be abrogated by carbohydrate interference. This finding opens an interesting scenario on the possibility to target specific galectins for therapeutic purposes [4,5].

In cancer, altered glycosylation is a common finding, both on the tumor cell surface and in the extracellular matrix, and the discovery of "tumor specific galectin signatures" opens a fashionable scenario. Many molecular interactions mediated by galectins, in fact, could be functionally relevant for tumor growth and progression. It is not surprising indeed that scientists are going to look "in the sugar box" in order to identify potential diagnostic and predictive tumor markers. At the same time, researchers are dissecting, at the structural and functional levels, those functionally relevant galectin-mediated molecular interactions that are critical for tumors and can be potentially targetable with tailored neo-glycoproteins, glycomimetics, or other biological tools. It is likely that new avenues for cancer therapy will be explored in the near future [3–7].

As aforementioned, cancer does not represent the only playground in which galectins works. Several physiological processes seem to be regulated by galectins. Regulation of immune response, bone cell differentiation, and macrophages activation just represent some of the examples. A large number of papers recently published in the literature also support the notion that the biological relevance of galectins is not restricted to cancer. Galectins have been discovered to be functionally active in many different diseases, some of which deserve special attention for the possibility to target these molecules for a therapeutic intervention [8–10].

In this special issue, focused on galectins in translational medicine, a total of 20 interesting papers, 15 of which are represented by comprehensive reviews and 5 original research studies, have been finally considered. These contributions are clustered into 4 groups: (1) Structural and functional studies on galectins and ligands; (2) Galectins and cancer; (3) Galectins in non- neoplastic diseases; and (4) Miscellanea, as detailed in Table 1.

Table 1. Contributions to the special issue "Galectins in Cancer and Translational Medicine: From Bench to Bedside".

Cluster	Paper	Reference
(1) Structural and Functional Studies on Galectins and Ligands	Dissecting the structure-activity relationship of galectin-ligand interactions	Chan Y.-C. et al. [11]
	Poly-N-acetyllactosamine neo-glycoproteins as nanomolar ligands of human galectin-3: binding kinetics and modeling	Bumba, L. et al. [12]
(2) Galectins and Cancer	Proteomic identification of the galectin-1-involved molecular pathways in urinary bladder urothelial carcinoma	Li C.F. et al. [13]
	Overall survival of ovarian cancer patients is determined by expression of galectins-8 and -9	Schulz H. et al. [14]
	Galectin-1, -3, and -7 are prognostic markers for survival of ovarian cancer patients	Schulz H. et al. [15]
	Galectin-3 performance in histologic and cytologic assessment of thyroid nodules: a systematic review and meta-analysis	Trimboli P. et al. [16]
	Galectins and carcinogenesis: their role in head and neck carcinoma and thyroid carcinomas	Kindt N. et al. [17]
	Galectin-3: the impact on the clinical management of patients with thyroid nodules and future perspectives	Bartolazzi A. et al. [18]
	Galectin-7 in epithelial homeostasis and carcinoma	Advedissian T. et al. [19]
	TrkB-target galectin-1 impairs immune activation and radiation responses in neuroblastoma: implication for tumor therapy	Batzke K. et al. [7]
	Role of galectins in multiple myeloma	Storti P. et al. [20]
	Galectin targeted therapy in Oncology: Current knowledge and perspectives	Wdowiak K. et al. [5]
	Role of galectins in tumors and in clinical immunotherapy	Chou F.-C. et al. [6]
	Galectins as molecular targets for therapeutic intervention	Dings R.P.M. et al. [4]
	Galectin-1 inhibitor OTX008 induces tumor vessel normalization and tumor growth inhibition in human head and neck squamous cell carcinoma models	Koonce N.A. et al. [21]

Table 1. *Cont.*

Cluster	Paper	Reference
(3) Galectins in Non-Neoplastic Diseases	Galectin-3: one molecule for an alphabet of diseases, from A to Z	Sciacchitano S. et al. [22]
	Galectin-3 in atrial fibrillation: mechanisms and therapeutic implications	Clementy N. et al. [8]
	Translational implication of galectin-9 in the pathogenesis and treatment of viral infection	Lai J.-H. et al. [9]
(4) Miscellanea	Role of galectin-3 in bone cell differentiation, bone pathophysiology and vascular osteogenesis	Iacobini C. et al. [10]
	Galectin-12 in cellular differentiation, apoptosis and polarization	Wan L. et al. [23]

The first two papers [11,12] belong to the structural cluster. Research on galectin glycobiology has drawn much attention for the multifunctional features of these molecules, which can be considered potential targets for selective therapeutic interventions in many pathological contexts. The elegant report by Chan at al. [11] provides an in-depth review on galectin inhibitors, focusing on the structure-activity relationship with the attempt to clarify how these potential therapeutic tools interact with galectins. Structural and functional studies on the galectin-ligand interaction represent an important and necessary step for developing new tools tailored to bind galectins and to eventually block galectin-mediate functions. Bumba et al. [12] report a molecular model focused on the galectin-3 carbohydrate recognition domain. This specific study analyzes the mode and kinetics of galectin-3 bindings to a panel of multivalent neo-glycoproteins, which present complex poly-LacNAc-based oligosaccharide ligands on albumin scaffold. Significant differences in the binding kinetics are observed among the different glycoproteins. A tetrasaccharide capped with N,N'-diacetillactosamine (LacdiNAc) showed the strongest ligand ability to galectin-3. This information may drive the development of neo-glycoproteins ingegnerized ad hoc to specifically interfere with galectin-3 mediated functions.

The cluster worded "Galectins and Cancer", as expected, is the larger one. The cluster includes several papers focused on cancer diagnosis, prognosis, and galectin-based therapeutic approaches [4–7,13–21]. The implication of galectins in tumor growth and progression, and more in general the impact that these molecules may have in oncology, is very well-known and has been extensively reviewed [3]. More specifically, for this paper collection, Li et al. [13] demonstrate the possibility to identify deregulated proteins, functionally related to galectin-1, by using an in vitro model of urinary bladder urothelial carcinoma and a proteomic approach. Modulation of these galectin-1 related molecules may have prognostic value for this tumor type. Similarly, a specific galectin signature, which can be used as prognostic marker, has been also identified in ovarian carcinoma [14,15].

Thyroid cancer represents the paradigmatic tumor model, in which experimental studies on galectins glycobiology, in particular on galectin-3, contributed greatly to improving cancer diagnosis. A validated galectin-3 test-method for the preoperative detection of thyroid cancer has been already translated in the clinical setting and is changing the clinical management of patients with thyroid nodules, reducing consistently unnecessary surgical procedures [24–29]. Three contributions in this specific field, showing different clinical experiences, are presented here [16–18]. A novel galectin-3-based immune-positron emission tomography (immune-PET) for imaging thyroid cacer in vivo, and its biological rationale is also presented and discussed [18,30,31]. One of the main functions of galectins in cancer is the regulation of tumor cell adhesion and migration. Modulation of these functions is likely relevant for tumor growth and progression. Galectin-7 is abnormally expressed in epithelial tumors and is involved in tumor progression and metastasis. Advedissian et al. [19] investigated the physiological function of this lectin in epithelial cells and demonstrated its aberrant expression in malignant epithelial counterparts. Studies on the functional role of galectins in neuroblastoma [7] and multiple myeloma [20], two relatively rare tumor conditions, are also presented in this special issue. Up-regulation of galectin-1 has been previously reported in aggressive neuroblastoma, the most common extracranial tumor of childhood. Galectin-1 is necessary for balancing immune response and angiogenic processes in physiological conditions, but cancer exploits these functions to escape from attacks of the host immune system and to better survive in hypoxic conditions. A strategic approach that combines radiotherapy and galectin-1 blockade is proposed and may be useful for reactivating an efficient tumor immune response [7]. Multiple myeloma are characterized by the tight adhesion between neoplastic plasma cells and a bone marrow microenvironment that allows for tumor cell survival and drug resistance. In this niche, neo-angiogenesis is induced as well as immunosuppression and osteolytic metastasis. In the review by Storti et al. [20], the authors analyze the expression profile of different galectin molecules and their functional activities in these specific tumors, with special focus on tumor progression, angiogenesis, and osteoclastogenesis. Several of the identified galectins may be attractive targets for multiple myeloma therapy. The expression of galectin-1 and -8 on tumor cells in fact, correlate with patients' survival.

Finally, three very interesting and comprehensive reviews exploring new potential therapeutic strategies for cancer treatment, based on galectin targeting, are considered in this book [4–6]. This paper represent a hot topic in oncology. Structural and functional studies on galectin-ligand interactions performed at molecular level, in fact, may produce important information for creating therapeutic interventions based on galectin-targeting. The proposed therapeutic approaches are supported by a strong biological rationale. New therapeutic strategies for cancer may involve the use of specific galectin inhibitors, such as competitive carbohydrates, small glycomimetic molecules, and monoclonal antibodies to specific galectins, that can be used alone or in combination with other therapeutic options.

An interesting issue is also represented by the fact that tumor-derived galectins may have bifunctional effects on tumor cells and immune cells. This has been extensively studied for galectin-1, galectin-3, and galectin-9 in different cancer types. In the review by Chou et al. [6], this issue is specifically discussed.

A short communication by Koonce et al. [21] reports the biological effects of a non-peptidic galectin-1 allosteric inhibitor, worded OTX008 and able to induce tumor vessel normalization and tumor growth inhibition in squamous cell carcinoma models. These effects are related to the improved tumor oxygenation registered after treatment with such a galectin-1 inhibitor.

This condition allows a more permissive tumor microenvironment status for improved radiation or chemotherapeutic treatment.

The cluster entitled "Galectins in Non-Neoplastic Diseases" includes three interesting reviews. The first by Sciacchitano and co-authors [22] is focused on galectin-3. The authors report in an elegant and original way a large number of studies published so far on the role of galectin-3 in different diseases, listed in alphabetical order. This original contribution provides the base for further investigations in each specific medical field.

It has been previously reported that galectin-3 is required for TGF-β mediated myofibroblast activation and matrix production. Increased galectin-3 expression seems to be a common feature in tissue fibrosis. Disruption of the galectin-3 gene blocks myofibroblast activation and procollagen I expression, in an experimental model of liver fibrosis in vitro and in vivo [23].

Clementy et al. described the role of galectin-3 in the pathogenesis of atrial fibrillation [8]. Considering the fact that galectin-3 is a well-recognized biomarker of fibrosis and tissue remodeling, the authors discuss the mechanisms of such an electrical alteration and its therapeutic implications.

A very interesting review published by Lai and co-authors [9] is focused on the translational implication of galectin-9 in the pathogenesis and treatment of viral infection. The interaction between galectin-9 and the ligand Tim-3 triggers signaling events that regulate immune response. Galectin-9 has been reported to be over-expressed in a variety of target cells infected by different viruses, among which include HCV, HBV, HSV, influenza virus, dengue virus, and HIV. The biological significance of galectin-9 expression in virus infected cells and its functional implications in regulating virus specific immune-response also represent important fields of research to be pursued. Understanding the biological role of galectin-9 in this specific context may help to identify better methods for monitoring and treating viral infections. The last cluster of contributions includes the work by Iacobini et al. [10], where they reported original data on galectin-3's biological role in all stages of bone biology, from development to remodeling. The regulatory role of galectin-3 in inflammatory bone and joint disorders is also discussed, together with the possibility to target galectin-3 for therapeutic purposes.

The special review issue on the recently discovered galectin-12 and its functional role in regulating important physiological mechanisms, like cellular differentiation, apoptosis, and macrophage polarization to M2 subtype [32], closes on galectins and translational medicine. Although galectin biology is attracting the interest of many scientists operating in different experimental contexts and clinical fields, from developmental biology to clinical oncology, many aspects concerning galectins are still incompletely understood and need further investigation. Among these, some deserve mention: elucidation of galectin-mediated intracellular mechanisms; galectin-dependent signaling; structural aspects of galectin functions (i.e., oligomerization, galectin-derived extracellular lattice); and the biological and functional significance of specific "galectin signatures" in benign and malignant tissue as well. Addressing these issues will open the way to new therapeutic strategies for different human diseases, stimulating research for developing glycomimetics, neo-glycoproteins, and/or monoclonal antibodies directed to specifically modulate or block the functions of these fascinating molecules.

Overall 20 important contributions published in this special issue illustrate recent advances in the field of Galectins, which may have potential clinical implications in cancer but also in other clinical contexts. It seems to me that an interesting chapter of translational glycobiology is on the horizon.

I would like to thank all the authors who contributed to this special issue and I remain hopeful that the increasing knowledge in the field will require, in the very near future, a new and more extensive effort for preparing the second volume entitled "Galectin in cancer and translational medicine II".

Acknowledgments: I would like to thank my research group including senior scientists and students for their support and suggestions, in particular: Salvatore Sciacchitano, Carlo Bellotti, Gian Paolo Di Francesco, Calogero D'Alessandria, Giorgia Scafetta, Claudia Cippitelli, Niccolò Noccioli.

Conflicts of Interest: The author has an ownership of a patent related to the use of radiolabeled mAbs to galectin-3 for tumor imaging in vivo (patent n. 1388763, registered on 20 February 2008, Rome, Italy).

Abbreviation

CRD	Carbohydrate recognition domain
LacdiNAc	N,N'-diacetillactosamine
HCV	Hepatitis C virus
HBV	Hepatitis B virus
HSV	Herpes simplex virus
HIV	Human immunodeficiency virus
TGF-β	Transforming growth factor-β
Tim-3	T-cell immunoglobulin domain, mucin domain-3

References

1. Hirabayashi, J.; Kasai, K.-I. The family of metazoal metal-independent β-galactoside-binding lectins: Structure, function and molecular evolution. *Glycobiology* **1993**, *3*, 297–304. [CrossRef] [PubMed]
2. Johannes, L.; Jacob, R.; Leffler, H. Galectins at a glance. *J. Cell Sci.* **2018**, *131*, jcs208884. [CrossRef] [PubMed]
3. Liu, F.; Rabinovich, G.A. Galectins as modulators of tumor progression. *Nat. Rev. Cancer* **2005**, *5*, 29–41. [CrossRef] [PubMed]
4. Dings, R.P.M.; Miller, M.C.; Griffin, R.J.; Mayo, K.H. Galectins as molecular targets for therapeutic intervention. *Int. J. Mol. Sci.* **2018**, *19*, 905. [CrossRef] [PubMed]

5. Wdoviak, K.; Frankuz, T.; Gallego-Colon, E.; Ruiz-Agamez, N.; Kubeczko, M.; Grochola, I.; Wojnar, J. Galectin targeted therapy in oncology: Current knowledge and perspectives. *Int. J. Mol. Sci.* **2018**, *19*, 210. [CrossRef] [PubMed]

6. Chou, F.-C.; Chen, H.-Y.; Kuo, C.-C.; Sytwu, H.-K. Role of galectins in tumors and in clinical immunotherapy. *Int. J. Mol. Sci.* **2018**, *19*, 430. [CrossRef] [PubMed]

7. Batzke, K.; Buchel, G.; Hansen, W.; Schramm, A. TrkB-target galectin-1 impairs immune activation and radiation responses in neuroblastoma: Implications for tumor therapy. *Int. J. Mol. Sci.* **2018**, *19*, 718. [CrossRef] [PubMed]

8. Clementy, N.; Piver, E.; Bisson, A.; Andre, C.; Bernard, A.; Pierre, B.; Fauchier, L.; Babuty, D. Galectin-3 in atrial fibrillation: Mechanisms and therapeutic implications. *Int. J. Mol. Sci.* **2018**, *19*, 976. [CrossRef] [PubMed]

9. Lai, J.-H.; Luo, S.-F.; Wang, M.-J.; Ho, L.-J. Translational implications of galectin-9 in the pathogenesis and treatment of viral infection. *Int. J. Mol. Sci.* **2017**, *18*, 2108. [CrossRef] [PubMed]

10. Iacobini, C.; Blasetti Fantauzzi, C.; Pugliese, G.; Menini, S. Role of galectin-3 in bone cell differentiation, bone pathophysiology and vascular osteogenesis. *Int. J. Mol. Sci.* **2017**, *18*, 2481. [CrossRef] [PubMed]

11. Chan, Y.-C.; Lin, H.-Y.; Tu, Z.; Kuo, Y.-H.; Hsu, S.-T.D.; Lin, C.-H. Dissecting the structure-activity relationship of galectin-ligand interactions. *Int. J. Mol. Sci.* **2018**, *19*, 392. [CrossRef] [PubMed]

12. Bumba, L.; Laaf, D.; Spiwok, V.; Elling, L.; Kren, V.; Bojarova, P. Poly-*N*-acetyllactosamine neo-glycoproteins as nanomolar ligands of human galectin-3: Binding kinetics and modeling. *Int. J. Mol. Sci.* **2018**, *19*, 372. [CrossRef] [PubMed]

13. Li, C.-F.; Shen, K.-H.; Chien, L.-H.; Huang, C.-H.; Wu, T.-F.; He, H.-L. Proteomic identification of the galectin-1-involved molecular pathways in urinary bladder urothelial carcinoma. *Int. J. Mol. Sci.* **2018**, *19*, 1242. [CrossRef] [PubMed]

14. Schulz, H.; Kuhn, C.; Hofmann, S.; Mayr, D.; Mahner, S.; Jeschke, U.; Schmoeckel, E. Overall survival of ovarian cancer patients is determined by expression of galectins-8 and -9. *Int. J. Mol. Sci.* **2018**, *19*, 323. [CrossRef] [PubMed]

15. Schulz, H.; Schmoeckel, E.; Kuhn, C.; Hofmann, S.; Mayr, D.; Mahner, S.; Jeschke, U. Galectin-1, -3, and-7 are prognostic markers for survival of ovarian cancer patients. *Int. J. Mol. Sci.* **2017**, *18*, 1230. [CrossRef] [PubMed]

16. Trimboli, P.; Virili, C.; Romanelli, F.; Crescenzi, A.; Giovanella, L. Galectin-3 performance in histologic and cytologic assessment of thyroid nodules: A systematic review and meta-analysis. *Int. J. Mol. Sci.* **2017**, *18*, 1756. [CrossRef] [PubMed]

17. Kindt, N.; Journe, F.; Ghanem, G.E.; Saussez, S. Galectins and carcinogenesis: Their role in head and neck carcinoma and thyroid carcinomas. *Int. J. Mol. Sci.* **2017**, *18*, 2745. [CrossRef] [PubMed]

18. Bartolazzi, A.; Sciacchitano, S.; D'Alessandria, C. Galectin-3: The impact on the clinical management of patients with thyroid nodules and future perspectives. *Int. J. Mol. Sci.* **2018**, *19*, 445. [CrossRef] [PubMed]

19. Advedissian, T.; Deshayes, F.; Viguier, M. Galectin-7 in epithelial homeostasis and carcinoma. *Int. J. Mol. Sci.* **2017**, *18*, 2760. [CrossRef] [PubMed]

20. Storti, P.; Marchica, V.; Giuliani, N. Role of galectins in multiple myeloma. *Int. J. Mol. Sci.* **2017**, *18*, 2740. [CrossRef] [PubMed]

21. Koonce, N.; Griffin, R.; Dings, R.P.M. Galectin-1 inhibitor OTX008 induces tumor vessel normalization and tumor growth inhibition in human head and neck squamous cell carcinoma models. *Int. J. Mol. Sci.* **2017**, *18*, 2671. [CrossRef] [PubMed]

22. Sciacchitano, S.; Lavra, L.; Morgante, A.; Ulivieri, A.; Magi, F.; De Francesco, G.P.; Bellotti, C.; Salehi, L.B.; Ricci, A. Galectin-3: One molecule for an alphabet of diseases, from A to Z. *Int. J. Mol. Sci.* **2018**, *19*, 379. [CrossRef] [PubMed]

23. Henderson, N.C.; Mackinnon, A.C.; Farnworth, S.L.; Poirier, F.; Russo, F.P.; Iredale, J.P.; Haslett, C.; Simpson, K.J.; Sethi, T. Galectin-3 regulates myofibroblast activation and hepatic fibrosis. *Proc. Natl. Acad. Sci. USA* **2006**, *103*, 5061–5065. [CrossRef] [PubMed]

24. Bartolazzi, A. Improving accuracy of cytology for nodular thyroid lesions. *Lancet* **2000**, *355*, 1661–1662. [CrossRef]

25. Bartolazzi, A.; Gasbarri, A.; Papotti, M.; Bussolati, G.; Lucante, T.; Khan, A.; Inohara, H.; Marandino, F.; Orlandi, F.; Nardi, F.; et al. Application of immunodiagnostic method for improving preoperative diagnosis of nodular thyroid lesions. *Lancet* **2001**, *357*, 1644–1650. [CrossRef]

26. Bartolazzi, A.; Orlandi, F.; Saggiorato, E.; Volante, M.; Arecco, F.; Rossetto, R.; Palestini, N.; Ghigo, E.; Papotti, M.; Bussolati, G.; et al. Galectin-3-expression analysis in the surgical selection of follicular thyroid nodules with indeterminate fine-needle aspiration cytology: A prospective multicenter study. *Lancet Oncol.* **2008**, *9*, 543–549. [CrossRef]

27. Carpi, A.; Naccarato, A.G.; Iervasi, G.; Nicolini, A.; Bevilacqua, G.; Viacava, P.; Collecchi, P.; Lavra, L.; Marchetti, C.; Sciacchitano, S.; et al. Large needle aspiration biopsy and galectin-3 determination in selected thyroid nodules with indeterminate FNA-cytology. *Br. J. Cancer* **2006**, *95*, 204–209. [CrossRef] [PubMed]

28. Carpi, A.; Rossi, G.; Di Coscio, G.D.; Iervasi, G.; Nicolini, A.; Carpi, F.; Mechanick, J.I.; Bartolazzi, A. Galectin-3 detection on large-needle aspiration biopsy improves preoperative selection of thyroid nodules: A prospective cohort study. *Ann. Med.* **2010**, *42*, 70–78. [CrossRef] [PubMed]

29. Saggiorato, E.; De Pompa, R.; Volante, M.; Cappia, S.; Arecco, F.; Dei Tos, A.P.; Orlandi, F.; Papotti, M. Characterization of thyroid 'follicular neoplasms' in fine-needle aspiration cytological specimens using a panel of immunohistochemical markers: A proposal for clinical application. *Endocr. Relat. Cancer* **2005**, *12*, 305–317. [CrossRef] [PubMed]

30. Bartolazzi, A.; D'Alessandria, C.; Parisella, M.G.; Signore, A.; Del Prete, F.; Lavra, L.; Braesch-Andersen, S.; Massari, R.; Trotta, C.; Soluri, A.; et al. Thyroid cancer imaging in vivo by targeting the anti-apoptotic molecule galectin-3. *PLoS ONE* **2008**, *3*, e3768. [CrossRef] [PubMed]

31. D'Alessandria, C.; Braesch-Andersen, S.; Bejo, K.; Reder, S.; Blechert, B.; Schwaiger, M.; Bartolazzi, A. Noninvasive In Vivo Imaging and Biologic Characterization of Thyroid Tumors by ImmunoPET Targeting of Galectin-3. *Cancer Res.* **2016**, *76*, 3583–3592. [CrossRef] [PubMed]

32. Wan, L.; Yang, R.-Y.; Liu, F.-T. Galectin-12 in cellular differentiation, apoptosis and polarization. *Int. J. Mol. Sci.* **2018**, *19*, 176. [CrossRef] [PubMed]

International Journal of
Molecular Sciences

MDPI

Review

Dissecting the Structure–Activity Relationship of Galectin–Ligand Interactions

Yi-Chen Chan [1,2,3,†], Hsien-Ya Lin [1,†], Zhijay Tu [1,†], Yen-Hsi Kuo [1], Shang-Te Danny Hsu [1,4] and Chun-Hung Lin [1,2,4,5,6,*]

[1] Institute of Biological Chemistry, Academia Sinica, No. 128, Academia Road Section 2, Nan-Kang, Taipei 11529, Taiwan; cyc1786@yahoo.com (Y.-C.C.); pipis_lsy@hotmail.com (H.-Y.L.); s42304@yahoo.com.tw (Z.T.); yenkuo1@gmail.com (Y.-H.K.); sthsu@gate.sinica.edu.tw (S.-T.D.H.)
[2] Taiwan International Graduate Program (TIGP), Sustainable Chemical Science and Technology (SCST), Academia Sinica, Taipei 11529, Taiwan
[3] Department of Applied Chemistry, National Chiao Tung University, Hsinchu 300, Taiwan
[4] Institute of Biochemical Sciences, National Taiwan University, Taipei 10617, Taiwan
[5] Department of Chemistry, National Taiwan University, Taipei 10617, Taiwan
[6] The Genomics Research Center, Academia Sinica, Taipei 11529, Taiwan
* Correspondence: chunhung@gate.sinica.edu.tw; Tel.: +886-227-890-110
† These authors have equal contribution to this work.

Received: 17 December 2017; Accepted: 24 January 2018; Published: 29 January 2018

Abstract: Galectins are β-galactoside-binding proteins. As carbohydrate-binding proteins, they participate in intracellular trafficking, cell adhesion, and cell–cell signaling. Accumulating evidence indicates that they play a pivotal role in numerous physiological and pathological activities, such as the regulation on cancer progression, inflammation, immune response, and bacterial and viral infections. Galectins have drawn much attention as targets for therapeutic interventions. Several molecules have been developed as galectin inhibitors. In particular, TD139, a thiodigalactoside derivative, is currently examined in clinical trials for the treatment of idiopathic pulmonary fibrosis. Herein, we provide an in-depth review on the development of galectin inhibitors, aiming at the dissection of the structure–activity relationship to demonstrate how inhibitors interact with galectin(s). We especially integrate the structural information established by X-ray crystallography with several biophysical methods to offer, not only in-depth understanding at the molecular level, but also insights to tackle the existing challenges.

Keywords: binding interaction; galectin; inhibitor; isothermal titration calorimetry; NMR (Nuclear Magnetic Resonance); thiodigalactoside; X-ray crystallography

---

1. Introduction

Lectins are carbohydrate-binding proteins that recognize a diversity of glycan structures existing in numerous glycoconjugates. As unique members of the lectin family, galectins are small, cytosolic proteins that bind β-galactosides. Albeit the lack of a signal sequence required for protein secretion, galectins usually exist both intra- and extracellularly and were thus proposed to follow a nonclassical non-Golgi/ER secretion pathway [1]. The β-galactoside-binding feature is attributed to the evolutionarily conserved carbohydrate recognition domain (CRD) where the glycan binding takes place [2,3]. So far, 15 members of the galectin family have been identified in mammalians and were named in accordance with their sequential order of discovery. These galectins (abbreviated as GALs) are categorized into three groups according to the organization of their CRD [4]. Prototype galectins (including GAL-1, -2, -5, -7, -10, -11, -13, -14, and -15) form homodimers in solution via a noncovalent interaction through their single CRD, while tandem-repeat galectins (GAL-4, -6, -8, -9, and -12) comprise two distinct CRDs at their N- and C-termini that are tethered by a linker of variable length. In a chimera structure, GAL-3 is the only member that has a CRD at the C-terminus and a short non-lectin peptide motif (Gly–Pro–Tyr-rich domain) at the N-terminus. GAL-3 can self-oligomerize into a pentamer or other forms to form an adhesive network when bridging carbohydrates on the cell surface through multivalent interactions [5,6].

The CRD consists of 130 amino acid residues that fold into two antiparallel β-sheets of six (S1–S6) and five (F1–F5) strands, jointly forming a β-sheet sandwich structure and named S-sheets and F-sheets, respectively. The S1–S6 β-strands make up a concave surface to which specific glycans (up to the length of tetrasaccharides) are bound (Figure 1). To discuss the binding interactions in detail, the binding groove is further subdivided into subsites A–E, with the principle subsite C that is highly conserved among all galectin members. Subsite C harbors β-galactose, whereas the less conserved subsite D accommodates the sugar residue (e.g., glucose in lactose) next to β-galactose. In contrast, subsites A, B and E are more variable and thus specific for individual galectins. These subsites are occupied by sugars or functional groups adjacent to β-galactose.

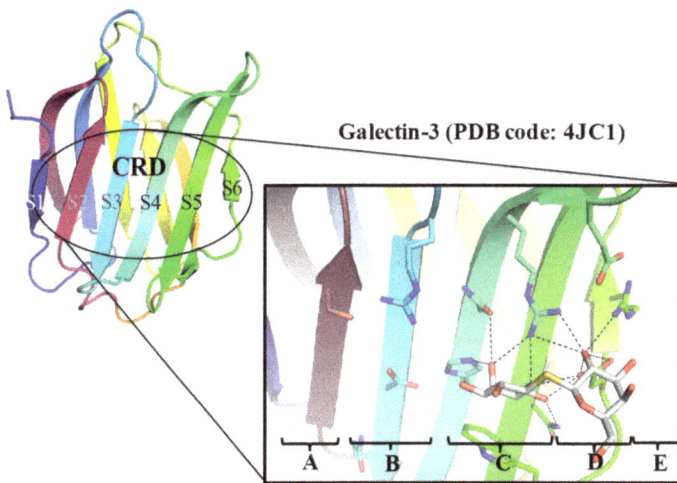

Figure 1. Galectin CRD contains five subsites (A–E; GAL-3 is shown here for instance) where thiodigalactosides bind, interacting with subsites C and D in a similar manner to lactose or *N*-acetyllactosamine.

2. Galectins in Cancer and Translational Medicine

Protein-carbohydrate interactions are essential for biomolecular communications within and between cells. Since galectins are capable of recognizing β-galactoside-containing glycoconjugates on the cell surface, in extracellular matrices, or in the lumen of intracellular vesicles, they play a pivotal role in apoptosis [7], cell proliferation [8], inflammation, immune response [9–12], and cell adhesion [13–15] and migration [8,16–18]. These activities, in turn, have significant roles in various diseases, such as cancer progression [5], HIV [19], autoimmune disorders, fibrosis, arthritis, obesity, cardiovascular diseases, allergies, and microbial infections [20].

Galectins are potential biomarkers in chronic or acute heart failure [21], as well as attractive targets for anticancer or anti-inflammatory agents. Cancer development is initiated when normal cells undergo neoplastic transformation that causes the dysregulation of cell growth or of regulatory mechanisms, resulting, for instance, in defective apoptosis and cell cycle alterations [5]. Several galectins are closely associated with these processes. Among them, GAL-1 and -3 are well studied for their roles in cancer progression. Previous studies demonstrated that GAL-3 is involved in processes linked to tumorigenesis, including the transformation to a malignant form, metastasis, and the increased invasiveness of tumor cells [22,23]. Particularly, GAL-3 is important for promoting cell–cell interactions and enabling cellular motility, which explains why cancer cells take advantage of GAL-3 to move and grow [24]. Furthermore, GAL-3 might also be secreted by tumor cells to sustain the microenvironment they favor, which subsequently helps them to escape from immune surveillance in two ways: (1) by inhibiting the afferent arm of the immune system, preventing the body from increasing the number of T cells in response to the tumor, and (2) by inhibiting the efferent arm, which is used to attack the tumors [22,25,26]. Accordingly, all this strongly suggests that blocking the function of GAL-3 is likely a feasible option to rescue the immune system's ability to attack cancer cells.

It is a challenging task to develop selective inhibitors for GAL-3, because the desirable inhibitors have to distinguish GAL-3 from the other members. Intracellular GAL-3, for instance, displays apoptosis-suppressing activity, but extracellular GAL-3 plays the opposite role and promotes apoptosis like GAL-1, -2, and -9 [5,27–29]. The high similarity in the CRD structures is another issue. The differential activities of specific galectins in normal and pathological processes [30,31] also explain the urgent need to develop potent and selective inhibitors. To date, there are three different categories of GAL-3 inhibitors to attenuate cancer progression. Two of them have been advanced to clinical trials, indicating that they could be available for therapeutic intervention in the near future. Firstly, G3-C12 is a peptide (the sequence: ANTPCGPYTHDCPVKR) discovered by phage display exhibiting an excellent inhibition (K_d = 72 nM) towards GAL-3 [32]. This peptide was shown to prevent the metastasis of breast cancer cells to the lung. However, several pieces of information still remain ambiguous, including its inhibition potency towards other galectins and its mode of action. Secondly, pectin is a complex polysaccharide rich in anhydrogalacturonic acid, galactose, and arabinose. This polysaccharide can bind to GAL-3 in a multivalent manner. GBC-590 (developed by Safescience, Inc., Boston, MA, USA.) is one of the modified citrus pectin derivatives [33,34]. It was shown to reduce colorectal carcinomas in Phase II trials [35]. Likewise, GCS-100 produced significant activity in Phase II clinical trials to treat patients suffering from relapsed chronic lymphocytic leukemia [36]. However, its binding to galectins has not been clearly demonstrated. Two additional polysaccharide-based multivalent inhibitors, GM-CT-01 (Davanat^{TM}, formerly invented by Pro-Pharmaceuticals, Inc.) and GR-MD-02 (Figure 2a,b), both developed by Galectin Therapeutics, showed moderate affinity with GAL-3 (K_d = 2.9 and 2.8 µM, respectively). GM-CT-01 is a natural galactomannan polysaccharide with an average molecular weight up to 60 kDa. Its polymannoside backbone is branched with galactose residues. GR-MD-02 is a galactoarabino-rhamnogalacturonan polysaccharide with a molecular weight of ~50 kDa. These molecules are currently examined under Phase I or Phase II clinical trials for several cancers [37–39]. Nonetheless, it was noted that both GM-CT-01 and GR-MD-02 display comparable inhibition of GAL-1 and -3 (K_d = 10 µM and 8 µM for GAL-1, respectively, determined by NMR studies) [38,40–42]. Because of their high water solubility and safe features in humans, these plant polysaccharides are good drug candidates. The use of these pectins as galectin inhibitors is so far based on studies in cell culture and animal models. It could be risky to correlate the clinical efficacy of pectins to GAL-3-mediated activities. On the other hand, there is no clear and satisfying structural explanation on how these pectins bind to galectins and how their affinities for GAL-1 and -3 are linked to their therapeutic efficacy.

Furthermore, TD139 (Figure 2c) [43], which is in clinical development by the Swedish startup Galecto Biotech [44], is a small-sized, monovalent inhibitor. Despite its low affinity for GAL-2, -4N, -4C, -7, -8N, and -9N, TD139 displays potent inhibition of GAL-1 (K_d = 10 nM, determined by fluorescence polarization (FP)) and GAL-3 (K_d = 14 nM, also by FP) [45], exhibiting a high selectivity for GAL-1 and -3. This inhibitor has completed Phase Ib/IIa clinical trials for the treatment of Idiopathic Pulmonary Fibrosis. TD139 was generated several years after the optimization of TDG-based inhibitors started in 2004, [46–51], representing the combined efforts of chemical synthesis, X-ray crystallography, and computational modeling. Since thiodigalactoside (TDG) and TD139 are symmetric saccharides, and TD139 represents a TDG derivative bearing two identical substituents (4-fluorophenyl-triazole) at the C3- and C3′-positions of TDG, we prepared TAZTDG (an asymmetrical derivative of TD139), containing one 4-fluorophenyl-triazole at C3 to understand how the inhibition potency is established by an extra binding interaction with the introduction of an additional substituent [52]. Meanwhile, in addition to the resolved X-ray crystal structures, we also relied on the use of several biophysical methods to obtain insights about the binding interactions.

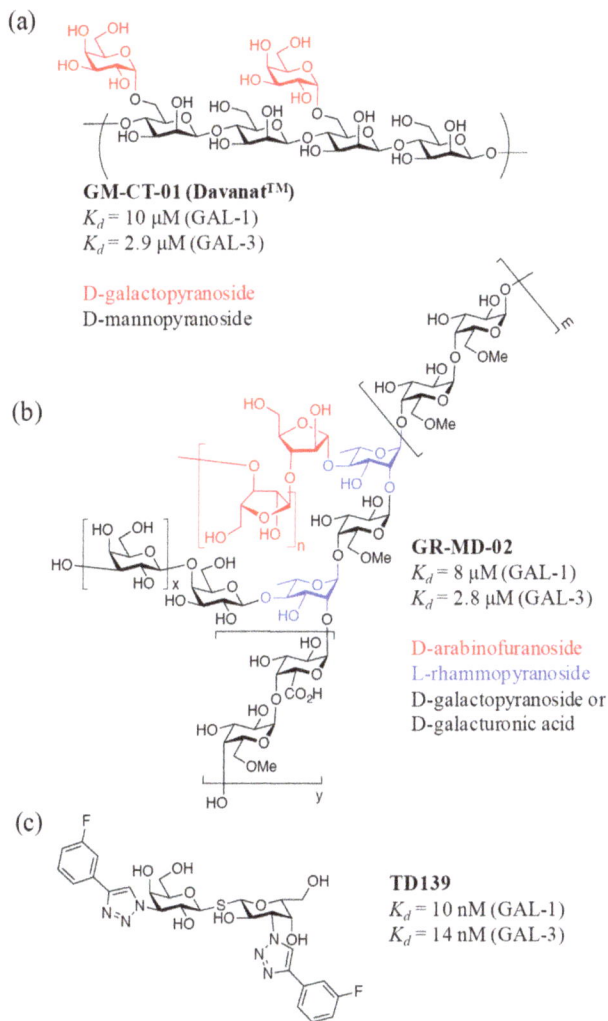

Figure 2. Structures of GM-CT-01, GR-MD-02, and TD139 that have been examined in clinical trials for GAL-3-related diseases.

3. Rationale for the Design of Anti-Galectin Agents

Since the majority of galectin activities is associated with their carbohydrate-binding features, the inhibition of the CRD by antagonists (or inhibitors) to compete with the natural ligand appears to be a feasible option, not only to disclose their exact functions, but also to develop molecules for therapeutic intervention. The glycotope interacting with galectin was outlined in 1986, and the first structural information came from the X-ray structure of galectin CRD in complex with lactose (PDB code: 1HLC) [53,54], which delineated the binding interactions at the molecular level. In accordance with the complex structure of GAL-3-Gal-β1,4-GlcNAc, several interactions were noticed between the galactose moiety, a series of conserved residues on S4–S6 β strands, and the loop connecting the S4 and S5 strands [55]. Taking GAL-3 as an example, the interactions included three hydrogen bonds (H-bonds) between the oxygen atom of galactose C4–OH and His158, Asn160, and Arg162, with the bond lengths of 2.8, 3.3, and 2.9 Å, respectively (Figure 3). Meanwhile, O5 was found to be H-bonded to Arg162 and Glu184, and C6–OH was found to interact with Asn174 and Glu184, so that seven H-bonds in total were observed with galactose. On the other hand, only three H-bonds were spotted on the GlcNAc moiety, between the C3–OH and the residues Arg162 and Glu184.

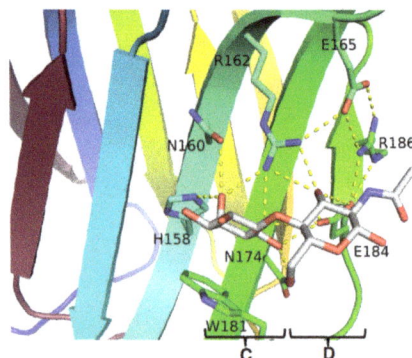

Figure 3. The complex structure of GAL-3/*N*-acetylactosamine (LacNAc) (PDB code: 3ZSJ). Residues in subsites C and D are highly conserved in galectins. The natural ligand LacNAc resides at subsites C and D forming several H-bonds that are shown in dotted yellow lines.

In addition to the H-bond network, H3, H4, and H5 of galactose were also found to interact with Trp181 via van der Waals forces. Taken together, the galactose of lactose/LacNAc is a paramount component for the recognition by galectins and is indispensable for the interactions between the CRD residues and a ligand or inhibitor.

3.1. Enhancement of the Binding Affinity

As the C4–OH, O5, and C6–OH of galactose are highly engaged in binding interactions, the remaining C2–OH and C3–OH of galactose do not participate in any interaction and can thus be available for modifications to gain extra binding [47,56]. Nilsson and coworkers developed the chemistry of thio-glycosides and generated a number of TDG derivatives [48,57], because TDG not only possesses a similar binding affinity as lactose/LacNAc, but also resists chemical or enzymatic hydrolysis. Like LacNAc or lactose, TDG situates at subsites C and D. Modifications of the LacNAc or TDG scaffold can introduce additional positive protein–ligands interactions (e.g., electrostatic interaction, van der Waals forces, and H-bond) at other subsites of the CRD, and this strategy has become a common approach to increase the binding affinity. Since different galectins have variations in their protein sequences at subsites A, B, and E, the introduced binding interactions at these sites would likely enhance the affinity and selectivity at the same time.

TD139 is an ideal example in which the introduction of 4-fluorophenyl-triazole to the C3- and C3' positions of TDG led to ~1000-fold higher affinity for GAL-3, as compared to TDG (K_d = 0.068 and 75 µM measured by isothermal titration calorimetry (ITC) for TD139 and TDG, respectively). Several interactions were identified to contribute to this dramatic enhancement. The 4-fluorophenyl-triazole substituent was stacked between Arg144 and Ala146 at subsite B of

GAL-3, forming arginine-π interactions (Figure 4a) [52]. The terminal fluorine atom of TD139 thus formed multiple, orthogonal polar interactions (namely, fluorine bondings) with the protein backbone carbonyls and backbone amide NHs, at an average distance <3.5 Å (Figure 4a), resulting in a characteristic fluorophilic microenvironment [58,59]. Meanwhile, at the other end of TD139, tandem arginine-π interactions were observed between the 4-fluorophenyl and triazole moieties and Arg186 at subsite E (Figure 5). We previously reported that the salt bridge between Asp/Glu and Arg (e.g., Arg162–Glu165–Glu184–Arg186 in GAL-3 [55,60]) is indispensable to orient the Arg residue that participates in the aforementioned arginine-π interactions.

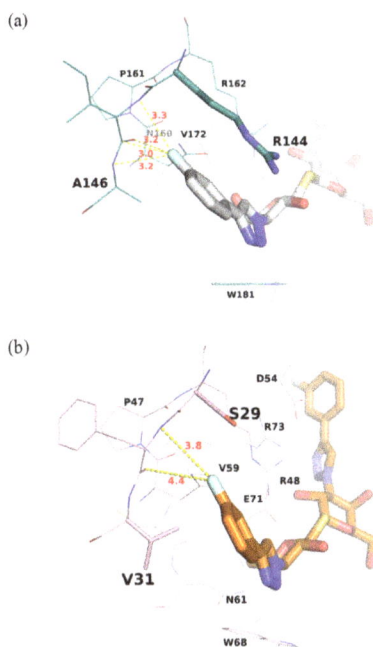

Figure 4. Close-up view of (**a**) GAL-3 and (**b**) GAL-1 subsite B in complex with TD139. The multiple fluorine bonds are shown as dotted yellow lines between the fluorine and the backbone peptide.

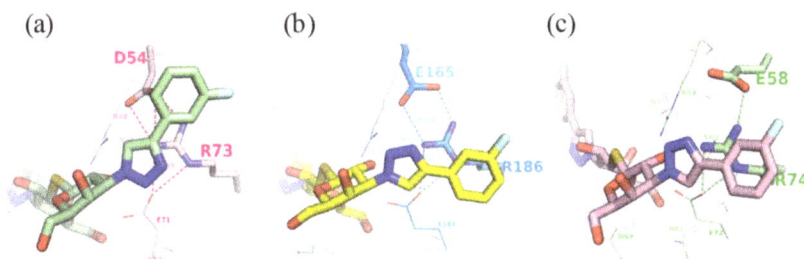

Figure 5. Structures of (**a**) GAL-1 (PDB code: 4Y24), (**b**) GAL-3 (5H9P), (**c**) GAL-7 (5H9Q) in complex with TD139 to display the ion-pair-π (**a**) or tandem arginine-π (**b,c**) interactions at subsite E. The salt bridges between Asp/Glu and Arg (in dotted lines) are essential for controlling the orientation of Asp and Arg involved in the aforementioned interactions.

Meanwhile, compound **1** is the analogue of TD139 containing two coumarylmethyl substituents at both positions C3 and C3′ of TDG (Figure 6) [49]. Its binding affinity for GAL-3 (K_d = 91 nM) is 176-fold higher than that for GAL-1 (K_d = 16 μM). It is suggested that the increment of the binding affinity is due to the two arginine-π interactions, as in TD139 where the coumarin substituents are proposed to stack on the guardinium groups of Arg144 and Arg186 in GAL-3, forming two face-to-face arene-arginine interactions. Neither of the GAL-3 mutants (R144K, R144S, R186K, or R186S) shows a prominent affinity for GAL-3. Additionally, water-mediated H-bonds were found to bridge between the coumaryl carbonyl oxygen and Lys76. Although the H-bonds contributed to the affinity, they were, however, less important than the two coumarine–arginine interactions.

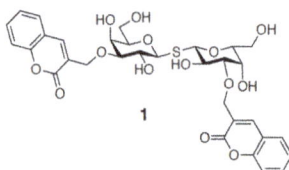

Figure 6. Molecular structure of di-substituted 3-*O*-coumarylmethyl thiodigalactoside (**1**).

3.2. Enhancement of the Binding Selectivity

To discuss the binding specificity, we will place a special emphasis on the binding interactions at subsites B and E. When comparing GAL-1 and -3, for instance, Ser29 and Val31 at subsite B of GAL-1 correspond to Arg144 and Ala146 in GAL-3, respectively. Since Val is a little bigger than Ala, the 4-fluorophenyl moiety is forced to rotate ~16° (Figure 7a) in the GAL-1 complex, leading to the formation of only two fluorine–protein interactions, in contrast to the four interactions observed in the GAL-3 complex (Figure 4a,b). Moreover, GAL-7 contains Arg31 and His33 in similar positions as those held by Arg144 and Ala146 in GAL-3, respectively, but Arg31[GAL-7] is remote from subsite B and is thus not close to the 4-fluorophenyl substituent of TD139. The 4-fluorophenyl moiety is found to turn ~50° away as compared to that in the GAL-3 complex (Figure 7b), likely because of the steric hindrance caused by the imidazole of His33[GAL-7], a much bulkier residue than its counterparts Val31[GAL-1] and Ala146[GAL-3]. Two water molecules are present in the resulting vacated space in subsite B of GAL-7 (Figure 7b). The different arrangement explains the lower affinity of TD139 for GAL-7 (K_d = 38 μM) than for GAL-3 (K_d = 0.068 μM, determined by ITC) [52].

Furthermore, Arg183[GAL-3] and Arg71[GAL-7] adopt a different rotametric position when binding with 1,2-fused (3,5-dimethoxyphenyl)-oxazoline-substituted LacNAc (compound **2**, Figure 8a). Arg183 extends outside subsite E (Figure 8b), whereas Arg71 is curved back towards the lactosamine-binding site (Figure 8c) [61]. The conformational difference causes a 60-fold higher preference of compound **2** for GAL-3 (K_d = 4.4 μM) than for GAL-7 (K_d = 249 μM). In comparison with the binding in GAL-3, the conformational change in subsite E of GAL-7 is too narrow to accommodate the 3,5-dimethoxybenzyl-oxazoline moiety. The inhibitor, thus, has to take a different conformation to be situated at the same subsite.

Additionally, compound **3** represents another analogue of TD139 containing diphenyl ether-linked triazole at positions C3 and C3′ of TDG (Figure 9). It shows a 230-fold higher selectivity for GAL-3 (K_d = 0.36 μM) than for GAL-1 (K_d = 84 μM) [62]. The augmented preference for GAL-3 was proposed to be due to the presence of three arginine-π interactions with Arg144[GAL-3], Arg168[GAL-3], and Arg186[GAL-3] in subsites B and E (Figure 9). On the other hand, only one interaction (with Arg73) can be observed in GAL-1.

We previously mentioned that C2–OH of galactose is not involved in any binding interaction. It is of note that a pocket mainly created by L4 (between S4 and S5) is on the top of the galactose ring (Figure 10). One approach taking advantage of this feature is to synthesize taloside derivatives (that are the C2-epimers of galactosides; see Figure 10a,b), and introduce a substituent at C2–OH. Interestingly, His52 of GAL-1 appears to be more interactive towards the O2-substituent of talosides, whereas Arg48[GAL-1] forms one edge of the pocket and Ser29[GAL-1] borders the other edge (Figure 10c,d). Since Ser29[GAL-1] is smaller than the equivalent Arg144[GAL-3], the pocket in GAL-1 apparently has a larger space than that in GAL-3 for the close contact with the O2-substituent. Indeed, 2-*O*-Toluoyl taloside (**II**) shows higher affinity for GAL-1 than 2-*O*-acetyl taloside (**I**) and forms the proposed histidine–aromatic stacking interaction (Figure 10b,d) [63]. Unfortunately, both **I** and **II** display neither satisfying potency (mM range inhibitory potential) nor selectivity for GAL-1 and -3. With a larger, optimized O2-substituent, it is still possible to obtain a more promising result.

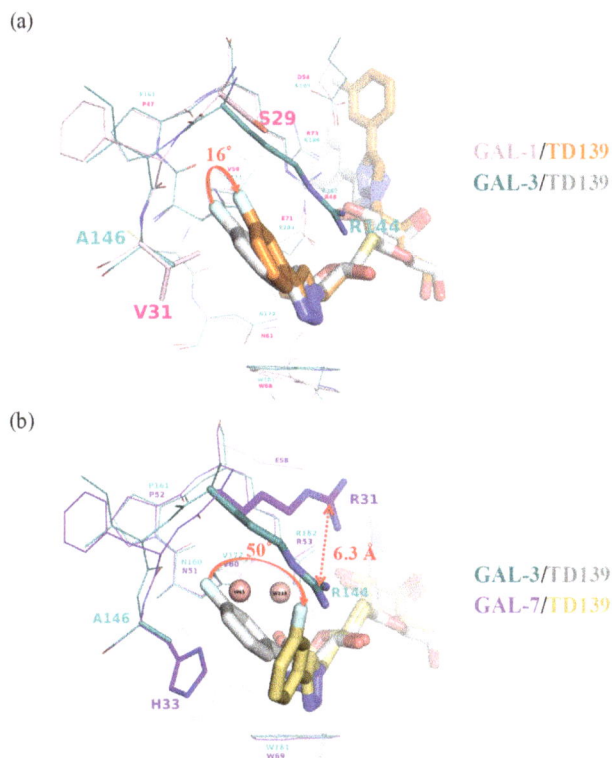

Figure 7. (**a**) Structural superimposition of the GAL-1 (pink)–TD139 (orange) complex and GAL-3 (deep teal)–TD139 (gray) complex. Val31 of GAL-1 is bulkier than the corresponding residue (Ala146) in GAL-3, and the 4-fluorophenyl moiety is forced to rotate approximately 16°. (**b**) When the structures of GAL-3 (deep teal)–TD139 (gray) and GAL-7 (purple blue)–TD139 (yellow orange) are superimposed, the fluorine atom of the 4-fluorophenyl moiety at GAL-7 B subsite turns approximately 50° away as compared to that in GAL-3. In addition, a large difference is observed between Arg144[GAL-3] and Arg31[GAL-7], that are 6.3 Å apart.

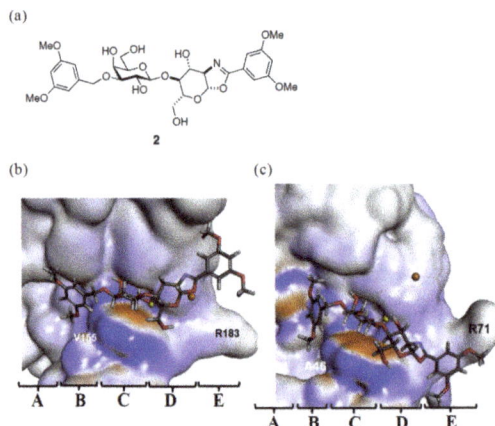

Figure 8. (a) Molecular structure of 1,2-fused (3,5-dimethoxyphenyl)-oxazoline-substituted LacNAc (compound **2**). (b,c) Postulated binding mode of **2** with GAL-3 and -7, respectively. The binding interactions were generated by the CDOCKER docking protocol using Discovery Studio version 4.1. Reprinted with permission from © 2017 Wiley-VCH Verlag GmbH & Co [61].

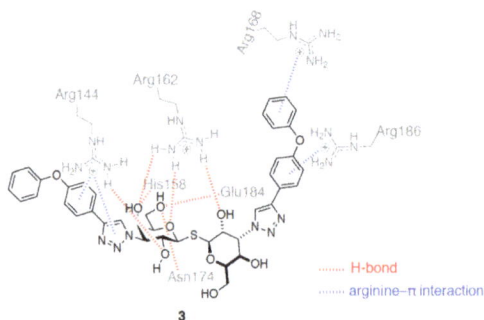

Figure 9. Representative binding mode of compound **3** with GAL-3 that contains H-bonds (shown in red dotted lines) and arginine-π interactions (blue dotted lines). The mentioned binding interactions were derived from the molecular docking study.

Furthermore, we previously identified an Arg-Asp–Glu-Glu-Arg salt-bridge network critical for the binding preference of Galβ1-3/4GlcNAc disaccharides for GAL-1, -3, and -7, though the salt bridge was not involved in the direct interactions with the bound ligand [55]. Notably, the salt bridge was found to be critical for the binding interaction with inhibitors. Asp54^{GAL-1}, Glu165^{GAL-3}, and Glu58^{GAL-7} that participate in the formation of the salt bridges (Figure 11), were found to orient the Arg residue in subsite E to interact with the 4-fluorophenyl substituent of TD139. Tandem arginine-π interactions were similar between the 4-fluorophenyl-triazole and Arg186^{GAL-3} or Arg74^{GAL-7} in the complex structures of GAL-3 and -7 with TD139. The 4-fluorophenyl-triazole moiety of TD139 was, nevertheless, found to flip over at subsite E of GAL-1. The resulting change made the aromatic substituent become close to Asp54^{GAL-1}, leading to the formation of an ion-pair π-interaction, which was favored by the fluorinated arene (i.e., an electron-deficient π system) (Figure 5) [52]. Taking this difference into consideration may help to design potent inhibitors against specific galectin(s).

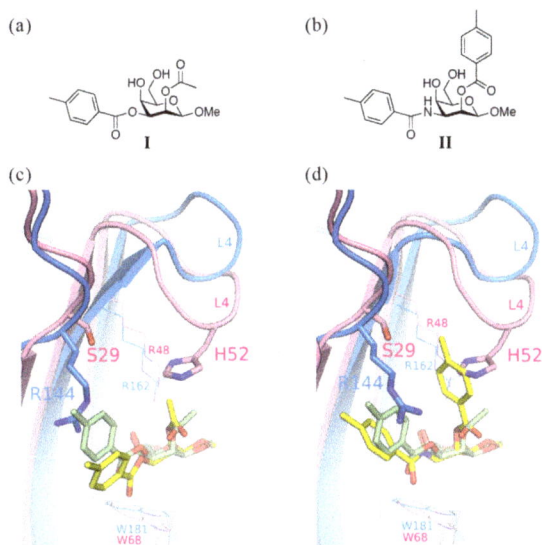

Figure 10. Molecular structures of the two talosides, including (**a**) methyl 2-*O*-acetyl-3-*O*-toluoyl-β-D-talopyranoside (**I**) and (**b**) methyl 3-deoxy-2-*O*-toluoyl-3-*N*-toluoyl-β-D-talopyranoside (**II**). Despite similar structures of **I** and **II**, their affinity with GAL-1 is different ($K_d > 4$ mM and $K_d = 2.4$ mM, respectively). (**c**) The complex structure of GAL-1 (pink)–**I** (pale green) is superimposed with that of GAL-3 (marine blue)–**I** (yellow). (**d**) The complex structure of GAL-1 (pink)–**I** (pale green) is superimposed with that of GAL-3 (marine blue)–**II** (yellow).

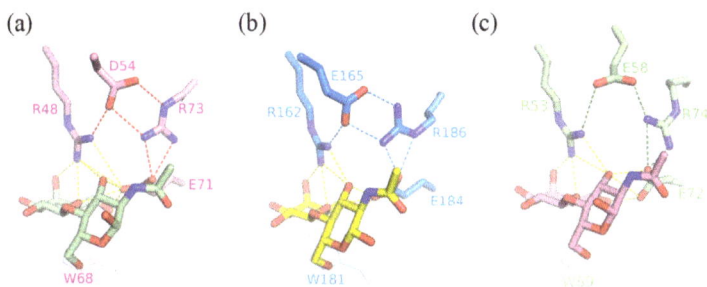

Figure 11. Structures of (**a**) GAL-1 (PDB ID: 1W6P), (**b**) GAL-3 (1KJL), and (**c**) GAL-7 (5GAL) in complex with LacNAc. The salt bridges between Asp/Glu and Arg in LacNAc complex structures are shown in dotted lines for GAL-1 (red), -3 (marine blue), and -7 (forest green), respectively, and all interactions related to galectin–LacNAc complexes formation are colored in yellow.

4. Tools to Characterize the Binding Interactions

X-ray crystallography is the most powerful tool to give precise insights on the arrangement of atoms of a protein–ligand complex in solid state. Nevertheless, the comprehensive understanding of protein–ligand interactions often requires collecting additional information about several pivotal aspects, such as the dynamic movement of the ligand and/or a certain part of the protein, and kinetic and thermodynamic properties of the binding process. This explains the reason why X-ray crystallography is not the only solution. To understand the binding process at a molecular level, X-ray crystal structure analysis has to be complemented with other methods to afford an insightful analysis during the stage of drug development and optimization. In the following sections, we will discuss the methods that have been employed in the development of galectin inhibitors.

4.1. Fluorescence Polarization

Fluorescence polarization (FP) allows the quantitative analysis of molecular interactions in solution by measuring the difference in the degree of polarized light of a bound and unbound fluorophore, which in turn is related to the size of the mobility and the size of the protein–probe complex. Because of its simple and rapid operation, FP has become popular for high-throughput screening in search of lead compounds via competitive binding assays. FP does not require the protein of interest to be labeled or immobilized [64,65]. In addition, the concentrations of all interacting components are known. The most beneficial credit is that the affinity of the ligands is determined when target proteins exist in solution, in a situation that mimics their biological environment. In the field of galectins, Leffler and coworkers were the first to establish an FP-based binding assay when evaluating the IC_{50} values of several 3'-O-benzyl ether-substituted LacNAc derivatives, in which fluorescein-tagged LacNAc served as the reference compound [46,66]. The derivative *p*-methoxy substituted benzyl ether (compound **4**, Figure 12) gave the highest inhibitory potency (IC_{50} = 13 μM) in comparison to the parent ligand, methyl glycoside of LacNAc (IC_{50} = 158 μM). FP was also applied more recently to determine the dissociation constants of a series of 3-3' disubstituted (4-aryl-1,2,3-triazolyl)thiodigalactoside-based derivatives (e.g., TD139, Figure 2c) in relation to GAL-1, -2, -3, -4N, -4C, -7, -8N, -9N, and -9C. Overall, the disubstituted (4-aryltriazolyl) thiodigalactosides were more potent inhibitors than TDG of all the aforementioned galectins, except for GAL-8. Among these galectins, GAL-1 and -3 were strongly inhibited [45].

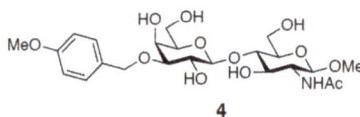

Figure 12. Molecular structure of 3'-O-*p*-methoxybenzyl ether-substituted LacNAc (compound **4**).

In general, a fluorescence probe of satisfying affinity is the prerequisite of FP analysis to avoid false positives. The groups of Tamara [45] and Johann [67] obtained up to 1000-fold different K_d values for GAL-1–TD139 because their fluorescent probes (**5** and **6**, respectively, Figure 13) displayed distinct affinities for GAL-1. Hence, it is noteworthy that potential lead compounds can be likely ignored at the beginning of the screening if a high-affinity probe is used. When using a probe of low affinity, it is necessary to study the target protein at high concentrations, which becomes inevitable to pick up weak-binding compounds.

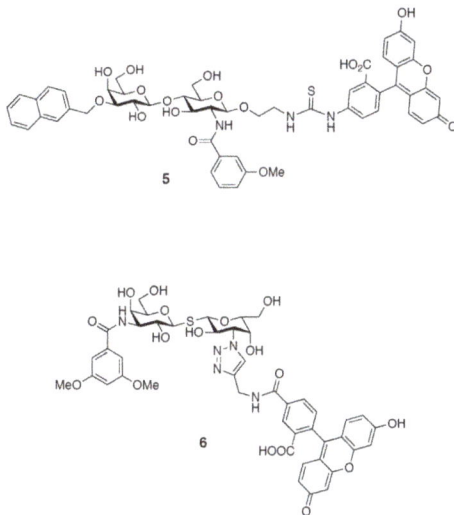

Figure 13. Molecular structures of 2-(fluorescein-5-ylthioureidoethyl) [3-*O*-(naphthalen-2-methyl)-β-D-galactopyranosyl]-(1→4)-2-(3-methoxybenzamido)-2-deoxy-β-D-glucopyranoside,   **5**   and 3,3′-dideoxy-3-[4-(fluorescein-5-yl-carbonylaminomethyl)-1H-1,2,3-triazol-1-yl]-3′-(3,5-di-methoxybenzamido)-1,1′-sulfanediyl-di-β-D-galactopyranoside, **6**.

4.2. Biolayer Interferometry

Biolayer Interferometry (BLI) is a relatively new method to measure biomolecular interactions. The proteins of interest have to be immobilized on the biosensor tip. Upon the binding with an analyte (ligand), the thickness of the biosensor tip is usually changed. BLI analyzes the optical interference pattern of a white light reflected from the immobilized surface, in comparison to an internal reference surface, thus allowing for direct real-time measurements of binding kinetics, including the rate of association (k_{on}), the rate of dissociation (k_{off}), and, concomitantly, the equilibrium constant (K_d) [68]. BLI has become recently popular to determine K_d values. Zhang et al. applied BLI for examining pectin binding of GAL-3 and comparing it with the other four methods, namely, surface plasmon resonance, FP, competitive fluorescence-linked immunosorbent assay, and cell-based hemagglutination assay [69]. Chien and coworkers measured the K_d values of lactose with GAL-1, -7, -8N, -8C, and the full-length GAL-8 [70].

The lifetime of a drug–protein target complex is highly associated with the k_{off} that is particularly important in rating the efficacy of a drug. A drug's k_{off} is not determined merely by its affinity, as a drug only takes effect when it is bound to the protein target [71–73]. Two distinct candidates may have the same level of affinity, but greatly different binding kinetics. TD139 was found to have very slow off rates (k_{off}) in the binding with GAL-1 and -3 (10^{-2} s^{-1} and 10^{-3} s^{-1}, respectively), as revealed by the BLI analysis. This result is consistent with the data measured by the ^{15}N-^{1}H correlation spectra. However, to our knowledge, no report has discussed how the binding kinetics of galectin inhibitors correlates with their complex half-life in solution or in a biological system.

4.3. Isothermal Titration Calorimetry

There have been more and more studies relying on thermodynamics measurements in drug discovery because they provides clues on the energy-driven binding interactions [74]. ITC is the ideal method to detect the heat released or absorbed during a biomolecular binding interaction down to the nanomolar range [75]. The binding affinity is determined by the change of free energy (ΔG) that is in turn affected by two parameters: the change of enthalpy (ΔH) and entropy (ΔS) (i.e., $\Delta G = \Delta H - T\Delta S$). Enthalpy reflects the heat differences as a result of the binding process, whereas entropy accounts for the degree of disorder in association with the same event [76]. A number of thermodynamic analyses focus on the binding of galactose-bearing glycans with galectins. The results are consistent with the structural studies. For example, Bachhawat-Sikder et al. performed the ITC analysis for the binding of LacNAc with GAL-3 [77]. The process was enthalpically driven ($\Delta H = -8.88$ kcal/mol) with a seven times higher affinity than that of the binding of lactose, which was equivalent to 4.08 kcal/mol of favorable enthalpy change. The preferential binding was realized by the C2-acetamide of the GlcNAc moiety that interacts with Glu165 via a water molecule, in addition to the aforementioned H-bond network established by Arg162^{GAL-3} and Glu184^{GAL-3} with O3 of GlcNAc.

The binding of TDG towards GAL-1 and -3 is also an enthalpy-driven, exothermic process with a heat release of -11.6 and -9.4 kcal/mol, respectively. The introduction of one 4-flurophenyl-triazole moiety at the C3 of TDG to produce TAZTDG resulted in a significant favorable change in enthalpy, with $\Delta\Delta H_{TAZTDG-TDG} = -4.5$ and -5.3 kcal/mol for GAL-1 and GAL-3, respectively. The lower energy level of TAZTDG is assumed to be a consequence of an arginine-π interaction at subsites B and E. On the other hand, the incorporation of a second 4-flurophenyl-triazole moiety did not have much effect on GAL-1 ($\Delta\Delta H_{TD139-TAZTDG} = -0.6$ kcal/mol), in contrast to GAL-3 ($\Delta\Delta H_{TD139-TAZTDG} = -3.5$ kcal/mol) [52], suggesting a differential binding at subsites B and E. This is consistent with the presence of two Arg residues at the CRD of GAL-3 (Arg144 at subsite B, Arg186 at subsite E) with the possibility of forming another arginine-π interaction. For GAL-1, however, there is only one Arg (i.e., Arg73) available, which is located at the subsite E of CRD. The additional 4-fluorophenyl-triazole moiety, does not, thus, provide much impact on the enthalpy. The effect of additional binding interactions became more obvious when comparing TD139 ($K_d = 68$ nM) and TDG (that contains no substituent, $K_d = 75$ μM). In addition to displaying the highest binding preference for GAL-3, TD139 possessed approximately a 1000-fold binding enhancement as compared to TDG. The great enhancement was due to a change in the free energy ($\Delta\Delta G_{TD139-TDG}$) of -4.3 kcal/mol, which was contributed by a huge enthalpy change ($\Delta\Delta H_{TD139-TDG} = -8.8$ kcal/mol) but offset by an unfavorable entropy change ($-T\Delta\Delta S_{TD139-TDG} = 4.6$ kcal/mol). It is noted that the ITC method is usually suitable for inhibitors of moderate affinity (K_d from sub-mM to sub-μM). When ITC is applied for the direct measurement of high-affinity inhibitors (K_d of low nM and better), it always generates very steep binding curves with few data points on the slope of the curves. Since the value of $K_d/\Delta G$ is extracted from the shape or slope of the titration curves, the measurement often results in large errors in the $K_d/\Delta G$ values. Nowadays, the direct ITC measurement has been substituted by an indirect procedure in which a medium-affinity ligand is competed and displaced by the desirable high-affinity inhibitor.

Subsequent modifications to overcome the enthalpy and entropy compensation are important to ameliorate the binding affinity of a ligand. In fact, one has to be aware that experimental ΔG, ΔH, and calculated ΔS only serve as a guideline, and other molecular studies need to be integrated. ΔS is not determined solely from a separate experiment, hence any error that occurr during the measurements of ΔG and ΔH will be reflected on the entropy change [78]. Meanwhile, the presumption that in a protein–ligand binding event the net enthalpy is predominantly contributed by polar interactions while entropy is dominated by desolvation effects, does not hold in all cases [79].

4.4. ^{19}F-NMR Spectroscopy

In the last two decades, the introduction of a fluorine atom to drug molecules has become more pronounced in medicinal chemistry [80,81] because of several key advantages, including the improvement on the metabolic stability, the regulation of the acidity and basicity, and the increase of the affinity by forming extra interactions with target proteins (e.g., H-bonds, hydrophobic, and polar interactions) [82,83]. A growing number of reports have incorporated fluorine to lead compounds in the initial stage of drug discovery. Moreover, ^{19}F-NMR spectroscopy provides a rapid and sensitive way to detect the presence of fluorine-containing compounds, since biomolecules usually do not contain fluorine, and thus there is almost no background in ^{19}F-NMR. This method is also useful to reveal binding interactions, because the ^{19}F signals of fluorine-containing ligands would shift either upfield or downfield under different chemical environments in the bound state, in comparison with those of the free ligands. A shielded fluorine atom is thought to be in the proximity to H-bond donors within a protein structure, while a deshielded one is predominantly found in close contact with hydrophobic side chains or with the carbonyl carbon in the protein backbone [84,85].

We previously employed ^{19}F-NMR to explore the binding features of TD139 in complex with GAL-1 and -3 [52]. The free-form ligand exhibited a single and sharp resonance at ~113 ppm as a consequence of rapid tumbling in solution (Figure 14a). Interestingly, the GAL-1 and -3 complexed with TD139 generated similar spectra that contained three distinct resonances of broader linewidths, suggesting the presence of a slow conformational interconversion of the fluorinated phenyl rings of TD139 on the NMR timescale. One unique ^{19}F resonance represents a fingerprint of one binding mode of the 4-fluorophenyl moiety. Therefore, the two upfield-shifted ^{19}F resonances were attributed to two different binding modes of the 4-fluorophenyl moiety that is densely packed in the subsites B of GAL-1 and -3. Meanwhile, the upfield shifts were due to the shielding effect of the multiple fluorine bonds on the fluorine atom at the subsite B. On the other hand, the most downfield-shifted resonance ($\delta < -113$ ppm, Figure 14a) in the ^{19}F spectrum was likely corresponding to the binding with the subsite E. The Arg and Asp/Glu side chains at the subsite E of both galectins established an extended π-electron surface at which the terminal fluorine was situated. The fluorine atom was deshielded, hence leading to the observed downfield shifts in the peak positions.

The asymmetrical TAZTDG in complex with GAL-1 and -3 also gave similar spectra as those of the symmetrical TD139. This indicated that the monosubstituted TAZTDG could bind with either subsites B or E in solution, in a similar manner as the di-substituted TD139, to concurrently occupy both subsites (Figure 14b). The corresponding heteronuclear ^{15}N-^{1}H correlation spectra further supported the dual binding modes of TAZTDG and GAL-1 or -3.

Additionally, in the ^{19}F-NMR study of TD139 and TAZTDG in complex with GAL-7, only one resonance was observed in each case. This corresponded to a weak binding and, hence, fast averaging of free and bound signals, and at the same time reflected an apparent population-weighted chemical shift.

Additionally, ^{19}F-NMR was also applied to examine the binding of TAZTDG and TD139 with GAL-8. This galectin is known to contain two different CRDs located at the N- and C-termini of GAL-8. We thus prepared three different forms of GAL-8, including the full length and the two individual domains (namely, N-CRD and C-CRD). Like the bound form of GAL-7 previously mentioned, the GAL-8 complexes exhibited a single broadened resonance, which reflected its low binding affinity for TAZTDG and TD139. In comparison with GAL-8^{NCRD}, the signal pattern of GAL-8^{CCRD} was similar to that of GAL-8^{full} (Figure 15), implying that the C-terminal domain might be predominant in the binding of GAL-8 to TAZTDG and TD139. The observation is consistent with the K_d values of TD139 (K_d = 45, 88 and 14 μM for GAL-8^{full}, GAL-8^{NCRD}, and GAL-8^{CCRD}, respectively). Additionally, the linewidths of the GAL-8^{NCRD} bound form were much broader than those of GAL-8^{full} and GAL-8^{CCRD}. In spite of the lower affinity with GAL-8^{NCRD}, the signals were not split even when the temperature was dropped to 278 K, which is likely explained by the slow conformational interconversion of the fluorinated phenyl rings, as mentioned previously in the study of GAL-1 and GAL-3.

There is no specific inhibitor for either N-CRD or C-CRD of GAL-8. It is known that N-CRD prefers binding to negative-charged glycans, while C-CRD binds better to neutral carbohydrates. This is in agreement with a previous study showing that 3′-sialyl LacNAc and 3′-sulfated lactose displayed higher affinity for GAL-8^{NCRD} than for GAL-8^{CCRD} [70,86–88]. In accordance with our ^{19}F-NMR study, TD139 appeared to bind better to GAL-8^{CCRD} than to GAL-8^{NCRD}, probably because the inhibitor has no charge.

(a)

: Subsite E binding mode
: Subsite B binding mode

(b)

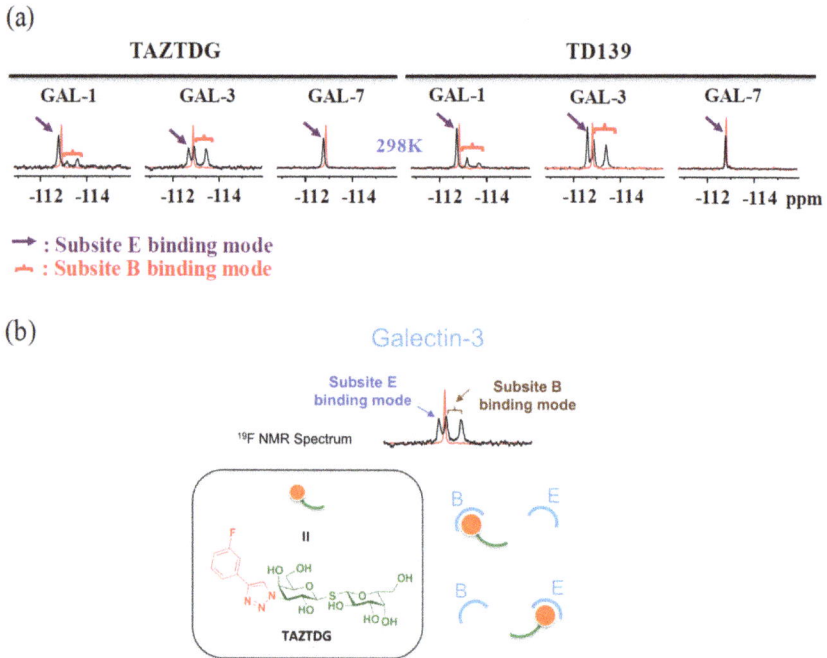

Figure 14. (a) ^{19}F-NMR spectra of TAZTDG and TD139 in complex with GAL-1, -3, and -7. (b) Dual binding modes of the galectin inhibitor TAZTDG. On the basis of the integrated analysis of isothermal titration calorimetry, X-ray crystallography, and NMR spectroscopy, the 4-fluorophenyl-triazole moiety of TAZTDG can bind with either subsites B or E, leading to the observed dual binding modes.

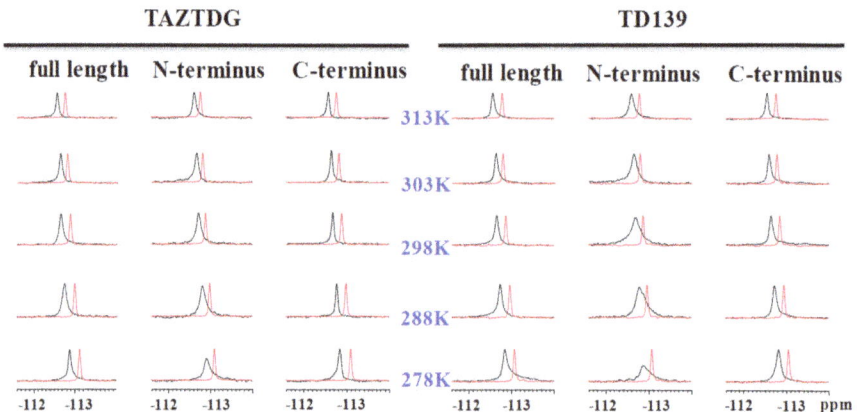

Figure 15. The ^{19}F-NMR spectra of TAZTDG (left panel) and TD139 (right panel) in complex with GAL-8^{full}, GAL-8^{NCRD}, and GAL-8^{CCRD} are shown in the order of decreasing temperatures (from top to bottom). The spectra of free TAZTDG and TD139 under the same conditions are shown in red as references.

5. Future Direction

Galectins are found primarily in the cytosol, nucleus, extracellular space or in the circulation. Their functions do not entirely take place outside the cells. Extra- and intracellular galectins often display opposite functions. Extracellular GAL-1, for instance, is responsible for promoting cancer cell proliferation, but intracellular GAL-1 plays an opposite role in inhibiting cell proliferation [89]. Another example comes from GAL-8 that was shown to play a specific role in bacterial or viral infection. After pathogens are engulfed by cells, they typically try to escape from the endosome to access nutrients in the cytosol. GAL-8 was found to recognize special glycosylation sites found within the endosomes and recruit the adapter machinery CALCOCOL2 (namely calcium-binding and coiled-coil domain-containing protein 2) that activates autophagy [90]. As a consequence, it is indispensable to develop inhibitors of high affinity to specifically target either extra- or intracellular galectins, or even selectively label their cellular locations. It is indeed a challenging issue to come up with molecules that mostly stay either outside or inside the cells.

Moreover, several selective inhibitors have been developed so far, but there are noticeable challenges remaining. One major problem is that only few galectins were examined, such as GAL-1, -3, -7, and -8. Great success has been achieved to distinguish between GAL-3 and other galectins. Most of the developed inhibitors display satisfying affinity for GAL-3 but not for weak-binding galectins (e.g., GAL-7). It is certainly a major breakthrough if the selectivity can be reversed (e.g., to develop inhibitors more selective for GAL-7 than for GAL-3).

Acknowledgments: The NMR spectra were measured at the Core Facility for Protein Structural Analysis, supported by the National Core Facility Program for Biotechnology, Taiwan. This work was supported by the Summit Project in Academia Sinica and Ministry of Science and Technology (MOST 106-0210-01-15-02, MOST 107-0210-01-19-01), Taiwan.

Author Contributions: Yi-Chen Chan, Yi-Chen Chan, Zhijay Tu, and Chun-Hung Lin organized and wrote the manuscript. Yi-Chen Chan, Zhijay Tu, Hsien-Ya Lin, and Yen-Hsi Kuo performed all the experiments and analyzed the data. Shang-Te Danny Hsu provided assistance and discussion in NMR experiments. All authors reviewed the manuscript and approved the manuscript for publication.

Conflicts of Interest: The authors declare no competing financial interests.

References

1. Pieters, R.J. Inhibition and detection of galectins. *ChemBioChem* **2006**, *7*, 721–728. [CrossRef] [PubMed]
2. Leffler, H.; Carlsson, S.; Hedlund, M.; Qian, Y.; Poirier, F. Introduction to galectins. *Glycoconj. J.* **2002**, *19*, 433–440. [CrossRef] [PubMed]
3. Ingrassia, L.; Camby, I.; Lefranc, F.; Mathieu, V.; Nshimyumukiza, P.; Darro, F.; Kiss, R. Anti-galectin compounds as potential anti-cancer drugs. *Curr. Med. Chem.* **2006**, *13*, 3513–3527. [CrossRef] [PubMed]
4. Masuyer, G.; Jabeen, T.; Oberg, C.T.; Leffler, H.; Nilsson, U.J.; Acharya, K.R. Inhibition mechanism of human galectin-7 by a novel galactose-benzylphosphate inhibitor. *FEBS J.* **2012**, *279*, 193–202. [CrossRef] [PubMed]
5. Liu, F.-T.; Rabinovich, G.A. Galectins as modulators of tumour progression. *Nat. Rev. Cancer* **2005**, *5*, 29–41. [CrossRef] [PubMed]
6. Lepur, A.; Salomonsson, E.; Nilsson, U.J.; Leffler, H. Ligand induced galectin-3 protein self-association. *J. Biol. Chem.* **2012**, *287*, 21751–21756. [CrossRef] [PubMed]
7. Ahmed, H.; AlSadek, D.M.M. Galectin-3 as a Potential target to prevent cancer metastasis. *Clin. Med. Insights Oncol.* **2015**, *9*, 113–121. [CrossRef] [PubMed]
8. Gendronneau, G.; Sidhu, S.S.; Delacour, D.; Dang, T.; Calonne, C.; Houzelstein, D.; Magnaldo, T.; Poirier, F. Galectin-7 in the control of epidermal homeostasis after injury. *Mol. Biol. Cell* **2008**, *19*, 5541–5549. [CrossRef] [PubMed]
9. Rabinovich, G.A.; Toscano, M.A. Turning 'sweet' on immunity: Galectin-glycan interactions in immune tolerance and inflammation. *Nat. Rev. Immunol.* **2009**, *9*, 338–352. [CrossRef] [PubMed]
10. Chen, H.-Y.; Weng, I.-C.; Hong, M.-H.; Liu, F.-T. Galectins as bacterial sensors in the host innate response. *Curr. Opin. Microbiol.* **2014**, *17*, 75–81. [CrossRef] [PubMed]

11. Rabinovich, G.A.; van Kooyk, Y.; Cobb, B.A. Glycobiology of immune responses. *Ann. N. Y. Acad. Sci.* **2012**, *1253*, 1–15. [CrossRef] [PubMed]

12. De Oliveira, F.L.; Gatto, M.; Bassi, N.; Luisetto, R.; Ghirardello, A.; Punzi, L.; Doria, A. Galectin-3 in autoimmunity and autoimmune diseases. *Exp. Biol. Med.* **2015**, *240*, 1019–1028. [CrossRef] [PubMed]

13. Boscher, C.; Zheng, Y.-Z.; Lakshminarayan, R.; Johannes, L.; Dennis, J.W.; Foster, L.J.; Nabi, I.R. Galectin-3 protein regulates mobility of N-cadherin and GM1 ganglioside at cell-cell junctions of mammary carcinoma cells. *J. Biol. Chem.* **2012**, *287*, 32940–32952. [CrossRef] [PubMed]

14. Hughes, R.C. Galectins as modulators of cell adhesion. *Biochimie* **2001**, *83*, 667–676. [CrossRef]

15. Yabuta, C.; Yano, F.; Fujii, A.; Shearer, T.R.; Azuma, M. Galectin-3 enhances epithelial cell adhesion and wound healing in rat cornea. *Ophthalmic Res.* **2014**, *51*, 96–103. [CrossRef] [PubMed]

16. Fortin, S.; Le Mercier, M.; Camby, I.; Spiegl-Kreinecker, S.; Berger, W.; Lefranc, F.; Kiss, R. Galectin-1 is implicated in the protein kinase C epsilon/vimentin-controlled trafficking of integrin-beta1 in glioblastoma cells. *Brain Pathol.* **2010**, *20*, 39–49. [CrossRef] [PubMed]

17. Panjwani, N. Role of galectins in re-epithelialization of wounds. *Ann. Transl. Med.* **2014**, *2*, 89. [PubMed]

18. Liu, W.; Hsu, D.-K.; Chen, H.-Y.; Yang, R.-Y.; Carraway, K.L., 3rd; Isseroff, R.R.; Liu, F.-T. Galectin-3 regulates intracellular trafficking of EGFR through Alix and promotes keratinocyte migration. *J. Investig. Dermatol.* **2012**, *132*, 2828–2837. [CrossRef] [PubMed]

19. Sato, S.; Ouellet, M.; St-Pierre, C.; Tremblay, M.J. Glycans, galectins, and HIV-1 infection. *Ann. N. Y. Acad. Sci.* **2012**, *1253*, 133–148. [CrossRef] [PubMed]

20. Klyosov, A.A.; Traber, P.G. Galectins in Disease and Potential Therapeutic Approaches. In Galectins and Disease Implications for Targeted Therapeutics. *ACS Symp. Ser.* **2012**, *1115*, 3–43.

21. Hrynchyshyn, N.; Jourdain, P.; Desnos, M.; Diebold, B.; Funck, F. Galectin-3: A new biomarker for the diagnosis, analysis and prognosis of acute and chronic heart failure. *Arch. Cardiovasc. Dis.* **2013**, *106*, 541–546. [CrossRef] [PubMed]

22. Radosavljevic, G.; Volarevic, V.; Jovanovic, I.; Milovanovic, M.; Pejnovic, N.; Arsenijevic, N.; Hsu, D.-K.; Lukic, M.L. The roles of Galectin-3 in autoimmunity and tumor progression. *Immunol. Res.* **2012**, *52*, 100–110. [CrossRef] [PubMed]

23. Reticker-Flynn, N.E.; Malta, D.F.; Winslow, M.M.; Lamar, J.M.; Xu, M.-J.; Underhill, G.H.; Hynes, R.O.; Jacks, T.E.; Bhatia, S.N. A combinatorial extracellular matrix platform identifies cell-extracellular matrix interactions that correlate with metastasis. *Nat. Commun.* **2012**, *3*, 1122. [CrossRef] [PubMed]

24. Barrow, H.; Rhodes, J.M.; Yu, L.-G. The role of galectins in colorectal cancer progression. International journal of cancer. *J. Int. Cancer* **2011**, *129*, 1–8. [CrossRef] [PubMed]

25. Kouo, T.; Huang, L.; Pucsek, A.B.; Cao, M.; Solt, S.; Armstrong, T.; Jaffee, E. Galectin-3 shapes antitumor immune responses by suppressing CD8+ T cells via LAG-3 and inhibiting expansion of plasmacytoid dendritic cells. *Cancer Immunol. Res.* **2015**, *3*, 412–423. [CrossRef] [PubMed]

26. Wieers, G.; Demotte, N.; Godelaine, D.; Van der Bruggen, P. Immune suppression in tumors as a surmountable obstacle to clinical efficacy of cancer vaccines. *Cancers* **2011**, *3*, 2904–2954. [CrossRef] [PubMed]

27. Yang, R.-Y.; Hsu, D.-K.; Liu, F.-T. Expression of galectin-3 modulates T-cell growth and apoptosis. *Proc. Natl. Acad. Sci. USA* **1996**, *93*, 6737–6742. [CrossRef] [PubMed]

28. Rabinovich, G.A.; Rubinstein, N.; Toscano, M.A. Role of galectins in inflammatory and immunomodulatory processes. *Biochim. Biophys. Acta* **2002**, *1572*, 274–284. [CrossRef]

29. Clark, M.C.; Pang, M.; Hsu, D.-K.; Liu, F.-T.; de Vos, S.; Gascoyne, R.D.; Said, J.; Baum, L.G. Galectin-3 binds to CD45 on diffuse large B-cell lymphoma cells to regulate susceptibility to cell death. *Blood* **2012**, *120*, 4635–4644. [CrossRef] [PubMed]

30. Nishi, N.; Abe, A.; Iwaki, J.; Yoshida, H.; Itoh, A.; Shoji, H.; Kamitori, S.; Hirabayashi, J.; Nakamura, T. Functional and structural bases of a cysteine-less mutant as a long-lasting substitute for galectin-1. *Glycobiology* **2008**, *18*, 1065–1073. [CrossRef] [PubMed]

31. Battig, P.; Saudan, P.; Gunde, T.; Bachmann, M.F. Enhanced apoptotic activity of a structurally optimized form of galectin-1. *Mol. Immunol.* **2004**, *41*, 9–18. [CrossRef] [PubMed]

32. Zou, J.; Glinsky, V.V.; Landon, L.A.; Matthews, L.; Deutscher, S.L. Peptides specific to the galectin-3 carbohydrate recognition domain inhibit metastasis-associated cancer cell adhesion. *Carcinogenesis* **2005**, *26*, 309–318. [CrossRef] [PubMed]

33. Platt, D.; Raz, A. Modulation of the lung colonization of B16-F1 melanoma cells by citrus pectin. *J. Natl. Cancer Inst.* **1992**, *84*, 438–442. [CrossRef] [PubMed]

34. Pienta, K.J.; Naik, H.; Akhtar, A.; Yamazaki, K.; Replogle, T.S.; Lehr, J.; Donat, T.L.; Tait, L.; Hogan, V.; Raz, A. Inhibition of spontaneous metastasis in a rat prostate cancer model by oral administration of modified citrus pectin. *J. Natl. Cancer Inst.* **1995**, *87*, 348–353. [CrossRef] [PubMed]

35. McAuliffe, J.; Hindsgaul, O. Carbohydrates in Medicine. In *Molecular and Cellular Glycobiology*; Fukuda, M., Hindsgaul, O., Eds.; Oxford University Press: Oxford, UK, 2000; pp. 249–285.

36. Cotter, F.; Smith, D.A.; Boyd, T.E.; Richards, D.A.; Alemany, C.; Loesch, D.; Salogub, G.; Tidmarsh, G.F.; Gammon, G.M.; Gribben, J. Single-agent activity of GCS-100, a first-in-class galectin-3 antagonist, in elderly patients with relapsed chronic lymphocytic leukemia. *J. Clin. Oncol.* **2009**, *27*, 7006.

37. Traber, P.G.; Zomer, E. Therapy of experimental NASH and fibrosis with galectin inhibitors. *PLoS ONE* **2013**, *8*, e83481. [CrossRef] [PubMed]

38. Traber, P.G.; Chou, H.; Zomer, E.; Hong, F.; Klyosov, A.; Fiel, M.I.; Friedman, S.L. Regression of fibrosis and reversal of cirrhosis in rats by galectin inhibitors in thioacetamide-induced liver disease. *PLoS ONE* **2013**, *8*, e75361. [CrossRef] [PubMed]

39. Klyosov, A.; Zomer, E.; Platt, D. Studies. In *Glycobiology and Drug Design*; American Chemical Society: New York, NY, USA, 2012; Volume 1102, pp. 89–130.

40. Miller, M.C.; Klyosov, A.; Mayo, K.H. The alpha-galactomannan Davanat binds galectin-1 at a site different from the conventional galectin carbohydrate binding domain. *Glycobiology* **2009**, *19*, 1034–1045. [CrossRef] [PubMed]

41. Miller, M.C.; Ribeiro, J.P.; Roldos, V.; Martin-Santamaria, S.; Canada, F.J.; Nesmelova, I.A.; Andre, S.; Pang, M.; Klyosov, A.A.; Baum, L.G.; et al. Structural aspects of binding of alpha-linked digalactosides to human galectin-1. *Glycobiology* **2011**, *21*, 1627–1641. [CrossRef] [PubMed]

42. Blanchard, H.; Yu, X.; Collins, P.M.; Bum-Erdene, K. Galectin-3 inhibitors: A patent review (2008-present). *Expert Opin. Ther. Patents* **2014**, *24*, 1053–1065. [CrossRef] [PubMed]

43. Mackinnon, A.C.; Gibbons, M.A.; Farnworth, S.L.; Leffler, H.; Nilsson, U.J.; Delaine, T.; Simpson, A.J.; Forbes, S.J.; Hirani, N.; Gauldie, J.; et al. Regulation of transforming growth factor-beta1-driven lung fibrosis by galectin-3. *Am. J. Respir. Crit. Care Med.* **2012**, *185*, 537–546. [CrossRef] [PubMed]

44. Garber, K. Galecto Biotech. *Nat. Biotechnol.* **2013**, *31*, 481. [CrossRef] [PubMed]

45. Delaine, T.; Collins, P.; MacKinnon, A.; Sharma, G.; Stegmayr, J.; Rajput, V.K.; Mandal, S.; Cumpstey, I.; Larumbe, A.; Salameh, B.A.; et al. Galectin-3-binding glycomimetics that strongly reduce bleomycin-induced lung fibrosis and modulate intracellular glycan recognition. *ChemBioChem* **2016**, *17*, 1759–1770. [CrossRef] [PubMed]

46. Sorme, P.; Kahl-Knutsson, B.; Huflejt, M.; Nilsson, U.J.; Leffler, H. Fluorescence polarization as an analytical tool to evaluate galectin-ligand interactions. *Anal. Biochem.* **2004**, *334*, 36–47. [CrossRef] [PubMed]

47. Sorme, P.; Arnoux, P.; Kahl-Knutsson, B.; Leffler, H.; Rini, J.M.; Nilsson, U.J. Structural and thermodynamic studies on cation-Pi interactions in lectin-ligand complexes: High-affinity galectin-3 inhibitors through fine-tuning of an arginine-arene interaction. *J. Am. Chem. Soc.* **2005**, *127*, 1737–1743. [CrossRef] [PubMed]

48. Cumpstey, I.; Sundin, A.; Leffler, H.; Nilsson, U.J. C2-symmetrical thiodigalactoside bis-benzamido derivatives as high-affinity inhibitors of galectin-3: Efficient lectin inhibition through double arginine-arene interactions. *Angew. Chem. Int. Ed.* **2005**, *44*, 5110–5112. [CrossRef] [PubMed]

49. Rajput, V.K.; MacKinnon, A.; Mandal, S.; Collins, P.; Blanchard, H.; Leffler, H.; Sethi, T.; Schambye, H.; Mukhopadhyay, B.; Nilsson, U.J. A selective galactose-coumarin-derived galectin-3 inhibitor demonstrates involvement of galectin-3-glycan interactions in a pulmonary fibrosis model. *J. Med. Chem.* **2016**, *59*, 8141–8147. [CrossRef] [PubMed]

50. Mackinnon, A.; Chen, W.-S.; Leffler, H.; Panjwani, N.; Schambye, H.; Sethi, T.; Nilsson, U.J. Design, Synthesis, and Applications of Galectin Modulators in Human Health. In *Carbohydrates as Drugs*; Seeberger, P.H., Rademacher, C., Eds.; Springer International Publishing: Cham, Switzerland, 2014; pp. 95–121.

51. Oberg, C.T.; Leffler, H.; Nilsson, U.J. Inhibition of galectins with small molecules. *Chimia* **2011**, *65*, 18–23. [CrossRef] [PubMed]

52. Hsieh, T.-J.; Lin, H.-Y.; Tu, Z.; Lin, T.-C.; Wu, S.-C.; Tseng, Y.-Y.; Liu, F.-T.; Hsu, S.-T.D.; Lin, C.-H. Dual thio-digalactoside-binding modes of human galectins as the structural basis for the design of potent and selective inhibitors. *Sci. Rep.* **2016**, 29457. [CrossRef] [PubMed]

53. Leffler, H.; Barondes, S.H. Specificity of binding of three soluble rat lung lectins to substituted and unsubstituted mammalian β-Galactoside. *J. Biol. Chem.* **1986**, *261*, 10119–10126. [PubMed]

54. Lobsanov, Y.D.; Gitt, M.A.; Leffler, H.; Barondes, S.H.; Rini, J.M. X-ray crystal structure of the human dimeric S-Lac lectin, L-14-II, in complex with lactose at 2.9 Å resolution. *J. Biol. Chem.* **1993**, *268*, 27034–27038. [PubMed]

55. Hsieh, T.-J.; Lin, H.-Y.; Tu, Z.; Huang, B.-S.; Wu, S.-C.; Lin, C.-H. Structural Basis Underlying the binding preference of human galectins-1, -3 and -7 for Galβ1-3/4GlcNAc. *PLoS ONE* **2015**, *10*, e0125946. [CrossRef] [PubMed]

56. Sorme, P.; Qian, Y.; Nyholm, Per-G.; Leffler, H.; Nilsson, U.J. Low micromolar inhibitors of galectin-3 based on 3′-derivatization of N-acetyllactosamine. *ChemBioChem* **2002**, *3*, 183–189. [CrossRef]

57. Cumpstey, I.; Salomonsson, E.; Sundin, A.; Leffler, H.; Nilsson, U.J. Double affinity amplification of galectin-ligand interactions through arginine-arene interactions: Synthetic, thermodynamic, and computational studies with aromatic diamido thiodigalactosides. *Chem. Eur. J.* **2008**, *14*, 4233–4245. [CrossRef] [PubMed]

58. Paulini, R.; Muller, K.; Diederich, F. Orthogonal multipolar interactions in structural chemistry and biology. *Angew. Chem. Int. Ed.* **2005**, *44*, 1788–1805. [CrossRef] [PubMed]

59. Pollock, J.; Borkin, D.; Lund, G.; Purohit, T.; Dyguda-Kazimierowicz, E.; Grembecka, J.; Cierpicki, T. Rational design of orthogonal multipolar interactions with fluorine in protein-ligand complexes. *J. Med. Chem.* **2015**, *58*, 7465–7474. [CrossRef] [PubMed]

60. Cumpstey, I.; Salomonsson, E.; Sundin, A.; Leffler, H.; Nilsson, U.J. Studies of arginine-arene interactions through synthesis and evaluation of a series of galectin-binding aromatic lactose esters. *ChemBioChem* **2007**, *8*, 1389–1398. [CrossRef] [PubMed]

61. Dion, J.; Advedissian, T.; Storozhylova, N.; Dahbi, S.; Lambert, A.; Deshayes, F.; Viguier, M.; Tellier, C.; Poirier, F.; Teletchea, S.; et al. Development of a sensitive microarray platform for the ranking of galectin inhibitors: Identification of a selective galectin-3 inhibitor. *ChemBioChem* **2017**, *18*, 1–14. [CrossRef] [PubMed]

62. Van Hattum, H.; Branderhorst, H.M.; Moret, E.E.; Nilsson, U.J.; Leffler, H.; Pieters, R.J. Tuning the preference of thiodigalactoside- and lactosamine-based ligands to galectin-3 over galectin-1. *J. Med. Chem.* **2013**, *56*, 1350–1354. [CrossRef] [PubMed]

63. Collins, P.M.; Oberg, C.T.; Leffler, H.; Nilsson, U.J.; Blanchard, H. Taloside inhibitors of galectin-1 and galectin-3. *Chem. Biol. Drug Des.* **2012**, *79*, 339–346. [CrossRef] [PubMed]

64. Sorme, P.; Kahl-Knutson, B.; Wellmar, U.; Nilsson, U.J.; Leffler, H. Fluorescence polarization to study galectin-ligand interactions. *Methods Enzymol.* **2003**, *362*, 504–512. [PubMed]

65. Lea, W.A.; Simeonov, A. Fluorescence polarization assays in small molecule screening. *Expert Opin. Drug Discov.* **2011**, *6*, 17–32. [CrossRef] [PubMed]

66. Sorme, P.; Kahl-Knutsson, B.; Wellmar, U.; Magnusson, B.G.; Leffler, H.; Nilsson, U.J. Design and synthesis of galectin inhibitors. *Methods Enzymol.* **2003**, *363*, 157–169. [PubMed]

67. Dion, J.; Deshayes, F.; Storozhylova, N.; Advedissian, T.; Lambert, A.; Viguier, M.; Tellier, C.; Dussouy, C.; Poirier, F.; Grandjean, C. Lactosamine-based derivatives as tools to delineate the biological functions of galectins: Application to skin tissue repair. *ChemBioChem* **2017**, *18*, 782–789. [CrossRef] [PubMed]

68. Petersen, R. Strategies Using Bio-Layer Interferometry biosensor technology for vaccine research and development. *Biosensors* **2017**, *7*, 49. [CrossRef] [PubMed]

69. Zhang, T.; Zheng, Y.; Zhao, D.; Yan, J.; Sun, C.; Zhou, Y.; Tai, G. Multiple approaches to assess pectin binding to galectin-3. *Intl. J. Biol. Macromol.* **2016**, *91*, 994–1001. [CrossRef] [PubMed]

70. Chien, C.-H.; Ho, M.-R.; Lin, C.-H.; Hsu, S.-T.D. Lactose binding induces opposing dynamics changes in human galectins revealed by NMR-based hydrogen-deuterium exchange. *Molecules* **2017**, *22*, 1357. [CrossRef] [PubMed]

71. Lu, H.; Tonge, P.J. Drug-target residence time: Critical information for lead optimization. *Curr. Opin. Chem. Biol.* **2010**, *14*, 467–474. [CrossRef] [PubMed]

72. Núñez, S.; Venhorst, J.; Kruse, C.G. Target–drug interactions: First principles and their application to drug discovery. *Drug Discov. Today* **2012**, *17*, 10–22. [CrossRef] [PubMed]

73. Vauquelin, G. Effects of target binding kinetics on in vivo drug efficacy: Koff, kon and rebinding. *Br. J. Pharmacol.* **2016**, *173*, 2319–2334. [CrossRef] [PubMed]

74. Garbett, N.C.; Chaires, J.B. Thermodynamic studies for drug design and screening. *Expert Opin. Drug Discov.* **2012**, *7*, 299–314. [CrossRef] [PubMed]

75. Velazquez-Campoy, A.; Freire, E. Isothermal titration calorimetry to determine association constants for high-affinity ligands. *Nat. Protoc.* **2006**, *1*, 186–191. [CrossRef] [PubMed]

76. Freire, E. Do enthalpy and entropy distinguish first in class from best in class? *Drug Discov. Today* **2008**, *13*, 869–874. [CrossRef] [PubMed]

77. Bachhawat-Sikder, K.; Thomas, C.J.; Surolia, A. Thermodynamic analysis of the binding of galactose and poly-*N*-acetyllactosamine derivatives to human galectin-3. *FEBS Lett.* **2001**, *500*, 75–79. [CrossRef]

78. Klebe, G. Applying thermodynamic profiling in lead finding and optimization. *Nat. Rev. Drug Discov.* **2015**, *14*, 95–110. [CrossRef] [PubMed]

79. Geschwindner, S.; Ulander, J.; Johansson, P. Ligand binding thermodynamics in drug discovery: Still a hot tip? *J. Med. Chem.* **2015**, *58*, 6321–6335. [CrossRef] [PubMed]

80. Kirk, K.L. Fluorine in medicinal chemistry: Recent therapeutic applications of fluorinated small molecules. *J. Fluorine Chem.* **2006**, *127*, 1013–1029. [CrossRef]

81. Zhou, Y.; Wang, J.; Gu, Z.; Wang, S.; Zhu, W.; Acena, J.L.; Soloshonok, V.A.; Izawa, K.; Liu, H. Next generation of fluorine-containing pharmaceuticals, compounds currently in phase II–III clinical trials of major pharmaceutical companies: New structural trends and therapeutic areas. *Chem. Rev.* **2016**, *116*, 422–518. [CrossRef] [PubMed]

82. Bohm, H.J.; Banner, D.; Bendels, S.; Kansy, M.; Kuhn, B.; Muller, K.; Obst-Sander, U.; Stahl, M. Fluorine in medicinal chemistry. *ChemBioChem* **2004**, *5*, 637–643. [CrossRef] [PubMed]

83. Shah, P.; Westwell, A.D. The role of fluorine in medicinal chemistry. *J. Enzyme Inhib. Med. Chem.* **2007**, *22*, 527–540. [CrossRef] [PubMed]

84. Dalvit, C.; Vulpetti, A. Fluorine–protein interactions and ^{19}F-NMR isotropic chemical shifts: An empirical correlation with implications for drug design. *ChemMedChem* **2011**, *6*, 104–114. [CrossRef] [PubMed]

85. Vulpetti, A.; Dalvit, C. Fluorine local environment: From screening to drug design. *Drug Discov. Today* **2012**, *17*, 890–897. [CrossRef] [PubMed]

86. Hirabayashi, J.; Hashidate, T.; Arata, Y.; Nishi, N.; Nakamura, T.; Hirashima, M.; Urashima, T.; Oka, T.; Futai, M.; Muller, W.E.; Yagi, F.; Kasai, K. Oligosaccharide specificity of galectins: A search by frontal affinity chromatography. *Biochim. Biophys. Acta* **2002**, *1572*, 232–254. [CrossRef]

87. Ideo, H.; Seko, A.; Ishizuka, I.; Yamashita, K. The N-terminal carbohydrate recognition domain of galectin-8 recognizes specific glycosphingolipids with high affinity. *Glycobiology* **2003**, *13*, 713–723. [CrossRef] [PubMed]

88. Ideo, H.; Matsuzaka, T.; Nonaka, T.; Seko, A.; Yamashita, K. Galectin-8-N-domain recognition mechanism for sialylated and sulfated glycans. *J. Biol. Chem.* **2011**, *286*, 11346–11355. [CrossRef] [PubMed]

89. Than, N.G.; Romero, R.; Balogh, A.; Karpati, E.; Mastrolia, S.A.; Staretz-Chacham, O.; Hahn, S.; Erez, O.; Papp, Z.; Kim, C.J. Galectins: Double-edged swords in the cross-roads of pregnancy complications and female reproductive tract inflammation and neoplasia. *J. Pathol. Transl. Med.* **2015**, *49*, 181–208. [CrossRef] [PubMed]

90. Thurston, T.L.; Wandel, M.P.; Von Muhlinen, N.; Foeglein, A.; Randow, F. Galectin 8 targets damaged vesicles for autophagy to defend cells against bacterial invasion. *Nature* **2012**, *482*, 414–418. [CrossRef] [PubMed]

International Journal of
Molecular Sciences

MDPI

Article

Poly-*N*-Acetyllactosamine Neo-Glycoproteins as Nanomolar Ligands of Human Galectin-3: Binding Kinetics and Modeling

Ladislav Bumba [1,†], Dominic Laaf [2,†], Vojtěch Spiwok [3], Lothar Elling [2], Vladimír Křen [1] and Pavla Bojarová [1,*]

[1]   Institute of Microbiology of the Czech Academy of Sciences, Vídeňská 1083, 14220 Prague, Czech Republic;
      bumba@biomed.cas.cz (L.B.); kren@biomed.cas.cz (V.K.)
[2]   Laboratory for Biomaterials, Institute for Biotechnology and Helmholtz-Institute for Biomedical Engineering,
      RWTH Aachen University, Pauwelsstrasse 20, 52074 Aachen, Germany; d.laaf@biotec.rwth-aachen.de (D.L.);
      l.elling@biotec.rwth-aachen.de (L.E.)
[3]   Department of Biochemistry and Microbiology, University of Chemistry and Technology Prague,
      Technická 3, 16628 Prague 6, Czech Republic; Vojtech.Spiwok@vscht.cz
*   Correspondence: bojarova@biomed.cas.cz; Tel.: +420-296-442-510
†   These authors contributed equally to this work.

Received: 15 January 2018; Accepted: 23 January 2018; Published: 26 January 2018

Abstract: Galectin-3 (Gal-3) is recognized as a prognostic marker in several cancer types. Its involvement in tumor development and proliferation makes this lectin a promising target for early cancer diagnosis and anti-cancer therapies. Gal-3 recognizes poly-*N*-acetyllactosamine (LacNAc)-based carbohydrate motifs of glycoproteins and glycolipids with a high specificity for internal LacNAc epitopes. This study analyzes the mode and kinetics of binding of Gal-3 to a series of multivalent neo-glycoproteins presenting complex poly-LacNAc-based oligosaccharide ligands on a scaffold of bovine serum albumin. These neo-glycoproteins rank among the strongest Gal-3 ligands reported, with K_d reaching sub-nanomolar values as determined by surface plasmon resonance. Significant differences in the binding kinetics were observed within the ligand series, showing the tetrasaccharide capped with *N*,*N'*-diacetyllactosamine (LacdiNAc) as the strongest ligand of Gal-3 in this study. A molecular model of the Gal-3 carbohydrate recognition domain with docked oligosaccharide ligands is presented that shows the relations in the binding site at the molecular level. The neo-glycoproteins presented herein may be applied for selective recognition of Gal-3 both on the cell surface and in blood serum.

Keywords: carbohydrate; galectin-3; galectins in diagnosis; galectins in therapy; glycosyltransferase; surface plasmon resonance; molecular modeling

1. Introduction

Galectin-3 (Gal-3) is the only member of the chimeric subgroup of galectins [1] found in vertebrate animals. It is overexpressed in many cancers, e.g., gastric, colorectal, breast tumors, hepatocellular and pancreatic carcinomas, melanomas or glioblastomas. It participates in crucial cancer-related processes: tumorigenesis, metastasis, and neoplasia, angiogenesis, cell adhesion, apoptosis and survival of tumor cells, as well as their immune escape from the host defense system [2–4].

In view of all the tumor promoting effects of Gal-3, the design of synthetic inhibitors of Gal-3 represents the principal strategy in the search for new antitumor drugs and precise cancer diagnostics. High-affinity carbohydrate ligands of Gal-3 are of potential interest in numerous biomedical applications [5]. Though β-galactosides are generally recognized as Gal-3 ligands, we recently found that the terminal *N*,*N'*-diacetyllactosamine (LacdiNAc; GalNAcβ1,4GlcNAc) epitope acts as a selective ligand of Gal-3 compared to galectin-1 [6,7]. The LacdiNAc disaccharide specifically occurs in some *O*-

and *N*-linked mammalian glycoproteins and has specialized functions [8,9]; it has also been identified as a specific glyco-biomarker in several types of cancers [10–12]. Otherwise, it is overexpressed in parasites [13] and other organisms [14]. We demonstrated that the tetrasaccharide composed of a terminal LacdiNAc epitope and an internal LacNAc (GalNAcβ1,4GlcNAcβ1,3Galβ1,4GlcNAc; LacdiNAc-LacNAc) is a superior ligand of Gal-3 when presented in multivalent mode on bovine serum albumin (BSA) as a protein scaffold [15]. In the present work, we evaluate eight oligosaccharide analogs of poly-LacNAc type by surface plasmon resonance studies, of which the LacdiNAc-LacNAc-decorated neo-glycoconjugate exhibits the highest binding affinity in sub-nanomolar range.

Gal-3 is composed of a C-terminal highly conserved carbohydrate recognition domain (CRD) which binds glycan ligands, and an N-terminal non-lectin domain rich in proline and glycine tandem repeats. The N-terminal domain contains a serine phosphorylation site responsible for the oligomerization and cross-linking activity of Gal-3 [16]. The CRD of Gal-3 comprises 135 amino acids, which harbor a specific binding groove of a total of five subsites (A–E) (Figure S1). This binding groove is able to accommodate up to a tetrasaccharide [17]. The galactose unit of lactose (Galβ1,4Glc) or LacNAc (Galβ1,4GlcNAc) binds to the most conserved subsite C whereas the reducing end Glc(NAc) occupies the second most important subsite D; together they form the conserved binding pocket of the binding site (β-strands S4–S6). Subsites A and B extend beyond the C-3 of galactose and constitute the less conserved binding pocket of the CRD (β-strands S1–S3). The non-conserved and generally less defined subsite E reaches beyond the reducing end of the tetrasaccharide and may interact with moieties attached to the C-1 of Glc(NAc) [18,19]. The eight conserved amino acid residues in the CRD of Gal-3 (Arg144, His158, Asn160, Arg162, Asn174, Trp181, Glu184 and Arg186), and Asp148 then coin the binding specificity for particular carbohydrate ligands and provide the main interactions with the bound ligand in the form of hydrogen bonds and van der Waals interactions [20]. In the present paper we show molecular dynamics simulations of LacdiNAc-LacNAc (**3**) and LacNAc-LacNAc (**4**) tetrasaccharides (the best ligands in the series) in the binding site of Gal-3 and compare them with LacdiNAc (**1**) and LacNAc (**2**) disaccharides (the worst ligands in the series).

Gal-3 is monomeric in solution, though small amounts of oligomeric species have been detected at high concentrations [21]. Several mechanisms have been proposed concerning the self-association and oligomerization of Gal-3 subunits or even lattice formation [22] and precipitation upon contact with multivalent ligands such as laminin [23], asialofetuin (ASF) [24], with synthetic multivalent carbohydrates [25] or even complex monovalent glycans such as lacto-*N*-neotetraose [26].

In sum, the present study reveals the binding kinetics of the strongest multivalent neo-glycoprotein ligands of Gal-3 ever reported and the prominent importance of the terminal LacdiNAc epitope in the tetrasaccharide glycan. The sub-nanomolar affinities of ligands to Gal-3 were determined by surface plasmon resonance (SPR). The novel SPR design with immobilized Gal-3-AVI construct (Gal-3 containing AviTag peptide sequence) shown here represents the optimum approach for measuring kinetics with Gal-3 thanks to its fully maintained flexibility. Moreover, the reversed SPR setup with immobilized neo-glycoprotein ligands provides clues pertaining to the Gal-3 oligomerization. Molecular docking of selected glycan ligands in the Gal-3 CRD discloses the relations in the Gal-3 binding site and relevant lectin-ligand interactions.

2. Results

2.1. Preparation of functionalized Poly-LacNAc glycans **1–8** and Neo-Glycoproteins **9–16**

Oligosaccharide glycans **1–8** (Figure 1) carrying a *t*-Boc-protected thioureido linker at the reducing end were prepared as described previously [7,15,27]. The sequential preparative reactions employed a library of tailored glycosyltransferases. The human β4-galactosyltransferase (β4GalT), and the *Helicobacter pylori* β3-*N*-acetylglucosaminyltransferase (β3GlcNAcT) were used for the synthesis of poly-LacNAc type 2 (Galβ4GlcNAc)$_n$, the mutant human β4-galactosaminyltransferase (β4GalTY284L) for the preparation of LacdiNAc (GalNAcβ4GlcNAc; **1**, **3**, **5**), the *E. coli* β3-galactosyltransferase (β3GalT) for the generation of LacNAc type 1 (Galβ3GlcNAc; **7**), and the murine α3-galactosyltransferase (α3GalT) for the synthesis of Galili (Galα3Galβ4GlcNAc; **8**) (Scheme S1). The structure and purity of glycans **1–8** (Figure 1) were confirmed by HPLC-ESI-MS and NMR (Figures S2 and S3) [7,15,27].

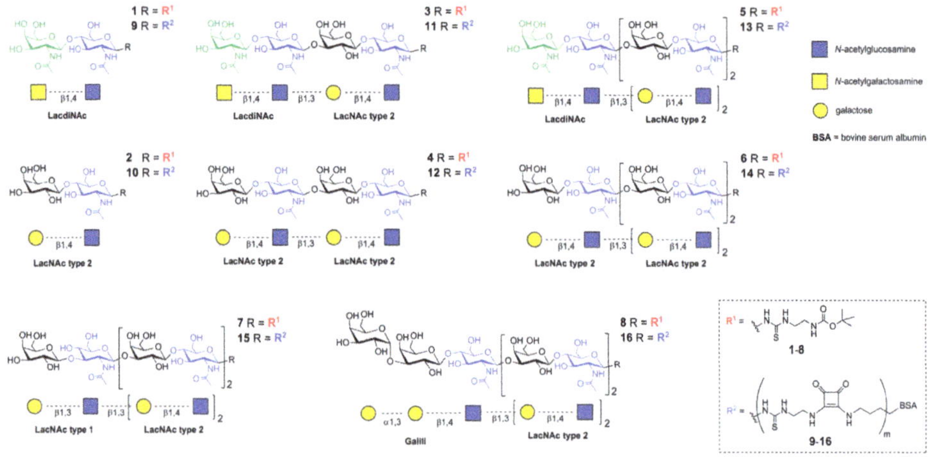

Figure 1. Schematic representation of di- (**1,2**), tetra- (**3,4**), hexa- (**5–7**), and heptasaccharide (**8**) glycans used for covalent modification of BSA to yield respective neo-glycoproteins **9–16** [7,15,27].

Neo-glycoproteins **9–16** were prepared by conjugating the respective amino-functionalized glycans (**1–8**) to free lysine residues of bovine serum albumin (BSA) via two-step amidation using diethyl squarate (3,4-diethoxy-3-cyclobutene-1,2-dione) as described previously [7,15,27] (Scheme S2). The integrity of neo-glycoproteins was checked by SDS-PAGE (Figure S4).

2.2. Binding Properties of Glycans **1–8** and Neo-Glycoproteins **9–16** to Gal-3 in ELISA Assay

A soluble His-tagged construct of human galectin-3 (Gal-3) was expressed in *E. coli* and purified by immobilized metal-ion affinity chromatography as described before [27]. The binding properties and inhibition parameters of glycans **1–8** and respective neo-glycoproteins **9–16** were compared using enzyme-linked immunosorbent assays (ELISA).

To determine the binding affinities between Gal-3 and neo-glycoproteins **9–16**, the binding of soluble Gal-3 to the neo-glycoproteins immobilized in microplate wells was quantified by colorimetric immunodetection using anti-His antibody conjugated to horseradish peroxidase (HRP) (Figure 2a) [7,27]. Gal-3 bound neo-glycoproteins in a concentration-dependent manner and the interaction was exclusively conferred by glycan moieties since no binding of Gal-3 to the glycan-free BSA was detected. Apparent dissociation constants (K_d) of the Gal-3/neo-glycoprotein complexes were calculated from a non-linear regression of binding curves. As documented in Table 1, the K_d values were found to range between 30 and 700 nM, except for the neo-glycoproteins **9** and **10**, whose binding affinities were in micromolar concentrations.

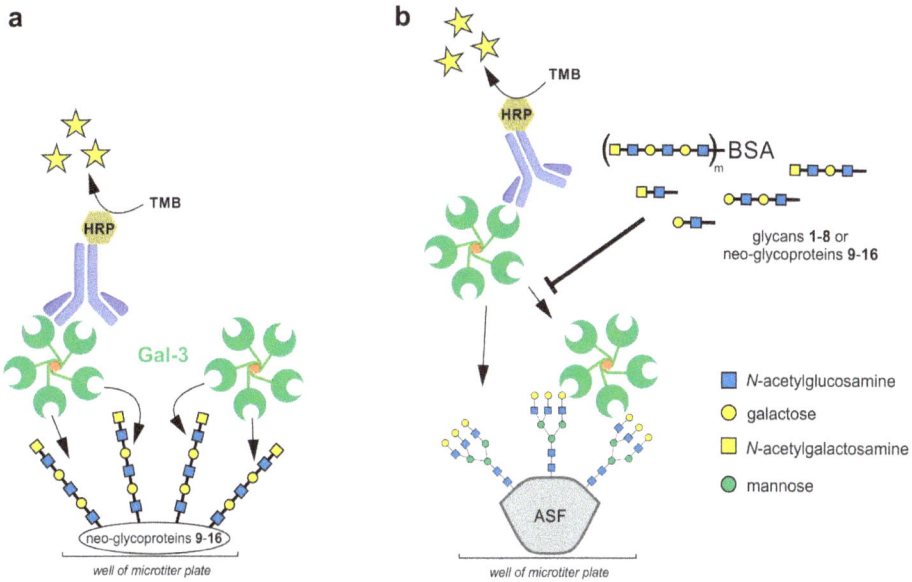

Figure 2. ELISA assays used in the study. (**a**) Direct ELISA assay with immobilized neo-glycoproteins **9–16**; (**b**) Competitive ELISA assay using glycans **1–8** or neo-glycoproteins **9–16** as competing ligands for the inhibition of binding of Gal-3 to immobilized asialofetuin (ASF). The proposed Gal-3 oligomer structure is based on previous reports [25]. Horseradish peroxidase (HRP)-conjugated antibody was used for the detection of bound Gal-3. The HRP converted the added 3,3′,5,5′-tetramethylbenzidine (TMB) to obtain a photometric signal.

Table 1. Binding properties of glycans **1–8** and respective neo-glycoproteins **9–16** in ELISA assay.

Neo-Glycoprotein (Respective Glycan) [a]	M_W (kDa)	m [b]	IC_{50} Glycan (µM) [c]	IC_{50} Neo-Glycoprotein (nM)	r_p [d]	r_p/m [e]	K_d Neo-Glycoprotein (nM)
9 (glycan 1)	76.4	17	42 ± 2	2026 ± 147	20.7	1.2	6290 ± 530
10 (glycan 2)	78.2	18	36 ± 1	344 ± 38	104.7	5.8	4780 ± 1240
11 (glycan 3)	87.4	21	7 ± 1 [f]	11 ± 2	642.9	30.6	30 ± 4 [f]
12 (glycan 4)	84.7	19	13 ± 3 [f]	31 ± 1	419.4	22.1	300 ± 60 [f]
13 (glycan 5)	86.6	16	6.20 ± 0.02 [g]	37 ± 7 [g]	169.9	10.6	76 ± 19 [g]
14 (glycan 6)	86.6	17	20.1 ± 0.1 [g]	76 ± 8 [g]	263.4	15.5	350 ± 110 [g]
15 (glycan 7)	85.9	17	12.5 ± 0.1 [g]	212 ± 22 [g]	58.9	3.5	700 ± 100 [g]
16 (glycan 8)	89.6	18	8.4 ± 0.1 [g]	65 ± 4 [g]	128.2	7.1	290 ± 90 [g]

[a] The numbers in parentheses indicate the compound numbers of respective glycans bound on the neo-glycoprotein; [b] Average number of glycans per BSA molecule; [c] *cf.* IC_{50} (lactose) = 137 ± 27 µM; [d] Relative potency, i.e., IC_{50} (monovalent glycan)/IC_{50} (multivalent neo-glycoprotein); [e] Relative potency per glycan bound on BSA; [f] The data were adopted from our previous study [15]; [g] The data were adopted from our previous study [27].

To assess the inhibitory potential of glycans **1–8** and the respective neo-glycoproteins **9–16** towards Gal-3, their capacity to inhibit the binding of Gal-3 to immobilized asialofetuin (ASF) was determined by competitive ELISA inhibition analyses (Figure 2b) [7,15,27]. ASF is a multivalent glycoprotein presenting three triantennary *N*-glycans terminated with LacNAc, which is reported to interact with Gal-3. Here, Gal-3 was incubated with increasing concentrations of the glycans or neo-glycoproteins as competing ligands and the inhibition of Gal-3 binding to immobilized ASF was quantified by colorimetric immunodetection using anti-His antibody conjugated to HRP. Lactose was utilized as a positive control. The respective inhibition constants (IC_{50}) were calculated from the non-linear regression of the sigmoidal inhibition curves and they are listed in Table 1.

As shown in Table 1, the inhibitory potencies of individual glycans to Gal-3, expressed as IC_{50}, range between 42 and 6.2 µM, with the Galili- (**8**) and especially LacdiNAc capped glycans (**3**, **5**) being the most potent monovalent glycan inhibitors. In comparison, lactose, used as a positive control, had an IC_{50} value of 137 µM. The presence of one (LacNAc, **2**) or two (LacdiNAc, **1**) acetamido groups increased the inhibitory potency up to 4 times compared to lactose. LacdiNAc disaccharide **1** was a slightly less efficient inhibitor than LacNAc (**2**; Figure S5, Table 1). However, LacdiNAc was previously identified as a highly selective inhibitor for Gal-3 compared to other galectins abundant in vivo, in particular galectin-1 [6]. This fact provides an advantage in the development of selective inhibitors for diagnostic and/or specific therapeutic applications.

Neo-glycoproteins **9** and **10** carrying the disaccharides LacdiNAc (**1**) and LacNAc (**2**), respectively, were identified as the least efficient inhibitors (Figure S6). In contrast, the IC_{50} values for neo-glycoproteins **11–16** were found to be in low nanomolar range, which reflects a significant increase in their capacity to inhibit Gal-3 binding to ASF. The best neo-glycoprotein ligands in this series are compounds **11** (with tetrasaccharides) and **13** (with hexasaccharides) carrying the terminal LacdiNAc-LacNAc epitope. The terminal LacNAc-LacNAc motif in compounds **12** and **14** is also a good ligand but the inhibition strength is reduced by a factor of 2-3 (*cf.* **11** vs. **12** and **13** vs. **14**), presumably due to the binding preference of the non-conserved pocket of Gal-3 CRD (β-sheets S1–S3) for LacdiNAc as shown further.

The effect of multivalent ligand presentation is reflected in the comparison of the inhibitory potencies of multivalent conjugates with the respective monovalent glycans under consideration of the individual glycan density (relative inhibitory potency per glycan, r_p/m, *cf.* Table 1). All neo-glycoproteins except for **1** were shown to exhibit a considerable cluster glycoside effect [28] due to multivalent glycan presentation. Notably, neo-glycoprotein **9** carrying LacdiNAc disaccharide (**1**) did not induce any multivalent effect at all. We conclude that LacdiNAc disaccharide alone, without the LacNAc attachment at the reducing end, is not suitable for a proper interaction with the Gal-3 CRD and for induction of Gal-3 oligomerization. On the contrary, neo-glycoprotein **10** (carrying LacNAc disaccharide) showed a relative inhibitory potency per glycan of 5.8, which confirms the binding specificity of the conserved binding pocket of Gal-3 CRD (β-sheets S4–S6) for LacNAc and the induction of multivalence effect.

2.3. Binding Kinetics of Neo-Glycoproteins **9–16** with Gal-3 Determined by Surface Plasmon Resonance

The kinetics of the interaction of neo-glycoproteins **9–16** with Gal-3 were studied by surface plasmon resonance (SPR). This technique measures biomolecular interactions in real-time in a label free environment, where one of the interactants is immobilized to the sensor surface, and the other passes free in solution over the surface as analyte. Several experimental approaches were examined to find optimal conditions for the evaluation of interactions of the tested neo-glycoproteins with Gal-3. To assess ASF binding to immobilized Gal-3, the recombinant His-tagged Gal-3 protein was either covalently immobilized to a carboxylated surface of the GLC (General Layer Chemistry) sensor chip by amine coupling chemistry through its lysine residues or captured by a Ni^{2+}-nitrilotriacetate (Ni-NTA) surface through its polyhistidine tag, and ten-fold dilutions of ASF (0.01–10 µM) were injected over the sensor surface. Surprisingly, no SPR response was observed after repeated injections of ASF on the Gal-3 surface in either case, indicating that the covalent immobilization of Gal-3 to the sensor chip completely abolishes its lectin activity. Moreover, the interaction of ASF with Gal-3 captured to the Ni-NTA surface was burdened by a high nonspecific binding of ASF to the sensor chip, which prevented a detailed characterization of the interaction between ASF and Gal-3. Further optimization of the experimental protocols did not improve the quality of the data, showing that the His-tagged Gal-3 protein could not be used as a ligand in the SPR interaction studies using either of these experimental approaches.

To overcome the difficulties with Gal-3 immobilization, we prepared a biotinylated version of Gal-3 through in vivo biotinylation of an AviTag peptide that was genetically fused to the

N-terminus of Gal-3 (Gal-3-AVI) (Figure 3). The AviTag is a specific 15-amino acid peptide sequence (GLNDIFEAQKIEWHE) that directs a highly targeted enzymatic conjugation of a single biotin molecule to the specific lysine (**K**) residue within the AviTag sequence using biotin ligase (BirA). In contrast to chemical biotinylation, which usually generates heterogeneous products with impaired function, the co-translational biotinylation of the AviTag peptide is site specific and provides a highly homogeneous protein preparation. Moreover, the N-terminal localization of the AviTag peptide provides a favorable orientation of Gal-3-AVI on a streptavidin-coated surface, leaving the C-terminal carbohydrate-binding domain of Gal-3 freely accessible for binding interactions. Hence, the Gal-3-AVI protein was co-expressed with BirA in *E. coli* and purified by using the Ni-chelating affinity chromatography. Western blot analysis of the purified Gal-3-AVI showed a single protein band recognized by anti-biotin antibody, indicating the covalent attachment of biotin to Gal-3-AVI (Figure 3). As shown in Table S1, the binding properties of the biotinylated Gal-3-AVI protein were comparable to those of the original Gal-3 construct, documenting that the biotinylated AviTag peptide does not interfere with the lectin activity of Gal-3.

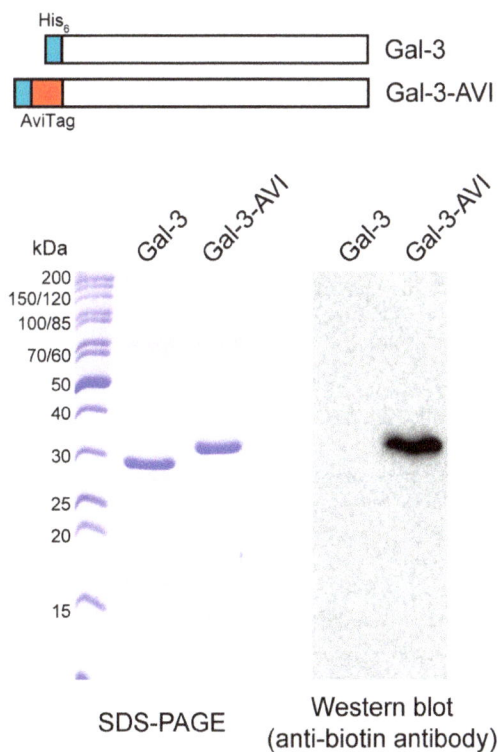

Figure 3. SDS-PAGE and Western blot analyses of the Gal-3-AVI construct.

For SPR analysis, the biotinylated Gal-3-AVI was captured on a neutravidin-coated sensor chip, and the binding of serially diluted neo-glycoproteins **9–16** to immobilized Gal-3-AVI was analyzed. Three coupling concentrations of Gal-3-AVI were tested, so that the effect of mass transfer and the non-specific binding were minimized. As a result, the coupling level of 250 relative units (RU) along with the flow rate of 30 μL/min gave the optimum conditions and these parameters were used for the subsequent experiments. Real time kinetics of the interactions of neo-glycoproteins with immobilized Gal-3-AVI are shown in Figure 4.

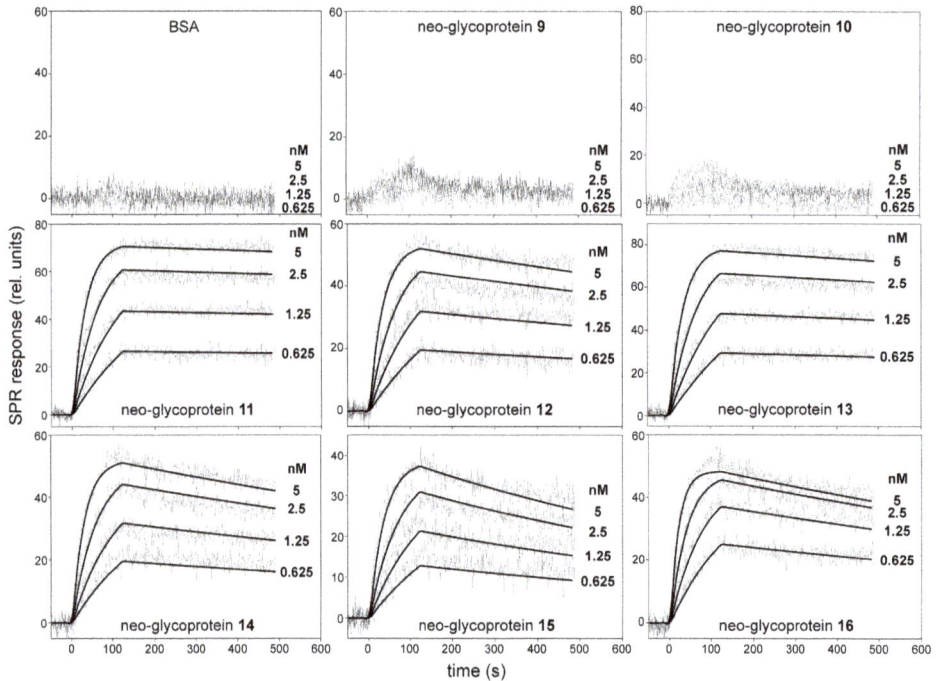

Figure 4. SPR kinetic binding analysis of the interactions between neo-glycoproteins **9–16** and immobilized Gal-3-AVI. The neo-glycoproteins at indicated concentrations were injected in parallel over the neutravidin-coated sensor chip coated with the biotinylated Gal-3-AVI at a flow rate of 30 μL/min. The kinetic data were globally fitted by using a 1:1 Langmuir binding model. The fitted curves are superimposed as thin black lines on top of the sensograms. BSA, bovine serum albumin; rel., relative.

The concentration-dependent binding curves revealed significant differences in the intensity of SPR responses between the neo-glycoproteins carrying the disaccharide (**9**, **10**) and tetra- to heptasaccharide glycans (**11–16**). The binding of all tested neo-glycoproteins was exclusively conferred by a glycan moiety since no binding of the glycan-free BSA to Gal-3-AVI was detected. As shown in Figure 4, injection of neo-glycoproteins **9** and **10** at nanomolar concentrations yielded a very low SPR response, indicating that these conjugates carrying disaccharide glycans are rather low-affinity Gal-3 ligands. In contrast, the interaction of neo-glycoproteins **11–16** carrying tetra- to heptasaccharide glycans at nanomolar concentrations exhibited typical concentration-dependent binding curves, maintaining both association and dissociation phase of the sensograms. The binding data fitted well to a simple 1:1 Langmuir binding model and the calculated association and dissociation rate constants (k_a and k_d, respectively) for the interactions between immobilized Gal-3-AVI and the tested neo-glycoproteins are listed in Table 2. The data show that binding affinities (K_D) of Gal-3 to **11–16** range in subnanomolar concentrations (10^{-10}–10^{-11} M). The LacdiNAc-capped neo-glycoproteins **11** (tetrasaccharide-decorated) and **13** (hexasaccharide-decorated) were identified as the strongest Gal-3 ligands in the series (K_D of 14 pM and 26 pM, respectively). Such a high binding affinity appears to be due to a very slow dissociation rate, particularly in the **11**/Gal-3 complex ($k_d = 8.5 \times 10^{-5} \cdot \text{s}^{-1}$) since the association rates of the interaction between the neo-glycoproteins and Gal-3 are comparable. In contrast, the interaction of Gal-3 with neo-glycoprotein **15** carrying a hexasaccharide capped with LacNAc type 1 epitope was characterized by a fast dissociation rate of the complex ($k_d = 1.3 \times 10^{-3} \cdot \text{s}^{-1}$), yielding a significant decrease of the K_D value to 270 pM. In comparison, the standard ligand ASF featured a

K_D of 8.3 nM (Table 2 and Figure S7), i.e., it had almost 600-times lower affinity to Gal-3-AVI than **11**. The found K_D values are by far the lowest values ever determined for Gal-3 by SPR [29,30] and are in the range of affinities of monoclonal antibodies. The K_D for ASF (positive control) correlates well with the value determined previously by isothermal titration calorimetry (7.14 nM) [31]. Therefore, we conclude that the new approach to immobilization of Gal-3 on the sensor chip via the AviTag-biotin-neutravidin spacer presented here is the most appropriate experimental design for measuring kinetics of the interaction of Gal-3 and carbohydrate ligands thanks to the fully maintained flexibility of Gal-3 on the chip surface. All other experimental approaches give underestimated results, probably due to steric hindrance and partial blocking of the Gal-3 binding site.

Table 2. Kinetic affinity constants for the interactions of neo-glycoconjugates **9–16** with immobilized Gal-3-AVI.

Compound	Attached Glycan [a]	k_a ($M^{-1} \cdot s^{-1}$)	k_d (s^{-1})	K_D (M)
9	LacdiNAc	N.D.	N.D.	N.D.
10	LacNAc	N.D.	N.D.	N.D.
11	LacdiNAc-LacNAc	$(5.9 \pm 1.3) \times 10^6$	$(8.5 \pm 2.5) \times 10^{-5}$	$(1.4 \pm 0.4) \times 10^{-11}$
12	LacNAc-LacNAc	$(6.2 \pm 2.1) \times 10^6$	$(4.2 \pm 3.1) \times 10^{-4}$	$(6.8 \pm 3.9) \times 10^{-11}$
13	LacdiNAc-LacNAc-LacNAc	$(5.8 \pm 2.2) \times 10^6$	$(1.5 \pm 0.4) \times 10^{-4}$	$(2.6 \pm 0.9) \times 10^{-11}$
14	LacNAc-LacNAc-LacNAc	$(7.8 \pm 2.9) \times 10^6$	$(5.1 \pm 2.8) \times 10^{-4}$	$(6.5 \pm 2.6) \times 10^{-11}$
15	LacNAc type 1-LacNAc-LacNAc	$(4.9 \pm 1.8) \times 10^6$	$(1.3 \pm 0.3) \times 10^{-3}$	$(2.7 \pm 1.0) \times 10^{-10}$
16	Galili-LacNAc-LacNAc	$(9.3 \pm 3.1) \times 10^6$	$(6.3 \pm 2.0) \times 10^{-4}$	$(6.8 \pm 3.5) \times 10^{-11}$
ASF	Positive control standard	$(4.8 \pm 1.9) \times 10^4$	$(4.0 \pm 1.7) \times 10^{-4}$	$(8.3 \pm 3.1) \times 10^{-9}$

[a] Structures of glycans attached to respective neo-glycoproteins are depicted in Figure 1. N.D., not determined.

To gain further information on the affinity of neo-glycoproteins **9–16** to Gal-3, the kinetics of their interaction was determined by SPR in the reversed setup. Neo-glycoproteins **9–16** were immobilized at a very low density (~300 RU) on a carboxylated surface of the sensor chip and serial dilutions of His$_6$-tag Gal-3, prepared in the same way as for the ELISA experiments, were injected over the sensor surface. Initial binding experiments revealed that low concentrations of Gal-3 (<10 nM) did not yield any SPR response, indicating that binding affinity of Gal-3 to the immobilized neo-glycoproteins is significantly decreased as compared to that obtained for binding of neo-glycoproteins to the immobilized Gal-3-AVI. Higher concentrations of Gal-3 (16–250 nM) resulted in typical concentration-dependent binding curves, except for interaction of Gal-3 with the immobilized disaccharide-carrying neo-glycoproteins **9** and **10** where no binding was detected even at high Gal-3 concentrations (10 µM) (Figure S8). Kinetic parameters of interactions between Gal-3 and neo-glycoproteins **11–16** were calculated from global fitting of the binding curves. The data were fitted to several kinetic models, such as 1:1 Langmuir-type, bivalent analyte, heterogeneous ligand and heterogeneous analyte binding models, but only the heterogeneous ligand model provided reasonable fits ("ligand" in this sense refers to the binding partner immobilized on the chip surface). Further optimization of the experimental layout (changing the ligand density, flow rates, etc.) did not improve the quality of the fits and the heterogeneous ligand model was used for final evaluation of the binding curves (Figure S8). This model accounts for binding of an analyte to two ligand species, which may represent either two different molecules or two different (types of) binding sites on the same ligand molecule. Thus, the interaction can be described by complex kinetics involving two separate binding events. The calculated association and dissociation rate constants (k_{a1}, k_{d1} and k_{a2}, k_{d2}) are listed in Table S2. The data showed that the interactions of immobilized neo-glycoproteins **11–16** with Gal-3 are characterized by a set of two binding affinity constants, where the former (K_{D1}) range in submicromolar (10^{-7}) and the latter (K_{D2}) in nanomolar (10^{-8}) concentrations. The difference lies mainly in the dissociation rate constants (cf. k_{d1} and k_{d2}) whereas the association rate constants are rather similar. The K_D values are very similar throughout the set of neo-glycoconjugates and generally follow the trends observed in ELISA assay and SPR setup with Gal-3-AVI (cf. Tables 1 and 2).

However, the binding affinities of immobilized neo-glycoproteins to Gal-3 are much lower than those obtained from the measurements with immobilized Gal-3-AVI, and, most likely, they do not reflect the real interaction data. Moreover, binding of Gal-3 to multivalent glyco-ligands is known to provoke spontaneous Gal-3 oligomerization [24,26], which complicates the use of Gal-3 as an analyte in the SPR studies. Taken together, these data clearly suggest that the use of biotinylated Gal-3-AVI is a much more suitable approach for probing the binding activity of Gal-3 by SPR.

2.4. Molecular Dynamics of Disaccharide (**1,2**) and Tetrasaccharide (**3,4**) Glycans in Gal-3 CRD

We performed molecular dynamics simulations in order to describe the differences in affinities between disaccharide ligands **1** and **2** carried by neo-glycoconjugates **9** and **10**, and ligands **3** and **4** carried by **11** and **12**, which represent the weakest and the strongest pair of ligands in the series, respectively. For this aim, we built a molecular model of Gal-3 CRD based on the experimentally determined crystal structure of Gal-3 CRD in complex with lacto-*N*-neotetraose, Galβ1,4GlcNAcβ1,3Galβ1,4Glc (PDB ID: 4LBN [17]). This is one of the only two known crystal structures of wild-type Gal-3 CRD complexed with an oligosaccharide ligand (PDB ID: 4LBN, 4LBM [17]) and out of the two ligands published, the structure of lacto-*N*-neotetraose is more closely related to glycans **3** and **4**, differing basically in the 2-acetamido group at the reducing-end GlcNAc. So far, all crystallization attempts of Gal-3 were limited only to the carbohydrate binding domain as the N-terminal domain is too flexible and therefore difficult to crystallize [20]. Moreover, the N-terminal domain is known to participate in the formation of the quaternary Gal-3 structure, contributing to the Gal-3 oligomerization and avidity [23], whereas it is less important for the very ligand binding [26].

All simulations were done in GROMACS 5.1.3 [32] and the system was simulated in Amber99SB-ILDN force field [33]. Glycan ligand topologies were generated using GLYCAM web server (Glycam 06) [34] and converted to GROMACS by ACPYPE [35]. For the modeling purposes, we used glycans **1**–**4** truncated to the saccharide part (without the C-1 linker) since general force fields used for modeling of organic compounds are not appropriate for carbohydrates, and it was necessary to use a carbohydrate-tailored force field for the Glycam-generated topologies. Generally, it is not excluded to combine a carbohydrate and general force field for one molecule but this procedure is not without risks of possible artefacts, and our preliminary study showed the disadvantages of this approach.

Our initial attempts for docking of glycans **1**–**4** by PLANTS 1.2 software [36] did not give satisfactory results for the terminal LacNAc domain in substrates **2** and **4** and also for lacto-*N*-neotetraose used as a control ligand, probably due to too many degrees of freedom for the position and rotation of the terminal non-reducing galactose unit. Thus, the protein-glycan complexes acquired by this approach were not suitable as a starting point for molecular dynamics simulations. Therefore, in order to test the binding modes of compounds **1**–**4**, we opted for manual docking of **1**–**4** into the Gal-3 CRD, followed by molecular dynamics simulation (30 ns) for each complex.

All four simulated complexes were stable during simulations. The main glycan-protein interactions, such as the stacking interaction to Trp181, and the hydrogen bond to His158, were retained throughout the whole simulations. Figure 5 shows snapshots at the end of respective simulations for glycans **1**–**4** with Gal-3 CRD (PDB ID: 4LBN), and for lacto-*N*-neotetraose for comparison. Both disaccharidic ligands **1** and **2** bind to the binding groove (conserved binding pocket of subsites C, D) in perfect agreement with the experimentally determined complexes with lactose (PDB ID: 3ZSJ [37]), lacto-*N*-neotetraose (PDB ID: 4LBN [17]), and lacto-*N*-tetraose (PDB ID: 4LBM [17]). The simulation shows that the *N*-acetamido moiety at the terminal galactoside residue in **1** (GalNAcβ1,4GlcNAc) does not form any direct hydrogen bonds with the protein and it rather protrudes in the direction out of the binding site (Figure 5). This may explain the fact that binding affinities of disaccharides **1** and **2** are comparable as determined in ELISA assays (Table 1). A higher affinity for LacNAc over lactose is caused by stacking of the reducing-end GlcNAc *N*-acetamido moiety against the side chain of Arg186.

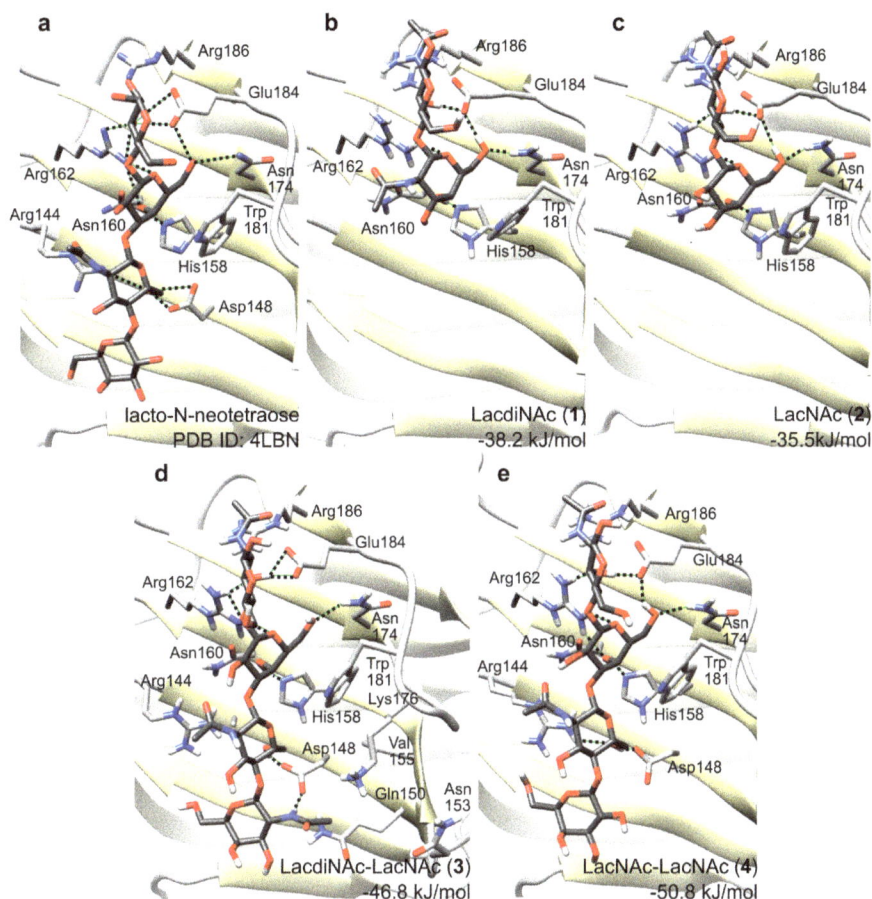

Figure 5. Complexes of glycans **1**–**4** in Gal-3 CRD after 30 ns molecular dynamics simulation. (**a**) Experimentally determined structure of the complex with lacto-*N*-neotetraose (PBD ID: 4LBN) [17]; (**b**) LacdiNAc disaccharide ligand **1**; (**c**) LacNAc disaccharide ligand **2**; (**d**) LacdiNAc-LacNAc tetrasaccharide ligand **3**; (**e**) LacNAc-LacNAc tetrasaccharide ligand **4**. The amino acid residues important for the glycan binding are depicted in stick representations, with oxygens in red and nitrogens in blue.

The binding free energies for glycan ligands **1**–**4** were predicted by Linear Interaction Energy method as −38.2, −35.5, −46.8 and −50.8 kJ/mol, respectively. These values tend to overestimate binding, and an accurate prediction would require tuning of the scaling factors of Linear Interaction Energy method using a large series of ligands; nevertheless, they follow the trend of binding free energies with approx. 4 kJ/mol accuracy. Binding of **3** and **4** involves additional interactions absent in the binding of **1** or **2**, which occur between the non-reducing end disaccharide GalNAcβ1,4GlcNAc bound in the second, less conserved binding pocket (subsites A, B), and mainly with Arg144, Asp148 as well as water-mediated interactions. This fact explains a significantly higher affinity of Gal-3 CRD for the tetrasaccharide ligands. This is also reflected in the binding free energies, which are significantly lower for tetrasaccharides **3** and **4** compared to disaccharides **1** and **2**. The *N*-acetamido moiety of GalNAc in **3** binds into the pocket formed by residues Gln150, Asn153, Val155 and Lys176 and it makes a hydrogen bond with one oxygen of Asp148 whereas the second oxygen binds the 6-hydroxyl group of the GlcNAc residue. These additional interactions may explain a higher affinity towards **3** relative to **4**, which does not contain this acetamido moiety.

3. Discussion

The present study on various structural aspects of multivalent poly-LacNAc-type glycans with respect to the binding and inhibition of Gal-3 clearly highlights the tetrasaccharide LacdiNAc-LacNAc (**3**) as the optimum ligand for Gal-3. For efficient Gal-3 binding, the interaction with longer oligosaccharides is required as suggested previously [17,38,39]. Not only are longer glycans sterically more accessible for lectin interaction [7] but, more importantly, the unique poly-LacNAc-type pattern specifically binds to both the conserved and non-conserved parts of the binding groove on the Gal-3 CRD. Thus, both the first (β-strands S4–S6) and second (β-strands S1–S3) binding pockets are occupied by LacNAc or related structures (Figure 1 and Table 1) as hypothesized previously [17,27,39,40]. The advantage of the tetrasaccharide ligand **4** is in the optimum combination of binding of the LacNAc part in the conserved first binding pocket (β-strands S4–S6) and of LacdiNAc part in the non-conserved second binding pocket of the CRD (β-strands S1–S3). The terminal GalNAc moiety of **3** exhibits additional hydrogen bonding with Asp148, which apparently stabilizes glycan **3** in the Gal-3 binding groove relative to **4**. Binding kinetics data clearly support this theory since they show that the difference between the LacdiNAc-capped glycan **3** and its LacNAc counterpart **4** is not given by a better recognition of the glycan but by its better stabilization in the binding site (k_a rate constants are similar for both neo-glycoproteins **11** and **12** whereas k_d is almost 5 times lower for **11**; *cf.* Table 2).

A detailed comparison of found K_D values (Table 2) reveals that the binding properties of the neo-glycoproteins carrying the same epitope are relatively similar for the tetrasaccharide- and hexasaccharide-carrying neo-glycoproteins (*cf.* Table 2, **11** vs. **13**, and **12** vs. **14**). Gal-3 is stated to primarily recognize the internal LacNAc motifs [41,42]. Therefore, the LacdiNAc-LacNAc-LacNAc hexasaccharide on neo-glycoprotein **13** can be bound in two modes, either involving the terminal LacdiNAc-LacNAc or only the internal LacNAc-LacNAc. We suggest that this fact accounts for the 1.8-times lower affinity (K_D) of Gal-3 to **13** compared to **11**, since **13** allows also the less efficient LacNAc-LacNAc binding mode.

Another noteworthy result is the huge difference in binding kinetics observed between the individual SPR setups. The binding kinetics using immobilized Gal-3-AVI is clearly described by 1:1 Langmuir-type model, correctly mirroring the fact that the immobilized Gal-3 maintained its monomeric form throughout the interaction. In contrast, the reversed setup with Gal-3 in solution states a much more complex binding kinetics described by heterogenous ligand model. When consulting the literature, a similar complex behavior was observed by SPR when Gal-3 interacted with some multivalent macromolecular ligands such as laminin [43]. Laminin is a complex glycoprotein carrying poly-LacNAc chains [44]. In contrast, SPR with immobilized monovalent low-molecular ligands such as lactosides fit the simple 1:1 binding model [45,46]. Taking into account that complex, especially multivalent glyco-ligands are known to incite Gal-3 oligomerization by several described mechanisms [24,26], we hypothesize that the heterogenous nature of the interaction of immobilized neo-glycoconjugates with Gal-3 in solution might reflect the oligomerization of Gal-3 monomers, triggered by binding to these multivalent ligands. A more detailed experimental analysis of this process is yet to be performed. The considerably weaker interaction found in the reversed setup (nevertheless, comparable with other SPR setups for Gal-3 found in the literature) may be caused by a partial steric obstruction of some glycans presented on the immobilized multivalent BSA scaffold, as a result of the non-directed (random) immobilization procedure. In contrast, our novel SPR setup using Gal-3-AVI exploits the advantage of the fully flexible attachment via biotin-AviTag.

4. Materials and Methods

4.1. Synthesis of Glycans **1–8** and Neo-Glycoproteins **9–16**

The recombinant glycosyltransferases used in this study were produced in *E. coli* and purified by affinity chromatography as described earlier [15,27,47–49]. Functionalized glycans **1–8** were synthesized from (*tert*-butoxycarbonylamino)ethylthioureidyl 2-acetamido-2-deoxy-β-D-glucopyranoside [50] in high-yielding sequential preparative reactions employing a series of glycosyltransferases as described previously [27,51]. The synthetic protocol (Scheme S1) for the preparation of **1–8** can be found in the Supplementary Materials. The synthesis of neo-glycoproteins **9–16** was performed as reported earlier [7,15,27]. Briefly, the amino-functionalized glycans **1–8** were deprotected and conjugated with 3,4-diethoxy-3-cyclobutene-1,2-dione (diethyl squarate) to yield squarate monoamide esters, which were isolated by preparative HPLC. Then, the amidation of free lysine residues of BSA with and subsequent purification by preparative HPLC yielded neo-glycoproteins **9–16**. Detailed information on syntheses including Scheme S2 can be found in the Supplementary Materials.

4.2. Protein Production

Asialofetuin was purchased from Sigma-Aldrich (Steinheim, Germany), dissolved in phosphate-buffered saline (PBS) at 1 mg/mL and stored at aliquots at $-20\,°C$. The recombinant human His_6-tagged galectin-3 protein (Gal-3) was produced and purified as described earlier [15,27,30,49]; the experimental details are included in the Supplementary Materials. For the expression of the biotinylated Gal-3-AVI protein, the AviTag sequence (GLNDIFEAQKIEWHE) was genetically introduced into the plasmid construct encoding the His_6-tagged Gal-3 (pETDuet1-Gal-3). A pair of synthetic oligonucleotides (5′-GATCCGGGTCTGAACGACATCTTCGAGGCTCAGAAAATCGAAT GGCACGAAGGTGGATCTGGTGGATCTGCG-3′ and 5′-AATTCGCAGATCCACCAGATCCACCTT CGTGCCATTCGATTTTCTGAGCCTCGAAGATGTCGTTCAGACCCG-3′) was annealed to give the double-stranded DNA fragment carrying the BamHI and EcoRI overhangs and cloned into the BamHI/EcoRI-cleaved pETDuet1-Gal-3 vector. The expression of the biotinylated Gal-3-AVI protein was performed in *E. coli* BL21 (λDE3) transformed with the pETDuet1-Gal-3 vector along with an IPTG inducible plasmid containing the *birA* gene. The cells were grown in mineral M9 medium supplemented with glycerol (20 g·L^{-1}), yeast extract (20 g·L^{-1}), 150 µg·mL^{-1} ampicillin and 10 µg·mL^{-1} chloramphenicol to an optical density of 0.6 at 600 nm, induced with 1 mM isopropyl 1-thio-β-D-galactopyranoside (IPTG) in the presence of 50 µM D-biotin (Sigma), and grown for additional 4 h at 37 °C. The in vivo biotinylated His_6-tagged Gal-3 protein was purified by Ni^{2+}-chelating affinity chromatography analogously to the original construct as described in the Supplementary Materials. Protein concentrations were determined by Bradford assay using a calibration with bovine serum albumin. The molecular weight of Gal-3 and Gal-3-AVI was determined by amino acid sequence to be 28,023 and 30,550 Da, respectively.

4.3. ELISA Assays with Gal-3

The affinity of Gal-3 for neo-glycoproteins **9–16** was determined using ELISA as reported previously [15,27,30,49]. For the immobilization of the respective neo-glycoproteins or non-modified BSA (negative control), we used F16 Maxisorp NUNC-Immuno Modules (Thermo Scientific, Roskilde, Denmark). Per well, an amount of 5 pmol was incubated overnight at a working concentration of 0.1 µM (PBS). Then the wells were blocked with bovine serum albumin (2% *w/v*) diluted in PBS (1 h, room temperature). Afterwards, recombinant Gal-3 in varying concentration (total volume 50 µL) was added and incubated for 1 h. Detection of bound Gal-3 was achieved using anti-His_6-IgG1 antibody from mouse conjugated with horseradish peroxidase (Roche Diagnostics, Mannheim, Germany) diluted in PBS (1:2000, 50 µL, 1 h, room temperature). TMB One (Kem-En-Tec, Taastrup, Denmark) substrate solution was utilized to initiate reaction of IgG-conjugated peroxidase. The reaction was stopped by adding 3 M hydrochloric acid (50 µL). The binding signal of bound galectin was measured with a spectrophotometer (Spectra Max Plus, Molecular Devices, Sunnyvale, CA, USA) at an optical density of 450 nm. Obtained data were analyzed using SigmaPlot 10 software (Systat Software GmbH, Erkrath, Germany).

In the competitive ELISA design, the F16 Maxisorp NUNC-Immuno Modules (Thermo Scientific, Roskilde, Denmark) were coated overnight with ASF (Sigma Aldrich, Steinheim, Germany; 0.1 µM in PBS, 50 µL, 5 pmol per well) and blocked with BSA (2% *w/v* diluted in PBS (1 h, room temperature). Afterwards, a mixture of the respective compound **1–16** in varying concentrations together with Gal-3 (total volume 50 µL; 5 µM final Gal-3 concentration) were added and incubated for 1 h. Detection of bound Gal-3 and data analysis were performed as described above.

4.4. Surface Plasmon Resonance Measurements

SPR measurements were performed at 25 °C using a set of sensor chips (GLC, NLC, HTE) mounted on a ProteOn XPR36 Protein Interaction Array System (Bio-Rad, Hercules, CA, USA) as described previously [52]. Briefly, Gal-3-AVI protein was diluted to a final concentration of 5 µg·mL^{-1} in PBS and captured on a neutravidin-coated NLC sensor chip (Bio-Rad) at a flow rate of 30 µL·min^{-1}. Neo-glycoproteins **9–16**, used here as analytes, were serially diluted in the running buffer (PBS supplemented with 0.005% Tween 20), and injected in parallel over the immobilized Gal-3-AVI at a flow rate of 30 µL·min^{-1}. The Gal-3-AVI surface was regenerated by a washing step with 50 mM HCl for 60 s. For the reversed experimental setup, the neo-glycoproteins **9–16** were immobilized on the GLC sensor chip by amine coupling chemistry and serial dilutions of Gal-3 were injected over the sensor surface. The correction of the sensograms for sensor background was performed by the interspot referencing procedure, which utilizes the sites on the 6 × 6 array without exposure to the immobilization of ligand but only to the flow of analyte). Furthermore, a "blank" injection was used for substracting the value for analyte. The global analysis of data was done by fitting the binding curves using 1:1 Langmuir-type and heterogeneous ligand binding models. The Langmuir-type model presumes that ligand (L) and analyte (A) interact together under the direct formation of the final complex (AL):

$$A + L \xrightarrow{ka,kd} AL$$

where k_a stands for the association rate constant, and k_d for the dissociation rate constant. Heterogeneous ligand models accounts for the presence of ligands that bind one analyte in two separate binding events:

$$A + L1 \xrightarrow{ka1,kd1} AL1$$

$$A + L2 \xrightarrow{ka2,kd2} AL2$$

where k_{a1} and k_{d1}, and k_{a2} and k_{d2} represent the association and dissociation rate constants of the former and the latter binding, respectively. The equilibrium dissociation constant, K_D, was determined as a ratio between the dissociation and association rate constants:

$$K_D = \frac{k_d}{k_a}$$

4.5. Molecular Dynamics Simulations

The molecular model of Gal-3 CRD, used for molecular dynamics simulations, was constructed from the experimentally determined structure of Gal-3 CRD in complex with lacto-*N*-neotetraose (PDB ID: 4LBN [17]). All simulations were done in GROMACS 5.1.3 [32]. The system was simulated in Amber99SB-ILDN force field [33]. Ligand topologies were generated using GLYCAM web server (Glycam 06) [34] and converted to GROMACS by ACPYPE [35]. Lengths of all covalent bonds were constrained in all simulations. Simulation step was set to 2 fs. Electrostatics was calculated using particle-mesh Ewald method [53] and temperature was controlled by Parrinello-Bussi thermostat [54].

Ligands were docked into the binding site by manual alignment of respective carbohydrate moieties and adjustment of ligand torsion in UCSF Chimera [55]. Each system was minimized in vacuum (approximately 15–25,000 steps). Then it was solvated and neutralized by replacing the randomly chosen water molecules by chloride anions. Each system contained a monomer of Gal-3 CRD, a glycan ligand, 8453–9046 TIP3P water molecules and 4 chloride counterions. Then, it was minimized again (approximately 10–15,000 steps), followed by restrained NVT (constant number of particles, volume and temperature) simulations at 10, 50 and 100 K; 10 ps each. Furthermore, restrained NPT (constant number of particles, pressure and temperature) simulation of 100 ps at 300 K and restrained NVT simulation of 1.1 ns at the same temperature were performed. Harmonic position restraints with a force constant of 1000 $kJ \cdot mol^{-1} \cdot nm^{-2}$ were used in all restrained simulations. They were followed by unrestrained simulations (30 ns) at 300 K.

Binding free energies were predicted by the Linear Interaction Energy method [56] based on the last 10 ns of each simulation and 10 ns simulations of unbound solvated ligands. Scaling factors for Lennard-Jones and electrostatic energies were 0.181 and 0.3, respectively.

5. Conclusions

The present paper discloses the binding kinetics of the strongest multivalent ligand of Gal-3 ever reported: a BSA-based neo-glycoprotein carrying LacdiNAc-LacNAc tetrasaccharides. This conjugate exhibits pM affinities to Gal-3 as determined by SPR in a novel setup with Gal-3 immobilized via Avi-biotin-neutravidin tag. This setup allows the determination of binding constants comparable to the values in solution as shown by isothermal titration calorimetry. Through molecular dynamics studies, we identified the interactions of this tetrasaccharide in the Gal-3 binding site and proposed possible explanations for its high affinity. The results herein shed more light on the interaction of poly-LacNAc-type glycoconjugates with Gal-3 and draft new pathways in the direction of developing specific efficient inhibitors of Gal-3 applicable in cancer-related diagnosis and therapies.

Supplementary Materials: Supplementary materials can be found at http://www.mdpi.com/2079-6382/19/2/372/s1.

Acknowledgments: The authors gratefully acknowledge the financial support by the twin collaborative projects German Research Foundation (DFG, project EL 135/12-1) & Czech Science Foundation (GAČR, project 15-02578J). The authors are thankful for financial and networking support from the EU-COST actions CM1102 and CM1303. Pavla Bojarová acknowledges support through LTC17005 (the Ministry of Education, Youth and Sports of the Czech Republic; COST Action CA15135). Ladislav Bumba acknowledges support from the project LM2015064 (Czech National Node to the European Infrastructure for Translational Medicine) of the Ministry of Education, Youth and Sports of the Czech Republic. Vojtěch Spiwok acknowledges support from the projects ELIXIR CZ (LM2015047), CESNET (LM2015042) and the CERIT Scientific Cloud (LM2015085) research infrastructure projects including the access to computing and storage facilities.

Author Contributions: Ladislav Bumba measured and evaluated SPR data; Dominic Laaf synthesized glycans and neo-glycoproteins and performed ELISAs; Vojtěch Spiwok performed molecular modeling; Lothar Elling and Vladimír Křen supervised the study and revised the manuscript for important intellectual content; Pavla Bojarová conceived the experiments, analyzed the data and wrote the manuscript. All authors participated in manuscript writing and reading and approved the final version of the manuscript.

Conflicts of Interest: The authors declare no conflict of interest.

Abbreviations

ASF	Asialofetuin
BSA	Bovine serum albumin
CRD	Carbohydrate recognition domain
ELISA	Enzyme-linked immunosorbent assay
ESI-MS	Electrospray ionization mass spectrometry
Gal-3	Galectin-3
Gal-3-AVI	Galectin-3 containing AviTag peptide sequence
HPLC	High-performance liquid chromatography

Int. J. Mol. Sci. **2018**, *19*, 372

IC$_{50}$	Half maximal inhibitory concentration
LacNAc	*N*-Acetyllactosamine (β-D-Gal-(1→4)-D-GlcNAc)
LacdiNAc	*N*,*N*′-Diacetyllactosamine (β-D-GalNAc-(1→4)-D-GlcNAc)
NMR	Nuclear magnetic resonance
SDS-PAGE	Sodium dodecyl sulfate polyacrylamide gel electrophoresis
SPR	Surface plasmon resonance
TMB	3,3′,5,5′-tetramethylbenzidine

References

1. Hirabayashi, J.; Kasai, K. The family of metazoan metal-independent β-galactoside-binding lectins: Structure, function and molecular evolution. *Glycobiology* **1993**, *3*, 297–304. [CrossRef] [PubMed]
2. Liu, F.T.; Rabinovich, G.A. Galectins as modulators of tumour progression. *Nat. Rev. Cancer* **2005**, *5*, 29–41. [CrossRef] [PubMed]
3. Iurisci, I.; Cumashi, A.; Sherman, A.A.; Tsvetkov, Y.E.; Tinari, N.; Piccolo, E. Synthetic inhibitors of galectin-1 and -3 selectively modulate homotypic cell aggregation and tumor cell apoptosis. *Anticancer Res.* **2009**, *29*, 403–410. [PubMed]
4. Ebrahim, A.H.; Alalawi, Z.; Mirandola, L.; Rakhshanda, R.; Dahlbeck, S.; Nguyen, D.; Jenkins, M.; Grizzi, F.; Cobos, E.; Figueroa, J.A.; et al. Galectins in cancer: Carcinogenesis, diagnosis and therapy. *Ann. Transl. Med.* **2014**, *2*, 88. [CrossRef] [PubMed]
5. Tellez-Sanz, R.; Garcia-Fuentes, L.; Vargas-Berenguel, A. Human galectin-3 selective and high affinity inhibitors. Present state and future perspectives. *Curr. Med. Chem.* **2013**, *20*, 2979–2990. [CrossRef] [PubMed]
6. Šimonová, A.; Kupper, C.E.; Böcker, S.; Müller, A.; Hofbauerová, K.; Pelantová, H.; Elling, L.; Křen, V.; Bojarová, P. Chemo-enzymatic synthesis of LacdiNAc dimers of varying length as novel galectin ligands. *J. Mol. Catal. B Enzym.* **2014**, *101*, 47–55. [CrossRef]
7. Laaf, D.; Bojarová, P.; Mikulová, B.; Pelantová, H.; Křen, V.; Elling, L. Two-step enzymatic synthesis of β-D-N-acetylgalactosamine-(1→4)-D-N-acetylglucosamine (LacdiNAc) chitooligomers for deciphering galectin binding behavior. *Adv. Synth. Catal.* **2017**, *359*, 2101–2108. [CrossRef]
8. Breloy, I.; Söte, S.; Ottis, P.; Bonar, D.; Grahn, A.; Hanisch, F.-G. *O*-Linked LacdiNAc-modified glycans in extracellular matrix glycoproteins are specifically phosphorylated at the subterminal GlcNAc. *J. Biol. Chem.* **2012**, *287*, 18275–18286. [CrossRef] [PubMed]
9. Jin, C.; Kenny, D.T.; Skoog, E.C.; Padra, M.; Adamczyk, B.; Vitizeva, V.; Thorell, A.; Venkatakrishnan, V.; Lindén, S.K.; Karlsson, N.G. Structural diversity of human gastric mucin glycans. *Mol. Cell. Proteom.* **2017**, *16*, 743–758. [CrossRef] [PubMed]
10. Hirano, K.; Matsuda, A.; Shirai, T.; Furukawa, K. Expression of LacdiNAc groups on *N*-glycans among human tumors is complex. *Biomed. Res. Int.* **2014**, *2014*, 981627. [CrossRef] [PubMed]
11. Haji-Ghassemi, O.; Gilbert, M.; Spence, J.; Schur, M.J.; Parker, M.J.; Jenkins, M.L.; Burke, J.E.; van Faassen, H.; Young, N.M.; Evans, S.V. Molecular basis for recognition of the cancer glycobiomarker, LacdiNAc (GalNAc[β1→4]GlcNAc), by *Wisteria floribunda* agglutinin. *J. Biol. Chem.* **2016**, *291*, 24085–24095. [CrossRef] [PubMed]
12. Anugraham, M.; Jacob, F.; Everest-Dass, A.V.; Schoetzau, A.; Nixdorf, S.; Hacker, N.; Fink, D.; Heinzelmann-Schwarz, V.; Packer, N.H. Tissue glycomics distinguish tumour sites in women with advanced serous adenocarcinoma. *Mol. Oncol.* **2017**, *11*, 1595–1615. [CrossRef] [PubMed]
13. Wuhrer, M.; Koeleman, C.A.M.; Deelder, A.M.; Hokke, C.H. Repeats of LacdiNAc and fucosylated LacdiNAc on *N*-glycans of the human parasite *Schistosoma mansoni*. *FEBS J.* **2006**, *273*, 347–361. [CrossRef] [PubMed]
14. Hanzawa, K.; Suzuki, N.; Natsuka, S. Structures and developmental alterations of *N*-glycans of zebrafish embryos. *Glycobiology* **2017**, *27*, 228–245. [CrossRef] [PubMed]
15. Böcker, S.; Laaf, D.; Elling, L. Galectin binding to neo-glycoproteins: LacDiNAc conjugated BSA as ligand for human galectin-3. *Biomolecules* **2015**, *5*, 1671–1696. [CrossRef] [PubMed]
16. Ahmed, H.; AlSadek, D.M. Galectin-3 as a potential target to prevent cancer metastasis. *Clin. Med. Insights Oncol.* **2015**, *25*, 113–121. [CrossRef] [PubMed]
17. Collins, P.M.; Bum-Erdene, K.; Yu, X.; Blanchard, H. Galectin-3 interactions with glycosphingolipids. *J. Mol. Biol.* **2014**, *426*, 1439–1451. [CrossRef] [PubMed]

18. Leffler, H.; Carlsson, S.; Hedlund, M.; Qian, Y.; Poirier, F. Introduction to galectins. *Glycoconj. J.* **2004**, *19*, 433–440. [CrossRef] [PubMed]

19. Hsieh, T.-J.; Lin, H.-Y.; Tu, Z.; Lin, T.-C.; Wu, S.-C.; Tseng, Y.-Y.; Liu, F.-T.; Hsu, S.-T.D.; Lin, C.-H. Dual thio-digalactoside-binding modes of human galectins as the structural basis for the design of potent and selective inhibitors. *Sci. Rep.* **2016**, *6*, 29457. [CrossRef] [PubMed]

20. Rapoport, E.M.; Kurmyshkina, O.V.; Bovin, N.V. Mammalian galectins: Structure, carbohydrate specificity, and functions. *Biochemistry* **2008**, *73*, 393–405. [CrossRef] [PubMed]

21. Morris, S.; Ahmad, N.; André, S.; Kaltner, H.; Gabius, H.-J.; Brenowitz, M.; Brewer, F. Quaternary solution structures of galectins-1, -3, and -7. *Glycobiology* **2004**, *14*, 293–300. [CrossRef] [PubMed]

22. Nieminen, J.; Kuno, A.; Hirabayashi, J.; Sato, S. Visualization of galectin-3 oligomerization on the surface of neutrophils and endothelial cells using fluorescence resonance energy transfer. *J. Biol. Chem.* **2007**, *282*, 1374–1383. [CrossRef] [PubMed]

23. Massa, S.M.; Cooper, D.N.W.; Leffler, H.; Barondes, S.H. L-29, an endogenous lectin, binds to glycoconjugate ligands with positive cooperativity. *Biochemistry* **1993**, *32*, 260–267. [CrossRef] [PubMed]

24. Lepur, A.; Salomonsson, E.; Nilsson, U.J.; Leffler, H. Ligand induced galectin-3 protein self-association. *J. Biol. Chem.* **2012**, *287*, 21751–21756. [CrossRef] [PubMed]

25. Ahmad, N.; Gabius, H.J.; André, S.; Kaltner, H.; Sabesan, S.; Roy, R.; Liu, B.; Macaluso, F.; Brewer, C.F. Galectin-3 precipitates as a pentamer with synthetic multivalent carbohydrates and forms heterogeneous cross-linked complexes. *J. Biol. Chem.* **2004**, *279*, 10841–10847. [CrossRef] [PubMed]

26. Halimi, H.; Rigato, A.; Byrne, D.; Ferracci, G.; Sebban-Kreuzer, C.; ElAntak, L.; Guerlesquin, F. Glycan dependence of galectin-3 self-association properties. *PLoS ONE* **2014**, *9*, e111836. [CrossRef] [PubMed]

27. Laaf, D.; Bojarová, P.; Pelantová, H.; Křen, V.; Elling, L. Tailored multivalent neo-glycoproteins: Synthesis, evaluation, and application of a library of galectin-3-binding glycan ligands. *Bioconjug. Chem.* **2017**, *28*, 2832–2840. [CrossRef] [PubMed]

28. Lundquist, J.J.; Toone, E.J. The cluster glycoside effect. *Chem. Rev.* **2002**, *102*, 555–578. [CrossRef] [PubMed]

29. Wang, H.; Huang, W.; Orwenyo, J.; Banerjee, A.; Vasta, G.R.; Wang, L.-X. Design and synthesis of glycoprotein-based multivalent glyco-ligands for influenza hemagglutinin and human galectin-3. *Bioorg. Med. Chem.* **2013**, *21*, 2037–2044. [CrossRef] [PubMed]

30. Böcker, S.; Elling, L. Biotinylated *N*-acetyllactosamine- and *N,N*-diacetyllactosamine-based oligosaccharides as novel ligands for human galectin-3. *Bioengineering* **2017**, *4*, 31. [CrossRef] [PubMed]

31. Dam, T.K.; Gabius, H.J.; André, S.; Kaltner, H.; Lensch, M.; Brewer, C.F. Galectins bind to the multivalent glycoprotein asialofetuin with enhanced affinities and a gradient of decreasing binding constants. *Biochemistry* **2005**, *44*, 12564–12571. [CrossRef] [PubMed]

32. Abraham, M.J.; Murtola, T.; Schulz, R.; Páll, S.; Smith, J.C.; Hess, B.; Lindahl, E. GROMACS: High performance molecular simulations through multi-level parallelism from laptops to supercomputers. *SoftwareX* **2015**, *1*, 19–25. [CrossRef]

33. Lindorff-Larsen, K.; Piana, S.; Palmo, K.; Maragakis, P.; Klepeis, J.L.; O'Dror, R.; Shaw, D.E. Improved side-chain torsion potentials for the Amber ff99SB protein force field. *Proteins* **2010**, *78*, 1950–1958. [CrossRef] [PubMed]

34. Kirschner, K.N.; Yongye, A.B.; Tschampel, S.M.; Daniels, C.R.; Foley, B.L.; Woods, R.J. GLYCAM06: A generalizable Biomolecular force field. Carbohydrates. *J. Comput. Chem.* **2008**, *29*, 622–655. [CrossRef] [PubMed]

35. Sousa da Silva, A.W.; Vranken, W.F. ACPYPE—AnteChamber PYthon Parser interfacE. *BMC Res. Notes* **2012**, *5*, 367. [CrossRef] [PubMed]

36. Korb, O.; Stützle, T.; Exner, T.E. Empirical scoring functions for advanced protein-ligand docking with PLANTS. *J. Chem. Inf. Model.* **2009**, *49*, 84–96. [CrossRef] [PubMed]

37. Saraboji, K.; Hakansson, M.; Genheden, S.; Diehl, C.; Qvist, J.; Weininger, U.; Nilsson, U.J.; Leffler, H.; Ryde, U.; Akke, M.; et al. The carbohydrate-binding site in galectin-3 is pre-organized to recognize a sugar-like framework of oxygens: Ultra-high resolution structures and water dynamics. *Biochemistry* **2012**, *51*, 296–306. [CrossRef] [PubMed]

38. Blanchard, H.; Yu, X.; Collins, P.M.; Bum-Erdene, K. Galectin-3 inhibitors: A patent review (2008–present). *Expert Opin. Ther. Pat.* **2014**, *24*, 1053–1065. [CrossRef] [PubMed]

39. Öberg, C.T.; Leffler, H.; Nilsson, U.J. Inhibition of galectins with small molecules. *CHIMIA Int. J. Chem.* **2011**, *65*, 18–23. [CrossRef]

40. Laaf, D.; Steffens, H.; Pelantová, H.; Bojarová, P.; Křen, V.; Elling, L. Chemo-enzymatic synthesis of branched *N*-acetyllactosamine glycan oligomers for galectin-3 inhibition. *Adv. Synth. Catal.* **2017**, *359*, 4015–4024. [CrossRef]

41. Kamili, N.A.; Arthur, C.M.; Gerner-Smidt, C.; Tafesse, E.; Blenda, A.; Dias-Baruffi, M.; Stowell, S.R. Key regulators of galectin–glycan interactions. *Proteomics* **2016**, *16*, 3111–3125. [CrossRef] [PubMed]

42. Stowell, S.R.; Arthur, C.M.; Mehta, P.; Slanina, K.A.; Blixt, O.; Leffler, H.; Smith, D.F.; Cummings, R.D. Galectin-1, -2, and -3 exhibit differential recognition of sialylated glycans and blood group antigens. *J. Biol. Chem.* **2008**, *283*, 10109–10123. [CrossRef] [PubMed]

43. Barboni, E.A.M.; Bawumia, S.; Colin Hughes, R. Kinetic measurements of binding of galectin 3 to a laminin substratum. *Glycoconjug. J.* **1999**, *16*, 365–373. [CrossRef]

44. Jin, F.; Chammas, R.; Engel, J.; Reinhold, V. Structure and function of laminin 1 glycans; glycan profiling. *Glycobiology* **1995**, *5*, 157–158. [CrossRef] [PubMed]

45. Yoshioka, K.; Sato, Y.; Murakami, T.; Tanaka, M.; Niwa, O. One-step detection of galectins on hybrid monolayer surface with protruding lactoside. *Anal. Chem.* **2010**, *82*, 1175–1178. [CrossRef] [PubMed]

46. Javier Muñoz, F.; Ignacio Santos, J.; Ardá, A.; André, S.; Gabius, H.-J.; Sinisterra, J.V.; Jiménez-Barbero, J.; Hernáiz, M.J. Binding studies of adhesion/growth-regulatory galectins with glycoconjugates monitored by surface plasmon resonance and NMR spectroscopy. *Org. Biomol. Chem.* **2010**, *8*, 2986–2992. [CrossRef] [PubMed]

47. Sauerzapfe, B.; Namdjou, D.J.; Schumacher, T.; Linden, N.; Křenek, K.; Křen, V.; Elling, L. Characterization of recombinant fusion constructs of human beta 1,4-galactosyltransferase 1 and the lipase pre-propeptide from *Staphylococcus hyicus*. *J. Mol. Catal. B Enzym.* **2008**, *50*, 128–140. [CrossRef]

48. Kupper, C.E.; Rosencrantz, R.R.; Henssen, B.; Pelantová, H.; Thönes, S.; Drozdová, A.; Křen, V.; Elling, L. Chemo-enzymatic modification of poly-*N*-acetyllactosamine (LacNAc) oligomers and *N,N*-diacetyllactosamine (LacDINAc) based on galactose oxidase treatment. *Beilstein J. Org. Chem.* **2012**, *8*, 712–725. [CrossRef] [PubMed]

49. Fischöder, T.; Laaf, D.; Dey, C.; Elling, L. Enzymatic synthesis of *N*-acetyllactosamine (LacNAc) type 1 oligomers and characterization as multivalent galectin ligands. *Molecules* **2017**, *22*, 1320. [CrossRef] [PubMed]

50. Sauerzapfe, B.; Křenek, K.; Schmiedel, J.; Wakarchuk, W.W.; Pelantová, H.; Křen, V.; Elling, L. Chemo-enzymatic synthesis of poly-*N*-acetyllactosamine (poly-LacNAc) structures and their characterization for CGL2-galectin-mediated binding of ECM glycoproteins to biomaterial surfaces. *Glycoconj. J.* **2009**, *26*, 141–159. [CrossRef] [PubMed]

51. Rech, C.; Rosencrantz, R.R.; Křenek, K.; Pelantová, H.; Bojarová, P.; Römer, C.E.; Hanisch, F.G.; Křen, V.; Elling, L. Combinatorial one-pot synthesis of poly-*N*-acetyllactosamine oligosaccharides with Leloir-glycosyltransferases. *Adv. Synth. Catal.* **2011**, *353*, 2492–2500. [CrossRef]

52. Bojarová, P.; Chytil, P.; Mikulová, B.; Bumba, L.; Konefał, R.; Pelantová, H.; Krejzová, J.; Slámová, K.; Petrásková, L.; Kotrchová, L.; et al. Glycan-decorated HPMA copolymers as high-affinity lectin ligands. *Polym. Chem.* **2017**, *8*, 2647–2658. [CrossRef]

53. Essmann, U.; Perera, L.; Berkowitz, M.L.; Darden, T.; Lee, H.; Pedersen, L.G. A smooth particle mesh Ewald method. *J. Chem. Phys.* **1995**, *103*, 8577–8593. [CrossRef]

54. Bussi, G.; Donadio, D.; Parrinello, M. Canonical sampling through velocity-rescaling. *J. Chem. Phys.* **2007**, *126*, 014101. [CrossRef] [PubMed]

55. Pettersen, E.F.; Goddard, T.D.; Huang, C.C.; Couch, G.S.; Greenblatt, D.M.; Meng, E.C.; Ferrin, T.E. UCSF Chimera—A visualization system for exploratory research and analysis. *J. Comput. Chem.* **2004**, *25*, 1605–1612. [CrossRef] [PubMed]

56. Hansson, T.; Marelius, J.; Åqvist, J. Ligand binding affinity prediction by linear interaction energy methods. *J. Comput. Aided Mol. Des.* **1998**, *12*, 27–35. [CrossRef] [PubMed]

International Journal of
Molecular Sciences

MDPI

Article

Proteomic Identification of the Galectin-1-Involved Molecular Pathways in Urinary Bladder Urothelial Carcinoma

Chien-Feng Li [1,2,3], Kun-Hung Shen [4], Lan-Hsiang Chien [5], Cheng-Hao Huang [1,†], Ting-Feng Wu [1,*] and Hong-Lin He [2]

[1]   Department of Biotechnology, Southern Taiwan University of Science and Technology, Tainan 710, Taiwan; angelo.p@yahoo.com.tw (C.-F.L.); plane32033@gmail.com (C.-H.H.)
[2]   Departments of Pathology, Chi Mei Medical Center, Tainan 710, Taiwan; baltic1023@gmail.com
[3]   National Institute of Cancer Research, National Health Research Institutes, Miaoli 350, Taiwan
[4]   Department of Urology, Chi Mei Medical Center, Tainan 710, Taiwan; robert.shen@msa.hinet.net
[5]   Department of Medical Research, Chi Mei Medical Center, Tainan 710, Taiwan; m96h0207@stust.edu.tw
*   Correspondence: wutingfe@stust.edu.tw; Tel.: +886-253-3131 (ext. 8394)
†   Current address: Nan Pao Resins Chemical Company Ltd., Tainan 710, Taiwan.

Received: 5 February 2018; Accepted: 16 April 2018; Published: 19 April 2018

Abstract: Among various heterogeneous types of bladder tumors, urothelial carcinoma is the most prevalent lesion. Some of the urinary bladder urothelial carcinomas (UBUCs) develop local recurrence and may cause distal invasion. Galectin-1 de-regulation significantly affects cell transformation, cell proliferation, angiogenesis, and cell invasiveness. In continuation of our previous investigation on the role of galectin-1 in UBUC tumorigenesis, in this study, proteomics strategies were implemented in order to find more galectin-1-associated signaling pathways. The results of this study showed that galectin-1 knockdown could induce 15 down-regulated proteins and two up-regulated proteins in T24 cells. These de-regulated proteins might participate in lipid/amino acid/energy metabolism, cytoskeleton, cell proliferation, cell-cell interaction, cell apoptosis, metastasis, and protein degradation. The aforementioned dys-regulated proteins were confirmed by western immunoblotting. Proteomics results were further translated to prognostic markers by analyses of biopsy samples. Results of cohort studies demonstrated that over-expressions of glutamine synthetase, alcohol dehydrogenase (NADP+), fatty acid binding protein 4, and toll interacting protein in clinical specimens were all significantly associated with galectin-1 up-regulation. Univariate analyses showed that de-regulations of glutamine synthetase and fatty acid binding protein 4 in clinical samples were respectively linked to disease-specific survival and metastasis-free survival.

Keywords: urinary bladder urothelial carcinoma; galectin-1; fatty acid binding protein 4; glutamine synthetase; two-dimensional gel electrophoresis

---

1. Introduction

The surfaces of bladders and ureters are mostly covered by urothelium. Among various heterogeneous types of bladder tumors, urothelial carcinoma is the most prevalent lesion. Urinary bladder urothelial carcinoma (UBUC) is often recognized as non-invasively papillary or superficially invasive tumors [1]. Nevertheless, some of the aforementioned tumors develop local recurrence and may cause distal invasion [1]. UBUC cells are identified as low-grade or less differentiated and high-grade cells.

Galectin-1 is one of the fifteen mammalian proteins belonging to the β-galactoside binding family. It is a 14-kDa monomer with one carbohydrate-binding domain and usually forms a non-covalent homodimer in the cells [2]. The galectin-1 protein expression is precisely controlled by the methylation degree of the *LGALS1* gene promoter located on chromosome 22q12 [3].

Galectin-1 de-regulation significantly affects cell transformation [4], cell proliferation [2], angiogenesis [5], and cell adhesion and invasiveness [6–8], as well as immunosuppression [9,10]. Up-regulated galectin-1 expression has been observed in UBUC [11], colorectal cancer [12], breast cancer [10], lung cancer [13], head/neck cancer [14], ovarian cancer [15], prostate carcinoma [16], glioma [17], Kaposi's sarcoma [18], and Hodgkin's lymphoma [19]. It is found that tumor stages, tumor invasiveness, and metastasis are associated with the increased galectin-1 expression in UBUC [11]. Furthermore, there is a positive correlation between poor prognosis and increased galectin-1 amount in lesions in patients with UBUC [11] and glioblastoma [20]. Comparative immunohistochemical results showed a higher galectin-1 protein amount in invaded areas than that in non-invaded areas of human U87 and U373 xenografted glioblastoma in nude mice [21]. The aforementioned phenomena may be attributed to the modification of actin through increasing the small RhoAGTPase expression. In surprising similarity, immunohistochemistry (IHC) analyses of oral squamous cell carcinoma (OSCC) specimens demonstrated that galectin-1 is over-expressed at the invasion front [22]. Further galectin-1 augments the expression and activities of matrix metalloproteinase proteins (MMP) 2 and 9 to provoke OSCC cell invasion. It can also induce cytoskeleton re-organization to promote invasiveness by regulating the activity of cell division cycle 42 (cdc42), a member of the RhoGTPase family [8]. The above results implicate that the galectin-1 protein is intimately associated with tumor invasiveness.

In addition to the link to invasion, the galectin-1 protein has been shown to bind to GTPase HRas proto-Oncogene (H-Ras) and cause the membrane anchorage of H-Ras. Increased galectin-1 expression in tumor cells eventually enhances H-Ras membrane localization and evokes the RAF proto-oncogene serine/threonine-protein kinase (Raf-1)/mitogen-activated protein kinase (MEK)/ extracellular signal–regulated kinases (Erk) pathway to strengthen the cell transformation [23]. In addition to the involvement in cell transformation, Rubinstein et al. (2004) found that melanoma cells can secret the galectin-1 protein to prevent cell-mediated immunity by provoking activated T cell apoptosis, thus giving rise to the immune privilege of tumor cells [9].

Our previous studies show that galectin-1 over-expression in tumor cells is correlated with tumor stages, grades, and invasion [11]. It can also significantly predict disease specific survival and metastasis-free survival at the univariate and multivariate levels. Cell signaling examination indicates that the galectin-1 protein takes part in UBUC cell invasion by regulating the MMP9 activity via the Ras–Ras-related C3 botulinum toxin substrate 1 (Rac1)–mitogen-activated protein kinase kinase kinase 4 (MEKK4)–c-Jun N-terminal kinase (JNK)–Activator protein 1 (AP1) signaling pathway [24]. In continuation of our previous investigation on the role of galectin-1 in UBUC tumorigenesis, in this study, proteomics strategies were implemented in order to find more key signaling pathways which are initiated by the galectin-1 protein in tumor cells further to our existing findings. Proteomics results demonstrated that de-regulated proteins in galectin-1 knockdown T24 cells might participate in lipid/amino acid/energy metabolism, cytoskeleton, cell proliferation, cell-cell interaction, cell apoptosis, metastasis, and protein degradation. Furthermore, the results of cohort studies showed that dys-regulations of glutamine synthetase and fatty acid binding protein 4 in clinical samples were respectively linked to disease-specific survival and metastasis-free survival in univariate analyses.

2. Results

2.1. Search for the De-Regulated Proteins in Sh-Gal(+120) T24 Cells by Two-Dimensional Gel Electrophoresis

In our previous studies, we found that the galectin-1 protein is correlated with UBUC cell invasive capability by regulating the MMP9 activity through the Ras–Rac1–MEKK4–JNK–AP1 signaling pathway [24]. In this study, we exploited proteomics to find more novel molecular pathways evoked by galectin-1 dysregulation in UBUC cells, which might be related to UBUC carcinogenesis.

The experimental design for this study is summarized in Figure 1. To obtain the above goal, Lavapurple™-stained two-dimensional gel electrophoresis (2-DE) was first carried out to acquire the protein profiles of Sh-Gal-1(+120) (galectin-1 knockdown stable cell line) and Sc-Gal-1(+120) (scrambled control) T24 cells, as described in Materials and Methods. Then, the 2-DE protein pictures of Sh-Gal-1(+120) T24 cells were compared to those of Sc-Gal-1(+120) T24 cells to search for the differentially expressed protein features which were recognized according to the definition, as described in Materials and Methods. To prevent the gel-to-gel variation, 11 replica gel pairs were obtained to find those de-regulated proteins provoked by galectin-1 knockdown. The typical 2-DE gel diagrams of Sh-Gal-1(+120) and Sc-Gal-1(+120) T24 cells are shown in Figure 1 and the other 10 pairs are provided in the supplementary materials (Supplementary Figure S1).

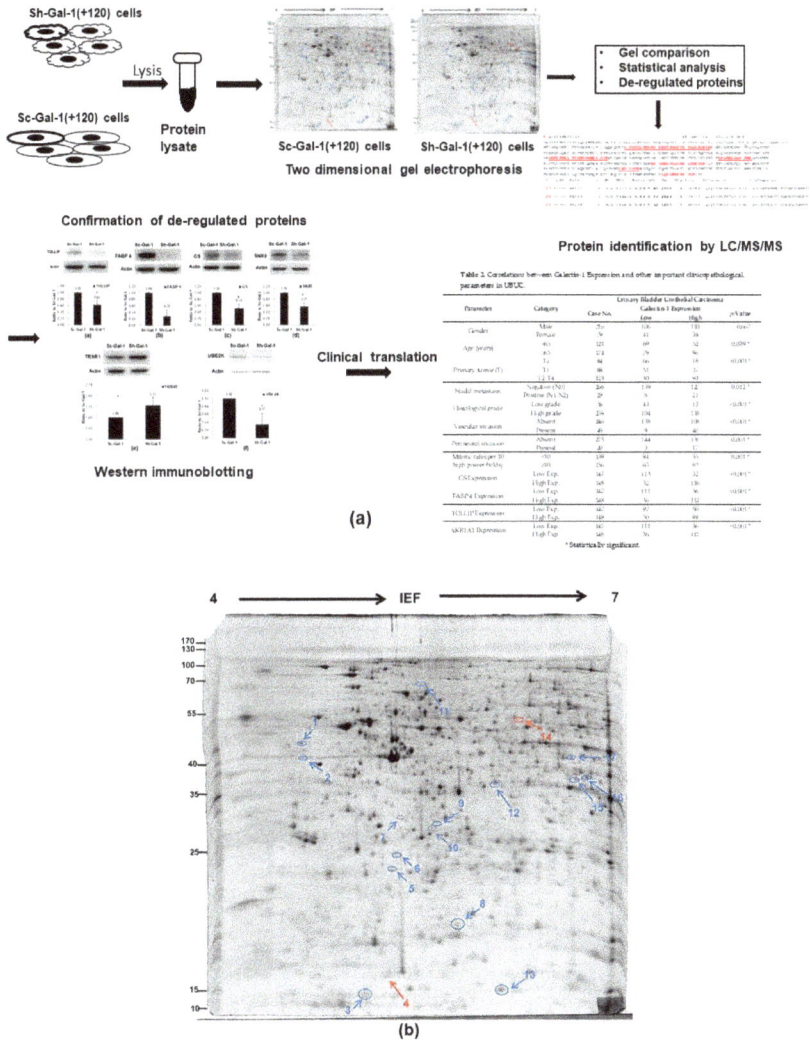

(a)

(b)

Figure 1. *Cont.*

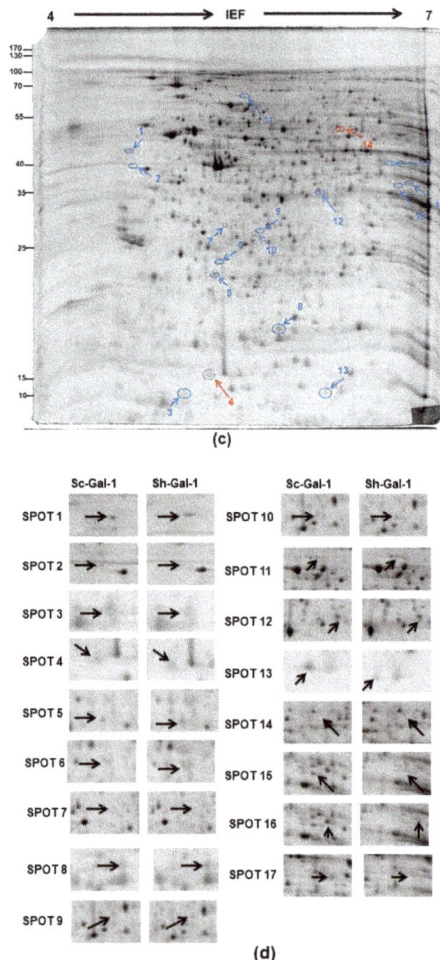

Figure 1. 2-DE protein profiles of Sc-Gal-1(+120) and Sh-Gal-1(+120) T24 cells. (**a**) The experimental design; (**b**) Sc-Gal-1 (+120) T24 cells; (**c**) Sh-Gal (+120) T24 cells. (**d**) The expanded images of de-regulated protein spots. A total of 200 µg of T24 lysate proteins was used in each 2-DE gel, carried out as described in Materials and Methods. Down-regulated protein spots were indicated by blue arrows, while up-regulated ones were indicated by red arrows.

The protein profiles of Sh-Gal-1(+120) T24 cells were very similar to those of Sc-Gal-1(+120) T24 cells, but the amount of some identical proteins in the knockdown cell line was more than twice as high as that in its scrambled counterpart. Fifteen under-expressed and two over-expressed protein spots were found, as indicated by the arrows in Figure 1. Spots 4 ($p = 0.07$) and 10 ($p = 0.05$) were near $p < 0.05$. Spot 7 ($p < 0.01$) was not listed because it was not identified by liquid chromatography/tandem mass spectrometry (LC-MS/MS) (Table 1 and Supplementary Table S1).

Table 1. Mass spectrometric data and identification of the galectin-1-associated de-regulated proteins.

Spot	Identity	Accession Number	Theo. [b] PI/Mr.	Exper. [c] PI/Mr.	Matched [d] Peptide Number	Coverage (%) [e]	Incidence	Fold [f] Difference	*p*-Value
1	ribonuclease/ angiogenin inhibitor 1 (RNH1) [a]	P13489	4.5/49	4.74/49.4	11	22.1	6/11	−2.77	0.003
2	reticulocalbin 1 (RCN1)	Q15293	4.6/42.7	4.86/38.9	3	11.8	6/11	−3.74	0.04
3	galectin 1 (LGALS1)	P09382	4.8/16.7	5.3/14.7	8	52.6	8/11	−56.5	0.01
4	tubulin specific chaperone A (TBCA)	O75347	5.3/12.9	4.9/17.8	8	63.8	6/11	+2.92	0.07
5	ubiquitin conjugating enzyme E2 K (UBE2K)	P61086	5/26.1	5.33/22.6	9	46.5	6/11	−3.71	0.0005
6	polyamine-modulated factor 1 (PMF1)	Q6P1K2	5/27.7	5.37/23.3	5	33.7	7/11	−5.1	0.00003
7	Unidentified	-	-	5.0/34.0	-	-	7/11	−5.33	0.0047
8	stathmin 1 (STMN1)	P16949	6.1/20.5	5.75/17.3	17	29.5	6/11	−1.88	0.008
9	toll interacting protein (TOLLIP)	Q9H0E2	5.8/33.1	5.68/30.3	8	23.4	6/11	−3.52	0.04
10	protein CWC15 homolog (CWC15)	Q9P013	5.6/26.6	5.7/32.3	5	19	6/11	−3.35	0.05
11	sorting nexin 9 (SNX9)	Q9Y5X1	5.5/83.4	5.4/66.6	11	23	6/11	−2.98	0.00004
12	scavenger mRNA-decapping enzyme (DCPS)	Q96C86	6.5/37.3	5.93/38.6	12	32.3	6/11	−2.82	0.028
13	fatty acid binding protein 4 (FABP 4)	P15090	6.5/16.6	7.93/15.2	5	41.9	10/11	−5.16	0.00003
14	thioredoxin reductase 1 (TRXR1)	Q16881	6.6/58.3	6.07/54.4	13	24.1	6/11	+4.32	0.00001
15	alcohol dehydrogenase [NADP+] (AKR1A1)	P14550	6.9/37.5	6.32/36.6	14	33.2	6/11	−2.24	0.005
16	*N*-acetylneuraminate synthase (sialic acid synthase) (NANS)	Q9NR45	6.9/37.4	6.29/40.3	7	22.6	7/11	−2.3	0.005
17	glutamine synthetase (GS)	P15104	6.8/41.4	6.43/42.1	7	14.7	9/11	−5.02	0.0001

[a] HUPO Gene Nomenclature Committee approved symbol; [b] Theo.: Theoretical; [c] Exper.: Experimental; [d] Matched peptide number: Number of peptides matched with protein in MS/MS query. The detailed data of MS/MS identification for each peptide is provided in Supplementary Table S1, [e] Coverage: Total percentage of amino acid sequence covered by peptides identified by MS/MS analyses; [f] Fold difference: the difference between mean of normalized volume of Sh-Gal-1(+120) and Sc-Gal-1(+120) cells.

2.2. Identification of the Differentially Expressed Proteins in Sh-Gal-1(+120) T24 Cells

After the recognition of dys-regulated proteins by 2-DE gel comparison, de-regulated proteins were identified using LC-MS/MS, as described in Materials and Methods. The spectrometric results of down- and up-regulated proteins are shown in Table 1. The experimental molecular weight and pI of each de-regulated protein were close to the theoretical numbers and most of the spectrometric protein coverages were near or over 20%, except for spots 2 and 17. The fragment ion spectra (MS2 spectra) are presented in Supplementary Figure S2. Literature reviews indicated that the identified de-regulated proteins were associated with amino acid/energy metabolism (GS) [25], lipid/energy metabolism (FABP 4) [26], cytoskeleton (STMN1, TBCA) [27,28], cell proliferation (STMN1, PMF1, RNH1) [29–31], cell-cell interaction (NANS) [32], cell apoptosis (TRXR1, RCN1) [33,34], metastasis (NANS, SNX9) [32,35], protein degradation (TOLLIP, UBE2K) [36], and glucose metabolism (AKR1A1).

2.3. Confirmation of De-Regulated Proteins Found in Sh-Gal-1(+120) T24 Cells by Western Immunoblotting

After the identification of de-regulated proteins, we validated our findings via proteomics examination using western immunoblotting. The expressions of TOLLIP, FABP 4, GS, SNX9, UBE2K, PMF1, AKR1A1, STMN1, and TRXR1 were investigated in Sh-Gal-1(+120) T24 cells. These nine proteins were selected from de-regulated proteins for validation due to their respective high-fold changes, except for STMN1, and possible association with carcinogenesis. Results of western immunoblotting demonstrated that the expressions of TOLLIP, FABP 4, GS, SNX9, TRXR1, UBE2K, STMN1, AKR1A1, and PMF1 were under-expressed in Sh-Gal-1(+120) T24 cells, while TRXR1 was over-expressed in Sh-Gal-1(+120) cells. The above results were in line with proteomics results. In this study, our results showed that galectin-1 knockdown in T24 cells could lower FABP 4 protein expression. Currently, no documented literature has reported the relationship between FABP 4 and galectin-1 in cancer cells. Boiteux et al. (2009) found that the peroxisome proliferator-activated receptor γ (PPAR-γ) can regulate FABP 4 protein expression [37]. Thus, we then explored the impacts of rosiglitazone, a PPAR-γ agonist, on the FABP 4 expression in Sh-Gal-1(+120) T24 cells. The results in Figure 2 demonstrated that rosiglitazone treatment could profoundly rescue FABP 4 protein expression in Sh-Gal-1(+120) T24 cells, but not galectin-1 expression, suggesting that galectin-1 was upstream from PPAR-γ and FABP 4 was downstream from PPAR-γ. Galectin-1 might regulate FABP 4 expression through PPAR-γ.

2.4. Confirmation of Proteomics Data Using Cohort Studies of GS, FABP 4, TOLLIP, and AKR1A1

The results from western immunoblotting validated the dys-regulated proteins provoked by galectin-1 knockdown in Sh-Gal-1(+120) T24 cells. Furthermore, we designed cohort studies to confirm the expressed levels of GS, FABP 4, TOLLIP, and AKR1A1 in clinical specimens. Results of cohort studies showed that galectin-1 expression in biopsy samples was significantly correlated with the primary tumor (pT) status, grade, nodal metastasis, vascular invasion, perineural invasion, and mitotic rate. More importantly, the expression levels of GS, FABP 4, TOLLIP, and AKR1A1 in clinical samples were associated with the galectin-1 amount (Table 2), respectively. This finding was consistent with that of proteomics studies.

The results of the univariate log-rank analysis on disease-specific survival (DSS) and metastasis-free survival (MFS) indicated that primary tumor status, histological grade, nodal metastasis, the presence of vascular invasion and perineural invasion, the high mitotic rate, galectin-1 (Figure 3a), GS (Figure 3b), and FABP 4 (Figure 3c) expression levels were significantly correlated with DSS (Table 3). However, the impacts on both TOLLIP and AKR1A1 abundance were not significant in predicting DSS. In line with our previous results [11] and the above proteomics data, a high galectin-1 expression level was predictive of a shorter metastasis-free survival time (Figure 3d). The same results were also observed in patients with high amounts of GS (Figure 3e) or the FABP 4 (Figure 3f) protein. However, the impacts on both of the TOLLIP and AKR1A1 amounts were not significant (Table 3). Further multivariate analyses demonstrated that factors of primary tumor status (T1, relative risk (R.R.) = 2.732; T2–T4, R. R. = 9.346), mitotic rate (R.R. = 2.048), and galectin-1 level (R.R. = 4.628) independently predicted MFS (Table 3), respectively. Moreover, the aforementioned three factors were also significantly predictive of worse MFS, including primary tumor status (T1, R.R. = 4.181; T2–T4, R.R. = 5.543), a high mitotic rate (R.R. = 1.885), and a high galectin-1 level (R.R. = 2.386) (Table 3).

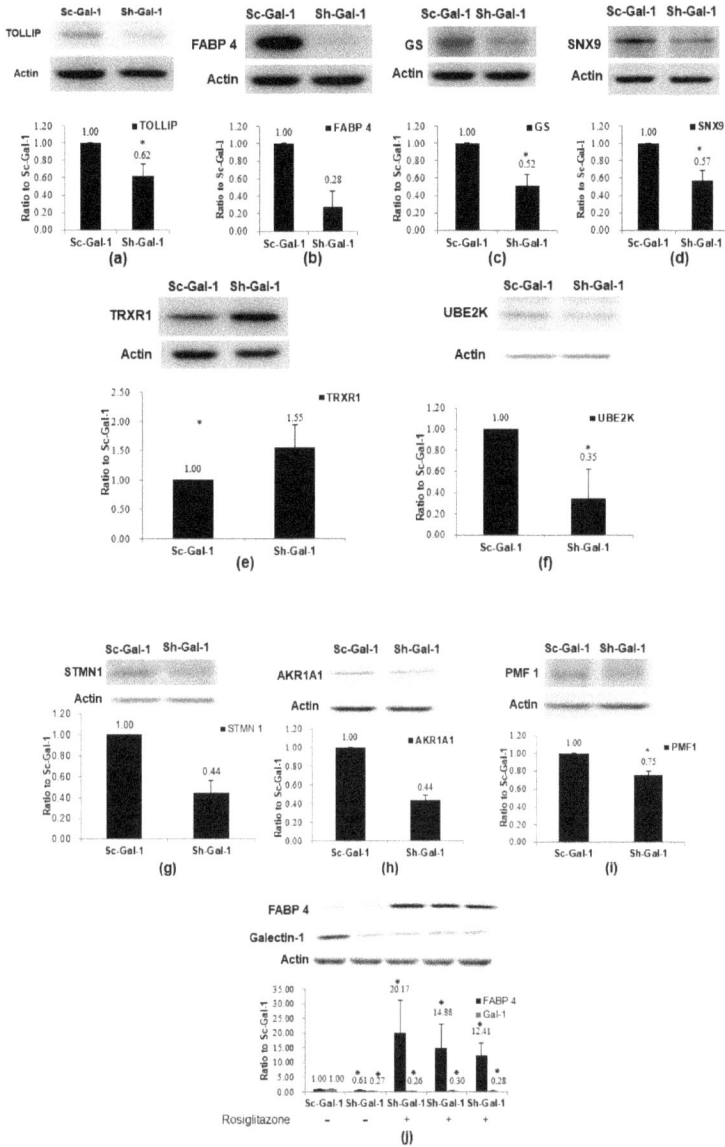

Figure 2. Confirmation of de-regulated proteins evoked by galectin-1 knockdown in T24 cells. (**a**) TOLLIP; (**b**) FABP 4; (**c**) GS; (**d**) SNX9; (**e**) TRXR1; (**f**) UBE2K; (**g**) STMN1; (**h**) AKR1A1; (**i**) PMF1; (**j**) Effects of rosiglitazone (PPAR-γ agonist) on the FABP 4 expression. Western immunoblotting was carried out as described in Materials and Methods. The blot was the representative result of three independent experiments. The protein expression fold (mean ± S.D.) was expressed as the ratio of normalized intensity of the protein of interest (observed protein/actin) in Sh-Gal-1(+120) T24 cells to that in Sc-Gal-1(+120) T24 cells. * $p < 0.05$.

Table 2. Correlations between Galectin-1 Expression and other important clinicopathological, parameters in UBUC.

Parameter	Category	Urinary Bladder Urothelial Carcinoma			
		Case No.	Galectin-1 Expression		*p*-Value
			Low	High	
Gender	Male	216	106	110	0.667
	Female	79	41	38	
Age (years)	<65	121	69	52	0.039 *
	≥65	174	78	96	
Primary tumor (T)	Ta	84	66	18	
	T1	88	51	37	<0.001 *
	T2–T4	123	30	93	
Nodal metastasis	Negative (N0)	266	139	127	0.012 *
	Positive (N1-N2)	29	8	21	
Histological grade	Low grade	56	43	13	<0.001 *
	High grade	239	104	135	
Vascular invasion	Absent	246	138	108	<0.001 *
	Present	49	9	40	
Perineural invasion	Absent	275	144	131	0.001 *
	Present	20	3	17	
Mitotic rate (per 10 high power fields)	<10	139	84	55	0.001 *
	≥10	156	63	93	
GS Expression	Low Exp.	147	115	32	<0.001 *
	High Exp.	148	32	116	
FABP 4 Expression	Low Exp.	147	111	36	<0.001 *
	High Exp.	148	36	112	
TOLLIP Expression	Low Exp.	147	97	50	<0.001 *
	High Exp.	148	50	98	
AKR1A1 Expression	Low Exp.	147	111	36	<0.001 *
	High Exp.	148	36	112	

* Statistically significant.

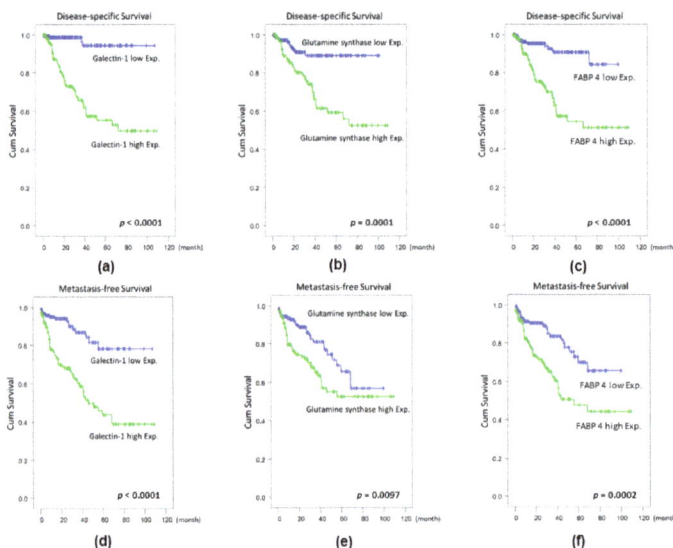

Figure 3. Survival analyses by log-rank tests. Disease-specific survival (DSS) and metastasis-free survival (MFS) of patients with urinary bladder urothelial carcinoma (UBUC) were significantly associated with (**a,d**) galectin-1 over-expression, (**b,e**) GS over-expression, (**c,f**), and FABP 4 over-expression.

Table 3. Univariate log-rank and multivariate analyses for Disease-specific and Metastasis-free Survivals in urinary bladder urothelial carcinoma.

Parameter	Category	Case No.	Disease-Specific Survival					Metastasis-Free Survival				
			Univariate Analysis		Multivariate Analysis			Univariate Analysis		Multivariate Analysis		
			No. of Event	p-Value	R.R.[a]	95% C.I.	p-Value	No. of Event	p-Value	R.R.	95% C.I.	p-Value
Gender	Male	216	39	0.5392	–	–	–	59	0.2999	–	–	–
	Female	79	11		–	–	–	16		–	–	–
Age (years)	<65	121	16	0.0972	–	–	–	30	0.5912	–	–	–
	≥65	174	34		–	–	–	45		–	–	–
Primary tumor (T)	Ta	84	1	<0.0001 *	1	–	0.003 *	4	<0.0001 *	1	–	0.033 *
	T1	88	9		2.732	1.232–6.061		23		4.181	1.996–2.591	
	T2–T4	123	40		9.346	1.027–83.333		48		5.543	1.961–23.810	
Nodal metastasis	Negative (N0)	266	41	0.0037 *	1	–	0.640	61	<0.0001 *	1	–	0.086
	Positive (N1-N2)	29	9		1.199	0.561–2.561		14		1.746	0.924–3.303	
Histological grade	Low grade	56	2	0.0017 *	1	–	0.703	5	0.0008 *	1	–	0.807
	High grade	239	48		1.359	0.282–6.558		75		0.873	0.293–2.598	
Vascular invasion	Absent	246	36	0.0052 *	1	–	0.209	54	0.0003 *	1	–	0.994
	Present	49	14		1.616	0.765–3.413		21		0.998	0.521–1.911	
Perineural invasion	Absent	275	44	0.0085 *	1	–	0.353	65	0.0006 *	1	–	0.202
	Present	20	6		1.586	0.599–4.196		10		1.646	0.765–3.542	
Mitotic rate (per 10 high power fields)	<10	139	12	<0.0001 *	1	–	0.044 *	22	<0.0001 *	1	–	0.021 *
	≥10	156	38		2.048	1.020–4.109		53		1.885	1.103–3.224	
Galectin 1 expression	High Exp.	147	4	<0.0001 *	1	–	0.007 *	16	<0.0001 *	1	–	0.012
	Low Exp.	148	46		4.628	1.5.5–14.225		59		2.386	1.215–4.685	
GS expression	Low Exp.	147	11	0.0001 *	1	–	0.867	27	0.0097 *	1	–	0.354
	High Exp.	148	39		1.066	0.504–2.254		48		0.763	0.431–1.351	
FABP 4 expression	Low Exp.	147	10	<0.0001 *	1	–	0.405	25	0.0002 *	1	–	0.904
	High Exp.	148	40		1.378	0.648–2.927		50		1.035	0.588–1.822	
TOLLIP expression	Low Exp.	147	28	0.3415	–	–	–	34	0.5027	–	–	–
	High Exp.	148	22		–	–	–	41		–	–	–
AKR1A1 expression	Low Exp.	147	21	0.2467	–	–	–	30	0.0779	–	–	–
	High Exp.	148	29		–	–	–	45		–	–	–

* Statistically significant. [a] R. R., relative risk.

3. Discussion

Our previous findings suggest that the galectin-1-increased expression in UBUC is intimately associated with the primary tumor status, grade, and invasion [11]. Univariate and multivariate analyses demonstrate that galectin-1 over-expression in UBUC can also independently predict DSS and MFS. Further signaling pathway investigation indicates that the galectin-1 protein participates in UBUC cell invasion by modulating the MMP9 activity through the Ras–Rac1–MEKK4–JNK–AP1 signaling pathway [24]. To expand our knowledge regarding the role of galectin-1 in UBUC carcinogenesis, in this study, proteomic approaches were exploited to find more key signaling pathways which were provoked by galectin-1 up-regulation in UBUC cells.

In this study, the proteome maps of Sh-Gal-1(+120), a galectin-1 knockdown stable cell line, were compared to those of scrambled control cells, Sc-Gal-1(+120) cells. Our results showed that fifteen under-expressed and two over-expressed proteins were attributed to galectin-1 depletion in T24 cells. These impaired-regulated proteins found by proteomics could be categorized into a wide range of protein types involved in amino acid/energy metabolism (GS), lipid/energy metabolism (FABP 4), cytoskeleton (STMN1, TBCA), cell proliferation (STMN1, PMF1, RNH1), cell-cell interaction (NANS), cell apoptosis (TRXR1), invasion (SNX9), protein degradation (UBE2K, TOLLIP), and glucose metabolism (AKR1A1). SNX9, UBE2K, PMF1, STMN1, and TRXR1 were confirmed by western immunoblotting in the next step, while FABP 4, GS, TOLLIP, and AKR1A1 were validated by western immunoblotting and cohort studies. The aforementioned four proteins are found to be correlated with galectin-1 expression, tumor progression, and DSS/MFS with substantial clinical significance. The FABP 4 protein may participate in the trafficking of lipids to specific compartments. Abnormal FABP 4 functions may disrupt the lipid-mediated biological processes and result in diseases such as diabetes, obesity, and cancer [26]. Recently, Nieman et al. (2011) reported that FABP 4 demonstrates a higher up-regulation level in human omental metastatic ovarian lesions than in primary clinical ovarian tumor specimens and is over-expressed at the adipocyte/cancer cell interface [38]. Their results suggested that adipocytes support the cancer cells to metastasize from ovary to omentum. In omental adipocytes, lipolysis is induced to produce fatty acid and FABP 4 participates in fatty acid trafficking to cancer cells. In tumor cells, fatty acid is broken down by β-oxidation to fuel cancer cells for fast cell proliferation [38]. In contrast to the ovarian cancer, FABP 4 augmented expression is found to be associated with pTa UBUC with subsequent progression [39]. Our present studies suggested that galectin-1 knockdown reduced FABP 4 expression, likely through modulating PPAR-γ. Taken together, galectin-1 might take part in UBUC progression and invasion through regulating PPAR-γ/FABP 4.

Rapidly growing tumor cells would increase nucleotide and protein synthesis. They rely on the continuous provision of amino acids. Glutamine contains the most abundant amino acid in the body and is also the richest source for the nontoxic form of ammonia. Thus, tumor cells are highly dependent upon glutamine. When tumors begin to grow in the body, they elicit the host to mobilize and augment circulating glutamine. In addition, tumor cells use aerobic glycolysis and glutaminolysis (for breaking down glutamine to α-ketoglutarate to enter tricarboxylic acid (TCA) cycle) to quickly obtain the energy and precursor molecules for the synthesis of macromolecules [25]. Glutamine synthetase catalyzes the conversion of glutamate to glutamine and is thus involved in glutamine metabolism. GS has been implied as an early maker for hepatocellular carcinoma [40] and its positive immunostaining is correlated with both specific and overall mortality at the multivariate level [41]. Our data and the above observations all lead to the implication that galectin-1 might be involved in metabolic transformation. TOLLIP and UBE2K are shown to be associated with protein degradation. TOLLIP is originally recognized as an effector in interleukine-1 and Toll-like receptor signaling pathways [36]. However, it is also considered to take part in the trafficking of ubiquitinated proteins [36]. The increased protein turnover contributes to carcinogenesis by giving anti-apoptotic protection to tumor cells and augmented elimination of abnormal proteins [42]. Our present data showed that galectin-1 knockdown reduced TOLLIP expression and galectin-1 expression was closely associated with TOLLIP expression in the findings of biopsy samples. It might be likely that galectin-1 favors UBUC development by increasing the protein turnover. This is the first observation that TOLLIP was correlated with galectin-1 expression and UBUC patient's DSS/MFS. More molecular and functional studies are required to uncover the role of TOLLIP in UBUC carcinogenesis. In this study, we found that AKR1A1 expression was significantly associated with galectin-1 expression. Since the above results were the first findings, further work should be carried out to elucidate the signaling route by which galectin-1 regulates AKR1A1 expression.

Among the confirmed de-regulated proteins by western immunoblotting, STMN1 is a regulatory protein for microtubule assembly, and thus modulates the cytoskeleton [27]. It also participates in cell cycle regulation [29]. Cytoskeleton changes and impaired cell cycle regulation are essential for carcinogenesis. STMN 1 has been shown to be up-regulated in multiple tumors and is significantly linked to a poor survival rate of the patients [43]. Results in this investigation showed that galectin-1 knockdown decreased STMN 1 expression, suggesting that galectin-1 might control UBUC progression, probably through modulating STMN 1 expression. SNX9 is shown to provoke breast cancer metastasis via modulating RhoGTPase [35]. SNX9 can control the activation of RhoA and cdc42GTPase. It also controls breast cancer cell invasion through the RhoA-ROCK pathway and N-WASP. Our previous findings suggested that galectin-1 may promote UBUC invasion via the Ras–Rac1–MEKK4–JNK pathway [24]. The present observation might uncover an alternative pathway for galectin-1 to contribute to UBUC invasion. PMF1 is recognized as one of the proteins participating in polyamine homeostasis. Polyamine homeostasis impacts the cellular transcription rate and the decreased expression of early growth-associated genes including c-jun, c-myc, and c-fos [30]. Besides, Aleman et al. (2008) reported that PMF1 methylation is significantly linked to UBUC progression [44]. Our present data provided the novel observation that galectin-1 could regulate PMF1 expression.

In conclusion, the results of this study implicated that multi-functional galectin-1 contributed to UBUC carcinogenesis, probably through regulating amino acid/lipid/glucose metabolism, cytoskeleton rearrangement, cellular transcription, cell invasion, and protein degradation.

4. Materials and Methods

4.1. Cell Lines

Human UBUC T24 cells (high grade/invasive) were bought from Bioresource Collection and Research Center, Hsinchu, Taiwan and cultured at 37 °C in McCoy's5A (GIBCO (Life Technologies), Grand Island, NY, USA) medium, supplemented with 10% (*v*/*v*) fetal bovine serum (FBS). Human UBUC J82 cells (high grade) were gifted by Chien-Feng Li from the Department of Pathology, Chi-Mei Medical Center, Tainan, Taiwan and cultured at 37 °C in Dulbecco's Modified Eagle Medium supplemented with 10% (*v*/*v*) FBS (GIBCO, Grand Island, NY, USA).

4.2. Construction of Galectin-1 Knockdown T24 Cells with Short-Hairpin RNA (shRNA)

Galectin-1 knockdown stable clones were established as described previously [24]. In brief, custom-made DNA precursor oligonucleotides of silencing RNA (siRNA) were annealed and the DNA duplex was cloned between *Bgl*II and *Hind*III restriction enzyme sites of the expression vector pSUPER-NEO (Oligoengine Corporate, Seattle, WA, USA), which has a neomycin resistance gene as the selection marker. Two complementary oligonucleotide strands of the siRNA DNA precursor, which would be transcribed into a shRNA processed into a 19-mer small interference RNA (siRNA), were designed with Oligoengine software (version 2.0, Oligoengine Corporate, Seattle, WA, USA). Oligonucleotide DNA sequences are listed in the supplementary File S1. T24 cells were transfected with the siRNA expression plasmid using PolyJet™ transfection reagents (SignaGen Laboratories, Ijamsville, MD, USA) according to the manufacturer's suggestions. Stable knockdown T24 cells were screened in the medium containing 150 μg/mL of G418 antibiotics. T24 cell clones with the best knockdown results were chosen using western immunoblotting and real-time reverse transcriptase polymerase chain reaction (RT-PCR) from 20 cell clones with targeting at +120 nucleotide (nt), named

Sh-Gal-1(+120). Control cells were constructed by the stable transfection of T24 cells with the scrambled siRNA DNA precursor and T24 cell clones with the least impacts on galectin-1 protein amount were chosen from 20 cell clones scrambled at +120 nt and recognized as Sc-Gal-1(+120).

4.3. Preparation of Protein Lysates for Two-Dimensional Gel Electrophoresis

In this study, 70–80% confluent Sh-Gal-1(+120) and Sc-Gal-1(+120) T24 cells were trypsinized and recovered by centrifugation at $1000\times g$ for 5 min at 25 °C and lysed with lysis solution (7 M urea (J.T Baker, Center Valley, PA, USA), 2 M thiourea (J.T Baker, Center Valley, PA, USA), 100 mM dithiothreitol (DTT) (USB Corcorporation, Cleveland, OH, USA), 4% (*v*/*v*) 3-[(3-cholamidopropyl) dimethylammonio]-1-propanesulfonate (CHAPS) (aMResco, Solon, OH, USA), 40 mM Tris-base (pH 10) (aMResco, Solon, OH, USA), 1 mM phenylmethanesulfonylfluoride (PMSF) (aMResco, Solon, OH, USA), and one Complete Mini protease inhibitor cocktail tablet (Roche, Diagnostics, Indianapolis, USA) per liter) by shaking at room temperature for 1 h. Then, the lysates were centrifuged at $349,000\times g$ for 2 h at 15 °C in Type 90 Ti rotor (Beckman Coulter, Fullerton, CA, USA) and proteins present in the supernatant were precipitated with the 2-D Clean-up Kit (GE Healthcare Bio-Sciences AB, Uppsala, Sweden) according to the manufacturer's procedure. The pellet from the 2D Cleanup kit was first re-suspended in MRB buffer (7M Urea, 4% (*v*/*v*) CHAPS) and then incubated at room temperature for 1 h with shaking to dissolve the pellet. Then, the protein concentration of the lysate was quantitated by the Bio-Rad DC protein assay. After measurement, an appropriate amount of thiourea, IPG buffer, and DTT were added into the lysate to re-constitute the rehydration buffer (7 M urea, 2 M thiourea, 0.5% (*v*/*v*) immobilized pH gradient (IPG) buffer (pH 4–7), 4% (*v*/*v*) CHAPS, 1 mM DTT and 0.1 (*w*/*v*) bromophenol blue (aMResco, Solon, OH, USA)) and incubated at room temperature with shaking for another 1 h. Following this, the protein lysates were stored at −80 °C until isoelectric focusing.

4.4. Isoelectric Focusing (IEF) and SDS-Polyacrylamide Gel Electrophoresis (SDS-PAGE)

Isoelectric focusing (IEF) and sodium dodecyl sulfate polyacrylamide gel electrophoresis (SDS-PAGE) were conducted as described previously [45]. The 18-cm immobibline dry strips (pH 4–7) (GE Healthcare Bio-Sciences AB, Uppsala, Sweden) were rehydrated within the BioRad Protean IEF Cell for 16 h at 20 °C with 300 μL rehydration buffer including 200 μg protein lysates respectively prepared from Sh-Gal-1(+120) and Sc-Gal-1(+120) T24 cells. Then, the proteins were concentrated at 20 °C at 50, 100, 200, 500, 1000, 5000, and 8000 V, respectively, with a total of 81,434 voltage-hours. After isoelectric focusing, the gel strips were equilibrated in the equilibration buffer (6 M urea, 30% (*v*/*v*) glycerol (Kanto Chemical, Portland, OR, USA), 2% (*w*/*v*) SDS (aMResco, Solon, OH, USA)) containing 2% (*w*/*v*) DTT for 15 min and then in the equilibration buffer containing 5% (*w*/*v*) iodoacetamide (aMResco, Solon, OH, USA) for a further 15 min. The equilibrated gel was loaded onto the top of a 12.5% (*w*/*v*) polyacrylamide gel and sealed with 0.5% (*w*/*v*) agarose (aMResco, Solon, OH, USA) and the proteins were separated at 420 V using BioRad Protean IIxi until bromophenol blue reached the bottom of the gel.

4.5. Lavapurple Staining, Image Analysis, and Statistical Analysis

Image analysis and statistical analysis were carried out as described formerly with some exceptions [45]. LavaPurple staining was carried out following the manufacturer's suggested protocol (Fluorotechnics, Sydney, Australia). In brief, the 2-DE gels were first soaked within solution I ((1% (*w*/*v*) citric acid and 15% (*v*/*v*) ethanol) for 1.5 h and then incubated in solution II (0.63% (*w*/*v*) boric acid, 0.385% (*w*/*v*) and 0.5% (*v*/*v*) Lavapurple) for 1.5 h in the dark. Lastly, the gels were put in solution III (15% (*v*/*v*) ethanol) for 30 min and then in solution I for 30 min. Lavapurple-stained 2-DE gels were scanned using a Typhoon 9400 fluorescence scanner (GE healthcare, Little Chalfont, UK) with a green laser (green laser PMT: 600 volt and emission filter: 580 BP) to obtain the digital images. To find the de-regulated proteins, a total of 11 pairs of well-separated gel images acquired from Sh-Gal-1(+120) and Sc-Gal-1(+120) T24 cells were compared with PDQuest 8.0.1 (BioRad, Hercules,

CA, USA) software. Differentially expressed protein spots recognized by computer analysis were further verified by visualization. The intensity (volume) of the spot was quantitated and normalized as a percentage (ppm) of the total intensities of all spots in a gel (total normalized volume). Normalized volumes of individual de-regulated protein spots across all the replica gels of Sh-Gal-1(+120) and Sc-Gal-1(+120) T24 cells were first analyzed by the normal distribution test and then the Student's *t*-Test (STATISTICA Version 10.0 MR1, StatSoft, Tulsa, OK, USA) was carried out when reaching normal distribution. However, when normal distribution was not obtained, log transformation was carried out, followed by the normal distribution test and then the Student's *t*-Test. The normalized volume of each spot in Sh-Gal-1(+120) cells was compared to that of the same spot in Sc-Gal-1(+120) cells. In all cases, statistical variance of the Sh-Gal-1(+120): Sc-Gal-1(+120) spot normalized volume ratio within 95% (Student's *t*-Test; $p < 0.05$) was deemed as significantly different. Furthermore, the differentially expressed proteins present in at least six out of 11 gel pairs were regarded as galectin-1-associated proteins.

4.6. In-Gel Digestion and Protein Identification Analysis Via Liquid Chromatography-Tandem Mass Spectrometry (LC-MS/MS)

Since the ideal identification of Lavapurple stained protein spot was not obtained, the silver stained protein spots of interest were picked up for in-gel digestion. The protocol of silver staining is listed in the supplementary File S1. The in-gel digestion and mass spectrometric protein identification were carried out as described previously [45]. Briefly, the peptides present in protein digest were resolved in the LTQ-Orbitrap hybrid tandem mass spectrometer (ThermoFisher, Waltham, MA, USA) in-line coupled with the Agilent 1200 nanoflow HPLC system mounted with LC Packing C18 PepMap 100 (length: 5 mm; internal diameter: 300 µm; bead size: 5 µm) as the trap column and Agilent ZORBAX XDB-C18 (length: 50 mm; internal diameter: 75 µm; bead size: 3.5 µm) as the resolving column. File Converter in the Xcalibur 2.0SR package (ThermoFisher, Waltham, MA, USA), as well as an in-house program, were exploited to obtain the MS/MS information and the charge and mass of each analyzed peptide were calculated. TurboSequest program (ver. 27, rev. 11; Thermo Fisher Scientific, Waltham, MA, USA) was then implemented to determine the best matched peptides from a non-redundant protein database whose FASTA sequences were downloaded from the National Center for Biotechnology Information (ftp://ftp.ncifcrf.gov/pub/nonredun/) on 12 October 2010 with 541,927 entries. Only the tryptic peptides with ≤2 missed cuts were recorded. During the database search, the mass ranges were 1 and 3.5 m/z for fragment and precursor ions, respectively. The protein identities were only confirmed when there were at least two peptides matched and search results had a high Xcore (i.e., ≥2.0 for doubly charged peptides and ≥3.0 for triply charged ones) and minimal differences between the observed and hypothetical masses (i.e., ΔM < 10 ppm). For each set of MS/MS analyses, 25 fmol of bovine serum albumin (BSA) (aMResco, Solon, OH, USA) in gel was analyzed in parallel for verification of the effectiveness of the entire protein identification procedure, including in-gel digestion, nanoflow HPLC, MS/MS, and informatics analyses. The experimental data were only taken into account when a 10 ppm mass accuracy and over 70% coverage were observed in the co-processed BSA samples.

4.7. Western Immunoblotting

Western immunoblotting was carried out as mentioned before [45]. Sh-Gal-1(+120) and Sc-Gal-1(+120) cells were collected and lysed in the lysis buffer (10 mM Tris (pH 8.0), 0.32 M sucrose (Avantor Performance Matericals, Center Valley, PA, USA), 1% (*v/v*) Triton X-100 (Merck Millipore, Darmstadt, Germany), 5 mM EDTA (Merck Millipore, Darmstadt, Germany), 2 mM DTT and 1 mM PMSF). After measurement of the protein concentration with the Bio-Rad DC protein assay kit, an equal volume of 2× sample buffer (0.1M Tris (pH 6.8), 2% (*w/v*) SDS, 0.2% (*v/v*) β-mercaptoethanol (aMResco, Solon, OH, USA), 10% (*v/v*) glycerol, and 0.0016% (*w/v*) bromophenol blue) was mixed with the protein lysate. Suitable amounts of the protein lysates were resolved using electrophoresis at 100 V with 10% (w/v) SDS-PAGE, and further transferred onto PVDF membranes (Strategene, La Jolla, CA, USA). After blocking for 1 h in 3% (w/v) bovine albumin serum (BSA) (aMResco, Solon, OH, USA) at room temperature, membranes were probed for 2 h at room temperature with primary antibodies (UBE2K (1:10,000), STMN1 (1:8000), TOLLIP (1:2000), FABP 4 (1:1000), AKR1A1(1:5000), GS (1:500) and actin (1:5000) from Merck Millipore, Darmstadt, Germany; PMF1 (1:1000), SNX9 (1:2000), TRXR1 (1:1000) from Santa Cruz Biotechnology, Dallas, TX, USA). The membranes were washed and then hybridized with appropriate secondary antibodies for 1 h at room temperature. Secondary antibodies binding on the membranes were detected by the chemiluminescence ECL detection system (GE Healthcare Bio-Sciences AB, Uppsala, Sweden) using the Fujifilm LAS-4000 Luminescent Image Analyzer (Fujifilm Corporation, Tokyo, Japan). The intensity of each protein band, normalized with the actin protein expression level, was determined by PDQUEST Quantity One software (Bio-Rad Laboratory, Hercules, CA, USA) and analyzed with the Student's *t*-Test (STATISTICA Ver 10.0 MR1, StatSoft, Tulsa, OK, USA).

4.8. Case Selection

For immunohistochemistry, the Institutional Review Board (IRB) of Chi Mei Medical Center admitted the retrospective retrieval of 295 available primary UBUC blocks (Identification code: IRB10102-004, approved by IRB of Chi Mei Medical Center, project started from 14 Feb. 2012), which were adopted from the patients undergoing surgical resection with curative intent between January 1998 and May 2004. However, those who received the palliative resection were excluded. Patients with validated or suspicious lymph node metastasis underwent regional lymph node dissection. Cisplatin-based postoperative adjuvant chemotherapy was carried out in patients with pT3-pT4 status or nodal linkage. The UBUC histologic diagnosis was validated in all specimens according to the recent World Health Organization classification. Histologic grading was identified based upon Edmonson–Steiner criteria, while tumor staging was recognized according to the seventh edition of the American Joint Committee on Cancer system. Medical charts were reviewed for each patient to ensure the accuracy of other related clinicopathologic data. Follow-up information was available in all cases with a median period of 42 months (ranging 3–176 months).

4.9. Immunohistochemistry (IHC) and Statistical Analysis

IHC was conducted on representative tissue sections dissected from formalin-fixed, paraffin-embedded tumor blocks at a 3-μm thickness. Xylene was first used to deparaffinize the slides. Then, the slides were rehydrated with ethanol and heated using a microwave in a 10 mM citrate buffer (pH 6) for 7 min to retrieve the antigen epitopes. A total of 3% H_2O_2 was used to quench endogenous peroxidase. After antigen retrieval, slides were washed in Tris-buffered saline for 15 min and then hybridized with a primary antibody again for 1 h, followed by antibody detection using a ChemMate EnVision[TM] kit (K5001; DAKO, Glostrup, Denmark). Two expert pathologists (CF Li and HL He) blinded to clinicopathological information and patient outcomes interpreted the immunostainings. The immune-intensities of membranous and cytosolic staining of Galectin 1, FABP 4, GS, TOLLIP, and AKRAK1 were respectively recorded by *H* score. Scoring of the immunoreactivity was measured based on a composite of both the percentage and intensity of the cytoplasm of positively stained tumor cells to generate an *H* score, using the equation:

$$H \text{ score} = \sum P_i(I + 1)$$

where *i* is the intensity of the stained tumor cells (0 to 3+), and Pi is the percentage of stained tumor cells for each intensity varying from 0% to 100%. The specimens that showed an H-score above the median value were defined as high expression, while the cases below the median value were defined as low expression. The median H-score of all specimens was regarded as the cutoff point to divide high and low expression.

Statistical analyses were carried out with SPSS 14.0. The immuno-expressions of Galectin 1 (Invitrogen, Carlsbad, CA, USA), FABP 4 (Epitoimcs, Burlingame, CA, USA), GS (BD Transduction Laboratories™-BD Bioscience, San Jose, CA, USA), TOLLIP (Novus Biologicals, Littleton Colorado, USA), and AKRAK1 (Abcam, Cambridge, MA, USA) were compared by various parameters with the chi-square test. The end points evaluated in the whole cohort were DSS and MFS measured based on the date of UBUC operation to disease-related mortality and metastasis. In stepwise forward fashion, parameters at univariate $p < 0.05$ were basically analyzed for their relative prognostic importance using the Cox regression model. Of all analyses, the two-sided test of significance with $p < 0.05$ was considered significant.

5. Conclusions

The results of this study implicated that multi-functional galectin-1 contributed to UBUC carcinogenesis, probably through regulating amino acid/lipid/glucose metabolism, cytoskeleton rearrangement, cellular transcription, cell invasion, and protein degradation.

Supplementary Materials: The following are available online at http://www.mdpi.com/1422-0067/19/4/1242/s1.

Acknowledgments: The authors would like to thank the Ministry of Science and Technology of Taiwan for offering the grants MOST 105-2320-B-218 -001-MY3 and NSC 99-2314-B-384-003-MY3 to support this research. We also would like to thank the technical services (supports) provided by Proteomics Research Center of the National Yang-Ming University, Taiwan.

Author Contributions: Ting-Feng Wu and Kun-Hung Shen conceived and designed the experiments; Chien-Feng Li and Hong-Lin He performed IHC, specimens scoring, and univariate/multivariate analyses; Lan-Hsiang Chien, Cheng-Hao Huang, and Ting-Feng Wu carried out all the western immunblotting; Ting-Feng Wu, Kun-Hung Shen, and Chien-Feng Li wrote the paper.

Conflicts of Interest: The authors declare no conflict of interest.

Abbreviations

UBUC	Urinary bladder urothelial carcinoma
IHC	Immunohistochemistry
MMP	Matrix metalloproteinase protein
OSCC	Oral squamous cell carcinoma
2-DE	Two-dimensional gel electrophoresis
LC-MS/MS	Liquid chromatography/tandem mass spectrometry
PPAR-γ	Peroxisome proliferator-activated receptor gamma
RNH1	Ribonuclease/angiogenin inhibitor 1
RCN1	Reticulocalbin 1
LGALS1	Galectin 1
TBCA	Tubulin specific chaperone A
UBE2K	Ubiquitin conjugating enzyme E2 K
PMF1	Polyamine-modulated factor 1
STMN1	Stathmin 1
TOLLIP	toll interacting protein
CWC15	protein CWC15 homolog
SNX9	sorting nexin 9
DCPS	Scavenger mRNA-decapping enzyme
FABP 4	fatty acid binding protein 4
TRXR1	thioredoxin reductase 1
AKR1A1	Alcohol dehydrogenase (NADP$^+$)
NANS	N-acetylneuraminate synthase (sialic acid synthase)
GS	glutamine synthetase
R. R.	relative risk

References

1. Stein, J.P.; Grossfeld, G.D.; Ginsberg, D.A.; Esrig, D.; Freeman, J.A.; Figueroa, A.J.; Skinner, D.G.; Cote, R.J. Prognostic markers in bladder cancer: A contemporary review of the literature. *J. Urol.* **1998**, *160*, 645–659. [CrossRef]
2. Camby, I.; Le Mercier, M.; Lefranc, F.; Kiss, R. Galectin-1: A small protein with major functions. *Glycobiology* **2006**, *16*, 137R–157R. [CrossRef] [PubMed]
3. Benvenuto, G.; Carpentieri, M.L.; Salvatore, P.; Cindolo, L.; Bruni, C.B.; Chiariotti, L. Cell-specific transcriptional regulation and reactivation of galectin-1 gene expression are controlled by DNA methylation of the promoter region. *Mol. Cell. Biol.* **1996**, *16*, 2736–2743. [CrossRef] [PubMed]
4. Paz, A.; Haklai, R.; Elad-Sfadia, G.; Ballan, E.; Kloog, Y. Galectin-1 binds oncogenic H-Ras to mediate Ras membrane anchorage and cell transformation. *Oncogene* **2001**, *20*, 7486–7493. [CrossRef] [PubMed]
5. Thijssen, V.L.; Barkan, B.; Shoji, H.; Aries, I.M.; Mathieu, V.; Deltour, L.; Hackeng, T.M.; Kiss, R.; Kloog, Y.; Poirier, F.; et al. Tumor cells secrete galectin-1 to enhance endothelial cell activity. *Cancer Res.* **2010**, *70*, 6216. [CrossRef] [PubMed]
6. Jung, T.Y.; Jung, S.; Ryu, H.H.; Jeong, Y.I.; Jin, Y.H.; Jin, S.G.; Kim, I.Y.; Kang, S.S.; Kim, H.S. Role of galectin-1 in migration and invasion of human glioblastoma multiforme cell lines. *J. Neurosurg.* **2008**, *109*, 273–284. [CrossRef] [PubMed]
7. Hsieh, S.H.; Ying, N.W.; Wu, M.H.; Chiang, W.F.; Hsu, C.L.; Wong, T.Y.; Jin, Y.T.; Hong, T.M.; Chen, Y.L. Galectin-1, a novel ligand of neuropilin-1, activates VEGFR-2 signaling and modulates the migration of vascular endothelial cells. *Oncogene* **2008**, *27*, 3746–3753. [CrossRef] [PubMed]
8. Wu, M.H.; Hong, T.M.; Cheng, H.W.; Pan, S.H.; Liang, Y.R.; Hong, H.C.; Chiang, W.F.; Wong, T.Y.; Shieh, D.B.; Shiau, A.L.; et al. Galectin-1-mediated tumor invasion and metastasis, up-regulated matrix metalloproteinase expression, and reorganized actin cytoskeletons. *Mol. Cancer Res.* **2009**, *7*, 311–318. [CrossRef] [PubMed]
9. Rubinstein, N.; Alvarez, M.; Zwirner, N.W.; Toscano, M.A.; Ilarregui, J.M.; Bravo, A.; Mordoh, J.; Fainboim, L.; Podhajcer, O.L.; Rabinovich, G.A. Targeted inhibition of *galectin-1* gene expression in tumor cells results in heightened T cell-mediated rejection; a potential mechanism of tumor immune privilege. *Cancer Cell* **2004**, *5*, 241–251. [CrossRef]
10. Dalotto-Moreno, T.; Croci, D.O.; Cerliani, J.P.; Martinez-Allo, V.C.; Dergan-Dylon, S.; Méndez-Huergo, S.P.; Stupirski, J.C.; Mazal, D.; Osinaga, E.; Toscano, M.A.; et al. Targeting galectin-1 overcomes breast cancer associated immunosuppression and prevents metastatic disease. *Cancer Res.* **2013**, *73*, 1107–1117. [CrossRef] [PubMed]
11. Wu, T.F.; Li, C.F.; Chien, L.H.; Shen, K.H.; Huang, H.Y.; Su, C.C.; Liao, A.C. Galectin-1 dysregulation independently predicts disease specific survival in bladder urothelial carcinoma. *J. Urol.* **2015**, *193*, 1002–1008. [CrossRef] [PubMed]
12. Barrow, H.; Rhodes, J.M.; Yu, L.G. The role of galectins in colorectal cancer progression. *Int. J. Cancer* **2011**, *129*, 1–8. [CrossRef] [PubMed]
13. Szoke, T.; Kayser, K.; Baumhakel, J.D.; Trojan, I.; Furak, J.; Tiszlavicz, L.; Horvath, A.; Szluha, K.; Gabius, H.J.; Andre, S. Prognostic significance of endogenous adhesion/growth-regulatory lectins in lung cancer. *Oncology* **2005**, *69*, 167–174. [CrossRef] [PubMed]
14. Saussez, S.; Camby, I.; Toubeau, G.; Kiss, R. Galectins as modulators of tumor progression in head and neck squamous cell carcinomas. *Head Neck* **2007**, *29*, 874–884. [CrossRef] [PubMed]
15. Chow, S.N.; Chen, R.J.; Chen, C.H.; Chang, T.C.; Chen, L.C.; Lee, W.J.; Shen, J.; Chow, L.P. Analysis of protein profiles in human epithelial ovarian cancer tissues by proteomic technology. *Eur. J. Gynaecol. Oncol.* **2010**, *31*, 55–62. [PubMed]
16. Laderach, D.J.; Gentilini, L.D.; Giribaldi, L.; Delgado, V.C.; Nugnes, L.; Croci, D.O.; Al Nakouzi, N.; Sacca, P.; Casas, G.; Mazza, O.; et al. A unique galectin signature in Human prostate cancer progression suggests galectin-1 as a key target for treatment of advanced disease. *Cancer Res.* **2013**, *73*, 86–96. [CrossRef] [PubMed]
17. Rorive, S.; Belot, N.; Decaestecker, C.; Lefranc, F.; Gordower, L.; Micik, S.; Maurage, C.A.; Kaltner, H.; Ruchoux, M.M.; Danguy, A.; et al. Galectin-1 is highly expressed in human gliomas with relevance for modulation of invasion of tumor astrocytes into the brain parenchyma. *Glia* **2001**, *33*, 241–255. [CrossRef]

18. Croci, D.O.; Salatino, M.; Rubinstein, N.; Cerliani, J.P.; Cavallin, L.E.; Leung, H.J.; Ouyang, J.; Ilarregui, J.M.; Toscano, M.A.; Domaica, C.I.; et al. Disrupting galectin-1 interactions with *N*-glycans suppresses hypoxia-driven angiogenesis and tumorigenesis in Kaposi's sarcoma. *J. Exp. Med.* **2012**, *209*, 1985–2000. [CrossRef] [PubMed]

19. D'Haene, N.; Maris, C.; Sandras, F.; Dehou, M.F.; Remmelink, M.; Decaestecker, C.; Salmon, I. The differential expression of galectin-1 and galectin-3 in normal lymphoid tissue and non-Hodgkins and Hodgkins lymphomas. *Int. J. Immunopathol. Pharmacol.* **2005**, *18*, 431–443. [CrossRef] [PubMed]

20. Puchades, M.; Nilsson, C.L.; Emmett, M.R.; Aldape, K.D.; Ji, Y.; Lang, F.F.; Liu, T.J.; Conrad, C.A. Proteomic investigation of glioblastoma cell lines treated with wild-type p53 and cytotoxic chemotherapy demonstrates an association between galectin-1 and p53 expression. *J. Proteome Res.* **2007**, *6*, 869–875. [CrossRef] [PubMed]

21. Camby, I.; Belot, N.; Lefranc, F.; Sadeghi, N.; de Launoit, Y.; Kaltner, H.; Musette, S.; Darro, F.; Danguy, A.; Salmon, I.; et al. Galectin-1 modulates human glioblastoma cell migration into the brain through modifications to the actin cytoskeleton and levels of expression of small GTPases. *J. Neuropathol. Exp. Neurol.* **2002**, *61*, 585–596. [CrossRef] [PubMed]

22. Chiang, W.F.; Liu, S.Y.; Fang, L.Y.; Lin, C.N.; Wu, M.H.; Chen, Y.C.; Chen, Y.L.; Jin, Y.T. Overexpression of galectin-1 at the tumor invasion front is associated with poor prognosis in early-stage oral squamous cell carcinoma. *Oral Oncol.* **2008**, *44*, 325–334. [CrossRef] [PubMed]

23. Blaževitš, O.; Mideksa, Y.G.; Šolman, M.; Ligabue, A.; Ariotti, N.; Nakhaeizadeh, H.; Fansa, E.K.; Papageorgiou, A.C.; Wittinghofer, A.; Ahmadian, M.R.; et al. Galectin-1 dimers can scaffold Raf-effectors to increase H-ras nanoclustering. *Sci. Rep.* **2016**, *6*, 24165. [CrossRef] [PubMed]

24. Shen, K.H.; Li, C.F.; Chien, L.H.; Huang, C.H.; Su, C.C.; Liao, A.C.; Wu, T.F. Role of galectin-1 in urinary bladder urothelial carcinoma cell invasion through the JNK pathway. *Cancer Sci.* **2016**, *107*, 1390–1398. [CrossRef] [PubMed]

25. Medina, M.A. Glutamine and cancer. *J. Nutr.* **2001**, *131* (Suppl. S9), 2539S–2542S. [CrossRef] [PubMed]

26. Furuhashi, M.; Hotamisligil, G.S. Fatty acid-binding proteins: Role in metabolic diseases and potential as drug targets. *Nat. Rev. Drug Discov.* **2008**, *7*, 489–503. [CrossRef] [PubMed]

27. Cassimeris, L. The oncoprotein 18/stathmin family of microtubule destabilizers. *Curr. Opin. Cell Biol.* **2002**, *14*, 18–24. [CrossRef]

28. Tian, G.; Cowan, N.J. Tubulin-specific chaperones: Components of a molecular machine that assembles the α/β heterodimer. *Methods Cell. Biol.* **2013**, *115*, 155–171. [CrossRef] [PubMed]

29. Rubin, C.I.; Atweh, G.F. The role of stathmin in the regulation of the cell cycle. *J. Cell. Biochem.* **2004**, *93*, 242–250. [CrossRef] [PubMed]

30. Wang, J.Y.; McCormack, S.A.; Viar, M.J.; Wang, H.; Tzen, C.Y.; Scott, R.E.; Johnson, L.R. Decreased expression of protooncogenes c-fos, c-myc, and c-jun following polyamine depletion in IEC-6 cells. *Am. J. Physiol.* **1993**, *265 Pt 1*, G331–G338. [CrossRef] [PubMed]

31. Pizzo, E.; Sarcinelli, C.; Sheng, J.; Fusco, S.; Formiggini, F.; Netti, P.; Yu, W.; D'Alessio, G.; Hu, G.F. Ribonuclease/angiogenin inhibitor 1 regulates stress-induced subcellular localization of angiogenin to control growth and survival. *J. Cell Sci.* **2013**, *126 Pt 18*, 4308–4319. [CrossRef] [PubMed]

32. Büll, C.; Stoel, M.A.; den Brok, M.H.; Adema, G.J. Sialic acids sweeten a tumor's life. *Cancer Res.* **2014**, *74*, 3199–31204. [CrossRef] [PubMed]

33. Chen, W.; Zou, P.; Zhao, Z.; Weng, Q.; Chen, X.; Ying, S.; Ye, Q.; Wang, Z.; Ji, J.; Liang, G. Selective killing of gastric cancer cells by a small molecule via targeting TrxR1 and ROS-mediated ER stress activation. *Oncotarget* **2016**, *7*, 16593–16609. [CrossRef] [PubMed]

34. Xu, S.; Xu, Y.; Chen, L.; Fang, Q.; Song, S.; Chen, J.; Teng, J. RCN1 suppresses ER stress-induced apoptosis via calcium homeostasis and PERK-CHOP signaling. *Oncogenesis* **2017**, *6*, e304. [CrossRef] [PubMed]

35. Bendris, N.; Williams, K.C.; Reis, C.R.; Welf, E.S.; Chen, P.H.; Lemmers, B.; Hahne, M.; Leong, H.S.; Schmid, S.L. SNX9 promotes metastasis by enhancing cancer cell invasion via differential regulation of RhoGTPases. *Mol. Biol. Cell* **2016**, *27*, 1409–1419. [CrossRef] [PubMed]

36. Zhu, L.; Wang, L.; Luo, X.; Zhang, Y.; Ding, Q.; Jiang, X.; Wang, X.; Pan, Y.; Chen, Y. Tollip, an intracellular trafficking protein, is a novel modulator of the transforming growth factor-β signaling pathway. *J. Biol. Chem.* **2012**, *287*, 39653–39663. [CrossRef] [PubMed]

37. Boiteux, G.; Lascombe, I.; Roche, E.; Plissonnier, M.L.; Clairotte, A.; Bittard, H.; Fauconnet, S. A-FABP, a candidate progression marker of human transitional cell carcinoma of the bladder, is differentially regulated by PPAR in urothelial cancer cells. *Int. J. Cancer* **2009**, *124*, 1820–1828. [CrossRef] [PubMed]

38. Nieman, K.M.; Kenny, H.A.; Penicka, C.V.; Ladanyi, A.; Buell-Gutbrod, R.; Zillhardt, M.R.; Romero, I.L.; Carey, M.S.; Mills, G.B.; Hotamisligil, G.S.; et al. Adipocytes promote ovarian cancer metastasis and provide energy for rapid tumor growth. *Nat. Med.* **2011**, *17*, 1498–14503. [CrossRef] [PubMed]

39. Wild, P.J.; Herr, A.; Wissmann, C.; Stoehr, R.; Rosenthal, A.; Zaak, D.; Simon, R.; Knuechel, R.; Pilarsky, C.; Hartmann, A. Gene expression profiling of progressive papillary noninvasive carcinomas of the urinary bladder. *Clin. Cancer Res.* **2005**, *11*, 4415–4429. [CrossRef] [PubMed]

40. Long, J.; Lang, Z.W.; Wang, H.G.; Wang, T.L.; Wang, B.E.; Liu, S.Q. Glutamine synthetase as an early marker for hepatocellular carcinoma based on proteomic analysis of resected small hepatocellular carcinomas. *Hepatobiliary Pancreat. Dis. Int.* **2001**, *9*, 296–305.

41. Dal Bello, B.; Rosa, L.; Campanini, N.; Tinelli, C.; Torello Viera, F.; D'Ambrosio, G.; Rossi, S.; Silini, E.M. Glutamine synthetase immunostaining correlates with pathologic features of hepatocellular carcinoma and better survival after radiofrequency thermal ablation. *Clin. Cancer Res.* **2010**, *16*, 2157–2166. [CrossRef] [PubMed]

42. Mani, A.; Gelmann, E.P. The ubiquitin-proteasome pathway and its role in cancer. *J. Clin. Oncol.* **2005**, *23*, 4776–4789. [CrossRef] [PubMed]

43. Belletti, B.; Baldassarre, G. Stathmin: A protein with many tasks. New biomarker and potential target in cancer. *Expert Opin. Ther. Targets* **2011**, *15*, 1249–1266. [CrossRef] [PubMed]

44. Aleman, A.; Cebrian, V.; Alvarez, M.; Lopez, V.; Orenes, E.; Lopez-Serra, L.; Algaba, F.; Bellmunt, J.; López-Beltrán, A.; Gonzalez-Peramato, P.; et al. Identification of PMF1 methylation in association with bladder cancer progression. *Clin. Cancer Res.* **2008**, *14*, 8236–8243. [CrossRef] [PubMed]

45. Wu, T.F.; Hsu, L.T.; Tsang, B.X.; Huang, L.C.; Shih, W.Y.; Chen, L.Y. Clarification of the molecular pathway of Taiwan local pomegranate fruit juice underlying the inhibition of urinary bladder urothelial carcinoma cell by proteomics strategy. *BMC Complement. Altern. Med.* **2016**, *16*, 96. [CrossRef] [PubMed]

International Journal of
Molecular Sciences

MDPI

Article

Overall Survival of Ovarian Cancer Patients Is Determined by Expression of Galectins-8 and -9

Heiko Schulz [1], Christina Kuhn [1], Simone Hofmann [1], Doris Mayr [2], Sven Mahner [1], Udo Jeschke [1,*] and Elisa Schmoeckel [2]

[1] LMU Munich, University Hospital, Department of Obstetrics and Gynecology, Maistrasse 11, 80337 Munich, Germany; heiko.schulz@med.uni-muenchen.de (H.S.); christina.kuhn@med.uni-muenchen.de (C.K.); simone.hofmann@med.uni-muenchen.de (S.H.); sven.mahner@med.uni-muenchen.de (S.M.)
[2] LMU Munich, Department of Pathology, Ludwig Maximilians University of Munich, Thalkirchner Str. 142, 80337 Munich, Germany; doris.mayr@med.uni-muenchen.de (D.M.); elisa.schmoeckel@med.uni-muenchen.de (E.S.)
*   Correspondence: udo.jeschke@med.uni-muenchen.de; Tel.: +49-89-4400-54-240

Received: 28 November 2017; Accepted: 18 January 2018; Published: 22 January 2018

Abstract: The evaluation of new prognostic factors that can be targeted in ovarian cancer diagnosis and therapy is of the utmost importance. Galectins are a family of carbohydrate binding proteins with various implications in cancer biology. In this study, the presence of galectin (Gal)-8 and -9 was investigated in 156 ovarian cancer samples using immunohistochemistry (IHC). Staining was evaluated using semi-quantitative immunoreactivity (IR) scores and correlated to clinical and pathological data. Different types of galectin expression were compared with respect to disease-free survival (DFS) and overall survival (OS). Gal-8 served as a new positive prognostic factor for the OS and DFS of ovarian cancer patients. Gal-9 expression determined the DFS and OS of ovarian cancer patients in two opposing ways—moderate Gal-9 expression was correlated with a reduced outcome as compared to Gal-9 negative cases, while patients with high Gal-9 expression showed the best outcome.

Keywords: galectin-8; galectin-9; immunochemistry; ovarian cancer; prognostic factor; disease-free survival; overall survival

---

1. Introduction

Ovarian cancer is the fifth leading cause of cancer death among women of all ages [1]. Due to its frequent diagnosis in advanced stages, characterized by a wide cancer dissemination into the peritoneum and the acquisition of chemo resistance after treatment [2], ovarian cancer displays 5-year relative survival rates of less than 50% [3]. Ovarian cancer management lacks effective screening methods and specific treatment options. As prognosticators in ovarian cancer, the histological subtype, disease stage at diagnosis, extent of residual disease after surgery, and volume of ascites can be used [4]. However, except for breast cancer gene (*BRCA*) status, no biological prognostic factor is commonly considered [4]. Various studies have attempted to introduce new prognostic factors in ovarian cancer, and for several proteins a prognostic value independent of clinical parameters has been detected. However, so far none of them can be applied in ovarian cancer therapy or diagnosis. Hence, there is a tremendous need for the evaluation of new prognostic factors that can be targeted in ovarian cancer.

In 1994 the galectin (Gal) family was described as group of proteins sharing a binding affinity for β-galactosides, with significant similarity in the carbohydrate- recognition domain (CRD) [5,6]. Since then, the galectin family has grown in members. In total, 10 different galectins (Gal-1–4, Gal-7–10, Gal-12, and Gal-13) are known to be present in humans [7]. According to the arrangement of CRDs, galectins can be subdivided into three groups. Prototype galectins contain a single CRD, often forming homodimers, while tandem-repeat galectins contain two CRDs connected by a linker chain, and chimeric galectins (a group containing only member Gal-3) have a second N-terminal domain connected to a single CRD [8]. In this study, we will focus on two tandem-repeat galectins, Gal-8 and -9. By binding β-galactosides on certain glyoproteins with their CRDs, galectins are known to modulate cell–cell and cell–matrix interactions as well as intracellular pathways [6]. Galectins have been discovered to play an important role in several diseases including cancer [9]. Several mechanisms of tumor biology, also referred to as "hallmarks of cancer", are known to be influenced by galectins: enhanced proliferation, resistance to cell death, and induction of angiogenesis, as well as tumor invasion and metastasis [7,10,11]. Therefore, galectin expression in cancer tissues of several malignancies has been found to affect patients' disease-free survival (DFS) or overall survival (OS). For this reason, several studies assessed different galectins as prognostic survival markers, but thus far, most efforts have been spent on Gal-1 and -3. However, Gal-8 and -9 have been evaluated as prognostic markers in few cancer types. In triple-negative breast cancer, for instance, patients displaying Gal-8 expression in nuclei had significantly better DFS and OS [12]. Higher Gal-9 expression, on the other hand, was associated with prolonged OS of gastric cancer patients [13].

In ovarian cancer, however, most previous studies focused on galectin-1, -3 and -7 as prognostic factors [14–19]. Our group recently published an article in the international journal of molecular sciences, presenting high tumor and stroma Gal-1 expression, as well as higher Gal-7 expression as negative prognostic markers for OS of ovarian cancer patients, while nuclear Gal-3 expression was correlated with a better OS [20]. In fact, to our knowledge there is only one very recent study on Gal-8 and Gal-9 in ovarian cancer [21]. In this study, high epithelial Gal-8 expression was associated with the acquisition of chemo resistance. However, no correlation with DFS or OS was observed. In the same study, the Gal-9 expression that was observed in "cytosolic or perinuclear puncta", was correlated with poor OS. However, this special Gal-9 expression was not associated with altered DFS. Cytoplasmic Gal-9 expression, however, showed no association to either DFS or OS. In general, with a 5-year follow-up time, all of the observations were limited to a rather short period of observation and the analysis was performed in a collective of only high-grade serous ovarian cancer samples, with their prognostic role in other than subtypes remaining elusive. Summing up, there are a limited number of studies on Gal-8 and -9 in ovarian cancer and several aspects of their prognostic features still remain to be elucidated.

Therefore, in this study, we evaluated the prognostic influence of Gal-8 and -9 in patients with epithelial ovarian cancer using immunohistochemistry and analyzed correlations to each other and to clinical and pathological parameters. We hypothesize that Gal-8 and -9 are prognostic for overall survival in ovarian cancer patients. Since it is known that galectin function and their effect on patients' survival can be determined by expression in the nucleus or cytoplasm of cancer cells as well as the peritumoral stroma, we paid attention to the specific location of galectin expression in our analysis.

2. Results

2.1. In Silico Analysis of Gal-8 and -9 Expression in Normal Ovarian Tissues and Ovarian Cancer

The human protein atlas (available at www.proteinatlas.org) was used to analyze Gal-8 and -9 expression in normal ovarian tissues as well as ovarian cancer tissues [22]. In ovarian stroma cells, Gal-8 (human Gal-8 gene, *LGALS8*) was not detected via antibody staining. However out of 12 ovarian cancer tissues, 8 showed medium Gal expression. For Gal-9 (human Gal-9 gene, LGALS9), out of 12 ovarian cancer patients, 3 showed medium Gal-9 expression and 5 patients showed low Gal-9 expression. In normal ovarian tissues, Gal-9 was found to have low expression in ovarian stromal cells. According to this, both Gal-8 and Gal-9 seem to be altered in ovarian cancer compared to normal ovarian tissues. This further motivated us to specify Gal-8 and -9 expression in ovarian cancer tissues using immunochemistry.

2.2. Gal-8 is a Positive Prognostic Factor for OS and DFS in Ovarian Cancer Patients

Galectin-8 staining could be evaluated in 143 ovarian cancer samples. Gal-8 expression occurred predominantly in the cytoplasm and nuclei of ovarian cancer cells but not in the peritumoral stroma (Figure 1). In total, 96 cases (67.1%) showed a high Gal-8 expression in the cytoplasm (immunoreactivity score, IRS > 1), while in 47 specimens (32.9%), only low Gal-8 expression was observed (IRS $\leq$ 1). The median IRS of Gal-8 staining in the cytoplasm was 3. According to chi-squared statistics, low Gal-8 expression in the cytoplasm correlated with lymph node metastasis as well as a higher International Federation of Gynecology and Obstetrics (FIGO) stage ($p = 0.019$, $p = 0.033$, respectively) and (Table 1). In 70 of the samples (51.4%) Gal-8 positive nuclei were observed, while 74 specimens (48.6%) did not present with nuclear Gal-8 staining. Positive nuclear Gal-8 staining was more often observed in lower FIGO stages ($p = 0.011$) and (Table 1). Besides, no other correlation of nuclear Gal-8 expression and clinical or pathological data was observed.

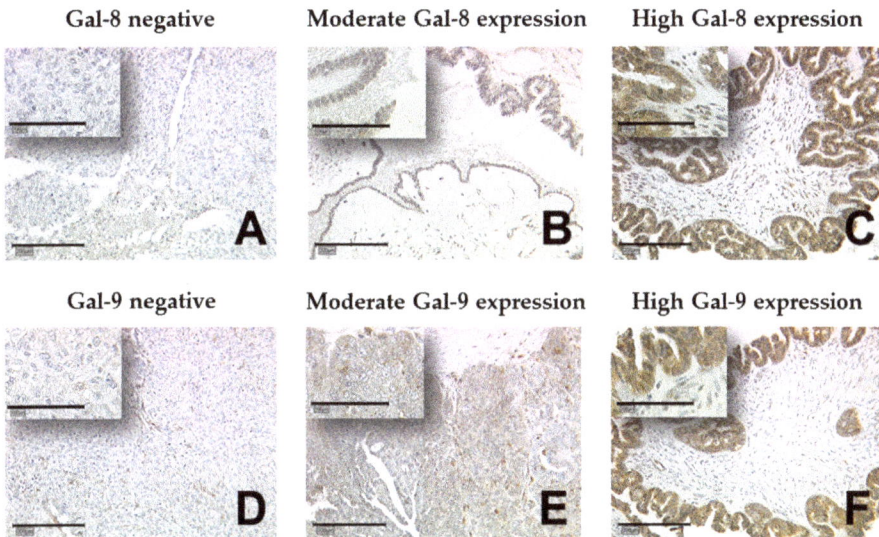

Figure 1. Detection of galectin-8 and -9 using immunohistochemistry (IHC). Gal-8 staining (**A–C**) and Gal-9 staining (**D–F**) was predominantly present in the cytoplasm of ovarian cancer cells but not in the peritumoral stroma. Representative photomicrographs are shown. There is a 10× magnification for the outer pictures and 25× for the inserts. Scale bar in (**A**) equals 200 µm (outer pictures) and 100 µm (inserts). Gal: galectin.

Table 1. Gal-8 and -9 staining correlated with clinical and pathological data.

	Gal-8 Expression (Cytoplasm)			Gal-8 Expression (Nucleus)			Gal-9 Expression (Cytoplasm)			
	Low	High	p-Value	Negative	Positive	p-Value	Negative	Moderate	High	p-Value
Histology										
Serous	40	62	NS	54	48	NS	24	71	9	0.024
Clear cell	2	9		3	8		1	10	0	
Endometrioid	2	17		8	11		6	10	5	
Mucinous	3	8		5	7		1	6	4	
Tumor Stage										
pT1	10	26	NS	14	23	NS	8	20	9	0.018
pT2+	37	69		55	51		24	77	8	
Lymph node										
pN0/pNX	25	70	0.019	43	53	NS	26	61	12	NS
pN1	22	26		27	21		6	36	6	
Distant Metastasis										
pM0/pMX	45	94	NS	68	72	NS	32	92	17	NS
pM1	2	2		2	2		0	5	1	
Grading										
G1	7	26	NS	13	21	NS	7	19	8	0.006
G2+	37	62		53	46		24	72	5	
FIGO										
I/II	8	33	0.033	14	28	0.011	6	25	11	0.002
III/IV	37	60		55	42		25	69	6	
Age										
≤60 years	24	51	NS	32	43	NS	10	52	17	<0.001
>60 years	23	45		38	31		22	45	1	

TNM staging was accomplished according to the Union for International Cancer Control (UICC); pT1 = tumor stage 1; pT2+ = tumor stage 2 or higher; pN0 = lymph node stage 0; pNX = lymph node stage not evaluated; pN1 = lymph node stage 1; pM0 = distant metastasis stage 0; pMX = distant metastasis not evaluated; pM1 = distant metastasis stage 1; G1 = grade 1; G2+ = grade 2 or higher; NS = Not significant ($p > 0.05$).

Different groups of Gal-8 expression were compared using Kaplan-Meier analysis (Figure 2). Patients presenting with high Gal-8 expression showed a significantly better overall survival and disease-free survival ($p = 0.024$, $p = 0.018$, respectively). Nuclear Gal-8 expression had no significant influence on overall or disease-free survival. In multivariate analysis, Gal-8 staining served as a prognostic factor independent of clinical and pathological variables (Table 2).

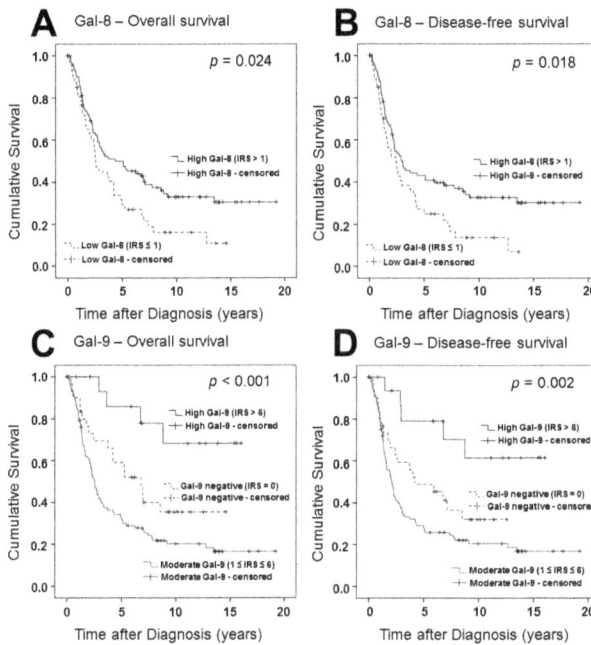

Figure 2. Survival times of patient groups with different galectin-8 and -9 expression levels were compared. Patients with high Gal-8 expression in the cytoplasm showed better progression-free (**A**) and overall survival (**B**) compared to patients without or with low Gal-8 expression. Cases with a moderate Gal-9 expression in the cytoplasm displayed a reduced progression-free (**C**) and overall survival (**D**) compared to Gal-9 negative cases. However, patients with high Gal-9 expression showed the best progression-free (**C**) and overall survival (**D**). Galectin-8 and -9 expression was determined in the cytoplasm of cancer cells using IHC and immunoreactivity (IR) scores. Survival times were plotted as Kaplan–Meier graphs. Graph shows the percentage of living patients (vertical axis) in dependence of time (horizontal axis). Patients without reported death who exited the study before the observation period ended were censored by the software. Censoring events have been marked in the graphs.

Table 2. Multivariate analysis.

Covariate	Coefficient (b_i)	HR Exp (b_i)	95% CI		p-Value
			Lower	Upper	
Histology	−0.005	0.995	0.989	1.002	0.135
Grading	0.614	1.848	1.342	2.544	<0.001
FIGO	0.763	2.144	1.503	3.058	<0.001
Patients' age (≤60 vs. >60 years)	0.737	2.089	1.265	3.447	0.004
Gal-8 staining (low vs. high)	−0.487	0.615	0.388	0.973	0.038
Gal-9 staining (neg. vs. low vs. high)	0.687	1.988	1.257	3.145	0.003

HR = hazard ratio; CI = confidence interval.

2.3. Gal-9 Expression Determines DFS and OS of Ovarian Cancer Patients in Two Different Ways

Staining for galectin-9 was assessed in 147 ovarian cancer samples using IR scores. Gal-9 staining was mostly present in the cytoplasm of ovarian cancer cells, but not the nuclei or the peritumoral stroma. Throughout the panel a median IRS of 3 was observed. In total, 32 cases (20.5%) were Gal-9 negative (IRS = 0), 79 cases (50.6%), however, presented with moderate Gal-9 staining ($1 \geq IRS \geq 6$) and in 36 cases (24.5%) high Gal-9 expression (IRS > 6) was observed. Gal-9 staining showed different distribution in different histological subtypes (Table 1). Cases with high Gal-9 expression presented more often with low tumor stage, lower grading, early FIGO stage, and younger age. The majority of Gal-9 negative cases, however, showed a high tumor stage, higher grading, advanced FIGO stage, and older age (Table 1).

Using Kaplan-Meier analysis, different groups of Gal-9 expression showed significant differences in overall and disease-free survival. Cases with a moderate Gal-9 expression ($1 \geq IRS \geq 6$) displayed a reduced progression-free and overall survival compared to Gal-9 negative cases (IRS = 0). However, the small group of patients presenting with a high Gal-9 expression (IRS > 6) showed the best progression-free (C) and overall survival (D). In multivariate analysis, this correlation proved to be independent of clinical and pathological variables, together with grading, FIGO, patients' age and Gal-8 expression.

2.4. Correlation Analysis

A correlation analysis between IR scores of Gal-8 and Gal-9 staining in the cytoplasm was performed. Results are shown in Table 3. We observed a rather weak, but highly significant correlation between cytoplasmic Gal-8 and Gal-9 staining ($p < 0.001$).

Table 3. Correlation analysis.

	Gal-9 Cytoplasm
Gal-8 cytoplasm	
cc	0.464
p	<0.001
n	142

IR scores for Gal-8 and -9 staining were correlated using Spearman's correlation analysis. cc = correlation coefficient, *p* = two-tailed significance, *n* = number of patients.

3. Discussion

According to our data, high Gal-8 expression in the cytoplasm of cancer cells is a novel positive prognostic factor for DFS and OS in ovarian cancer patients. Cytoplasmic Gal-9 expression, however, determines the DFS and OS of ovarian cancer patients in two opposing ways: On one hand, moderate Gal-9 expression correlates with a reduced overall and disease-free survival, compared to Gal-9-negative cancers, while high Gal-9 expression correlated with the best outcome. Stromal Gal-8 or -9 was not observed in ovarian cancer samples and nuclear expression does not seem to play an important role for survival of ovarian cancer patients.

In 1995, Gal-8 was cloned for the first time from a rat liver cDNA expression library [23]. Later, a homolog gene was detected in the human prostate adenocarcinoma cell line LNCaP, that was identified as prostate carcinoma tumor antigen-1 (*PCTA-1*). Also, altered Gal-8 expressed was found in prostate carcinomas compared to normal prostate and benign prostatic hypertrophy [24]. Several alternative splicing variants have been reported in Gal-8 mRNA processing [25]. In total, seven different isoforms of Gal-8 are encoded by the human Gal-8 gene (*LGALS8*). Three of them belong to the tandem-repeat galectin group and four to the prototype group with only one CRD. However, prototype isoforms of Gal-8 were not found at the protein level [26]. Nevertheless, rather than as a single protein, galectin-8 should be regarded as a discrete subfamily among all galectins. To our knowledge, there are no antibodies available to target specific isoforms of galectin-8. Hence, immunohistochemistry is limited to the observation of the total expression level of all galectin-8 isoforms. Whether different anti-Gal-8 antibodies have a higher affinity for certain Gal-8 isoforms remains elusive as well. However, this could be a reason for different results evaluating Gal-8 as a prognostic factor using different antibodies in immunohistochemistry [21]. This problem should be addressed in further experiments, e.g., using Western blot analysis to determine the individual Gal-8 subtype expressed in ovarian cancer tissues.

Gal-8 has been found to contribute to several mechanism of tumor biology. Endothelial cell migration and tube formation in vitro as well as angiogenesis in vivo has been demonstrated to be induced by Gal-8 [27]. Cell adhesion in human non-small cell lung carcinoma cells (H1299) and rat hepatoma cells (Fao) as well as Chinese hamster ovary (CHO-P) cells was affected by the presence of Gal-8, either positively or negatively dependent on its concentration [28]. In glioblastoma cell line U87, Gal-8 was shown to promote cell migration and proliferation and has been observed to prevent tumor cell apoptosis [29]. However, none of these effects have been examined in ovarian cancer, and further studies are required to explain the role of Gal-8 in ovarian cancer biology.

One of the first descriptions of Galectin-9 was in 1997, after which it was cloned and identified as a tumor antigen in Hodgkin's lymphoma [30]. Since then, many implications of Gal-9 in cancer have been reported [31]. In melanoma cells, galectin-9 was able to induce cell aggregation and apoptosis, and down-regulation of Gal-9 was associated with distant metastasis [32]. Similarly, in breast cancer, Gal-9 negative tumors were more likely to show distant metastasis and therefore correlated with an unfavorable prognosis [33]. In both, melanoma and breast cancer, tumor cell adhesion has been found to be influenced by Gal-9 expression [32,34]. Furthermore, changing Gal-9 expression was discovered during endothelial cell activation, implying a function in angiogenesis [35]. However, best studied role of Gal-9 is in immunity and inflammation. Most prominent mechanism here is the binding of Gal-9 to TIM3, a T cell-specific surface molecule, leading to intracellular calcium flux, aggregation, and apoptosis of T-helper type 1 cells [36]. Similarly, in CD8+ cytotoxic T-cells, Gal-9 was able to induce apoptosis in vitro and vivo, inhibiting the immune response to alloantigen of a skin graft [37]. The same mechanism can be implicated in the acquaintance of tumor immunity. Furthermore, Gal-9 induced the differentiation of naive T cells to T regulatory (T reg) cells, decreasing the number of CD4(+) TIM3(+) T cells and increasing the number of T reg cells in the peripheral blood of a mouse model [38]. T reg cells, however, are known to suppress the antitumor immune response and therefore enable the tumor immune escape [39]. In line with this, in ovarian cancer, a higher number of T reg cells in lymphoid aggregates surrounding the tumor were associated with significantly reduced patient survival [40]. Summing up, the role of Gal-9 in cancer immunity implicates a reduced survival of patients with Gal-9 expressing cancers, while its functions in apoptosis, cancer cell adhesion, and metastasis could explain a better outcome in Gal-9 expressing ovarian cancers. Both taken together could serve as an explanation for the two opposing ways in which Gal-9 determined the survival of ovarian cancer patients in this study.

Similar to Gal-8, several splice variants have been reported for Gal-9 [31]. Again, varying antibody affinity to different Gal-9 isoforms could explain contradictory results in different studies [21]. Furthermore, heterogeneity in Gal-9 splice variants could explain the complex effects of Gal-9 expression on patients' survival, which is described in literature, but was also observed in this study. Assuming different Gal-9 isoforms can realize different functions in cancer biology, patient survival could be affected by different Gal-9 isoforms in opposing ways. However, since these considerations are rather speculative, further studies are required to address this problem.

4. Materials and Methods

4.1. Patients

Tissue micro arrays (TMAs) were constructed from a collective of formalin-fixed, paraffin-embedded (FFPE) ovarian cancer samples obtained from a collective of 156 female patients who underwent surgery at the Department of Obstetrics and Gynecology, University Hospital, LMU Munich, Germany between 1990 and 2002. No patient had received chemotherapy before surgery. Four histological subtypes were included into the panel (serous (n = 110), endometrioid (n = 21), clear cell (n = 12), and mucinous (n = 13)). Experienced gynecological pathologists (E.S., D.M.) performed tumor grading (G1 (n = 38), G2 (n = 53), G3 (n = 53)) according to WHO. TNM classification (T = tumor, N = lymph nodes, M = metastasis) was performed according to the Union for International Cancer Control (UICC). Extent of the primary tumor (T1 (n = 40), T2 (n = 18), T3 (n = 93), T4 (n = 4)), lymph node involvement (N0 (n = 43), N1 (n = 52) and distant metastasis (M0 (n = 3), M1 (n = 6) was evaluated. FIGO stage was determined (I (n = 35), II (n = 10), III (n = 103), IV (n = 3)) according to the criteria of the International Federation of Gynecology and Obstetrics (FIGO). Patient follow up data was received from the Munich Cancer Registry. Median patients' age was 62 ± 12 years with a range between 31 and 88 years. During the study 104 deaths have been observed with a mean overall survival of 3.2 ± 3.0 years.

4.2. Immunohistochemistry

TMA slides were stained using immunohistochemistry as previously described [16]. All sections were dewaxed in xylol for 20 min, before endogenous peroxidase was quenched with 3% hydrogen peroxide (Merck, Darmstadt, Germany). Next, slides were rehydrated in a descending series of alcohol (100%, 75%, and 50%) and heat-induced antigen retrieval was performed by cooking in sodium citrate buffer (0.1 mol/L citric acid/0.1 mol/L sodium citrate, pH 6.0) in a pressure cooker for 5 min. Tissues were blocked with Blocking Solution (Reagent 1; ZytoChem Plus HRP Polymer System (Mouse/Rabbit); Zytomed Systems GmbH, Berlin, Germany) for 5 min at room temperature (RT). Then, specimens were incubated with Anti-Gal-8 (rabbit, monoclonal, Abcam, Cambridge, UK) at a final concentration of 10 μg/mL (1:100 dilution) in phosphate buffered saline (PBS) for 1 h at RT, and Anti-Gal-9 (rabbit, polyclonal, Abcam, Cambridge, UK) at a final concentration of 3.34 μg/mL (1:300 dilution) in PBS overnight (16 h) at 4 °C. Afterwards, slides were incubated with post-block reagent (Reagent 2) (Zytomed Systems GmbH, Berlin, Germany) for 20 min at RT and HRP-Polymer (Reagent 3) (Zytomed Systems GmbH, Berlin, Germany) for 30 min at RT. After each incubation, slides were washed in PBS twice for 4 min. Visualization reaction was performed with 3,3'-diaminobenzidine chromagen (DAB; Dako, Glostrup, Denmark) and stopped after 2 min in tap water. Counterstaining was performed with Mayer acidic hematoxylin. Specimens were dehydrated in an ascending series of alcohol (50%, 75%, and 100%) followed by xylol. Tissue sections, incubated with PBS instead of a primary antibody, were used as a negative control. Tissue samples of colon mucosa served as a positive control. Staining results were received using a semi-quantitative method analog to the immunoreactivity score. Staining for Gal-8 and -9 was evaluated in the cytoplasm of ovarian cancer cells. The predominant staining intensity (0 = negative, 1 = low, 2 = moderate, and 3 = strong) and the percentage of stained cells (0 = 0%, 1 = 1–10%, 2 = 11–50%, 3 = 51–80%, and 4 = 81–100% stained cells) were assessed and multiplied resulting in values of the IRS. For survival analysis, Gal-8 was grouped into low (IRS ≤ 1) and high expression cases (IRS > 1). Gal-9 was divided into negative (IRS = 0), moderate (1 ≥ IRS ≥ 6) and high (IRS > 6) expression.

4.3. Statistical Analysis

Statistical data was processed using SPSS 23.0 (v23, IBM, Armonk, New York, NY, USA) statistic software. Chi-squared statistics were used to test for correlation to clinical and pathological variables. Correlations between staining results were calculated using spearman's correlation analysis. Kaplan–Meier curves and the log-rank test (Mantel–Cox) were used for survival analysis. Data are presented with the mean ± standard deviation. Significance was assumed for $p < 0.05$.

4.4. Ethics Statement

The current study was approved by the Ethics Committee of the Ludwig Maximilians University, Munich, Germany (approval number 227-09) on 30 September 2009. All tissue samples used for this study were obtained from left-over material from the archives of LMU Munich, Department

Gynecology and Obstetrics, Ludwig-Maximilians University, Munich, Germany, initially used for pathological diagnostics. The diagnostic procedures were completed before the current study was performed. During the analysis, the observers were fully blinded for patients' data. The study was approved by the Ethics Committee of LMU Munich. All experiments were performed according to the standards of the Declaration of Helsinki (1975).

5. Conclusions

We were able to show that Gal-8 expression is a positive prognostic factor for overall and disease-free survival of ovarian cancer patients, while Gal-9 expression determines overall and disease-free survival in two different ways: Moderate Gal-9 expression correlates with a reduced survival, compared to Gal-9 negative cases, while patients with high Gal-9 expression showed the best outcome.

Acknowledgments: This study was funded by the FöFoLe program of the Ludwig-Maximilians University of Munich for Heiko Schulz.

Author Contributions: Heiko Schulz, Christina Kuhn and Simone Hofmann performed the experiments; and Heiko Schulz analyzed the data and wrote the paper. Elisa Schmoeckel, Doris Mayr and Sven Mahner revised the manuscript for important intellectual content. Udo Jeschke initiated and supervised the study and designed the experiments. All authors read and approved the final version of the manuscript.

Conflicts of Interest: The authors declare no conflict of interest.

Abbreviations

Gal	Galectin
IHC	Immunohistochemistry
IRS	Immunoreactivity score
DFS	Disease-free survival
OS	Overall survival
BRCA	Breast cancer gene
CRD	Carbohydrate-recognition domain
LGALS8	Human Gal-8 gene
LGALS9	Human Gal-9 gene
FIGO	Fédération Internationale de Gynécologie et d'Obstétrique
UICC	Union for International Cancer Control
PCTA-1	Prostate carcinoma tumor antigen-1
CD8	cluster of differentiation 8
CD4	cluster of differentiation 4
TIM3	T-cell immunoglobulin and mucin-domain containing-3
T reg	T-regulatory
TMA	Tissue micro array
FFPE	Formalin-fixed paraffin-embedded
WHO	World Health Organization
TNM	T = tumor, N = lymph nodes, M = metastasis
PBS	phosphate buffered saline

References

1. Siegel, R.L.; Miller, K.D.; Jemal, A. Cancer Statistics. *CA Cancer J. Clin.* **2017**, *67*, 7–30. [CrossRef] [PubMed]
2. Thibault, B.; Castells, M.; Delord, J.P.; Couderc, B. Ovarian cancer microenvironment: Implications for cancer dissemination and chemoresistance acquisition. *Cancer Metastasis Rev.* **2014**, *33*, 17–39. [CrossRef] [PubMed]
3. Baldwin, L.A.; Huang, B.; Miller, R.W.; Tucker, T.; Goodrich, S.T.; Podzielinski, I.; DeSimone, C.P.; Ueland, F.R.; van Nagell, J.R.; Seamon, L.G. Ten-year relative survival for epithelial ovarian cancer. *Obstet. Gynecol.* **2012**, *120*, 612–618. [CrossRef] [PubMed]

4. Davidson, B.; Trope, C.G. Ovarian cancer: Diagnostic, biological and prognostic aspects. *Womens Health (Lond.)* **2014**, *10*, 519–533. [CrossRef] [PubMed]

5. Barondes, S.H.; Castronovo, V.; Cooper, D.N.; Cummings, R.D.; Drickamer, K.; Feizi, T.; Gitt, M.A.; Hirabayashi, J.; Hughes, C.; Kasai, K.; et al. Galectins: A family of animal beta-galactoside-binding lectins. *Cell* **1994**, *76*, 597–598. [CrossRef]

6. Barondes, S.H.; Cooper, D.N.; Gitt, M.A.; Leffler, H. Galectins. Structure and function of a large family of animal lectins. *J. Biol. Chem.* **1994**, *269*, 20807–20810. [PubMed]

7. Ebrahim, A.H.; Alalawi, Z.; Mirandola, L.; Rakhshanda, R.; Dahlbeck, S.; Nguyen, D.; Jenkins, M.; Grizzi, F.; Cobos, E.; Figueroa, J.A.; et al. Galectins in cancer: Carcinogenesis, diagnosis and therapy. *Ann. Transl. Med.* **2014**, *2*, 88. [PubMed]

8. Leffler, H.; Carlsson, S.; Hedlund, M.; Qian, Y.; Poirier, F. Introduction to galectins. *Glycoconj. J.* **2002**, *19*, 433–440. [CrossRef] [PubMed]

9. Yang, R.Y.; Rabinovich, G.A.; Liu, F.T. Galectins: Structure, function and therapeutic potential. *Expert Rev. Mol. Med.* **2008**, *10*, e17. [CrossRef] [PubMed]

10. Hanahan, D.; Weinberg, R.A. Hallmarks of cancer: The next generation. *Cell* **2011**, *144*, 646–674. [CrossRef] [PubMed]

11. Liu, F.T.; Rabinovich, G.A. Galectins as modulators of tumour progression. *Nat. Rev. Cancer* **2005**, *5*, 29–41. [CrossRef] [PubMed]

12. Grosset, A.A.; Labrie, M.; Vladoiu, M.C.; Yousef, E.M.; Gaboury, L.; St-Pierre, Y. Galectin signatures contribute to the heterogeneity of breast cancer and provide new prognostic information and therapeutic targets. *Oncotarget* **2016**, *7*, 18183–18203. [CrossRef] [PubMed]

13. Jiang, J.; Jin, M.S.; Kong, F.; Cao, D.; Ma, H.X.; Jia, Z.; Wang, Y.P.; Suo, J.; Cao, X. Decreased galectin-9 and increased tim-3 expression are related to poor prognosis in gastric cancer. *PLoS ONE* **2013**, *8*, e81799. [CrossRef] [PubMed]

14. Zhang, P.; Shi, B.; Zhou, M.; Jiang, H.; Zhang, H.; Pan, X.; Gao, H.; Sun, H.; Li, Z. Galectin-1 overexpression promotes progression and chemoresistance to cisplatin in epithelial ovarian cancer. *Cell Death Dis.* **2014**, *5*, e991. [CrossRef] [PubMed]

15. Kim, H.J.; Jeon, H.K.; Cho, Y.J.; Park, Y.A.; Choi, J.J.; Do, I.G.; Song, S.Y.; Lee, Y.Y.; Choi, C.H.; Kim, T.J.; et al. High galectin-1 expression correlates with poor prognosis and is involved in epithelial ovarian cancer proliferation and invasion. *Eur. J. Cancer* **2012**, *48*, 1914–1921. [CrossRef] [PubMed]

16. Mirandola, L.; Yu, Y.; Cannon, M.J.; Jenkins, M.R.; Rahman, R.L.; Nguyen, D.D.; Grizzi, F.; Cobos, E.; Figueroa, J.A.; Chiriva-Internati, M. Galectin-3 inhibition suppresses drug resistance, motility, invasion and angiogenic potential in ovarian cancer. *Gynecol. Oncol.* **2014**, *135*, 573–579. [CrossRef] [PubMed]

17. Brustmann, H. Epidermal growth factor receptor expression in serous ovarian carcinoma: An immunohistochemical study with galectin-3 and cyclin d1 and outcome. *Int. J. Gynecol. Pathol.* **2008**, *27*, 380–389. [CrossRef] [PubMed]

18. Kim, H.J.; Jeon, H.K.; Lee, J.K.; Sung, C.O.; Do, I.G.; Choi, C.H.; Kim, T.J.; Kim, B.G.; Bae, D.S.; Lee, J.W. Clinical significance of galectin-7 in epithelial ovarian cancer. *Anticancer Res.* **2013**, *33*, 1555–1561. [PubMed]

19. Labrie, M.; Vladoiu, M.C.; Grosset, A.A.; Gaboury, L.; St-Pierre, Y. Expression and functions of galectin-7 in ovarian cancer. *Oncotarget* **2014**, *5*, 7705–7721. [CrossRef] [PubMed]

20. Schulz, H.; Schmoeckel, E.; Kuhn, C.; Hofmann, S.; Mayr, D.; Mahner, S.; Jeschke, U. Galectins-1, -3, and -7 are prognostic markers for survival of ovarian cancer patients. *Int. J. Mol. Sci.* **2017**, *18*, 1230. [CrossRef] [PubMed]

21. Labrie, M.; De Araujo, L.O.F.; Communal, L.; Mes-Masson, A.M.; St-Pierre, Y. Tissue and plasma levels of galectins in patients with high grade serous ovarian carcinoma as new predictive biomarkers. *Sci. Rep.* **2017**, *7*, 13244. [CrossRef] [PubMed]

22. Uhlen, M.; Fagerberg, L.; Hallstrom, B.M.; Lindskog, C.; Oksvold, P.; Mardinoglu, A.; Sivertsson, A.; Kampf, C.; Sjostedt, E.; Asplund, A.; et al. Proteomics. Tissue-based map of the human proteome. *Science* **2015**, *347*, 1260419. [CrossRef] [PubMed]

23. Hadari, Y.R.; Paz, K.; Dekel, R.; Mestrovic, T.; Accili, D.; Zick, Y. Galectin-8. A new rat lectin, related to galectin-4. *J. Biol. Chem.* **1995**, *270*, 3447–3453. [CrossRef] [PubMed]

24. Su, Z.Z.; Lin, J.; Shen, R.; Fisher, P.E.; Goldstein, N.I.; Fisher, P.B. Surface-epitope masking and expression cloning identifies the human prostate carcinoma tumor antigen gene pcta-1 a member of the galectin gene family. *Proc. Natl. Acad. Sci. USA* **1996**, *93*, 7252–7257. [CrossRef] [PubMed]

25. Nishi, N.; Itoh, A.; Shoji, H.; Miyanaka, H.; Nakamura, T. Galectin-8 and galectin-9 are novel substrates for thrombin. *Glycobiology* **2006**, *16*, 15C–20C. [CrossRef] [PubMed]
26. Troncoso, M.F.; Ferragut, F.; Bacigalupo, M.L.; Cardenas Delgado, V.M.; Nugnes, L.G.; Gentilini, L.; Laderach, D.; Wolfenstein-Todel, C.; Compagno, D.; Rabinovich, G.A.; et al. Galectin-8: A matricellular lectin with key roles in angiogenesis. *Glycobiology* **2014**, *24*, 907–914. [CrossRef] [PubMed]
27. Delgado, V.M.; Nugnes, L.G.; Colombo, L.L.; Troncoso, M.F.; Fernandez, M.M.; Malchiodi, E.L.; Frahm, I.; Croci, D.O.; Compagno, D.; Rabinovich, G.A.; et al. Modulation of endothelial cell migration and angiogenesis: A novel function for the "tandem-repeat" lectin galectin-8. *FASEB J.* **2011**, *25*, 242–254. [CrossRef] [PubMed]
28. Levy, Y.; Arbel-Goren, R.; Hadari, Y.R.; Eshhar, S.; Ronen, D.; Elhanany, E.; Geiger, B.; Zick, Y. Galectin-8 functions as a matricellular modulator of cell adhesion. *J. Biol. Chem.* **2001**, *276*, 31285–31295. [CrossRef] [PubMed]
29. Metz, C.; Doger, R.; Riquelme, E.; Cortes, P.; Holmes, C.; Shaughnessy, R.; Oyanadel, C.; Grabowski, C.; Gonzalez, A.; Soza, A. Galectin-8 promotes migration and proliferation and prevents apoptosis in u87 glioblastoma cells. *Biol. Res.* **2016**, *49*, 33. [CrossRef] [PubMed]
30. Tureci, O.; Schmitt, H.; Fadle, N.; Pfreundschuh, M.; Sahin, U. Molecular definition of a novel human galectin which is immunogenic in patients with hodgkin's disease. *J. Biol. Chem.* **1997**, *272*, 6416–6422. [CrossRef] [PubMed]
31. Heusschen, R.; Griffioen, A.W.; Thijssen, V.L. Galectin-9 in tumor biology: A jack of multiple trades. *Biochim. Biophys. Acta* **2013**, *1836*, 177–185. [CrossRef] [PubMed]
32. Kageshita, T.; Kashio, Y.; Yamauchi, A.; Seki, M.; Abedin, M.J.; Nishi, N.; Shoji, H.; Nakamura, T.; Ono, T.; Hirashima, M. Possible role of galectin-9 in cell aggregation and apoptosis of human melanoma cell lines and its clinical significance. *Int. J. Cancer* **2002**, *99*, 809–816. [CrossRef] [PubMed]
33. Yamauchi, A.; Kontani, K.; Kihara, M.; Nishi, N.; Yokomise, H.; Hirashima, M. Galectin-9, a novel prognostic factor with antimetastatic potential in breast cancer. *Breast J.* **2006**, *12*, S196–S200. [CrossRef] [PubMed]
34. Irie, A.; Yamauchi, A.; Kontani, K.; Kihara, M.; Liu, D.; Shirato, Y.; Seki, M.; Nishi, N.; Nakamura, T.; Yokomise, H.; et al. Galectin-9 as a prognostic factor with antimetastatic potential in breast cancer. *Clin. Cancer Res.* **2005**, *11*, 2962–2968. [CrossRef] [PubMed]
35. Thijssen, V.L.; Hulsmans, S.; Griffioen, A.W. The galectin profile of the endothelium: Altered expression and localization in activated and tumor endothelial cells. *Am. J. Pathol.* **2008**, *172*, 545–553. [CrossRef] [PubMed]
36. Zhu, C.; Anderson, A.C.; Schubart, A.; Xiong, H.; Imitola, J.; Khoury, S.J.; Zheng, X.X.; Strom, T.B.; Kuchroo, V.K. The tim-3 ligand galectin-9 negatively regulates t helper type 1 immunity. *Nat. Immunol.* **2005**, *6*, 1245–1252. [CrossRef] [PubMed]
37. Wang, F.; He, W.; Zhou, H.; Yuan, J.; Wu, K.; Xu, L.; Chen, Z.K. The tim-3 ligand galectin-9 negatively regulates cd8+ alloreactive t cell and prolongs survival of skin graft. *Cell Immunol.* **2007**, *250*, 68–74. [CrossRef] [PubMed]
38. Seki, M.; Oomizu, S.; Sakata, K.M.; Sakata, A.; Arikawa, T.; Watanabe, K.; Ito, K.; Takeshita, K.; Niki, T.; Saita, N.; et al. Galectin-9 suppresses the generation of th17, promotes the induction of regulatory T cells, and regulates experimental autoimmune arthritis. *Clin. Immunol.* **2008**, *127*, 78–88. [CrossRef] [PubMed]
39. Facciabene, A.; Motz, G.T.; Coukos, G. T-regulatory cells: Key players in tumor immune escape and angiogenesis. *Cancer Res.* **2012**, *72*, 2162–2171. [CrossRef] [PubMed]
40. Hermans, C.; Anz, D.; Engel, J.; Kirchner, T.; Endres, S.; Mayr, D. Analysis of foxp3+ t-regulatory cells and cd8+ T-cells in ovarian carcinoma: Location and tumor infiltration patterns are key prognostic markers. *PLoS ONE* **2014**, *9*, e111757. [CrossRef] [PubMed]

International Journal of
Molecular Sciences

MDPI

Article

Galectins-1, -3, and -7 Are Prognostic Markers for Survival of Ovarian Cancer Patients †

Heiko Schulz [1], Elisa Schmoeckel [2], Christina Kuhn [1], Simone Hofmann [1], Doris Mayr [2], Sven Mahner [1] and Udo Jeschke [1,*]

[1] Department of Gynaecology and Obstetrics, Ludwig-Maximilians University of Munich, Campus Großhadern: Marchioninistr. 15, 81377 Munich and Campus Innenstadt: Maistr. 11, Munich 80337, Germany; Heiko.Schulz@med.uni-muenchen.de (H.S.); Christina.Kuhn@med.uni-muenchen.de (C.K.); Simone.Hofmann@med.uni-muenchen.de (S.H.); Sven.Mahner@med.uni-muenchen.de (S.M.)
[2] LMU Munich, Department of Pathology, Ludwig Maximilians University of Munich, Thalkirchner Str. 142, Munich 80337, Germany; Elisa.Schmoeckel@med.uni-muenchen.de (E.S.); Doris.Mayr@med.uni-muenchen.de (D.M.)
*   Correspondence: udo.jeschke@med.uni-muenchen.de; Tel.: +49-89-4400-54240
†   This study is dedicated to the memory of Susanne Kunze (1948–2015).

Academic Editor: Armando Bartolazzi
Received: 9 May 2017; Accepted: 5 June 2017; Published: 8 June 2017

Abstract: There is a tremendous need for developing new useful prognostic factors in ovarian cancer. Galectins are a family of carbohydrate binding proteins which have been suggested to serve as prognostic factors for various cancer types. In this study, the presence of Galectin-1, -3, and -7 was investigated in 156 ovarian cancer specimens by immunochemical staining. Staining was evaluated in the cytoplasm and nucleus of cancer cells as well as the peritumoral stroma using a semi quantitative score (Remmele (IR) score). Patients' overall survival was compared between different groups of Galectin expression. Galectin (Gal)-1 and -3 staining was observed in the peritumoral stroma as well as the nucleus and cytoplasm of tumor cells, while Gal-7 was only present in the cytoplasm of tumor cells. Patients with Gal-1 expression in the cytoplasm or high Gal-1 expression in the peritumoral stroma showed reduced overall survival. Nuclear Gal-3 staining correlated with a better outcome. We observed a significantly reduced overall survival for cases with high Gal-7 expression and a better survival for Gal-7 negative cases, when compared to cases with low expression of Gal-7. We were able to show that both tumor and stroma staining of Gal-1 could serve as negative prognostic factors for ovarian cancer. We were able to confirm cytoplasmic Gal-7 as a negative prognostic factor. Gal-3 staining in the nucleus could be a new positive prognosticator for ovarian cancer.

Keywords: Galectin-1; Galectin-3; Galectin-7; ovarian cancer; overall survival

1. Introduction

Ovarian cancer is the most lethal gynecological malignancy, ranking fifth in estimated cancer deaths among women in the USA [1]. First-line treatment consists of primary debulking surgery followed by platinum and paclitaxel chemotherapy [2]. Still, the 5-year relative survival rate for epithelial ovarian cancer patients is less than 50% [3]. A lack of screening methods and the frequent presentation with advanced stage disease are considered as the main reasons for the poor outcome of ovarian cancer patients. Disease stage at diagnosis, extent of residual disease after surgery, histological subtype, and a high volume of ascites can be used as prognosticators in ovarian cancer [4]. Numerous studies have aimed to introduce new biological prognostic factors in ovarian cancer. Recently, carbohydrate stem cell marker TF1 has been proposed as negative prognostic marker in ovarian cancer displaying wildtype p53, while estrogen receptor promoter methylation could predict overall survival in low-grade ovarian carcinoma patients [5,6]. Although for these and various other molecules the prognostic value independently of clinical parameters has been proven, until today, except for breast cancer gene (BRCA)-status, no biological marker is commonly accepted [4]. Further specification of anti-cancer therapy necessarily requires an improvement of biological prognostic markers in ovarian cancer.

Galectins have been defined as a family of proteins sharing two main characteristics: a binding affinity for β-galactosides and a significant similarity in the carbohydrate-recognition domain (CRD) [7]. The first member of this family described was Galectin-1, which is isolated as homodimers composed of two identical CRD subunits [8]. Since then, the Galectin family has had a growing number of members, but only Galectin (Gal)-1–4, Gal-7–10, Gal-12, and Gal-13 are known to be present in humans [9]. Similar to Gal-1, Gal-7 typically occurs in homodimers, while Gal-3 is the only Galectin characterized chimeric protein known to form higher order oligomers [10,11]. In several cancer types, Galectins are known to affect tumor growth, metastasis, angiogenesis, cell migration, as well as tumor invasiveness and progression, and are therefore very likely to show a prognostic value for patients' survival [9,12].

The role of Galectin-1 in cancer has been studied by various groups, and several papers already exist on this topic. For patients' sera and ovarian cancer tissue, it has been shown that a combination of CA-125 and Galectin-1 serves as a possible two-marker combination for preoperative discrimination of benign and malignant ovarian masses [13]. Also, patients suffering from metastatic epithelial ovarian cancer were observed to show higher serum Gal-1 levels than those with non- metastatic type. Elevated peritumoral stroma staining of Gal-1 was shown to occur in advanced stages of epithelial ovarian cancer and is also connected with poorer progression-free survival in univariate analysis [14]. However, these results have not yet been reproduced for overall survival or confirmed by multivariate analysis [15]. Due to this, the possibility of Gal-1 as an independent prognostic marker in ovarian cancer still needs to be further investigated.

High cytoplasmic Galectin-3 expression has been suggested as a negative prognostic factor, as it was shown to correlate with shorter progression-free survival in ovarian cancer [16]. However, in another study, Gal-3 expression did not correlate to reduced overall survival, but a cytoplasmic staining pattern was associated with poor outcome when compared to patterns including nuclear staining [17]. Although Gal-3 staining in nucleus and stroma has been observed, their influence on overall survival still maintains elusive. Galectin-7 has been proposed to serve as negative prognostic factor in ovarian cancer by two independent groups. In both studies, its influence on progression-free survival and overall survival has been confirmed by univariate and multivariate analysis [16,18]. Yet, there is further disagreement whether Gal-7 staining occurs predominantly in the nucleus or the cytoplasm. Also, it remains unknown if there is a correlation between expressions of different Galectins in ovarian cancer, and there is a desperate need for comprehensive studies of various Galectins on a representative ovarian cancer panel. Therefore, in this study, we investigated the prognostic influence of Gal-1, -3, and -7 in patients with epithelial ovarian cancer and analyzed correlations to each other and to clinical and pathological parameters. We hypothesize that Gal-1, -3, and -7 are prognostic for overall survival in ovarian cancer patients, dependent of the localization of the Galectin expression.

2. Results

2.1. Gal-1 Tumor and Stroma Staining Is Negative Prognostic for Overall Survival

Galectin-1 staining was successfully performed on 150 ovarian cancer specimens. Gal-1 was present in the cytoplasm and the nuclei of ovarian cancer cells, as well as the peritumoral stroma (Figure 1). In 102 cases (68.0%), the cytoplasms of tumor cells were positive for Gal-1, with a median Remmele score (IRS) of 3. Peritumoral stroma was positive for Gal-1 in 148 cases (98.0%), with a median IRS of 8. Gal-1 expression significantly correlated with several clinical and pathological data (Table 1).

Gal-1 tumor staining	Gal-1 stroma staining	Gal-3 tumor staining

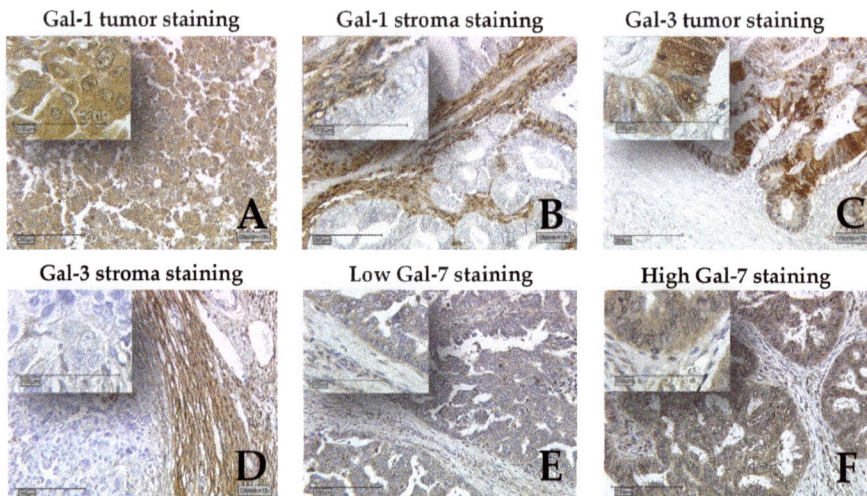

Gal-3 stroma staining	Low Gal-7 staining	High Gal-7 staining

Figure 1. Detection of Galectins by immunohistochemistry. Representative photomicrographs are shown. Galectin (Gal)-1 was present in the cytoplasm and the nuclei of ovarian cancer cells (**A**) as well as the peritumoral stroma (**B**); Gal-3 staining was observed in the nuclei, cytoplasm (**C**), and stroma (**D**); Staining for Galectin-7 was mainly observed in the cytoplasm (**E**); only a few individual cases showed nuclear staining (**F**); 10× magnification was used for the outer pictures and 50× magnification for the inserts. The scale bars in in the outer pictures equal 200 μm (10× magnification) and the scale bars in the inserts equal 100 μm (50× magnification).

Table 1. Gal-1 staining correlated with clinical and pathological data.

Clinical and Pathological Variables	Gal-1 Expression Cytoplasm		p	Gal-1 Expression Stroma		p	Gal-1 Expression Nucleus		p
	negative	positive		low	high		negative	positive	
Histology									
Serous	26	79	0.008	34	71	NS	27	78	0.002
Clear cell	5	7		6	6		3	9	
Endometrioid	8	12		7	13		11	9	
Mucinous	9	4		3	10		9	4	
Tumor Stage									
pT1	22	17	<0.001	20	19	0.006	19	20	0.020
pT2+	26	84		30	80		31	79	
Lymph node									
pN0/pNX	36	65	NS	34	67	NS	43	58	0.001
pN1	12	37		16	33		7	42	
Distant Metastasis									
pM0/pMX	47	97	NS	49	95	NS	49	95	NS
pM1	1	5		1	5		1	5	
Grading									
G1	20	16	<0.001	13	23	NS	14	22	NS
G2+	22	80		31	71		31	71	
FIGO									
I/II	22	21	0.001	17	26	NS	21	22	0.013
III/IV	24	78		31	71		28	74	
Age									
≤60 years	27	52	NS	28	51	NS	24	55	NS
≤60 years	21	50		22	49		26	45	

TNM staging was accomplished according to actual standards of Union for International Cancer Control (UICC); pT1 = tumor stage 1; pT2+ = tumor stage 2 or higher; pN0 = lymph node stage 0; pNX = lymph node stage not evaluated; pN1 = lymph node stage 1; pM0 = distant metastasis stage 0; pMX = distant metastasis not evaluated; pM1 = distant metastasis stage 1; G1 = grade 1; G2+ = grade 2 or higher; FIGO = Fédération Internationale de Gynécologie et d'Obstétrique; NS = Not significant ($p > 0.05$).

Gal-1 staining in cytoplasm and nucleus showed differences for several histological subtypes ($p = 0.008$, $p = 0.002$, respectively). Cytoplasmic Gal-1 staining was significantly stronger in serous, clear cell, or endometrioid subtypes, while for mucinous subtype we found more negative cases. Also, more cases showed Gal-1 positive nuclei for serous and clear cell subtypes, while endometrioid and mucinous subtypes had weaker nuclear Gal-1 stainings. Furthermore, Gal-1 staining in nucleus, cytoplasm, and stroma were significantly higher in cases with advanced tumor stage ($p < 0.001$, $p = 0.006$, $p = 0.02$, respectively). Gal-1 expression in the cytoplasm was significantly higher in cases with higher grading ($p < 0.001$) and advanced FIGO (Fédération Internationale de Gynécologie et d'Obstétrique) stage ($p = 0.001$). Gal-1 staining in the nucleus showed higher IR scores in lymph node positive cases ($p = 0.001$) and cases with advanced FIGO stage ($p = 0.013$).

Survival times of different groups of Gal-1 expression in nucleus, cytoplasm, and stroma have been compared (Figure 2). Cases with Gal-1 expression in the cytoplasm showed significantly reduced overall survival compared to cases without any Gal-1 expression in the cytoplasm ($p = 0.029$) Moreover, cases displaying high Gal-1 expression in the stroma showed a significantly reduced outcome compared to cases with low Gal-1 expression in the stroma ($p = 0.045$). Comparing negative versus positive cases of Gal-1 expression in the nucleus did not show any differences with regard to overall survival. However, based on considering a multivariate analysis, only Gal-1 stroma staining would serve as an independent prognostic factor (Table 2).

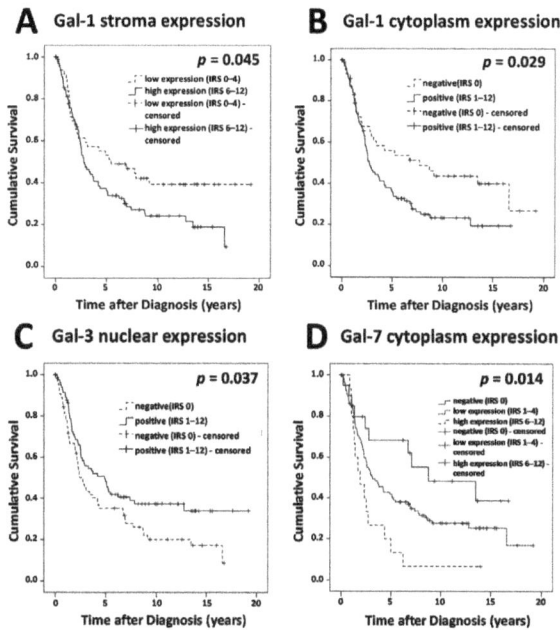

Figure 2. Survival times were plotted as Kaplan-Meier graphs. Percentage of living patients (vertical axis) was plotted in dependence of time (horizontal axis). Patients without an observed event (death) who exited the study before the observation period ended have been censored. Censoring has been marked in the graphs. Survival times of different groups of Galectin expression have been compared. Cases displaying high Gal-1 expression in the stroma showed a significantly reduced outcome compared to cases with low Gal-1 expression in the stroma. (**A**) Cases with Gal-1 expression in the cytoplasm showed significantly reduced overall survival compared to cases without any Gal-1 expression in cytoplasm; (**B**) Cases without Gal-3 expression in nuclei showed significantly reduced overall survival compared to cases with nuclear Gal-3 expression; (**C**) Cases with high Gal-7 expression showed a significantly reduced overall survival and Gal-7 negative cases showed better overall survival, when compared to cases with low expression of Gal-7; (**D**) Galectin expression was determined in cytoplasm, nucleus, and stroma using Remmele (IR) scores.

Table 2. Multivariate analysis.

Covariate	Coefficient (b₁)	HR Exp (b₁)	95% CI		p-Value
			Lower	Upper	
Histology (serous vs. other)	0.211	1.235	0.658	2.317	0.511
Grade (G1 vs. G2, G3)	0.942	2.565	1.290	5.100	0.007
FIGO (I, II vs. III, IV)	1.140	3.126	1.537	6.357	0.002
Patients' age (≤60 vs. >60 years)	0.312	1.367	0.861	2.169	0.185
Gal-1 stroma (low vs. high)	0.571	1.770	1.044	2.999	0.034
Gal-1 cytoplasm (neg. vs. pos.)	−0.187	0.830	0.423	1.626	0.586
Gal-3 nucleus (neg. vs. pos.)	−0.265	0.767	0.480	1.227	0.269
Gal-7 cytoplasm (neg. vs. pos.)	0.636	1.889	1.160	3.077	0.011

HR = *hazard ratio*; CI = *confidence interval*.

2.2. Presence of Gal-3 in Nuclei Is A Positive Prognosticator in Ovarian Cancer

Gal-3 positive nuclei were observed in 83 (55%) out of 151 cases, while 96 cases (63.6%) showed cytoplasmic Gal-3 staining and 85 cases (56.3%) presented with Gal-3 positive peritumoral stroma (Figure 1). Median IR scores for Gal-3 in nuclei, cytoplasm, and stroma were 1, 2, and 1, respectively. Gal-3 staining showed correlations with clinical and pathological data (Table 3). Gal-3 expression in stroma and nucleus was different for several histological subtypes ($p = 0.008$, $p = 0.013$, respectively). Gal-3 stroma staining was stronger in serous and clear cell subtypes but weaker in endometrioid and mucinous subtypes, while nuclear Gal-3 staining was stronger in serous, clear cell, and mucinous subtypes but weaker in endometrioid subtype. Tumors rated as pT1 presented with significantly stronger nuclear Gal-3 staining than pT2 or higher staged cases ($p = 0.042$). We observed a correlation of Gal-3 in nucleus and cytoplasm with patients' age ($p = 0.022$, $p = 0.013$, respectively), with higher IR scores for patients younger than 60 years. For our study panel, Gal-3 overexpression in the cytoplasm was not correlated with poorer outcome of ovarian cancer patients. Also, Gal-3 staining in the peritumoral stroma could not serve as a prognostic factor. However, nuclear Gal-3 expression could serve as a positive prognostic factor (Figure 2). Cases without Gal-3 expression in nuclei showed significantly reduced overall survival compared to cases with nuclear Gal-3 expression ($p = 0.034$). According to the results of a multivariate analysis, nuclear Gal-3 staining could not serve as an independent prognostic factor, probably due to its strong correlations with patients' age, tumor stage, and histology (Table 2).

Table 3. Gal-3 staining correlated with clinical and pathological data.

Clinical and Pathological Variables	Gal-3 Expression Cytoplasm		p	Gal-3 Expression Stroma		p	Gal-3 Expression Nucleus		p
	neg.	pos.		neg.	pos.		neg.	pos.	
Histology									
Serous	37	69	NS	42	64	0.008	44	62	0.013
Clear cell	3	9		2	10		3	9	
Endometrioid	12	9		13	8		16	5	
Mucinous	3	9		9	3		5	7	
Tumor Stage									
pT1	12	27	NS	21	18	NS	12	27	0.042
pT2+	43	68		44	67		55	56	
Lymph node									
pN0/pNX	39	62	NS	47	54	NS	48	53	NS
pN1	16	34		19	31		20	30	
Distant Metastasis									
pM0/pMX	53	92	NS	64	81	NS	65	80	NS
pM1	2	4		2	4		3	3	
Grading									
G1	9	28	NS	16	21	NS	13	24	NS
G2+	40	62		44	58		51	51	
FIGO									
I/II	13	30	NS	21	22	NS	15	28	NS
III/IV	41	62		43	60		51	52	
Age									
≤60 years	22	57	0.022	33	46	NS	28	51	0.013
>60 years	33	39		33	39		40	32	

TNM staging was accomplished according to actual standards of UICC; pT1 = tumor stage 1; pT2+ = tumor stage 2 or higher; pN0 = lymph node stage 0; pNX = lymph node stage not evaluated; pN1 = lymph node stage 1; pM0 = distant metastasis stage 0; pMX = distant metastasis not evaluated; pM1 = distant metastasis stage 1; G1 = grade 1; G2+ = grade 2 or higher; NS = Not significant ($p > 0.05$).

2.3. Gal-7 Expression Level Predicts Shortened Overall Survival in Ovarian Cancer

Staining for Galecin-7 was mainly observed in the cytoplasm; only few individual cases showed nuclear staining (Figure 1). Cytoplasmic Gal-7 staining was present in 129 (86.6%) out of 149 specimens, with a median IR score of 3. In total, 20 cases presented negative for Gal-7, while 114 cases showed low and 15 cases showed high expression of Gal-7. Gal-7 expression appeared to show differences for several histological subtypes ($p = 0.026$). The strongest Gal-7 staining was found in serous subtype, and the weakest was in endometrioid subtype (Table 4). No other correlation of Gal-7 with pathological data was found. Survival times of Gal-7 negative cases and cases displaying a high Gal-7 expression were compared to cases with low Gal-7 expression (Figure 2). We observed a significantly reduced overall survival for cases with high Gal-7 expression and a better survival for Gal-7 negative cases, when compared to cases with low expression of Gal-7 ($p = 0.014$). Also, according to the results of a multivariate analysis, higher Gal-7 expression can be confirmed as an independent prognostic factor for overall survival in ovarian cancer (Table 2).

Table 4. Gal-7 staining correlated with clinical and pathological data.

Clinical and Pathological Variables	Gal-7 Expression Cytoplasm			p
	neg.	low	high	
Histology				
Serous	10	83	12	0.026
Clear cell	0	10	2	
Endometrioid	7	13	0	
Mucinous	3	8	1	
Tumor Stage				
pT1	4	29	5	NS
pT2+	15	85	10	
Lymph node				
pN0/pNX	15	75	8	NS
pN1	5	39	7	
Distant Metastasis				
pM0/pMX	19	110	14	NS
pM1	1	4	1	
Grading				
G1	6	25	3	NS
G2+	12	80	11	
FIGO				
I/II	8	29	4	NS
III/IV	11	81	11	
Age				
≤60 years	12	59	8	NS
>60 years	8	55	7	

TNM staging was accomplished according to actual standards of UICC; pT1 = tumor stage 1; pT2+ = tumor stage 2 or higher; pN0 = lymph node stage 0; pNX = lymph node stage not evaluated; pN1 = lymph node stage 1; pM0 = distant metastasis stage 0; pMX = distant metastasis not evaluated; pM1 = distant metastasis stage 1; G1 = grade 1; G2+ = grade 2 or higher; NS = Not significant ($p > 0.05$).

2.4. Correlation Analysis

A correlation analysis is shown in Table 5. For Gal-1 staining, we observed positive correlations between staining in cytoplasm, nucleus, and stroma. Also, staining results of Gal-3 in cytoplasm, nucleus, and stroma were positively correlated among each other. Furthermore, we found correlations between Galectin-1 and -3 staining in nucleus, cytoplasm, and stroma. Gal-7 staining showed positive correlations with Gal-1 in cytoplasm and nucleus and all types of Gal-3 staining.

Table 5. Correlation analysis.

Staining	Gal-1 Cytoplasm	Gal-1 Stroma	Gal-1 Nucleus	Gal-3 Cytoplasm	Gal-3 Stroma	Gal-3 Nucleus	Gal-7 Cytoplasm
Gal-1 cytoplasm							
cc	1.000	0.382	0.748	0.356	0.263	0.282	0.272
p	.	<0.001	<0.001	<0.001	0.001	<0.001	0.001
n	150	150	150	149	149	149	146
Gal-1 stroma							
cc	0.382	1.000	0.231	0.123	0.280	−0.006	−0.040
p	<0.001	.	0.004	0.135	0.001	0.937	0.633
n	150	150	150	149	149	149	146
Gal-1 nucleus							
cc	0.748	0.231	1.000	0.302	0.315	0.329	0.249
p	<0.001	0.004	.	<0.001	<0.001	<0.001	0.002
n	150	150	150	149	149	149	146
Gal-3 cytoplasm							
cc	0.356	0.123	0.302	1.000	0.293	0.839	0.276
p	<0.001	0.135	<0.001	.	<0.001	<0.001	0.001
n	149	149	149	151	151	151	146
Gal-3 stroma							
cc	0.263	0.280	0.315	0.293	1.000	0.267	0.231
p	0.001	0.001	<0.001	<0.001	.	0.001	0.005
n	149	149	149	151	151	151	146
Gal-3 nucleus							
cc	0.282	−0.006	0.329	0.839	0.267	1.000	0.335
p	<0.001	0.937	<0.001	<0.001	0.001	.	<0.001
n	149	149	149	151	151	151	146
Gal-7 cytoplasm							
cc	0.272	−0.040	0.249	0.276	0.231	0.335	1.000
p	0.001	0.633	0.002	0.001	0.005	<0.001	.
n	146	146	146	146	146	146	149

IR scores of Gal-1, -3, and -7 staining in different compartments were correlated with each other using Spearman's correlation analysis. cc = correlation coefficient, *p* = two-tailed significance, *n* = number of patients.

3. Discussion

According to our data, Gal-1 staining in cytoplasm and stroma share a negative prognostic impact on overall survival in ovarian cancer. In accordance, in vitro experiments showed that overexpression of Galectin-1 significantly increases migrative and invasive behavior of ovarian cancer cells [19]. Furthermore, Gal-1 knockdown experiments in ovarian cancer cells displayed a reduction in cell growth, migration, and invasion. Gal-1 interaction with H-Ras and activation of the Raf/extracellular signal-regulated kinase (ERK) pathway as well as the downregulation of matrix metalloproteinase-9 (MMP-9) and c-Jun could have been explored as possible mechanisms. Moreover, Gal-1 overexpression could significantly decrease the sensitivities of ovarian cancer cells to cisplatin, illustrating a possible explanation for decreased survival of ovarian cancer patients with increased Gal-1 expression [14]. Thus, Gal-1 is a promising new target for ovarian cancer therapy. For this purpose, several compounds targeting Gal-1 have been introduced [20]. OTX008, for instance, a new compound binding non-covalently to Gal-1 on the side back face, was able to inhibit proliferation and invasion of various cancer cells lines [21]. Anti-proliferative effects of OTX008 correlated with Gal-1 expression across a large panel of cell lines. Moreover, OTX008 efficiently inhibited the growth of ovarian cancer xenografts in vivo [22]. According to the results of a multivariate analysis, only Gal-1 stroma staining could serve as independent prognostic factor. Accumulation of Gal-1 in peritumoral stroma has been described for various other tumor entities [23–25]. Some groups tried to investigate the mechanisms responsible for this phenomenon. In situ hybridization experiments were able to show that fibroblast cells, adjacent to malignant cells, express Gal-1 mRNA, illustrating a possible explanation for peritumoral Gal-1 accumulation. Also, it was demonstrated that ovarian cancer cells produce Gal-1 and release it to the medium. Furthermore, conditioned medium obtained from ovarian carcinoma cells is able to induce increased gal-1 expression in fibroblast cells. Both experiments suggest that primarily the ovarian cancer cells might be responsible for stromal Gal-1 expression [26]. Our exploration of the positive correlation between Gal-1 staining in peritumoral stroma and malignant cells is consistent with this hypothesis. However, it requires further investigations to explain cases without Gal-1 expression in cancer cells but in the stroma or vice versa.

Several groups have suggested that higher Gal-3 expression is associated with reduced progression-free survival in ovarian cancer [17,27]. However, in these studies, observation of Gal-3 expression was limited to the cytoplasm, while the prognostic value of nuclear Gal-3 staining has not been further studied. We could not confirm a negative influence of cytoplasmic Gal-3 overexpression on overall survival for our study panel. On the contrary, nuclear Gal-3 staining served as a positive prognostic factor, although not independent of clinical and pathological parameters. Apparently, it is the nuclear and not cytoplasmic Gal-3 expression that has a major influence on patients' outcome. In line with this, Gal-3 has been observed to play an important role in nuclear cell physiology, as it is involved in the mechanisms of pre-mRNA-splicing or mRNA transport [28,29]. Furthermore, cell culture experiments using human cervix adenocarcinoma HeLa-cells showed a delayed DNA damage repair response activation and a decrease in the G2/M cell cycle checkpoint arrest in the absence of Gal-3 [30]. A similar mechanism could be conceivable in ovarian cancer, predisposing cells for further mutations in the absence of nuclear Gal-3. To our knowledge, reduced Gal-3 expression as an indicator of poorer prognosis has only been observed in gastric cancer so far [31]. In cholangiocarcinoma, Gal-3 expression was associated with a poorly-differentiated type, while in vitro experiments showed significantly increased cell migration and invasion after suppression of Gal-3 expression [32]. However, for ovarian cancer, in vitro experiments showed knockdown of Gal-3 inhibits migration and invasion of cancer cells, while apoptosis and sensitivity to carboplatin increases [33]. Moreover, paclitaxel and additional Gal-3 inhibitor treatment showed synergistic cytotoxic effects and increased apoptosis in an on ovarian cancer cell line [34]. Since there are disagreements in previous research and our data is neither consistent with previous studies on progression-free survival nor with recent results of in vitro research, further investigation on the prognostic role of Gal-3 in ovarian cancer is definitely required.

As recently proposed by other groups, we were able to confirm Gal-7 as negative prognosticator for overall survival in ovarian cancer in uni- and multivariate analysis. Further cell culture experiments were able to prove that Gal-7 expression is induced by a mutant form of p53. Also, gal-7 was shown to increase proliferation [16], invasiveness, and motility of ovarian cancer cells, while interacting immunosuppressive by killing Jurkat T-cells and human peripheral T-cells [18]. All in all, these investigations confirm Gal-7 as a new promising target for specific therapeutic option in epithelial ovarian cancer. We observed various positive correlations between Gal-1, -3, and -7. This observation, and the fact that Galectins share binding affinities and have similarities in protein structure, suggests the assumption that Galectins might also share common functions in ovarian cancer molecular biology. However, since this observation is rather descriptive, further investigations are required to explore the biological characteristics and functions of different Galectins to determine the manner(s) in which they are similar or different in specific regards to their role(s) in ovarian cancer.

4. Materials and Methods

4.1. Patients

Formalin-fixed, paraffin-embedded (FFPE) ovarian cancer samples from 156 female patients who underwent surgery at the Department of Obstetrics and Gynecology, Ludwig-Maximilians- University of Munich, Germany between 1990 and 2002 were analyzed in this study. Women diagnosed for benign or for borderline tumors of the ovary were excluded and no patient had received neo-adjuvant chemotherapy. Tumor grading (G1 ($n = 38$), G2 ($n = 53$), G3 ($n = 53$)), and histological characterization (serous ($n = 110$), endometrioid ($n = 21$), clear cell ($n = 12$), mucinous ($n = 13$)) were performed by a gynecological pathologist. Tumor staging was accomplished using FIGO classification (I ($n = 35$), II ($n = 10$), III ($n = 103$), IV ($n = 3$)). TNM classification was performed according to UICC. Data on the extension of the primary tumor was available in 155 cases (T1 ($n = 40$), T2 ($n = 18$), T3 ($n = 93$), T4 ($n = 4$)), data on lymph node involvement was available in 95 cases (N0 ($n = 43$), N1 ($n = 52$) and data on the presence of distant metastasis was available in 9 cases (M0 (*n* = 3), M1 (*n* = 6). Clinical data was retrieved from patients' charts and follow up data was requested from the Munich Cancer Registry. Patients' age at surgery ranged between 31 and 88 years, with a median age of 62 ±12 years. Mean overall survival was 3.2 ± 3.0 years and 104 deaths were observed in total. The mean follow up time was 5.1 ± 4.8 years.

4.2. Immunochemistry

Resected ovarian cancer tissue samples were fixed in formalin and embedded in paraffin after surgery. For histopathological investigations, sections were dewaxed in Xylol for 20 minutes and immersed in 3% hydrogen peroxide (Merck, Darmstadt, Germany) to quench endogenous peroxidase. Then slides were rehydrated in a descending series of alcohol (100%, 75%, and 50%), and cooked for 5 minutes in sodium citrate buffer (0.1 mol/L citric acid/0.1 mol/L sodium citrate, pH 6.0) in a pressure cooker to ensure epitope retrieval. Afterwards, slides were washed in distilled water and phosphate-buffered saline (PBS), followed by a specific procedure for each Galectin staining. In particular, for Galectin-1 (Gal-1) staining, slides were blocked using power block (BioGenex, San Ramon, CA, USA) for 3 min at room temperature and incubated with Anti-Galectin 1 primary antibody (goat, polyclonal; R&D Systems, Minneapolis, MN, USA) at a final concentration of 0.033 μg/mL in power block (BioGenex, San Ramon, CA, USA) for 16 h at 4 °C. Galectin-3 (Gal-3) staining was performed by blocking specimens with 1.5% horse serum (Vector Laboratories, Burlingame, CA, USA) for 30 min at room temperature and incubating with Anti-Galectin 3 primary antibody (mouse, monoclonal, Novocastra Reagents, Leica Biosystems, Wetzlar, Germany) at a final concentration of 4.6 μg/mL in PBS for 16 h at 4 °C. For Galectin-7 (Gal-7) staining, specimens were blocked with Blocking Solution (Reagent 1; ZytoChem Plus HRP Polymer System (Mouse/Rabbit); Zytomed Systems GmbH, Berlin, Germany) for 5 minutes at room temperature. Slides were then incubated with Anti-Gal-7 (rabbit, polyclonal; Abcam, Cambridge, UK) at a final concentration of 2.5 μg/mL in PBS for 16 h at 4 °C. Afterwards, for Gal-1 and -3 staining, slides were incubated with isotype-matching anti-goat/mouse-IgG secondary antibody and avidin-biotin-peroxidase complex both for 30 min at room temperature, according to ABC Vectastain kit (Vector Laboratories, Burlingame, CA, USA). For Gal-7 staining, specimens were incubated in post-block reagent (Reagent 2) (Zytomed Systems GmbH, Berlin, Germany) and HRP-Polymer (Reagent 3) (Zytomed Systems GmbH, Berlin, Germany) for 30 min at room temperature, according to the manufacturer's protocol (ZytoChem Plus HRP Polymer System (Mouse/Rabbit). All slides were washed twice in PBS for 2 min after every incubation step. For visualization reaction, every specimen was stained with 3,3'-diaminobenzidine chromogen (DAB; Dako, Glostrup, Denmark), stopped after 30 s to 2 min with tap water, counterstained in Mayer acidic hematoxylin, dehydrated in an ascending series of alcohol followed by xylol, and covered with Consul Mount (Thermo Shandon, Pittsburgh, PA, USA). Tissue sections that had been previously incubated with isotype-matched rabbit-/mouse-/goat- IgG (Dako, Hamburg, Germany) instead of the primary antibody served as negative controls. For positive control, tissue slides of placental tissue (Gal-1, -3) or breast cancer (Gal-7) were used. Primary antibodies were chosen due to high expected staining specificities according to the results of positive control staining, description, and example pictures on the manufacturer's homepages. A semi-quantitative method (IR score; Remmele score) was performed by two independent observers in consensus to obtain staining results. For this purpose, the predominant staining intensity (0 = negative, 1 = low, 2 = moderate, and 3 = strong) and the percentage of stained cells (0 = 0%, 1 = 1–10%, 2 = 11–50%, 3 = 51–80%, and 4 = 81–100% stained cells) has to be multiplied, resulting in values from 0 to 12. Staining intensity was measured in the cytoplasm and the nucleus of the cancer cells, and in the peritumoral stroma. Cut-off points for IR scores were chosen specifically for each staining with regard to the distribution pattern of IR scores in the collective. For Gal-1 staining in cytoplasm and nucleus of cancer cells, IRS = 0 was considered as negative and an IRS ≥ 1 as positive. For stroma staining, Gal-1 groups of low expression (IRS < 5) and high expression (IRS ≥ 5) were compared. For analysis of Gal-3 staining, negative cases with an

IRS = 0 were compared to positive cases with an IRS $\geq$ 1. Gal-7 expression was grouped as negative (IRS = 0), low (1 $\geq$ IRS $\geq$ 4), and high (IRS $\geq$ 6).

4.3. Statistical Analysis

Statistical data was obtained using SPSS 23.0 (v23, IBM, Armonk, NY, USA) statistic software. Distribution of clinicopathological variables was tested with Chi-Square Statistics. Mann-Whitney *U*-test was used to compare IR scores of Galectins between different clinical and pathological subgroups. Correlations between immunochemical staining results were calculated using Spearman's correlation analysis. Kaplan-Meier curves and Log-rank test (Mantel-Cox) were used to compare survival times between different groups. Data are presented with the mean $\pm$ standard deviation. Values of $p < 0.05$ were considered as significant.

4.4. Ethics Statement

All tissue samples used for this study were left-over material from the archives of LMU Munich, Department Gynecology and Obstetrics, Ludwig-Maximilians-University, Munich, Germany, that had initially been collected for histopathological diagnostics. All diagnostic procedures had already been fully completed at the time the histopathological investigations for the current study were performed. Patients' data have been fully anonymized. The study was approved by the Ethics Committee of LMU Munich. All experiments were performed according to the standards set in the declaration of Helsinki 1975.

5. Conclusions

We were able to show that Galectin expression and its impact on overall survival of ovarian cancer patients is strongly dependent of its localization, whether it is in the nucleus or cytoplasm of tumor cells or the peritumoral stroma. We elaborated that Gal-1 tumor and stroma staining, and Gal-7 staining in the cytoplasm serves as a negative prognostic factor for overall survival in ovarian cancer, while nuclear Gal-3 staining could serve as a positive prognostic factor. According to the results of a multivariate analysis, Gal-1 stroma staining and Gal-7 staining are prognostic factors, independent of clinical and pathological parameters.

Acknowledgments: This study was funded by the FöFoLe program of the Ludwig-Maximilians-University of Munich for Heiko Schulz.

Author Contributions: Udo Jeschke conceived and designed the experiments; Christina Kuhn and Simone Hofmann performed the experiments; Heiko Schulz analyzed the data and wrote the paper. Elisa Schmoeckel and Doris Mayr revised the manuscript for important intellectual content. Sven Mahner and Udo Jeschke initiated and supervised the study. All authors read and approved the final version of the manuscript.

Conflicts of Interest: The authors declare no conflict of interest.

Abbreviations

Gal	Galectin
BRCA	Breast cancer gene
CRD	Carbohydrate-recognition domain
UICC	Union for International Cancer Control
FIGO	Fédération Internationale de Gynécologie et d'Obstétrique
IRS	Remmele score

1. Siegel, R.L.; Miller, K.D.; Jemal, A. Cancer statistics, 2016. *CA Cancer J. Clin.* **2016**, *66*, 7–30. [CrossRef] [PubMed]
2. Du Bois, A.; Quinn, M.; Thigpen, T.; Vermorken, J.; Avall-Lundqvist, E.; Bookman, M.; Bowtell, D.; Brady, M.; Casado, A.; Cervantes, A.; et al. 2004 consensus statements on the management of ovarian cancer: Final document of the 3rd International Gynecologic Cancer Intergroup Ovarian Cancer Consensus Conference (GCIG OCCC 2004). *Ann. Oncol.* **2005**, *16*, viii7–viii12. [CrossRef] [PubMed]
3. Baldwin, L.A.; Huang, B.; Miller, R.W.; Tucker, T.; Goodrich, S.T.; Podzielinski, I.; DeSimone, C.P.; Ueland, F.R.; van Nagell, J.R.; Seamon, L.G. Ten-year relative survival for epithelial ovarian cancer. *Obstet. Gynecol.* **2012**, *120*, 612–618. [CrossRef] [PubMed]
4. Davidson, B.; Trope, C.G. Ovarian cancer: Diagnostic, biological and prognostic aspects. *Womens Health* **2014**, *10*, 519–533. [CrossRef] [PubMed]
5. Heublein, S.; Sabina, K.P.; Doris, M.; Nina, D.; Udo, J. p53 determines prognostic significance of the carbohydrate stem cell marker TF1 (CD176) in ovarian cancer. *J. Cancer Res. Clin. Oncol.* **2016**, *142*, 1163–1170. [CrossRef] [PubMed]
6. Kirn, V.; Heublein, S.; Knabl, J.; Guenthner-Biller, M.; Andergassen, U.; Fridrich, C.; Malter, W.; Harder, J.; Friese, K.; Mayr, D. Estrogen receptor promoter methylation predicts survival in low-grade ovarian carcinoma patients. *J. Cancer Res. Clin. Oncol.* **2014**, *140*, 1681–1687. [CrossRef] [PubMed]
7. Barondes, S.H.; Robbins, B.A.; Liu, F.T. Galectins: A family of animal β-galactoside-binding lectins. *Cell* **1994**, *76*, 597–608. [CrossRef]
8. Barondes, S.H.; Cooper, D.N.; Gitt, M.A.; Leffler, H. Galectins. Structure and function of a large family of animal lectins. *J. Biol. Chem.* **1994**, *269*, 20807–20810. [PubMed]
9. Ebrahim, A.H.; Alalawi, Z.; Mirandola, L.; Rakhshanda, R.; Dahlbeck, S.; Nguyen, D.; Jenkins, M.; Grizzi, F.; Cobos, E.; Figueroa, J.A.; et al. Galectins in cancer: Carcinogenesis, diagnosis and therapy. *Ann. Transl Med.* **2014**, *2*, 88. [PubMed]
10. Cummings, R.D.; Liu, F.T. Galectins. In *Essentials of Glycobiology*, 2nd ed.; Varki, A., Cummings, R.D., Eds.; Cold Spring Harbor: New York, NY, USA, 2009.
11. Leffler, H.; Carlsson, S.; Hedlund, M.; Qian, Y.; Poirier, F. Introduction to galectins. *Glycoconj. J.* **2004**, *19*, 433–440. [CrossRef] [PubMed]
12. Danguy, A.; Camby, I.; Kiss, R. Galectins and cancer. *Biochim. Biophys. Acta* **2002**, *1572*, 285–293. [CrossRef]
13. Freydanck, M.K.; Laubender, R.P.; Rack, B.; Schuhmacher, L.; Jeschke, U.; Scholz, C. Two-marker combinations for preoperative discrimination of benign and malignant ovarian masses. *Anticancer Res.* **2012**, *32*, 2003–2008. [PubMed]
14. Zhang, P.; Shi, B.; Zhou, M.; Jiang, H.; Zhang, H.; Pan, X.; Gao, H.; Sun, H.; Li, Z. Galectin-1 overexpression promotes progression and chemoresistance to cisplatin in epithelial ovarian cancer. *Cell Death Dis.* **2014**, *5*, e991. [CrossRef] [PubMed]
15. Kim, H.J.; Jeon, H.K.; Cho, Y.J.; Park, Y.A.; Choi, J.J.; Do, I.G.; Song, S.Y.; Lee, Y.Y.; Choi, C.H.; Kim, T.J.; et al. High galectin-1 expression correlates with poor prognosis and is involved in epithelial ovarian cancer proliferation and invasion. *Eur. J. Cancer* **2012**, *48*, 1914–1921. [CrossRef] [PubMed]
16. Kim, H.J.; Jeon, H.K.; Lee, J.K.; Sung, C.O.; Do, I.G.; Choi, C.H.; Kim, T.J.; Kim, B.G.; Bae, D.S.; Lee, J.W. Clinical significance of galectin-7 in epithelial ovarian cancer. *Anticancer Res.* **2013**, *33*, 1555–1561. [PubMed]
17. Brustmann, H. Epidermal growth factor receptor expression in serous ovarian carcinoma: An immunohistochemical study with galectin-3 and cyclin D1 and outcome. *Int. J. Gynecol. Pathol.* **2008**, *7*, 380–389. [CrossRef] [PubMed]
18. Labrie, M.; Vladoiu, M.C.; Grosset, A.A.; Gaboury, L.; St-Pierre, Y. Expression and functions of galectin-7 in ovarian cancer. *Oncotarget* **2014**, *5*, 7705–7721. [CrossRef] [PubMed]
19. Chen, L.; Yao, Y.; Sun, L.; Zhou, J.; Liu, J.; Wang, J.; Li, J.; Tang, J. Clinical implication of the serum galectin-1 expression in epithelial ovarian cancer patients. *J. Ovarian Res.* **2015**, *8*, 78. [CrossRef] [PubMed]
20. Astorgues-Xerri, L.; Riveiro, M.E.; Tijeras-Raballand, A.; Serova, M.; Neuzillet, C.; Albert, S.; Raymond, E.; Faivre, S. Unraveling galectin-1 as a novel therapeutic target for cancer. *Cancer Treatment Rev.* **2014**, *40*, 307–319. [CrossRef] [PubMed]

21. Dings, R.P.M.; Miller, M.C.; Nesmelova, I.; Astorgues-Xerri, L.; Kumar, N.; Serova, M.; Chen, X.; Raymond, E.; Hoye, T.R.; Mayo, K.H. Anti-tumor agent calixarene 0118 targets human galectin-1 as an allosteric inhibitor of carbohydrate binding. *J. Med. Chem.* **2012**, *5*, 5121–5129. [CrossRef] [PubMed]

22. Astorgues-Xerri, L.; Riveiro, M.E.; Tijeras-Raballand, A.; Serova, M.; Rabinovich, G.A.; Bieche, I.; Vidaud, M.; de Gramont, A.; Martinet, M.; Cvitkovic, E.; et al. OTX008, a selective small-molecule inhibitor of galectin-1, downregulates cancer cell proliferation, invasion and tumour angiogenesis. *Eur. J. Cancer* **2014**, *50*, 2463–2477. [CrossRef] [PubMed]

23. Xu, X.C.; El -Naggar, A.K.; Lotan, R. Differential expression of galectin-1 and galectin-3 in thyroid tumors. Potential diagnostic implications. *Am. J. Pathol.* **1995**, *147*, 815–822. [PubMed]

24. Gillenwater, A.; Xu, X.C.; El-Naggar, A.K.; Clayman, G.L.; Lotan, R. Expression of galectins in head and neck squamous cell carcinoma. *Head Neck* **1996**, *18*, 422–432. [CrossRef]

25. Sanjuan, X.; Fernandez, P.L.; Castells, A.; Castronovo, V.; Van den Brule, F.; Liu, F.T.; Cardesa, A.; Campo, E. Differential expression of galectin 3 and galectin 1 in colorectal cancer progression. *Gastroenterology* **1997**, *113*, 1906–1915. [CrossRef]

26. Van den Brule, F.; Califice, S.; Garnier, F.; Fernandez, P.L.; Berchuck, A.; Castronovo, V. Galectin-1 accumulation in the ovary carcinoma peritumoral stroma is induced by ovary carcinoma cells and affects both cancer cell proliferation and adhesion to laminin-1 and fibronectin. *Lab. Investig.* **2003**, *83*, 377–386. [CrossRef] [PubMed]

27. Kim, M.K.; Sung, C.O.; Do, I.G.; Jeon, H.K.; Song, T.J.; Park, H.S.; Lee, Y.Y.; Kim, B.G.; Lee, J.W.; Bae, D.S. Overexpression of Galectin-3 and its clinical significance in ovarian carcinoma. *Int. J. Clin. Oncol.* **2011**, *16*, 352–358. [CrossRef] [PubMed]

28. Patterson, R.J.; Wang, W.; Wang, J.L. Understanding the biochemical activities of galectin-1 and galectin-3 in the nucleus. *Glycoconj. J.* **2004**, *19*, 499–506. [CrossRef] [PubMed]

29. Patterson, R.J.; Haudek, K.C.; Voss, P.G.; Wang, J.R. Examination of the role of galectins in pre-mRNA splicing. In *Galectin*; Stowel, S.R., Cummings, R.D., Eds.; Springer: New York, NY, USA, 2015; pp. 431–449.

30. Carvalho, R.S.; Fernandes, V.C.; Nepomuceno, T.C.; Rodrigues, D.C.; Woods, N.T.; Suarez-Kurtz, G.; Chammas, R.; Monteiro, A.N.; Carvalho, M.A. Characterization of LGALS3 (galectin-3) as a player in DNA damage response. *Cancer Biol. Ther.* **2014**, *15*, 840–850. [CrossRef] [PubMed]

31. Okada, K.; Shimura, T.; Suehiro, T.; Mochiki, E.; Kuwano, H. Reduced galectin-3 expression is an indicator of unfavorable prognosis in gastric cancer. *Anticancer Res.* **2006**, *26*, 1369–1376. [PubMed]

32. Kramer, M.W.; Kuczyk, M.A.; Hennenlotter, J.; Serth, J.; Schilling, D.; Stenzl, A.; Merseburger, A.S. Decreased expression of galectin-3 predicts tumour recurrence in pTa bladder cancer. *Oncol. Rep.* **2008**, *20*, 1403–1408. [CrossRef]

33. Lu, H.; Liu, Y.; Wang, D.; Wang, L.; Zhou, H.; Xu, G.; Xie, L.; Wu, M.; Lin, Z.; Yu, Y.; et al. Galectin-3 regulates metastatic capabilities and chemotherapy sensitivity in epithelial ovarian carcinoma via NF-κB pathway. *Tumour Biol.* **2016**, *37*, 11469–11477. [CrossRef] [PubMed]

34. Hossein, G.; Keshavarz, M.; Ahmadi, S.; Naderi, N. Synergistic effects of PectaSol-C modified citrus pectin an inhibitor of Galectin-3 and paclitaxel on apoptosis of human SKOV-3 ovarian cancer cells. *Asian Pac. J. Cancer Prev.* **2013**, *14*, 7561–7568. [CrossRef] [PubMed]

International Journal of
Molecular Sciences

MDPI

Review

Galectin-3 Performance in Histologic and Cytologic Assessment of Thyroid Nodules: A Systematic Review and Meta-Analysis

Pierpaolo Trimboli [1,*], Camilla Virili [2], Francesco Romanelli [3], Anna Crescenzi [4] and Luca Giovanella [1]

[1] Department of Nuclear Medicine and Thyroid Centre, Oncology Institute of Southern Switzerland, Via Ospedale, 6500 Bellinzona, Switzerland; luca.giovanella@eoc.ch

[2] Department of Medico-surgical Sciences and Biotechnologies, Sapienza University of Rome, 04100 Latina, Italy; camillavirili@libero.it

[3] Department of Experimental Medicine, Sapienza University of Rome, 00161 Rome, Italy; francesco.romanelli@uniroma1.it

[4] Pathology Unit, University Hospital Campus Bio Medico, 00128 Rome, Italy; a.crescenzi@unicampus.it

* Correspondence: pierpaolo.trimboli@eoc.ch; Tel.: +41-91-811-6446

Received: 15 July 2017; Accepted: 10 August 2017; Published: 11 August 2017

Abstract: The literature on Galectin-3 (Gal-3) was systematically reviewed to achieve more robust information on its histologic reliability in identifying thyroid cancers and on the concordance between Gal-3 test in histologic and cytologic samples. A computer search of the PubMed and Scopus databases was conducted by combinations of the terms thyroid and Gal-3. Initially, 545 articles were found and, after their critical review, 52 original papers were finally included. They reported 8172 nodules with histologic evaluation of Gal-3, of which 358 with also preoperative FNAC Gal-3 assessment. At histology, Gal-3 sensitivity was 87% (95% confidence intervals [CI] from 86% to 88%), and specificity 87% (95% CI from 86% to 88%); in both cases, we found heterogeneity (I_2 85% and 93%, respectively) and significant publication bias ($p < 0.001$). The pooled rate of positive Gal-3 at fine needle aspiration (FNAC) among cancers with histologically proven Gal-3 positivity was 94% (95% CI from 89% to 97%), with neither heterogeneity (I_2 14.5%) nor bias ($p = 0.086$). These data show high reliability of Gal-3 for thyroid cancer at histology, while its sensitivity on FNAC samples is lower. The limits of cytologic preparations and interpretation of Gal-3 results have to be solved.

Keywords: thyroid cancer; meta-analysis; cytology; Galectin-3; fine needle aspiration (FNAC)

1. Introduction

The mammalian galectins are a family of soluble sugar binding proteins characterized by a carbohydrate recognition domain that shows a high affinity for β-galactosides [1]. Among the galectin family, Galectin-3 (Gal-3) is the only chimeric type. It is composed of a proline- and glycin-rich amino-terminal domain fused to a carboxy-terminal carbohydrate recognition domain. This domain recognizes and binds β-galactosides on cell glycoproteins and glycolipids. Gal-3 may be observed in the cytoplasm and in the nucleus as well as extracellular matrix [2]. The expression and distribution of Gal-3 between the nucleus and cytosol changes during cell differentiation and cancer development [3]. This protein is involved in a large number of physiological and pathological processes such as cell proliferation, differentiation, survival, apoptosis, intracellular trafficking and tumor progression [4]. Then, Gal-3 expression was reported in tumors of different organs: in thyroid, liver, stomach, and central nervous system, the protein is up-regulated, whereas, in cancers of the breast, ovary, uterus and prostate, it is down-regulated [5]. Nuclear Gal-3 is involved in various functions such as regulation of gene transcription; within thyroid cells, it modulates gene transcription by the interaction with the nuclear thyroid-specific transcription factor TTF-1 [6]. Gal-3 has received significant attention for its utility as a diagnostic marker for thyroid cancer, being differentially expressed in thyroid carcinoma compared with benign and normal thyroid specimens [7].

In thyroid nodules, Gal-3 expression was demonstrated by immunohistochemistry (IHC) utilizing histologic or cytologic specimens; numerous studies investigated Gal-3 in thyroid malignancy either as a single marker or as a part of a molecular panel. Variability in the reported performances of this marker, however, was frequently seen; a standardized protocol has been proposed and published several years ago by experts in the fields for widespread use of Gal-3 [8]. Recent studies have also demonstrated improved methodological reliability for Gal-3 expression in thyroid fine-needle aspiration (FNAC) material by performing the IHC evaluation on cell-block preparation so avoiding problems related to antigen accessibility in cell smears. The use of cell-block for performing IHC for diagnostic marker is the only recommended by American Thyroid Association (ATA) [9].

Following the above open problems, the actual reliability of Gal-3 in detecting thyroid cancers remains not definitely assessed. Thus, here we aimed to systematically review the literature to achieve more robust information about: (1) the histologic sensitivity and specificity of Gal-3 in the identification of thyroid malignancy; and (2) the concordance between histologic results and preoperative data obtained in FNAC samples. The latter was performed to emphasize the reliability of Gal-3 test in the pre-surgical diagnostic workup. According to these objectives, we designed a very careful selection of papers for meta-analysis searching those paper reporting Gal-3 evaluation in both FNAC and histologic specimens in the same patient.

2. Results

2.1. Eligible Articles and Description of the Studies

The comprehensive computer literature search revealed 545 articles. Once duplicates articles were excluded, the papers initially included were 357. Abstracts of these articles were screened and 305 were excluded according to the abovementioned criteria; specifically, 41 were excluded due to unclear data and another seven due to incomplete data on Gal-3 (i.e., Gal-3 performed only in FNAC samples with no correlated data on histologic specimens). Two papers were excluded after contacting authors due to potential overlapping data with other studies. Finally, the systematic review included 52 original articles reporting Gal-3 IHC in histologic samples [10–61], of which five with both FNAC and histologic examination of Gal-3 in the same series of lesions [13,15,26,38,43]. Figure 1 illustrates the diagram of flow to retrieve the final series of papers.

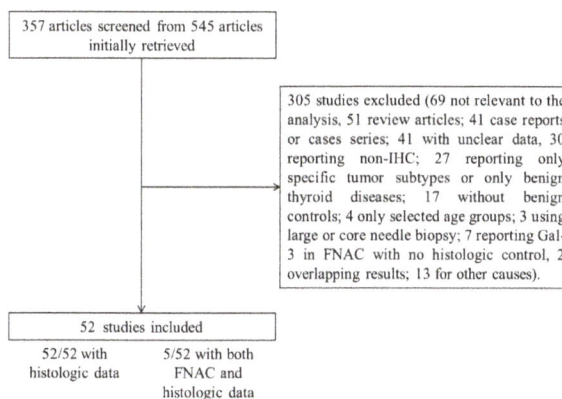

357 articles screened from 545 articles initially retrieved

305 studies excluded (69 not relevant to the analysis, 51 review articles; 41 case reports or cases series; 41 with unclear data, 30 reporting non-IHC; 27 reporting only specific tumor subtypes or only benign thyroid diseases; 17 without benign controls; 4 only selected age groups; 3 using large or core needle biopsy; 7 reporting Gal-3 in FNAC with no histologic control, 2 overlapping results; 13 for other causes).

52 studies included

52/52 with histologic data	5/52 with both FNAC and histologic data

Figure 1. Diagram of flow of search strategy and results.

2.2. Qualitative Analysis

The fifty-two papers included [10–61] were published from 1995 to 2016. Authors were European in 25 cases, while 15 were from Asia, seven from USA, four from Africa, and one from Latin America. All studies had a retrospective design. Overall, more than eight thousand lesions were submitted to Gal-3 test at histology, of which half were malignant. As mentioned above, five papers reported Gal-3 results obtained in the same patients in both FNAC and histology; one of these [38] was excluded from FNAC analysis because Gal-3 was evaluated on conventional smears from FNAC which are not the gold standard for ancillary examinations such as IHC. The remaining four studies [13,15,26,43] described 358 cases, of which 142 cancers and 216 benign nodules. Table 1 summarizes the data from the 52 studies included for the meta-analysis. Notably, the large majority of benign lesions with positivity of Gal-3 were represented by adenomas and thyroiditis, probably because Gal-3 is expressed in the cytosol of thyrocytes and blocks the apoptotic pathway (this condition may be present in these benign lesions).

Table 1. Summary of results of the 52 studies included for the meta-analysis.

Samples	Total Lesions	Carcinomas	Carcinomas with Gal-3+ (%)	Benign Lesions	Benign Lesions with Gal-3– (%)
Histologic data	8172	4237	3654 (86%)	3935	3341 (85%)
Cytologic data (cell-block preparation from FNAC samples)	358	142	129 (91%)	216	194 (90%)

2.3. Quantitative Analysis (Meta-Analysis) of Gal-3 on Histology

The histologic performance of Gal-3 in histologic specimens was evaluated by all 52 studies. The pooled rate of Gal-3 positive results among carcinomas (i.e., Gal-3 sensitivity for thyroid malignancy histology) was 87% (95% CI from 86% to 88%), ranging from 53% to 100%. The series of cancers was heterogeneous (I_2 85.1%, 95% CI from 81.7% to 87.8%) and showed significant publication bias (Egger test: −2.53, 95% CI from −3.59 to −1.47, $p < 0.001$). The pooled percentage of benign lesions with negative Gal-3 (i.e., Gal-3 specificity for thyroid benignancy histology) was 87% (95% CI from 86% to 88%), ranging from 20% to 100%. This series was heterogeneous (I_2 93.5%, 95% CI from 92.5% to 94.3%) and showed significant publication bias (Egger test: −3.59, 95% CI from −4.85 to −2.32, $p < 0.001$). Figure 2 summarizes the performance of Gal-3 at histology.

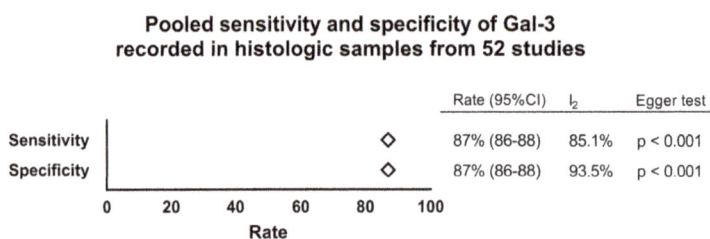

Pooled sensitivity and specificity of Gal-3 recorded in histologic samples from 52 studies

		Rate (95%CI)	I_2	Egger test
Sensitivity	◇	87% (86-88)	85.1%	p < 0.001
Specificity	◇	87% (86-88)	93.5%	p < 0.001

0   20   40   60   80   100
Rate

Figure 2. Forest plot of pooled sensitivity and specificity (fixed-effect) of Gal-3 test at histology of 4237 thyroid cancers and 3935 thyroid benign nodules from 52 studies. High heterogeneity (I_2) and significant publication bias (Egger test) were observed.

In an attempt to avoid the above heterogeneity and risk of bias, we meta-analyzed only the larger series and selected papers with at least 200 cases with histologic follow-up [16,27,30,37,41,44,47,48,55]; nevertheless, heterogeneity and publication bias were still present (sensitivity and specificity were 89% and 92%, respectively). On the other hand, when we evaluated a subseries of papers using biotin-free method for Gal-3 [18,21,26–28,34,35,37,40,43,47,50,54,57,59], interestingly we found absence of publication bias in the calculation of sensitivity (86% sensitivity, I_2 88.3%, 95% CI from 82.7% to 91.4%, Egger test: -2.9, 95% CI from -6.17 to 0.37, $p = 0.078$; 84% specificity, I_2 91.8%, 95% CI from 88.7% to 93.8%, Egger test: $-4,09$, 95% CI from -7.22 to -0.97, $p = 0.014$).

2.4. Quantitative Analysis (Meta-Analysis) of Gal-3 on FNAC, and Concordance of Cytologic Results with Histologic Ones

The reliability of Gal-3 when performed in cell-block from FNAC was investigated and reported in four papers [13,15,26,43].

Initially, we analyzed the histologic results in this subgroup: the pooled histologic sensitivity of Gal-3 was 96% (95% CI from 92% to 99%), and neither heterogeneity (I_2 19.1%, 95% CI from 0% to 73.6%) nor significant bias (Egger test: -0.55, 95% CI from -6.55 to 5.44, $p = 0.73$) was found; in addition, histologic specificity of Gal-3 was 89% (95% CI from 85% to 93%), and there was neither heterogeneity (I_2 0%, 95% CI from 0% to 67.9%) nor significant publication bias (Egger test: -0.6, 95% CI from -13.59 to 12.38, $p = 0.86$).

Then, we evaluated the pooled rate of preoperative Gal-3 positivity between cancers (i.e., Gal-3 sensitivity for thyroid malignancy at FNAC): the sensitivity of Gal-3 was 90% (95% CI from 85% to 94%), ranging from 80% to 94%; the series of cancers was not heterogeneous (I_2 0%, 95% CI from 0% to 67.9%) and showed no significant publication bias (Egger test: -0.87 (95% CI from -4.29 to 2.55, $p = 0.389$).

Finally, we calculated the concordance between Gal-3 results at histology with Gal-3 at FNAC. The pooled rate of positive Gal-3 at FNAC within the series of cancers with positive Gal-3 test at histology was 94% (95% CI from 89% to 97%); the series was not heterogeneous (I_2 14.5%, 95% CI from 0% to 72.3%) and there was no publication bias (Egger test: -1.25 (95% CI from -2.95 to 0.44, $p = 0.086$) (Figure 3). Perfect concordance (100%) was found between negative Gal-3 results at FNAC ($n = 216$) and histology ($n = 216$).

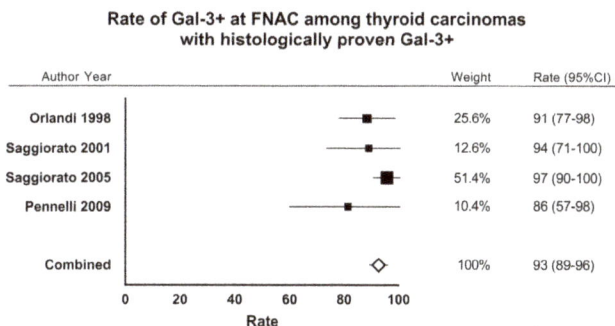

Rate of Gal-3+ at FNAC among thyroid carcinomas with histologically proven Gal-3+

Author Year		Weight	Rate (95%CI)
Orlandi 1998		25.6%	91 (77-98)
Saggiorato 2001		12.6%	94 (71-100)
Saggiorato 2005		51.4%	97 (90-100)
Pennelli 2009		10.4%	86 (57-98)
Combined		100%	93 (89-96)

0   20   40   60   80   100
Rate

Figure 3. Forest plot of meta-analysis (fixed-effect) of proportion of FNAC samples (cell block preparation) with positivity of Gal-3 test among thyroid cancers with histologically proven positivity of Gal-3.

3. Discussion

The major result of the present meta-analysis is that Gal-3 IHC is positive in 87% of thyroid cancers, as proven at analysis of histologic findings. This is pivotal information because it demonstrates that many thyroid carcinomas have over-expression of this marker. The weakness of this result is that we observed very relevant heterogeneity between the studies and significant publication bias. However, it is reasonable to consider this percentage as quite reliable due to the short 95% CI of sensitivity. In addition, 87% of benign lesions are proven to be Gal-3 negative at histology. High heterogeneity, significant publication bias, and short 95% CI were also recorded in specificity. Interestingly, we found a significant improvement of results of sensitivity of Gal-3 when we meta-analyzed only those papers using a biotin-free method. This indirectly demonstrates the Gal-3 has to be evaluated by biotin-free approach. An interesting study on Gal-3 in preoperative samples and definitive histologic specimens was published by Bartolazzi et al. [7]. Unfortunately, this study was excluded from our meta-analysis due to high risk of overlapping data. In that study, 465 nodules were tested for Gal-3 (biotin-free method), 134 of these were positive and 101/134 were carcinomas (the remaining false positive cases were largely represented by benign neoplasia such as adenomas). Table 2 reports the most significant results from this study.

Table 2. Results obtained in the multicenter study by Bartolazzi et al. [7]: Galectin-3 immunocytochemical evaluation on 465 follicular thyroid proliferations compared to the final histology.

Samples	Malignant	Benign	Borderline	Positive Predictive Value
Galectin-3 positive	101	22	11	82%
Galectin-3 negative	280	29	22	91%

Interestingly, this study [7] is the most relevant in the literature and was conducted according to a specific protocol [8]. Thus, our present study strongly corroborates those findings and demonstrates, with evidence-based data, that the vast majority of thyroid malignancy is Gal-3 positive.

The most relevant application of Gal-3 is obviously in the assessment of thyroid nodules, especially when the conventional cytologic report is non-conclusive/indeterminate. The present meta-analysis shows that Gal-3 test on FNAC samples (cell-block preparation) has a sensitivity lower than that recorded at histology. We have to underline that some studies of this subgroup were published by the same authors; there were no overlapping data (we contacted the authors), but it might represent a potential risk of bias. However, many problems might affect these data. In fact, among the different methodologies used for Gal-3 evaluation in thyroid specimens, there is great technical variation in antibody clones and characteristics, antibody dilution, antigen retrieval methods and biotin free protocols for reaction visualization. Moreover, relevant differences exist in staining interpretation: discordance among pathologists occurs for marker localization (nuclear, cytoplasm and cell membrane) and there is still lack of fixed cutoff for positive classification (i.e., qualitative [weak, moderate, and intense] and quantitative [5%, 10%, 50%, etc.]) [62]. The scoring criterion is a crucial step to improve the reliability of molecular markers for clinical use and then carry out their sensitivity and specificity. As for other markers used as oncological diagnostic, prognostic and predictive tools, Gal-3 strongly needs for a standardized protocol and laboratory validation to establish its actual utility and reliability. Importantly, the laboratory validation of Gal-3 test-method for thyroid cancer diagnosis has already been extensively performed, and two papers have adopted it [7,16]. This strongly encourages using Gal-3 protocol [8]. Recent studies about other markers reported improved accuracy in protein expression analysis by the use of RNA in situ hybridization especially when there is no suitable antibody or the target molecule is a secreted protein [63]. Moreover, the RNAscope 2.0 assay system (Advanced Cell Diagnostics, Hayward, CA, USA) is a relatively new technology accomplished with recommended scoring that allows reliable mRNA assessment on a glass slide [64]. After studies in thyroid pathology, mRNA in situ evaluation approach could integrate or replace IHC for Gal-3 for its use in clinical practice. As a potential weakness of this evaluation, it should be underlined that a scantly (i.e., undetectable at IHC) Gal-3 expression is normal in nucleosome of normal thyroid cell because there Gal-3 plays a physiological function as transcriptional regulator. Thus, mRNA in situ evaluation might carry a risk of potential false positive results. As mentioned above, one study evaluated Gal-3 in FNAC smears and was excluded from meta-analysis of cytologic reliability of Gal-3 [38]; however, regardless this cytologic preparation is not the gold standard for IHC, the authors reported good sensitivity.

A discussion for clinical practice may be addressed. After defining the above issues of the preoperative Gal-3 use, we could have a significant improvement in the management of those patients with indeterminate FNAC reports; in fact, thyroid cancers are not frequent in this class of patients (i.e., up to 25%) and the prognosis of malignant lesions with this preoperative assessment is really good [65,66]. This encourages reducing diagnostic thyroidectomy in patients with indeterminate FNAC and negative Gal-3 result [67]. The use of Gal-3 test should have a favorable impact on the costs for management of our patients especially when compared with other molecular testing [68], even if this has to be proven in a prospective study, if possible.

Finally, other preoperative uses of Gal-3 test were reported in literature. Specifically, high reliability of Gal-3 was recorded in microhistologic specimens from core needle biopsy [69–71]; the latter results have to be confirmed but this technique has not been diffused. In addition, Gal-3 was investigated as a serum marker; unfortunately a few studies on small series have been reported in this topic and no significant difference between cancers and benign lesion were found. These findings are probably due to the sub- intra-cellular localization of Gal-3 and the lack of its secretory pathway. To date, the role of serum Gal-3 is excluded [40,45,72–74]. Two papers investigated an imaging Gal-3 approach [75,76], but until now no data on application of these radiolabelled antibodies in humans exist. As these studies come from the same group, an independent validation should be desirable before proceeding with phase I clinical studies. Notably, other widely employed nuclear medicine procedures (i.e., 99mTc-sestaMIBI scan and 18FDG-PET/CT) already show good accuracy in discriminating cytologically indeterminate nodules [77–81].

In conclusion, the present meta-analysis demonstrates that a high rate of thyroid cancers has Gal-3 over-expression at histology and elevated percentage of thyroid benign nodules has negative Gal-3 test. However, high heterogeneity and significant publication bias were evident. In addition, the sensitivity of Gal-3 test on FNAC samples is lower than that recorded at histology, prompting to solve limits of cytologic preparations and interpretation of results.

4. Materials and Methods

4.1. Search Strategy

Initially, we searched studies investigating the role of Galectin-3 in diagnosing thyroid nodules. A comprehensive computer literature search of the PubMed/MEDLINE and Scopus databases was conducted to find published articles on this topic. The search algorithm was based on the combinations of the terms "thyroid" and "galectin" (i.e., "thyro*" AND "galect*" on PubMed/MEDLINE: by this search we could find all combinations of terms beginning with thyro and those beginning with galect). A beginning date limit was not used, and the search was updated until 14 May 2017. The research was restricted to English language studies. To identify additional studies and expand our search, references lists of the retrieved articles could be also screened. Three investigators (Pierpaolo Trimboli, Camilla Virili, and Francesco Romanelli) independently searched articles applying the above strategy.

4.2. Study Selection

Original articles reporting experience on the use of Gal-3 in humans were eligible for inclusion. In particular, we selected studies encompassing analysis of Gal-3 IHC on malignant and benign thyroid lesions; we also selected papers about Gal-3 usefulness in preoperatively made FNAC in ameliorating cytologic diagnosis of thyroid tumors. Specifically, only studies using formalin fixed and paraffin embedded histologic samples and cell-block cytologic samples were considered as eligible for the final series to be meta-analyzed. Studies using non-IHC technique for Gal-3 evaluation, review articles, case reports, cases series, articles with unclear data, and series with overlapping results were also excluded. The same three authors (PT, CV, FR) independently screened titles and abstracts of the retrieved articles according to the above criteria, reviewed the full-texts, and selected articles for their inclusion.

4.3. Data Extraction

For each included study, information was extracted concerning study data: authors, year of publication, journal, study aim, study design, main results, type of Gal-3 evaluation (i.e., cytologic and/or histologic specimens, conventional or cell-block preparation) and country of patient's origin. Number, gender, and age of patients enrolled in the studies were also collected, when available.

4.4. Statistical Analysis

For statistical pooling of the data, the fixed-effects model was used. Pooled data were presented with 95% confidence intervals (95% CI) and displayed using a forest plot. I-square index was used to quantify the heterogeneity among the studies, and a significant heterogeneity was defined as an I-square value > 50%. Egger test was carried out to evaluate the possible presence of a significant publication bias. Statistical significance was set at $p < 0.01$. All analyses were performed using the StatsDirect statistical software version (StatsDirect Ltd., Altrincham, UK).

Author Contributions: Pierpaolo Trimboli and Luca Giovanella conceived and designed the study; Camilla Virili, Pierpaolo Trimboli and Francesco Romanelli performed the literature search; Pierpaolo Trimboli analyzed the data; and Anna Crescenzi and Luca Giovanella critically reviewed the results. All five authors contributed to write the manuscript.

Conflicts of Interest: Authors declare no conflict of interest.

References

1. Leffler, H.; Carlsson, S.; Hedlund, M.; Qian, Y.; Poirier, F. Introduction to galectins. *Glycoconj. J.* **2004**, *19*, 433–440. [CrossRef] [PubMed]
2. Wang, J.L.; Gray, R.M.; Haudek, K.C.; Patterson, R.J. Nucleocytoplasmic lectins. *Biochim. Biophys. Acta* **2004**, *673*, 75–93. [CrossRef] [PubMed]
3. Fritsch, K.; Mernberger, M.; Nist, A.; Stiewe, T.; Brehm, A.; Jacob, R. Galectin-3 interacts with components of the nuclear ribonucleoprotein complex. *BMC Cancer* **2016**, *16*, 502. [CrossRef] [PubMed]
4. Liu, F.T.; Rabinovich, G. Galectins as modulators of tumour progression. *Nat. Rev. Cancer* **2005**, *5*, 29–41. [CrossRef] [PubMed]
5. Haudek, K.C.; Spronk, K.J.; Voss, P.G.; Patterson, R.J.; Wang, J.L.; Arnoys, E.J. Dynamics of Galectin-3 in the Nucleus and Cytoplasm. *Biochim. Biophys. Acta* **2010**, *1800*, 181–189. [CrossRef] [PubMed]
6. Paron, I.; Scaloni, A.; Pines, A.; Bachi, A.; Liu, F.T.; Puppin, C.; Pandolfi, M.; Ledda, L.; Di Loreto, C.; Damante, G.; et al. Nuclear localization of Galectin-3 in transformed thyroid cells: A role in transcriptional regulation. *Biochem. Biophys. Res. Commun.* **2003**, *302*, 545–553. [CrossRef]
7. Bartolazzi, A.; Orlandi, F.; Saggiorato, E.; Volante, M.; Arecco, F.; Rossetto, R.; Palestini, N.; Ghigo, E.; Papotti, M.; Bussolati, G.; et al. Italian Thyroid Cancer Study Group (ITCSG). Galectin-3-expression analysis in the surgical selection of follicular thyroid nodules with indeterminate fine-needle aspiration cytology: A prospective multicentre study. *Lancet Oncol.* **2008**, *9*, 543–549. [CrossRef]
8. Bartolazzi, A.; Bellotti, C.; Sciacchitano, S. Methodology and technical requirements of the galectin-3 test for the preoperative characterization of thyroid nodules. *Appl. Immunohistochem. Mol. Morphol.* **2012**, *20*, 2–7. [CrossRef] [PubMed]
9. Haugen, B.R.; Alexander, E.K.; Bible, K.C.; Doherty, G.M.; Mandel, S.J.; Nikiforov, Y.E.; Pacini, F.; Randolph, G.W.; Sawka, A.M.; Schlumberger, M.; et al. 2015 American Thyroid Association Management Guidelines for Adult Patients with Thyroid Nodules and Differentiated Thyroid Cancer: The American Thyroid Association Guidelines Task Force on Thyroid Nodules and Differentiated Thyroid Cancer. *Thyroid* **2016**, *26*, 1–133. [CrossRef] [PubMed]
10. Xu, X.C.; el-Naggar, A.K.; Lotan, R. Differential expression of galectin-1 and galectin-3 in thyroid tumors. Potential diagnostic implications. *Am. J. Pathol.* **1995**, *147*, 815–822. [PubMed]
11. Fernández, P.L.; Merino, M.J.; Gómez, M.; Campo, E.; Medina, T.; Castronovo, V.; Sanjuán, X.; Cardesa, A.; Liu, F.T.; Sobel, M.E. Galectin-3 and laminin expression in neoplastic and non-neoplastic thyroid tissue. *J. Pathol.* **1997**, *181*, 80–86. [CrossRef]
12. Cvejic, D.; Savin, S.; Paunovic, I.; Tatic, S.; Havelka, M.; Sinadinovic, J. Immunohistochemical localization of galectin-3 in malignant and benign human thyroid tissue. *Anticancer Res.* **1998**, *18*, 2637–2641. [PubMed]
13. Orlandi, F.; Saggiorato, E.; Pivano, G.; Puligheddu, B.; Termine, A.; Cappia, S.; De Giuli, P.; Angeli, A. Galectin-3 is a presurgical marker of human thyroid carcinoma. *Cancer Res.* **1998**, *58*, 3015–3020. [PubMed]

14. Inohara, H.; Honjo, Y.; Yoshii, T.; Akahani, S.; Yoshida, J.; Hattori, K.; Okamoto, S.; Sawada, T.; Raz, A.; Kubo, T. Expression of galectin-3 in fine-needle aspirates as a diagnostic marker differentiating benign from malignant thyroid neoplasms. *Cancer* **1999**, *85*, 2475–2484. [CrossRef]

15. Saggiorato, E.; Cappia, S.; De Giuli, P.; Mussa, A.; Pancani, G.; Caraci, P.; Angeli, A.; Orlandi, F. Galectin-3 as a presurgical immunocytodiagnostic marker of minimally invasive follicular thyroid carcinoma. *J. Clin. Endocrinol. Metab.* **2001**, *86*, 5152–5158. [CrossRef] [PubMed]

16. Bartolazzi, A.; Gasbarri, A.; Papotti, M.; Bussolati, G.; Lucante, T.; Khan, A.; Inohara, H.; Marandino, F.; Orlandi, F.; Nardi, F.; et al. Thyroid Cancer Study Group. Application of an immunodiagnostic method for improving preoperative diagnosis of nodular thyroid lesions. *Lancet* **2001**, *357*, 1644–1650. [CrossRef]

17. Beesley, M.F.; McLaren, K.M. Cytokeratin 19 and galectin-3 immunohistochemistry in the differential diagnosis of solitary thyroid nodules. *Histopathology* **2002**, *41*, 236–243. [CrossRef] [PubMed]

18. Herrmann, M.E.; LiVolsi, V.A.; Pasha, T.L.; Roberts, S.A.; Wojcik, E.M.; Baloch, Z.W. Immunohistochemical expression of galectin-3 in benign and malignant thyroid lesions. *Arch. Pathol. Lab. Med.* **2002**, *126*, 710–713. [PubMed]

19. Cvejić, D.; Savin, S.; Petrović, I.; Paunović, I.; Tatić, S.; Havelka, M. Differential expression of galectin-3 in papillary projections of malignant and non-malignant hyperplastic thyroid lesions. *Acta. Chir. Iugosl.* **2003**, *50*, 67–70. [CrossRef] [PubMed]

20. Giannini, R.; Faviana, P.; Cavinato, T.; Elisei, R.; Pacini, F.; Berti, P.; Fontanini, G.; Ugolini, C.; Camacci, T.; De Ieso, K.; et al. Galectin-3 and oncofetal-fibronectin expression in thyroid neoplasia as assessed by reverse transcription-polymerase chain reaction and immunochemistry in cytologic and pathologic specimens. *Thyroid* **2003**, *13*, 765–770. [CrossRef] [PubMed]

21. Jakubiak-Wielganowicz, M.; Kubiak, R.; Sygut, J.; Pomorski, L.; Kordek, R. Usefulness of galectin-3 immunohistochemistry in differential diagnosis between thyroid follicular carcinoma and follicular adenoma. *Pol. J. Pathol.* **2003**, *54*, 111–115. [PubMed]

22. Kovács, R.B.; Földes, J.; Winkler, G.; Bodó, M.; Sápi, Z. The investigation of galectin-3 in diseases of the thyroid gland. *Eur. J. Endocrinol.* **2003**, *149*, 449–453. [CrossRef] [PubMed]

23. Mehrotra, P.; Okpokam, A.; Bouhaidar, R.; Johnson, S.J.; Wilson, J.A.; Davies, B.R.; Lennard, T.W. Galectin-3 does not reliably distinguish benign from malignant thyroid neoplasms. *Histopathology* **2004**, *45*, 493–500. [CrossRef] [PubMed]

24. Oestreicher-Kedem, Y.; Halpern, M.; Roizman, P.; Hardy, B.; Sulkes, J.; Feinmesser, R.; Stern, Y. Diagnostic value of galectin-3 as a marker for malignancy in follicular patterned thyroid lesions. *Head Neck* **2004**, *26*, 960–966. [CrossRef] [PubMed]

25. Weber, K.B.; Shroyer, K.R.; Heinz, D.E.; Nawaz, S.; Said, M.S.; Haugen, B.R. The use of a combination of galectin-3 and thyroid peroxidase for the diagnosis and prognosis of thyroid cancer. *Am. J. Clin. Pathol.* **2004**, *122*, 524–531. [CrossRef] [PubMed]

26. Saggiorato, E.; De Pompa, R.; Volante, M.; Cappia, S.; Arecco, F.; Dei Tos, A.P.; Orlandi, F.; Papotti, M. Characterization of thyroid 'follicular neoplasms' in fine-needle aspiration cytological specimens using a panel of immunohistochemical markers: A proposal for clinical application. *Endocr. Relat. Cancer* **2005**, *12*, 305–317. [CrossRef] [PubMed]

27. De Matos, P.S.; Ferreira, A.P.; de Oliveira Facuri, F.; Assumpção, L.V.; Metze, K.; Ward, L.S. Usefulness of HBME-1, cytokeratin 19 and galectin-3 immunostaining in the diagnosis of thyroid malignancy. *Histopathology* **2005**, *47*, 391–401. [CrossRef] [PubMed]

28. Galusca, B.; Dumollard, J.M.; Lassandre, S.; Niveleau, A.; Prades, J.M.; Estour, B.; Peoc'h, M. Global DNA methylation evaluation: Potential complementary marker in differential diagnosis of thyroid neoplasia. *Virchows Arch.* **2005**, *447*, 18–23. [CrossRef] [PubMed]

29. Nucera, C.; Mazzon, E.; Caillou, B.; Violi, M.A.; Moleti, M.; Priolo, C.; Sturniolo, G.; Puzzolo, D.; Cavallari, V.; Trimarchi, F.; et al. Human galectin-3 immunoexpression in thyroid follicular adenomas with cell atypia. *J. Endocrinol. Investig.* **2005**, *28*, 106–112. [CrossRef]

30. Prasad, M.L.; Pellegata, N.S.; Huang, Y.; Nagaraja, H.N.; de la Chapelle, A.; Kloos, R.T. Galectin-3, fibronectin-1, CITED-1, HBME1 and cytokeratin-19 immunohistochemistry is useful for the differential diagnosis of thyroid tumors. *Mod. Pathol.* **2005**, *18*, 48–57. [CrossRef] [PubMed]

31. Jo, Y.S.; Li, S.; Song, J.H.; Kwon, K.H.; Lee, J.C.; Rha, S.Y.; Lee, H.J.; Sul, J.Y.; Kweon, G.R.; Ro, H.K.; et al. Influence of the BRAF V600E mutation on expression of vascular endothelial growth factor in papillary thyroid cancer. *J. Clin. Endocrinol. Metab.* **2006**, *91*, 3667–3670. [CrossRef] [PubMed]

32. Nakamura, N.; Erickson, L.A.; Jin, L.; Kajita, S.; Zhang, H.; Qian, X.; Rumilla, K.; Lloyd, R.V. Immunohistochemical separation of follicular variant of papillary thyroid carcinoma from follicular adenoma. *Endocr. Pathol.* **2006**, *17*, 213–223. [CrossRef] [PubMed]

33. Rossi, E.D.; Raffaelli, M.; Mule', A.; Miraglia, A.; Lombardi, C.P.; Vecchio, F.M.; Fadda, G. Simultaneous immunohistochemical expression of HBME-1 and galectin-3 differentiates papillary carcinomas from hyperfunctioning lesions of the thyroid. *Histopathology* **2006**, *48*, 795–800. [CrossRef] [PubMed]

34. Scognamiglio, T.; Hyjek, E.; Kao, J.; Chen, Y.T. Diagnostic usefulness of HBME1, galectin-3, CK19, and CITED1 and evaluation of their expression in encapsulated lesions with questionable features of papillary thyroid carcinoma. *Am. J. Clin. Pathol.* **2006**, *126*, 700–708. [CrossRef] [PubMed]

35. Sapio, M.R.; Guerra, A.; Posca, D.; Limone, P.P.; Deandrea, M.; Motta, M.; Troncone, G.; Caleo, A.; Vallefuoco, P.; Rossi, G.; et al. Combined analysis of galectin-3 and BRAFV600E improves the accuracy of fine-needle aspiration biopsy with cytological findings suspicious for papillary thyroid carcinoma. *Endocr. Relat. Cancer* **2007**, *14*, 1089–1097. [CrossRef] [PubMed]

36. Coli, A.; Bigotti, G.; Parente, P.; Federico, F.; Castri, F.; Massi, G. Atypical thyroid nodules express both HBME-1 and Galectin-3, two phenotypic markers of papillary thyroid carcinoma. *J. Exp. Clin. Cancer Res.* **2007**, *26*, 221–227. [PubMed]

37. Park, Y.J.; Kwak, S.H.; Kim, D.C.; Kim, H.; Choe, G.; Park, D.J.; Jang, H.C.; Park, S.H.; Cho, B.Y.; Park, S.Y. Diagnostic value of galectin-3, HBME-1, cytokeratin 19, high molecular weight cytokeratin, cyclin D1 and p27(kip1) in the differential diagnosis of thyroid nodules. *J. Korean Med. Sci.* **2007**, *22*, 621–628. [CrossRef] [PubMed]

38. Aiad, H.A.; Kandil, M.A.; Asaad, N.Y.; El-Kased, A.M.; El-Goday, S.F. Galectin-3 immunostaining in cytological and histopathological diagnosis of thyroid lesions. *J. Egypt. Natl. Cancer Inst.* **2008**, *20*, 36–46.

39. Hooft, L.; van der Veldt, A.A.; Hoekstra, O.S.; Boers, M.; Molthoff, C.F.; van Diest, P.J. Hexokinase III, cyclin A and galectin-3 are overexpressed in malignant follicular thyroid nodules. *Clin. Endocrinol. Oxf.* **2008**, *68*, 252–257. [CrossRef] [PubMed]

40. Inohara, H.; Segawa, T.; Miyauchi, A.; Yoshii, T.; Nakahara, S.; Raz, A.; Maeda, M.; Miyoshi, E.; Kinoshita, N.; Yoshida, H.; et al. Cytoplasmic and serum galectin-3 in diagnosis of thyroid malignancies. *Biochem. Biophys. Res. Commun.* **2008**, *376*, 605–610. [CrossRef] [PubMed]

41. Savin, S.; Cvejic, D.; Isic, T.; Paunovic, I.; Tatic, S.; Havelka, M. Thyroid peroxidase and galectin-3 immunostaining in differentiated thyroid carcinoma with clinicopathologic correlation. *Hum. Pathol.* **2008**, *39*, 1656–1663. [CrossRef] [PubMed]

42. Than, T.H.; Swethadri, G.K.; Wong, J.; Ahmad, T.; Jamil, D.; Maganlal, R.K.; Hamdi, M.M.; Abdullah, M.S. Expression of Galectin-3 and Galectin-7 in thyroid malignancy as potential diagnostic indicators. *Singap. Med. J.* **2008**, *49*, 333–338.

43. Pennelli, G.; Mian, C.; Pelizzo, M.R.; Naccamulli, D.; Piotto, A.; Girelli, M.E.; Mescoli, C.; Rugge, M. Galectin-3 cytotest in thyroid follicular neoplasia: A prospective, monoinstitutional study. *Acta Cytol.* **2009**, *53*, 533–539. [CrossRef] [PubMed]

44. Barut, F.; Onak Kandemir, N.; Bektas, S.; Bahadir, B.; Keser, S.; Ozdamar, S.O. Universal markers of thyroid malignancies: Galectin-3, HBME-1, and cytokeratin-19. *Endocr. Pathol.* **2010**, *21*, 80–89. [CrossRef] [PubMed]

45. Išić, T.; Savin, S.; Cvejić, D.; Marečko, I.; Tatić, S.; Havelka, M.; Paunović, I. Serum Cyfra 21.1 and galectin-3 protein levels in relation to immunohistochemical cytokeratin 19 and galectin-3 expression in patients with thyroid tumors. *J. Cancer Res. Clin. Oncol.* **2010**, *136*, 1805–1812. [CrossRef] [PubMed]

46. Saleh, H.A.; Jin, B.; Barnwell, J.; Alzohaili, O. Utility of immunohistochemical markers in differentiating benign from malignant follicular-derived thyroid nodules. *Diagn. Pathol.* **2010**, *5*, 9. [CrossRef] [PubMed]

47. Zhu, X.; Sun, T.; Lu, H.; Zhou, X.; Lu, Y.; Cai, X.; Zhu, X. Diagnostic significance of CK19, RET, galectin-3 and HBME-1 expression for papillary thyroid carcinoma. *J. Clin. Pathol.* **2010**, *63*, 786–989. [CrossRef] [PubMed]

48. Song, Q.; Wang, D.; Lou, Y.; Li, C.; Fang, C.; He, X.; Li, J. Diagnostic significance of CK19, TG, Ki67 and galectin-3 expression for papillary thyroid carcinoma in the northeastern region of China. *Diagn. Pathol.* **2011**, *6*, 126. [CrossRef] [PubMed]

49. Abulkheir, I.L.; Mohammad, D.B. Value of immunohistochemical expression of p27 and galectin-3 in differentiation between follicular adenoma and follicular carcinoma. *Appl. Immunohistochem. Mol. Morphol.* **2012**, *20*, 131–140. [CrossRef] [PubMed]

50. Cui, W.; Sang, W.; Zheng, S.; Ma, Y.; Liu, X.; Zhang, W. Usefulness of cytokeratin-19, galectin-3, and Hector Battifora mesothelial-1 in the diagnosis of benign and malignant thyroid nodules. *Clin. Lab.* **2012**, *58*, 673–680. [PubMed]

51. Gong, L.; Chen, P.; Liu, X.; Han, Y.; Zhou, Y.; Zhang, W.; Li, H.; Li, C.; Xie, J. Expressions of D2-40, CK19, galectin-3, VEGF and EGFR in papillary thyroid carcinoma. *Gland Surg.* **2012**, *1*, 25–32. [PubMed]

52. Manivannan, P.; Siddaraju, N.; Jatiya, L.; Verma, S.K. Role of pro-angiogenic marker galectin-3 in follicular neoplasms of thyroid. *Indian J. Biochem. Biophys.* **2012**, *49*, 392–394. [PubMed]

53. Mataraci, E.A.; Ozgüven, B.Y.; Kabukçuoglu, F. Expression of cytokeratin 19, HBME-1 and galectin-3 in neoplastic and nonneoplastic thyroid lesions. *Pol. J. Pathol.* **2012**, *63*, 58–64. [PubMed]

54. Guerra, A.; Marotta, V.; Deandrea, M.; Motta, M.; Limone, P.P.; Caleo, A.; Zeppa, P.; Esposito, S.; Fulciniti, F.; Vitale, M. BRAF (V600E) associates with cytoplasmatic localization of p27kip1 and higher cytokeratin 19 expression in papillary thyroid carcinoma. *Endocrine* **2013**, *44*, 165–171. [CrossRef] [PubMed]

55. Wu, G.; Wang, J.; Zhou, Z.; Li, T.; Tang, F. Combined staining for immunohistochemical markers in the diagnosis of papillary thyroid carcinoma: Improvement in the sensitivity or specificity? *J. Int. Med. Res.* **2013**, *41*, 975–983. [CrossRef] [PubMed]

56. Abd-El Raouf, S.M.; Ibrahim, T.R. Immunohistochemical expression of HBME-1 and galectin-3 in the differential diagnosis of follicular-derived thyroid nodules. *Pathol. Res. Pract.* **2014**, *210*, 971–978. [CrossRef] [PubMed]

57. Ma, H.; Xu, S.; Yan, J.; Zhang, C.; Qin, S.; Wang, X.; Li, N. The value of tumor markers in the diagnosis of papillary thyroid carcinoma alone and in combination. *Pol. J. Pathol.* **2014**, *65*, 202–209. [CrossRef] [PubMed]

58. Ceyran, A.B.; Şenol, S.; Şimşek, B.Ç.; Sağıroğlu, J.; Aydın, A. Role of cd56 and e-cadherin expression in the differential diagnosis of papillary thyroid carcinoma and suspected follicular-patterned lesions of the thyroid: The prognostic importance of e-cadherin. *Int. J. Clin. Exp. Pathol.* **2015**, *8*, 3670–3680. [PubMed]

59. Sumana, B.S.; Shashidhar, S.; Shivarudrappa, A.S. Galectin-3 Immunohistochemical Expression in Thyroid Neoplasms. *J. Clin. Diagn. Res.* **2015**, *9*, EC07–EC11. [CrossRef] [PubMed]

60. Al-Sharaky, D.R.; Younes, S.F. Sensitivity and Specificity of Galectin-3 and Glypican-3 in Follicular-Patterned and Other Thyroid Neoplasms. *J. Clin. Diagn. Res.* **2016**, *10*, EC06–EC10. [CrossRef] [PubMed]

61. Chao, T.T.; Maa, H.C.; Wang, C.Y.; Pei, D.; Liang, Y.J.; Yang, Y.F.; Chou, S.J.; Chen, Y.L. CIP2A is a poor prognostic factor and can be a diagnostic marker in papillary thyroid carcinoma. *APMIS* **2016**, *124*, 1031–1037. [CrossRef] [PubMed]

62. Chiu, C.G.; Strugnell, S.S.; Griffith, O.L.; Jones, S.J.; Gown, A.M.; Walker, B.; Nabi, I.R.; Wiseman, S.M. Diagnostic utility of galectin-3 in thyroid cancer. *Am. J. Pathol.* **2010**, *176*, 2067–2081. [CrossRef] [PubMed]

63. Choi, J.; Lee, H.E.; Kim, M.A.; Jang, B.G.; Lee, H.S.; Kim, W.H. Analysis of MET mRNA Expression in Gastric Cancers Using RNA In situ hybridization assay: Its clinical implication and comparison with immunohistochemistry and silver in situ hybridization. *PLoS ONE* **2014**, *9*, e111658. [CrossRef] [PubMed]

64. Yu, H.; Batenchuk, C.; Badzio, A.; Boyle, T.A.; Czapiewski, P.; Chan, D.C.; Lu, X.; Gao, D.; Ellison, K.; Kowalewski, A.A.; et al. Expression by two complementary diagnostic assays and mRNA in situ hybridization in small cell lung cancer. *J. Thorac. Oncol.* **2017**, *12*, 110–120. [CrossRef] [PubMed]

65. Rago, T.; Scutari, M.; Latrofa, F.; Loiacono, V.; Piaggi, P.; Marchetti, I.; Romani, R.; Basolo, F.; Miccoli, P.; Tonacchera, M.; et al. The large majority of 1520 patients with indeterminate thyroid nodule at cytology have a favorable outcome, and a clinical risk score has a high negative predictive value for a more cumbersome cancer disease. *J. Clin. Endocrinol. Metab.* **2014**, *99*, 3700–3707. [CrossRef] [PubMed]

66. Trimboli, P.; Bongiovanni, M.; Rossi, F.; Guidobaldi, L.; Crescenzi, A.; Ceriani, L.; Nigri, G.; Valabrega, S.; Romanelli, F.; Giovanella, L. Differentiated thyroid cancer patients with a previous indeterminate (Thy 3) cytology have a better prognosis than those with suspicious or malignant FNAC reports. *Endocrine* **2015**, *49*, 191–195. [CrossRef] [PubMed]

67. Sciacchitano, S.; Lavra, L.; Ulivieri, A.; Magi, F.; Porcelli, T.; Amendola, S.; De Francesco, G.P.; Bellotti, C.; Trovato, M.C.; Salehi, L.B.; et al. Combined clinical and ultrasound follow-up assists in malignancy detection in Galectin-3 negative Thy-3 thyroid nodules. *Endocrine* **2016**, *54*, 139–147. [CrossRef] [PubMed]

68. Sciacchitano, S.; Lavra, L.; Ulivieri, A.; Magi, F.; Paolo De Francesco, G.; Bellotti, C.; Salehi, L.B.; Trovato, M.; Drago, C.; Bartolazzi, A. Comparative analysis of diagnostic performance, feasibility and cost of different test-methods for thyroid nodules with indeterminate cytology. *Oncotarget* **2017**. [CrossRef] [PubMed]

69. Carpi, A.; Naccarato, A.G.; Iervasi, G.; Nicolini, A.; Bevilacqua, G.; Viacava, P.; Collecchi, P.; Lavra, L.; Marchetti, C.; Sciacchitano, S.; et al. Large needle aspiration biopsy and galectin-3 determination in selected thyroid nodules with indeterminate FNA-cytology. *Br. J. Cancer* **2006**, *95*, 204–209. [CrossRef] [PubMed]

70. Carpi, A.; Rossi, G.; Coscio, G.D.; Iervasi, G.; Nicolini, A.; Carpi, F.; Mechanick, J.I.; Bartolazzi, A. Galectin-3 detection on large-needle aspiration biopsy improves preoperative selection of thyroid nodules: A prospective cohort study. *Ann. Med.* **2010**, *42*, 70–78. [CrossRef] [PubMed]

71. Trimboli, P.; Guidobaldi, L.; Amendola, S.; Nasrollah, N.; Romanelli, F.; Attanasio, D.; Ramacciato, G.; Saggiorato, E.; Valabrega, S.; Crescenzi, A. Galectin-3 and HBME-1 improve the accuracy of core biopsy in indeterminate thyroid nodules. *Endocrine* **2016**, *52*, 39–45. [CrossRef] [PubMed]

72. Makki, F.M.; Taylor, S.M.; Shahnavaz, A.; Leslie, A.; Gallant, J.; Douglas, S.; The, E.; Trites, J.; Bullock, M.; Inglis, K.; et al. Serum biomarkers of papillary thyroid cancer. *J. Otolaryngol. Head Neck Surg.* **2013**, *42*, 16. [CrossRef] [PubMed]

73. Saussez, S.; Glinoer, D.; Chantrain, G.; Pattou, F.; Carnaille, B.; André, S.; Gabius, H.J.; Laurent, G. Serum galectin-1 and galectin-3 levels in benign and malignant nodular thyroid disease. *Thyroid* **2008**, *18*, 705–712. [CrossRef] [PubMed]

74. Yılmaz, E.; Karşıdağ, T.; Tatar, C.; Tüzün, S. Serum Galectin-3: Diagnostic value for papillary thyroid carcinoma. *Ulus. Cerrahi Derg.* **2015**, *31*, 192–196. [CrossRef] [PubMed]

75. D'Alessandria, C.; Braesch-Andersen, S.; Bejo, K.; Reder, S.; Blechert, B.; Schwaiger, M.; Bartolazzi, A. Noninvasive In Vivo Imaging and Biologic Characterization of Thyroid Tumors by ImmunoPET Targeting of Galectin-3. *Cancer Res.* **2016**, *76*, 3583–3592. [CrossRef] [PubMed]

76. Bartolazzi, A.; D'Alessandria, C.; Parisella, M.G.; Signore, A.; Del Prete, F.; Lavra, L.; Braesch-Andersen, S.; Massari, R.; Trotta, C.; Soluri, A.; et al. Thyroid cancer imaging in vivo by targeting the anti-apoptotic molecule galectin-3. *PLoS ONE* **2008**, *3*, e3768. [CrossRef] [PubMed]

77. Campennì, A.; Siracusa, M.; Ruggeri, R.M.; Laudicella, R.; Pignata, S.A.; Baldari, S.; Giovanella, L. Differentiating malignant from benign thyroid nodules with indeterminate cytology by 99Tc-MIBI scan: A new quantitative method for improving diagnostic accuracy. *Sci. Rep.* **2017**, *7*, 6147. [CrossRef] [PubMed]

78. Campennì, A.; Giovanella, L.; Siracusa, M.; Alibrandi, A.; Pignata, S.A.; Giovinazzo, S.; Trimarchi, F.; Ruggeri, R.M.; Baldari, S. (99 m)Tc-Methoxy-Isobutyl-Isonitrile Scintigraphy Is a Useful Tool for Assessing the Risk of Malignancy in Thyroid Nodules with Indeterminate Fine-Needle Cytology. *Thyroid* **2016**, *26*, 1101–1109. [CrossRef] [PubMed]

79. Piccardo, A.; Puntoni, M.; Treglia, G.; Foppiani, L.; Bertagna, F.; Paparo, F.; Massollo, M.; Dib, B.; Paone, G.; Arlandini, A.; et al. Thyroid nodules with indeterminate cytology: Prospective comparison between 18F-FDG-PET/CT, multiparametric neck ultrasonography, 99 mTc-MIBI scintigraphy and histology. *Eur. J. Endocrinol.* **2016**, *174*, 693–703. [CrossRef] [PubMed]

80. Heinzel, A.; Müller, D.; Behrendt, F.F.; Giovanella, L.; Mottaghy, F.M.; Verburg, F.A. Thyroid nodules with indeterminate cytology: Molecular imaging with ^{9}9mTc-methoxyisobutylisonitrile (MIBI) is more cost-effective than the Afirma gene expression classifier. *Eur. J. Nucl. Med. Mol. Imaging* **2014**, *41*, 1497–1500. [CrossRef] [PubMed]

81. Giovanella, L.; Campenni, A.; Treglia, G.; Verburg, F.A.; Trimboli, P.; Ceriani, L.; Bongiovanni, M. Molecular imaging with (99 m)Tc-MIBI and molecular testing for mutations in differentiating benign from malignant follicular neoplasm: A prospective comparison. *Eur. J. Nucl. Med. Mol. Imaging* **2016**, *43*, 1018–1026. [CrossRef] [PubMed]

International Journal of
Molecular Sciences

MDPI

Review

Galectins and Carcinogenesis: Their Role in Head and Neck Carcinomas and Thyroid Carcinomas

Nadège Kindt [1], Fabrice Journe [1,2], Ghanem E. Ghanem [2] and Sven Saussez [1,3,*]

[1] Laboratory of Anatomy, Department of Human Anatomy and Experimental Oncology,
Faculty of Medicine and Pharmacy, University of Mons (UMons), Pentagone 2A,
6 Ave du Champ de Mars, B-7000 Mons, Belgium; nadege.kindt@umons.ac.be (N.K.);
fabrice.journe@umons.ac.be (F.J.)

[2] Laboratory of Oncology and Experimental Surgery, Institut Jules Bordet,
Université Libre de Bruxelles (ULB), 1000 Brussels, Belgium; gghanem@ulb.ac.be

[3] Department of Oto-Rhino-Laryngology, Université Libre de Bruxelles (ULB), CHU Saint-Pierre,
1000 Brussels, Belgium

* Correspondence: sven.saussez@umons.ac.be; Tel.: +32-65-373-584

Received: 15 November 2017; Accepted: 15 December 2017; Published: 18 December 2017

Abstract: Head and neck cancers are among the most frequently occurring cancers worldwide. Of the molecular drivers described for these tumors, galectins play an important role via their interaction with several intracellular pathways. In this review, we will detail and discuss this role with specific reference to galectins-1, -3, and -7 in angiogenesis, cell proliferation, and invasion as well as in cell transformation and cancer progression. Furthermore, we will evaluate the prognostic value of galectin expression in head and neck cancers including those with oral cavity, salivary gland, and nasopharyngeal pathologies. In addition, we will discuss the involvement of these galectins in thyroid cancers where their altered expression is proposed as a new diagnostic biomarker.

Keywords: galectins; head and neck cancer; thyroid cancer

---

1. Introduction

Head and neck cancers are among the most frequently occurring cancers worldwide, and squamous cell carcinoma is the predominant histological type [1]. Related major risk factors include tobacco and alcohol use; but in recent years, HPV infection has appeared as an additional risk factor that contributes to head and neck squamous cell carcinoma (HNSCC), although contradictory prognostic values have been reported [2,3]. This contradiction can be explained by the choice of the studied population. Indeed, two HPV-positive cancer populations may be distinguished. The first group is mainly composed of non-smoking younger adults with a favorable prognosis, while the second includes smoking and drinking patients with a poor prognosis [4,5].

The thyroid is an endocrine gland located in the lower part of the neck, and its malignancies are the most frequent endocrine cancers, with an increasing incidence over the last several years. Several types of thyroid carcinomas exist, including papillary carcinoma and follicular carcinoma, which are both well-differentiated and the most frequent histological thyroid cancers, as well as medullary carcinoma and anaplastic carcinoma, the latter being the most undifferentiated and aggressive. All differ in their incidence, prognosis and response to treatment [6]. The major risk factors of thyroid cancers are ionizing radiation exposure and pre-existent benign thyroid pathology, and recently obesity has been suggested as a risk factor [7].

Galectins (gal) are proteins that are encoded by LGALS genes. They are members of the lectin family, of which 14 mammalian galectins have been identified. Galectins belong to the family of glycan-binding proteins and present three main structures (Figure 1). The first structure is the prototype galectin which is composed of one carbohydrate recognition domain (CRD) and can dimerize; this group is composed of gal-1, -2, -5, -7, -10, -11, -13, -14, and -15. The second structure is the tandem-repeat galectin which is composed of two CRD domains connected by a peptide bond; this group contains gal-4, -6, -8, -9, and -12. The third structure is a chimeric galectin with the unique gal-3 for which the CRD domain contains an N-terminal sequence that allows the oligomerization to form a pentamer. The galectins have an affinity for carbohydrates, including *N*-acetyl-lactosamine disaccharide (LacNac) or β-galactose and are found both in the intracellular and extracellular compartments. Moreover, galectins display various roles depending on their location. Indeed, in the intracellular compartment, gal-1 and -3 regulate some signaling pathways involved in cell survival by direct interaction with, notably, H-Ras and K-Ras pathways, respectively [8,9]. In the extracellular matrix, gal-3 contributes to the stimulation of endothelial cell migration and proliferation leading to angiogenesis promotion [10]. Thus, galectins are involved in several mechanisms leading and promoting cancer. The present review will focus on their implications in head and neck, as well as thyroid cancers.

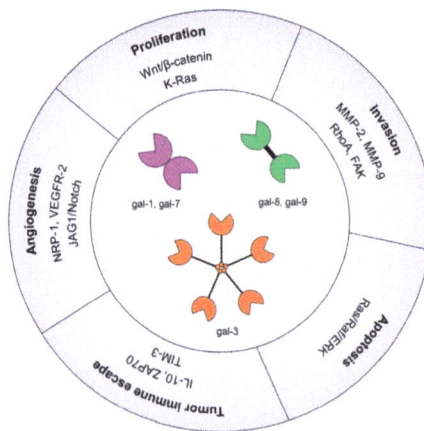

Figure 1. Hallmark of galectins in head and neck cancers. Gal-1, -3, -7, -8, and -9 in cancer progression affecting several pathways including proliferation, apoptosis, invasion, angiogenesis, and tumor-immune escape. Matrix Metalloproteinase (MMP); Interleukin (IL); Focal adhesion kinsase (FAK).

2. Galectins in Head and Neck Cancer and Thyroid Cancer

Galectins contribute in many ways to carcinogenesis and tumor progression and are involved in biological processes such as cell growth and differentiation, cell adhesion, migration, invasion, and angiogenesis [10–12] (Figure 1). Gal-1 is more frequently studied in HNSCC whereas gal-3 is more extensively examined in thyroid cancer. In their respective cancers, these galectins are reported to be powerful diagnostic markers and are indicative of a good prognosis [13,14].

In our review based on the most relevant data, gal-1, -3, -7, -8, and -9 are discussed in relation to head and neck and/or thyroid cancers. Genomic alterations of these galectins were investigated thanks to The Cancer Genome Atlas. The analysis of mRNA expression, copy number variation and mutation in 279 cases of head and neck squamous cell carcinomas using the cBioPortal for Cancer Genomics [15] revealed molecular alterations in only 7% of cases for gal-1, 6% for gal-3, 5% for gal-7, 9% for gal-8, and 4% for gal-9, mainly including mRNA up-regulation and a few gene amplification [16] (Figure 2A). The same analysis was done for 507 papillary thyroid carcinomas, and genomic alterations were reported in only 2% of cases for gal-1, 2.4% for gal-3, 2.4% for gal-7, 6% for gal-8, and 6% for gal-9, mainly involving mRNA up-regulation [17] (Figure 2B). Of note, no molecular alteration in these galectins was reported in the evaluation of 117 poorly-differentiated and anaplastic thyroid cancers [18].

(A)

(B)

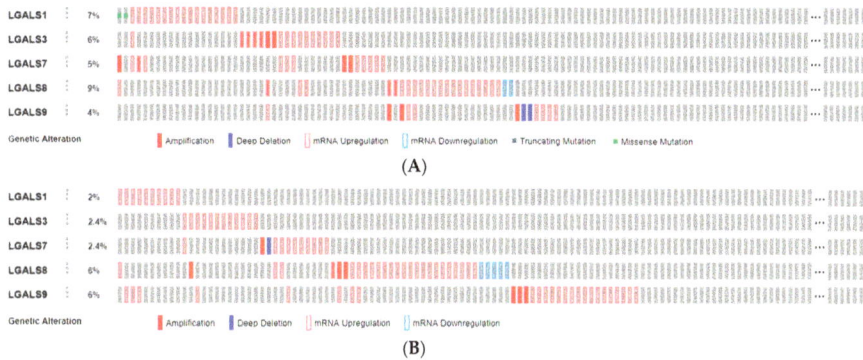

Figure 2. Alterations of galectin-1, -3, -7, -8, and -9 genes in head and neck squamous cell carcinoma (**A**) and thyroid cancer (**B**). Molecular alterations were investigated with the cBioPortal for Cancer Genomics [15], evaluating RNA expression, copy number variation, and mutation using The Cancer Genome Atlas (TCGA) datasets. Alteration events are displayed by samples ((**A**), *n* = 279; (**B**), *n* = 507).

2.1. Galectins in Carcinogenesis

HNSCCs are characterized by a field of cancerization around the main tumor. This tumor develops from normal epithelium to low and high-grade dysplasia which finally progresses to carcinoma. It is well known that most patients present several areas of dysplasia around the field of the main tumor. These other pre-neoplastic lesions could be the origin of additional head and neck carcinomas seen in the follow-up of the patient. Therefore, the study of the contribution of galectins in HNSCC carcinogenesis is very interesting. Regarding salivary gland transformation, it was demonstrated that the expression of gal-3 was strong in the cytoplasm of normal inter- and intra-lobular ductal cells and in benign lesions, while its expression was decreased in high grade carcinomas [19,20]. By contrast, expression of gal-1 was weaker than gal-3 in normal ductal cells with stronger labelling of myoepithelial cells, and in carcinomas gal-1 was expressed in the cytoplasm and/or the nucleus of the majority of tumors (Table 1) [19,20]. Gal-7 was expressed in the cytoplasm and the nuclei of normal ductal cells with a stronger immunoreactivity in the basal layer. In carcinoma, the expression of gal-7 decreased significantly compared to adenoma (Table 1) [20]. Finally, gal-8 was exclusively detected at high level in the cytoplasm of normal ductal cells while its expression decreased in cancer cells, with a labelling mainly localized in the cytoplasm and in some cases in the nucleus (Table 1) [20]. Concerning carcinoma of the oral cavity, gal-1 protein expression and mRNA level significantly increased during transformation (normal epithelium < dysplasia < carcinoma) [21–23]. Poorly differentiated carcinoma showed a stronger gal-1 expression compared to the well-differentiated one [21]. The expression of gal-3 protein and mRNA were also increased in carcinoma compared to normal epithelium (Table 1) [23,24]. Moreover, gal-1 and gal-3 serum levels were 8 and 3 fold higher, respectively, in oral cavity cancer patients compared to healthy volunteers [23]. Regarding laryngeal and hypopharyngeal carcinoma, it was established that gal-1 and gal-7 increased during carcinogenesis (Table 1), with a shift of gal-1 localization from the nucleus towards the cytoplasm during the progression of high grade dysplasia to carcinoma, and a shift of gal-7 from the cytoplasm to the nucleus between normal epithelium and dysplasia [25]. In nasopharyngeal cancer, it was reported that the gal-1 protein and mRNA expression increased in cancer tissue compared to normal tissue (Table 1) [26]. Additionally, Duray et al. observed that the expression of gal-9 increased in naso-sinusal carcinoma compared to non-malignant naso-sinusal diseases. In contrast, gal-8 expression decreased in naso-sinusal cancer compared to non-malignant naso-sinusal diseases (Table 1) [27].

Regarding thyroid cancer, multicenter studies reported that gal-3 expression increased in thyroid cancer tissues, definitively demonstrating its diagnostic value [28,29]. More recently, Arcolia et al.

investigated gal-1 and gal-3 immunohistochemical expression in normal thyroid tissues, benign thyroid lesions and thyroid cancers, and reported that both galectin cytoplasmic immunostainings were significantly higher in cancer cells of malignant thyroid lesions compared to epithelial cells in benign lesions as well as in normal tissues [14]. These data complete a previous study showing that gal-3 expression was stronger in cancer tissue compared to non-malignant tissue (nodular goiter), while no difference was observed for gal-7 [30]. Additionally, gal-1 and -3 serum levels increased in carcinoma compared to healthy volunteers (Table 1) [31].

Table 1. Galectin-1, -3, -7, -8, and -9 expression during carcinogenesis of head and neck and thyroid carcinomas.

Type of Tissues	Gal-1	Gal-3	Gal-7	Gal-8	Gal-9	Ref.
Salivary gland carcinomas	* ↗	** ↘	↘	↘		[19,20]
Oral cavity carcinoma	↗	↗				[21–24]
Laryngeal carcinoma	↗		↗			[25]
Hypopharyngeal carcinoma	↗		↗			[25]
Naso-pharyngeal carcinoma	↗			↘	↗	[26,27]
Thyroid carcinomas	↗	↗	No change			[14,28–31]

* Increase; ** Decrease.

2.2. Galectins and Patient Prognosis

With a five-year relative survival rate of approximately 50%, patients with HNSCC have poor prognoses. The assumption that galectins could be used as biomarkers to identify HNSCC individuals with a better or a poorer prognosis is based on many arguments. Indeed, overexpression of gal-1 in HNSCC is associated with a poor prognosis in terms of recurrence-free survival and overall survival [32–35]. Moreover, it was demonstrated that patients with tongue carcinoma expressing low level of gal-3 had a favorable five-year disease-free survival (73.9%) compared to patients expressing high levels (40.8%) [36]. In adenoid cystic carcinoma, patients presenting an increase of gal-3 expression tended to have a shorter disease-free survival [37]. In addition, the combination of low gal-1 and gal-3 serum levels in HNSCC patients was significantly associated with a longer survival compared to HNSCC patients with high gal-1 or gal-3 [38]. Regarding gal-7, it was also demonstrated in stage IV hypopharyngeal carcinomas that high expression of gal-7 correlated with poor prognosis [39]. However, in laryngeal carcinoma, researchers have observed that a gal-3 positive tumor correlated with a better prognosis in terms of recurrence-free survival and overall survival and that gal-3 expression was positively associated with the histopathological grade and tumor keratinization [40,41]. Of note, molecular alterations, as evaluated using the cBioPortal for Cancer Genomics [15] in gal-1, -3, -7, -8, and -9, were not significantly associated with patient survival [16].

In thyroid cancer, well-differentiated thyroid carcinomas, including papillary and follicular carcinomas, are associated with an excellent prognosis, with 85% to 90% cure rates thanks to early detection and appropriate treatment. In contrast, undifferentiated anaplastic thyroid carcinoma is highly aggressive with a five-year survival rate of approximately 7% [42]. In this context, Kim et al. reported that the absence of gal-3 expression in papillary thyroid carcinomas was associated with pathological parameters considered less aggressive, including negative lymph node involvement [43]. However, another study indicated that gal-3 expression was not significantly associated with extra-thyroidal extension or lymph node metastases [44]. Additionally, genomic alterations in these galectins were not correlated with patient survival [17]. Thus, large-scale clinical studies are needed to verify the potential prognosis values of galectin expression in cancer associated with patient prognosis.

On the other hand, gal-3 and gal-1 have been suggested as diagnostic markers. Indeed, higher gal-3 expression correlated with the diagnosis of thyroid carcinoma [45,46], leading to the translation of gal-3 assessment in clinical settings [28,29], as well as the use of this lectin for thyroid cancer imaging in vivo by using an immunoPET targeting gal-3 to improve the diagnosis [47]. Our own recent findings demonstrated that gal-1 is also a useful immunohistochemical marker to discriminate malignant tumors from benign thyroid nodules [14]. Furthermore, we also validated that gal-3 is a sensitive marker for the diagnosis of thyroid malignancy, and we added support for its combination with CK19 and HBME-1 with the highest performance for the diagnosis of well-differentiated thyroid cancer [48]. By contrast, gal-7 and gal-8 were not differentially expressed between thyroid adenoma and cancer, suggesting that such galectins are not mainly involved in the carcinogenesis and tumor progression in HNSCC [48].

2.3. Galectins and Cell Proliferation/Survival

Galectins contribute to tumor growth by affecting cell proliferation, survival and apoptosis. They are present in the intracellular and extracellular compartments of the cells. Extracellularly, galectins can bind glycoproteins such as laminin, elastin, and fibronectin. Additionally, they can cross-link cell-surface glycoconjugates, which can activate signaling pathways within the cell leading to the modulation of cell cycle progression and apoptosis [49]. In the intracellular compartment, galectins also modulate cell cycle progression and apoptosis through various interactions with intracellular proteins implicated in these processes [49].

Gal-3 is the most studied galectin as to its involvement in tumor cell survival. Indeed, it shows anti-apoptotic properties; its translocation from the cytosol to mitochondria following exposure to pro-apoptotic stimuli blocks changes of the mitochondrial membrane potential, thereby preventing apoptosis [50]. Additionally, the mutation of gal-3 (serine^6 to alanine or glutamic acid) prevents phosphorylation, resulting in a decrease of its anti-apoptotic activity [51]. Furthermore, gal-3 has a role in cell proliferation, as the inhibition of this protein by knockdown lead to the decrease of esophageal and renal cancer cell proliferation in vitro [52,53]. In oral tongue squamous cell carcinoma, an in vitro study using Tca8113 cell line established that the overexpression of gal-3 increased cell proliferation. Inversely, the knockdown of gal-3 in Tca8113 cells exhibited a decrease of cell proliferation. The latter study also showed that gal-3 acts on cell proliferation through the activation of the Wnt/β-catenin pathway (Figure 3) [24].

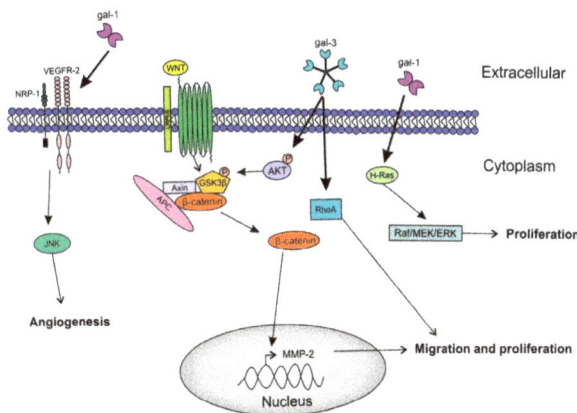

Figure 3. Involvement of galectins during carcinogenesis. Gal-1 interacts with Neuropilin-1/Vascular endothelial growth factor-2 (NPR-1/VEGFR-2) complex to promote endothelial cells migration through the c-Jun N-terminal kinase (JNK) pathway. Also, gal-1 interacts with H-Ras to stimulate cell proliferation by the stimulation of the Extracellular signal-regulated kinase (ERK) pathway. Gal-3 promotes cell proliferation and cell migration by the activation of β-catenin through the Phosphoinositide 3-kinase (PI3K)/Akt pathway, but also through Ras homolog family member A (RhoA) signaling. Glycogen synthase kinase 3β (GSK3β); Adenomatous polyposis coli (APC).

Int. J. Mol. Sci. **2017**, *18*, 2745

In thyroid cancer, it was demonstrated in vitro that gal-3 plays an anti-apoptotic role by its interaction with the pro-apoptotic Bax factor, which belongs to the Bcl-2 protein family [54]. Moreover, it is known that gal-3 interacts with K-Ras in cancer cells, leading to the activation of important signaling cascades such as PI3K/Akt [55]. Inhibition of gal-3 with Td131_1 demonstrated a synergistic effect with doxorubicin in a human ex vivo papillary thyroid cancer model by the activation of caspase 3 and PARP cleavage [56]. Furthermore, gal-3 inhibition with modified citrus pectin (MCP) combined with the inhibition of Ras with S-trans, trans-farnesyl thiosalicylic acid (FTS) not only drove the decrease of anaplastic cell proliferation both in vitro and in vivo, but also induced cell cycle arrest at the G1 phase and apoptosis of these cells [57].

Concerning gal-1, it was demonstrated that its knockdown caused a decrease in cell proliferation in anaplastic, but not in papillary thyroid cancer cells. In vivo, the orthotopic mouse model of anaplastic gal-1KD cells displayed a significant decrease in tumor growth [14]. Similar evidence was reported with ovarian cancer, and explained its effect on cell growth by the activation of the H-Ras/Raf/ERK pathway (Figure 3) [58]. Finally, the addition of recombinant gal-1 or gal-3 to the medium of three oral squamous cell carcinoma cell lines increased tumor cell proliferation [23].

2.4. Galectins and Cell Migration/Invasion

Tumor cell invasion contributes to the development of metastases. Several studies have brought to light the role of galectins in different metastatic processes, notably in cell adhesion, cell migration and invasiveness. Among all galectins, gal-3 is the most extensively studied regarding its role in metastasis occurrence. It interacts with adhesion molecules (CAM), such as integrins and cadherins, at the cell surface to promote cancer cell invasion [59]. For example, in pancreatic cancer cell lines, gal-3 silencing lead to a two-fold decreased in cell migration and invasion [60]. The mechanism by which gal-3 affects cell migration proceeds through an inhibition of the phosphorylation of Akt and GSK-3β occurring after 12 h, followed by a decrease of β-catenin expression after 48 hours. Indeed, the down-regulation of β-catenin leads to a lower expression of matrix metalloproteinase MMP-2, meaning that gal-3 may regulate the Wnt/β-catenin signaling pathway in cancer cells (Figure 3) [60]. Similar results have been published for HNSCC, where silencing of Wnt reduced the ability of gal-3 to stimulate cell migration and invasion of oral tongue squamous cell carcinoma [24].

In thyroid cancer, Shankar et al. observed that gal-3 and caveolin-1 (known to promote tumor cell migration [61]), are both up-regulated in differentiated thyroid cancer compared to benign lesions, and cause RhoA activation and FAK stabilization in focal adhesions, ultimately leading to cell migration [62]. Interestingly, the knockdown of RhoA in tongue squamous cell carcinoma (TSCC) cells negatively affects cell migration and invasion [63]. Moreover, RhoA protein was expressed at higher levels in TSCC xenografts in mice with lymph node metastases. Silencing of RhoA in such tumors downregulates the expression of gal-3, β-catenin and MMP-9 (Figure 3) [63].

Concerning gal-1, it was demonstrated that in cancer-associated fibroblasts (CAF), which play an important role in cell migration and invasion, under-expression of gal-1 significantly decreased breast cancer cell migration. Additionally, MMP-9 expression levels were attenuated in CAF silenced for gal-1, thus limiting the degradation of the extracellular matrix [64]. Similarly, in oral squamous cell carcinoma (OSCC), the silencing of gal-1 in CAF negatively affected cell migration and invasion by reducing the amount of monocyte chemotactic protein-1 (MCP-1/CCL2) [65]. Moreover, grafting OSCC cells mixed with CAF in mice allowed a faster tumor growth than OSCC cells alone, but also a fewer number of circulating tumor cells when gal-1 is under-expressed in CAF [65]. Another study by Wu et al. reported that overexpression of gal-1 in an OSCC cell line leads to an increase of MMP-2 and MMP-9 mRNA in vitro [66]. Additional data suggest that gal-1 highly affects actin filament localization, with modulation of filopodia formation, and stimulates tumor cell adhesion, migration, and invasiveness [66,67]. In vivo, Wu et al. demonstrated that gal-1 overexpression in oral cancer cells was associated with a higher number of lung metastases [66]. In thyroid cancer, our group recently demonstrated that the knockdown of gal-1 in papillary and anaplastic thyroid cancer cell lines induced decreased cell motility and invasion [14].

2.5. Galectins and Angiogenesis

Angiogenesis is an important event in cancer progression, promoting tumor growth and metastasis. Gal-1 and gal-3 are the two main galectins studied in angiogenesis in relation to cell proliferation and invasion. It has been reported that activation of endothelial cells in vitro induced an increase of gal-1 expression in the extracellular compartment [68]. In vivo, gal-1 null mice showing no expression of gal-1 in endothelial cells had a lower number of vessels in the tumor site compared to wild type mice expressing high gal-1 levels [69]. In oral cavity cancers, gal-1 is overexpressed in cancer associated-endothelial cells compared to normal endothelial cells [70]. Moreover, human umbilical vein endothelial cell (HUVECs) adhesion to laminin was significantly increased by stimulating the expression of extracellular gal-1, and dramatically decreased by its silencing [70].

Additionally, gal-1 binds the neuropilin-1 (NRP-1) on endothelial cells. NRP-1 is an important mediator of angiogenesis that can facilitate the binding of the vascular endothelial growth factor A (VEGF-A) to VEGFR-2, enhancing the phosphorylation of the receptor and the activation of angiogenesis. VEGFR-2 silencing leads to a reduction in gal-1-induced HUVECs migration. This activation of endothelial cell migration by gal-1 via the complex of VEGFR-2/NRP-1 required the participation of the JNK pathway (Figure 3) [70]. D'Haene et al. also showed that gal-1 and gal-3 might mediate angiogenesis by causing increased VEGF receptor density on endothelial cells, making them more accessible to low levels of endogenous VEGF [71].

Concerning gal-3, Chen et al. demonstrated that gal-3 was able to induce the release of pro-inflammatory cytokine such as IL-6, G-CSF, GM-CSF, and sICAM-1 from endothelial cells, which lead to the promotion of metastasis by increasing endothelial cell surface adhesion molecules, such as integrin $\alpha v \beta 1$. Cytokine release might be regulated by the binding of gal-3 to the transmembrane protein MUC-1, because MUC-1 negative cells do not abundantly release such cytokines [72]. Additionally, endogenous gal-3 silencing significantly reduced VEGF- and bFGF-mediated endothelial cell migration, as well as capillary tubule formation in vitro, with a reversible effect when exogenous gal-3 was added to the cells [73]. Finally, gal-3 also induced angiogenesis through JAG1/Notch pathway activation, leading to endothelial cell sprouting [74].

However, regarding head and neck and thyroid cancers, there is a lack of data concerning the role of galectins in angiogenesis.

2.6. Galectins and Tumor Immune Escape

Some members of the glycan-binding protein family, such as gal-1, -3, and -9, are promoters of tumor immune evasion. Indeed, gal-1 is reported to induce growth arrest and apoptosis of activated T cells [75]. The mechanism of apoptosis induced by tumor cell-derived gal-1 could be due to the interaction of gal-1 with several surface glycoproteins of T cells such as CD45, CD43, CD7, and CD3 [76,77]. Furthermore, Kovacs et al. showed that gal-1 leads to T cell apoptosis through the stimulation of the tyrosine kinases p56lck and ZAP70, depolarization of the mitochondrial membrane, and activation of the caspase cascade [78]. Additionally, gal-1 is able to promote the production of IL-10 and to decrease the secretion of IFN-γ, together contributing to the immune-tolerance of tumor cells [79]. In head and neck cancers, more particularly in gingival carcinoma, it was demonstrated that cancer tissue highly expressing gal-1 correlated with a decrease in T cell infiltration and an increase in CD3+ and CD8+ T cell apoptosis [35].

On the other hand, it was previously documented that extracellular gal-3 was able to induce apoptosis of lymphoid cells, human PBMCs, and activated mouse T cells after binding to the cell surface [80]. Peng et al. also demonstrated that soluble gal-3 inhibited T cell responses in vivo [81]. Furthermore, several studies showed that gal-3 drove to the loss of natural killer (NK) cell action. Indeed, binding of gal-3 to poly-N-acetyllactosamine residue of cell surface mucin (MUC1) attenuated the interaction of tumor cells with NK cells, leading to tumor cell evasion from NK cell immunity [82]. Also, the interaction of gal-3 with NKp30 receptors on NK cells inhibited NK-mediated cytolysis. Moreover, the down-regulation of gal-3 in HeLa cells increased the sensitivity of tumor cells to lysis by NK cells [83].

Once again, there is no specific data concerning gal-3 in tumor immune evasion in head and neck or thyroid cancers.

Finally, gal-9 was identified, by mass spectrometry in 2005, as a ligand of immunoglobulins and mucin-domain-containing molecule-3 (TIM-3) that is specifically expressed in IFN-γ-producing T cells [84]. It was demonstrated that this binding can lead to the death of Th1 cells but can also inhibit Th17 polarization and advance the expansion of FoxP3+ Tregs [84,85]. In thyroid cancer, Severson et al. showed that gal-9 is expressed by tumor-infiltrating lymph nodes, and thus may contribute to immune dysfunction, suggesting that immune-modulating therapies that inhibit TIM-3/gal-9 may be viable options for patients with advanced diseases [86].

2.7. Clinical Potential of Galectin Inhibition

Several preclinical and clinical studies in phase I and II have used different kinds of galectin inhibitors, such as GM-CT-01 and GR-MD-02 in lymphoma, breast cancer, and melanoma. GM-CT-01 (DANAVAT®) is a galactomannan derived from plants which binds to gal-1. In vivo data revealed that injection of GM-CT-01 to colon tumor-bearing mice potentiated the anti-tumor activity of 5-fluorouracil (5-FU) [87]. In metastatic colon cancer patients, phase I and II demonstrated that the combination of GM-CT-01 with 5-FU increased the longevity of patients by 46% and decreased the serious adverse effects by 41% [87]. GM-CT-01 was also studied as a booster of tumor-infiltrating lymphocyte (TIL) function; this action is due to an increase in the release of IFNγ, allowing tumor regression [88]. This property is used in a phase I/II study in which GM-CT-01 is associated with a peptide vaccination to induce a more efficient and long-lasting anti-tumor immune response in metastatic melanoma patients (NCT01723813). Another inhibitor tested in pre-clinical and clinical studies is GR-MD-02, which binds to the carbohydrate-binding domain of gal-3. This inhibitor was developed by Galectins Therapeutics (Norcross, GA, USA) and was used in mice models of melanoma, breast cancer, prostate cancer, and sarcoma. The researchers observed in these models that GR-MD-02 potentiated the effect of immune modulators such as anti-PD1, anti-CTLA4, and anti-OX40 antibodies by reducing tumor size and improving mice survival [88]. Finally, clinical studies combined GR-MD-02 with other therapeutic molecules, such as ipilimumab (anti-CTLA4 antibody) in metastatic melanoma patients (NCT02117362) and pembrolizumab (anti-PD-1 antibody) in patients with metastatic melanoma, non-small cell lung cancer, and head and neck cancer (NCT02575404). In the latter study, one partial response has been observed showing a marked reduction in tumor size at week 12 of therapy, after 3 doses of GR-MD-02 and pembrolizumab combination [89].

3. Conclusions

Galectins, especially gal-1 and gal-3, have been largely studied in head and neck carcinoma and thyroid cancer (Table 2). They participate in carcinogenesis and tumor progression by stimulating cancer cell proliferation and survival, migration and invasion, and angiogenesis, as well as immune system escape through multiple mechanisms. Regarding their clinical value, the most recent studies revealed that galectins may be used as markers for prognosis in head and neck cancer and for diagnosis of thyroid carcinoma. Indeed, the use of gal-3 thyrotest preoperatively can help to characterize benign from malignant lesions of the thyroid, which could prevent unnecessary thyroid surgical procedures. However, larger clinical trials are required to validate these findings in head and neck cancer. Of note, the number of genomic alterations was very low in galectins and did not correlate with patient survival. Globally, these data also support galectins as potential targets for anti-cancer therapy. Several in vitro studies are promising and a few clinical trials are ongoing. Similarly, galectin inhibitors/antagonists can also be good options, and deserve to be tested in head and neck carcinoma and thyroid cancers, most likely in combination with various other therapies.

Table 2. Main references supporting a role for galectins in head and neck and thyroid cancer cell proliferation, migration/invasion, angiogenesis, and tumor immune escape.

	Gal-1		Gal-3		Gal-9	
	HNSCC	TC	HNSCC	TC	HNSCC	TC
Cell proliferation	[22]	[14]	[23]	[54,56]		
Apoptosis				[53,55,56]		
Cell migration/invasion	[64,65]	[14]	[23,62]	[61]		
Angiogenesis	[69]					
Tumor immune escape	[34]					[85]

HNSCC: Head and neck squamous cell carcinoma; TC: Thyroid carcinoma.

Acknowledgments: We thank financial support from the "Fonds pour l'Action de Recherche Concertée" (ARC-14/19 UMONS 3) (Belgium).

Author Contributions: Nadège Kindt and Fabrice Journe have written the review and Fabrice Journe, Ghanem E. Ghanem, and Sven Saussez have reviewed and edited this review.

Conflicts of Interest: The authors declare no conflict of interest.

References

1. Bose, P.; Brockton, N.T.; Dort, J.C. Head and neck cancer: From anatomy to biology. *Int. J. Cancer* **2013**, *133*, 2013–2023. [CrossRef] [PubMed]

2. Ang, K.K.; Harris, J.; Wheeler, R.; Weber, R.; Rosenthal, D.I.; Nguyen-Tân, P.F.; Westra, W.H.; Chung, C.H.; Jordan, R.C.; Lu, C.; et al. Human papillomavirus and survival of patients with oropharyngeal cancer. *N. Engl. J. Med.* **2010**, *363*, 24–35. [CrossRef] [PubMed]

3. Duray, A.; Descamps, G.; Decaestecker, C.; Remmelink, M.; Sirtaine, N.; Lechien, J.; Ernoux-Neufcoeur, P.; Bletard, N.; Somja, J.; Depuydt, C.E.; et al. Human papillomavirus DNA strongly correlates with a poorer prognosis in oral cavity carcinoma. *Laryngoscope* **2012**, *122*, 1558–1565. [CrossRef] [PubMed]

4. Buckley, L.; Gupta, R.; Ashford, B.; Jabbour, J.; Clark, J.R. Oropharyngeal cancer and human papilloma virus: Evolving diagnostic and management paradigms. *ANZ J. Surg.* **2015**. [CrossRef] [PubMed]

5. Descamps, G.; Karaca, Y.; Lechien, J.R.; Kindt, N.; Decaestecker, C.; Remmelink, M.; Larsimont, D.; Andry, G.; Hassid, S.; Rodriguez, A.; et al. Classical risk factors, but not HPV status, predict survival after chemoradiotherapy in advanced head and neck cancer patients. *J. Cancer Res. Clin. Oncol.* **2016**, *142*, 2185–2196. [CrossRef] [PubMed]

6. Sipos, J.A.; Mazzaferri, E.L. Thyroid cancer epidemiology and prognostic variables. *Clin. Oncol. R. Coll. Radiol.* **2010**, *22*, 395–404. [CrossRef] [PubMed]

7. Pazaitou-Panayiotou, K.; Polyzos, S.A.; Mantzoros, C.S. Obesity and thyroid cancer: Epidemiologic associations and underlying mechanisms. *Obes. Rev. Off. J. Int. Assoc. Study Obes.* **2013**, *14*, 1006–1022. [CrossRef] [PubMed]

8. Paz, A.; Haklai, R.; Elad-Sfadia, G.; Ballan, E.; Kloog, Y. Galectin-1 binds oncogenic H-Ras to mediate Ras membrane anchorage and cell transformation. *Oncogene* **2001**, *20*, 7486–7493. [CrossRef] [PubMed]

9. Levy, R.; Grafi-Cohen, M.; Kraiem, Z.; Kloog, Y. Galectin-3 promotes chronic activation of K-Ras and differentiation block in malignant thyroid carcinomas. *Mol. Cancer Ther.* **2010**, *9*, 2208–2219. [CrossRef] [PubMed]

10. Funasaka, T.; Raz, A.; Nangia-Makker, P. Galectin-3 in angiogenesis and metastasis. *Glycobiology* **2014**, *24*, 886–891. [CrossRef] [PubMed]

11. Thijssen, V.L.; Heusschen, R.; Caers, J.; Griffioen, A.W. Galectin expression in cancer diagnosis and prognosis: A systematic review. *Biochim. Biophys. Acta* **2015**, *1855*, 235–247. [CrossRef] [PubMed]

12. Liu, F.-T.; Patterson, R.J.; Wang, J.L. Intracellular functions of galectins. *Biochim. Biophys. Acta* **2002**, *1572*, 263–273. [CrossRef]

13. Tang, W.; Huang, C.; Tang, C.; Xu, J.; Wang, H. Galectin-3 may serve as a potential marker for diagnosis and prognosis in papillary thyroid carcinoma: A meta-analysis. *OncoTargets Ther.* **2016**, *9*, 455–460. [CrossRef] [PubMed]

14. Arcolia, V.; Journe, F.; Wattier, A.; Leteurtre, E.; Renaud, F.; Gabius, H.-J.; Remmelink, M.; Decaestecker, C.; Rodriguez, A.; Boutry, S.; et al. Galectin-1 is a diagnostic marker involved in thyroid cancer progression. *Int. J. Oncol.* **2017**, *51*, 760–770. [CrossRef] [PubMed]

15. cBioPortal for Cancer Genomics. Available online: http://www.cbioportal.org/ (accessed on 15 December 2017).

16. Cancer Genome Atlas Network Comprehensive genomic characterization of head and neck squamous cell carcinomas. *Nature* **2015**, *517*, 576–582. [CrossRef]

17. Agrawal, N.; Akbani, R.; Aksoy, B.A.; Ally, A.; Arachchi, H.; Asa, S.L.; Auman, J.T.; Balasundaram, M.; Balu, S.; Baylin, S.B.; et al. Integrated Genomic Characterization of Papillary Thyroid Carcinoma. *Cell* **2014**, *159*, 676–690. [CrossRef] [PubMed]

18. Landa, I.; Ibrahimpasic, T.; Boucai, L.; Sinha, R.; Knauf, J.A.; Shah, R.H.; Dogan, S.; Ricarte-Filho, J.C.; Krishnamoorthy, G.P.; Xu, B.; et al. Genomic and transcriptomic hallmarks of poorly differentiated and anaplastic thyroid cancers. *J. Clin. Investig.* **2016**, *126*, 1052–1066. [CrossRef] [PubMed]

19. Xu, X.C.; Sola Gallego, J.J.; Lotan, R.; El-Naggar, A.K. Differential expression of galectin-1 and galectin-3 in benign and malignant salivary gland neoplasms. *Int. J. Oncol.* **2000**, *17*, 271–276. [CrossRef] [PubMed]

20. Remmelink, M.; de Leval, L.; Decaestecker, C.; Duray, A.; Crompot, E.; Sirtaine, N.; André, S.; Kaltner, H.; Leroy, X.; Gabius, H.-J.; Saussez, S. Quantitative immunohistochemical fingerprinting of adhesion/growth-regulatory galectins in salivary gland tumours: Divergent profiles with diagnostic potential. *Histopathology* **2011**, *58*, 543–556. [CrossRef] [PubMed]

21. Ding, Y.-M.; Dong, J.-H.; Chen, L.-L.; Zhang, H.-D. Increased expression of galectin-1 is associated with human oral squamous cell carcinoma development. *Oncol. Rep.* **2009**, *21*, 983–987. [PubMed]

22. De Vasconcelos Carvalho, M.; Pereira, J.D.S.; Alves, P.M.; da Silveira, E.J.D.; de Souza, L.B.; Queiroz, L.M.G. Alterations in the immunoexpression of galectins-1, -3 and -7 between different grades of oral epithelial dysplasia. *J. Oral Pathol. Med.* **2013**, *42*, 174–179. [CrossRef] [PubMed]

23. Aggarwal, S.; Sharma, S.C.; Das, S.N. Galectin-1 and galectin-3: Plausible tumour markers for oral squamous cell carcinoma and suitable targets for screening high-risk population. *Clin. Chim. Acta Int. J. Clin. Chem.* **2015**, *442*, 13–21. [CrossRef] [PubMed]

24. Wang, L.-P.; Chen, S.-W.; Zhuang, S.-M.; Li, H.; Song, M. Galectin-3 accelerates the progression of oral tongue squamous cell carcinoma via a Wnt/β-catenin-dependent pathway. *Pathol. Oncol. Res.* **2013**, *19*, 461–474. [CrossRef] [PubMed]

25. Saussez, S.; Decaestecker, C.; Lorfevre, F.; Chevalier, D.; Mortuaire, G.; Kaltner, H.; André, S.; Toubeau, G.; Gabius, H.-J.; Leroy, X. Increased expression and altered intracellular distribution of adhesion/growth-regulatory lectins galectins-1 and -7 during tumour progression in hypopharyngeal and laryngeal squamous cell carcinomas. *Histopathology* **2008**, *52*, 483–493. [CrossRef] [PubMed]

26. Tang, C.-E.; Tan, T.; Li, C.; Chen, Z.-C.; Ruan, L.; Wang, H.-H.; Su, T.; Zhang, P.-F.; Xiao, Z.-Q. Identification of Galectin-1 as a novel biomarker in nasopharyngeal carcinoma by proteomic analysis. *Oncol. Rep.* **2010**, *24*, 495–500. [PubMed]

27. Duray, A.; De Maesschalck, T.; Decaestecker, C.; Remmelink, M.; Chantrain, G.; Neiveyans, J.; Horoi, M.; Leroy, X.; Gabius, H.-J.; Saussez, S. Galectin fingerprinting in naso-sinusal diseases. *Oncol. Rep.* **2014**, *32*, 23–32. [CrossRef] [PubMed]

28. Bartolazzi, A.; Gasbarri, A.; Papotti, M.; Bussolati, G.; Lucante, T.; Khan, A.; Inohara, H.; Marandino, F.; Orlandi, F.; Nardi, F.; et al. Thyroid Cancer Study Group Application of an immunodiagnostic method for improving preoperative diagnosis of nodular thyroid lesions. *Lancet* **2001**, *357*, 1644–1650. [CrossRef]

29. Bartolazzi, A.; Orlandi, F.; Saggiorato, E.; Volante, M.; Arecco, F.; Rossetto, R.; Palestini, N.; Ghigo, E.; Papotti, M.; Bussolati, G.; et al. Italian Thyroid Cancer Study Group (ITCSG) Galectin-3-expression analysis in the surgical selection of follicular thyroid nodules with indeterminate fine-needle aspiration cytology: A prospective multicentre study. *Lancet Oncol.* **2008**, *9*, 543–549. [CrossRef]

30. Than, T.H.; Swethadri, G.K.; Wong, J.; Ahmad, T.; Jamil, D.; Maganlal, R.K.; Hamdi, M.M.; Abdullah, M.S. Expression of Galectin-3 and Galectin-7 in thyroid malignancy as potential diagnostic indicators. *Singap. Med. J.* **2008**, *49*, 333–338. [PubMed]

31. Saussez, S.; Glinoer, D.; Chantrain, G.; Pattou, F.; Carnaille, B.; André, S.; Gabius, H.-J.; Laurent, G. Serum galectin-1 and galectin-3 levels in benign and malignant nodular thyroid disease. *Thyroid Off. J. Am. Thyroid Assoc.* **2008**, *18*, 705–712. [CrossRef] [PubMed]

32. Le, Q.-T.; Shi, G.; Cao, H.; Nelson, D.W.; Wang, Y.; Chen, E.Y.; Zhao, S.; Kong, C.; Richardson, D.; O'Byrne, K.J.; et al. Galectin-1: A link between tumor hypoxia and tumor immune privilege. *J. Clin. Oncol. Off. J. Am. Soc. Clin. Oncol.* **2005**, *23*, 8932–8941. [CrossRef] [PubMed]

33. Saussez, S.; Decaestecker, C.; Lorfevre, F.; Cucu, D.-R.; Mortuaire, G.; Chevalier, D.; Wacreniez, A.; Kaltner, H.; André, S.; Toubeau, G.; Camby, I.; et al. High level of galectin-1 expression is a negative prognostic predictor of recurrence in laryngeal squamous cell carcinomas. *Int. J. Oncol.* **2007**, *30*, 1109–1117. [CrossRef] [PubMed]

34. Chang, S.-L.; Li, C.-F.; Lin, C.; Lin, Y.-S. Galectin-1 overexpression in nasopharyngeal carcinoma: Effect on survival. *Acta Otolaryngol. (Stockh.)* **2014**, *134*, 536–542. [CrossRef] [PubMed]

35. Noda, Y.; Kishino, M.; Sato, S.; Hirose, K.; Sakai, M.; Fukuda, Y.; Murakami, S.; Toyosawa, S. Galectin-1 expression is associated with tumour immunity and prognosis in gingival squamous cell carcinoma. *J. Clin. Pathol.* **2017**, *70*, 126–133. [CrossRef] [PubMed]

36. Honjo, Y.; Inohara, H.; Akahani, S.; Yoshii, T.; Takenaka, Y.; Yoshida, J.; Hattori, K.; Tomiyama, Y.; Raz, A.; Kubo, T. Expression of cytoplasmic galectin-3 as a prognostic marker in tongue carcinoma. *Clin. Cancer Res. Off. J. Am. Assoc. Cancer Res.* **2000**, *6*, 4635–4640.

37. Teymoortash, A.; Pientka, A.; Schrader, C.; Tiemann, M.; Werner, J.A. Expression of galectin-3 in adenoid cystic carcinoma of the head and neck and its relationship with distant metastasis. *J. Cancer Res. Clin. Oncol.* **2006**, *132*, 51–56. [CrossRef] [PubMed]

38. Saussez, S.; Lorfevre, F.; Lequeux, T.; Laurent, G.; Chantrain, G.; Vertongen, F.; Toubeau, G.; Decaestecker, C.; Kiss, R. The determination of the levels of circulating galectin-1 and -3 in HNSCC patients could be used to monitor tumor progression and/or responses to therapy. *Oral Oncol.* **2008**, *44*, 86–93. [CrossRef] [PubMed]

39. Saussez, S.; Cucu, D.-R.; Decaestecker, C.; Chevalier, D.; Kaltner, H.; André, S.; Wacreniez, A.; Toubeau, G.; Camby, I.; Gabius, H.-J.; Kiss, R. Galectin 7 (p53-induced gene 1): A new prognostic predictor of recurrence and survival in stage IV hypopharyngeal cancer. *Ann. Surg. Oncol.* **2006**, *13*, 999–1009. [CrossRef] [PubMed]

40. Piantelli, M.; Iacobelli, S.; Almadori, G.; Iezzi, M.; Tinari, N.; Natoli, C.; Cadoni, G.; Lauriola, L.; Ranelletti, F.O. Lack of expression of galectin-3 is associated with a poor outcome in node-negative patients with laryngeal squamous-cell carcinoma. *J. Clin. Oncol.* **2002**, *20*, 3850–3856. [CrossRef] [PubMed]

41. Plzák, J.; Betka, J.; Smetana, K.; Chovanec, M.; Kaltner, H.; André, S.; Kodet, R.; Gabius, H.-J. Galectin-3—An emerging prognostic indicator in advanced head and neck carcinoma. *Eur. J. Cancer* **2004**, *40*, 2324–2330. [CrossRef] [PubMed]

42. Thyroid Cancer Survival Rates, by Type and Stage. Available online: https://www.cancer.org/cancer/thyroid-cancer/detection-diagnosis-staging/survival-rates.html (accessed on 27 October 2017).

43. Kim, E.S.; Lim, D.J.; Lee, K.; Jung, C.K.; Bae, J.S.; Jung, S.L.; Baek, K.H.; Lee, J.M.; Moon, S.D.; Kang, M.I.; et al. Absence of galectin-3 immunostaining in fine-needle aspiration cytology specimens from papillary thyroid carcinoma is associated with favorable pathological indices. *Thyroid Off. J. Am. Thyroid Assoc.* **2012**, *22*, 1244–1250. [CrossRef] [PubMed]

44. Lee, Y.M.; Lee, J.B. Prognostic value of epidermal growth factor receptor, p53 and galectin-3 expression in papillary thyroid carcinoma. *J. Int. Med. Res.* **2013**, *41*, 825–834. [CrossRef] [PubMed]

45. Carpi, A.; Rossi, G.; Coscio, G.D.; Iervasi, G.; Nicolini, A.; Carpi, F.; Mechanick, J.I.; Bartolazzi, A. Galectin-3 detection on large-needle aspiration biopsy improves preoperative selection of thyroid nodules: A prospective cohort study. *Ann. Med.* **2010**, *42*, 70–78. [CrossRef] [PubMed]

46. Sumana, B.S.; Shashidhar, S.; Shivarudrappa, A.S. Galectin-3 Immunohistochemical Expression in Thyroid Neoplasms. *J. Clin. Diagn. Res.* **2015**, *9*, EC07-11. [CrossRef] [PubMed]

47. D'Alessandria, C.; Braesch-Andersen, S.; Bejo, K.; Reder, S.; Blechert, B.; Schwaiger, M.; Bartolazzi, A. Noninvasive In Vivo Imaging and Biologic Characterization of Thyroid Tumors by ImmunoPET Targeting of Galectin-3. *Cancer Res.* **2016**, *76*, 3583–3592. [CrossRef] [PubMed]

48. Arcolia, V.; Journe, F.; Renaud, F.; Leteurtre, E.; Gabius, H.-J.; Remmelink, M.; Saussez, S. Combination of galectin-3, CK19 and HBME-1 immunostaining improves the diagnosis of thyroid cancer. *Oncol. Lett.* **2017**, *14*, 4183–4189. [CrossRef] [PubMed]

49. Liu, F.-T.; Rabinovich, G.A. Galectins as modulators of tumour progression. *Nat. Rev. Cancer* **2005**, *5*, 29–41. [CrossRef] [PubMed]

50. Matarrese, P.; Tinari, N.; Semeraro, M.L.; Natoli, C.; Iacobelli, S.; Malorni, W. Galectin-3 overexpression protects from cell damage and death by influencing mitochondrial homeostasis. *FEBS Lett.* **2000**, *473*, 311–315. [CrossRef]

51. Yoshii, T.; Fukumori, T.; Honjo, Y.; Inohara, H.; Kim, H.-R.C.; Raz, A. Galectin-3 phosphorylation is required for its anti-apoptotic function and cell cycle arrest. *J. Biol. Chem.* **2002**, *277*, 6852–6857. [CrossRef] [PubMed]

52. Qiao, L.; Liang, N.; Xie, J.; Luo, H.; Zhang, J.; Deng, G.; Li, Y.; Zhang, J. Gene silencing of galectin-3 changes the biological behavior of Eca109 human esophageal cancer cells. *Mol. Med. Rep.* **2016**, *13*, 160–166. [CrossRef] [PubMed]

53. Xu, Y.; Li, C.; Sun, J.; Li, J.; Gu, X.; Xu, W. Antitumor effects of galectin-3 inhibition in human renal carcinoma cells. *Exp. Biol. Med.* **2016**, *241*, 1365–1373. [CrossRef] [PubMed]

54. Harazono, Y.; Kho, D.H.; Balan, V.; Nakajima, K.; Zhang, T.; Hogan, V.; Raz, A. Galectin-3 leads to attenuation of apoptosis through Bax heterodimerization in human thyroid carcinoma cells. *Oncotarget* **2014**, *5*, 9992–10001. [CrossRef] [PubMed]

55. Elad-Sfadia, G.; Haklai, R.; Balan, E.; Kloog, Y. Galectin-3 augments K-Ras activation and triggers a Ras signal that attenuates ERK but not phosphoinositide 3-kinase activity. *J. Biol. Chem.* **2004**, *279*, 34922–34930. [CrossRef] [PubMed]

56. Lin, C.-I.; Whang, E.E.; Donner, D.B.; Jiang, X.; Price, B.D.; Carothers, A.M.; Delaine, T.; Leffler, H.; Nilsson, U.J.; Nose, V.; et al. Galectin-3 targeted therapy with a small molecule inhibitor activates apoptosis and enhances both chemosensitivity and radiosensitivity in papillary thyroid cancer. *Mol. Cancer Res.* **2009**, *7*, 1655–1662. [CrossRef] [PubMed]

57. Menachem, A.; Bodner, O.; Pastor, J.; Raz, A.; Kloog, Y. Inhibition of malignant thyroid carcinoma cell proliferation by Ras and galectin-3 inhibitors. *Cell Death Discov.* **2015**, *1*, 15047. [CrossRef] [PubMed]

58. Zhang, P.; Zhang, P.; Shi, B.; Zhou, M.; Jiang, H.; Zhang, H.; Pan, X.; Gao, H.; Sun, H.; Li, Z. Galectin-1 overexpression promotes progression and chemoresistance to cisplatin in epithelial ovarian cancer. *Cell Death Dis.* **2014**, *5*, e991. [CrossRef] [PubMed]

59. Xin, M.; Dong, X.-W.; Guo, X.-L. Role of the interaction between galectin-3 and cell adhesion molecules in cancer metastasis. *Biomed. Pharmacother.* **2015**, *69*, 179–185. [CrossRef] [PubMed]

60. Kobayashi, T.; Shimura, T.; Yajima, T.; Kubo, N.; Araki, K.; Tsutsumi, S.; Suzuki, H.; Kuwano, H.; Raz, A. Transient gene silencing of galectin-3 suppresses pancreatic cancer cell migration and invasion through degradation of β-catenin. *Int. J. Cancer* **2011**, *129*, 2775–2786. [CrossRef] [PubMed]

61. Díaz, J.; Mendoza, P.; Silva, P.; Quest, A.F.; Torres, V.A. A novel caveolin-1/p85α/Rab5/Tiam1/Rac1 signaling axis in tumor cell migration and invasion. *Commun. Integr. Biol.* **2014**, *7*. [CrossRef] [PubMed]

62. Shankar, J.; Wiseman, S.M.; Meng, F.; Kasaian, K.; Strugnell, S.; Mofid, A.; Gown, A.; Jones, S.J.M.; Nabi, I.R. Coordinated expression of galectin-3 and caveolin-1 in thyroid cancer. *J. Pathol.* **2012**, *228*, 56–66. [CrossRef] [PubMed]

63. Yan, G.; Zou, R.; Chen, Z.; Fan, B.; Wang, Z.; Wang, Y.; Yin, X.; Zhang, D.; Tong, L.; Yang, F.; et al. Silencing RhoA inhibits migration and invasion through Wnt/β-catenin pathway and growth through cell cycle regulation in human tongue cancer. *Acta Biochim. Biophys. Sin.* **2014**, *46*, 682–690. [CrossRef] [PubMed]

64. Zhu, X.; Wang, K.; Zhang, K.; Xu, F.; Yin, Y.; Zhu, L.; Zhou, F. Galectin-1 knockdown in carcinoma-associated fibroblasts inhibits migration and invasion of human MDA-MB-231 breast cancer cells by modulating MMP-9 expression. *Acta Biochim. Biophys. Sin.* **2016**, *48*, 462–467. [CrossRef] [PubMed]

65. Wu, M.-H.; Hong, H.-C.; Hong, T.-M.; Chiang, W.-F.; Jin, Y.-T.; Chen, Y.-L. Targeting galectin-1 in carcinoma-associated fibroblasts inhibits oral squamous cell carcinoma metastasis by downregulating MCP-1/CCL2 expression. *Clin. Cancer Res.* **2011**, *17*, 1306–1316. [CrossRef] [PubMed]

66. Wu, M.-H.; Hong, T.-M.; Cheng, H.-W.; Pan, S.-H.; Liang, Y.-R.; Hong, H.-C.; Chiang, W.-F.; Wong, T.-Y.; Shieh, D.-B.; Shiau, A.-L.; et al. Galectin-1-mediated tumor invasion and metastasis, up-regulated matrix metalloproteinase expression, and reorganized actin cytoskeletons. *Mol. Cancer Res.* **2009**, *7*, 311–318. [CrossRef] [PubMed]

67. Rizqiawan, A.; Tobiume, K.; Okui, G.; Yamamoto, K.; Shigeishi, H.; Ono, S.; Shimasue, H.; Takechi, M.; Higashikawa, K.; Kamata, N. Autocrine galectin-1 promotes collective cell migration of squamous cell carcinoma cells through up-regulation of distinct integrins. *Biochem. Biophys. Res. Commun.* **2013**, *441*, 904–910. [CrossRef] [PubMed]

68. Baum, L.G.; Seilhamer, J.J.; Pang, M.; Levine, W.B.; Beynon, D.; Berliner, J.A. Synthesis of an endogenous lectin, galectin-1, by human endothelial cells is up-regulated by endothelial cell activation. *Glycoconj. J.* **1995**, *12*, 63–68. [CrossRef] [PubMed]

69. Thijssen, V.L.J.L.; Postel, R.; Brandwijk, R.J.M.G.E.; Dings, R.P.M.; Nesmelova, I.; Satijn, S.; Verhofstad, N.; Nakabeppu, Y.; Baum, L.G.; Bakkers, J.; et al. Galectin-1 is essential in tumor angiogenesis and is a target for antiangiogenesis therapy. *Proc. Natl. Acad. Sci. USA* **2006**, *103*, 15975–15980. [CrossRef] [PubMed]

70. Hsieh, S.H.; Ying, N.W.; Wu, M.H.; Chiang, W.F.; Hsu, C.L.; Wong, T.Y.; Jin, Y.T.; Hong, T.M.; Chen, Y.L. Galectin-1, a novel ligand of neuropilin-1, activates VEGFR-2 signaling and modulates the migration of vascular endothelial cells. *Oncogene* **2008**, *27*, 3746–3753. [CrossRef] [PubMed]

71. D'Haene, N.; Sauvage, S.; Maris, C.; Adanja, I.; Le Mercier, M.; Decaestecker, C.; Baum, L.; Salmon, I. VEGFR1 and VEGFR2 involvement in extracellular galectin-1- and galectin-3-induced angiogenesis. *PLoS ONE* **2013**, *8*, e67029. [CrossRef] [PubMed]

72. Chen, C.; Duckworth, C.A.; Zhao, Q.; Pritchard, D.M.; Rhodes, J.M.; Yu, L.-G. Increased circulation of galectin-3 in cancer induces secretion of metastasis-promoting cytokines from blood vascular endothelium. *Clin. Cancer Res. Off. J. Am. Assoc. Cancer Res.* **2013**, *19*, 1693–1704. [CrossRef] [PubMed]

73. Markowska, A.I.; Liu, F.-T.; Panjwani, N. Galectin-3 is an important mediator of VEGF- and bFGF-mediated angiogenic response. *J. Exp. Med.* **2010**, *207*, 1981–1993. [CrossRef] [PubMed]

74. Dos Santos, S.N.; Sheldon, H.; Pereira, J.X.; Paluch, C.; Bridges, E.M.; El-Cheikh, M.C.; Harris, A.L.; Bernardes, E.S. Galectin-3 acts as an angiogenic switch to induce tumor angiogenesis via Jagged-1/Notch activation. *Oncotarget* **2017**, *8*, 49484–49501. [CrossRef] [PubMed]

75. Rabinovich, G.A.; Ramhorst, R.E.; Rubinstein, N.; Corigliano, A.; Daroqui, M.C.; Kier-Joffé, E.B.; Fainboim, L. Induction of allogenic T-cell hyporesponsiveness by galectin-1-mediated apoptotic and non-apoptotic mechanisms. *Cell Death Differ.* **2002**, *9*, 661–670. [CrossRef] [PubMed]

76. Camby, I.; Le Mercier, M.; Lefranc, F.; Kiss, R. Galectin-1: A small protein with major functions. *Glycobiology* **2006**, *16*, 137R–157R. [CrossRef] [PubMed]

77. Rubinstein, N.; Alvarez, M.; Zwirner, N.W.; Toscano, M.A.; Ilarregui, J.M.; Bravo, A.; Mordoh, J.; Fainboim, L.; Podhajcer, O.L.; Rabinovich, G.A. Targeted inhibition of galectin-1 gene expression in tumor cells results in heightened T cell-mediated rejection; A potential mechanism of tumor-immune privilege. *Cancer Cell* **2004**, *5*, 241–251. [CrossRef]

78. Kovács-Sólyom, F.; Blaskó, A.; Fajka-Boja, R.; Katona, R.L.; Végh, L.; Novák, J.; Szebeni, G.J.; Krenács, L.; Uher, F.; Tubak, V.; Kiss, R.; Monostori, E. Mechanism of tumor cell-induced T-cell apoptosis mediated by galectin-1. *Immunol. Lett.* **2010**, *127*, 108–118. [CrossRef] [PubMed]

79. Cedeno-Laurent, F.; Opperman, M.J.; Barthel, S.R.; Hays, D.; Schatton, T.; Zhan, Q.; He, X.; Matta, K.L.; Supko, J.G.; Frank, M.H.; et al. Metabolic inhibition of galectin-1-binding carbohydrates accentuates antitumor immunity. *J. Investig. Dermatol.* **2012**, *132*, 410–420. [CrossRef] [PubMed]

80. Fukumori, T.; Takenaka, Y.; Yoshii, T.; Kim, H.-R.C.; Hogan, V.; Inohara, H.; Kagawa, S.; Raz, A. CD29 and CD7 mediate galectin-3-induced type II T-cell apoptosis. *Cancer Res.* **2003**, *63*, 8302–8311. [PubMed]

81. Peng, W.; Wang, H.Y.; Miyahara, Y.; Peng, G.; Wang, R.-F. Tumor-associated galectin-3 modulates the function of tumor-reactive T cells. *Cancer Res.* **2008**, *68*, 7228–7236. [CrossRef] [PubMed]

82. Suzuki, Y.; Sutoh, M.; Hatakeyama, S.; Mori, K.; Yamamoto, H.; Koie, T.; Saitoh, H.; Yamaya, K.; Funyu, T.; Habuchi, T.; et al. MUC1 carrying core 2 O-glycans functions as a molecular shield against NK cell attack, promoting bladder tumor metastasis. *Int. J. Oncol.* **2012**, *40*, 1831–1838. [CrossRef] [PubMed]

83. Wang, W.; Guo, H.; Geng, J.; Zheng, X.; Wei, H.; Sun, R.; Tian, Z. Tumor-released Galectin-3, a soluble inhibitory ligand of human NKp30, plays an important role in tumor escape from NK cell attack. *J. Biol. Chem.* **2014**, *289*, 33311–33319. [CrossRef] [PubMed]

84. Zhu, C.; Anderson, A.C.; Schubart, A.; Xiong, H.; Imitola, J.; Khoury, S.J.; Zheng, X.X.; Strom, T.B.; Kuchroo, V.K. The Tim-3 ligand galectin-9 negatively regulates T helper type 1 immunity. *Nat. Immunol.* **2005**, *6*, 1245–1252. [CrossRef] [PubMed]

85. Seki, M.; Oomizu, S.; Sakata, K.-M.; Sakata, A.; Arikawa, T.; Watanabe, K.; Ito, K.; Takeshita, K.; Niki, T.; Saita, N.; et al. Galectin-9 suppresses the generation of Th17, promotes the induction of regulatory T cells, and regulates experimental autoimmune arthritis. *Clin. Immunol.* **2008**, *127*, 78–88. [CrossRef] [PubMed]

86. Severson, J.J.; Serracino, H.S.; Mateescu, V.; Raeburn, C.D.; McIntyre, R.C.; Sams, S.B.; Haugen, B.R.; French, J.D. PD-1+Tim-3+ CD8+ T Lymphocytes Display Varied Degrees of Functional Exhaustion in Patients with Regionally Metastatic Differentiated Thyroid Cancer. *Cancer Immunol. Res.* **2015**, *3*, 620–630. [CrossRef] [PubMed]

87. Klyosov, A.; Zomer, E.; Platt, D. DAVANAT®(GM-CT-01) and Colon Cancer: Preclinical and Clinical (Phase I and II) Studies. In *Glycobiology and Drug Design*; ACS Symposium Series; American Chemical Society: Washington, DC, USA, 2012; Volume 1102, pp. 89–130. ISBN 978-0-8412-2765-1.

88. Demotte, N.; Bigirimana, R.; Wieërs, G.; Stroobant, V.; Squifflet, J.-L.; Carrasco, J.; Thielemans, K.; Baurain, J.-F.; Van Der Smissen, P.; Courtoy, P.J.; van der Bruggen, P. A short treatment with galactomannan GM-CT-01 corrects the functions of freshly isolated human tumor-infiltrating lymphocytes. *Clin. Cancer Res. Off. J. Am. Assoc. Cancer Res.* **2014**, *20*, 1823–1833. [CrossRef] [PubMed]

89. Redmond, W. Immunotherapy plus a galectin-3 inhibitor improves anti-tumor immunity: Insights from mice and a first-in-human phase I clinical trial. In Proceedings of the GTCbio 9th Immunotherapeutics & Immunomonitoring Conference, San Diego, CA, USA, 2 February 2017.

International Journal of
Molecular Sciences

MDPI

Review

Galectin-3: The Impact on the Clinical Management of Patients with Thyroid Nodules and Future Perspectives

Armando Bartolazzi [1],*, Salvatore Sciacchitano [2,3] and Calogero D'Alessandria [4]

[1]   Pathology Research Laboratory, Saint Andrea University Hospital, via di Grottarossa 1035, 00189 Rome, Italy
[2]   Department of Clinical and Molecular Medicine, Sapienza University, Policlinico Umberto I viale Regina Elena 324, 00161 Rome, Italy; Salvatore.Sciacchitano@uniroma1.it
[3]   Laboratory of Biomedical Research, Niccolò Cusano University Foundation, via Don Carlo Gnocchi 3, 00166 Rome, Italy
[4]   Nuklearmedizinische Klinik und Poliklinik, Klinikum rechts der Isar, Technische Universität München, Ismaninger Strasse 22, 81675 München, Germany; Calogero.dalessandria@tum.de
*   Correspondence: Armando.Bartolazzi@ki.se; Tel.: +39-06-3377-5321; Fax: +39-06-3377-5032

Received: 21 December 2017; Accepted: 29 January 2018; Published: 2 February 2018

Abstract: Galectins (S-type lectins) are an evolutionarily-conserved family of lectin molecules, which can be expressed intracellularly and in the extracellular matrix, as well. Galectins bind β-galactose-containing glycoconjugates and are functionally active in converting glycan-related information into cell biological programs. Altered glycosylation notably occurring in cancer cells and expression of specific galectins provide, indeed, a fashionable mechanism of molecular interactions able to regulate several tumor relevant functions, among which are cell adhesion and migration, cell differentiation, gene transcription and RNA splicing, cell cycle and apoptosis. Furthermore, several galectin molecules also play a role in regulating the immune response. These functions are strongly dependent on the cell context, in which specific galectins and related glyco-ligands are expressed. Thyroid cancer likely represents the paradigmatic tumor model in which experimental studies on galectins' glycobiology, in particular on galectin-3 expression and function, contributed greatly to the improvement of cancer diagnosis. The discovery of a restricted expression of galectin-3 in well-differentiated thyroid carcinomas (WDTC), compared to normal and benign thyroid conditions, contributed also to promoting preclinical studies aimed at exploring new strategies for imaging thyroid cancer in vivo based on galectin-3 immuno-targeting. Results derived from these recent experimental studies promise a further improvement of both thyroid cancer diagnosis and therapy in the near future. In this review, the biological role of galectin-3 expression in thyroid cancer, the validation and translation to a clinical setting of a galectin-3 test method for the preoperative characterization of thyroid nodules and a galectin-3-based immuno-positron emission tomography (immuno-PET) imaging of thyroid cancer in vivo are presented and discussed.

Keywords: galectin-3; thyroid cancer; thyroid FNA-cytology, tumor imaging in vivo; immuno-PET

---

1. The Clinical Problem of the Preoperative Characterization of Follicular Thyroid Nodules

It has been estimated that as many as 5–6% of the adult population in the USA (about 15 million people) have clinically-evident thyroid nodules, but this number is much higher (up to 65%) if sub-clinical nodules incidentally discovered during a thyroid echo-scan are also counted [1,2]. Distinguishing among benign and malignant thyroid nodules is a challenge. In fact, it is very difficult to detect preoperatively a relatively rare thyroid cancer (about 12–15% of total thyroid nodules) among a multitude of benign thyroid nodules [3–12]. The wide use of thyroid fine-needle-aspiration (FNA) cytology contributed to improving the preoperative characterization of thyroid nodules, allowing a better surgical selection of these lesions and increasing, at the same time, the number of thyroid cancers diagnosed at histology.

Thyroid FNA cytology was first proposed at Radiumhemmet, Karolinska Hospital, Solna (Sweden), during the year 1950 with the main purpose of distinguishing preoperatively among benign and malignant thyroid proliferations. The first report was published in Acta Medica Scandinava in 1952 [13]. Actually, thyroid FNA cytology is widely accepted as the most effective procedure for the preoperative diagnosis of thyroid cancer. However, since the first experience with the method, two major limitations were observed: the first is linked to the adequacy of the sampling, although it can be presently improved by performing FNA with an ultrasonographic guidance; the second limitation, which occurs in 15–30% of the cases (depending on the diagnostic thyroid center considered), represents, instead, an intrinsic limit of thyroid-cytology in providing a correct morphological diagnosis of benign (hyperplasia and adenoma) and malignant follicular thyroid lesions (follicular carcinoma, oncocytic follicular carcinoma and follicular variant of papillary carcinoma). This task is practically impossible on cytological bases alone due to the overlapping cyto-morphological features in benign and malignant follicular thyroid lesions [14–22]. In particular, the diagnosis of follicular carcinoma requires unequivocal demonstration of capsular and/or vascular invasion. These specific morphological hallmarks cannot be observed on cytological bases (Figure 1).

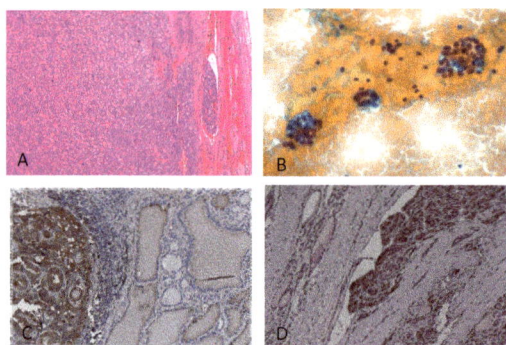

Figure 1. Follicular thyroid lesions and galectin-3 expression on histological samples. (**A**) Histological features of follicular carcinoma showing capsular and vascular infiltration. (**B**) This features cannot be detected on cytological bases. Fine-needle-aspiration-derived thyrocytes arranged in follicular structures remain indeterminate (defined as category Thy-3f according to the British Thyroid Association). (**C**) Galectin-3 expression on a histological sample of the follicular variant of papillary thyroid carcinoma. (**D**) Galectin-3 expression on follicular thyroid carcinoma with capsular infiltration. Magnification: (**A**) ×200; (**B**–**D**) ×250. (**A**,**B**) Conventional haematoxylin/eosin staining; (**C**,**D**) direct immunoperoxidase staining by using a horseradish-peroxidase-conjugated (HRP) monoclonal antibody to galectin-3.

Follicular thyroid nodules at conventional FNA cytology remain, indeed, indeterminate and are classified as Thy-3 according to the British Thyroid Association, or Diagnostic Category III: AUS/FLUS (atypia of undetermined significance/follicular lesion of undetermined significance) according to the Bethesda System for thyroid cytology classification [23,24]. As a consequence, patients bearing cytologically-indeterminate follicular nodules are frequently referred to surgery for a partial or complete thyroidectomy, more for diagnosis rather than for a real therapeutic necessity. As expected, 85–90% of these lesions will be classified as benign at the final histology, and a large proportion of patients will be over-treated [3,15–17,25–28].

Although the current clinical management of patients with thyroid nodules seems to be "methodologically efficient", considering the relatively low incidence of thyroid cancer and its low mortality rate, the surgical overtreatment of such a large number of patients with benign nodules represents a social problem. In Germany, for example, a country with a high prevalence of follicular thyroid nodules, around 100,000 thyroid surgical procedures for benign thyroid lesions are registered each year [29]. The large majority of these lesions are likely benign follicular nodules that were not fully-characterized preoperatively.

In addition to the relevant costs for the public health system and patients' distress, people undergoing surgery for thyroid nodules have a low, but significant risk of permanent injury of the parathyroid glands and recurrent nerves. Life-long substitutive thyroid hormone therapy is always necessary, and calcium supplementation may be also required for patients with impaired (iatrogenic) parathyroid function.

In the last two decades, many efforts have been directed toward developing new potential diagnostic tools for improving thyroid cancer diagnosis preoperatively, and galectin-3 is probably one of the most extensively-studied marker so far. The expression analysis of galectin-3 in thyroid lesions, applied preoperatively for improving the diagnostic performance of conventional FNA cytology, has been finally validated for translation to the clinical setting. Two large multi-institutional studies performed at the international level and involving university hospitals and specialized thyroid centers contributed to this achievement [30,31]. Since the year 1995 when a preliminary report on galectin-3 and thyroid cancer was published [32], despite a great variance in the methodology used, more than 300 papers published in the English literature confirmed the restricted expression of galectin-3 in thyroid cancer, compared to normal and benign thyroid conditions [33–41]. Moreover, an independent study of the largest thyroid cancer diagnostic marker panel reported to date showed that galectin-3 was the most accurate stand-alone marker for well-differentiated thyroid cancer, compared to 56 different candidate molecules [42,43]. Altogether, these data strongly support the potential role of galectin-3 as a reliable diagnostic marker for thyroid cancer diagnosis.

2. The Biological Rationale of Galectin-3 Expression in Transformed Thyroid Cells

Galectin-3 is almost invariably expressed in well-differentiated thyroid carcinomas and can be promptly detected in the cytoplasm of malignant thyroid cells by using immunohistochemical procedures. On the other side, galectin-3 is undetectable in the cytoplasm of normal thyroid follicular cells. This diagnostically-relevant finding has been extensively confirmed in the literature in experimental tumor models in vitro, as well as on cyto-histological substrates ex vivo [30–43].

By using an in vitro model of normal thyroid cells named TAD-2, Takenaka et al. demonstrated that a forced expression of galectin-3 via cDNA transfection induced a transformed phenotype [44]. Stable galectin-3-positive TAD-2 transfectants, in fact, acquired the phenotype of serum-independent growth, clonogenicity in soft agar and loss of contact inhibition. Moreover, a gene expression profile performed on galectin-3 transfectants revealed activation of genes involved in tumor growth and progression, among which were PCNA (proliferating cell nuclear antigen), replication factor C and *Rb* retinoblastoma gene. The latter's protein product plays a significant role in G1–S transition. Conversely, in a different set of experiments, which used a thyroid cancer and a breast carcinoma cell line, inhibition of galectin-3 expression by using mRNA interference reverted the transformed phenotype [45,46].

These experimental findings clearly demonstrate that galectin-3 likely plays a relevant biological role in thyroid cancer. The aberrant expression of galectin-3 in normal thyroid cells, in fact, blocks the apoptotic program, allowing accumulation of DNA mutations and molecular alterations, which in turn promote the development of cancer.

The galectin-3 COOH-terminal domain contains an NWGR amino acid motif highly conserved in the BH1 domain of the Bcl-2 family of anti-apoptotic molecules. The NWGR amino acid sequence is critical for regulating apoptosis as demonstrated by experimental studies in vitro, which used cell transfectants carrying glycine to alanine substitution in the NWGR motif, exposed to *cis*-platinum

(CDDP), a potent anticancer compound that produces an interstrand DNA cross-link and induces apoptosis. Galectin-3 mutant transfectants in the NWGR motif showed high sensitivity to CDDP exposure in vitro compared to the control cell lines expressing wild-type galectin-3 that remain largely viable [47]. More recently, it has been reported that galectin-3 is a physiological target of p53 transcriptional activity. A p53-dependent down-regulation of galectin-3 expression, occurring at transcriptional level, is required for triggering the p53-mediated apoptotic program in different cell systems [48]. This means that following DNA damage, wild-type p53 does not work properly in activating the apoptotic program in a cell context in which galectin-3 remains upregulated. Indeed, in well-differentiated thyroid carcinoma (WDTC) that notably express wt-p53, an unexplained paradoxical concomitant expression of galectin-3 seems to occur. Interestingly, a loss of p53 activator HIPK2 (homeodomain interacting protein kinase-2), a critical molecule that is necessary for p53 phosphorylation on serine 46, has been finally demonstrated in WDTC and was found responsible for p53 loss of function, galectin-3 overexpression and block of apoptosis [49]. In line with these findings, genetic studies also show that a hypomethylation state of 5 CpG sites in the galectin-3 gene correlated with thyroid malignancies [50].

All together, these findings provide a strong biological rationale for the restricted expression of galectin-3 in malignant thyroid cells compared to normal and benign thyroid conditions. Furthermore, a plethora of experimental data published in the literature definitively demonstrates that WDTC almost invariably expresses galectin-3, while normal thyroid tissue, follicular nodular hyperplasia (multinodular goiters) and the large majority of thyroid follicular adenomas do not [33–43,51].

3. Validation of a Galectin-3 Test Method for Clinical Use

With this biological background, the potential diagnostic value of galectin-3 expression analysis in distinguishing among benign and malignant thyroid nodules has been deeply investigated in a large retrospective international multicenter study, which included institutions from Italy, Sweden, the United States and Japan [30]. In this study, as many as 1006 retrospective and histologically well-characterized thyroid lesions were independently analyzed at the immunohistochemical level for galectin-3 expression. The analysis used a purified and well-characterized mAb to galectin-3. Sensitivity, specificity, positive predictive value and diagnostic accuracy of galectin-3 expression in distinguishing among benign and malignant thyroid lesions were 99%, 98%, 91% and 97%, respectively, demonstrating that galectin-3 expression analysis is a potent and reliable diagnostic tool for thyroid cancer detection ex vivo [30].

A galectin-3 test-method optimized for clinical use was applied, indeed, on cytological substrates in a prospective large multicenter study, which involved 11 thyroid institutions and cancer centers. In this study, carried out on 466 patients bearing Thy-3 follicular thyroid proliferations as candidates for surgery, galectin-3 expression analysis was applied preoperatively on FNA-derived cellblock preparations by using immunocyto-histochemistry [31].

The final centralized histological characterization of the resected follicular thyroid lesions performed in blind by two independent and eminent pathologists confirmed that galectin-3 expression analysis is an inexpensive and useful diagnostic tool, which can be easily used in clinical practice for improving the diagnostic performance of conventional thyroid FNA cytology. The sensitivity and specificity of the test-method applied preoperatively were 78% and 93%, respectively. Most importantly, 88% of the Thy-3 follicular thyroid proliferations referred for surgery were correctly classified preoperatively by using the galectin-3 test method alone [31].

Results of these studies are very exciting from the clinical point of view, considering the fact that the correct preoperative application of the test method allows one to avoid up to 71% of unnecessary thyroid surgical procedures, in a clinical scenario in which almost all of the follicular thyroid proliferations (with indeterminate cytology) are still referred for surgery. Moreover, when the galectin-3 test method is correctly used in a clinical-pathological multidisciplinary context its diagnostic performance is further improved [52].

Presently, an optimized galectin-3 test method for cyto-histological use has been translated to the clinical setting and is routinely used in many thyroid institutions worldwide (Figure 2) [53].

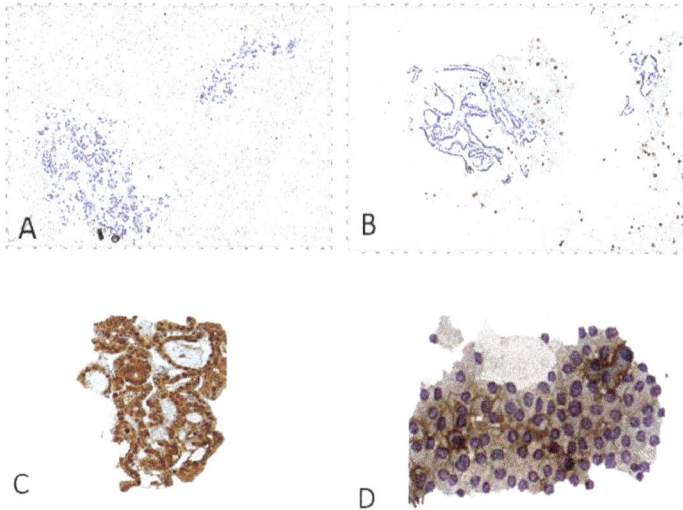

Figure 2. Cellblocks' preparations from thyroid FNA cytology stained with an HRP-conjugated mAb to galectin-3. Benign follicular thyroid proliferations do not express galectin-3: (**A**) follicular adenoma; (**B**) nodular hyperplasia (gal-3-positive scattered foamy macrophages serve as the internal positive control). Thyroid malignancies expressing galectin-3: (**C**) papillary carcinoma follicular variant; (**D**) follicular carcinoma; (**A–D**) direct immunoperoxidase staining by using an HRP-conjugated mAb to gal-3. Magnification: ×250.

Since the year 2003, galectin-3 expression analysis has been mentioned in the American Thyroid Association Guidelines for the clinical management of thyroid nodules [54] and more recently in the revised work [55]. As highlighted by Trimboli et al. in their contribution to this special issue [56], the galectin-3 test method performs better on histological substrates compared to FNA-derived cytological preparations. This is clearly expected if we consider the diagnostic and technical variability that typically affects cytology and immunocytochemistry. In our experience, there is still space for improving FNA-derived cytological preparations, immunocyto-histochemical staining and interpretation of galectin-3 results on FNA-derived thyroid cellblocks. Basic technical and operative guidelines for the optimal diagnostic performance of the galectin-3 test method have been published in order to avoid the occurrence of false negative and false positive results [54]. The most important technical requirements for a reliable galectin-3 test method are summarized in Table 1.

When galectin-3 expression analysis is used for diagnostic purposes, the specific demonstration of galectin-3 accumulation in the cytoplasm of transformed follicular thyroid cells is imperative, independently of the presence or not of positive nuclear staining. This finding is biologically relevant because it is the accumulation of galectin-3 in the cytoplasm that triggers the anti-apoptotic function. With this in mind, we can write with confidence that a correct and reliable biological characterization of a thyroid lesion always requires a combined morphological and immuno-phenotypical approach. For this reason, galectin-3 expression analysis does not replace the conventional thyroid FNA cytology, but integrates this method, with the final result of a consistent improvement of the diagnostic performance [31].

Table 1.   Technical requirements for galectin-3 expression analysis to be applied on fine-needle-aspiration-derived thyroid cells and conventional histological substrates.

A: A purified and well-characterized mAb to human galectin-3 (concentration ranging from 5–10 µg/mL) must be used in immunohisto-cytochemistry (direct or indirect immunoperoxidase) with a biotin-free detection system.
B: Galectin-3 immunostaining must be applied on formalin-fixed and paraffin embedded cyto-histological substrates (i.e., FNA-derived cellblocks).
C: Antigen retrieval microwave treatment with 0.01 M citrate buffer, pH6 for three cycles of 3–5 min each at 750 W is necessary.
D: Follicular thyroid cells showing galectin-3 accumulation in the cytoplasm, with or without nuclear staining, are considered positive. Scattered foamy macrophages serve as the internal positive control.
E: In Hashimoto's thyroiditis (HT) and less frequently in chronic lymphocytic thyroiditis, false positive immunostaining for galectin-3 may occur in follicular thyroid cells within inflammatory follicles. This may generate false positive results (nodular lesions in HT always require a multidisciplinary clinical-pathological evaluation for a better therapeutic decision).
F: The surgical option for galectin-3-positive cases is advisable, also in the presence of a few galectin-3-positive thyroid follicular cells (cytoplasm +).

(Modified from Bartolazzi et al. [53].)

The clinical problem of the preoperative characterization of thyroid nodules is a challenge, and since the publication of the first multicenter study in which galectin-3 expression analysis was applied on thyroid histological samples [30], several molecular approaches including gene expression profile [57] and mutational analysis [58] were also proposed for resolving this important clinical problem. Recently, a comprehensive comparative study on the diagnostic performance, feasibility and cost effectiveness of several diagnostic test methods, including the molecular methods proposed for thyroid cancer diagnosis, has been published [59].

Surprisingly galectin-3 expression analysis applied at the immunocyto-histochemical level on paraffin-embedded cyto-histological substrates showed a better diagnostic performance compared to the GEC and Afirma molecular tests. Galectin-3 serves well both as an efficient rule-out and rule-in test method, with a good likelihood ratio and diagnostic accuracy. Furthermore, and very importantly, galectin-3 expression analysis does not require centralization in a single specialized laboratory, but can be easily performed in each conventional laboratory of histology [59].

At this point, objective considerations concerning the cost of these diagnostic test methods are also necessary. The cost of galectin-3 immunohistochemical analysis is about 113 US Dollars/test and indeed is very competitive if compared with the cost estimated for the molecular genetic tests (20-times more expensive). For this reason, the galectin-3 test-method has a potential screening role [58].

Considering the prevalence of indeterminate thyroid nodules at conventional cytology (about 15–30% of FNA specimens), the cost savings offered by a galectin-3-based immunocyto-histological assay would result in being significant for both low income countries and industrialized countries where specialized thyroid hospitals examine thousands of patients per year.

4. Thyroid Cancer Imaging In Vivo by Targeting Galectin-3

As previously mentioned, the high prevalence of benign thyroid nodules in the adult population and the relatively rare occurrence of thyroid cancer make the preoperative identification of malignant nodules very difficult. Several imaging approaches are currently used in clinical practice with the attempt to detect thyroid cancer, but unfortunately, these methods are not specific enough.

Thyroid scintigraphy with radioiodine, for example, is widely used preoperatively in patients with one or more thyroid nodules. The method provides functional information on iodine uptake (cold or hot nodules), but it fails to distinguish among benign and malignant thyroid nodules.

On the other side, positron emission tomography (PET) in combination with different PET tracers like [^{18}F]-2-fluoro-2-deoxy-D-glucose (^{18}F-FDG), ^{18}F-DOPA and ^{68}Ga-somatostatin analogues has been also used for the same purpose [60]. In particular ^{18}F-FDG PET has been proposed as a preoperative diagnostic procedure for detecting thyroid cancer. Although a cancer diagnosis is generally ruled out in the presence of negative ^{18}F-FDG PET, the sensitivity and specificity of the method are not optimal [61–63]. Considering the plethora of data derived by the extensive analysis of galectin-3 expression in normal, benign and malignant thyroid tissue and the aforementioned biological rationale of galectin-3 expression in transformed thyroid cells, the idea to image thyroid cancer in vivo by targeting galectin-3 seems to be promising, coherent and clinically relevant.

Our group first proposed a novel approach for imaging thyroid cancer in vivo based on galectin-3 immunotargeting [64]. A preliminary set of experiments was performed on human thyroid cancer cell lines xenografted in a murine experimental model. A galectin-3-based thyroid immunoscintigraphy, which used as tracer a ^{99m}Tc-labeled mAb to galectin-3 and as detector a prototype of a mini-γ camera for small animal imaging, was used ad hoc for imaging thyroid cancer xenografts in vivo. Results from this preliminary study showed a good and reliable imaging of galectin-3-positive thyroid tumors between 6 and 9 h from injection of 100 µCi of radiotracer in the tail vein, opening a new avenue in thyroid cancer diagnosis [64]. These preliminary results clearly show the real possibility of detecting thyroid cancer in vivo by targeting galectin-3.

Recently, a galectin-3-based immuno-positron emission tomography (immuno-PET) for imaging thyroid cancer in vivo has been developed and used in preclinical experimental models of thyroid cancer xenografts. The method used a thyroid cancer-specific probe obtained by radiolabeling a purified and well-characterized mAb to galectin-3 amino-terminal epitope, with the long half-life positron emitter zirconium-89 (^{89}Zr; $t_{1/2}$ = 74.8 h, β+ = 22.6%). Xenografted athymic Nude-Foxn1$^{nu/nu}$ mice received 1.5 MBq (40 µCi) of ^{89}Zr-labeled mAb to gal-3 in the tail vein and were subjected to imaging sessions at 48 h post-injection.

Static PET/computer tomography (PET/CT) images showed high tumor binding specificity of the radiotracer and a reliable imaging of thyroid cancer in vivo [64] (Figure 3).

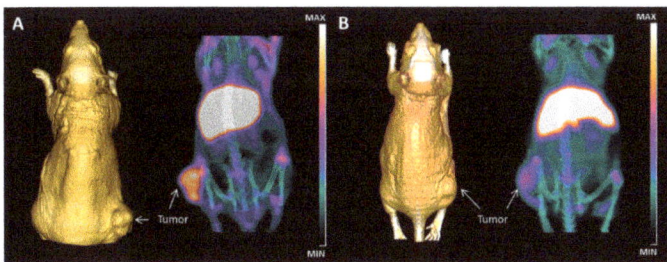

Figure 3. In vivo immuno-positron emission tomography (immuno-PET) imaging of thyroid tumor by targeting galectin-3. (**A**) Maximum intensity projection (MIP) of µ-PET images acquired at 48 h post-injection of 40 µCi of radiotracer showing accumulation of ^{89}Zr-labelled mAb to gal-3 in subcutaneous tumor growing in the right thigh. 3D volume rendering (3D VR), based on CT acquisition performed using OsiriX Image Software 5.0 (Pixmeo, Geneva, Switzerland), allows a tri-dimensional reconstruction of the anatomy of the mouse and visualization of the tumor shape. The activity in the liver is due to residualization of antibody metabolites. (**B**) A strong reduction of uptake is visualized in mice pre-injected with 100-fold excess of unlabeled mAb to gal-3 and imaged 48 h post-injection of the radiotracer, confirming the binding specificity. Normal thyroid gland does not express detectable galectin-3, and as expected, no accumulation of the radiotracer was visible in the neck region.

The reliability of the proposed imaging approach has been confirmed in three different animal models of human thyroid cancer xenografts including a follicular carcinoma and a poorly-differentiated thyroid carcinoma.

The specificity of galectin-3 immuno-PET targeting for imaging thyroid cancer has been further confirmed by an extensive ex vivo biodistribution analysis, measuring the amount of ^{89}Zr-labeled probe accumulated in tumors and normal tissues explanted from the experimental animals [65].

Concluding, galectin-3 immuno-PET targeting represents a new potential diagnostic method for in vivo detection and biological characterization of thyroid nodules, which deserves to be further improved for clinical translation. Preclinical studies on orthotopic models of thyroid cancer are ongoing to confirm the preliminary findings. Galectin-3 immuno-PET strategy for imaging thyroid cancer in vivo is supported by a strong clinical and biological rationale and has the potential to improve, in the near future, the clinical management of patients bearing thyroid nodules, reducing unnecessary surgery and social costs [66–68].

Galectin-3 immuno-PET may potentially improve thyroid cancer diagnosis in the following conditions: (a) in the presence of multiple thyroid nodules, which cannot be easily analyzed preoperatively by using conventional FNA cytology; (b) in the presence of small suspicious sub-centimetric lesions (3–4 mm) discovered in a deep mediastinal position or intimately associated with vascular structures, for which fine-needle-aspiration-biopsy evaluation can be difficult or harmful; (c) the method may be also applied to distinguish among thyroid cancer infiltration/residual tumor nests and minimal normal thyroid tissue residues after thyroid surgery. This represents a common clinical problem during follow-up of patients after a thyroid cancer surgery. Although many of these conditions are presently detected with radioiodine, they remain largely uncharacterized at biological level.

(d) Moreover, a fraction of poorly-differentiated thyroid carcinomas and anaplastic thyroid carcinomas (rare lesions) generally express galectin-3, but lose the ability to uptake radioiodine.

In these specific cases, immuno-PET targeting of galectin-3 might be useful for detecting residual disease and to provide information for a better clinical management and therapeutic intervention of each specific case.

Although the preclinical experimental work performed with galectin-3 immuno-PET has been focused on thyroid cancer detection in vivo, the proposed technology could be useful in different tumor conditions. Galectin-3, in fact, is expressed in different primary and metastatic tumors (i.e., melanoma, breast carcinoma, prostatic carcinoma) [51]. The possibility to image these tumors in vivo or to detect metastasis (i.e., metastatic sentinel lymph nodes) represents an interesting field of research to be investigated.

At least theoretically, increasing the LET (linear energy transfer) of the radionuclide linked to the galectin-3-specific probe will generate a tool for immuno-radiotherapy and/or radio-ablation of the so-called "occult thyroid carcinomas" of millimetric dimension. These lesions are generally undetectable preoperatively. Further, preclinical studies in the field are ongoing, as well as humanization of galectin-3-specific mAbs and/or adequate structural modifications of these probes (i.e., the creation of radiolabeled F(ab)-fragments or chimeric molecules). This work will be necessary for translating the method to the clinical setting.

5. Conclusions

In the present Special Issue of the International Journal of Molecular Sciences focused on galectins in cancer and translational medicine, other experiences with galectin-1, galectin-3 and thyroid cancer are reported by different research groups [56,69]. Very interestingly, some papers published in this issue also show the possibility to inhibit specific galectins in vivo by using both natural and synthetic inhibitors [70]. We are confident that research in glycobiology of galectins will contribute in the near future to improve both cancer diagnosis and treatment of several tumor conditions, discovering targetable galectin-mediated functions that are critical for cancer [51,71].

Int. J. Mol. Sci. **2018**, *19*, 445

Concluding: i) glycobiology of galectin-3; ii) the biological effects induced by galectin-3 in tumor cells in vitro; iii) the extensive characterization of galectin-3 expression performed on human thyroid tissues ex vivo; iv) the possibility to imaging thyroid cancer in vivo by using a galectin-3 immuno-PET represent all together a fashionable journey from the bench to the bed-side, which has already provided important improvements in clinical practice.

Acknowledgments: This work was partially supported by A.I.R.C. (Italian Association for Cancer Research) and Compagnia di San Paolo Progetto Oncologia granted to Armando Bartolazzi and by the Deutsche Forschungsgemeinschaft (DFG Project: DA 1552/2-1) granted to Calogero D'Alessandria.

Conflicts of Interest: Armando Bartolazzi has an ownership of a patent related to the use of radiolabeled mAbs to galectin-3 for tumor imaging in vivo (Patent No. 1388763, registered on 20 February 2008, Rome, Italy). Other co-authors declare no conflict of interest.

References

1. Rosai, J.; Carcangiu, M.L.; De Lellis, R.A. *Atlas of Tumor Pathology: Tumors of the Thyroid Gland*, 3rd ed.; Armed Force Institute of Pathology: Washington, DC, USA, 1992; pp. 1–343.
2. Tan, G.H.; Gharib, H. Thyroid incidentalomas: Management approaches to no palpable nodules discovered incidentally on thyroid imaging. *Ann. Intern. Med.* **1997**, *126*, 226–231. [CrossRef] [PubMed]
3. Bartolazzi, A. Improving accuracy of cytology for nodular thyroid lesions. *Lancet* **2000**, *355*, 1661–1662. [CrossRef]
4. Castro, M.R.; Gharib, H. Continuing controversies in the management of thyroid nodules. *Ann. Intern. Med.* **2005**, *142*, 926–931. [CrossRef] [PubMed]
5. Schlumberger, M.J.; Pacini, F. Thyroid nodule. In *Thyroid Tumors*, 2nd ed.; Editions Nucleon: Paris, France, 2003; pp. 11–31.
6. Alexander, E.K. Approach to the patient with cytologically indeterminate thyroid nodule. *J. Clin. Endocrinol. Metabol.* **2008**, *93*, 4175–4182. [CrossRef] [PubMed]
7. Davis, N.L.; Gordon, M.; Germann, E.; Robins, R.E.; Mc Gregor, G.I. Clinical parameters predictive of malignancy of thyroid follicular neoplasms. *Am. J. Surg.* **1991**, *161*, 567–569. [CrossRef]
8. Tuttle, R.M.; Lemar, H.; Burch, H.B. Clinical features associated with an increased risk of thyroid malignancy in patients with follicular neoplasia by fine-needle aspiration. *Thyroid* **1998**, *8*, 377–383. [CrossRef] [PubMed]
9. Raber, W.; Kaserer, K.; Niederle, B.; Vierhapper, H. Risk factors for malignancy of thyroid nodules initially identified as follicular neoplasia by fine-needle aspiration: Results of a prospective study of one hundred twenty patients. *Thyroid* **2000**, *10*, 709–712. [CrossRef] [PubMed]
10. Yassa, L.; Cibas, E.S.; Benson, C.B.; Frates, M.C.; Doubilet, P.M.; Gawande, A.A.; Moore, F.D., Jr.; Kim, B.W.; Nosè, W.; Marqusee, E.; et al. Long-term assessment of a multidisciplinary approach to thyroid nodule diagnostic evaluation. *Cancer* **2007**, *111*, 508–516. [CrossRef] [PubMed]
11. Rago, T.; Scutari, M.; Latrofa, F.; Loiacono, V.; Piaggi, P.; Marchetti, I.; Romani, R.; Basolo, F.; Miccoli, P.; Tonacchera, M.; et al. The large majority of 1520 patients with indeterminate thyroid nodule at cytology have a favorable outcome and a clinical risk score has a high negative predictive value for a more cumbersome cancer disease. *J. Clin. Endocrinol. Metab.* **2014**, *99*, 3700–3707. [CrossRef] [PubMed]
12. Ianni, F.; Campanella, P.; Rota, C.A.; Prete, A.; Castellino, L.; Pontecorvi, A.; Corsello, S.M. A meta-analysis-derived proposal for a clinical, ultrasonographic and cytological scoring system to evaluate thyroid nodules: The "CUT" score. *Endocrine* **2016**, *52*, 313–321. [CrossRef] [PubMed]
13. Soderstrom, N. Puncture of goiters for aspiration biopsy. *Acta Med. Scand.* **1952**, *144*, 237–244. [CrossRef] [PubMed]
14. Kini, S.R. Thyroid. In *Guide to Clinical Aspiration Biopsy*; Kline, T.S., Ed.; Igaku Shoin: New York, NY, USA, 1987; Volume 3, pp. 1–380.
15. Baloch, Z.W.; Sack, M.J.; Yu, G.H.; Li Volsi, V.A.; Gupta, P.K. Fine-needle aspiration of thyroid: An institutional experience. *Thyroid* **1998**, *8*, 565–569. [CrossRef] [PubMed]
16. Hall, T.L.; Layfield, L.J.; Philippe, A.; Rosenthal, D.L. Sources of diagnostic error in fine needle aspiration of the thyroid. *Cancer* **1989**, *63*, 718–725. [CrossRef]

17. Lloyd, R.V.; Erickson, L.A.; Casey, M.B.; Lam, K.Y.; Lohse, C.M.; Asa, S.L.; Chan, J.K.; De Lellis, R.A.; Harach, H.L.; Kakudo, K.; et al. Observer variation in the diagnosis of follicular variant of papillary thyroid carcinoma. *Am. J. Surg. Pathol.* **2004**, *28*, 1336–1340. [CrossRef] [PubMed]

18. Wu, H.H.; Jones, J.N.; Qsman, J. Fine-needle aspiration cytology of the thyroid: Ten years experience in a community teaching hospital. *Diagn. Cytopathol.* **2006**, *34*, 93–96. [CrossRef] [PubMed]

19. Yang, J.; Schnadig, V.; Logrono, R.; Wasserman, P.G. Fine-needle aspiration of thyroid nodules: A study of 4703 patients with histologic and clinical correlations. *Cancer* **2007**, *111*, 306–315. [CrossRef] [PubMed]

20. Oertel, Y.C.; Miyahara-Felipe, L.; Mendoza, M.G.; Yu, K. Value of repeated fine needle aspirations of the thyroid: An analysis of over ten thousand FNAs. *Thyroid* **2007**, *17*, 1061–1066. [CrossRef] [PubMed]

21. Mihai, R.; Parker, A.J.; Roskell, D.; Sadler, G.P. One in four patients with follicular thyroid cytology (THY3) has a thyroid carcinoma. *Thyroid* **2009**, *19*, 33–37. [CrossRef] [PubMed]

22. Gharib, H. Fine-needle aspiration biopsy of thyroid nodules: Advantages, limitations and effects. *Mayo Clin. Proc.* **1994**, *69*, 44–50. [CrossRef]

23. Cibas, E.S.; Ali, S.Z. The Bethesda System for Reporting Thyroid Cytopathology. *Thyroid* **2009**, *19*, 1159–1165. [CrossRef] [PubMed]

24. Perros, P.; Boalert, K.; Colley, S.; Evans, C.; Evans, R.M.; Gerrard Ba, G.; Gilbert, J.; Harrison, B.; Johnson, S.J.; Giles, T.E.; et al. British Thyroid Association. Guidelines for the management of thyroid cancer. *Clin. Endocrinol.* **2014**, *61*, 1–122. [CrossRef] [PubMed]

25. Kim, E.S.; Nam-Goong, I.S.; Gong, G.; Hong, S.J.; Kim, W.B.; Shong, Y.K. Postoperative findings and risk for malignancy in thyroid nodules with cytological diagnosis of the so-called "follicular neoplasms". *Korean J. Intern. Med.* **2003**, *18*, 94–97. [CrossRef] [PubMed]

26. Sclabas, G.M.; Staekel, G.A.; Shapiro, S.E.; Fornage, B.D.; Sherman, S.L.; Vassillopoulou-Sellin, R.; Lee, J.E.; Evans, D.B. Fine-needle aspiration of the thyroid and correlation with histopathology in a contemporary serie of 240 patients. *Am. J. Surg.* **2003**, *186*, 702–709. [CrossRef] [PubMed]

27. Sorrenti, S.; Trimboli, P.; Catania, A.; Ulisse, S.; De Antoni, E.; D'Armiento, M. Comparison of malignancy rate in thyroid nodules with cytology of indeterminate follicular or indeterminate Hurthle cell neoplasm. *Thyroid* **2009**, *19*, 355–360. [CrossRef] [PubMed]

28. Asari, R.; Niederle, B.E.; Scheuba, C.; Riss, P.; Koperek, O.; Kaserer, K.; Niederle, B. Indeterminate thyroid nodules: A challenge for the surgical strategy. *Surgery* **2010**, *148*, 516–525. [CrossRef] [PubMed]

29. Musholt, T.J.; Clerici, T.; Dralle, H.; Frilling, A.; Goretzki, P.E.; Hermann, M.M.; Kußmann, J.; Lorenz, K.; Nies, C.; Schabram, J.; et al. German Association of Endocrine Surgeons practice guidelines for the surgical treatment of benign thyroid disease. *Langenbecks Arch. Surg.* **2011**, *396*, 639–649. [CrossRef] [PubMed]

30. Bartolazzi, A.; Gasbarri, A.; Papotti, M.; Bussolati, G.; Lucante, T.; Khan, A.; Inohara, H.; Marandino, F.; Orlandi, F.; Nardi, F.; et al. Application of immunodiagnostic method for improving preoperative diagnosis of nodular thyroid lesions. *Lancet* **2001**, *357*, 1644–1650. [CrossRef]

31. Bartolazzi, A.; Orlandi, F.; Saggiorato, E.; Volante, M.; Arecco, F.; Rossetto, R.; Palestini, N.; Ghigo, E.; Papotti, M.; Bussolati, G.; et al. Galectin-3-expression analysis in the surgical selection of follicular thyroid nodules with indeterminate fine-needle aspiration cytology: A prospective multicenter study. *Lancet Oncol.* **2008**, *9*, 543–549. [CrossRef]

32. Xu, X.C.; el-Naggar, A.K.; Lotan, R. Differential expression of Galectin-1 and Galectin-3 in thyroid tumors. Potential diagnostic implications. *Am. J. Pathol.* **1995**, *147*, 815–822. [PubMed]

33. Gasbarri, A.; Martegani, M.P.; Del Prete, F.; Lucante, T.; Natali, P.G.; Bartolazzi, A. Galectin-3 and CD44v6 isoforms in the pre-operative evaluation of thyroid nodules. *J. Clin. Oncol.* **1999**, *17*, 3494–3502. [CrossRef] [PubMed]

34. Saggiorato, E.; Cappia, S.; De Giuli, P.; Mussa, A.; Pancani, G.; Caraci, P.; Angeli, A.; Orlandi, F. Galectin-3 as a presurgical immunocytodiagnostic marker of minimally invasive follicular thyroid carcinoma. *J. Clin. Endocrinol. Metab.* **2001**, *86*, 5152–5158. [CrossRef] [PubMed]

35. Papotti, M.; Rodriguez, J.; De Pompa, R.; Bartolazzi, A.; Rosai, J. Galectin-3 and HBME-1 expression in well-differentiated thyroid tumors with follicular architecture of uncertain malignant potential. *Mod. Pathol.* **2005**, *18*, 541–546. [CrossRef] [PubMed]

36. Carpi, A.; Naccarato, A.G.; Iervasi, G.; Nicolini, A.; Bevilacqua, G.; Viacava, P.; Collecchi, P.; Lavra, L.; Marchetti, C.; Sciacchitano, S.; et al. Large needle aspiration biopsy and galectin-3 determination in selected thyroid nodules with indeterminate FNA cytology. *Br. J. Cancer* **2006**, *95*, 204–209. [CrossRef] [PubMed]

37. De Matos, L.L.; Del Giglio, A.B.; Matsubayashi, C.O.; de Lima Farah, M.; Del Giglio, A.; da Silva Pinhal, M.A. Expression of CK-19, galectin-3 and HBME-1 in the differentiation of thyroid lesions: Systematic review and diagnostic meta-analysis. *Diagn. Pathol.* **2012**, *7*, 97. [CrossRef] [PubMed]

38. Herrmann, M.E.; LiVolsi, V.A.; Pasha, T.L.; Roberts, S.A.; Wojcik, E.M.; Baloch, Z.W. Immunohistochemical Expression of Galectin-3 in Benign and Malignant Thyroid Lesions. *Arch. Pathol. Lab. Med.* **2002**, *126*, 710–713.

39. Carpi, A.; Rossi, G.; Di Coscio, G.D.; Iervasi, G.; Nicolini, A.; Carpi, F.; Mechanick, J.I.; Bartolazzi, A. Galectin-3 detection on large-needle aspiration biopsy improves preoperative selection of thyroid nodules: A prospective cohort study. *Ann. Med.* **2010**, *42*, 70–78. [CrossRef] [PubMed]

40. Saggiorato, E.; De Pompa, R.; Volante, M.; Cappia, S.; Arecco, F.; Dei Tos, A.P.; Orlandi, F.; Papotti, M. Characterization of thyroid 'follicular neoplasms' in fine-needle aspiration cytological specimens using a panel of immunohistochemical markers: A proposal for clinical application. *Endocr.-Relat. Cancer* **2005**, *12*, 305–317. [CrossRef] [PubMed]

41. Weber, K.B.; Shroyer, K.R.; Heinz, D.E.; Nawaz, S.; Said, M.S.; Haugen, B.R. The use of combination of galectin-3 and thyroid peroxidase for the diagnosis and prognosis of thyroid cancer. *Am. J. Clin. Pathol.* **2004**, *122*, 524–531. [CrossRef] [PubMed]

42. Griffith, O.L.; Chiu, C.G.; Gown, A.M.; Jones, S.J.; Wiseman, S.M. Biomarker panel diagnosis of thyroid cancer: A critical review. *Expert Rev. Anticancer Ther.* **2008**, *8*, 1399–1413. [CrossRef] [PubMed]

43. Chiu, C.G.; Strugnell, S.S.; Griffith, O.L.; Jones, S.J.; Gown, A.M.; Walker, B.; Nabi, I.R.; Wiseman, S.M. Diagnostic utility of galectin-3 in thyroid cancer. *Am. J. Pathol.* **2010**, *176*, 2067–2081. [CrossRef] [PubMed]

44. Takenaka, Y.; Inohara, H.; Yoshii, T.; Oshima, K.; Nakahara, S.; Akahani, S.; Honjo, Y.; Yamamoto, Y.; Raz, A.; Kubo, T. Malignant transformation of thyroid follicular cells by galectin-3. *Cancer Lett.* **2003**, *195*, 111–119. [CrossRef]

45. Yoshii, T.; Inohara, H.; Takenaka, Y.; Honjo, Y.; Akahani, S.; Nomura, T.; Raz, A.; Kubo, T. Galectin-3 maintains the transformed phenotype of thyroid papillary carcinoma cells. *Int. J. Oncol.* **2001**, *18*, 787–792. [CrossRef] [PubMed]

46. Honjo, Y.; Nangia-Makker, P.; Inohara, H.; Raz, A. Down-regulation of galectin-3 suppresses tumorigenicity of human breast carcinoma cells. *Clin. Cancer Res.* **2001**, *7*, 661–668. [PubMed]

47. Akahani, S.; Nangia-Makker, P.; Inohara, H.; Kim, H.R.; Raz, A. Galectin-3: A novel antiapoptotic molecule with a functional BH1 (NWGR) domain of Bcl-2 family. *Cancer Res.* **1997**, *57*, 5272–5276. [PubMed]

48. Cecchinelli, B.; Lavra, L.; Rinaldo, C.; Iacovelli, S.; Gurtner, A.; Gasbarri, A.; Ulivieri, A.; Del Prete, F.; Trovato, M.; Piaggio, G.; et al. Repression of the antiapoptotic molecule galectin-3 by homeodomain-interacting protein kinase 2-activated p53 is required for p53-induced apoptosis. *Mol. Cell. Biol.* **2006**, *26*, 4746–4757. [CrossRef] [PubMed]

49. Lavra, L.; Rinaldo, C.; Ulivieri, A.; Luciani, E.; Fidanza, P.; Giacomelli, L.; Bellotti, C.; Ricci, A.; Trovato, M.; Soddu, S.; et al. The loss of the p53 activator HIPK2 is responsible for galectin-3 overexpression in well differentiated thyroid carcinomas. *PLoS ONE* **2011**, *6*, e20665. [CrossRef] [PubMed]

50. Keller, S.; Angrisano, T.; Florio, E.; Pero, R.; Decaussin-Petrucci, M.; Troncone, G.; Capasso, M.; Lembo, F.; Fusco, A.; Chiarotti, L. DNA methylation state of the galectin-3 gene represents a potential new marker of thyroid malignancy. *Oncol. Lett.* **2013**, *6*, 86–90. [CrossRef] [PubMed]

51. Liu, F.; Rabinovich, G.A. Galectins as modulators of tumor progression. *Nat. Rev. Cancer* **2005**, *5*, 29–41. [CrossRef] [PubMed]

52. Sciacchitano, S.; Lavra, L.; Ulivieri, A.; Magi, F.; Porcelli, T.; Amendola, S.; De Francesco, G.P.; Bellotti, C.; Trovato, M.C.; Salehi, L.B.; et al. Combined clinical and ultrasound follow-up assists in malignancy detection in Galectin-3 negative Thy-3 thyroid nodules. *Endocrine* **2016**, *54*, 139–147. [CrossRef] [PubMed]

53. Bartolazzi, A.; Bellotti, C.; Sciacchitano, S. Methodology and technical requirements of the galectin-3 test for the preoperative characterization of thyroid nodules. *Appl. Immunohistochem. Mol. Morphol.* **2012**, *20*, 2–7. [CrossRef] [PubMed]

54. The American Thyroid Association. Consensus Guidelines for Thyroid Testing in the New Millennium: In Laboratory Medicine Practice Guidelines. Laboratory support for the diagnosis and monitoring of thyroid disease. (Section H): Thyroid fine needle aspiration (FNA) and cytology. *Thyroid* **2003**, *13*, 80–86.

55. Cooper, D.S.; Doherty, G.M.; Haugen, B.R.; Kloos, R.T.; Lee, S.L.; Mandel, S.J.; Mazzaferri, E.L.; McIver, B.; Pacini, F.; Schlumberger, M.; et al. Revised American Thyroid Association Management Guidelines for patients with thyroid nodules and differentiated thyroid cancer. *Thyroid* **2009**, *19*, 1167–1214. [CrossRef] [PubMed]

56. Trimboli, P.; Virili, C.; Romanelli, F.; Crescenzi, A.; Giovanella, L. Galectin-3 Performance in Histologic a Cytologic Assessment of Thyroid Nodules: A Systematic Review and Meta-Analysis. *Int. J. Mol. Sci.* **2017**, *18*, 1756. [CrossRef] [PubMed]

57. Alexander, E.K.; Kennedy, G.C.; Baloch, Z.W.; Cibas, E.S.; Chudova, D.; Diggans, J.; Friedman, L.; Kloos, R.T.; LiVolsi, V.A.; Mandel, S.J.; et al. Preoperative diagnosis of benign thyroid nodules with indeterminate cytology. *N. Engl. J. Med.* **2012**, *367*, 705–715. [CrossRef] [PubMed]

58. Nikiforov, Y.E.; Carty, S.E.; Chiosea, S.I.; Coyne, C.; Duvvuri, U.; Ferris, R.L.; Gooding, W.E.; LeBeau, S.O.; Ohri, N.P.; Seethala, R.R.; et al. Impact of the Multi-Gene Thyroseq Next-Generation Sequencing Assay on Cancer Diagnosis in Thyroid nodules with atypia of underdeterminate significance/Follicular lesion of underdeterminate significance cytology. *Thyroid* **2015**, *25*, 1217–1223. [CrossRef] [PubMed]

59. Sciacchitano, S.; Lavra, L.; Ulivieri, A.; Magi, F.; De Francesco, G.P.; Bellotti, C.; Salehi, L.B.; Trovato, M.; Drago, C.; Bartolazzi, A. Comparative analysis of diagnostic performance, feasibility and cost of different test-methods for thyroid nodules with indeterminate cytology. *Oncotarget* **2017**, *8*, 49421–49442. [CrossRef] [PubMed]

60. Treglia, G.; Castaldi, P.; Villani, M.F.; Perotti, G.; de Waure, C.; Filice, A.; Ambrosini, V.; Cremonini, N.; Santimaria, M.; Versari, A.; et al. Comparison of 18F-DOPA, 18F-FDG and 68Ga-somatostatin analogue PET/CT in patients with recurrent medullary thyroid carcinoma. *Eur. J. Nucl. Med. Mol. Imaging* **2012**, *39*, 569–580. [CrossRef] [PubMed]

61. Traugott, A.L.; Dehdashti, F.; Trinkaus, K.; Cohen, M.; Fialkowski, E.; Quayle, F.; Hussain, H.; Davila, R.; Ylagan, L.; Moley, J.F. Exclusion of malignancy in thyroid nodules with indeterminate fine-needle aspiration cytology after negative 18F-fluorodeoxyglucose positron emission tomography: Interim analysis. *World J. Surg.* **2010**, *34*, 1247–1253. [CrossRef] [PubMed]

62. Deandreis, D.; Al Ghuzlan, A.; Auperin, A.; Vielh, P.; Caillou, B.; Chami, L.; Lumbroso, J.; Travagli, J.P.; Hartl, D.; Baudin, E.; et al. Is (18)F-fluorodeoxyglucose-PET/CT useful for the presurgical characterization of thyroid nodules with indeterminate fine needle aspiration cytology? *Thyroid* **2012**, *22*, 165–172. [CrossRef] [PubMed]

63. Vriens, D.; de Wilt, J.H.; van der Wilt, G.I.; Netea-Maier, R.T.; Oyen, W.J.; de Geus-Oei, L.F. The role of [18F]-2-fluoro-2-deoxy-d-glucose-positron emission tomography in thyroid nodules with indeterminate fine-needle aspiration biopsy: Systematic review and meta-analysis of the literature. *Cancer* **2011**, *117*, 4582–4594. [CrossRef] [PubMed]

64. Bartolazzi, A.; D'Alessandria, C.; Parisella, M.G.; Signore, A.; Del Prete, F.; Lavra, L.; Braesch-Andersen, S.; Massari, R.; Trotta, C.; Soluri, A.; et al. Thyroid cancer imaging in vivo by targeting the anti-apoptotic molecule galectin-3. *PLoS ONE* **2008**, *3*, e3768. [CrossRef] [PubMed]

65. D'Alessandria, C.; Braesch-Andersen, S.; Bejo, K.; Reder, S.; Blechert, B.; Schwaiger, M.; Bartolazzi, A. Noninvasive in Vivo Imaging and Biologic Characterization of Thyroid Tumors by ImmunoPET Targeting of Galectin-3. *Cancer Res.* **2016**, *76*, 3583–3592. [CrossRef] [PubMed]

66. Hodak, S.P.; Rosenthal, D.S.; American Thyroid Association Clinical Affairs Committee. Information for clinicians: Commercially available molecular diagnosis testing in the evaluation of thyroid nodule fine-needle aspiration specimens. *Thyroid* **2013**, *23*, 131–134. [CrossRef] [PubMed]

67. Facey, K.; Bradbury, I.; Laking, G.; Payne, E. Overview of the clinical effectiveness of positron emission tomography imaging in selected cancers. *Health Technol. Assess.* **2007**, *11*. [CrossRef]

68. Buck, A.K.; Herrmann, K.; Stargardt, T.; Dechow, T.; Krause, B.J.; Schreyogg, J. Economic evaluation of PET and PET/CT in oncology: Evidence and methodologic approaches. *J. Nucl. Med. Technol.* **2010**, *38*, 6–17. [CrossRef] [PubMed]

69. Kindt, N.; Journe, F.; Ghanem, G.E.; Saussez, S. Galectins and carcinogenesis: Their role in head and neck carcinomas and thyroid carcinomas. *Int. J. Mol. Sci.* **2017**, *18*, 2745. [CrossRef] [PubMed]

70. Wdowiak, K.; Francuz, T.; Gallego-Colon, E.; Ruiz-Agamez, N.; Kubeczko, M.; Grochola, I.; Wojnar, J. Galectin targeted therapy in oncology: Current knowledge and perspectives. *Int. J. Mol. Sci.* **2018**, *19*, 210. [CrossRef] [PubMed]

71. Levy, R.; Biran, A.; Poirier, F.; Raz, A.; Kloog, Y. Galectin-3 mediates cross-talk between k-Ras and Let-7c tumor suppressor microRNA. *PLoS ONE* **2011**, *6*, e27490. [CrossRef] [PubMed]

International Journal of
Molecular Sciences

MDPI

Review

Galectin-7 in Epithelial Homeostasis and Carcinomas

Tamara Advedissian, Frédérique Deshayes and Mireille Viguier *

Team Morphogenesis, Homeostasis and Pathologies, Institut Jacques Monod, UMR 7592 CNRS—University Paris Diderot, Sorbonne Paris Cité, 15 rue Hélène Brion, 75013 Paris, France; tamara.advedissian@ijm.fr (T.A.); frederique.deshayes@ijm.fr (F.D.)
* Correspondence: mireille.viguier@univ-paris-diderot.fr

Received: 23 November 2017; Accepted: 14 December 2017; Published: 19 December 2017

Abstract: Galectins are small unglycosylated soluble lectins distributed both inside and outside the cells. They share a conserved domain for the recognition of carbohydrates (CRD). Although galectins have a common affinity for β-galatosides, they exhibit different binding preferences for complex glycans. First described twenty years ago, galectin-7 is a prototypic galectin, with a single CRD, able to form divalent homodimers. This lectin, which is mainly expressed in stratified epithelia, has been described in epithelial tissues as being involved in apoptotic responses, in proliferation and differentiation but also in cell adhesion and migration. Most members of the galectins family have been associated with cancer biology. One of the main functions of galectins in cancer is their immunomodulating potential and anti-angiogenic activity. Indeed, galectin-1 and -3, are already targeted in clinical trials. Another relevant function of galectins in tumour progression is their ability to regulate cell migration and cell adhesion. Among these galectins, galectin-7 is abnormally expressed in various cancers, most prominently in carcinomas, and is involved in cancer progression and metastasis but its precise functions in tumour biology remain poorly understood. In this issue, we will focus on the physiological functions of galectin-7 in epithelia and present the alterations of galectin-7 expression in carcinomas with the aim to describe its possible functions in tumour progression.

Keywords: galectins; galectin-7; epithelia; carcinoma

---

1. Introduction

Galectins are a family of soluble lectins, which possess a large variety of ligands and functions. Among the galectins, galectin-7 presents a unique tissue-specific expression pattern and participates in diverse biological processes, notably in the regulation of epithelial homeostasis. In this review, after discussing general information about the galectin family, we will present galectin-7 expression profile and structure. Then, we will focus on the role of galectin-7 in epithelial homeostasis, in cell adhesion and migration by presenting the results obtained both in animal models and cell lines. Finally, we will address the association of galectin-7 with carcinoma and its putative function in cancer progression.

1.1. Galectins

Galectins were identified in the 1970s [1,2] and formerly named S-type lectins, due to their solubility and the sulfhydryl-dependency of the first galectins discovered [3], but their nomenclature became systematic in 1994. Since then, they were ranked according to their order of discovery [4]. Several galectins are expressed in the same species with up to 16 galectins identified in mammals and 12 in humans [5–7]. Galectins are a family of proteins characterized by a common affinity for β-galactoside containing carbohydrates and an evolutionary conserved Carbohydrate Recognition Domain (CRD) [4]. The different galectins do not possess any signal peptide or any anchoring domain and are synthesized by the free polysomes in the cytosol [7]. The unique exception is galectin-3, which possess a NES (Nuclear Export Signal) [8] and a NLS (Nuclear Localization Signal)-like motif with similarities with the NLS of p53 and c-Myc [9]. The galectins are secreted by an unconventional pathway and thus can be localized in the extracellular compartment [10,11]. However, they are also found in the cytosol, in the nucleus, or even in the mitochondria [11,12].

Moreover, while some galectins such as galectin-1 and galectin-3 are widely expressed, other family members have a more restricted tissue localisation. Hence, galectin-2 expression is limited to digestive epithelia [13] and galectin-7 is preferentially expressed in stratified epithelia [14].

Galectin sequences are similar from the lower invertebrates to mammals. The common basic structure of the galectin domain is composed of about 130 amino acids organised in two β-sheets containing five (F1–F5) and six (S1–S6) anti-parallel β-strands forming a jellyroll topology (see galectin-7 structure in Figure 1a) [4,15]. Seven carbohydrate-binding amino acids in strands S4, S5 and S6 are essential for the specific binding of β-galatosides [16] and are highly conserved among galectins [17,18]. These amino acids are encoded by three consecutive exons in mammalian galectins and form a characteristic sequence of galectins called the CRD [4].

According to the structural organization of their CRD, galectins can be classified into 3 subtypes. Hence, "proto-type" galectins, are composed of a single galectin domain that is able to dimerize (galectin-1, -2, -5, -7, -10, -11, -13, -14, and -15) whereas "tandem repeat-type" galectins possess a single polypeptide chain with two CRDs connected by a linker peptide (galectin-4, -6, -8, -9, and -12). The "chimera-type" subtype, with galectin-3 being the unique member, is constituted of one C-terminal CRD linked to a N-terminal non-lectinic domain [19,20].

Another classification of galectins, which is based on the determination of two CRD subtypes, also exists and refers to the evolution of galectins. The CRDs are thus defined according to the relative position of intron/exon corresponding to the sequence of the F4 or F3 β-sheet. Indeed, among the 3 exons encoding the CRD, there are two subtypes of the second exon, also called "W" exon because of the presence of a highly conserved tryptophan residue. One of the "W" exon ends within the sequence encoding the F4 β-strand and the other ends within the sequence encoding the F3 β-strand. These two subtypes have been called respectively F4-CRD and F3-CRD. Prototype galectins can belong to the F4-CRD subtype (e.g., galectin-7, galectin-10) or to the F3-CRD subtype (e.g., galectin-1, galectin-2) and galectin-3 contains a F3-CRD. The tandem-repeat galectins contain both a F4-CRD and a F3-CRD subtypes [7,20].

Thanks to their CRD, galectins recognize oligosaccharides present in proteins, lipids or microbial molecules. The minimal ligand recognized by galectins is *N*-Acetyl-Lactosamine (LacNAc), a disaccharide found on both *N*- and *O*-glycans [7,21]. However, galectins have a selective affinity for sugars with complex organisation according to their structure and composition: amount of branching, of LacNAc repeats (poly-LacNAc) or the presence of terminal saccharides such as sialic acid or fucose [21–23]. These differences enable the specific affinity of a given galectin for its ligand. In general, the affinity of galectins for complex carbohydrates increases with the number of LacNAc repeats and the number of branches. Hence the major ligands of galectins are *N*-glycans [21,24].

As a consequence, galectins can bind multiple glycosylated partners, either glycoproteins or glycolipids. Due to their multivalence, they form networks of molecules termed "lattice " [25,26]. In addition to binding to glycans on glycoconjugates, galectins interact with unglycosylated intracellular but also extracellular ligands. As an illustration, galectin-1 has been shown to interact directly with the pseudo-light chain λ5 of the pre-BCR (B Cell Receptor) [27] and galectin-7 with Bcl-2 [12] or E-cadherin [28].

Due to their diversity of localisation and their various partners, the different members of the galectin family display a striking functional diversification. In particular, they are involved in intracellular trafficking as well as in cell adhesion and cell migration, in the regulation of the immune system or even in mRNA splicing [11]. Galectins can also affect cell signalling and impact development and tissue homeostasis leading to the emergence of pathologies such as cancer [7,25,29].

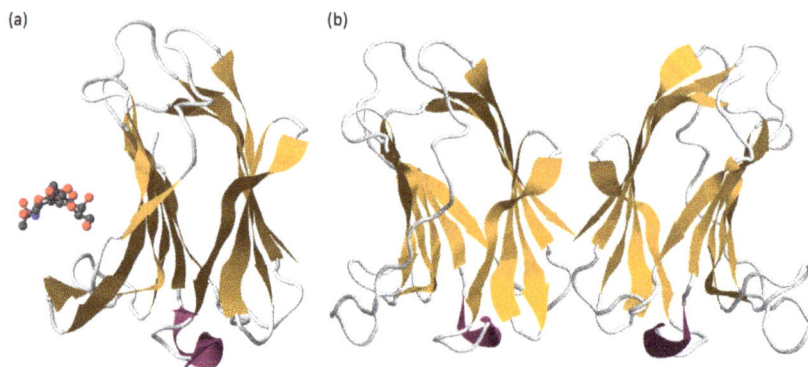

Figure 1. (**a**) Representation of the crystal structure of galectin-7 in complex with LacNAc (PBD 4XBQ) [30]. Carbohydrates are recognized by the residues H49, N51, R53, N62, W69, E72 and R74 forming the CRD. (**b**) Crystal structure of homodimeric galectin-7 (PBD 1BKZ) [18] illustrating the "back-to-back" arrangement of galectin-7 dimers with the two CRD orientated in the opposite direction. Structures obtained from www.rcsb.org.

1.2. Galectin-7

The galectin-7 gene was simultaneously discovered by Madsen and colleagues [31] and Magnaldo and colleagues [32] to be expressed in the epidermis, respectively as a protein repressed during Simian Virus-40 (SV-40)-mediated transformation of keratinocytes and as a protein differentially expressed in normal keratinocytes and in squamous cancer cells which failed to complete terminal keratinocyte differentiation. This lectin is mostly expressed in stratified epithelia notably in the epidermis (where galectin-7 is found both in interfollicular region and in hair follicles), the oesophagus, the tongue, the anus, the lips and the cornea [14,33,34]. However, galectin-7 expression has also been described in thymic Hassall's corpuscles [14], in sebaceous glands [33] and in myoepithelial cells from the mammary epithelia [35]. Galectin-7 has also recently been detected in the gingiva [36]. Importantly, its expression is induced by p53 and Ultra-Violet B (UVB) light [37]. Galectin-7 is secreted by keratinocytes into their culture medium despite the fact that, as all the galectins, it does not possess a typical secretion signal peptide [31]. However, galectin-7 is also found in the cytosol, in mitochondria and the nucleus, but its function in the nucleus is largely unknown. Diverse studies suggest an intracellular function of galectin-7 as for example in the regulation of keratinocyte proliferation and differentiation through the c-Jun N-terminal Kinase (JNK1)–miR-203-p63 pathway (see below) [38]. Galectin-7 expression can be induced via p53 or TNFα activation pathways, and both wild type and p53 mutants harbouring "hot spot" point mutation bind the galectin-7 promoter [39]. In addition, Nuclear Factor-kappa B (NF-κB) also binds to the galectin-7 promoter [39].

Galectin-7 is widely present in mammals and only one orthologue has been described outside of the mammalian lineage in anol lizards [40]. Interestingly, a copy-number variation has been pointed out for galectin-7 for which a single copy of two genes has been identified notably in the human, cow and dog genome [40]. The *LGALS7* and *LGALS7B* genes have been duplicated in tandem but in opposite direction and are found in chromosome 19 in humans [40]. Both genes encode identical galectin-7 protein but exhibit different putative transcription factors binding sites in their promoter sequence suggesting differences in expression regulation [40]. It has been hypothesized that galectin-7 could come from a duplication of galectin-4 [20] which is present in its neighbourhoods as a single copy.

Galectin-7, as other prototypic galectins, is able to form homodimers but with a different topology. Indeed, despite sequence homologies with other prototypic galectins such as galectin-1 or galectin-2 that associate in dimer in a "side-by-side" organisation, galectin-7 form homodimer through a "back-to-back" arrangement giving rise to a larger dimer interface compared to other prototypic galectins (Figure 1b) [18,41]. This difference in structural arrangement suggests that the glycoconjugate bridging activity of galectin-7 may differ from other prototypic galectins. The substitution at position 74 of an arginine by a serine inhibits the carbohydrates-binding activity of galectin-7 but does not alter the capacity of galectin-7 to form homodimers in solution [42]. This indicates that binding to oligosaccharides is not required for galectin-7 to form homodimers even if it can slightly modify its conformation and influence the dimers' stability [43,44]. Regarding carbohydrate binding, galectin-7 displays preferential binding to internal or terminal LacNAc repeat carried by *N*-glycan (Figure 1a) [21,22].

Multiple cellular functions have been attributed to galectin-7, most of which are related to epithelial integrity maintenance and will be described below. However, the established and putative cellular functions of galectin-7 are summarized in Figure 2.

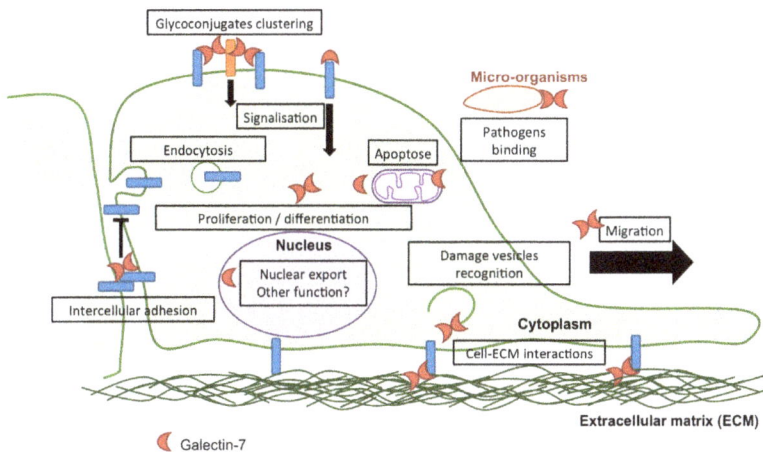

Figure 2. Schematic representation of known and putative functions of galectin-7. In addition to the functions of galectin-7 in cell proliferation, apoptosis, differentiation, migration and adhesion described in this issue, few evidence highlights other functions of galectin-7 in epithelia. As an illustration, galectin-7 has been shown to interfere with Transforming Growth factor β (TGFβ) signalling in response to Hepatocyte Growth Factor (HGF) by promoting smad3 export from the nucleus and thus preventing liver fibrosis occurrence [45]. In addition, the commensal bacteria *Finegoldia magna* has been described to adhere to the upper layers of the epidermis through binding of the adhesion bacterial protein *F. magna* Adhesion Factor (FAF) to galectin-7, indicating that galectin-7 can bind to ligands from microbial origin [46].

2. Galectin-7 in Epithelial Homeostasis

Galectin-7 participates in epithelial maintenance by regulating at least three key aspects of epithelia homeostasis: cell growth, cell differentiation and apoptosis. Nevertheless, the precise mechanisms by which galectin-7 participate in the regulation of these processes still remain to be decrypted more deeply.

2.1. Apoptosis

Several studies have revealed a role of galectin-7 in the apoptotic response (Figure 2) [34]. However, depending on experimental conditions, galectin-7 has been shown to be either a pro-apoptotic factor or an anti-apoptotic factor, indicating that galectin-7 activity in apoptosis varies according to the cellular context and/or the apoptotic stimulus. First, it was discovered, in the epidermis, that UVB-induced sunburns increase galectin-7 expression in keratinocytes from human skin ex vivo [37]. Remarkably, overexpression of galectin-7 occurs in apoptotic keratinocytes, highlighting a possible association [37]. This has been demonstrated using mouse models lacking or overexpressing galectin-7 in the epidermis in which both absence and excess of galectin-7 modify the kinetics of the apoptosis response to UVB irradiation and induce premature apoptotic response [47,48], pointing out the involvement of galectin-7 in the apoptosis process in vivo.

Addition of recombinant galectin-7 in absence of apoptotic stimuli is sufficient to induce apoptosis in the T lymphocyte Jurkat cell line [41,49,50] and in freshly isolated human T cells [50], as previously shown for other galectins [51]. Apoptosis induction by galectin-7 in Jurkat cells can be inhibited by lactose addition, indicating that this function of galectin-7 relies on its lectin activity [50]. However, in other cell types, addition of recombinant galectin-7 [50] or alterations of galectin-7 expression levels alone [52,53] are not sufficient to induce apoptosis indicating that direct induction of apoptosis by galectin-7 is restricted to T lymphocytes.

To investigate the function of galectin-7 in apoptosis, most researchers induce the ectopic expression of galectin-7 in diverse cancer cell lines and examine the sensitivity of the cells to apoptotic stimuli. As an illustration, de novo expression of galectin-7 in the cervical cancer HeLa cells and in the colorectal adenocarcinoma DLD-1 cell line makes these cells more sensitive to the induction of apoptosis by UVB irradiation or diverse chemical apoptotic stimuli [12,54–56]. Similarly, overexpression of galectin-7 in ST88-14 cells, a sarcoma-derived cell line, in the cervical carcinoma siHa cells or in the prostate cancer cells DU-145 results in an increased susceptibility of the cells to apoptotic stimuli [56–58]. Accordingly, galectin-7 downregulation in the cervical squamous carcinoma cells SiHa and C-33A increases cell viability in response to the apoptosis-inducing chemotherapeutic agent paclitaxel [59]. All these studies indicate that galectin-7 has a pro-apoptotic effect in many cell types. However, in the breast MCF-7 cancer cells or in the B16F1 melanoma cell line, ectopic expression of galectin-7 decreases the cell sensitivity to apoptotic stimuli [42,60], indicating that galectin-7 can also, contrastingly, have an anti-apoptotic effect.

Interestingly, the function of galectin-7 in apoptosis does not rely on its lectin activity and is predominantly intracellular. Indeed, St-Pierre and colleagues have shown that the de novo expression of a CRD-defective galectin-7 mutant harbouring a substitution of an arginine by a serine at position 74 (R74S mutant) had a similar effect to the expression of the wild type galectin-7 on apoptosis susceptibility. These results were observed in both DU-145 prostate cancer cells where galectin-7 has a pro-apoptotic effect [58] and in MCF-7 breast cancer cells where galectin-7 had an anti-apoptotic function [42]. Nevertheless, how galectin-7 regulates apoptosis is still unclear. Kuwabara et al., reported that galectin-7-overexpressing cells exhibit increased cytochrome c release and amplified JNK activation after apoptosis stimulation indicating that galectin-7 acts upstream of these two pathways [54]. The mechanism by which galectin-7 participates in apoptosis could also be linked to its interaction with the anti-apoptotic factor Bcl-2. Indeed, galectin-7 directly interacts with Bcl-2 in a carbohydrate-independent manner [12]. In accordance, overexpression of galectin-7 increases cell apoptosis in response to a specific Bcl-2 inhibitor [56]. Remarkably, the R74S galectin-7 mutant, which localizes far less efficiently to the mitochondria, still contributes to apoptosis regulation [58], suggesting that galectin-7 may also function outside of the mitochondria.

2.2. Proliferation

Studies performed on diverse cell types, mostly cancerous cell lines, have demonstrated that galectin-7 has a suppressive effect on cell proliferation (Figure 1). Indeed, ectopic expression or addition of exogenous galectin-7 in the DLD-1 human colon carcinoma cell line [55] and the neuroblastoma cells SK-N-MC, respectively [53], drastically reduced tumour cell proliferation. Consistently, galectin-7 knockdown in human keratinocytes results in a hyperproliferative phenotype [38]. However, ectopic expression of galectin-7 in the B16F1 melanoma cell line did not affect cell growth [60] indicating that cell context might be important for galectin-7 to modulate cell proliferation. Evidence obtained in vivo in galectin-7-null mice indicates that galectin-7 is also involved in the regulation of cell growth during stress responses. Indeed, galectin-7 deficiency induces enhanced cell proliferation after epidermal injury or UVB irradiation of the skin [47].

The molecular mechanism by which galectin-7 participates in apoptosis and cell proliferation remain to be clarified. However, galectin-7 could be an effector of the tumour suppressor gene *p53*. Strikingly, galectin-7 expression is strongly induced by p53 [61] and lack of wild type p53 in human keratinocytes cell lines prevents galectin-7 expression induction in response to UVB irradiation [37].

2.3. Differentiation

Both proliferating basal and quiescent differentiated suprabasal keratinocytes express and secrete galectin-7 [14,31,32]. As a consequence, galectin-7 was described as a marker of stratified epithelia but not as a marker of differentiation. However, some evidence has suggested a role of galectin-7 in keratinocyte differentiation (Figure 2) such as its reduced expression after addition of retinoic acid in cultured keratinocytes [32]. In addition, in keratinocytes cultured in vitro, galectin-7 mRNA expression increases with the cell density, suggesting a potential link with epidermal differentiation [62]. Regarding tumour biology, galectin-7 downregulation correlates with poor tumour differentiation in bladder squamous cell carcinomas [63] and in vulvar squamous cell carcinoma [64]. Recently, Liu and colleagues produced the first mechanistic evidence of a function of galectin-7 in keratinocyte differentiation in vitro [38]. In fact, in the keratinocyte cell line HaCaT, galectin-7 knockdown reduces cell differentiation as assessed by the expression of keratins. Moreover, they found that intracellular galectin-7 regulates keratinocyte differentiation through the JNK–miR-203-p63 pathway [38]. Indeed, their results indicate that galectin-7 interacts with JNK1 and prevents its ubiquitination and subsequent degradation by the proteasome. Both galectin-7 and JNK1 induce miR-203 expression and the subsequent inhibition of the transcription factor p63, an important regulator of keratinocyte proliferation and differentiation [65].

3. Galectin-7 in Adhesion and Migration

In this section, we will describe current understanding of galectin-7 function in two interconnected mechanisms: cell migration and adhesion (Figure 2). Both processes are required for tissue maintenance and are central in tumour biology. As a consequence, modification of cell migration and adhesion characteristics can promote cancer progression.

3.1. Adhesion

Several galectins have been implicated in cell–cell or cell–ECM (Extra-Cellular Matrix) adhesion and thus, galectins are considered as a family of adhesion-modulating proteins [66]. However, depending on conditions (i.e., galectin considered, galectin concentration or cell types) galectins can either favour or prevent interactions with either the substrate or the neighbouring cells. Regarding galectin-7, its subcellular localisation is enriched at cell-cell contacts in the suprabasal layers of human and mouse epidermis [31,47]. However, the potential role of galectin-7 in intercellular adhesion is poorly documented. Recently, a few studies came out indicating that galectin-7 mediates cell-cell adhesion. Indeed, in the uterus, Menkhorst and colleagues showed that galectin-7 is expressed in the endometrium and influenced trophoblast-endometrial epithelia intercellular adhesion in vitro [67]. This function of galectin-7 in cell-cell adhesion may have a crucial impact during embryo implantation.

In addition, our team previously showed that galectin-7 interacts with the adherent junctions-component E-cadherin in keratinocytes [48]. Interestingly, we recently demonstrated that galectin-7 directly binds to the extracellular domain of E-cadherin in a glycosylation-independent manner [28]. This interaction has a functional consequence on intercellular adhesion as galectin-7 knockdown in HaCaT keratinocytes importantly reduces adherent junction-mediated adhesion [28]. Focusing on the underlying mechanisms, we demonstrated that galectin-7 stabilises E-cadherin at the plasma membrane, preventing its endocytosis [28]. Interestingly, both galectin-7-null mice and galectin-7-overexpressing mice show intercellular adhesion defects in the epidermis [48]. These results are compatible with the current model for the regulation of adhesion by galectins. In this model, low concentration of galectins will promote bridging of molecules by bi- or multivalent galectins and favour adhesion. On the contrary, high amount of galectins will reduce their crosslinking properties by decreasing the probability of simultaneous binding to two or more ligands and thus decreasing adhesion [66]. As a consequence, absence or excess of galectins could have the same consequences.

Regarding the regulation of the adhesion to the ECM, a possible interaction between galectin-7 and β1-integrin has been suggested in polarised MDCK cells [68]. Consistently, the enhanced endometrial wound repair induced by addition of exogenous galectin-7 is prevented by the blockade of integrin–fibronectin interaction in vitro [69]. However, further investigations are needed to specify the potential link between galectin-7 and integrins. Finally, galectin-7 could also participate in cell-ECM adhesion by influencing Matrix Metallo-Proteinase proteins (MMP) expression. Indeed, in lymphoma cells or in HeLa cells, exogenously added galectin-7 was able to enhance MMP-9 expression, suggesting a potential role for galectin-7 in the regulation of cell–ECM adhesion during cancer dissemination [70,71]. Galectin-7 and MMP-9 also showed positive expression correlation in human hypopharyngeal and laryngeal squamous cell carcinomas [72].

3.2. Migration

Galectin-7 has first been found to be involved in cell migration during epithelial wound healing in mouse corneas where addition of exogenous galectin-7 accelerated re-epithelialisation after corneal injury [73,74]. Simultaneous addition of lactose with exogenous galectin-7 prevented the increase of healing rate due to galectin-7 supplement [73,74], suggesting that this function of galectin-7 in corneal healing might be dependent on its binding to extracellular glycoconjugates. In mice corneas and porcine skin epidermis, galectin-7 expression is increased following injury [73–76], indicating that galectin-7 expression can be induced under stress conditions such as injury occurrence. Then, studies performed in galectin-7-null mice revealed that galectin-7 is similarly involved in keratinocyte migration during epidermal wound healing [47]. These mice displayed re-epithelialisation delay when compared to Wild Type (WT) mice after tail superficial scratch. This role of galectin-7 in epidermal wound healing is independent of its function in cell growth regulation but is related to a reduced migratory potential of keratinocytes. Indeed, the delay in wound healing was observed as soon as 24h h after injury, whereas no difference in cell proliferation was observed at this period in galectin-7-null mice compared to WT mice. In addition, in the presence of the cell division inhibitor mitomycin, the delay in wound healing was still observed in newborns' skin explants from galectin-7-deficient mice compared to WT mice [47]. Surprisingly, overexpression of galectin-7 in mice epidermis also delayed wound closure after superficial epidermal injury [48], indicating that an optimal amount of galectin-7 is required for proper keratinocyte migration. Focusing on the underlying mechanisms, we have recently showed in vitro that galectin-7-depleted keratinocytes (HaCaT cells) have a reduced cell migration speed but also an impaired collective behaviour resulting in a decreased migration efficiency [28]. These alterations have been hypothesized to be related to the defective adherent junction functioning after galectin-7 silencing [28]. A role of galectin-7 in endometrial epithelial wound repair has also been highlighted by in vitro assays [69].

The function of galectin-7 in collective cell migration is relevant in pathological conditions such as cancer progression [34]. Indeed, in epithelial cancer, invasion processes of the surrounding healthy tissue by tumour cells frequently exhibit collective invasion reminiscent of regenerative migration of epithelial cells [77,78]. Accordingly, several groups reported an association between galectin-7 expression levels and cancer aggressiveness as we will discuss in the following section.

4. Galectin-7 and Carcinomas

Galectins are widely studied in cancer with several hundred references. Most members have been implicated as being associated either as markers, diagnostic cues, candidate effectors in cancer progression or modulators of treatment responses [29,79–84]. Interestingly, several galectins are usually co-expressed by a single cell or tumour and may exert ill-defined synergistic or compensatory properties.

4.1. Galectin-7 as a Tumour Progression Marker

Among these galectins, galectin-7, being expressed mostly in stratified epithelia, has been mainly studied in carcinomas and its implications in cancer were referenced only in a few tens of published articles. Among those, even fewer studies explored the ectopic expression of galectin-7 in tumours from non-epithelial origins such as lymphoid or melanomas tumours.

In lymphomas, ectopic expression of galectin-7 was shown to correlate with the metastatic potential of transplanted lymphomas cell lines [83,85–87]. Further observations in human lymphoid diseases suggested correlations between tumour progression and accumulation of galectin-7 [86] while no expression was detected in normal tissues.

Studies were also conducted to investigate the role of galectin-7 in melanoma. Indeed, tissue analysis on human melanomas and nevi first showed an expression of galectin-7 [88] but in situ labelling of nevi biopsies suggested that they were mostly the keratinocytes subpopulation of the biopsies that expressed galectin-7 [60]. In addition, it has been found that when the melanoma cells B16F1 are injected subcutaneously into mice, galectin-7 can be expressed by the resulting primary tumour as well as in lung metastasis [60]. However, even if galectin-7 increased the resistance of melanoma cells to apoptosis, studies on the melanoma cell line B16F1 showed that galectin-7 ectopic expression did not impact tumour growth and metastasis occurrence when cells are injected into mice [60].

Contrastingly, one of the major skin cancers apart from melanomas, basal cell carcinoma, appeared to be devoid of galectin-7 expression [89] while galectin-7 is expressed in normal tissues. Other carcinomas are also associated with decreased galectin-7 expression such as stomach [90], urothelial [91,92] and cervix [59] cancer. On the contrary, squamous epithelial and mucous tumours from head and neck [93], oesophagus [94] and thyroid [95] cancer do express higher galectin-7 levels than normal tissues.

In breast cancer, the earliest report on galectin-7 being associated to oncogenesis is from a chemically induced mammary tumour in rat which was found to overexpress galectin-7 contrastingly to chemically induced colon carcinomas which express a reduced level of galectin-7 [96]. Later on, galectin-7 was scored among the differentially expressed gene on human breast cancer [97]. Both galectin-3 and galectin-7 were associated with mammary tumour progression with galectin-7 being correlated to pejorative diagnosis when accumulated while reduced galectin-3 was of better prognosis [98]. Interestingly, these two galectins are mutually exclusive in mammary epithelia with galectin-7 being a marker of myoepithelial cells and luminal cells being galectin-3 positive [35]. The myoepithelial cells exhibit a basal type phenotype with a proliferative capacity [99,100]. Most importantly, galectin-7 is overexpressed in basal-like breast tumours [97,101]. Recently, using a mouse model for ErbB2 mammary tumours, galectin-7 expression was reported to accelerate tumour progression and the formation of tumour nodules in ErbB2-positive tumours [102]. On the contrary, using mouse model knocked-out for galectin-3 expression, no correlation was found with breast tumour progression or metastasis induced by transgenic expression of Mouse Mammary Tumour Virus- Polyoma Middle T (MMTV-PyMT) oncogene [103].

Furthermore, overexpression of galectin-7 was associated to metastatic potential of various tumour tissues, notably in breast cancer [101]. In oral squamous cell carcinoma (OSCC), galectin-7 has been associated with the histological malignancy grading system [104,105]. Consistently, in OSCC cell lines, galectin-7 downregulation reduces cell migration and invasion whereas its overexpression enhances cell migration and invasion of cancer cells [52]. Similarly to OSCC, in epithelial ovarian cancer, various studies described a correlation between galectin-7 ectopic expression in this monostratified epithelia and overall patient survival [50,106]. In addition, de novo galectin-7 expression was associated with progression to high-grade tumours and was enriched in metastatic samples compare to low-grade tumours [50]. On the contrary, in colon cancer, galectin-7 ectopic overexpression prevented metastatic dissemination [55] and promoted apoptosis after apoptosis induction by stimuli [54].

4.1. Galectin-7 as a Therapeutic Tool

The association of galectins with cancer [29,107] may be related to the well-known alterations in glycosylation signatures of cancer cells that may have incidental implications on galectins reactivity [108]. Indeed, during tumour progression, different factors affect protein glycosylation leading to altered glycan structure and composition as the appearance of truncated glycosylation motifs or enrichment of polyLacNAc [108–110]. This modification of galectin ligands will necessarily affect galectin activity and functions and can impact cell growth, adhesion, immunomodulation and cell migration [109]. As an illustration, modification of the surface glycome regulates galectin-1 binding to endothelial cells, and more importantly to VEGFR2, thus affecting angiogenesis and response to anti-VEGF treatments [111]. Among carcinomas, galectins-1 and -3 are the most studied and are already targeted in clinical trials [107,111,112]. Interestingly, immunomodulating potential is one of the most powerful actions of these galectins in cancer. Galectin-1 also has a relevant anti-angiogenic potential [113]. Another relevant function of galectins in cancer biology is cell migration and particularly adhesion to the MEC and to other cells (including surrounding or endothelial cells) [79]. As a consequence, galectin-7 appeared in the dedicated studies as a potential target for clinical approaches either as a prognostic marker, a direct modulating protein in cancer or a therapeutic target for inhibitors.

Indeed, galectin-7 has been proposed to serve as a marker in patients with certain type of cancer. Interestingly, galectin-7 overexpression on tumour tissues could be correlated to patient survival notably in ovarian [114] or in cervical [59,115] carcinomas. Furthermore, galectin-7 expression has been associated with better survival after chemo- or radio-therapies [91,115] and has been proposed to serve as a predictive marker of chemo- and radio-therapy resistance [116].

Administration of pure recombinant galectin-7 could be efficient in immunosuppressive approaches targeting T lymphocytes but contrastingly it could enhance invasive potential of tumour cells [50]. Soluble exogenously added galectin-7 has been used successfully to repair corneal injury in mice cornea ex vivo [73,74]. In a recent article, St-Pierre and colleagues demonstrated that recombinant galectin-7 added on tumour cells may rapidly traffic through intracellular compartments including endocytic vesicle and mitochondria and could induce ectopic expression of galectin-7 [117]. These new coming results reinforce the rationale of using galectin-7 on tumour that would benefit from its induced expression, as for example colon cancer [54,55]. On the contrary, using inhibitors of galectin-7 has to be tested in tumours where its expression is associated with pejorative prognosis.

Moreover, development of specific galectin-7 inhibitors that will selectively target the intracellular or extracellular functions of galectin-7 could be a strategy to inhibit not all but specific galectin-7-mediated processes [118]. Interestingly, inhibition of the intracellular lectin-activity of galectin-7 could be easily assessed. As previously demonstrated for galectins-1, -3, -8 and -9 [119], galectin-7 is able to detect damages to vesicles (Figures 2 and 3). Indeed, addition of a lysosomal-damaging agent induces intracellular recruitment of galectin-7 to permeabilized lysosomes indicating that this lectin recognizes altered vesicles (Figure 3). This feature shared by some galectins allows to identify galectin inhibitors that penetrate the cells and target the CRD domain of a given galectin, providing a functional assay to assess drug penetration and specificity [120,121]. Hence, some galectins are efficient to detect damaged vesicular compartment by binding to glycoproteins abnormally exposed in the cytoplasm due to vesicle integrity failure [119]. This approach is particularly relevant in the biomedical area where knowing the precise way of action of a chemical compound is a benefit.

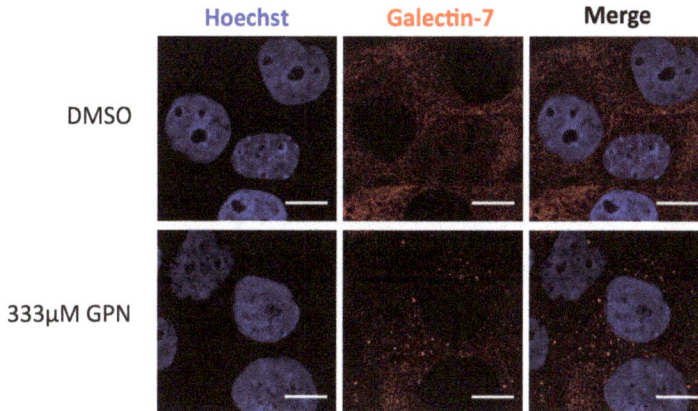

Figure 3. Galectin-7 is recruited at damaged lysosomes. Twelve minutes incubation with the lysosome-damaging agent GPN (glycyl-L-phenylalanine 2-naphthylamide) induces intracellular accumulation of galectin-7 at damaged lysosomes in HaCaT cells. Scale bar = 10 µm.

5. Conclusions

Galectin-7 is a prototypic galectin, which is preferentially found in stratified epithelia where it favours epithelial homeostasis. Though galectin-7-null mice or mice overexpressing galectin-7 in the epidermis are viable and fertile, they present defective responses to stress conditions. Indeed, acting intra- or extracellularly, galectin-7 participates in diverse processes such as the susceptibility to apoptosis, cell migration and cell-adhesion. However, the detailed mechanism of action of this tissue-specific galectin and its partners are mostly unknown. Thus, further work is required to uncover galectin-7 functions and ways of action. This is particularly important regarding the biomedical field because galectin-7 could influence tumour progression. Hence, specifying how galectin-7 influences a given process could help to design and predict the effects of galectin-7 inhibitors targeting diverse regions of the protein or generated recombinant mutant galectin-7 that retain only some of the functions of galectin-7.

Acknowledgments: This work was supported by FRM (Fondation pour la Recherche Médicale; DCM20121225750) and Fondation ARC (Association pour la Recherche contre le Cancer; PJA 20161204938) grants. We are thankful to Juliette Delafosse for English correction.

Conflicts of Interest: The authors declare no conflict of interest.

Abbreviations

BCR	B-Cell Receptor
CRD	Carbohydrate Recognition Domain
DMSO	Dimethyl Sulfoxide
FAF	*F. magna* Adhesion Factor
ECM	Extracellular Matrix
HGF	Hepatocyte Growth Factor
JNK	c-Jun N-terminal Kinase
LacNAc	*N*-Acetyl-Lactosamine
MMP	Matrix Metalloproteinase
MMTV-PyMT	Mouse Mammary Tumor Virus – Polyoma Middle T
NES	Nuclear Export Signal
NF-κB	Nuclear Factor kappa B
NLS	Nuclear Localisation Signal
OSCC	Oral Squamous Cell Carcinoma

SV-40	Simian Virus-40
TGFβ	Transforming Growth Factor β
UVB	Ultraviolet B
WT	Wild Type

References

1. Teichberg, V.I.; Silman, I.; Beitsch, D.D.; Resheff, G. A β-D-galactoside binding protein from electric organ tissue of Electrophorus electricus. *Proc. Natl. Acad. Sci. USA* **1975**, *72*, 1383–1387. [CrossRef] [PubMed]

2. De Waard, A.; Hickman, S.; Kornfeld, S. Isolation and properties of β-galactoside binding lectins of calf heart and lung. *J. Biol. Chem.* **1976**, *251*, 7581–7587. [PubMed]

3. Drickamer, K. Two distinct classes of carbohydrate-recognition domains in animal lectins. *J. Biol. Chem.* **1988**, *263*, 9557–9560. [PubMed]

4. Barondes, S.H.; Cooper, D.N.; Gitt, M.A.; Leffler, H. Galectins. Structure and function of a large family of animal lectins. *J. Biol. Chem.* **1994**, *269*, 20807–20810. [PubMed]

5. Than, N.G.; Romero, R.; Goodman, M.; Weckle, A.; Xing, J.; Dong, Z.; Xu, Y.; Tarquini, F.; Szilagyi, A.; Gal, P.; et al. A primate subfamily of galectins expressed at the maternal-fetal interface that promote immune cell death. *Proc. Natl. Acad. Sci. USA* **2009**, *106*, 9731–9736. [CrossRef] [PubMed]

6. Than, N.G.; Romero, R.; Xu, Y.; Erez, O.; Xu, Z.; Bhatti, G.; Leavitt, R.; Chung, T.H.; El-Azzamy, H.; LaJeunesse, C.; et al. Evolutionary origins of the placental expression of chromosome 19 cluster galectins and their complex dysregulation in preeclampsia. *Placenta* **2014**, *35*, 855–865. [CrossRef] [PubMed]

7. Cummings, R.D.; Liu, F.-T.; Vasta, G.R. Galectins. In *Essentials of Glycobiology*; Varki, A., Cummings, R.D., Esko, J.D., Stanley, P., Hart, G.W., Aebi, M., Darvill, A.G., Kinoshita, T., Packer, N.H., Prestegard, J.H., et al., Eds.; Cold Spring Harbor Laboratory Press: Cold Spring Harbor, NY, USA, 2015.

8. Li, S.-Y.; Davidson, P.J.; Lin, N.Y.; Patterson, R.J.; Wang, J.L.; Arnoys, E.J. Transport of galectin-3 between the nucleus and cytoplasm. II. Identification of the signal for nuclear export. *Glycobiology* **2006**, *16*, 612–622. [CrossRef] [PubMed]

9. Nakahara, S.; Hogan, V.; Inohara, H.; Raz, A. Importin-mediated nuclear translocation of galectin-3. *J. Biol. Chem.* **2006**, *281*, 39649–39659. [CrossRef] [PubMed]

10. Delacour, D.; Koch, A.; Jacob, R. The role of galectins in protein trafficking. *Traffic* **2009**, *10*, 1405–1413. [CrossRef] [PubMed]

11. Viguier, M.; Advedissian, T.; Delacour, D.; Poirier, F.; Deshayes, F. Galectins in epithelial functions. *Tissue Barriers* **2014**, *2*, e29103. [CrossRef] [PubMed]

12. Villeneuve, C.; Baricault, L.; Canelle, L.; Barboule, N.; Racca, C.; Monsarrat, B.; Magnaldo, T.; Larminat, F. Mitochondrial proteomic approach reveals galectin-7 as a novel BCL-2 binding protein in human cells. *Mol. Biol. Cell* **2011**, *22*, 999–1013. [CrossRef] [PubMed]

13. Oka, T.; Murakami, S.; Arata, Y.; Hirabayashi, J.; Kasai, K.; Wada, Y.; Futai, M. Identification and cloning of rat galectin-2: Expression is predominantly in epithelial cells of the stomach. *Arch. Biochem. Biophys.* **1999**, *361*, 195–201. [CrossRef] [PubMed]

14. Magnaldo, T.; Fowlis, D.; Darmon, M. Galectin-7, a marker of all types of stratified epithelia. *Differ. Res. Biol. Divers.* **1998**, *63*, 159–168. [CrossRef] [PubMed]

15. Rabinovich, G.A. Galectins: An evolutionarily conserved family of animal lectins with multifunctional properties; a trip from the gene to clinical therapy. *Cell Death Differ.* **1999**, *6*, 711–721. [CrossRef] [PubMed]

16. Lobsanov, Y.D.; Gitt, M.A.; Leffler, H.; Barondes, S.H.; Rini, J.M. X-ray crystal structure of the human dimeric S-Lac lectin, L-14-II, in complex with lactose at 2.9-A resolution. *J. Biol. Chem.* **1993**, *268*, 27034–27038. [PubMed]

17. Cooper, D.N.W. Galectinomics: Finding themes in complexity. *Biochim. Biophys. Acta* **2002**, *1572*, 209–231. [CrossRef]

18. Leonidas, D.D.; Vatzaki, E.H.; Vorum, H.; Celis, J.E.; Madsen, P.; Acharya, K.R. Structural basis for the recognition of carbohydrates by human galectin-7. *Biochemistry* **1998**, *37*, 13930–13940. [CrossRef] [PubMed]

19. Di Lella, S.; Sundblad, V.; Cerliani, J.P.; Guardia, C.M.; Estrin, D.A.; Vasta, G.R.; Rabinovich, G.A. When galectins recognize glycans: From biochemistry to physiology and back again. *Biochemistry (Moscow)* **2011**, *50*, 7842–7857. [CrossRef] [PubMed]

20. Houzelstein, D.; Gonçalves, I.R.; Fadden, A.J.; Sidhu, S.S.; Cooper, D.N.W.; Drickamer, K.; Leffler, H.; Poirier, F. Phylogenetic analysis of the vertebrate galectin family. *Mol. Biol. Evol.* **2004**, *21*, 1177–1187. [CrossRef] [PubMed]

21. Rabinovich, G.A.; Toscano, M.A. Turning "sweet" on immunity: Galectin-Glycan interactions in immune tolerance and inflammation. *Nat. Rev. Immunol.* **2009**, *9*, 338–352. [CrossRef] [PubMed]

22. Hirabayashi, J.; Hashidate, T.; Arata, Y.; Nishi, N.; Nakamura, T.; Hirashima, M.; Urashima, T.; Oka, T.; Futai, M.; Muller, W.E.G.; et al. Oligosaccharide specificity of galectins: A search by frontal affinity chromatography. *Biochim. Biophys. Acta* **2002**, *1572*, 232–254. [CrossRef]

23. Stowell, S.R.; Arthur, C.M.; Mehta, P.; Slanina, K.A.; Blixt, O.; Leffler, H.; Smith, D.F.; Cummings, R.D. Galectin-1, -2, and -3 exhibit differential recognition of sialylated glycans and blood group antigens. *J. Biol. Chem.* **2008**, *283*, 10109–10123. [CrossRef] [PubMed]

24. Boscher, C.; Dennis, J.W.; Nabi, I.R. Glycosylation, galectins and cellular signaling. *Curr. Opin. Cell Biol.* **2011**, *23*, 383–392. [CrossRef] [PubMed]

25. Rabinovich, G.A.; Toscano, M.A.; Jackson, S.S.; Vasta, G.R. Functions of cell surface galectin-glycoprotein lattices. *Curr. Opin. Struct. Biol.* **2007**, *17*, 513–520. [CrossRef] [PubMed]

26. Nabi, I.R.; Shankar, J.; Dennis, J.W. The galectin lattice at a glance. *J. Cell Sci.* **2015**, *128*, 2213–2219. [CrossRef] [PubMed]

27. Elantak, L.; Espeli, M.; Boned, A.; Bornet, O.; Bonzi, J.; Gauthier, L.; Feracci, M.; Roche, P.; Guerlesquin, F.; Schiff, C. Structural basis for galectin-1-dependent pre-B cell receptor (pre-BCR) activation. *J. Biol. Chem.* **2012**, *287*, 44703–44713. [CrossRef] [PubMed]

28. Advedissian, T.; Proux-Gillardeaux, V.; Nkosi, R.; Peyret, G.; Nguyen, T.; Poirier, F.; Viguier, M.; Deshayes, F. E-cadherin dynamics is regulated by galectin-7 at epithelial cell surface. *Sci. Rep.* **2017**, *7*, 17086. [CrossRef] [PubMed]

29. Liu, F.-T.; Rabinovich, G.A. Galectins as modulators of tumour progression. *Nat. Rev. Cancer* **2005**, *5*, 29–41. [CrossRef] [PubMed]

30. Hsieh, T.-J.; Lin, H.-Y.; Tu, Z.; Huang, B.-S.; Wu, S.-C.; Lin, C.-H. Structural Basis Underlying the Binding Preference of Human Galectins-1, -3 and -7 for Galβ1-3/4GlcNAc. *PLoS ONE* **2015**, *10*, e0125946. [CrossRef] [PubMed]

31. Madsen, P.; Rasmussen, H.H.; Flint, T.; Gromov, P.; Kruse, T.A.; Honoré, B.; Vorum, H.; Celis, J.E. Cloning, expression, and chromosome mapping of human galectin-7. *J. Biol. Chem.* **1995**, *270*, 5823–5829. [CrossRef] [PubMed]

32. Magnaldo, T.; Bernerd, F.; Darmon, M. Galectin-7, a human 14-kDa S-lectin, specifically expressed in keratinocytes and sensitive to retinoic acid. *Dev. Biol.* **1995**, *168*, 259–271. [CrossRef] [PubMed]

33. Nio-Kobayashi, J. Tissue- and cell-specific localization of galectins, β-galactose-binding animal lectins, and their potential functions in health and disease. *Anat. Sci. Int.* **2017**, *92*, 25–36. [CrossRef] [PubMed]

34. Saussez, S.; Kiss, R. Galectin-7. *Cell. Mol. Life Sci.* **2006**, *63*, 686–697. [CrossRef] [PubMed]

35. Jones, C.; Mackay, A.; Grigoriadis, A.; Cossu, A.; Reis-Filho, J.S.; Fulford, L.; Dexter, T.; Davies, S.; Bulmer, K.; Ford, E.; et al. Expression profiling of purified normal human luminal and myoepithelial breast cells: Identification of novel prognostic markers for breast cancer. *Cancer Res.* **2004**, *64*, 3037–3045. [CrossRef] [PubMed]

36. Yaprak, E.; Kasap, M.; Akpınar, G.; Kayaaltı-Yüksek, S.; Sinanoğlu, A.; Guzel, N.; Demirturk Kocasarac, H. The prominent proteins expressed in healthy gingiva: A pilot exploratory tissue proteomics study. *Odontology* **2017**. [CrossRef] [PubMed]

37. Bernerd, F.; Sarasin, A.; Magnaldo, T. Galectin-7 overexpression is associated with the apoptotic process in UVB-induced sunburn keratinocytes. *Proc. Natl. Acad. Sci. USA* **1999**, *96*, 11329–11334. [CrossRef] [PubMed]

38. Chen, H.-L.; Chiang, P.-C.; Lo, C.-H.; Lo, Y.-H.; Hsu, D.K.; Chen, H.-Y.; Liu, F.-T. Galectin-7 Regulates Keratinocyte Proliferation and Differentiation through JNK-miR-203-p63 Signaling. *J. Investig. Dermatol.* **2016**, *136*, 182–191. [CrossRef] [PubMed]

39. Campion, C.G.; Labrie, M.; Lavoie, G.; St-Pierre, Y. Expression of galectin-7 is induced in breast cancer cells by mutant p53. *PLoS ONE* **2013**, *8*, e72468. [CrossRef] [PubMed]

40. Kaltner, H.; Raschta, A.-S.; Manning, J.C.; Gabius, H.-J. Copy-number variation of functional galectin genes: Studying animal galectin-7 (p53-induced gene 1 in man) and tandem-repeat-type galectins-4 and -9. *Glycobiology* **2013**, *23*, 1152–1163. [CrossRef] [PubMed]

41. Vladoiu, M.C.; Labrie, M.; Létourneau, M.; Egesborg, P.; Gagné, D.; Billard, É.; Grosset, A.-A.; Doucet, N.; Chatenet, D.; St-Pierre, Y. Design of a peptidic inhibitor that targets the dimer interface of a prototypic galectin. *Oncotarget* **2015**, *6*, 40970–40980. [CrossRef] [PubMed]

42. Grosset, A.-A.; Labrie, M.; Gagné, D.; Vladoiu, M.-C.; Gaboury, L.; Doucet, N.; St-Pierre, Y. Cytosolic galectin-7 impairs p53 functions and induces chemoresistance in breast cancer cells. *BMC Cancer* **2014**, *14*, 801. [CrossRef] [PubMed]

43. Ramaswamy, S.; Sleiman, M.H.; Masuyer, G.; Arbez-Gindre, C.; Micha-Screttas, M.; Calogeropoulou, T.; Steele, B.R.; Acharya, K.R. Structural basis of multivalent galactose-based dendrimer recognition by human galectin-7. *FEBS J.* **2015**, *282*, 372–387. [CrossRef] [PubMed]

44. Ermakova, E.; Miller, M.C.; Nesmelova, I.V.; López-Merino, L.; Berbís, M.A.; Nesmelov, Y.; Tkachev, Y.V.; Lagartera, L.; Daragan, V.A.; André, S.; et al. Lactose binding to human galectin-7 (p53-induced gene 1) induces long-range effects through the protein resulting in increased dimer stability and evidence for positive cooperativity. *Glycobiology* **2013**, *23*, 508–523. [CrossRef] [PubMed]

45. Inagaki, Y.; Higashi, K.; Kushida, M.; Hong, Y.Y.; Nakao, S.; Higashiyama, R.; Moro, T.; Itoh, J.; Mikami, T.; Kimura, T.; et al. Hepatocyte growth factor suppresses profibrogenic signal transduction via nuclear export of Smad3 with galectin-7. *Gastroenterology* **2008**, *134*, 1180–1190. [CrossRef] [PubMed]

46. Murphy, E.C.; Mörgelin, M.; Reinhardt, D.P.; Olin, A.I.; Björck, L.; Frick, I.-M. Identification of molecular mechanisms used by Finegoldia magna to penetrate and colonize human skin. *Mol. Microbiol.* **2014**, *94*, 403–417. [CrossRef] [PubMed]

47. Gendronneau, G.; Sidhu, S.S.; Delacour, D.; Dang, T.; Calonne, C.; Houzelstein, D.; Magnaldo, T.; Poirier, F. Galectin-7 in the control of epidermal homeostasis after injury. *Mol. Biol. Cell* **2008**, *19*, 5541–5549. [CrossRef] [PubMed]

48. Gendronneau, G.; Sanii, S.; Dang, T.; Deshayes, F.; Delacour, D.; Pichard, E.; Advedissian, T.; Sidhu, S.S.; Viguier, M.; Magnaldo, T.; et al. Overexpression of galectin-7 in mouse epidermis leads to loss of cell junctions and defective skin repair. *PLoS ONE* **2015**, *10*, e0119031. [CrossRef] [PubMed]

49. Yamaguchi, T.; Hiromasa, K.; Kabashima-Kubo, R.; Yoshioka, M.; Nakamura, M. Galectin-7, induced by cis-urocanic acid and ultraviolet B irradiation, down-modulates cytokine production by T lymphocytes. *Exp. Dermatol.* **2013**, *22*, 840–842. [CrossRef] [PubMed]

50. Labrie, M.; Vladoiu, M.C.; Grosset, A.-A.; Gaboury, L.; St-Pierre, Y. Expression and functions of galectin-7 in ovarian cancer. *Oncotarget* **2014**, *5*, 7705–7721. [CrossRef] [PubMed]

51. Yang, R.-Y.; Rabinovich, G.A.; Liu, F.-T. Galectins: Structure, function and therapeutic potential. *Expert Rev. Mol. Med.* **2008**, *10*, e17. [CrossRef] [PubMed]

52. Guo, J.-P.; Li, X.-G. Galectin-7 promotes the invasiveness of human oral squamous cell carcinoma cells via activation of ERK and JNK signaling. *Oncol. Lett.* **2017**, *13*, 1919–1924. [CrossRef] [PubMed]

53. Kopitz, J.; André, S.; von Reitzenstein, C.; Versluis, K.; Kaltner, H.; Pieters, R.J.; Wasano, K.; Kuwabara, I.; Liu, F.-T.; Cantz, M.; et al. Homodimeric galectin-7 (p53-induced gene 1) is a negative growth regulator for human neuroblastoma cells. *Oncogene* **2003**, *22*, 6277–6288. [CrossRef] [PubMed]

54. Kuwabara, I.; Kuwabara, Y.; Yang, R.-Y.; Schuler, M.; Green, D.R.; Zuraw, B.L.; Hsu, D.K.; Liu, F.-T. Galectin-7 (PIG1) exhibits pro-apoptotic function through JNK activation and mitochondrial cytochrome c release. *J. Biol. Chem.* **2002**, *277*, 3487–3497. [CrossRef] [PubMed]

55. Ueda, S.; Kuwabara, I.; Liu, F.-T. Suppression of tumor growth by galectin-7 gene transfer. *Cancer Res.* **2004**, *64*, 5672–5676. [CrossRef] [PubMed]

56. Higareda-Almaraz, J.C.; Ruiz-Moreno, J.S.; Klimentova, J.; Barbieri, D.; Salvador-Gallego, R.; Ly, R.; Valtierra-Gutierrez, I.A.; Dinsart, C.; Rabinovich, G.A.; Stulik, J.; et al. Systems-level effects of ectopic galectin-7 reconstitution in cervical cancer and its microenvironment. *BMC Cancer* **2016**, *16*, 680. [CrossRef] [PubMed]

57. Barkan, B.; Cox, A.D.; Kloog, Y. Ras inhibition boosts galectin-7 at the expense of galectin-1 to sensitize cells to apoptosis. *Oncotarget* **2013**, *4*, 256–268. [CrossRef] [PubMed]

58. Labrie, M.; Vladoiu, M.; Leclerc, B.G.; Grosset, A.-A.; Gaboury, L.; Stagg, J.; St-Pierre, Y. A Mutation in the Carbohydrate Recognition Domain Drives a Phenotypic Switch in the Role of Galectin-7 in Prostate Cancer. *PLoS ONE* **2015**, *10*, e0131307. [CrossRef] [PubMed]

59. Zhu, H.; Wu, T.-C.; Chen, W.-Q.; Zhou, L.-J.; Wu, Y.; Zeng, L.; Pei, H.-P. Roles of galectin-7 and S100A9 in cervical squamous carcinoma: Clinicopathological and in vitro evidence. *Int. J. Cancer* **2013**, *132*, 1051–1059. [CrossRef] [PubMed]

60. Biron-Pain, K.; Grosset, A.-A.; Poirier, F.; Gaboury, L.; St-Pierre, Y. Expression and functions of galectin-7 in human and murine melanomas. *PLoS ONE* **2013**, *8*, e63307. [CrossRef] [PubMed]

61. Polyak, K.; Xia, Y.; Zweier, J.L.; Kinzler, K.W.; Vogelstein, B. A model for p53-induced apoptosis. *Nature* **1997**, *389*, 300–305. [CrossRef] [PubMed]

62. Sarafian, V.; Jans, R.; Poumay, Y. Expression of lysosome-associated membrane protein 1 (Lamp-1) and galectins in human keratinocytes is regulated by differentiation. *Arch. Dermatol. Res.* **2006**, *298*, 73–81. [CrossRef] [PubMed]

63. Ostergaard, M.; Rasmussen, H.H.; Nielsen, H.V.; Vorum, H.; Orntoft, T.F.; Wolf, H.; Celis, J.E. Proteome profiling of bladder squamous cell carcinomas: Identification of markers that define their degree of differentiation. *Cancer Res.* **1997**, *57*, 4111–4117. [PubMed]

64. Jiang, Y.; Tian, R.; Yu, S.; Zhao, Y.I.; Chen, Y.; Li, H.; Qiao, Y.; Wu, X. Clinical significance of galectin-7 in vulvar squamous cell carcinoma. *Oncol. Lett.* **2015**, *10*, 3826–3831. [CrossRef] [PubMed]

65. Truong, A.B.; Khavari, P.A. Control of keratinocyte proliferation and differentiation by p63. *Cell Cycle* **2007**, *6*, 295–299. [CrossRef] [PubMed]

66. Elola, M.T.; Wolfenstein-Todel, C.; Troncoso, M.F.; Vasta, G.R.; Rabinovich, G.A. Galectins: Matricellular glycan-binding proteins linking cell adhesion, migration, and survival. *Cell. Mol. Life Sci.* **2007**, *64*, 1679–1700. [CrossRef] [PubMed]

67. Menkhorst, E.M.; Gamage, T.; Cuman, C.; Kaitu'u-Lino, T.J.; Tong, S.; Dimitriadis, E. Galectin-7 acts as an adhesion molecule during implantation and increased expression is associated with miscarriage. *Placenta* **2014**, *35*, 195–201. [CrossRef] [PubMed]

68. Rondanino, C.; Poland, P.A.; Kinlough, C.L.; Li, H.; Rbaibi, Y.; Myerburg, M.M.; Al-bataineh, M.M.; Kashlan, O.B.; Pastor-Soler, N.M.; Hallows, K.R.; et al. Galectin-7 modulates the length of the primary cilia and wound repair in polarized kidney epithelial cells. *Am. J. Physiol. Ren. Physiol.* **2011**, *301*, F622–F633. [CrossRef] [PubMed]

69. Evans, J.; Yap, J.; Gamage, T.; Salamonsen, L.; Dimitriadis, E.; Menkhorst, E. Galectin-7 is important for normal uterine repair following menstruation. *Mol. Hum. Reprod.* **2014**, *20*, 787–798. [CrossRef] [PubMed]

70. Demers, M.; Magnaldo, T.; St-Pierre, Y. A novel function for galectin-7: Promoting tumorigenesis by up-regulating *MMP-9* gene expression. *Cancer Res.* **2005**, *65*, 5205–5210. [CrossRef] [PubMed]

71. Park, J.E.; Chang, W.Y.; Cho, M. Induction of matrix metalloproteinase-9 by galectin-7 through p38 MAPK signaling in HeLa human cervical epithelial adenocarcinoma cells. *Oncol. Rep.* **2009**, *22*, 1373–1379. [PubMed]

72. Saussez, S.; Cludts, S.; Capouillez, A.; Mortuaire, G.; Smetana, K.; Kaltner, H.; André, S.; Leroy, X.; Gabius, H.-J.; Decaestecker, C. Identification of matrix metalloproteinase-9 as an independent prognostic marker in laryngeal and hypopharyngeal cancer with opposite correlations to adhesion/growth-regulatory galectins-1 and -7. *Int. J. Oncol.* **2009**, *34*, 433–439. [CrossRef] [PubMed]

73. Cao, Z.; Said, N.; Amin, S.; Wu, H.K.; Bruce, A.; Garate, M.; Hsu, D.K.; Kuwabara, I.; Liu, F.-T.; Panjwani, N. Galectins-3 and -7, but not galectin-1, play a role in re-epithelialization of wounds. *J. Biol. Chem.* **2002**, *277*, 42299–42305. [CrossRef] [PubMed]

74. Cao, Z.; Said, N.; Wu, H.K.; Kuwabara, I.; Liu, F.-T.; Panjwani, N. Galectin-7 as a potential mediator of corneal epithelial cell migration. *Arch. Ophthalmol.* **2003**, *121*, 82–86. [CrossRef] [PubMed]

75. Cao, Z.; Wu, H.K.; Bruce, A.; Wollenberg, K.; Panjwani, N. Detection of differentially expressed genes in healing mouse corneas, using cDNA microarrays. *Investig. Ophthalmol. Vis. Sci.* **2002**, *43*, 2897–2904.

76. Klíma, J.; Lacina, L.; Dvoránková, B.; Herrmann, D.; Carnwath, J.W.; Niemann, H.; Kaltner, H.; André, S.; Motlík, J.; Gabius, H.-J.; et al. Differential regulation of galectin expression/reactivity during wound healing in porcine skin and in cultures of epidermal cells with functional impact on migration. *Physiol. Res.* **2009**, *58*, 873–884. [PubMed]

77. Friedl, P.; Gilmour, D. Collective cell migration in morphogenesis, regeneration and cancer. *Nat. Rev. Mol. Cell Biol.* **2009**, *10*, 445–457. [CrossRef] [PubMed]

78. Christiansen, J.J.; Rajasekaran, A.K. Reassessing epithelial to mesenchymal transition as a prerequisite for carcinoma invasion and metastasis. *Cancer Res.* **2006**, *66*, 8319–8326. [CrossRef] [PubMed]

79. Vladoiu, M.C.; Labrie, M.; St-Pierre, Y. Intracellular galectins in cancer cells: Potential new targets for therapy (Review). *Int. J. Oncol.* **2014**, *44*, 1001–1014. [CrossRef] [PubMed]
80. Thijssen, V.L.; Heusschen, R.; Caers, J.; Griffioen, A.W. Galectin expression in cancer diagnosis and prognosis: A systematic review. *Biochim. Biophys. Acta* **2015**, *1855*, 235–247. [CrossRef] [PubMed]
81. Satelli, A.; Rao, P.S.; Gupta, P.K.; Lockman, P.R.; Srivenugopal, K.S.; Rao, U.S. Varied expression and localization of multiple galectins in different cancer cell lines. *Oncol. Rep.* **2008**, *19*, 587–594. [CrossRef] [PubMed]
82. Balan, V.; Nangia-Makker, P.; Raz, A. Galectins as cancer biomarkers. *Cancers* **2010**, *2*, 592–610. [CrossRef] [PubMed]
83. St-Pierre, Y.; Campion, C.G.; Grosset, A.-A. A distinctive role for galectin-7 in cancer? *Front. Biosci.* **2012**, *17*, 438–450. [CrossRef]
84. Kaur, M.; Kaur, T.; Kamboj, S.S.; Singh, J. Roles of Galectin-7 in Cancer. *Asian Pac. J. Cancer Prev.* **2016**, *17*, 455–461. [CrossRef] [PubMed]
85. Moisan, S.; Demers, M.; Mercier, J.; Magnaldo, T.; Potworowski, E.F.; St-Pierre, Y. Upregulation of galectin-7 in murine lymphoma cells is associated with progression toward an aggressive phenotype. *Leukemia* **2003**, *17*, 751–759. [CrossRef] [PubMed]
86. Demers, M.; Biron-Pain, K.; Hébert, J.; Lamarre, A.; Magnaldo, T.; St-Pierre, Y. Galectin-7 in lymphoma: Elevated expression in human lymphoid malignancies and decreased lymphoma dissemination by antisense strategies in experimental model. *Cancer Res.* **2007**, *67*, 2824–2829. [CrossRef] [PubMed]
87. Demers, M.; Couillard, J.; Giglia-Mari, G.; Magnaldo, T.; St-Pierre, Y. Increased galectin-7 gene expression in lymphoma cells is under the control of DNA methylation. *Biochem. Biophys. Res. Commun.* **2009**, *387*, 425–429. [CrossRef] [PubMed]
88. Talantov, D.; Mazumder, A.; Yu, J.X.; Briggs, T.; Jiang, Y.; Backus, J.; Atkins, D.; Wang, Y. Novel genes associated with malignant melanoma but not benign melanocytic lesions. *Clin. Cancer Res. Off. J. Am. Assoc. Cancer Res.* **2005**, *11*, 7234–7242. [CrossRef] [PubMed]
89. Cada, Z.; Chovanec, M.; Smetana, K.; Betka, J.; Lacina, L.; Plzák, J.; Kodet, R.; Stork, J.; Lensch, M.; Kaltner, H.; et al. Galectin-7: Will the lectin's activity establish clinical correlations in head and neck squamous cell and basal cell carcinomas? *Histol. Histopathol.* **2009**, *24*, 41–48. [CrossRef] [PubMed]
90. Kim, S.-J.; Hwang, J.-A.; Ro, J.Y.; Lee, Y.-S.; Chun, K.-H. Galectin-7 is epigenetically-regulated tumor suppressor in gastric cancer. *Oncotarget* **2013**, *4*, 1461–1471. [CrossRef] [PubMed]
91. Matsui, Y.; Ueda, S.; Watanabe, J.; Kuwabara, I.; Ogawa, O.; Nishiyama, H. Sensitizing effect of galectin-7 in urothelial cancer to cisplatin through the accumulation of intracellular reactive oxygen species. *Cancer Res.* **2007**, *67*, 1212–1220. [CrossRef] [PubMed]
92. Langbein, S.; Brade, J.; Badawi, J.K.; Hatzinger, M.; Kaltner, H.; Lensch, M.; Specht, K.; André, S.; Brinck, U.; Alken, P.; et al. Gene-expression signature of adhesion/growth-regulatory tissue lectins (galectins) in transitional cell cancer and its prognostic relevance. *Histopathology* **2007**, *51*, 681–690. [CrossRef] [PubMed]
93. Chen, J.; He, Q.-Y.; Yuen, A.P.-W.; Chiu, J.-F. Proteomics of buccal squamous cell carcinoma: The involvement of multiple pathways in tumorigenesis. *Proteomics* **2004**, *4*, 2465–2475. [CrossRef] [PubMed]
94. Zhu, X.; Ding, M.; Yu, M.-L.; Feng, M.-X.; Tan, L.-J.; Zhao, F.-K. Identification of galectin-7 as a potential biomarker for esophageal squamous cell carcinoma by proteomic analysis. *BMC Cancer* **2010**, *10*, 290. [CrossRef] [PubMed]
95. Rorive, S.; Eddafali, B.; Fernandez, S.; Decaestecker, C.; André, S.; Kaltner, H.; Kuwabara, I.; Liu, F.-T.; Gabius, H.-J.; Kiss, R.; et al. Changes in galectin-7 and cytokeratin-19 expression during the progression of malignancy in thyroid tumors: Diagnostic and biological implications. *Mod. Pathol.* **2002**, *15*, 1294–1301. [CrossRef] [PubMed]
96. Lu, J.; Pei, H.; Kaeck, M.; Thompson, H.J. Gene expression changes associated with chemically induced rat mammary carcinogenesis. *Mol. Carcinog.* **1997**, *20*, 204–215. [CrossRef]
97. Perou, C.M.; Sørlie, T.; Eisen, M.B.; van de Rijn, M.; Jeffrey, S.S.; Rees, C.A.; Pollack, J.R.; Ross, D.T.; Johnsen, H.; Akslen, L.A.; et al. Molecular portraits of human breast tumours. *Nature* **2000**, *406*, 747–752. [CrossRef] [PubMed]
98. Castronovo, V.; van den Brûle, F.A.; Jackers, P.; Clausse, N.; Liu, F.T.; Gillet, C.; Sobel, M.E. Decreased expression of galectin-3 is associated with progression of human breast cancer. *J. Pathol.* **1996**, *179*, 43–48. [CrossRef]

99. Moumen, M.; Chiche, A.; Cagnet, S.; Petit, V.; Raymond, K.; Faraldo, M.M.; Deugnier, M.-A.; Glukhova, M.A. The mammary myoepithelial cell. *Int. J. Dev. Biol.* **2011**, *55*, 763–771. [CrossRef] [PubMed]

100. Prater, M.D.; Petit, V.; Alasdair Russell, I.; Giraddi, R.R.; Shehata, M.; Menon, S.; Schulte, R.; Kalajzic, I.; Rath, N.; Olson, M.F.; et al. Mammary stem cells have myoepithelial cell properties. *Nat. Cell Biol.* **2014**, *16*, 942–950. [CrossRef] [PubMed]

101. Demers, M.; Rose, A.A.N.; Grosset, A.-A.; Biron-Pain, K.; Gaboury, L.; Siegel, P.M.; St-Pierre, Y. Overexpression of galectin-7, a myoepithelial cell marker, enhances spontaneous metastasis of breast cancer cells. *Am. J. Pathol.* **2010**, *176*, 3023–3031. [CrossRef] [PubMed]

102. Grosset, A.-A.; Poirier, F.; Gaboury, L.; St-Pierre, Y. Galectin-7 Expression Potentiates HER-2-Positive Phenotype in Breast Cancer. *PLoS ONE* **2016**, *11*, e0166731. [CrossRef] [PubMed]

103. Eude-Le Parco, I.; Gendronneau, G.; Dang, T.; Delacour, D.; Thijssen, V.L.; Edelmann, W.; Peuchmaur, M.; Poirier, F. Genetic assessment of the importance of galectin-3 in cancer initiation, progression, and dissemination in mice. *Glycobiology* **2009**, *19*, 68–75. [CrossRef] [PubMed]

104. Mesquita, J.A.; Queiroz, L.M.G.; Silveira, É.J.D.; Gordon-Nunez, M.A.; Godoy, G.P.; Nonaka, C.F.W.; Alves, P.M. Association of immunoexpression of the galectins-3 and -7 with histopathological and clinical parameters in oral squamous cell carcinoma in young patients. *Eur. Arch. Oto-Rhino-Laryngol.* **2016**, *273*, 237–243. [CrossRef] [PubMed]

105. Alves, P.M.; Godoy, G.P.; Gomes, D.Q.; Medeiros, A.M.C.; de Souza, L.B.; da Silveira, E.J.D.; Vasconcelos, M.G.; Queiroz, L.M.G. Significance of galectins-1, -3, -4 and -7 in the progression of squamous cell carcinoma of the tongue. *Pathol. Res. Pract.* **2011**, *207*, 236–240. [CrossRef] [PubMed]

106. Kim, H.-J.; Jeon, H.-K.; Lee, J.-K.; Sung, C.O.; Do, I.-G.; Choi, C.H.; Kim, T.-J.; Kim, B.-G.; Bae, D.-S.; Lee, J.-W. Clinical significance of galectin-7 in epithelial ovarian cancer. *Anticancer Res.* **2013**, *33*, 1555–1561. [PubMed]

107. Cagnoni, A.J.; Pérez Sáez, J.M.; Rabinovich, G.A.; Mariño, K.V. Turning-Off Signaling by Siglecs, Selectins, and Galectins: Chemical Inhibition of Glycan-Dependent Interactions in Cancer. *Front. Oncol.* **2016**, *6*, 109. [CrossRef] [PubMed]

108. Pinho, S.S.; Reis, C.A. Glycosylation in cancer: Mechanisms and clinical implications. *Nat. Rev. Cancer* **2015**, *15*, 540–555. [CrossRef] [PubMed]

109. Stowell, S.R.; Ju, T.; Cummings, R.D. Protein glycosylation in cancer. *Annu. Rev. Pathol.* **2015**, *10*, 473–510. [CrossRef] [PubMed]

110. Ishida, H.; Togayachi, A.; Sakai, T.; Iwai, T.; Hiruma, T.; Sato, T.; Okubo, R.; Inaba, N.; Kudo, T.; Gotoh, M.; et al. A novel β1,3-N-acetylglucosaminyltransferase (β3Gn-T8), which synthesizes poly-N-acetyllactosamine, is dramatically upregulated in colon cancer. *FEBS Lett.* **2005**, *579*, 71–78. [CrossRef] [PubMed]

111. Croci, D.O.; Cerliani, J.P.; Dalotto-Moreno, T.; Méndez-Huergo, S.P.; Mascanfroni, I.D.; Dergan-Dylon, S.; Toscano, M.A.; Caramelo, J.J.; García-Vallejo, J.J.; Ouyang, J.; et al. Glycosylation-dependent lectin-receptor interactions preserve angiogenesis in anti-VEGF refractory tumors. *Cell* **2014**, *156*, 744–758. [CrossRef] [PubMed]

112. Mirandola, L.; Nguyen, D.D.; Rahman, R.L.; Grizzi, F.; Yuefei, Y.; Figueroa, J.A.; Jenkins, M.R.; Cobos, E.; Chiriva-Internati, M. Anti-galectin-3 therapy: A new chance for multiple myeloma and ovarian cancer? *Int. Rev. Immunol.* **2014**, *33*, 417–427. [CrossRef] [PubMed]

113. Thijssen, V.L.; Griffioen, A.W. Galectin-1 and -9 in angiogenesis: A sweet couple. *Glycobiology* **2014**, *24*, 915–920. [CrossRef] [PubMed]

114. Schulz, H.; Schmoeckel, E.; Kuhn, C.; Hofmann, S.; Mayr, D.; Mahner, S.; Jeschke, U. Galectins-1, -3, and -7 Are Prognostic Markers for Survival of Ovarian Cancer Patients. *Int. J. Mol. Sci.* **2017**, *18*, 1230. [CrossRef] [PubMed]

115. Tsai, C.J.; Sulman, E.P.; Eifel, P.J.; Jhingran, A.; Allen, P.K.; Deavers, M.T.; Klopp, A.H. Galectin-7 levels predict radiation response in squamous cell carcinoma of the cervix. *Gynecol. Oncol.* **2013**, *131*, 645–649. [CrossRef] [PubMed]

116. Matsukawa, S.; Morita, K.; Negishi, A.; Harada, H.; Nakajima, Y.; Shimamoto, H.; Tomioka, H.; Tanaka, K.; Ono, M.; Yamada, T.; et al. Galectin-7 as a potential predictive marker of chemo- and/or radio-therapy resistance in oral squamous cell carcinoma. *Cancer Med.* **2014**, *3*, 349–361. [CrossRef] [PubMed]

117. Bibens-Laulan, N.; St-Pierre, Y. Intracellular galectin-7 expression in cancer cells results from an autocrine transcriptional mechanism and endocytosis of extracellular galectin-7. *PLoS ONE* **2017**, *12*, e0187194. [CrossRef] [PubMed]

118. Advedissian, T.; Deshayes, F.; Poirier, F.; Grandjean, C.; Viguier, M. Galectins, a class of unconventional lectins. *Med. Sci.* **2015**, *31*, 499–505. [CrossRef]

119. Thurston, T.L.M.; Wandel, M.P.; von Muhlinen, N.; Foeglein, A.; Randow, F. Galectin 8 targets damaged vesicles for autophagy to defend cells against bacterial invasion. *Nature* **2012**, *482*, 414–418. [CrossRef] [PubMed]

120. Stegmayr, J.; Lepur, A.; Kahl-Knutson, B.; Aguilar-Moncayo, M.; Klyosov, A.A.; Field, R.A.; Oredsson, S.; Nilsson, U.J.; Leffler, H. Low or No Inhibitory Potency of the Canonical Galectin Carbohydrate-binding Site by Pectins and Galactomannans. *J. Biol. Chem.* **2016**, *291*, 13318–13334. [CrossRef] [PubMed]

121. Dion, J.; Advedissian, T.; Storozhylova, N.; Dahbi, S.; Lambert, A.; Deshayes, F.; Viguier, M.; Tellier, C.; Poirier, F.; Téletchéa, S.; et al. Development of a Sensitive Microarray Platform for the Ranking of Galectin Inhibitors: Identification of a Selective Galectin-3 Inhibitor. *ChemBioChem* **2017**. [CrossRef] [PubMed]

International Journal of
Molecular Sciences

MDPI

Review

TrkB-Target Galectin-1 Impairs Immune Activation and Radiation Responses in Neuroblastoma: Implications for Tumour Therapy

Katharina Batzke [1], Gabriele Büchel [2], Wiebke Hansen [3] and Alexander Schramm [1,*]

[1] Department of Medical Oncology, West German Cancer Center, University Hospital Essen, University of Duisburg-Essen, 45122 Essen, Germany; katharina.batzke@uk-essen.de

[2] Theodor Boveri Institute and Comprehensive Cancer Center Mainfranken, Biocenter, University of Würzburg, Am Hubland, 97074 Würzburg, Germany; gabriele.buechel@uni-wuerzburg.de

[3] Institute of Medical Microbiology, University Hospital Essen, University of Duisburg-Essen, 45122 Essen, Germany; wiebke.hansen@uk-essen.de

* Correspondence: alexander.schramm@uni-due.de; Tel.: +49-201-7232506

Received: 29 January 2018; Accepted: 1 March 2018; Published: 2 March 2018

Abstract: Galectin-1 (Gal-1) has been described to promote tumour growth by inducing angiogenesis and to contribute to the tumour immune escape. We had previously identified up-regulation of Gal-1 in preclinical models of aggressive neuroblastoma (NB), the most common extracranial tumour of childhood. While Gal-1 did not confer a survival advantage in the absence of exogenous stressors, Gal-1 contributed to enhanced cell migratory and invasive properties. Here, we review these findings and extend them by analyzing Gal-1 mediated effects on immune cell regulation and radiation resistance. In line with previous results, cell autonomous effects as well as paracrine functions contribute to Gal-1 mediated pro-tumourigenic functions. Interfering with Gal-1 functions in vivo will add to a better understanding of the role of the Gal-1 axis in the complex tumour-host interaction during immune-, chemo- and radiotherapy of neuroblastoma.

Keywords: Galectin-1; radiation response; neuroblastoma

1. Biology of Galectins—Physiology and Pathophysiology

Glycan-binding proteins were first described in the 1960s by Ashwell and Morrell [1], and this paved the way for identifying the family of Galectins. Galectins are highly conserved in the animal kingdom and differ from other lectins by their affinity for β-galactoside sugars. To date, 15 different Galectins have been found in mammals, but still many biological functions of these proteins are unknown [2,3]. They share a consensus carbohydrate recognition domain (CRD) comprising approximately 130 amino acids. This domain determines affinity for β-galactosides and allows for formation of galectin-glycan lattices. Based on CRD organisation, Galectins can be classified into three different groups: Galectin-1 is a member of the prototypical group, which is characterised by a single CRD that links monomers to form multimeric Galectin lattices. A second group, encompassing Galectin-4 and -5, among others, contains tandem-repeats with two non-identical CRDs connected by an amino acid linker. A third type of Galectins is referred to as "chimaera-type", because the CRD is here fused to proline- and glycine-rich stretches. The only member of this group found to date is galectin-3 [2,4]. Galectins are mainly cytosolic proteins, but they are also localised in the extracellular space while lacking a typical secretion signal peptide. Thus, Galectins are secreted by a non-classical and yet poorly understood mechanism [5,6].

As a member of the prototypical group, Galectin-1 (Gal-1) is divalent and its ability to cross-link carbohydrate chains on cell surfaces has been recognised early on [4]. Gal-1 is a multifunctional protein involved in development and differentiation. It also plays a role in cell-cell adhesion and cell-matrix interaction and thus regulates intercellular communication processes [3]. However, Gal-1 is dispensable for normal development in mice, as Gal-1$^{-/-}$ mice are healthy and fertile. Interestingly, targeted infection of Gal-1$^{-/-}$ mice results in severe autoimmune disease, pointing to a role of Gal-1 in regulating and terminating inflammatory responses. This process has been attributed to increased Th1 and Th17 responses in Gal-1$^{-/-}$ mice. Interestingly, Th2 cells ... seemed to be protected from the anti-proliferative effects of Gal-1 as a consequence of different glycosylation patterns [7]. It has been discussed that Gal-1 is also critical for regulating inflammation-mediated neurodegeneration in experimental autoimmune encephalomyelitis (EAE), which is the murine equivalent to multiple sclerosis (MS). This hypothesis was supported by the observation that progressive neurodegeneration occurs, when Gal-1 is depleted in the acute phase of EAE, while adoptive transfer of Gal1-secreting astrocytes or administration of recombinant Gal-1 suppressed EAE [8]. Thus, Gal-1 seems to exert a critical role especially in regulating neuroinflammatory processes.

Another important physiological function of Gal-1 has been described in mediating feto-maternal tolerance [4]. It has been suggested that suppression of maternal immune responses against placental alloantigens and pregnancy maintenance are both affected by Gal-1. Interestingly, decreased Gal-1 levels were observed in women suffering a miscarriage [9]. Similarly, preeclampsia, a pregnancy-specific disorder characterized by sudden onset of hypertension and proteinuria, seems to be correlated with Gal-1 levels [10]. Taken together, those findings contribute to the notion that Gal-1 is required for healthy gestation.

In tumours, mainly the intracellular functions of Gal-1 have been linked to pro-tumourigenic process, as Gal-1 can interact with oncogenic H-RAS to promote angiogenesis [11], while Galectin 3 (Gal-3) and K-RAS proteins are found to team up in an unholy alliance in cancer cells. These findings seem to apply to many K-RAS driven malignancies including pancreatic cancer [12] and colon tumours [13]. Here, Gal-3 has been identified as an amplifier of K-RAS signalling via PI-3 kinase [14], which is in part mediated by Gal-3 facilitated formation of K-RAS nanoclusters [15]. Similarly, Gal-1 dimers were reported to scaffold Raf-effectors to increase H-RAS nanoclusters [16,17]. This conformation regulates the magnitude of RAS-driven MAPK-signal output [18]. While these functions refer to intracellular activity of Galectins, Gal-1 is also a paracrine mediator of tumour aggressiveness by acting as a pro-angiogenic factor [19,20]. Paracrine functions of Gal-3 include up-regulation of IL-6 in the tumour microenvironment via Gal-3BP (Galectin-3 binding protein), which promotes angiogenesis and inflammation [21]. However, Gal-3 can also exert autocrine effects in neuroblastoma cells by increasing phenotypic differentiation and by impairing MYCN-primed apoptosis via modulation of HIPK2 [22–24]. As both, Gal-1 and Gal-3, are highly expressed in the majority of neuroblastoma, it is necessary to better understand their interplay in regulating neuroblastoma proliferation and interaction with the tumour microenvironment.

However, accumulating evidence suggest that both Gal-1 is linked to many hallmarks of cancer (Figure 1) and that the expression of Gal-1 correlates with tumour aggressiveness in different human cancer types including oral squamous cell carcinoma, ovarian and breast cancer [25]. Several strategies have thus been proposed to interfere with Gal-1 functions, including neutralising peptides [20], small molecule inhibitors, and supply with metabolic inhibitors of N-acetyllactosamine biosynthesis [26,27]. Several clinical trials evaluate safety and efficacy of inhibiting Gal-3 functions, mostly in combination with immune therapies. On the other hand, it still poses a major challenge to translate current knowledge into the design and development of effective galectin-1 inhibitors in cancer therapy.

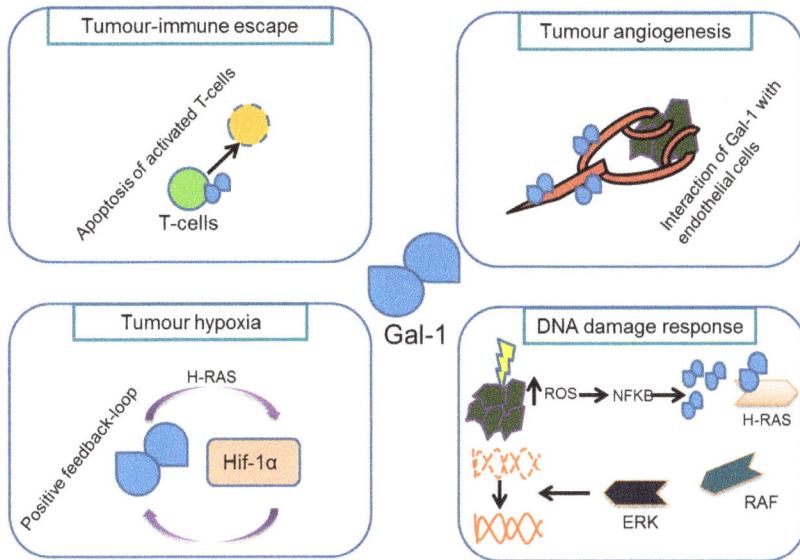

Figure 1. Schematic illustration of the multiple functions of Gal-1 in tumor-host interactions. Gal-1 contributes to the tumour-immune escape and promotes tumor angiogenesis by stimulating endothelial cell proliferation. Furthermore, Gal-1 is induced by (tumour) hypoxia and could contribute to radioresistance in hypoxic conditions by enabling a H-RAS dependent positive feedback-loop, which in turn induces Hif1α. Gal-1 was also reported to modulate DNA damage responses through activation of ROS and the H-RAS-RAF pathway (modified from [28]).

2. Gal-1 Expression in Neuroblastoma Is Linked to TrkB Activation

Neuroblastoma (NB) is the most common extracranial solid tumour of childhood and arises from the sympathetic nervous system [29]. Interestingly, neuroblastoma is characterized by a wide variety of clinical outcomes ranging from spontaneous regression to rapid progression and aggressive growth in other patients. Clinical observations and molecular analyses led to a classification of NB into two types according to patient outcome, cytogenetics and gene and protein expression profiles [29]. The type 1 neuroblastoma is the non-aggressive, favourable subtype, presenting with a hyperploid karyotype and expression of the TrkA neurotrophin receptor. By contrast, type 2 NB are characterised by a diploid karyotype with structural aberrations including loss of chromosome 1p and amplification of the MYCN oncogene [30]. Type 2 NB also highly express the receptor tyrosine kinase TrkB, but not TrkA. In normal tissue, activation of TrkB by its ligand BDNF (brain-derived neurotrophic factor) triggers neuronal development and survival, but also mediates synaptic reorganization processes through different signalling cascades involving protein-protein interaction [31]. There are three main pathways induced upon TrkB activation. First, PLC-γ activation causes release of inositol-1,4,5-triphosphat (IP3) and diacylglycerol (DAG) to increase intracellular calcium concentrations, which in turn activate Ca^{2+}-dependent protein kinases to stimulate cell growth. Secondly, TrkB induces the PI-3 kinase signalling pathway to activate AKT. Additionally, induction of the MAP/ERK pathway downstream of Ras/B-Raf promotes differentiation or cell survival depending on signal strength and duration [31,32]. The latter seems to be important for mediating the different effects exerted by the structurally highly related neurotrophin receptors, TrkA and TrkB. Analyses of the subtle differences between both receptors is deemed crucial for understanding the divergent biological outcomes of neurotrophin signalling in different cell types.

To further understand the biological consequences of the differently expressed neurotrophin receptors, TrkA and TrkB, we previously selected a Trk-negative human neuroblastoma cell line, SH-SY5Y, and designed subclones with either high TrkA or TrkB expression. Analysis of transcriptome and proteome changes revealed that Gal-1 is up-regulated only in SY5Y cells with ectopic TrkB expression (SY5Y-TrkB) [33,34]. As stimulation of TrkB by its ligand, BDNF, increases cell proliferation and resistance to chemotherapy in SY5Y-TrkB cells [35], we hypothesized that Gal-1 could contribute to TrkB mediated aggressiveness. This was supported by RNA profiling analysis of primary NB, which revealed a strong positive correlation between the expression of Gal-1 and TrkB, while expression of TrkA and Gal-1 were anti-correlated. Moreover, activation of TrkA or TrkB by their specific ligands in vitro confirmed up-regulation of Gal-1 only upon BDNF-mediated TrkB stimulation [36]. BDNF-mediated invasiveness of SY5Y-TrkB cells could be significantly reduced by neutralizing Gal-1 function, while recombinant Gal-1 could not fully recover the invasive phenotype of SY5Y-TrkB cells in the absence of BDNF. We concluded that Gal-1 is an important but not the only effector of BDNF-induced invasiveness of aggressive neuroblastoma cells [36].

3. Gal-1 Gene Dose Alters the Immune Phenotype and Tumour Angiogenesis in a Mouse Model of Neuroblastoma

Several mouse models have been developed to study neuroblastoma development in vivo as a consequence of MYCN or ALK overexpression in neural crest cell progenitors. While recent models used Cre-Lox technology to induce MYCN or ALK [37,38], an extensively characterized model has been developed by Weiss et al. more than 20 years ago. These transgenic mice (TH-MYCN) express the MYCN oncogene under control of the Tyrosine Hydroxylase (TH) promotor. They develop neuroblastic tumours comparable to those seen in humans concerning localisation, additional genomic aberrations and histological characteristics [39]. To investigate the function and biological role of Gal-1 in vivo, a Gal-1$^{-/-}$ mouse strain was generated by Poirier and co-workers back in 1993. They used homologous recombination techniques in embryonic stem cells to disrupt and inactivate the L14 lectin gene (the original name of the mouse homologue for LGALS1, the Gal-1 encoding gene). Animals with a global knock-out of Gal-1 were indistinguishable from control littermates. The mice were fertile and developed normally with an unimpaired lifespan indicating that other highly related lectins may be able to compensate for the function of Gal-1 in these animals [40]. As pointed out above, Gal-1 is crucial for maintaining feto-maternal tolerance in allogeneic matings, but did not affect survival of new-born mice from syngeneic breeding [41]. Impaired tolerance in the allogeneic setting might be due to enhanced cytotoxic T cell activity in the absence of Gal-1, while Gal-1 derived from NK cells was sufficient to induce apoptosis in activated T cells [42]. A tolerogenic role of Gal-1 has also been described in the context of bacterial infections [43] and in *T. cruzi* infected mice, in which Gal-1 is produced and secreted by B cells [44]. While Gal-1 is dispensable for normal development, at least in mice, it has important functions in modulating innate and adaptive immune mechanisms.

To further investigate the impact of Gal-1 on neuroblastoma aggressiveness, we crossbred TH-MYCN mice to Gal-1$^{-/-}$ mice. We demonstrated that the Gal-1 gene dosage did not affect tumour incidence, growth rate, or survival probability, but was correlated with alterations in the immune phenotype exemplified by a reduction of CD4^{+} T cell infiltration in tumours of Gal-1$^{-/-}$ mice [45]. Tumour infiltration by macrophages and NK cells was not affected by Gal-1 gene dosage in TH-MYCN mice. These findings are in line with the previously described role of Gal-1 in the tumour immune escape [46]. Here, Gal-1 was not only linked to induction of apoptosis in activated T cells [47] requiring tumour-immune cell contact [48] but also to induction of regulatory T cells [49]. Additionally, Gal-1 released by endothelial cells can induce apoptosis of activated but not resting T cells as a direct consequence of Gal-1 binding [47]. Hence, these findings confirm that T cells are main effectors of Gal-1 mediated immunomodulation. In addition to transgenic mouse models syngeneic transplantation models present one step further towards in vivo studies and are widely accepted as in vivo cell culture models. In the absence of a conditional Gal-1 ko mouse model, syngenic transplantation helped to dissect the effects of tumor-derived and host-derived Gal-1. Splenocytes from A/J mice receiving syngeneic NXS2 neuroblastoma cells with down-regulated Gal-1 (Gal-1 low) secreted higher amounts of IFN-γ and displayed enhanced cytotoxic T-cell function compared to NXS2 control cells or NXS2 cells overexpressing Gal-1 (Gal-1 high) [50]. Consequently, supernatants of NXS2 Gal-1 high or NXS2 control cells suppressed dendritic cell (DC) maturation and induced T cell apoptosis, while DCs and T cells exposed to supernatants from NXS2 Gal-1 low cells were largely unaffected. Moreover, shRNA-mediated downregulation of Gal-1 in murine cell lines leads to a significant decrease of tumour growth when transplanted subcutaneously into immunocompetent mice. [45]. Remarkably, low Gal-1 expressing murine NB cells also induced significantly lower numbers of liver metastases. Thus, Gal-1 levels affect aggressiveness and immune responses in both, transplanted and genetic models or murine neuroblastoma.

In addition to the altered immune phenotype of TH-MYCN/Gal-1$^{-/-}$ mice, these animals suffer from splenomegalies indicated by a significantly higher spleen weight observed in tumour bearing Gal-1$^{-/-}$ mice. These splenomegalies could be explained by a lower migratory capacity of Gal-1 deficient CD4^{+} T cells, while proliferation and apoptosis of CD4^{+} T cells was unaltered [45]. TH-MYCN/Gal-1$^{-/-}$ mice present with higher infiltration of CD11^{+} dendritic cells compared to Gal-1$^{+/-}$ or Gal-1 wt animals suggesting an enhanced tumour antigen presentation as a response to Gal-1 knock-out. However, tumours derived from mice lacking Gal-1 showed a reduction in CD31 endothelial cell staining confirming a role for Gal-1 in tumour angiogenesis depending on the Gal-1 gene dosage [45]. These findings are in line with a previously described role of Gal-1 in tumour angiogenesis, since Gal-1 was found to be overexpressed in both, tumour cells and tumour-associated endothelial cells, respectively [20]. It has been shown that Gal-1 enhances angiogenesis by interacting with N-glycans on the surface of endothelial cells in the extracellular space, and by augmenting VEGF signalling [27]. Furthermore, blocking Gal-1 functions using the small molecular inhibitor OTX008 reduced tumour growth and angiogenesis by targeting VEGFR-2 expression [51].

Standard therapies for aggressive neuroblastoma include surgery as well as chemo- and radiotherapy. Response to the latter therapeutic modalities is strongly modulated by the tumour microenvironment and development of resistance with subsequent disease relapse is frequently observed in aggressive type 2 neuroblastoma [52]. As both angiogenesis and the immune response have a strong impact on the outcome of radiotherapy in cancer, it remains an important task to understand the role of modifying factors, including Gal-1.

4. Gal-1 Expression and Its Impact on Radiotherapy

For neuroblastoma, standard of care includes surgery, chemotherapy and also radiotherapy. Since Gal-1 is overexpressed in the chemo- and radioresistant subtype of neuroblastoma, it can be considered as a potential therapeutic target for neuroblastoma therapy. Still, tumour irradiation is mainly performed by indirectly ionizing electromagnetic photons generated by an X- or γ-ray source, while only specialised centres make use of charged particles including protons, which are directly ionizing. Radiotherapeutic approaches mainly aim to induce DNA damage to kill malignant cells, as unrepaired DNA double-strand breaks subsequently induce apoptosis, mitotic catastrophes or autophagy. Thus, DNA damage repair pathways play a crucial role in determining the clinical outcome of radiotherapy, rendering it essential to find predictive biomarkers and novel targets for radiotherapy [53,54]. Gal-1 is induced by low dose ionizing radiation (0.5 Gy) in the tumour vasculature as well as in tumour-associated endothelial cells. Furthermore, specific enrichment of Gal-1 on the surface of tumour cells and stromal cells in the tumour microenvironment in vivo could be demonstrated using an isotope-labeled peptide, Anginex, which specifically binds to and inhibits Gal-1 functions [20]. Furthermore, co-cultivation of tumour cells and endothelial cells enhanced expression of Galectin-1 upon ionizing radiation, which might protect the tumour from radiation-induced cell death. It has been described that clinical or even subclinical doses sufficient to induce Gal-1 in the tumour microenvironment could be used to target Gal-1 by Anginex-loaded particles containing additional drugs [55]. Besides its role in direct induction of cell death as a consequence of unrepaired DNA damage and DNA double strand breaks, radiation also causes abscopal effects, defined as an action at a distance from the irradiated volume but within the same organism [56]. These systemic responses might be a consequence of radiation-induced release of immunogenic factors via a process termed "immunogenic cell death" (ICD). While several mediators of these abscopal effects have been identified, including pro-inflammatory cytokines, infrequency of these effects have been attributed to counterbalancing recruitment of myeloid-derived suppressor cells, enrichment of regulatory T cells and increased TGFβ-levels (reviewed in [57]). It is tempting to speculate that the immune-suppressive functions of Gal-1 also help tumours to dampen immune responses upon ionizing radiation. This hypothesis could be tested e.g., in the TH-MYCN model, in which differences in T cells and antigen-presenting cells between tumours with or without functional Gal-1 could be analysed. Hence, Gal-1 blockade might be beneficial in combination with radiation therapy by increasing immune responses towards the tumour.

Additionally, Gal-1 is not only an important player in the tumour microenvironment by modifying the immune response and angiogenesis, but it is also affected by tumour hypoxia. Several reports indicate a positive feedback loop expression between Hif1α, the major transcription factor involved in adaptation to hypoxic condition, and Gal-1 via the H-RAS oncogenic pathway [28]. Gal-1 expression is thus induced as a consequence of oxygen limitation in hypoxic environments. Hypoxic conditions limit the efficacy of radiotherapy as the formation of DNA-damaging H_2O_2 hydroxyl radicals requires oxygen. Thus, hypoxia within tumours causes a three-fold increase in the radiation dose required to generate the same amount of DNA damage when compared to normoxic conditions [28]. In addition to hypoxia, Gal-1 expression is induced by ionizing radiation of different cancer cells including breast, cervical and glioma cells [28,58]. Our own preliminary data point to Gal-1 upregulation on mRNA and protein level in Kelly and SY5Y human neuroblastoma cell lines upon ionizing radiation. However, shRNA mediated downregulation of Gal-1 in murine neuroblastoma cell lines did not alter the tumour cell responses to radiation with respect to clonogenic growth and cell cycle distribution. These findings confirm our previous results, which indicate a paracrine rather than an autocrine role for Gal-1 in neuroblastoma. Upon ionizing radiation, cells undergo autophagy or apoptosis as a consequence of DNA double-strand and single-strand breaks [28]. While it remains to be determined if Gal-1 mechanistically contributes to radiation responses, it can be hypothesized that upregulation of Gal-1 expression in a hypoxic tumour microenvironment additionally helps to stifle the immune responses to radiation.

5. Conclusions

Gal-1 is a multifaceted protein regulating different aspects of tumour biology. While necessary for balancing immune responses and angiogenic processes in physiological settings, tumours exploit these functions to escape from attacks of the host immune systems and to better cope with hypoxic conditions. Several strategies are currently evaluated to identify the best method to interfere with Gal-1 functions in therapeutic settings. One promising approach is to combine radiotherapy and Gal-1 blockade with the aim to reactivate immune responses towards the tumour and to limit angiogenic responses. Combining immune checkpoint inhibitors with blocking Gal-1 functions might also be an option to boost immunogenic anti-tumour responses. On the other hand, radiation-induced upregulation of Gal-1 could be used to selectively deliver drugs to Gal-1 positive cells in the tumour microenvironment. The results of these studies are anticipated to inform us about the optimal use of Gal-1 directed strategies in tumour therapy.

Conflicts of Interest: The authors declare no conflicts of interest.

References

1. Morell, A.G.; Gregoriadis, G.; Scheinberg, I.H.; Hickman, J.; Ashwell, G. The role of sialic acid in determining the survival of glycoproteins in the circulation. *J. Biol. Chem.* **1971**, *246*, 1461–1467. [PubMed]

2. Stowell, S.R.; Cummings, R.D. (Eds.) *Galectins: Methods and Protocols*; Humana Press: New York, NY, USA, 2015.

3. Danguy, A.; Camby, I.; Kiss, R. Galectins and cancer. *Biochim. Biophys. Acta* **2002**, *1572*, 285–293. [CrossRef]

4. Rabinovich, G.A. Galectin-1 as a potential cancer target. *Br. J. Cancer* **2005**, *92*, 1188–1192. [CrossRef] [PubMed]

5. Barondes, S.H.; Castronovo, V.; Cooper, D.N.; Cummings, R.D.; Drickamer, K.; Feizi, T.; Gitt, M.A.; Hirabayashi, J.; Hughes, C.; Kasai, K. Galectins: A family of animal beta-galactoside-binding lectins. *Cell* **1994**, *76*, 597–598. [CrossRef]

6. Paz, A.; Haklai, R.; Elad-Sfadia, G.; Ballan, E.; Kloog, Y. Galectin-1 binds oncogenic H-Ras to mediate Ras membrane anchorage and cell transformation. *Oncogene* **2001**, *20*, 7486–7493. [CrossRef] [PubMed]

7. Toscano, M.A.; Bianco, G.A.; Ilarregui, J.M.; Croci, D.O.; Correale, J.; Hernandez, J.D.; Zwirner, N.W.; Poirier, F.; Riley, E.M.; Baum, L.G.; et al. Differential glycosylation of TH1, TH2 and TH-17 effector cells selectively regulates susceptibility to cell death. *Nat. Immunol.* **2007**, *8*, 825–834. [CrossRef] [PubMed]

8. Starossom, S.C.; Mascanfroni, I.D.; Imitola, J.; Cao, L.; Raddassi, K.; Hernandez, S.F.; Bassil, R.; Croci, D.O.; Cerliani, J.P.; Delacour, D.; et al. Galectin-1 deactivates classically activated microglia and protects from inflammation-induced neurodegeneration. *Immunity* **2012**, *37*, 249–263. [CrossRef] [PubMed]

9. Tirado-González, I.; Freitag, N.; Barrientos, G.; Shaikly, V.; Nagaeva, O.; Strand, M.; Kjellberg, L.; Klapp, B.F.; Mincheva-Nilsson, L.; Cohen, M.; et al. Galectin-1 influences trophoblast immune evasion and emerges as a predictive factor for the outcome of pregnancy. *Mol. Hum. Reprod.* **2013**, *19*, 43–53. [CrossRef] [PubMed]

10. Freitag, N.; Tirado-González, I.; Barrientos, G.; Herse, F.; Thijssen, V.L.; Weedon-Fekjær, S.M.; Schulz, H.; Wallukat, G.; Klapp, B.F.; Nevers, T.; et al. Interfering with Gal-1-mediated angiogenesis contributes to the pathogenesis of preeclampsia. *Proc. Natl. Acad. Sci. USA* **2013**, *110*, 11451–11456. [CrossRef] [PubMed]

11. Ito, K.; Stannard, K.; Gabutero, E.; Clark, A.M.; Neo, S.-Y.; Onturk, S.; Blanchard, H.; Ralph, S.J. Galectin-1 as a potent target for cancer therapy: Role in the tumor microenvironment. *Cancer Metastasis Rev.* **2012**, *31*, 763–778. [CrossRef] [PubMed]

12. Song, S.; Ji, B.; Ramachandran, V.; Wang, H.; Hafley, M.; Logsdon, C.; Bresalier, R.S. Overexpressed galectin-3 in pancreatic cancer induces cell proliferation and invasion by binding Ras and activating Ras signaling. *PLoS ONE* **2012**, *7*, e42699. [CrossRef] [PubMed]

13. Wu, K.-L.; Huang, E.-Y.; Jhu, E.-W.; Huang, Y.-H.; Su, W.-H.; Chuang, P.-C.; Yang, K.D. Overexpression of galectin-3 enhances migration of colon cancer cells related to activation of the K-Ras-Raf-Erk1/2 pathway. *J. Gastroenterol.* **2013**, *48*, 350–359. [CrossRef] [PubMed]

14. Elad-Sfadia, G.; Haklai, R.; Balan, E.; Kloog, Y. Galectin-3 augments K-Ras activation and triggers a Ras signal that attenuates ERK but not phosphoinositide 3-kinase activity. *J. Biol. Chem.* **2004**, *279*, 34922–34930. [CrossRef] [PubMed]

15. Shalom-Feuerstein, R.; Plowman, S.J.; Rotblat, B.; Ariotti, N.; Tian, T.; Hancock, J.F.; Kloog, Y. K-ras nanoclustering is subverted by overexpression of the scaffold protein galectin-3. *Cancer Res.* **2008**, *68*, 6608–6616. [CrossRef] [PubMed]

16. Blaževitš, O.; Mideksa, Y.G.; Šolman, M.; Ligabue, A.; Ariotti, N.; Nakhaeizadeh, H.; Fansa, E.K.; Papageorgiou, A.C.; Wittinghofer, A.; Ahmadian, M.R.; et al. Galectin-1 dimers can scaffold Raf-effectors to increase H-ras nanoclustering. *Sci. Rep.* **2016**, *6*, 24165. [CrossRef] [PubMed]

17. Belanis, L.; Plowman, S.J.; Rotblat, B.; Hancock, J.F.; Kloog, Y. Galectin-1 is a novel structural component and a major regulator of H-ras nanoclusters. *Mol. Biol. Cell* **2008**, *19*, 1404–1414. [CrossRef] [PubMed]

18. Rotblat, B.; Belanis, L.; Liang, H.; Haklai, R.; Elad-Zefadia, G.; Hancock, J.F.; Kloog, Y.; Plowman, S.J. H-Ras nanocluster stability regulates the magnitude of MAPK signal output. *PLoS ONE* **2010**, *5*, e11991. [CrossRef] [PubMed]

19. Thijssen, V.L.; Barkan, B.; Shoji, H.; Aries, I.M.; Mathieu, V.; Deltour, L.; Hackeng, T.M.; Kiss, R.; Kloog, Y.; Poirier, F.; et al. Tumor cells secrete galectin-1 to enhance endothelial cell activity. *Cancer Res.* **2010**, *70*, 6216–6224. [CrossRef] [PubMed]

20. Thijssen, V.L.; Postel, R.; Brandwijk, R.J.M.G.E.; Dings, R.P.M.; Nesmelova, I.; Satijn, S.; Verhofstad, N.; Nakabeppu, Y.; Baum, L.G.; Bakkers, J.; et al. Galectin-1 is essential in tumor angiogenesis and is a target for antiangiogenesis therapy. *Proc. Natl. Acad. Sci. USA* **2006**, *103*, 15975–15980. [CrossRef] [PubMed]

21. Silverman, A.M.; Nakata, R.; Shimada, H.; Sposto, R.; DeClerck, Y.A. A galectin-3-dependent pathway upregulates interleukin-6 in the microenvironment of human neuroblastoma. *Cancer Res.* **2012**, *72*, 2228–2238. [CrossRef] [PubMed]

22. Petroni, M.; Veschi, V.; Prodosmo, A.; Rinaldo, C.; Massimi, I.; Carbonari, M.; Dominici, C.; McDowell, H.P.; Rinaldi, C.; Screpanti, I.; et al. MYCN sensitizes human neuroblastoma to apoptosis by HIPK2 activation through a DNA damage response. *Mol. Cancer Res.* **2011**, *9*, 67–77. [CrossRef] [PubMed]

23. Veschi, V.; Petroni, M.; Cardinali, B.; Dominici, C.; Screpanti, I.; Frati, L.; Bartolazzi, A.; Gulino, A.; Giannini, G. Galectin-3 impairment of MYCN-dependent apoptosis-sensitive phenotype is antagonized by nutlin-3 in neuroblastoma cells. *PLoS ONE* **2012**, *7*, e49139. [CrossRef] [PubMed]

24. Veschi, V.; Petroni, M.; Bartolazzi, A.; Altavista, P.; Dominici, C.; Capalbo, C.; Boldrini, R.; Castellano, A.; McDowell, H.P.; Pizer, B.; et al. Galectin-3 is a marker of favorable prognosis and a biologically relevant molecule in neuroblastic tumors. *Cell Death Dis.* **2014**, *5*, e1100. [CrossRef] [PubMed]

25. White, N.M.A.; Masui, O.; Newsted, D.; Scorilas, A.; Romaschin, A.D.; Bjarnason, G.A.; Siu, K.W.M.; Yousef, G.M. Galectin-1 has potential prognostic significance and is implicated in clear cell renal cell carcinoma progression through the HIF/mTOR signaling axis. *Br. J. Cancer* **2017**, *116*, e3. [CrossRef] [PubMed]

26. Cedeno-Laurent, F.; Opperman, M.J.; Barthel, S.R.; Hays, D.; Schatton, T.; Zhan, Q.; He, X.; Matta, K.L.; Supko, J.G.; Frank, M.H.; et al. Metabolic inhibition of galectin-1-binding carbohydrates accentuates antitumor immunity. *J. Investig. Dermatol.* **2012**, *132*, 410–420. [CrossRef] [PubMed]

27. Astorgues-Xerri, L.; Riveiro, M.E.; Tijeras-Raballand, A.; Serova, M.; Neuzillet, C.; Albert, S.; Raymond, E.; Faivre, S. Unraveling galectin-1 as a novel therapeutic target for cancer. *Cancer Treat. Rev.* **2014**, *40*, 307–319. [CrossRef] [PubMed]

28. Kuo, P.; Le, Q.-T. Galectin-1 links tumor hypoxia and radiotherapy. *Glycobiology* **2014**, *24*, 921–925. [CrossRef] [PubMed]

29. Brodeur, G.M. Neuroblastoma: Biological insights into a clinical enigma. *Nat. Rev. Cancer* **2003**, *3*, 203–216. [CrossRef] [PubMed]

30. Schulte, J.H.; Kuhfittig-Kulle, S.; Klein-Hitpass, L.; Schramm, A.; Biard, D.S.F.; Pfeiffer, P.; Eggert, A. Expression of the TrkA or TrkB receptor tyrosine kinase alters the double-strand break (DSB) repair capacity of SY5Y neuroblastoma cells. *DNA Repair* **2008**, *7*, 1757–1764. [CrossRef] [PubMed]

31. Baydyuk, M.; Xu, B. BDNF signaling and survival of striatal neurons. *Front. Cell. Neurosci.* **2014**, *8*, 254. [CrossRef] [PubMed]

32. Chao, M.V. Neurotrophins and their receptors: A convergence point for many signalling pathways. *Nat. Rev. Neurosci.* **2003**, *4*, 299–309. [CrossRef] [PubMed]

33. Schramm, A.; Schulte, J.H.; Klein-Hitpass, L.; Havers, W.; Sieverts, H.; Berwanger, B.; Christiansen, H.; Warnat, P.; Brors, B.; Eils, J.; et al. Prediction of clinical outcome and biological characterization of neuroblastoma by expression profiling. *Oncogene* **2005**, *24*, 7902–7912. [CrossRef] [PubMed]

34. Sitek, B.; Apostolov, O.; Stühler, K.; Pfeiffer, K.; Meyer, H.E.; Eggert, A.; Schramm, A. Identification of dynamic proteome changes upon ligand activation of Trk-receptors using two-dimensional fluorescence difference gel electrophoresis and mass spectrometry. *Mol. Cell. Proteom.* **2005**, *4*, 291–299. [CrossRef] [PubMed]

35. Ho, R.; Eggert, A.; Hishiki, T.; Minturn, J.E.; Ikegaki, N.; Foster, P.; Camoratto, A.M.; Evans, A.E.; Brodeur, G.M. Resistance to chemotherapy mediated by TrkB in neuroblastomas. *Cancer Res.* **2002**, *62*, 6462–6466. [PubMed]

36. Cimmino, F.; Schulte, J.H.; Zollo, M.; Koster, J.; Versteeg, R.; Iolascon, A.; Eggert, A.; Schramm, A. Galectin-1 is a major effector of TrkB-mediated neuroblastoma aggressiveness. *Oncogene* **2009**, *28*, 2015–2023. [CrossRef] [PubMed]

37. Althoff, K.; Beckers, A.; Bell, E.; Nortmeyer, M.; Thor, T.; Sprüssel, A.; Lindner, S.; de Preter, K.; Florin, A.; Heukamp, L.C.; et al. A Cre-conditional MYCN-driven neuroblastoma mouse model as an improved tool for preclinical studies. *Oncogene* **2015**, *34*, 3357–3368. [CrossRef] [PubMed]

38. Cazes, A.; Lopez-Delisle, L.; Tsarovina, K.; Pierre-Eugène, C.; de Preter, K.; Peuchmaur, M.; Nicolas, A.; Provost, C.; Louis-Brennetot, C.; Daveau, R.; et al. Activated Alk triggers prolonged neurogenesis and Ret upregulation providing a therapeutic target in ALK-mutated neuroblastoma. *Oncotarget* **2014**, *5*, 2688–2702. [CrossRef] [PubMed]

39. Weiss, W.A.; Aldape, K.; Mohapatra, G.; Feuerstein, B.G.; Bishop, J.M. Targeted expression of MYCN causes neuroblastoma in transgenic mice. *EMBO J.* **1997**, *16*, 2985–2995. [CrossRef] [PubMed]

40. Poirier, F.; Robertson, E.J. Normal development of mice carrying a null mutation in the gene encoding the L14 S-type lectin. *Development* **1993**, *119*, 1229–1236. [PubMed]

41. Blois, S.M.; Ilarregui, J.M.; Tometten, M.; Garcia, M.; Orsal, A.S.; Cordo-Russo, R.; Toscano, M.A.; Bianco, G.A.; Kobelt, P.; Handjiski, B.; et al. A pivotal role for galectin-1 in fetomaternal tolerance. *Nat. Med.* **2007**, *13*, 1450–1457. [CrossRef] [PubMed]

42. Kopcow, H.D.; Rosetti, F.; Leung, Y.; Allan, D.S.J.; Kutok, J.L.; Strominger, J.L. T cell apoptosis at the maternal-fetal interface in early human pregnancy, involvement of galectin-1. *Proc. Natl. Acad. Sci. USA* **2008**, *105*, 18472–18477. [CrossRef] [PubMed]

43. Davicino, R.C.; Méndez-Huergo, S.P.; Eliçabe, R.J.; Stupirski, J.C.; Autenrieth, I.; Di Genaro, M.S.; Rabinovich, G.A. Galectin-1-Driven Tolerogenic Programs Aggravate Yersinia enterocolitica Infection by Repressing Antibacterial Immunity. *J. Immunol.* **2017**, *199*, 1382–1392. [CrossRef] [PubMed]

44. Zuñiga, E.; Rabinovich, G.A.; Iglesias, M.M.; Gruppi, A. Regulated expression of galectin-1 during B-cell activation and implications for T-cell apoptosis. *J. Leukoc. Biol.* **2001**, *70*, 73–79. [PubMed]

45. Büchel, G.; Schulte, J.H.; Harrison, L.; Batzke, K.; Schüller, U.; Hansen, W.; Schramm, A. Immune response modulation by Galectin-1 in a transgenic model of neuroblastoma. *Oncoimmunology* **2016**, *5*, e1131378. [CrossRef] [PubMed]

46. Rubinstein, N.; Alvarez, M.; Zwirner, N.W.; Toscano, M.A.; Ilarregui, J.M.; Bravo, A.; Mordoh, J.; Fainboim, L.; Podhajcer, O.L.; Rabinovich, G.A. Targeted inhibition of galectin-1 gene expression in tumor cells results in heightened T cell-mediated rejection; A potential mechanism of tumor-immune privilege. *Cancer Cell* **2004**, *5*, 241–251. [CrossRef]

47. Perillo, N.L.; Pace, K.E.; Seilhamer, J.J.; Baum, L.G. Apoptosis of T cells mediated by galectin-1. *Nature* **1995**, *378*, 736–739. [CrossRef] [PubMed]

48. Kovács-Sólyom, F.; Blaskó, A.; Fajka-Boja, R.; Katona, R.L.; Végh, L.; Novák, J.; Szebeni, G.J.; Krenács, L.; Uher, F.; Tubak, V.; et al. Mechanism of tumor cell-induced T-cell apoptosis mediated by galectin-1. *Immunol. Lett.* **2010**, *127*, 108–118. [CrossRef] [PubMed]

49. Dalotto-Moreno, T.; Croci, D.O.; Cerliani, J.P.; Martinez-Allo, V.C.; Dergan-Dylon, S.; Méndez-Huergo, S.P.; Stupirski, J.C.; Mazal, D.; Osinaga, E.; Toscano, M.A.; et al. Targeting galectin-1 overcomes breast cancer-associated immunosuppression and prevents metastatic disease. *Cancer Res.* **2013**, *73*, 1107–1117. [CrossRef] [PubMed]

50. Soldati, R.; Berger, E.; Zenclussen, A.C.; Jorch, G.; Lode, H.N.; Salatino, M.; Rabinovich, G.A.; Fest, S. Neuroblastoma triggers an immunoevasive program involving galectin-1-dependent modulation of T cell and dendritic cell compartments. *Int. J. Cancer* **2012**, *131*, 1131–1141. [CrossRef] [PubMed]

51. Astorgues-Xerri, L.; Riveiro, M.E.; Tijeras-Raballand, A.; Serova, M.; Rabinovich, G.A.; Bieche, I.; Vidaud, M.; de Gramont, A.; Martinet, M.; Cvitkovic, E.; et al. OTX008, a selective small-molecule inhibitor of galectin-1, downregulates cancer cell proliferation, invasion and tumour angiogenesis. *Eur. J. Cancer* **2014**, *50*, 2463–2477. [CrossRef] [PubMed]

52. Brodeur, G.M.; Bagatell, R. Mechanisms of neuroblastoma regression. *Nat. Rev. Clin. Oncol.* **2014**, *11*, 704–713. [CrossRef] [PubMed]

53. Mahaney, B.L.; Meek, K.; Lees-Miller, S.P. Repair of ionizing radiation-induced DNA double-strand breaks by non-homologous end-joining. *Biochem. J.* **2009**, *417*, 639–650. [CrossRef] [PubMed]

54. Kastan, M.B.; Bartek, J. Cell-cycle checkpoints and cancer. *Nature* **2004**, *432*, 316–323. [CrossRef] [PubMed]

55. Upreti, M.; Jamshidi-Parsian, A.; Apana, S.; Berridge, M.; Fologea, D.A.; Koonce, N.A.; Henry, R.L.; Griffin, R.J. Radiation-induced galectin-1 by endothelial cells: A promising molecular target for preferential drug delivery to the tumor vasculature. *J. Mol. Med.* **2013**, *91*, 497–506. [CrossRef] [PubMed]

56. Mole, R.H. Whole body irradiation; radiobiology or medicine? *Br. J. Radiol.* **1953**, *26*, 234–241. [CrossRef] [PubMed]

57. Hu, Z.I.; McArthur, H.L.; Ho, A.Y. The Abscopal Effect of Radiation Therapy: What Is It and How Can We Use It in Breast Cancer? *Curr. Breast Cancer Rep.* **2017**, *9*, 45–51. [CrossRef] [PubMed]

58. Strik, H.M.; Schmidt, K.; Lingor, P.; Tönges, L.; Kugler, W.; Nitsche, M.; Rabinovich, G.A.; Bähr, M. Galectin-1 expression in human glioma cells: Modulation by ionizing radiation and effects on tumor cell proliferation and migration. *Oncol. Rep.* **2007**, *18*, 483–488. [CrossRef] [PubMed]

International Journal of
Molecular Sciences

MDPI

Review

Role of Galectins in Multiple Myeloma

Paola Storti [1], Valentina Marchica [1] and Nicola Giuliani [1,2,*

[1]   Department of Medicine and Surgery, University of Parma, Via Gramsci, 14, 43126 Parma, Italy;
      paola.storti@unipr.it (P.S.); valentina.marchica@studenti.unipr.it (V.M.)
[2]   Hematology, "Azienda Ospedaliero-Universitaria di Parma", Via Gramsci, 14, 43126 Parma, Italy
*   Correspondence: nicola.giuliani@unipr.it; Tel.: +39-052-103-3299

Received: 29 November 2017; Accepted: 12 December 2017; Published: 17 December 2017

Abstract: Galectins are a family of lectins that bind β-galactose-containing glycoconjugates and are characterized by carbohydrate-recognition domains (CRDs). Galectins exploit several biological functions, including angiogenesis, regulation of immune cell activities and cell adhesion, in both physiological and pathological processes, as tumor progression. Multiple myeloma (MM) is a plasma cell (PC) malignancy characterized by the tight adhesion between tumoral PCs and bone marrow (BM) microenvironment, leading to the increase of PC survival and drug resistance, MM-induced neo-angiogenesis, immunosuppression and osteolytic bone lesions. In this review, we explore the expression profiles and the roles of galectin-1, galectin-3, galectin-8 and galectin-9 in the pathophysiology of MM. We focus on the role of these lectins in the interplay between MM and BM microenvironment cells showing their involvement in MM progression mainly through the regulation of PC survival and MM-induced angiogenesis and osteoclastogenesis. The translational impact of these pre-clinical pieces of evidence is supported by recent data that indicate galectins could be new attractive targets to block MM cell growth in vivo and by the evidence that the expression levels of *LGALS1* and *LGALS8*, genes encoding for galectin-1 and galectin-8 respectively, correlate to MM patients' survival.

Keywords: galectins; myeloma; galectin-1; galectin-3; galectin-8; galectin-9

---

1. Biological and Pathophysiological Functions of Galectins

1.1. Galectin Family

Galectins are a family of lectins, evolutionarily conserved and with the ability to bind glycans [1]. All galectins contain one or more carbohydrate-recognition domains (CRDs), sequences of about 130 amino acids, responsible for the binding to carbohydrate [1,2]. To date, 15 mammalian galectins have been identified and some seem to be species specific, such as galectin-5 and galectin-6 found only in rodents, and galectin-11 and galectin-15 only in caprine and ovine [2,3]. Based on the number and the structure of CRD, galectins are divided in three groups: prototype (galectin-1, -2, -5, -7, -10, -11, -13, -14 and -15) that have only a CRD that can associate in dimer, tandem repeat-type (galectin-4, -6, -8 and -9) that carry two CRDs linked by a short peptide linker, and chimera-type (galectin-3) that has a CRD connected to a non-lectin amino-terminal region that allow the oligomerization into pentamers [1,2,4].

Galectins can be found in the intracellular compartment and in the cell nuclei; moreover, several galectins are secreted by cells through a non-classical secretory mechanism, lacking the signal sequence for classical secretion, and they are detected in the extracellular space [1,5]. Extracellularly, galectins can bind to cell surface glycoconjugates, bearing the *N*-acetyl-lactosamine (Galβ(1-4)GlcNAc; LacNAc) disaccharide, forming a galectin–glycan structure called lattice, and mediating an intracellular signal transduction [1,6–8]. Moreover, they can bind to some glycoproteins of the extracellular matrix, such as laminin, fibronectin and elastin [9]. In fact, each member of the galectin family exhibits preferences in different glycan binding [10], which could explain their differences in biological and pathophysiological functions and the wide range of identified receptors on cell surfaces [2,11]. In particular, the tandem repeat type galectins, galectin-8 and -9, carry two CRDs (N and C terminal) that, despite their structural similarity, recognize different oligosaccharides (i.e., sulfated or sialylated glycans or biantennary oligosaccharide) due to different affinity [12–14]. Moreover, the modification of the short peptide linker of tandem repeat-type galectins could also modify their biological functions [15]. Indeed, galectins could crosslink different glycoconjugates and trigger a cascade of transmembrane signaling pathways or they can cause the clustering of multiple glycoconjugates on cell surfaces [7].

The expression and the function of the galectins involved in bone marrow (BM) microenvironment are summarized in Table 1.

Table 1. Sources and functions of the mains galectins.

Galectin	Gene	Sources	Functions	References
Galectin-1	*LGALS1*	Pre-B cells, PCs, ECs, BMSCs, OBs, T cells, NK cells, DCs	Survival, Angiogenesis, Adhesion and migration, Immunosuppression, Invasion metastasis, Drug resistance	[16–29]
Galectin-3	*LGALS3*	OBs, OCs, BMSCs, PCs, T cells, ECs	Adhesion and migration, Angiogenesis, Anti-apoptotic, Invasion metastasis, Regulation bone homeostasis, Drug resistance	[30–37]
Galectin-8	*LGALS8*	ECs, PCs	Angiogenesis Adhesion and migration	[18,38,39]
Galectin-9	*LGALS9*	DCs, ECs, T cells, OCs	Pro-apoptotic, Adhesion, OC differentiation, Immunosuppression	[40–43]
Galectin-10	*CLC*	eosinophils and basophils	Immunosuppression	[44,45]

PCs, plasma cells; ECs, endothelial cells; BMSCs, bone marrow stromal cells; OBs, osteoblasts; NK, natural killer; DCs, dendritic cells; OCs, osteoclasts

1.2. Galectins in Hematopoiesis and Immunity

Galectins have a role in hematopoietic differentiation, in particular creating specific galectin–glycan interactions between hematopoietic and stromal cells, sustaining the formation of a microenvironmental niche [46].

During the T cell development, in the thymus, galectin-1, -3, -8 and -9 induce the apoptosis of the double positive (CD4+CD8+) or double negative (CD4−CD8−) developing thymocytes, favoring the interaction between thymocytes and thymic epithelial cells [46]. Once in the periphery, galectins fine regulate T cells homeostasis. Galectin-1 prolongs the survival of T naive cells, induces the apoptosis of T helper (Th) 1 and Th17 differentiated cells and protects Th2 cells, promoting the release of anti-inflammatory cytokines (interleukin (IL)-4, IL-5 and IL-10) [11]. Galectin-9 interacts with its receptor T-cell immunoglobulin and mucin-domain containing-3 (TIM-3) on Th1 cell surface, leading to their apoptosis [47]. Galectin-1 and -9 induce the expansion of FoxP3+ T regulatory (Treg) cells.

In B cell differentiation, galectin-1 creates a synapse between pre-B cell receptor (pre-BCR) on preB-cells and BM stromal cells, leading pre-BCR clustering and signaling [48]. In the periphery, galectin-1 is upregulated in PCs by the activation of B-lymphocyte-induced maturation protein (BLIMP)-1 and, with galectin-8, promotes the immunoglobulin production [11,18]. On the other hand, intracellular galectin-3 favors the differentiation toward memory B cells [46].

Thereafter, galectins regulate antigen presenting cell functions; in particular, galectin-1 induces the dendritic cell (DC) tolerogenic phenotype and the macrophage switch to M2 phenotype [49] and galectin-9 stimulates DCs and activates innate immunity [50].

Finally, galectin-3 is also involved in bone homeostasis. Galectin-3 deficient mice display increased osteoclast (OC) activity and, in vitro, galectin-3 interferes with the receptor activator of nuclear factor kappa-B ligand (RANKL) signaling on OCs, reducing their differentiation [32].

1.3. Galectins and Tumor Progression

Galectins are involved, since the first phases of tumor transformation, in the survival of tumoral cells [51]. Based on literature data, the inhibition of galectin-1 and -3 expression reverts the transformed phenotype into normal in glioma [52], breast [53] and thyroid papillary carcinoma cells [33]. Moreover, galectin-1 and -3 can interact with the oncogenic Ras proteins that are anchored to the cellular membrane [54], and sustain the activation of their downstream effectors, such as extracellular signal-regulated kinases (ERK), Raf-1 proto-oncogene, serine/threonine kinase (RAF1) and phosphoinositide 3-kinase (PI3K) [51,54,55].

Furthermore, galectins have a role in the regulation of cancer cells apoptosis and cell cycle. Different studies highlighted that galectin-1 could have different effects on cancer cell proliferation, depending on the localization of this lectin (intracellular or extracellular) and on the tumor type [29].

Moreover, galectin-1, -3 and -8 can support tumor cells migration and attachment to extracellular matrix (ECM) [56,57] but the major role of galectin in the tumor progression is their interaction with the microenvironment, promoting neo-angiogenesis and inducing immune escape. Galectin-1 promotes endothelial cell (EC) proliferation and migration, thus angiogenesis, binding to the vascular endothelial growth factor receptor (VEGFR) 2 and neuropilin-1. In Kaposi's sarcoma, prostate cancer, lung cancer and melanoma [17,58,59], it mimics the effect of vascular endothelial growth factor A (VEGFA) and confers resistance to anti-VEGF therapy [58]. Galectin-3 and -8 support the vascularization process in the tumor microenvironment binding to integrin $\alpha_v \beta_3$ and activated leucocyte cell adhesion molecule (ALCAM), respectively [39,60].

1.4. Galectins and Tumoral Immune Microenvironment

Tumor immune microenvironment is heavily shaped by galectins. One of the effects of galectins is to expand the regulatory myeloid cells; in fact, galectin-1 promotes differentiation of IL-27- and IL-10-producing tolerogenic DCs [61], and contributes to M2 macrophage polarization [49]. Galectin-1 and -9 also support the recruitment of myeloid-derived suppressor cells (MDSCs) and increase their regulatory capacity [16,43]. Secondly, galectin-1, -8, -9- and -10 enhance the expansion of Tregs and increase the immunosuppressive activity of CD4+CD25+Foxp3+ Tregs [62,63]. Further, galectin-1 and -9 selectively induce apoptosis in Th1 and Th17 effector cells [11]. On the other hand, galectin-3 and -9 participate in the immune inhibitory checkpoints cascade, including interactions with lymphocyte activation gene (LAG)-3 or TIM-3 [34,47], leading to an inhibition of tumor infiltrating T cells. Finally, in glioblastoma, galectin-1 induces a natural killer (NK) cell inhibition, mediated by the increased activity of tumor related immunosuppressive MDSCs [64]. In addition, galectin-3 and -9 reduce NK cell activity and cytokine production [63]. All the mechanisms described above decrease the potential of anti-tumor immune cells and boost the tumor immune evasion [63].

1.5. Galectins and Hematological Malignancies

In hematological malignancies, increased galectin-1 serum levels are correlated with increased tumor burden in Hodgkin lymphoma patients [65] and this lectin is highly expressed in cutaneous T-cell lymphomas cells [21] and chronic lymphocytic leukemia patients [66]. In Hodgkin lymphomas, galectin-1 reduces anti-tumor T cell activity, promoting the expansion of CD4+CD25+Foxp3+ Tregs [67]. Moreover, in lymphoma, galectin-1 expression is strongly correlated with resistance to immunotherapy; in fact, galectin-1-overexpressing lymphoma cells blunt antibody-dependent tumor phagocytosis in vitro, and, in vivo, galectin-1 reduces lymphoma cells sensitivity to CD20 immunotherapy [20].

Galectin-3 is overexpressed in diffuse large B-cell lymphoma [68] and chronic myelogenous leukemia [31]. This lectin increases the proliferative and chemotactic capacity of lymphoma and leukemia cells and enhances their resistance to chemotherapy. Finally, in acute myelogenous leukemia, galectin-3 expression is an independent unfavorable prognostic factor for patients' overall survival (OS) [69] and, in the same disease, galectin-9 induces, through TIM-3, T cell dysregulation [40].

2. Multiple Myeloma (MM) Pathophysiology

Multiple myeloma (MM) is a hematological malignancy characterized by an accumulation of malignant plasma cells (PCs) with a tight dependence to the BM microenvironment [70]. The active stage of MM is preceded by indolent stages as monoclonal gammopathy of undetermined significance (MGUS) and smoldering myeloma (SMM) [71]. The progression from an indolent stage to the symptomatic one is supported by sequential genetic events in the malignant clones and by microenvironment alterations that support the growth of malignant PCs [72], as the angiogenic switch and the development of osteolytic lesions [73].

2.1. Deregulated Pathways in Malignant Plasma Cells (PCs)

The MM PCs accumulate several genetic lesions, such as translocations, mutations, deletions or amplifications, which lead to a deregulation of different proliferative pathways [72]. In MM, the major translocations involve the immunoglobulins loci, putting under control of a strong enhancer different oncogenes such as cyclin (CCN) D1, CCND3, FGFR3, multiple myeloma SET domain (MMSET), MYC, MAF and MAFB [74]. A later genetic event is the mutation and the monoallelic deletion of chromosome the locus 17p13, carrying the onco-suppressor gene p53 [71]. Other recurrent mutations involve genes of ERK pathway (NRAS, BRAF and KRAS) [72] and of nuclear factor- kappa B (NF-κB) pathway (CYLD, TRAF3 and BIRC2/3) [75].

The malignant PCs are dependent on the BM microenvironment and the interaction between MM PCs and BM cells occurs mainly through adhesion molecules, chemokines and growth factors that support the survival and the proliferation of the malignant clones [73]. The PCs adhere to ECM by molecules, such as fibronectin and type I collagen, and the bone marrow stromal cells (BMSCs) interact with the malignant PCs using different molecular complexes as very late antigen-4 (VLA-4)/vascular cell adhesion molecule (VCAM)-1, CD38/CD31, lymphocyte function associate antigen (LFA)-1/intercellular adhesion molecule (ICAM)-1 and the homotypic binding of CD56 [76,77]; all the interactions described above support the production of soluble factors that sustain the growth of MM PCs. The major growth factor of MM cells is IL-6, produced either from PCs or BMSCs or osteoblasts (OBs); subsequently, the stimulation of NF-κB pathway activates the mitogen-activated protein kinase kinase 1 (MEK)/ERK, Janus kinase/signal transducers and activators of transcription (JAK/STAT3) and PI3K/ Protein kinase B (AKT) signaling pathways that promote cell survival and apoptosis resistance [78,79]. Besides IL-6, other important pro-survival and proliferation cytokines are VEGFA, insuline-like growth factor (IGF)-1, tumor necrosis factor α (TNFα), transforming growth factor β (TGFβ) and IL-1β [77,80].

2.2. Microenvironment Alterations in MM: Role of Angiogenesis and Bone Destruction

The MM progression is characterized by an avascular phase, corresponding to the two indolent stages SMM and MGUS, followed by an angiogenic switch, leading to the active MM [81]. Moreover, the BM microenvironment is hypoxic, and the MM cells overexpress the main factor involved in the cell adjustment to hypoxic stress, the hypoxia inducible factor (HIF)-1α [82,83]. Indeed, the MM angiogenic switch is supported by BM hypoxia, HIF-1α and the production of pro-angiogenic molecules either by MM PCs or by BMSCs, such as VEGFA, fibroblast growth factor (FGF), hepatocyte growth factor (HGF), angiopoietin (ANG)-1 and osteopontin (OPN), which act as chemoattractants or bind to receptor on endothelial cells (ECs), such as tyrosine kinase with immunoglobulin-like and EGF-like domains (TIE)-2 and VEGFR2, and promote their proliferation [81,84,85]. Moreover, BMSCs and MM cells secrete the proteolytic enzymes metalloprotease (MMP)-1, MMP-2 and MMP-9 that help to reshape the extracellular compartment and the migration of ECs [86]. The interaction between MM cells and

ECs, besides supports the angiogenic process, favors the homing of MM PCs into the BM supporting their survival [87].

One of the features of active MM is the presence of bone lesions, due to an imbalance between OC and OB formation and activity [88]. In the MM BM microenvironment, the RANKL/osteoprotegerin (OPG) ratio is altered; in particular, MM cells upregulate the production of RANKL and downregulate OPG expression by BMSCs and T lymphocytes [89,90]. Moreover, MM PCs sustain the osteoclastogenesis and the OC activity overexpressing chemokine (C-C motif) ligand (CCL)-3, IL-3, and IL-7 [91] [92]. Furthermore, activin A, secreted by BMSCs and monocytes, promotes OC differentiation and OB inhibition [93]. CCL20 is an additional factor involved in MM-induced OC activity [94]. Thereafter, MM cells interact with BMSCs and pre-OBs, through VLA-4/VCAM-1 and CD56/CD56 binding, and, in the latter ones, suppress the activity of the main pro-osteoblastogenic transcription factor, Runt-related transcription factor (Runx)-2, leading to an inhibition of OB differentiation [95]. Runx2 activity is also reduced by other soluble cytokines secreted by MM cells, as IL-7, IL-3 and HGF [96]. Finally, MM cells also inhibit the receptor for Wnt family member (Wnt)-5a, receptor tyrosine kinase-like orphan receptor (ROR)-2, on OB precursors and, indeed, block the activation of non-canonical Wnt signaling, important for OB differentiation [97].

2.3. The Immune Microenvironment in MM

MM patients share also several alterations in the immune system, due to a deficit of humoral immunity, immunoparesis and alterations in the activity of effectors cells [98]. In MM patients' peripheral blood, it is reported a reduction of the CD4+/CD8+ T cells ratio and an imbalance between Th1 and Th2, due to an overproduction of pro-Th2 cytokines, such as IL-4 and IL-10 [99]. In addition, an increase of Th17 cells is reported in MM BM; these cells secrete IL-17 that suppresses cytotoxic T cell activity, supports PC growth and is a key mediator of MM bone disease [100,101]. Moreover, MM cells express programmed cell death-ligand (PD-L)-1 that binds its receptor programmed cell death (PD)-1 expressed by T and NK cells and is deeply involved in the immunosuppression that characterize the MM immune microenvironment [102]. DCs are also dysfunctional for a deficit of co-stimulatory molecules, such as CD80 and CD86, needed for the activation of T cells [103]. The lack of DC functional activity is due to an accumulation of cytokines in the MM BM microenvironment, such as IL-6, IL-10, TGFβ and VEGFA [104]. Moreover, the interaction between MM PCs and DCs increases the production of indoleamine-pyrrole 2,3-dioxygenase (IDO) that promotes the T anergic phenotype and Tregs differentiation [105]. Finally, MM patients have elevated levels of MDSCs, compared to healthy subjects [106], and they are involved in the immune escape of MM cells.

3. Galectins and Multiple Myeloma

3.1. Galectin-1 and MM

In MM cells, galectin-1 is expressed at high level, at both mRNA and protein levels, maintaining inter-patient and inter-human myeloma cell line (HMCLs) variability [23,24]. Moreover, analyses of gene expression datasets of MM primary CD138+ cells revealed that *LGALS1* (galectin-1 gene) levels are significantly higher in newly diagnosed MM (MMD) patients, but not in MGUS, SMM and MM relapsed, compared to healthy donors [24,107]. Recently, it has been published that, in peripheral blood sera, galectin-1 protein level was borderline significantly higher in MMD compared to healthy controls and that the levels of this lectin in peripheral blood are not associated with OS, response to treatment and clinical pathological parameters [108]; on the other hand, this study shows only a positive correlation between galectin-1 and soluble (s)CD163, a macrophage activation marker, and sCD138 [108]. Moreover, galectin-1 has been identified as a ECM-associated protein that characterizes only the MM BM and not the ECM of MGUS patients or healthy controls [107].

Thereafter, Panero et al. highlighted that overexpression of *LGALS1* is associated with high mRNA expression of telomerase (*hTERT*) MM cells, ascribing a role of these lectin in MM cell proliferation [109].

In line with these data, in the first evidence of the role of galectin-1 in MM cells, Abroun et al. reported that galectin-1 binds β1-integrin and supports the proliferation of CD45RA(−) HMCLs, increasing the phosphorylation of ERK, AKT and IkB, and it has an opposite effect on CD45RA(+) MM cells, reducing their proliferation [23] (Figure 1).

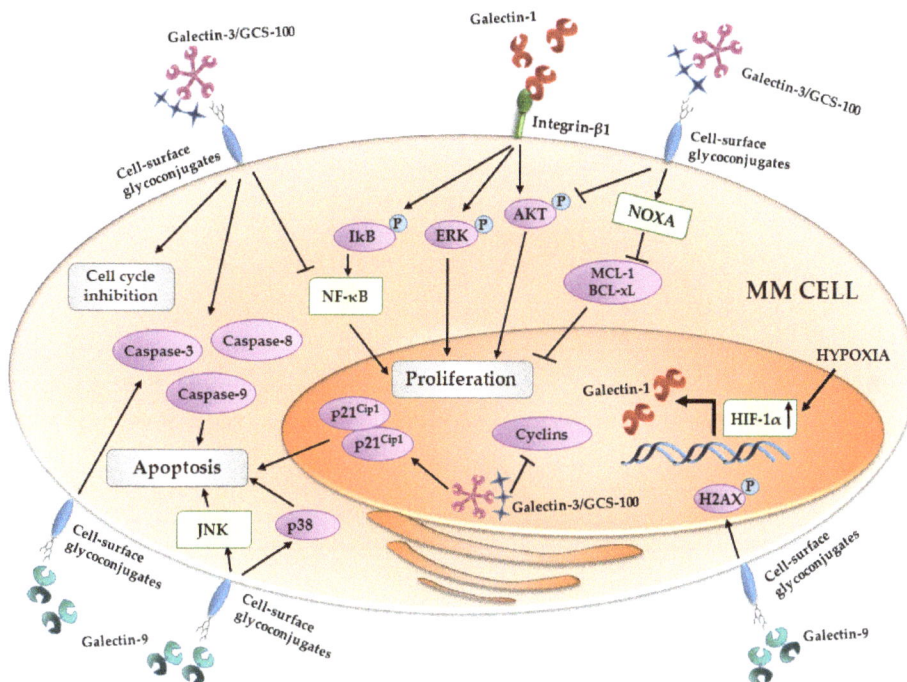

Figure 1. Effect of galectins on MM cells proliferation and apoptosis induction. Galectin-1 is regulated by HIF-1α and the galectin-1/integrin β1 binding induces pro-survival cascades. The inhibition of galectin-3 by GCS-100 (reported in figure as Galectin-3/GCS-100) leads to cell cycle arrest and an activation of different pro-apoptotic signals. Galectin-9 exerts its anti-proliferative activity inducing pro-apoptotic pathways. MM, multiple myeloma; IkB, inhibitor of kappa B; ERK, Extracellular signal-regulated kinases; AKT, Protein kinase B; NOXA, Phorbol-12-myristate-13-acetate-induced protein 1; MCL-1, Induced myeloid leukemia cell differentiation protein Mcl-1; Bcl-xL, B-cell lymphoma-extralarge; NF-κB, nuclear factor-kappaB; HIF-1α, Hypoxia inducible factor-1α; H2AX, H2A histone family, member X; JNK, c-Jun N-terminal kinases.

Recently, our group demonstrated that, in MM cells, galectin-1 is upregulated by hypoxia and its expression is controlled by HIF-1α [24]. Moreover, in HMCLs, the suppression of *LGALS1* by a shRNA lentiviral vector does not induce an alteration of cell proliferation or survival, but, on the other hand, galectin-1 inhibition leads to downregulation of pro-angiogenic molecules, such as monocyte chemoattractant protein (MCP)-1 and MMP-9, and an up-regulation of anti-angiogenic ones, such as Semaphorin-3A [24]. Indeed, galectin-1 suppression reduces the pro-angiogenic properties of HMCLs conditioned media (CM) in in vitro vessels formation [24] (Figure 2).

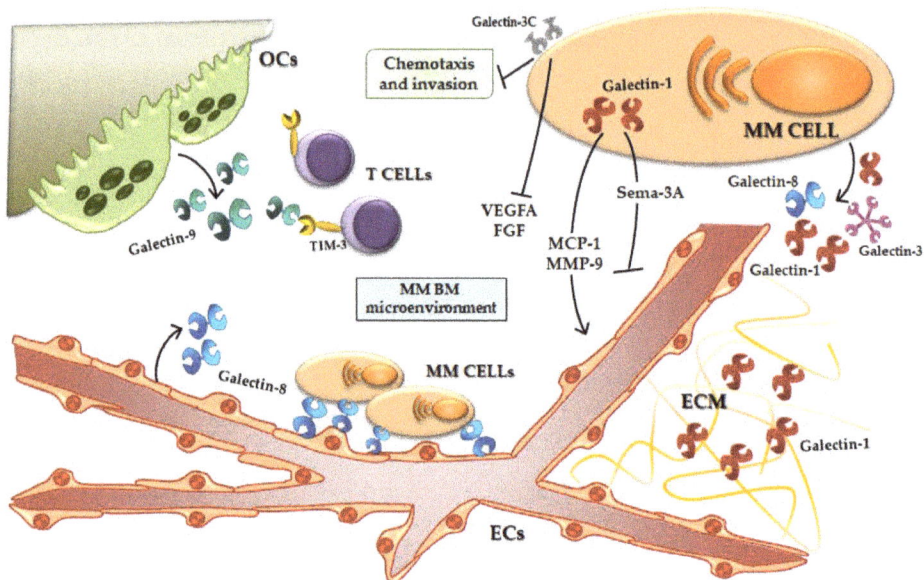

Figure 2. Galectins in the MM BM microenvironment. Galectin-1 induces the secretion of pro-angiogenic molecules and inhibits the production of anti-angiogenic ones. Galectin-1 is present in the ECM of MM patients. Galectin-3C reduces MM cell chemotaxis and invasion and the secretion of pro-angiogenic proteins. Galectin-8 is produced by ECs and mediates the adhesion between MM PCs and ECs. Galectin-9 is secreted by OCs and could mediate the T cell activity inhibition. OCs, Osteoclasts; TIM-3, T-cell immunoglobulin and mucin-domain containing-3; MM, multiple myeloma; VEGFA, Vascular endothelial growth factor *A*; FGF, Fibroblast growth factor; MCP-1, Monocyte chemoattractant protein-1; MMP-9, Metalloprotease-9; Sema-3A, Semaphorin -3A; ECM, Extracellular matrix; ECs, Endothelial cells; BM, bone marrow.

Finally, two different in vivo mouse models demonstrated the role of galectin-1 as a putative target in MM [24]. In both models, galectin-1 inhibition in MM cells significantly reduces tumor masses, tumor angiogenesis, and, in the intratibial model, the formation of bone lesions [24].

3.2. Galectin-3 and MM

Galectin-3 is variable expressed in HMCLs and in about 25% of primary MM CD138+ cells [30,36,110]. A citrus-derived polysaccharide inhibitor of galectin-3, GCS-100, induces apoptosis in primary MM cells and HMCLs, reduces MM cell proliferation supported by adhesion to BMSCs and blocks HMCLs migration induced by VEGFA [36]. In particular, GCS-100 modifies the MM cell cycle, leading to an accumulation of cells in sub-G_1 and G_1 phases, and induces an upregulation of the cell cycler inhibitor p21^{Cip1} and a reduction of different cyclins [35]. Thereafter, GCS-100-induced MM cell apoptosis is mediated by a reduction of MCL-1 and Bcl-xL (B-cell lymphoma-extralarge) proteins, an induction of NOXA (Phorbol-12-myristate-13-acetate-induced protein 1) and the activation of caspase-3, -8 and -9, associated with lower levels of activated AKT and NF-κB pathways [35] (Figure 1). Moreover, in MM cells, GCS-100 overcomes resistance to the proteasome inhibitor, bortezomib, and increases the apoptosis induced by dexamethasone treatment [36].

Mirandola et al. studied another galectin-3 negative dominant inhibitor, galectin-3C, a N-terminally truncated form of galectin-3, as a MM putative treatment [110]. Galectin-3C has a mild direct anti-proliferative effect on MM cells but shows an inhibition of chemotaxis and invasion stromal derived factor (SDF)-1α-mediated of U266, a HMCL [110]. Moreover, galectin-3C acts synergistically with bortezomib, reducing the migration of ECs mediated by MM cell CM and lowering secretion of VEGFA and FGF by HMCLs [110] (Figure 2). Finally, in a non-obese diabetic/severe combined immunodeficiency (NOD/SCID) murine model, MM tumor growth was inhibited by galectin-3C administration and, also, galectin-3C treatment enhances the anti-tumoral effect of bortezomib in vivo [110].

3.3. Galectin-8 and MM

Recently, Friedel et al. demonstrated that ECs produce galectin-8, in particular the splicing variants Gal-8S and Gal-8L, and HMCLs only the Gal-8S isoform; moreover, they reported that galectin-8 is present in the sera of about 45% of MM patients [38]. Both galectin-8 isoforms bind to MM cells and Gal-8L induces MM cell adhesion to ECs stronger than Gal-8S both in static tests and under dynamic shear stress test [38]. Finally, galectin-8 exploits its role in MM cell adhesion on ECs even after pro-inflammatory stimulus (TNF-α) [38] (Figure 2).

3.4. Galectin-9 and MM

Galectin-9, in contrast, has an anti-proliferative effect on cancer cells. In fact, recombinant protease resistant galectin-9 (hGal-9) exerts an anti-proliferative effect on HMCLs, with its efficiency positively correlated with the affinity for hGal-9 of each HMCLs [41]. In MM cells, the apoptotic induction by hGal-9 is mediated by the activation of caspase-3, -8 and -9 and the loss of mitochondrial outer membrane potential [41]. Moreover, the HMCL treatment with hGal-9 causes an up-regulation of c-Jun and JunD, the phosforilation of H2AX protein and the activation of JNK and p38/MAPK pathways that, taken together, support the pro-apoptotic process [41] (Figure 1). hGal-9 exploits its pro-apoptotic effect also on primary CD138+ from MM patients with poor cytogenetics factors and on bortezomib resistant HMCLs [41]. Finally, in in vivo mouse model, MM tumor growth is also inhibited with hGal-9 treatment [41].

Besides, An et al. explored the role of galectin-9 in osteoclastogenesis and MM BM immune suppression [42]. They have shown that galectin-9 protein is expressed by mature OCs but not by OC-precursors (monocytes) or by MM cells, that it is overexpressed during osteoclastogenesis and further induced by interferon-γ (IFN-γ) OC treatment [42]. Thereafter, BM plasma from MM patients show significantly higher levels of galectin-9, compared to healthy controls [42]. The authors concluded indicating a putative role of OC-secreted galectin-9 in Th1 cells negative regulation, through the interaction with its receptor TIM3 on T cell surface [42] (Figure 2).

3.5. Translational Implications in MM

As previously discussed, galectins are known to be involved in multiple biological processes and they could be an attractive therapeutic target in MM, blocking tumoral growth. However, there are some evidences that MM cells may change their glycome during the malignant transformation from normal PCs; this aspect deserve to be considered in a possible therapeutic context because changes on some tumor glycoproteins on MM cells surfaces, such as mucin-1, could lead to different affinity to galectins, resulting in possible alterations in their signals activation [111,112].

Targeting galectin-1 by specific inhibitors, e.g., OTX-008 tested in solid tumors [113], could be a strategy to block the process of neo-angiogenesis induced by MM cells, the expansion of the malignant clone and indeed the formation of lytic lesions [23,24]. Another way to reduce the cross talk between MM cell and ECs in the BM, and consequently the angiogenic process, is to disrupt the functions of galectin-8 [38]; however, this mechanism needs further studies since there are few literature data in MM setting. Moreover, the two compounds GCS-100 and Galectin-3C, which target galectin-3, showed promising data in vitro and in vivo, inhibiting MM cells proliferation and overcoming drug resistance [35,36,110]. In addition, galectin-9, that shows a pro-apoptotic effect on MM cells [41], could be targeted through the development of stable galectin-9, to be delivered on MM cells. Finally, because galectin-1, -3 and -9 have a role in tumor immune escape [11], we can suppose that targeting these galectins may restore the immune control of the disease. These aspects should be expanded and the combination of galectin-inhibition with the blocking of immune inhibitory checkpoints cascade by monoclonal antibodies (e.g., Pembrolizumab, Nivolumab or Ipilimumab) or with immunomodulatory drugs (IMiDs) such as lenalidomide and pomalidomide, might deserve further studies.

3.6. MM Patients' Overall Survival and Galectins

Galectins could be also associated with patients' outcome. In fact, *LGALS1* gene is included in the 70 genes PCs signature of Shaughnessy et al. that defines MM high risk [114]. Moreover, analyses of two independent gene expression datasets reveled that high level of *LGALS1* expression in MM CD138+ cells is correlated with reduced OS in MM patients compared to those with lower level of *LGALS1* [24,107]. In addition, *LGALS1* expression is also involved in patients' drug resistance; in fact, this gene is included in a 23 genes signature that distinguishes bortezomib-resistant MM mouse cell line from bortezomib-sensible cell line and significantly predicts differences in patient's outcomes after treatment with bortezomib [115].

Finally, galectin-8 could also be correlated with patients' outcome: OS and event free survival of MM patients of the total therapy 2 (TT2) and TT3 studies are longer in patients expressing low levels of *LGALS8* (galectin-8 gene) as compared to the *LGALS8* high group [38].

4. Conclusions

As we have explored in this review, galectins have pleiotropic effects in MM BM microenvironment, playing a role in survival, apoptosis, angiogenic properties of MM cell and in OC-mediated immunosuppression. Based on these literature data, the galectin-driven processes, which support MM cells survival and disease progression, deserve further studies and specific galectin inhibitors or activators (in the case of galectin-9) should be designed and fine-tuned. Moreover, different studies have demonstrated the relationship between the gene expression profile of *LGALS1* and *LGALS8* with the survival of MM patients indicating a possible prognostic role of these genes. Clearly further prospective studies should be necessary to confirm these observations.

Acknowledgments: This work was supported in part by a grant from the Associazione Italiana per la Ricerca sul Cancro (AIRC) IG2014 n.15531 (N.G.). We thank the "Associazione Italiana Contro le Leucemie, i Linfomi ed i Mielomi"" (AIL) Parma section for the support.

Author Contributions: Paola Storti wrote the manuscript; Valentina Marchica revised the manuscript and designed the figures; and Nicola Giuliani revised the manuscript.

Conflicts of Interest: The authors declare no conflict of interest.

Abbreviations

ALCAM	Activated Leucocyte Cell Adhesion Molecule
AKT	Protein kinase B
ANG-1	Angiopoietin-1
Bcl-XL	B-cell lymphoma-extralarge
BLIMP-1	B-Lymphocyte-Induced Maturation Protein 1
BM	Bone Marrow
BMSC	Bone Marrow Stromal Cell
CCL	Chemokine (C–C motif) Ligand
CCN	Cyclin
CM	Conditioned Media
CRD	Carbohydrate-Recognition Domains
DC	Dendritic Cell
EC	Endothelial Cell
ECM	Extracellular Matrix
ERK	Extracellular Signal-Regulated Kinases

FGF	Fibroblast Growth Factor
hGal-9	Recombinant Protease Resistant Galectin-9
H2AX	H2A histone family, member X
HIF-1α	Hypoxia inducible factor-1α
HGF	Hepatocyte Growth Factor
HMCL	Human Myeloma Cell Line
hTERT	Telomerase
ICAM-1	Intercellular Adhesion Molecule 1
IDO	Indoleamine-Pyrrole 2,3-Dioxygenase
IFN-γ	Interferon-γ
IGF-1	Insuline-Like Growth Factor 1
IκB	inhibitor of kappa B
IL	Interleukin
JAK	Janus Kinase
JNK	c-Jun N-terminal kinases
LAG-3	Lymphocyte-Activation Gene 3
LFA-1	Lymphocyte Function Associate Antigen 1
MCL-1	Induced myeloid leukemia cell differentiation protein Mcl-1
MCP-1	Monocyte Chemoattractant Protein-1
MDSC	Myeloid-Derived Suppressor Cells
MEK	Mitogen-Activated Protein Kinase Kinase 1
MGUS	Monoclonal Gammopathy of Undetermined Significance
MM	Multiple Myeloma
MMD	Newly Diagnosed Myeloma
MMP	Metalloprotease
MMSET	Multiple Myeloma SET Domain
NF-κB	Nuclear Factor-kappa-B
NK	Natural Killer
NOXA	Phorbol-12-myristate-13-acetate-induced protein 1
OB	Osteoblast
OC	Osteoclast
OPG	Osteoprotegerin
OPN	Osteopontin
OS	Overall Survival
PC	Plasma Cells
PD-1	Programmed Cell Death-1
PD-L1	Programmed Cell Death Ligand -1
PI3K	Phosphoinositide 3-Kinase
pre-BCR	Pre-B Cell Receptor
RAF1	Raf-1 Proto-Oncogene, Serine/Threonine Kinase
RANKL	Receptor Activator of Nuclear Factor kappa-B Ligand
ROR2	Receptor Tyrosine Kinase-Like Orphan Receptor 2
RUNX2	Runt-Related Transcription Factor 2
Sema-3A	Semaphorin-3A
SMM	Smoldering Myeloma
STAT	Signal Transducers and Activators of Transcription
TGFβ	Transforming Growth Factor β
Th	T Helper
TIE-2	Immunoglobulin-like and EGF-like Domains
TIM-3	T-cell Immunoglobulin and Mucin-Domain Containing-3
TNFα	Tumor Necrosis Factor α
Treg	T Regulatory
TT	Total Therapy
VCAM-1	Vascular Cell Adhesion Molecule 1
VEGFA	Vascular Endothelial Growth Factor *A*
VEGFR2	Vascular Endothelial Growth Factor Receptor 2
VLA-4	Very Late Antigen-4
WNT-5a	Wnt Family Member 5a

References

1. Yang, R.Y.; Rabinovich, G.A.; Liu, F.T. Galectins: Structure, function and therapeutic potential. *Expert Rev. Mol. Med.* **2008**, *10*, e17. [CrossRef] [PubMed]
2. Cummings, R.D.; Liu, F.T.; Vasta, G.R. Galectins. In *Essentials of Glycobiology*, 3rd ed.; Varki, A., Cummings, R.D., Esko, J.D., Stanley, P., Hart, G.W., Aebi, M., Darvill, A.G., Kinoshita, T., Packer, N.H., Prestegard, J.H., et al., Eds.; Cold Spring Harbor: New York, NY, USA, 2015.
3. Cooper, D.N. Galectinomics: Finding themes in complexity. *Biochim. Biophys. Acta* **2002**, *1572*, 209–231. [CrossRef]
4. Ahmad, N.; Gabius, H.J.; Andre, S.; Kaltner, H.; Sabesan, S.; Roy, R.; Liu, B.; Macaluso, F.; Brewer, C.F. Galectin-3 precipitates as a pentamer with synthetic multivalent carbohydrates and forms heterogeneous cross-linked complexes. *J. Biol. Chem.* **2004**, *279*, 10841–10847. [CrossRef] [PubMed]
5. Hughes, R.C. Secretion of the galectin family of mammalian carbohydrate-binding proteins. *Biochim. Biophys. Acta* **1999**, *1473*, 172–185. [CrossRef]
6. Rabinovich, G.A.; Toscano, M.A.; Jackson, S.S.; Vasta, G.R. Functions of cell surface galectin-glycoprotein lattices. *Curr. Opin. Struct. Biol.* **2007**, *17*, 513–520. [CrossRef] [PubMed]
7. Brewer, C.F.; Miceli, M.C.; Baum, L.G. Clusters, bundles, arrays and lattices: Novel mechanisms for lectin-saccharide-mediated cellular interactions. *Curr. Opin. Struct. Biol.* **2002**, *12*, 616–623. [CrossRef]
8. Nabi, I.R.; Shankar, J.; Dennis, J.W. The galectin lattice at a glance. *J. Cell Sci.* **2015**, *128*, 2213–2219. [CrossRef] [PubMed]
9. Hughes, R.C. Galectins as modulators of cell adhesion. *Biochimie* **2001**, *83*, 667–676. [CrossRef]
10. Hirabayashi, J.; Hashidate, T.; Arata, Y.; Nishi, N.; Nakamura, T.; Hirashima, M.; Urashima, T.; Oka, T.; Futai, M.; Muller, W.E.; et al. Oligosaccharide specificity of galectins: A search by frontal affinity chromatography. *Biochim. Biophys. Acta* **2002**, *1572*, 232–254. [CrossRef]
11. Rabinovich, G.A.; Toscano, M.A. Turning 'sweet' on immunity: Galectin-glycan interactions in immune tolerance and inflammation. *Nat. Rev. Immunol.* **2009**, *9*, 338–352. [CrossRef] [PubMed]
12. Ideo, H.; Seko, A.; Ishizuka, I.; Yamashita, K. The N-terminal carbohydrate recognition domain of galectin-8 recognizes specific glycosphingolipids with high affinity. *Glycobiology* **2003**, *13*, 713–723. [CrossRef] [PubMed]
13. Yoshida, H.; Teraoka, M.; Nishi, N.; Nakakita, S.; Nakamura, T.; Hirashima, M.; Kamitori, S. X-ray structures of human galectin-9 C-terminal domain in complexes with a biantennary oligosaccharide and sialyllactose. *J. Biol. Chem.* **2010**, *285*, 36969–36976. [CrossRef] [PubMed]
14. Kumar, S.; Frank, M.; Schwartz-Albiez, R. Understanding the specificity of human galectin-8C domain interactions with its glycan ligands based on molecular dynamics simulations. *PLoS ONE* **2013**, *8*, e59761. [CrossRef] [PubMed]
15. Levy, Y.; Auslender, S.; Eisenstein, M.; Vidavski, R.R.; Ronen, D.; Bershadsky, A.D.; Zick, Y. It depends on the hinge: A structure-functional analysis of galectin-8, a tandem-repeat type lectin. *Glycobiology* **2006**, *16*, 463–476. [CrossRef] [PubMed]
16. Verschuere, T.; van Woensel, M.; Fieuws, S.; Lefranc, F.; Mathieu, V.; Kiss, R.; van Gool, S.W.; de Vleeschouwer, S. Altered galectin-1 serum levels in patients diagnosed with high-grade glioma. *J. Neurooncol.* **2013**, *115*, 9–17. [CrossRef] [PubMed]
17. Croci, D.O.; Salatino, M.; Rubinstein, N.; Cerliani, J.P.; Cavallin, L.E.; Leung, H.J.; Ouyang, J.; Ilarregui, J.M.; Toscano, M.A.; Domaica, C.I.; et al. Disrupting galectin-1 interactions with *N*-glycans suppresses hypoxia-driven angiogenesis and tumorigenesis in Kaposi's sarcoma. *J. Exp. Med.* **2012**, *209*, 1985–2000. [CrossRef] [PubMed]
18. Tsai, C.M.; Guan, C.H.; Hsieh, H.W.; Hsu, T.L.; Tu, Z.; Wu, K.J.; Lin, C.H.; Lin, K.I. Galectin-1 and galectin-8 have redundant roles in promoting plasma cell formation. *J. Immunol.* **2011**, *187*, 1643–1652. [CrossRef] [PubMed]
19. Ito, K.; Stannard, K.; Gabutero, E.; Clark, A.M.; Neo, S.Y.; Onturk, S.; Blanchard, H.; Ralph, S.J. Galectin-1 as a potent target for cancer therapy: Role in the tumor microenvironment. *Cancer Metastasis Rev.* **2012**, *31*, 763–778. [CrossRef] [PubMed]
20. Lykken, J.M.; Horikawa, M.; Minard-Colin, V.; Kamata, M.; Miyagaki, T.; Poe, J.C.; Tedder, T.F. Galectin-1 drives lymphoma CD20 immunotherapy resistance: Validation of a preclinical system to identify resistance mechanisms. *Blood* **2016**, *127*, 1886–1895. [CrossRef] [PubMed]

21. Cedeno-Laurent, F.; Watanabe, R.; Teague, J.E.; Kupper, T.S.; Clark, R.A.; Dimitroff, C.J. Galectin-1 inhibits the viability, proliferation, and Th1 cytokine production of nonmalignant T cells in patients with leukemic cutaneous T-cell lymphoma. *Blood* **2012**, *119*, 3534–3538. [CrossRef] [PubMed]

22. Thijssen, V.L.; Postel, R.; Brandwijk, R.J.; Dings, R.P.; Nesmelova, I.; Satijn, S.; Verhofstad, N.; Nakabeppu, Y.; Baum, L.G.; Bakkers, J.; et al. Galectin-1 is essential in tumor angiogenesis and is a target for antiangiogenesis therapy. *Proc. Natl. Acad. Sci. USA* **2006**, *103*, 15975–15980. [CrossRef] [PubMed]

23. Abroun, S.; Otsuyama, K.; Shamsasenjan, K.; Islam, A.; Amin, J.; Iqbal, M.S.; Gondo, T.; Asaoku, H.; Kawano, M.M. Galectin-1 supports the survival of CD45RA(−) primary myeloma cells in vitro. *Br. J. Haematol.* **2008**, *142*, 754–765. [CrossRef] [PubMed]

24. Storti, P.; Marchica, V.; Airoldi, I.; Donofrio, G.; Fiorini, E.; Ferri, V.; Guasco, D.; Todoerti, K.; Silbermann, R.; Anderson, J.L.; et al. Galectin-1 suppression delineates a new strategy to inhibit myeloma-induced angiogenesis and tumoral growth in vivo. *Leukemia* **2016**. [CrossRef] [PubMed]

25. Le, Q.T.; Shi, G.; Cao, H.; Nelson, D.W.; Wang, Y.; Chen, E.Y.; Zhao, S.; Kong, C.; Richardson, D.; O'Byrne, K.J.; et al. Galectin-1: A link between tumor hypoxia and tumor immune privilege. *J. Clin. Oncol.* **2005**, *23*, 8932–8941. [CrossRef] [PubMed]

26. Camby, I.; Le Mercier, M.; Lefranc, F.; Kiss, R. Galectin-1: A small protein with major functions. *Glycobiology* **2006**, *16*, 137R–157R. [CrossRef] [PubMed]

27. Wu, M.H.; Hong, T.M.; Cheng, H.W.; Pan, S.H.; Liang, Y.R.; Hong, H.C.; Chiang, W.F.; Wong, T.Y.; Shieh, D.B.; Shiau, A.L.; et al. Galectin-1-mediated tumor invasion and metastasis, up-regulated matrix metalloproteinase expression, and reorganized actin cytoskeletons. *Mol. Cancer Res. MCR* **2009**, *7*, 311–318. [CrossRef] [PubMed]

28. Gieseke, F.; Bohringer, J.; Bussolari, R.; Dominici, M.; Handgretinger, R.; Muller, I. Human multipotent mesenchymal stromal cells use galectin-1 to inhibit immune effector cells. *Blood* **2010**, *116*, 3770–3779. [CrossRef] [PubMed]

29. Astorgues-Xerri, L.; Riveiro, M.E.; Tijeras-Raballand, A.; Serova, M.; Neuzillet, C.; Albert, S.; Raymond, E.; Faivre, S. Unraveling galectin-1 as a novel therapeutic target for cancer. *Cancer Treat. Rev.* **2014**, *40*, 307–319. [CrossRef] [PubMed]

30. Hoyer, K.K.; Pang, M.; Gui, D.; Shintaku, I.P.; Kuwabara, I.; Liu, F.T.; Said, J.W.; Baum, L.G.; Teitell, M.A. An anti-apoptotic role for galectin-3 in diffuse large B-cell lymphomas. *Am. J. Pathol.* **2004**, *164*, 893–902. [CrossRef]

31. Yamamoto-Sugitani, M.; Kuroda, J.; Ashihara, E.; Nagoshi, H.; Kobayashi, T.; Matsumoto, Y.; Sasaki, N.; Shimura, Y.; Kiyota, M.; Nakayama, R.; et al. Galectin-3 (Gal-3) induced by leukemia microenvironment promotes drug resistance and bone marrow lodgment in chronic myelogenous leukemia. *Proc. Natl. Acad. Sci. USA* **2011**, *108*, 17468–17473. [CrossRef] [PubMed]

32. Simon, D.; Derer, A.; Andes, F.T.; Lezuo, P.; Bozec, A.; Schett, G.; Herrmann, M.; Harre, U. Galectin-3 as a novel regulator of osteoblast-osteoclast interaction and bone homeostasis. *Bone* **2017**, *105*, 35–41. [CrossRef] [PubMed]

33. Yoshii, T.; Inohara, H.; Takenaka, Y.; Honjo, Y.; Akahani, S.; Nomura, T.; Raz, A.; Kubo, T. Galectin-3 maintains the transformed phenotype of thyroid papillary carcinoma cells. *Int. J. Oncol.* **2001**, *18*, 787–792. [CrossRef] [PubMed]

34. Kouo, T.; Huang, L.; Pucsek, A.B.; Cao, M.; Solt, S.; Armstrong, T.; Jaffee, E. Galectin-3 shapes antitumor immune responses by suppressing CD8+ T cells via LAG-3 and inhibiting expansion of plasmacytoid dendritic cells. *Cancer Immunol. Res.* **2015**, *3*, 412–423. [CrossRef] [PubMed]

35. Streetly, M.J.; Maharaj, L.; Joel, S.; Schey, S.A.; Gribben, J.G.; Cotter, F.E. GCS-100, a novel galectin-3 antagonist, modulates MCL-1, NOXA, and cell cycle to induce myeloma cell death. *Blood* **2010**, *115*, 3939–3948. [CrossRef] [PubMed]

36. Chauhan, D.; Li, G.; Podar, K.; Hideshima, T.; Neri, P.; He, D.; Mitsiades, N.; Richardson, P.; Chang, Y.; Schindler, J.; et al. A novel carbohydrate-based therapeutic GCS-100 overcomes bortezomib resistance and enhances dexamethasone-induced apoptosis in multiple myeloma cells. *Cancer Res.* **2005**, *65*, 8350–8358. [CrossRef] [PubMed]

37. D'Haene, N.; Sauvage, S.; Maris, C.; Adanja, I.; Le Mercier, M.; Decaestecker, C.; Baum, L.; Salmon, I. VEGFR1 and VEGFR2 involvement in extracellular galectin-1- and galectin-3-induced angiogenesis. *PLoS ONE* **2013**, *8*, e67029. [CrossRef] [PubMed]

38. Friedel, M.; Andre, S.; Goldschmidt, H.; Gabius, H.J.; Schwartz-Albiez, R. Galectin-8 enhances adhesion of multiple myeloma cells to vascular endothelium and is an adverse prognostic factor. *Glycobiology* **2016**, *26*, 1048–1058. [CrossRef] [PubMed]

39. Delgado, V.M.; Nugnes, L.G.; Colombo, L.L.; Troncoso, M.F.; Fernandez, M.M.; Malchiodi, E.L.; Frahm, I.; Croci, D.O.; Compagno, D.; Rabinovich, G.A.; et al. Modulation of endothelial cell migration and angiogenesis: A novel function for the "tandem-repeat" lectin galectin-8. *FASEB J.* **2011**, *25*, 242–254. [CrossRef] [PubMed]

40. Zhou, Q.; Munger, M.E.; Veenstra, R.G.; Weigel, B.J.; Hirashima, M.; Munn, D.H.; Murphy, W.J.; Azuma, M.; Anderson, A.C.; Kuchroo, V.K.; et al. Coexpression of Tim-3 and PD-1 identifies a CD8+ T-cell exhaustion phenotype in mice with disseminated acute myelogenous leukemia. *Blood* **2011**, *117*, 4501–4510. [CrossRef] [PubMed]

41. Kobayashi, T.; Kuroda, J.; Ashihara, E.; Oomizu, S.; Terui, Y.; Taniyama, A.; Adachi, S.; Takagi, T.; Yamamoto, M.; Sasaki, N.; et al. Galectin-9 exhibits anti-myeloma activity through JNK and p38 MAP kinase pathways. *Leukemia* **2010**, *24*, 843–850. [CrossRef] [PubMed]

42. An, G.; Acharya, C.; Feng, X.; Wen, K.; Zhong, M.; Zhang, L.; Munshi, N.C.; Qiu, L.; Tai, Y.T.; Anderson, K.C. Osteoclasts promote immune suppressive microenvironment in multiple myeloma: Therapeutic implication. *Blood* **2016**. [CrossRef] [PubMed]

43. Dardalhon, V.; Anderson, A.C.; Karman, J.; Apetoh, L.; Chandwaskar, R.; Lee, D.H.; Cornejo, M.; Nishi, N.; Yamauchi, A.; Quintana, F.J.; et al. Tim-3/galectin-9 pathway: Regulation of Th1 immunity through promotion of CD11b+Ly-6G+ myeloid cells. *J. Immunol.* **2010**, *185*, 1383–1392. [CrossRef] [PubMed]

44. Kubach, J.; Lutter, P.; Bopp, T.; Stoll, S.; Becker, C.; Huter, E.; Richter, C.; Weingarten, P.; Warger, T.; Knop, J.; et al. Human CD4+CD25+ regulatory T cells: Proteome analysis identifies galectin-10 as a novel marker essential for their anergy and suppressive function. *Blood* **2007**, *110*, 1550–1558. [CrossRef] [PubMed]

45. Lingblom, C.; Andersson, J.; Andersson, K.; Wenneras, C. Regulatory eosinophils suppress T cells partly through galectin-10. *J. Immunol.* **2017**, *198*, 4672–4681. [CrossRef] [PubMed]

46. Rabinovich, G.A.; Vidal, M. Galectins and microenvironmental niches during hematopoiesis. *Curr. Opin. Hematol.* **2011**, *18*, 443–451. [CrossRef] [PubMed]

47. Zhu, C.; Anderson, A.C.; Schubart, A.; Xiong, H.; Imitola, J.; Khoury, S.J.; Zheng, X.X.; Strom, T.B.; Kuchroo, V.K. The Tim-3 ligand galectin-9 negatively regulates T helper type 1 immunity. *Nat. Immunol.* **2005**, *6*, 1245–1252. [CrossRef] [PubMed]

48. Elantak, L.; Espeli, M.; Boned, A.; Bornet, O.; Bonzi, J.; Gauthier, L.; Feracci, M.; Roche, P.; Guerlesquin, F.; Schiff, C. Structural basis for galectin-1-dependent pre-B cell receptor (pre-BCR) activation. *J. Biol. Chem.* **2012**, *287*, 44703–44713. [CrossRef] [PubMed]

49. Starossom, S.C.; Mascanfroni, I.D.; Imitola, J.; Cao, L.; Raddassi, K.; Hernandez, S.F.; Bassil, R.; Croci, D.O.; Cerliani, J.P.; Delacour, D.; et al. Galectin-1 deactivates classically activated microglia and protects from inflammation-induced neurodegeneration. *Immunity* **2012**, *37*, 249–263. [CrossRef] [PubMed]

50. Anderson, D.E. Tim-3 as a therapeutic target in human inflammatory diseases. *Expert Opin. Ther. Targets* **2007**, *11*, 1005–1009. [CrossRef] [PubMed]

51. Liu, F.T.; Rabinovich, G.A. Galectins as modulators of tumour progression. *Nat. Rev. Cancer* **2005**, *5*, 29–41. [CrossRef] [PubMed]

52. Yamaoka, K.; Mishima, K.; Nagashima, Y.; Asai, A.; Sanai, Y.; Kirino, T. Expression of galectin-1 mRNA correlates with the malignant potential of human gliomas and expression of antisense galectin-1 inhibits the growth of 9 glioma cells. *J. Neurosci. Res.* **2000**, *59*, 722–730. [CrossRef]

53. Honjo, Y.; Nangia-Makker, P.; Inohara, H.; Raz, A. Down-regulation of galectin-3 suppresses tumorigenicity of human breast carcinoma cells. *Clin. Cancer Res.* **2001**, *7*, 661–668. [PubMed]

54. Paz, A.; Haklai, R.; Elad-Sfadia, G.; Ballan, E.; Kloog, Y. Galectin-1 binds oncogenic H-ras to mediate Ras membrane anchorage and cell transformation. *Oncogene* **2001**, *20*, 7486–7493. [CrossRef] [PubMed]

55. Elad-Sfadia, G.; Haklai, R.; Balan, E.; Kloog, Y. Galectin-3 augments K-Ras activation and triggers a Ras signal that attenuates ERK but not phosphoinositide 3-kinase activity. *J. Biol. Chem.* **2004**, *279*, 34922–34930. [CrossRef] [PubMed]

56. Nishi, N.; Shoji, H.; Seki, M.; Itoh, A.; Miyanaka, H.; Yuube, K.; Hirashima, M.; Nakamura, T. Galectin-8 modulates neutrophil function via interaction with integrin alpham. *Glycobiology* **2003**, *13*, 755–763. [CrossRef] [PubMed]

57. Hittelet, A.; Legendre, H.; Nagy, N.; Bronckart, Y.; Pector, J.C.; Salmon, I.; Yeaton, P.; Gabius, H.J.; Kiss, R.; Camby, I. Upregulation of galectins-1 and -3 in human colon cancer and their role in regulating cell migration. *Int. J. Cancer* **2003**, *103*, 370–379. [CrossRef] [PubMed]

58. Croci, D.O.; Cerliani, J.P.; Dalotto-Moreno, T.; Mendez-Huergo, S.P.; Mascanfroni, I.D.; Dergan-Dylon, S.; Toscano, M.A.; Caramelo, J.J.; Garcia-Vallejo, J.J.; Ouyang, J.; et al. Glycosylation-dependent lectin-receptor interactions preserve angiogenesis in anti-VEGF refractory tumors. *Cell* **2014**, *156*, 744–758. [CrossRef] [PubMed]

59. Thijssen, V.L.; Barkan, B.; Shoji, H.; Aries, I.M.; Mathieu, V.; Deltour, L.; Hackeng, T.M.; Kiss, R.; Kloog, Y.; Poirier, F.; et al. Tumor cells secrete galectin-1 to enhance endothelial cell activity. *Cancer Res.* **2010**, *70*, 6216–6224. [CrossRef] [PubMed]

60. Markowska, A.I.; Jefferies, K.C.; Panjwani, N. Galectin-3 protein modulates cell surface expression and activation of vascular endothelial growth factor receptor 2 in human endothelial cells. *J. Biol. Chem.* **2011**, *286*, 29913–29921. [CrossRef] [PubMed]

61. Ilarregui, J.M.; Croci, D.O.; Bianco, G.A.; Toscano, M.A.; Salatino, M.; Vermeulen, M.E.; Geffner, J.R.; Rabinovich, G.A. Tolerogenic signals delivered by dendritic cells to T cells through a galectin-1-driven immunoregulatory circuit involving interleukin 27 and interleukin 10. *Nat. Immunol.* **2009**, *10*, 981–991. [CrossRef] [PubMed]

62. Dalotto-Moreno, T.; Croci, D.O.; Cerliani, J.P.; Martinez-Allo, V.C.; Dergan-Dylon, S.; Mendez-Huergo, S.P.; Stupirski, J.C.; Mazal, D.; Osinaga, E.; Toscano, M.A.; et al. Targeting galectin-1 overcomes breast cancer-associated immunosuppression and prevents metastatic disease. *Cancer Res.* **2013**, *73*, 1107–1117. [CrossRef] [PubMed]

63. Mendez-Huergo, S.P.; Blidner, A.G.; Rabinovich, G.A. Galectins: Emerging regulatory checkpoints linking tumor immunity and angiogenesis. *Curr. Opin. Immunol.* **2017**, *45*, 8–15. [CrossRef] [PubMed]

64. Baker, G.J.; Chockley, P.; Zamler, D.; Castro, M.G.; Lowenstein, P.R. Natural killer cells require monocytic Gr-1(+)/CD11b(+) myeloid cells to eradicate orthotopically engrafted glioma cells. *Oncoimmunology* **2016**, *5*, e1163461. [CrossRef] [PubMed]

65. Ouyang, J.; Plutschow, A.; Pogge von Strandmann, E.; Reiners, K.S.; Ponader, S.; Rabinovich, G.A.; Neuberg, D.; Engert, A.; Shipp, M.A. Galectin-1 serum levels reflect tumor burden and adverse clinical features in classical Hodgkin lymphoma. *Blood* **2013**, *121*, 3431–3433. [CrossRef] [PubMed]

66. Pena, C.; Mirandola, L.; Figueroa, J.A.; Hosiriluck, N.; Suvorava, N.; Trotter, K.; Reidy, A.; Rakhshanda, R.; Payne, D.; Jenkins, M.; et al. Galectins as therapeutic targets for hematological malignancies: A hopeful sweetness. *Ann. Transl. Med.* **2014**, *2*, 87. [PubMed]

67. Juszczynski, P.; Ouyang, J.; Monti, S.; Rodig, S.J.; Takeyama, K.; Abramson, J.; Chen, W.; Kutok, J.L.; Rabinovich, G.A.; Shipp, M.A. The AP1-dependent secretion of galectin-1 by Reed Sternberg cells fosters immune privilege in classical Hodgkin lymphoma. *Proc. Natl. Acad. Sci. USA* **2007**, *104*, 13134–13139. [CrossRef] [PubMed]

68. Clark, M.C.; Pang, M.; Hsu, D.K.; Liu, F.T.; de Vos, S.; Gascoyne, R.D.; Said, J.; Baum, L.G. Galectin-3 binds to CD45 on diffuse large B-cell lymphoma cells to regulate susceptibility to cell death. *Blood* **2012**, *120*, 4635–4644. [CrossRef] [PubMed]

69. Cheng, C.L.; Hou, H.A.; Lee, M.C.; Liu, C.Y.; Jhuang, J.Y.; Lai, Y.J.; Lin, C.W.; Chen, H.Y.; Liu, F.T.; Chou, W.C.; et al. Higher bone marrow LGALS3 expression is an independent unfavorable prognostic factor for overall survival in patients with acute myeloid leukemia. *Blood* **2013**, *121*, 3172–3180. [CrossRef] [PubMed]

70. Palumbo, A.; Anderson, K. Multiple myeloma. *N. Engl. J. Med.* **2011**, *364*, 1046–1060. [CrossRef] [PubMed]

71. Kuehl, W.M.; Bergsagel, P.L. Molecular pathogenesis of multiple myeloma and its premalignant precursor. *J. Clin. Investig.* **2012**, *122*, 3456–3463. [CrossRef] [PubMed]

72. Morgan, G.J.; Walker, B.A.; Davies, F.E. The genetic architecture of multiple myeloma. *Nat. Rev. Cancer* **2012**, *12*, 335–348. [CrossRef] [PubMed]

73. Podar, K.; Richardson, P.G.; Hideshima, T.; Chauhan, D.; Anderson, K.C. The malignant clone and the bone-marrow environment. *Best Pract. Res. Clin. Haematol.* **2007**, *20*, 597–612. [CrossRef] [PubMed]

74. Gonzalez, D.; van der Burg, M.; Garcia-Sanz, R.; Fenton, J.A.; Langerak, A.W.; Gonzalez, M.; van Dongen, J.J.; San Miguel, J.F.; Morgan, G.J. Immunoglobulin gene rearrangements and the pathogenesis of multiple myeloma. *Blood* **2007**, *110*, 3112–3121. [CrossRef] [PubMed]

75. Annunziata, C.M.; Davis, R.E.; Demchenko, Y.; Bellamy, W.; Gabrea, A.; Zhan, F.; Lenz, G.; Hanamura, I.; Wright, G.; Xiao, W.; et al. Frequent engagement of the classical and alternative NF-κB pathways by diverse genetic abnormalities in multiple myeloma. *Cancer Cell* **2007**, *12*, 115–130. [CrossRef] [PubMed]

76. Quarona, V.; Ferri, V.; Chillemi, A.; Bolzoni, M.; Mancini, C.; Zaccarello, G.; Roato, I.; Morandi, F.; Marimpietri, D.; Faccani, G.; et al. Unraveling the contribution of ectoenzymes to myeloma life and survival in the bone marrow niche. *Ann. N. Y. Acad. Sci.* **2015**, *1335*, 10–22. [CrossRef] [PubMed]

77. Podar, K.; Chauhan, D.; Anderson, K.C. Bone marrow microenvironment and the identification of new targets for myeloma therapy. *Leukemia* **2009**, *23*, 10–24. [CrossRef] [PubMed]

78. Kawano, M.; Hirano, T.; Matsuda, T.; Taga, T.; Horii, Y.; Iwato, K.; Asaoku, H.; Tang, B.; Tanabe, O.; Tanaka, H.; et al. Autocrine generation and requirement of BSF-2/IL-6 for human multiple myelomas. *Nature* **1988**, *332*, 83–85. [CrossRef] [PubMed]

79. Klein, B.; Zhang, X.G.; Lu, Z.Y.; Bataille, R. Interleukin-6 in human multiple myeloma. *Blood* **1995**, *85*, 863–872. [PubMed]

80. Menu, E.; Kooijman, R.; van Valckenborgh, E.; Asosingh, K.; Bakkus, M.; van Camp, B.; Vanderkerken, K. Specific roles for the PI3K and the MEK-ERK pathway in IGF-1-stimulated chemotaxis, VEGF secretion and proliferation of multiple myeloma cells: Study in the 5T33MM model. *Br. J. Cancer* **2004**, *90*, 1076–1083. [CrossRef] [PubMed]

81. Giuliani, N.; Storti, P.; Bolzoni, M.; Palma, B.D.; Bonomini, S. Angiogenesis and multiple myeloma. *Cancer Microenviron.* **2011**, *4*, 325–337. [CrossRef] [PubMed]

82. Storti, P.; Bolzoni, M.; Donofrio, G.; Airoldi, I.; Guasco, D.; Toscani, D.; Martella, E.; Lazzaretti, M.; Mancini, C.; Agnelli, L.; et al. Hypoxia-inducible factor (HIF)-1α suppression in myeloma cells blocks tumoral growth in vivo inhibiting angiogenesis and bone destruction. *Leukemia* **2013**, *27*, 1697–1706. [CrossRef] [PubMed]

83. Colla, S.; Storti, P.; Donofrio, G.; Todoerti, K.; Bolzoni, M.; Lazzaretti, M.; Abeltino, M.; Ippolito, L.; Neri, A.; Ribatti, D.; et al. Low bone marrow oxygen tension and hypoxia-inducible factor-1α overexpression characterize patients with multiple myeloma: Role on the transcriptional and proangiogenic profiles of CD138(+) cells. *Leukemia* **2010**, *24*, 1967–1970. [CrossRef] [PubMed]

84. Giuliani, N.; Colla, S.; Lazzaretti, M.; Sala, R.; Roti, G.; Mancini, C.; Bonomini, S.; Lunghi, P.; Hojden, M.; Genestreti, G.; et al. Proangiogenic properties of human myeloma cells: Production of angiopoietin-1 and its potential relationship to myeloma-induced angiogenesis. *Blood* **2003**, *102*, 638–645. [CrossRef] [PubMed]

85. Ria, R.; Roccaro, A.M.; Merchionne, F.; Vacca, A.; Dammacco, F.; Ribatti, D. Vascular endothelial growth factor and its receptors in multiple myeloma. *Leukemia* **2003**, *17*, 1961–1966. [CrossRef] [PubMed]

86. Barille, S.; Akhoundi, C.; Collette, M.; Mellerin, M.P.; Rapp, M.J.; Harousseau, J.L.; Bataille, R.; Amiot, M. Metalloproteinases in multiple myeloma: Production of matrix metalloproteinase-9 (MMP-9), activation of proMMP-2, and induction of MMP-1 by myeloma cells. *Blood* **1997**, *90*, 1649–1655. [PubMed]

87. De Bruyne, E.; Andersen, T.L.; de Raeve, H.; van Valckenborgh, E.; Caers, J.; van Camp, B.; Delaisse, J.M.; van Riet, I.; Vanderkerken, K. Endothelial cell-driven regulation of CD9 or motility-related protein-1 expression in multiple myeloma cells within the murine 5T33MM model and myeloma patients. *Leukemia* **2006**, *20*, 1870–1879. [CrossRef] [PubMed]

88. Roodman, G.D. Pathogenesis of myeloma bone disease. *Leukemia* **2009**, *23*, 435–441. [CrossRef] [PubMed]

89. Giuliani, N.; Colla, S.; Sala, R.; Moroni, M.; Lazzaretti, M.; La Monica, S.; Bonomini, S.; Hojden, M.; Sammarelli, G.; Barille, S.; et al. Human myeloma cells stimulate the receptor activator of nuclear factor-kappa B ligand (RANKL) in T lymphocytes: A potential role in multiple myeloma bone disease. *Blood* **2002**, *100*, 4615–4621. [CrossRef] [PubMed]

90. Giuliani, N.; Bataille, R.; Mancini, C.; Lazzaretti, M.; Barille, S. Myeloma cells induce imbalance in the osteoprotegerin/osteoprotegerin ligand system in the human bone marrow environment. *Blood* **2001**, *98*, 3527–3533. [CrossRef] [PubMed]

91. Choi, S.J.; Cruz, J.C.; Craig, F.; Chung, H.; Devlin, R.D.; Roodman, G.D.; Alsina, M. Macrophage inflammatory protein 1-alpha is a potential osteoclast stimulatory factor in multiple myeloma. *Blood* **2000**, *96*, 671–675. [PubMed]

92. Lee, J.W.; Chung, H.Y.; Ehrlich, L.A.; Jelinek, D.F.; Callander, N.S.; Roodman, G.D.; Choi, S.J. Il-3 expression by myeloma cells increases both osteoclast formation and growth of myeloma cells. *Blood* **2004**, *103*, 2308–2315. [CrossRef] [PubMed]

93. Silbermann, R.; Bolzoni, M.; Storti, P.; Guasco, D.; Bonomini, S.; Zhou, D.; Wu, J.; Anderson, J.L.; Windle, J.J.; Aversa, F.; et al. Bone marrow monocyte-/macrophage-derived activin a mediates the osteoclastogenic effect of IL-3 in multiple myeloma. *Leukemia* **2014**, *28*, 951–954. [CrossRef] [PubMed]

94. Giuliani, N.; Lisignoli, G.; Colla, S.; Lazzaretti, M.; Storti, P.; Mancini, C.; Bonomini, S.; Manferdini, C.; Codeluppi, K.; Facchini, A.; et al. CC-chemokine ligand 20/macrophage inflammatory protein-3α and CC-chemokine receptor 6 are overexpressed in myeloma microenvironment related to osteolytic bone lesions. *Cancer Res.* **2008**, *68*, 6840–6850. [CrossRef] [PubMed]

95. Giuliani, N.; Colla, S.; Morandi, F.; Lazzaretti, M.; Sala, R.; Bonomini, S.; Grano, M.; Colucci, S.; Svaldi, M.; Rizzoli, V. Myeloma cells block RUNX2/CBFA1 activity in human bone marrow osteoblast progenitors and inhibit osteoblast formation and differentiation. *Blood* **2005**, *106*, 2472–2483. [CrossRef] [PubMed]

96. Giuliani, N.; Rizzoli, V.; Roodman, G.D. Multiple myeloma bone disease: Pathophysiology of osteoblast inhibition. *Blood* **2006**, *108*, 3992–3996. [CrossRef] [PubMed]

97. Bolzoni, M.; Donofrio, G.; Storti, P.; Guasco, D.; Toscani, D.; Lazzaretti, M.; Bonomini, S.; Agnelli, L.; Capocefalo, A.; Dalla Palma, B.; et al. Myeloma cells inhibit non-canonical Wnt co-receptor ROR2 expression in human bone marrow osteoprogenitor cells: Effect of Wnt5a/ROR2 pathway activation on the osteogenic differentiation impairment induced by myeloma cells. *Leukemia* **2013**, *27*, 451–463. [CrossRef] [PubMed]

98. Tete, S.M.; Bijl, M.; Sahota, S.S.; Bos, N.A. Immune defects in the risk of infection and response to vaccination in monoclonal gammopathy of undetermined significance and multiple myeloma. *Front. Immunol.* **2014**, *5*, 257. [CrossRef] [PubMed]

99. Frassanito, M.A.; Cusmai, A.; Dammacco, F. Deregulated cytokine network and defective Th1 immune response in multiple myeloma. *Clin. Exp. Immunol.* **2001**, *125*, 190–197. [CrossRef] [PubMed]

100. Noonan, K.; Marchionni, L.; Anderson, J.; Pardoll, D.; Roodman, G.D.; Borrello, I. A novel role of IL-17-producing lymphocytes in mediating lytic bone disease in multiple myeloma. *Blood* **2010**, *116*, 3554–3563. [CrossRef] [PubMed]

101. Prabhala, R.H.; Pelluru, D.; Fulciniti, M.; Prabhala, H.K.; Nanjappa, P.; Song, W.; Pai, C.; Amin, S.; Tai, Y.T.; Richardson, P.G.; et al. Elevated IL-17 produced by Th17 cells promotes myeloma cell growth and inhibits immune function in multiple myeloma. *Blood* **2010**, *115*, 5385–5392. [CrossRef] [PubMed]

102. Atanackovic, D.; Luetkens, T.; Kroger, N. Coinhibitory molecule PD-1 as a potential target for the immunotherapy of multiple myeloma. *Leukemia* **2014**, *28*, 993–1000. [CrossRef] [PubMed]

103. Brown, R.D.; Pope, B.; Murray, A.; Esdale, W.; Sze, D.M.; Gibson, J.; Ho, P.J.; Hart, D.; Joshua, D. Dendritic cells from patients with myeloma are numerically normal but functionally defective as they fail to up-regulate CD80 (B7–1) expression after huCD40LT stimulation because of inhibition by transforming growth factor-beta1 and interleukin-10. *Blood* **2001**, *98*, 2992–2998. [CrossRef] [PubMed]

104. Ratta, M.; Fagnoni, F.; Curti, A.; Vescovini, R.; Sansoni, P.; Oliviero, B.; Fogli, M.; Ferri, E.; Della Cuna, G.R.; Tura, S.; et al. Dendritic cells are functionally defective in multiple myeloma: The role of interleukin-6. *Blood* **2002**, *100*, 230–237. [CrossRef] [PubMed]

105. Beyer, M.; Kochanek, M.; Giese, T.; Endl, E.; Weihrauch, M.R.; Knolle, P.A.; Classen, S.; Schultze, J.L. In vivo peripheral expansion of naive CD4+CD25high FoxP3+ regulatory T cells in patients with multiple myeloma. *Blood* **2006**, *107*, 3940–3949. [CrossRef] [PubMed]

106. Gorgun, G.T.; Whitehill, G.; Anderson, J.L.; Hideshima, T.; Maguire, C.; Laubach, J.; Raje, N.; Munshi, N.C.; Richardson, P.G.; Anderson, K.C. Tumor-promoting immune-suppressive myeloid-derived suppressor cells in the multiple myeloma microenvironment in humans. *Blood* **2013**, *121*, 2975–2987. [CrossRef] [PubMed]

107. Glavey, S.V.; Naba, A.; Manier, S.; Clauser, K.; Tahri, S.; Park, J.; Reagan, M.R.; Moschetta, M.; Mishima, Y.; Gambella, M.; et al. Proteomic characterization of human multiple myeloma bone marrow extracellular matrix. *Leukemia* **2017**, *31*, 2426–2434. [CrossRef] [PubMed]

108. Andersen, M.N.; Ludvigsen, M.; Abildgaard, N.; Petruskevicius, I.; Hjortebjerg, R.; Bjerre, M.; Honore, B.; Moller, H.J.; Andersen, N.F. Serum galectin-1 in patients with multiple myeloma: Associations with survival, angiogenesis, and biomarkers of macrophage activation. *Onco Targets Ther.* **2017**, *10*, 1977–1982. [CrossRef] [PubMed]

109. Panero, J.; Stanganelli, C.; Arbelbide, J.; Fantl, D.B.; Kohan, D.; Garcia Rivello, H.; Rabinovich, G.A.; Slavutsky, I. Expression profile of shelterin components in plasma cell disorders. Clinical significance of POT1 overexpression. *Blood Cells Mol. Dis.* **2014**, *52*, 134–139. [CrossRef] [PubMed]

110. Mirandola, L.; Yu, Y.; Chui, K.; Jenkins, M.R.; Cobos, E.; John, C.M.; Chiriva-Internati, M. Galectin-3C inhibits tumor growth and increases the anticancer activity of bortezomib in a murine model of human multiple myeloma. *PLoS ONE* **2011**, *6*, e21811. [CrossRef] [PubMed]

111. Moehler, T.M.; Seckinger, A.; Hose, D.; Andrulis, M.; Moreaux, J.; Hielscher, T.; Willhauck-Fleckenstein, M.; Merling, A.; Bertsch, U.; Jauch, A.; et al. The glycome of normal and malignant plasma cells. *PLoS ONE* **2013**, *8*, e83719. [CrossRef] [PubMed]

112. Andrulis, M.; Ellert, E.; Mandel, U.; Clausen, H.; Lehners, N.; Raab, M.S.; Goldschmidt, H.; Schwartz-Albiez, R. Expression of mucin-1 in multiple myeloma and its precursors: Correlation with glycosylation and subcellular localization. *Histopathology* **2014**, *64*, 799–806. [CrossRef] [PubMed]

113. Astorgues-Xerri, L.; Riveiro, M.E.; Tijeras-Raballand, A.; Serova, M.; Rabinovich, G.A.; Bieche, I.; Vidaud, M.; de Gramont, A.; Martinet, M.; Cvitkovic, E.; et al. OTX008, a selective small-molecule inhibitor of galectin-1, downregulates cancer cell proliferation, invasion and tumour angiogenesis. *Eur. J. Cancer* **2014**, *50*, 2463–2477. [CrossRef] [PubMed]

114. Shaughnessy, J.D., Jr.; Zhan, F.; Burington, B.E.; Huang, Y.; Colla, S.; Hanamura, I.; Stewart, J.P.; Kordsmeier, B.; Randolph, C.; Williams, D.R.; et al. A validated gene expression model of high-risk multiple myeloma is defined by deregulated expression of genes mapping to chromosome 1. *Blood* **2007**, *109*, 2276–2284. [CrossRef] [PubMed]

115. Stessman, H.A.; Mansoor, A.; Zhan, F.; Janz, S.; Linden, M.A.; Baughn, L.B.; van Ness, B. Reduced CXCR4 expression is associated with extramedullary disease in a mouse model of myeloma and predicts poor survival in multiple myeloma patients treated with bortezomib. *Leukemia* **2013**, *27*, 2075–2077. [CrossRef] [PubMed]

International Journal of
Molecular Sciences

MDPI

Review

Galectin Targeted Therapy in Oncology: Current Knowledge and Perspectives

Kamil Wdowiak [1], Tomasz Francuz [1,2,*], Enrique Gallego-Colon [2], Natalia Ruiz-Agamez [2], Marcin Kubeczko [1,3], Iga Grochoła [1] and Jerzy Wojnar [1]

[1] Department of Internal Medicine and Oncology, Silesian Medical University, Katowice 40-027, Poland; wdowiak.kamil@op.pl (K.W.); marcin.kubeczko@gmail.com (M.K.); groc.iga@gmail.com (I.G.); jwojnar@sum.edu.pl (J.W.)
[2] Department of Biochemistry, Silesian Medical University, Katowice 40-752, Poland; enrique.gce@gmail.com (E.G.-C.); natalia.ruiz.agamez@gmail.com (N.R.-A.)
[3] Clinical and Experimental Oncology Department, Maria Skłodowska-Curie Memorial Cancer Center and Institute of Oncology, Gliwice Branch, Gliwice 44-101, Poland
* Correspondence: tfrancuz@sum.edu.pl; Tel.: +48-032-252-5088

Received: 31 October 2017; Accepted: 28 December 2017; Published: 10 January 2018

Abstract: The incidence and mortality of cancer have increased over the past decades. Significant progress has been made in understanding the underpinnings of this disease and developing therapies. Despite this, cancer still remains a major therapeutic challenge. Current therapeutic research has targeted several aspects of the disease such as cancer development, growth, angiogenesis and metastases. Many molecular and cellular mechanisms remain unknown and current therapies have so far failed to meet their intended potential. Recent studies show that glycans, especially oligosaccharide chains, may play a role in carcinogenesis as recognition patterns for galectins. Galectins are members of the lectin family, which show high affinity for β-galactosides. The galectin–glycan conjugate plays a fundamental role in metastasis, angiogenesis, tumor immunity, proliferation and apoptosis. Galectins' action is mediated by a structure containing at least one carbohydrate recognition domain (CRD). The potential prognostic value of galectins has been described in several neoplasms and helps clinicians predict disease outcome and determine therapeutic interventions. Currently, new therapeutic strategies involve the use of inhibitors such as competitive carbohydrates, small non-carbohydrate binding molecules and antibodies. This review outlines our current knowledge regarding the mechanism of action and potential therapy implications of galectins in cancer.

Keywords: galectins; cancer; diagnosis; galectins in therapy

---

1. Introduction

The carbohydrate alphabet acts as second genetic code containing necessary information to carry out many of cellular processes. The "sugar code", in the case of glycans has become immensely complex and creates a vast number of "word" combinations, which translate into bioactive information that triggers specific effects. This "sugar code" is translated inside the cell by sugar-binding proteins called lectins. Galectins are a subfamily of lectin proteins with high affinity for β-galactosides. In normal tissue and blood, galectins are expressed at low levels, but they are increased in serum, plasma and urine in neoplastic diseases [1]. Interestingly, galectins also play an important role in other chronic diseases such as cardiac insufficiency, diabetes, rheumatoid arthritis, asthma and liver cirrhosis [2]. The basic domain of galectins contains a carbohydrate recognition domain (CRD) through which they can bind with numerous carbohydrate ligands. To date, up to 16 members of the galectin family have been discovered in mammals, 12 of which have been identified in humans.

Depending on their structure, galectins may be divided into three groups: prototype, tandem repeats and chimeric galectins (Table 1) [3].

Table 1. Galectin characteristics according to molecular structure.

Subtype	Galectins	Model	Structure
Prototype	1, 2, 5, 7, 10, 11, 13, 14, 15, 16		Each monomer of homodimer contains CRD
Tandem Repeats	4, 6, 8, 9, 12		Two CRD domains connected with linker
Chimeric	3		Multimeric structures with one CRD in C terminus and N-terminus.

Galectins are present in numerous locations within the cell, such as nucleus, cytoplasm and plasma membrane, but also extracellularly. The distinct glycosylation of glycoproteins allows binding of galectins to β-galactosides in different areas of the protein, leading to prolonged receptor activation at the plasma membrane. Given that glycosylation is the most frequent post-translational modification, galectin–proteoglycan interactions might be an important phenomenon. In cancer, continuous stimulation of VEGFR2 promotes the formation of new blood vessels and thus facilitating cancer progression and metastasis [4]. Current therapies involve the use of vascular endothelial growth factor (VEGF) inhibitors such as Bevacizumab. Patients treated with VEGF-targeted therapies showed varying efficacies and tumor regrowth. Croci and collaborators found that Gal-1 can recognize *N*-glycans on VEGFR2 and trigger a VEGF-like signaling response thereby promoting vascular regrowth in absence of VEGF [4]. Additionally, galectins are involved in cancer-promoting processes such as proliferation, apoptosis and immune modulation. Galectins produced by tumor cells bind to T-cell glycoprotein receptors like CD45 and CD71 [5]. In particular, extracellular Gal-1 and extracellular Galectin-3 (Gal-3) have been implicated in promoting T-cell suppression and apoptosis, while intracellular Gal-3 promotes activation of anti-apoptotic pathways in T-lymphocytes [4,5].

The design of selective inhibitors for galectins is challenging because of the shared homology of the CRDs among lectins, which can range from 20% to 50% [6]. Additionally, tumor cells can generate multiple isoforms via alternative splicing and this can result in inhibitor-resistant galectins [1]. Furthermore, even if selective inhibition is achieved, other galectins can compensate for the inhibited type. This effect was observed for Gal-1 and Gal-3 in pancreatic cancer cells. The compensatory mechanism involves p16^{INK4a}, a tumor suppressor that inhibits cyclin-dependent kinases; p16^{INK4a} modulates and affects the reactivity and expression of lectins by downregulating Gal-3 levels. In this case, compensatory increased Gal-1 extracellular levels were observed. Intracellular Gal-3 downregulation caused reduction of the anti-apoptotic effect [7]. Potential adverse effects of galectin inhibition were also observed in human breast carcinoma. In breast cancer cells, Gal-1 and Gal-3 compete for cell surface receptors while generating opposite functions. Gal-3 binds with K-Ras and activates the MEK-ERK signaling pathway, while Gal-1 binds with H-Ras and activates PI3K/AKT cascade hence modulating rather distinct cellular functions [8].

2. Galectin-1

In healthy tissues, Gal-1 is located inside the cell, the cytoplasm, or nucleus [9] and remains there until cell activation [10]. Gal-1 secretion into the extracellular matrix (ECM) also occurs to a lesser extent. Increased expression of Gal-1 is observed in numerous neoplasms, including colorectal [11], lung [12], breast [13], pancreas [14,15], liver [16], thyroid [17] and hematological malignancies [18,19]. Additionally, an increase in Gal-1 blood concentration was observed in lung cancer [12], thyroid cancer [20], T cell lymphoma [21] and glioma [22].

Gal-1 can act at intracellular level as an effector of pre-mRNA splicing, or extracellularly as a binding protein to numerous glycoproteins, glycolipids and elements of the extracellular matrix (ECM) [23]. Consequently, Gal-1 has the potential to affect adhesion and aggregation of cells, especially in neoplastic cells where it can influence metastatic processes [24]. Gal-1 binding proteins have been identified including integrins, laminins, fibronectin, thrombospondin, vitronectin, osteopontin, neuropilin-1 (NRP-1), CD44, CD146 and CD326 [24]. Paz and colleagues suggested additional roles for Gal-1. Intracellular Gal-1 reacts with the active form of oncogenic H-Ras (H-Ras-GTP), thereby increasing its membrane anchorage, a crucial step in malignant transformation of certain cancers [25]. Gal-1 has also been associated with immunosuppression, stimulating apoptosis in activated T CD4+ and CD8+ lymphocytes [26]. Indeed, Gal-1 targeted therapy may contribute to reduce the dissemination of tumor cells and inhibit angiogenesis and tumor growth.

Currently, anti-angiogenic treatments have therapeutic limitations such as varying degrees of response and resistance. This phenomenon is thought to occur due to VEGF-independent mechanisms. In hypoxic areas, tumor cells survive oxygen-depleted environment by up-regulating the expression of hypoxia-inducible factor-1 (HIF-1α) [27]. Studies show that colorectal cancer cell lines cultured in hypoxic environment produce larger amounts of Gal-1, which correlated with increased hypoxic factors such as hypoxia induced factor α (HIF-1α) as well as carbonic anhydrase IX (CAIX) [28,29]. The studies described in this section suggest that inactivation of Gal-1 in tumor cells may result in an increased sensitivity to chemotherapeutic agents.

Several Gal-1 inhibitors have been designed with potential clinical application in cancer therapy.

2.1. Thiodigalactoside

Thiodigalactoside or TDG is a synthetic disaccharide with affinity for Gal-1. TDG non-selectively blocks Gal-1 action during angiogenesis and immune response and protects against oxidative stress (Table 2). Intra-tumoral treatment with TDG suppresses growth of breast cancer and melanoma in preclinical models [30]. Interestingly, the influence of TDG in blocking tumor progression was not observed in Gal-1 knock-out mice, indicating that Gal-1 is a TDG target. In a preclinical study, Gal-1 knock-out mice showed an increase in T CD4+ and CD8+ lymphocytes in the tumor milieu, in blood and in immunocompetent organs. An effect of TDG on angiogenesis was evidenced by the reduction in number of endothelial cells (CD31+) and in new vessel formation [30]. In subsequent studies, Ito and colleagues observed that after TDG administration, the number and size of lung metastases of mice carrying breast or colon tumors was decreased. The mechanism of action of TDG is by preventing binding of Gal-1 to CD44 and CD326 receptors on the surface of cancer stem cells (CSC) [30–32].

Table 2. Galectin-1 inhibitors.

Inhibitor	Target	Effect	Refs.
Thiodigalactoside (TDG)	Melanoma and breast cancer xenografts; Colon and breast cancer xenografts	Induction of apoptosis; Inhibition of angiogenesis, proliferation and tumor growth; Reduction of lung metastases; Restore T cells surveillance	[30,31]
Anginex (β pep-25)	Ovarian, melanoma and breast cancer xenografts	Inhibition of tumor growth, angiogenesis and migration; Increased sensitivity to radiotherapy and chemotherapy; Synergistic effects with bevacizumab	[33–36]

Table 2. *Cont.*

Inhibitor	Target	Effect	Refs.
6DBF7; DB16; DB21	Lung and ovarian cancer and melanoma xenografts	Inhibition of tumor growth; Inhibition of angiogenesis	[37,38]
OTX008 (0018)	Ovarian cancer xenografts; Head and neck and ovarian cancer cell lines; Clinical trial in patients with advanced solid tumors	Downregulation of cancer cell proliferation; Inhibition of tumor growth, angiogenesis and migration; Synergic effects with chemo- and immunotherapy	[24,39–42]
F8.G7	Endothelial cells; Kaposi's sarcoma xenografts	Inhibition of tumor growth, angiogenesis, migration	[43,44]
GM-CT-01 (DAVANAT®) oraz GR-MD-02	Colon cancer xenografts; Clinical trials in patients with colon cancer and melanoma	Inhibition of tumor growth; Restore the T cells surveillance	[45,46]

2.2. Anginex (β Pep-25)

Anginex is an antiangiogenic peptide involved in tumor growth (Table 2) [33,47]. The drug contains short sequences of known antiangiogenic factors such as platelet factor-4 (PF4), interleukin 8 (IL-8) and bactericidal-permeability increasing protein-1 (BPI-1) [48]. Anginex, specifically binds to the β-sheet motif of Gal-1, inhibiting neoplastic proliferation, migration and inducing apoptosis, thereby inhibiting tumor growth [49]. This drug also blocks Gal-1 uptake by endothelial cells, thereby preventing the translocation of H-Ras-GTP and phosphorylation of the Raf/MEK/ERK kinase cascade [50].

Numerous clinical studies evaluated the effect of Anginex in combination with radiotherapy and/or chemotherapy. The results showed that Anginex sensitizes tumor-associated endothelial cells to radiotherapy, thus strengthening the therapeutic effect [34,49–53]. In a human ovarian carcinoma mouse model, Anginex showed synergistic effect with a suboptimal dose of Carboplatin and boosted tumor regression [54]. Amano and colleagues showed that Anginex is able to prolong radiation-induced tumor regression in a squamous cell xenograft model [51]. Furthermore, several studies have focused on increasing Anginex bioavailability by modifying the structure and/or by conjugation with carrier proteins to increase treatment efficacy [36,55–57]. A very innovative and interesting study by Upreti and colleagues showed that Gal-1 is overexpressed in triple negative breast cancer (TNBC) relative to patients with normal tissue or benign breast lesions. They developed a murine model of TNBC, with radiation-induced Gal-1 expression in stromal tissue. Complexes of Anginex and arsenic trioxide, as well as Cisplatin-loaded liposomes were tested this model, leading to decreased tumor growth by ~80% (vs. 20% in non-irradiated mice treated with non-conjugated liposomes) [58]. Indeed, Anginex nanotherapy is a well-tolerated, very effective therapy with potential application as Gal-1 overexpression occurs in approximately eight to ten samples of ductal breast carcinoma human tissue. Interestingly, new Anginex analogues such as Dibenzofuran (6DBF7), DB16 and DB21 also showed similar results with an 80% reduction in tumor growth following administration in mice [37]. In particular, DB21 appears to inhibit angiogenesis and tumor growth very effectively [49,59]. Anginex's beneficial effects on tumor suppression are antagonized by its reduced stability and half-life, difficulty in manufacture and that it only sensitizes endothelial cells to radiotherapy in newly forming tumor vessels, not in tumor cells. Anginex interacts with other galectins, such as Gal-2, -7, -8N and 9N, but with lower affinity [38]. To conclude, Anginex therapy shows minimal side effects in Anginex-treated animal models, either alone or in combination with chemotherapy or radiotherapy [39,51,58].

2.3. OTX008 (0018)

OTX008 is chemically more stable and resistant to hydrolysis when compared to other Gal-1 inhibitors (Table 2). The low molecular weight (937 Da) and the fact that it is neither a protein nor saccharide, but a phenyl-based molecule greatly increases its bioavailability [39]. From a mechanistic point of view, OTX008 binds Gal-1 at a more distant location within the CRD as compared to Anginex. Additionally, OTX008 has both direct and indirect effects on cell survival, cell cycle and angiogenesis. Research has shown that administration of OTX008 in vivo and in vitro is effective, both in single and combination therapies [40–42]. Astorgues-Xerri et al. evaluated the efficiency of OTX008 in several cancer cell lines and in a murine ovarian carcinoma model [60]. Ovarian cancer cells of epithelial origin were shown to be more sensitive to OTX008 than cells of mesenchymal origin. Moreover, OTX008 exposure inhibited p-ERK 1/2 and p-AKT survival signaling pathways. OTX008 also caused G_2/M cell cycle arrest by modulating the activity of CDK1 via G_2/M checkpoint-regulators CDC25 and WEE1. In vivo experiments showed OTX008 inhibits tumor growth, accompanied by a decreased in Gal-1, Ki67 and VEGFR2 expression. Synergistic activity with other chemo- or immunotherapies was also achieved in in vitro therapies using drugs such as Cisplatin, Oxaliplatin, Docetaxel, 5-fluorouracil, Regorafenib, Sunitinib and Everolimus [55]. Combination treatment of mTOR inhibitor Rapamycin and OTX008 was more effective than Rapamycin alone in limiting tumor volume and reducing the number of cells with HRAS mutation [61]. In 2012, a phase I clinical trial aimed at evaluating the effects of subcutaneous administration of OTX008 for the treatment of advanced solid tumors (ClinicalTrials.gov: NCT01724320) [24]. However, so far, no results regarding outcome of treatment have been released. Recently, a Calixarene-based topomimetic of OTX008, PTX013, showed improved efficiency and greater potency than OTX008. Preliminary data indicates that PTX013 actions are not directed at Gal-1 and the molecular target is yet to be found [62]. Additionally, the possibility of combinatorial treatments with chemotherapeutics is an intriguing option that is being explored [43].

2.4. F8.G7

Interaction of Gal-1 and VEGFR2 leads to prolonged presence of the receptor in the cell membrane of endothelial cells thereby promoting tumor regrowth, which may limit the efficacy of anti-VEGF treatment [4]. Croci Do et al. showed that monoclonal anti-Gal-1 (F8.G7) (Table 2) based therapy inhibited tumor growth and angiogenesis, including pathways associated with VEGFR2/Gal-1 in mice with Kaposi's sarcoma [44]. Vessels of treated tumors decreased in size and number, were less dispersed and covered with mature pericytes. Additionally, increased T lymphocyte infiltration and production of IFN-γ and IL-17 was observed. Importantly, F8.G7 therapy only targets the non-canonical VEGF pathway and the canonical pathway can still contribute to tumor angiogenesis. As concluded by these researchers, further research is required and personalized therapy should be the aim of treatment.

2.5. GM-CT-01 (DAVANAT®) oraz GR-MD-02

GM-CT-01 (DAVANAT®) is a modified vegetal galactomannan oligomer extracted from Guar seeds (*Cyamopsis tetranoglonoloba*) (Table 2). Davanat shows affinity to the dimer interface rather than the CRDs in Gal-1 and Gal-3 [45]. Demotte et al. reported improved tumor infiltrating lymphocyte (TIL) function induced by GM-CT-01 [63]. Extracellular Gal-1 and Gal-3 are responsible for blockade of glycosylated receptors on the surface of TILs leading to reduced T-cell motility and overall function. Galactomannan treatment promotes IFN-γ secretion by T-cells, which promotes an antitumor response. GM-CT-01 therapy progressed into phase I and II clinical trials for the treatment of solid tumors. Unfortunately, the trials were prematurely terminated due to financial reasons, nevertheless a certain degree of therapeutic effect was observed in patients suffering from metastatic colorectal cancer (mCRC). In the DAVANAT® trial (NCT: NCT00054977), out of 20 subjects enrolled, one had a partial response to the drug while six other patients had stable disease. Moreover, lower frequency of 5-Fluorouracil (5-FU) side effects for grades 3–4 (G3–G4) was seen in combined treatment with

GM-CT-01 [46]. At present, an ongoing phase II clinical trial is being conducted using a GM-CT-01 vaccine in patients suffering from diffuse melanoma (NCT: NCT01723813). In preclinical models, Gal-1 facilitates the escape of melanoma cells from immune surveillance by reducing the number of helper T-cells and cytolytic T-cells [64]. Downregulation of Gal-1 by siRNA knockdown in B16F10 cell lines resulted in an increase in response rates to Temozolamide and increased survival time of B16F10 melanoma-bearing mice [65]. In a recent study, Wu and colleagues observed that patients treated with Bevacizumab (anti-VEGF antibody) and Ipilimumab (anti-CTLA-4 antibody) that also received anti-Gal-1 antibody had a longer overall survival (OS). In contrast patients with higher Gal-1 levels had shorter OS [66]. Inhibition of Gal-1 functions may enhance the activity of checkpoint inhibitors and restore T-cell activity.

Additionally, a modified version of the DAVANAT® drug, GR-MD-02, proved to be effective in the treatment of non-alcoholic steatohepatitis (NASH) in mice [67]. Reduction of inflammation, fat accumulation, fibrosis and hepatocellular damage were observed. In the randomized phase I study, no serious adverse events were observed with GR-MD-02 at doses of 2, 4 and 8 mg/kg [68]. In advanced stages of melanoma Gal-3 is overexpressed and its serum concentration increases [69–71]. Currently, two more clinical trials are being conducted using GR-MD-02 in combination with Ipilimumab or Pembrolizumab in patients suffering from melanoma (NCT: NCT02117362 and NCT02575404).

3. Galectin-3

Galectin-3 (Gal-3) is the only representative of the chimeric galectin group. Gal-3 is composed of a collagen-like sequence, a C-terminal domain (CTD) with a CRD, an N-terminal domain (NTD) with a serine phosphorylation site. The CRD of Gal-3 contains 110–130 amino acids with NWGR motifs which are important for interaction with anti-apoptotic proteins of the Bcl-2 family [72]. The C-terminus is responsible, among other functions, for binding saccharides such as *N*-acetyllactosamine (LacNAc) and lactose. Furthermore, Gal-3 has a higher affinity to polysaccharides terminating in galactose than to monosaccharides. The CRD contains five subunits (A–E) among which the C subunit is responsible for recognizing glycans containing β-galactosides [73]. The NTD facilitates multimerization and pentamer formation of galectin-3, which is necessary for extracellular secretion and nuclear translocation [74]. Based on the Gal-3 crystallographic structure, a number of low molecular weight and high affinity inhibitors have been developed [75]. A fraction of these compounds such as TDG and derivatives are currently being tested.

Gal-3 protein, present in both healthy tissues and neoplastic tissues, is involved in processes such as inflammation, neoplasia, cancer cell adhesion, angiogenesis, cell growth, proliferation and apoptosis [76]. A correlation between Gal-3 and such processes was shown for thyroid, stomach, large intestine, kidney, lung, prostate, breast and pancreatic cancers [1]. Importantly, immunohistochemical (IHC) staining of Gal-3 protein can provide a useful diagnostic tool for the differentiation of benign and malignant thyroid nodules, as demonstrated in several studies [77–79]. Based on a meta-analysis of 52 studies, the sensitivity and specificity of Gal-3 IHC expression was 87% and 87%, respectively [80]. Depending on its intracellular location, Gal-3 may have pro- or anti-apoptotic effects. In the nucleus, Gal-3 is responsible for gene expression regulation through transcription factors such as SP1 and β-catenin. It also plays a role in micro-RNA expression and splicing, as well as in transport of nuclear proteins [81–83]. In the cytoplasm, Gal-3 modulates numerous signaling pathways involved in cancer such as RAS, BCL-2 and MYC [84–86]. Moreover, this lectin is responsible for dampening the immune response through suppression of T-cells and natural killer (NK) cells and to induce apoptosis of T-cells by binding CD45 [87,88]. Additionally, endogenous Gal-3 may inhibit Cisplatin- or Etoposide-induced mitochondrial apoptosis pathway in prostate and breast cancer cells [84,89]. The role of Gal-3 in antineoplastic-resistant treatment is noteworthy and therefore its inhibition may be key in overcoming resistance and increase susceptibility of neoplastic cells to drugs [90]. Recently, Harazono et al. showed that extracellular Gal-3 takes part in a previously unknown chemoresistance mechanism [91] by which Gal-3 increases activity of Na/K ATPase. Following an application of Gal-3 inhibitor GCS-100, an increase in sensitivity to Doxorubicin was observed in tumor cells [92].

3.1. G3–C12

G3–C12 is an oligopeptide that binds Gal-3 at the CRD region. In a mouse model of breast cancer, mice subjected to G3–C12, had decreased metastasis formation (Table 3) [93]. Yang et al. were the first to use a conjugate therapy composed of a G3–C12 and 5-Fluorouracil, P-(G3–C12)-FU, on a mouse model of prostate cancer [94]. In this research, G3–C12 was also complexed with *N*-(2-hydroxypropyl) methacrylamide (HPMA) as a carrier molecule. Enhanced drug delivery was observed due to the low molecular weight of the HMPA compound, which facilitates delivery inside the cell. This complex yielded far better therapeutic results than 5-Fluorouracil (5-FU) therapy alone. Furthermore, G3–C12–HPMA conjugate shows better pharmacokinetics and bioavailability than with the chemotherapeutic agent [95]. Subsequently, Doxorubicin (DOX) or 5-FU was added to the HPMA/G3–C12 complex [96]. In vivo P-(G3–C12)-DOX-FU showed the strongest effect among other combinations and inhibited tumor growth in mice by 81.6%, whereas the other agents were less effective (P-DOX-FU—71.2%, P-DOX—63%, DOX-HCl—40.5%, P-FU—32%, 5-FU—14.6%). Current studies aim to identify how the copolymer binds the cell [97]. Researchers hypothesized that facilitated by G3–C12, the drug conjugate can bind Gal-3 and become internalized. Besides, the presence of DOX also leads to translocation of Gal-3 into mitochondria triggering antiapoptotic effects. However, progressive inflow of Gal-3 promotes drug accumulation in the mitochondria. With time, Gal-3 function is inhibited, while mitochondria dysfunction is exacerbated by the activity of DOX. Consequently, G3–C12 holds enormous potential, however, the mechanism of action is unknown.

Table 3. Galectin-3 inhibitors.

Inhibitor	Target	Effect	Ref.
G3–C12	Breast, colon and prostate cancer xenografts	Reduction of lung metastasis; Induction of apoptosis; Inhibition of tumor growth; Synergic effect with chemotherapy	[93–96]
Modified citrus pectin (MCP)	Breast and colon cancer xenografts; Prostate cancer cell lines; Patients with advance solid tumors	Inhibition of tumor growth, angiogenesis and metastasis; Induction of apoptosis; Cell cycle arrest; Increase sensitivity to chemotherapy; Rebalance the T cells surveillance	[98–102]
PectaSol-C Modified citrus pectin	Prostate and ovarian cancer cell lines	Induction of apoptosis; Inhibition of proliferation; Synergic effect with chemotherapy	[98,103,104]
GCS-100	Multiple myeloma, DLBCL cell lines; Prostate cancer cell lines; Patients with Chronic lymphocytic leukemia (CLL)	Inhibition of cell growth; Induction of apoptosis; Synergic effect with chemotherapy; Increased sesnsitivity to immunochemotherapy	[92,105–109]

3.2. Modified Citrus Pectins (MCP)

Modified citrus pectins (MCP) are a group of polysaccharides derived from citrus fruits, which have been subject to chemical or thermal modification. MCPs such as Pecta-Sol and GCS-100, inhibit Gal-3 function and ligands such as cytokines or type C lectins [110,111]. MCP antineoplastic actions include tumor growth suppression by halting cell cycle, apoptosis activation, sensitization of tumor cells to chemotherapy, reduction of metastatic and angiogenesis potential and restoration of immune function (Table 3).

In a study that included 26 patients with various solid tumors, a hydrolyzed form of MCPs was orally administered at a dosage of 5 grams three times a day. After two cycles (eight weeks) of treatment, 11 patients (42.3%) had achieved stable disease and six patients (23.0%) maintained stable disease status for at least 24 months [101]. These results should be weighted considering that the subjects had advanced and aggressive tumors. Additionally, the purpose of the study was to assess tolerability, quality of life and clinical benefit response. Other citrus pectins with higher affinity for Gal-3 are being investigated and already existing pectins are being modified with the aim of increasing their antineoplastic potential.

3.2.1. PectaSol-C

Yan and collaborators observed that the MCP family member PectaSol-C, inhibited tumor growth by blocking the MAPK cascade in prostate cancer cell lines. Effectiveness was assessed in prostate cancer cell lines, which showed a halt in proliferation and induction of apoptosis [98]. Later studies showed synergism of PectaSol-C with Doxorubicin in prostate cancer and with Paclitaxel in ovarian cancer cell lines [103,104]. In both cases tumor size was decreased (Table 3). Currently, a Phase III (NCT: NCT01681823) clinical trial is being carried out to test whether oral administration of PectaSol-C can improve prostate-specific antigen (PSA) kinetics in men with relapsed prostate cancer. In another study, MCP-treated HUVEC cells lost motility and cellular organization [99], tumors decreased in size and angiogenesis and growth of metastases were reduced [100].

3.2.2. GCS-100

GCS-100 is a branched polysaccharide, synthesized from modified MCPs. This Gal-3 inhibitor induces apoptosis in multiple myeloma cells including resistant myeloma cells to Doxorubicin, Melfalan, Dexamethason [92] and Bortezomib [105] (Table 3). A similar effect was observed in prostate cancer cells, where Gal-3 inhibition by siRNA or administration of GCS-100 increased Cisplatin-induced apoptosis [106]. Downregulation of Gal-3 expression on the surface of diffuse large B-Cell lymphoma (DLBCL) cells sensitized them to immunochemotherapy [107].

On the basis of the latest reports, when combined with a BH-mimetic, GCS-100 induces apoptosis of acute myeloid leukemia (AML) cells, especially in cases with predominant negative prognostic factors, such as FLT3 ITD mutations. The effect of GCS-100 appears to be related to induction of p53, because cases where its expression was not induced or p53 was otherwise inactive, resulted in no response to treatment [108]. Based on these studies, it can be concluded that GCS-100 is a good candidate, however, more research and larger studies are required to determine its efficacy. In a phase II clinical trial, 24 patients with recurrent chronic lymphocytic leukemia, were treated with GCS-100 intravenously at a dose of 160 mg/m^2 in a 5-day regiment, every 21 days. In 6 patients (25%) partial response was observed and the disease was stable in 12 patients (50%) [109]. Good overall tolerance was observed, with only minor complications such as nausea, skin rash and low to moderate fatigue. In addition to the antineoplastic roles of GCS-100, Gal-3 inhibition modulates immune system response. Demotte et al. explored the role of Gal-3 in immune system function. Inhibition of Gal-3 with GCS-100 in mice resulted in restoration of CD8+ and CD4+ T cell function. An increase in IFN-γ secretion by TILs was also observed along with tumor regression [102]. In a review by Zhang et al., the mechanism of action and anti-cancer properties of MCPs are discussed [112]. Unfortunately, limited reports on the application of MCPs in patients suffering from neoplasms are available.

4. Galectin-4

Galectin-4 (Gal-4) contains C- and N-terminal CRDs that share 38% amino acid sequence similarity. Two Gal-4 CRDs with different binding partners are connected by a linker region. Gal-4 is a tandem-repeat galectin expressed in epithelial cells of gastrointestinal tract [113]. To date, our knowledge on Gal-4 is restricted to the differences in expression observed in healthy versus cancerous tissues. Gal-4 is considered a risk factor for lymph node involvement in lung cancer [114]. Moreover, high levels of Gal-4 are seen in sera of patients suffering from colorectal cancer, especially in metastatic cases [115]. In two other reports, however, low levels of Gal-4 were associated with an advanced form of colorectal cancer [116,117], while stimulation of Gal-4 expression caused colorectal cancer cells to become sensitized to Camptothecin [117]. Conflicting results may be due to the presence of different Gal-4 isoforms that are not yet known. Evidently, further research is needed to clarify the function of Gal-4 and its role in cancer. Recently, there have been several studies which shed some light on the composition and function of Gal-4 [118–120]. Bum-Erdene et al. described, based on crystallography, the structure of the CRDs in relation with numerous ligands [119,120].

5. Galectin-7

Galectin-7 (Gal-7), was described for the first time in 1995 by Magnaldo et al. [121]. Initially considered as keratinocyte differentiation marker, Gal-7 is a prototype galectin capable of forming homodimers. Increased expression of Gal-7 was observed in numerous neoplasms, such as breast [122], thyroid [123] and throat [124] cancers, as well as in indolent lymphoproliferative diseases [125]. In more than one case, a correlation was identified between Gal-7 and the progression of neoplasms into more aggressive phenotypes [122,125]. Conversely, low expression of Gal-7 was observed in the case of gastric cancer [126], colon cancer [127], squamous cell carcinoma of the cervix [128] and urothelial bladder cancer [129]. Higher expression of Gal-7 in patients with squamous cell cervical cancer was associated with a better outcome after radiotherapy [130]. Labrie et al. observed increased Gal-7 expression when p53 was mutated [131]. Additionally, epithelial ovarian cancer cells secrete Gal-7, which through matrix metalloproteinase 9 (MMP-9) promotes invasiveness. In another study, mouse lymphoma cells transfected with an antisense Gal-7 plasmid showed reduced survival time [132]. We hypothesize that Gal-7 modulates *MMP-9* gene expression to some extent, since lymphoma cells transfected with Gal-7 antisense RNA, also showed a reduction in *MMP-9*. Based on crystallography of the Gal-7 molecule, a 2-*O*-galactoside benzyl phosphorane was synthesized. The new compound showed a 60-fold increased affinity for Gal-7 compared to galactoside [133]. Promising results were presented by Vladoiu et al., who used a selective inhibitor, hGal-7 to disrupt dimerization of Gal-7 and inhibit apoptosis of Jurkat T-cells [133]. This compound targets the dimer interface of Gal-7, but not at CRD. High concentrations of the drug were necessary to observe results and, thus, further studies are needed to improve on the molecule. A review article by Kaur and collaborators summarizes Gal-7 findings in cancer [134].

6. Galectin 8

The role of galectin 8 (Gal-8) in oncogenesis is not well understood. Gal-8 is a type of tandem-repeat galectin with two CRDs one at C- and another at N-terminal region joined by a polypeptide linker. The terminal domains are responsible for recognizing and binding ligands whereas the linking peptide regulates biological functions and it has a multimerization function [135]. Alternative splicing of the linker region results in the formation of a peptide of different length, which determines the formation of the various Gal-8 isoforms: Gal-8S (short liner region), Gal-8M (medium linker region) and Gal-8L (long linker region) [136,137]. These isoforms have different biological functions and can activate different signaling pathways, hence limiting the design of targeted therapies [138]. Gal-8 has been suggested to function in modulating angiogenesis [137]. Recent studies show that C-terminal CRD preferentially binds blood cell antigens A and B, as well as poly-LacNAc saccharides, while N-terminal CRDs have high affinity for sulfated and sialylated glycans [139,140]. In normal endothelial cells, Gal 8 binds CD166 [137] and CD44 [141]. Additionally, Gal-8 may be a useful marker of papillary thyroid cancer, where it is strongly expressed, unlike normal tissue where there is undetectable expression [142]. Loss of Gal-8 expression is associated with increased risk in urinary bladder cancer recurrence but not of tumor progression [143]. The expression of Gal-8 may be a potential predictor of early recurrence after nephrectomy in patients with localized pT1 clear cell renal cell carcinoma [144]. Additionally, Gal-8 is responsible for the progression of prostate cancer and initiation of metastatic phenotype [145]. Given that Gal-8 does is not expressed in healthy prostate tissue, it may be a potential therapeutic target in the future. Increased serum concentration of Gal-8 has been observed in breast cancer as well as colorectal cancer [146]. In breast cancer, Gal-8 expression was observed both intracellularly and extracellularly [147,148]. Satelli and colleagues hypothesized that intracellular Gal-8 undergoes post-translational processing representing the half-weight of the extracellular Gal-8 [148]. A recently published study presents the interactions between activated leukocyte cell adhesion molecule (ALCAM/CD166) and Gal-8, which may be important in the biology of breast cancer cells [149].

7. Galectin 9

Galectin 9 (Gal-9) was discovered and described for the first time in 1997 in patients suffering from Hodgkin's lymphoma (HL) [150]. Gal-9 is a type of tandem-repeat galectin with 2 CRDs, a 148-amino acid-long N-terminus and a 149-amino acid-long C-terminus. Between domains there is a connecting sequence, whose length determines the three isoforms of Gal-9: short Gal-9S, medium Gal-9M and long Gal-9L. The three isoforms exhibit varying degrees of chemotactic effects on eosinophils [151]. Gal-9 is present in both intracellular and extracellular compartments [11] and several ligands of Gal-9 have been described. Extracellular Gal-9 can bind to TIM-3, CD44 and Glut-2, while intracellular Gal-9 binds transcription factor NF-IL6 [152–155]. Gal-9 has been described to play an important role in numerous biological processes such as adhesion [156], aggregation of cancer cells [157], apoptosis [158,159], immunomodulation [160,161] and chemotaxis [162]. In most cases, Gal-9 expression in healthy tissues is higher than in neoplastic cells as observed in breast [157], liver [162], lung [163], prostate [164], kidney cancers [163] and melanoma [156]. Increased Gal-9 expression reflects progression and aggressiveness of the neoplasm. In a few cases, high Gal-9 expression was described in Hodgkin's lymphoma [150], colorectal [163], oral [165] and pancreatic cancer [166]. Differences in Gal-9 expression are caused by differential mRNA splicing and generation of different isoforms.

Gal-9 is a good prognostic factor in patients who suffer from renal cell carcinoma (RCC) [167]. High Gal-9 expression is associated with decreased overall survival (OS) time and decreased recurrence free survival (RFS). Patients with high Gal-9 expression showed more advanced progression of the disease with larger tumor size and necrosis [168]. Interestingly, Gal-9 proved ineffective in the stratification of patients with advanced disease (TNM III/IV, Fuhrman 3/4). Moreover, patients suffering from metastatic RCC, who had responded to IL-2 and IFN-γ therapy, showed high Gal-9 expression [168]. Unlike other galectins, Gal-9 predominantly functions as tumor suppressor. Studies performed in hepatocellular carcinoma (HCC) cell models indicated that silencing of Gal-9, by small interfering RNAs (siRNA), resulted in increased proliferation and migration [162]. Additionally, patients with positive Gal-9 expression had longer survival times than those with negative lesions [135].

A study by Wiersma and collaborators, showed that a recombinant form of Gal-9 was shown to promote cell death in colorectal carcinoma cells (CRCC) with KRAS mutations. CRCCs are commonly resistant to chemotherapy and immunotherapy when EGFR inhibitors are used. Following administration, recombinant soluble Gal-9 (rLGALS9) rapidly entered cells by endocytosis and accumulated in lysosomes. Internalization of rLGALS9 resulted in autophagosome–lysosome fusion failure, lysosome swelling, accumulation of autophagosomes and ultimately cell death [169]. Inhibition of the autophagosome-lysosome fusion was described earlier following the application of the lysosomal inhibitor, chloroquine [170]. Conversely, rLGALS9 therapy on CRC cells with the BRAF mutation caused no effects [169].

In another study, chronic myeloid leukemia (CML) cell lines resistant to tyrosine kinase inhibitors (TKIs) were treated with modified human Gal-9 (hGal-9). Resistance of cells to treatment was overcome and a synergistic activity of hGal-9 with TKIs was noted. The process of apoptosis occurred through the activation of activating transcription factor-3–Noxa proapoptotoic pathway (ATF3–Noxa) and was independent of p53 expression. In this study, hGal-9 was also reported to activate caspase-4 and caspase-8 through a TKI-independent pathway [171]. Interestingly, the Tim-3/Gal-9 signaling pathway is described as a one of the immune checkpoints responsible for T-cell exhaustion [172]. Inhibition of this signaling pathway may be an important therapeutic option in cancer patients. However, we now know that Gal-9 also has other membrane receptors on the surface of T cells [173] and that Tim-3 receptor has non-Gal-9 ligands [174]. A recent excellent review by Riayo Yang and Mien-Chie Hung summarizes the current research on the role of Tim-3 and Gal-9 in antitumor immunity [172].

8. Other Galectins

There are 16 types of galectins, however, the role of other Galectins has not been fully understood nor documented. In a recent study by Peng and collaborators, IHC staining showed increased levels of Gal-10 protein in all stages of colorectal carcinoma progression [175]. In another study, Gal-12 protein levels were found to be increased in a public data set of 526 acute myeloid leukemia (AML) samples of various FAB subtypes especially the M3 subtype [176]. Current research and publications knowledge is expected to be broadened soon.

9. Conclusions

In this review, the scientific evidence obtained through extensive study of galectins suggests that inhibition of galectin activity may contribute to better antineoplastic drugs. Preclinical and clinical studies indicate that inhibiting galectin action results in tumor growth arrest, inhibition of angiogenesis and occurrence of metastases. Additionally, administration of galectin inhibitors in combination with chemotherapy and radiotherapy improves efficacy of treatment of various neoplasms. Furthermore, inhibition of galectins has been shown to interfere with multidrug resistance mechanisms (MDR) enhancing sensitivity of chemotherapeutics. An important aspect to take into account is that tumor cells can express more than one galectin, and therefore, tumor-specific therapy is crucial for therapeutic benefit. Unfortunately, most research is restricted to cell lines and animal models and results from early phase human trials have been inconclusive. In colon cancer, tumor cells have increased expression of Gal-1, -3, -7, and -10, which correlated with increased blood levels of Gal-1, -2, -3, -4, -8, and -9 [1]. Another level of complexity in the design of novel therapies relies on the fact that galectins can present in various isoforms, of which only some affect the function of cancer cells. Additionally, tumor-clonal expansion may lead to the production of a tumor which may not express the expected galectin, hindering the discovery of targeted therapy. Another aspect to take into account in the design of therapeutic inhibitors is the balance between intracellular and extracellular localization of galectins. Current inhibitors only block extracellular functions of a given galectin and neglect intracellular functions as seen in Gal-1 and Gal-3. Consequently, further research is warranted to assess the role of galectins in cancer therapy. Nowadays, several clinical trials are being conducted focused on the use of galectins in the treatment of neoplasms (Table 4). The use of an inhibitor for a single galectin could be effective, provided that selectivity is adequate and that the above-mentioned problems are taken into account before clinical trials commence. However, as research progresses, galectin targeting therapy may increase the efficacy of cancer patient treatment.

Table 4. Ongoing clinical trials with galectin inhibitors in oncology.

NCT Number	Inhibitor	Target	Phase	Title of the Study
NCT01723813	GM-CT-01	Gal-3	I/II	Peptide Vaccinations Plus GM-CT-01 in Melanoma
NCT01724320	OTX008	Gal-1	I	A Phase I, First-in-man Study of OTX008 Given Subcutaneously as a Single Agent to Patients with Advanced Solid Tumors
NCT02117362	GR-MD-02	Gal-3	I	Galectin Inhibitor (GR-MD-02) and Ipilimumab in Patients with Metastatic Melanoma
NCT02575404	GR-MD-02	Gal-3	I	GR-MD-02 Plus Pembrolizumab in Melanoma Patients
NCT01681823	PectaSol-C	Gal-3	III	Effect of Modified Citrus Pectin on PSA Kinetics in Biochemical Relapsed PC with Serial Increases in PSA

Data were collected from: https://clinicaltrials.gov.

Conflicts of Interest: The authors declare no conflict of interest.

References

1. Thijssen, V.L.; Heusschen, R.; Caers, J.; Griffioen, A.W. Galectin expression in cancer diagnosis and prognosis: A systematic review. *Biochim. Biophys. Acta Rev. Cancer* **2015**, *1855*, 235–247. [CrossRef] [PubMed]

2. De Oliveira, F.L.; Gatto, M.; Bassi, N.; Luisetto, R.; Ghirardello, A.; Punzi, L.; Doria, A. Galectin-3 in autoimmunity and autoimmune diseases. *Exp. Biol. Med.* **2015**, *240*, 1019–1028. [CrossRef] [PubMed]

3. Barondes, S.H.; Cooper, D.N.; Gitt, M.A.; Leffler, H. Galectins. Structure and function of a large family of animal lectins. *J. Biol. Chem.* **1994**, *269*, 20807–20810. [PubMed]

4. Croci, D.O.; Rabinovich, G.A. Linking tumor hypoxia with VEGFR2 signaling and compensatory angiogenesis. *Oncoimmunology* **2014**, *3*, e29380. [CrossRef] [PubMed]

5. Stillman, B.N.; Hsu, D.K.; Pang, M.; Brewer, C.F.; Johnson, P.; Liu, F.-T.; Baum, L.G. Galectin-3 and Galectin-1 Bind Distinct Cell Surface Glycoprotein Receptors to Induce T Cell Death. *J. Immunol.* **2006**, *176*, 778–789. [CrossRef] [PubMed]

6. Fukumori, T.; Takenaka, Y.; Yoshii, T.; Kim, H.-R.C.; Hogan, V.; Inohara, H.; Kagawa, S.; Raz, A. CD29 and CD7 mediate galectin-3-induced type II T-cell apoptosis. *Cancer Res.* **2003**, *63*, 8302–8311. [PubMed]

7. Hirabayashi, J.; Hashidate, T.; Arata, Y.; Nishi, N.; Nakamura, T.; Hirashima, M.; Urashima, T.; Oka, T.; Futai, M.; Muller, W.E.G.; et al. Oligosaccharide specificity of galectins: A search by frontal affinity chromatography. *Biochim. Biophys. Acta Gen. Subj.* **2002**, *1572*, 232–254. [CrossRef]

8. Sanchez-Ruderisch, H.; Fischer, C.; Detjen, K.M.; Welzel, M.; Wimmel, A.; Manning, J.C.; André, S.; Gabius, H.-J. Tumor suppressor p16INK4a: Downregulation of galectin-3, an endogenous competitor of the pro-anoikis effector galectin-1, in a pancreatic carcinoma model. *FEBS J.* **2010**, *277*, 3552–3563. [CrossRef] [PubMed]

9. Shalom-Feuerstein, R.; Cooks, T.; Raz, A.; Kloog, Y. Galectin-3 regulates a molecular switch from N-Ras to K-Ras usage in human breast carcinoma cells. *Cancer Res.* **2005**, *65*, 7292–7300. [CrossRef] [PubMed]

10. Cho, M.; Cummings, R.D. Galectin-1, a β-galactoside-binding lectin in Chinese hamster ovary cells: II. Localization and biosynthesis. *J. Biol. Chem.* **1995**, *270*, 5207–5212. [CrossRef] [PubMed]

11. Thijssen, V.L.; Hulsmans, S.; Griffioen, A.W. The galectin profile of the endothelium. *Am. J. Pathol.* **2008**, *172*, 545–553. [CrossRef] [PubMed]

12. Hittelet, A.; Legendre, H.; Nagy, N.; Bronckart, Y.; Pector, J.C.; Salmon, I.; Yeaton, P.; Gabius, H.J.; Kiss, R.; Camby, I. Upregulation of galectins-1 and -3 in human colon cancer and their role in regulating cell migration. *Int. J. Cancer* **2003**, *103*, 370–379. [CrossRef] [PubMed]

13. Kuo, P.-L.; Hung, J.-Y.; Huang, S.-K.; Chou, S.-H.; Cheng, D.-E.; Jong, Y.-J.; Hung, C.; Yang, C.-J.; Tsai, Y.-M.; Hsu, Y.-L.; et al. Lung Cancer-Derived galectin-1 Mediates Dendritic Cell Anergy through Inhibitor of DNA Binding 3/IL-10 Signaling Pathway. *J. Immunol.* **2011**, *186*, 1521–1530. [CrossRef] [PubMed]

14. Dalotto-Moreno, T.; Croci, D.O.; Cerliani, J.P.; Martinez-Allo, V.C.; Dergan-Dylon, S.; Méndez-Huergo, S.P.; Stupirski, J.C.; Mazal, D.; Osinaga, E.; Toscano, M.A.; et al. Targeting galectin-1 overcomes breast cancer-associated immunosuppression and prevents metastatic disease. *Cancer Res.* **2013**, *73*, 1107–1117. [CrossRef] [PubMed]

15. Chen, R.; Pan, S.; Ottenhof, N.A.; de Wilde, R.F.; Wolfgang, C.L.; Lane, Z.; Post, J.; Bronner, M.P.; Willmann, J.K.; Maitra, A.; et al. Stromal galectin-1 expression is associated with long-term survival in resectable pancreatic ductal adenocarcinoma. *Cancer Biol. Ther.* **2012**, *13*, 899–907. [CrossRef] [PubMed]

16. Tang, D.; Yuan, Z.; Xue, X.; Lu, Z.; Zhang, Y.; Wang, H.; Chen, M.; An, Y.; Wei, J.; Zhu, Y.; et al. High expression of galectin-1 in pancreatic stellate cells plays a role in the development and maintenance of an immunosuppressive microenvironment in pancreatic cancer. *Int. J. Cancer* **2012**, *130*, 2337–2348. [CrossRef] [PubMed]

17. Spano, D.; Russo, R.; di Maso, V.; Rosso, N.; Terracciano, L.M.; Roncalli, M.; Tornillo, L.; Capasso, M.; Tiribelli, C.; Iolascon, A. Galectin-1 and its involvement in hepatocellular carcinoma aggressiveness. *Mol. Med.* **2010**, *16*, 102–115. [CrossRef] [PubMed]

18. Torres-Cabala, C.; Bibbo, M.; Panizo-Santos, A.; Barazi, H.; Krutzsch, H.; Roberts, D.D.; Merino, M.J. Proteomic identification of new biomarkers and application in thyroid cytology. *Acta Cytol.* **2006**, *50*, 518–528. [CrossRef] [PubMed]

19. Juszczynski, P.; Rodig, S.J.; Ouyang, J.; O'Donnell, E.; Takeyama, K.; Mlynarski, W.; Mycko, K.; Szczepanski, T.; Gaworczyk, A.; Krivtsov, A.; et al. MLL-Rearranged B Lymphoblastic Leukemias selectively express the immunoregulatory carbohydrate-binding protein galectin-1. *Clin. Cancer Res.* **2010**, *16*, 2122–2130. [CrossRef] [PubMed]

20. Koopmans, S.M.; Bot, F.J.; Schouten, H.C.; Janssen, J.; van Marion, A.M. The involvement of galectins in the modulation of the JAK/STAT pathway in myeloproliferative neoplasia. *Am. J. Blood Res.* **2012**, *2*, 119–127. [PubMed]

21. Saussez, S.; Glinoer, D.; Chantrain, G.; Pattou, F.; Carnaille, B.; André, S.; Gabius, H.-J.; Laurent, G. Serum galectin-1 and galectin-3 levels in benign and malignant nodular thyroid disease. *Thyroid* **2008**, *18*, 705–712. [CrossRef] [PubMed]

22. Cedeno-Laurent, F.; Watanabe, R.; Teague, J.E.; Kupper, T.S.; Clark, R.A.; Dimitroff, C.J. Galectin-1 inhibits the viability, proliferation, and Th1 cytokine production of nonmalignant T cells in patients with leukemic cutaneous T-cell lymphoma. *Blood* **2012**, *119*, 3534–3538. [CrossRef] [PubMed]

23. Verschuere, T.; van Woensel, M.; Fieuws, S.; Lefranc, F.; Mathieu, V.; Kiss, R.; van Gool, S.W.; de Vleeschouwer, S. Altered galectin-1 serum levels in patients diagnosed with high-grade glioma. *J. Neurooncol.* **2013**, *115*, 9–17. [CrossRef] [PubMed]

24. Vyakarnam, A.; Dagher, S.F.; Wang, J.L.; Patterson, R.J. Evidence for a role for galectin-1 in pre-mRNA splicing. *Mol. Cell. Biol.* **1997**, *17*, 4730–4737. [CrossRef] [PubMed]

25. Astorgues-Xerri, L.; Riveiro, M.E.; Tijeras-Raballand, A.; Serova, M.; Neuzillet, C.; Albert, S.; Raymond, E.; Faivre, S. Unraveling galectin-1 as a novel therapeutic target for cancer. In *Cancer Treatment Reviews*; Elsevier Ltd.: Amsterdam, the Netherlands, 2014; pp. 307–319.

26. Paz, A.; Haklai, R.; Elad-Sfadia, G.; Ballan, E.; Kloog, Y. Galectin-1 binds oncogenic H-Ras to mediate Ras membrane anchorage and cell transformation. *Oncogene* **2001**, *20*, 7486–7493. [CrossRef] [PubMed]

27. Gamrekelashvili, J.; Krüger, C.; von Wasielewski, R.; Hoffmann, M.; Huster, K.M.; Busch, D.H.; Manns, M.P.; Korangy, F.; Greten, T.F. Necrotic tumor cell death in vivo impairs tumor-specific immune responses. *J. Immunol.* **2007**, *178*, 1573–1580. [CrossRef] [PubMed]

28. Vaupel, P. Hypoxia and aggressive tumor phenotype: Implications for therapy and prognosis. *Oncologist* **2008**, *13*, 21–26. [CrossRef] [PubMed]

29. Zhao, X.-Y.; Chen, T.-T.; Xia, L.; Guo, M.; Xu, Y.; Yue, F.; Jiang, Y.; Chen, G.-Q.; Zhao, K.-W. Hypoxia inducible factor-1 mediates expression of galectin-1: The potential role in migration/invasion of colorectal cancer cells. *Carcinogenesis* **2010**, *31*, 1367–1375. [CrossRef] [PubMed]

30. Le, Q.-T.; Shi, G.; Cao, H.; Nelson, D.W.; Wang, Y.; Chen, E.Y.; Zhao, S.; Kong, C.; Richardson, D.; O'Byrne, K.J.; et al. Galectin-1: A link between tumor hypoxia and tumor immune privilege. *J. Clin. Oncol.* **2005**, *23*, 8932–8941. [CrossRef] [PubMed]

31. Ito, K.; Scott, S.A.; Cutler, S.; Dong, L.-F.; Neuzil, J.; Blanchard, H.; Ralph, S.J. Thiodigalactoside inhibits murine cancers by concurrently blocking effects of galectin-1 on immune dysregulation, angiogenesis and protection against oxidative stress. *Angiogenesis* **2011**, *14*, 293–307. [CrossRef] [PubMed]

32. Ito, K.; Ralph, S.J. Inhibiting galectin-1 reduces murine lung metastasis with increased CD4+ and CD8+ T cells and reduced cancer cell adherence. *Clin. Exp. Metastasis* **2012**, *29*, 561–572. [CrossRef] [PubMed]

33. Ito, K.; Stannard, K.; Gabutero, E.; Clark, A.M.; Neo, S.-Y.; Onturk, S.; Blanchard, H.; Ralph, S.J. Galectin-1 as a potent target for cancer therapy: Role in the tumor microenvironment. *Cancer Metastasis Rev.* **2012**, *31*, 763–778. [CrossRef] [PubMed]

34. Abdollahi, A.; Lipson, K.E.; Sckell, A.; Zieher, H.; Klenke, F.; Poerschke, D.; Roth, A.; Han, X.; Krix, M.; Bischof, M.; et al. Combined therapy with direct and indirect angiogenesis inhibition results in enhanced antiangiogenic and antitumor effects. *Cancer Res.* **2003**, *63*, 8890–8898. [PubMed]

35. Dings, R.P.M.; Loren, M.; Heun, H.; McNiel, E.; Griffioen, A.W.; Mayo, K.H.; Griffin, R.J. Scheduling of radiation with angiogenesis inhibitors anginex and avastin improves therapeutic outcome via vessel normalization. *Clin. Cancer Res.* **2007**, *13*, 3395–3402. [CrossRef] [PubMed]

36. Dong, D.F.; Li, E.X.; Wang, J.B.; Wu, Y.Y.; Shi, F.; Guo, J.J.; Wu, Y.; Liu, J.P.; Liu, S.X.; Yang, G.X. Anti-angiogenesis and anti-tumor effects of AdNT4-anginex. *Cancer Lett.* **2017**, *285*, 218–224. [CrossRef] [PubMed]

37. Dings, R.P.M.; Kumar, N.; Miller, M.C.; Loren, M.; Rangwala, H.; Hoye, T.R.; Mayo, K.H. Structure-based optimization of angiostatic agent 6DBF7, an allosteric antagonist of galectin-1. *J. Pharmacol. Exp. Ther.* **2013**, *344*, 589–599. [CrossRef] [PubMed]

38. Salomonsson, E.; Thijssen, V.L.; Griffioen, A.W.; Nilsson, U.J.; Leffler, H. The anti-angiogenic peptide anginex greatly enhances galectin-1 binding affinity for glycoproteins. *J. Biol. Chem.* **2011**, *286*, 13801–13804. [CrossRef] [PubMed]

39. Dings, R.P.M.; Miller, M.C.; Nesmelova, I.; Astorgues-Xerri, L.; Kumar, N.; Serova, M.; Chen, X.; Raymond, E.; Hoye, T.R.; Mayo, K.H. Antitumor agent calixarene 0118 targets human galectin-1 as an allosteric inhibitor of carbohydrate binding. *J. Med. Chem.* **2012**, *55*, 5121–5129. [CrossRef] [PubMed]

40. Dings, R.P.M.; Chen, X.; Hellebrekers, D.M.E.I.; van Eijk, L.I.; Zhang, Y.; Hoye, T.R.; Griffioen, A.W.; Mayo, K.H. Design of nonpeptidic topomimetics of antiangiogenic proteins with antitumor activities. *J. Natl. Cancer Inst.* **2006**, *98*, 932–936. [CrossRef] [PubMed]

41. Dings, R.P.M.; Van Laar, E.S.; Webber, J.; Zhang, Y.; Griffin, R.J.; Waters, S.J.; MacDonald, J.R.; Mayo, K.H. Ovarian tumor growth regression using a combination of vascular targeting agents anginex or topomimetic 0118 and the chemotherapeutic irofulven. *Cancer Lett.* **2008**, *265*, 270–280. [CrossRef] [PubMed]

42. Zucchetti, M.; Bonezzi, K.; Frapolli, R.; Sala, F.; Borsotti, P.; Zangarini, M.; Cvitkovic, E.; Noel, K.; Ubezio, P.; Giavazzi, R.; et al. Pharmacokinetics and antineoplastic activity of galectin-1-targeting OTX008 in combination with sunitinib. *Cancer Chemother. Pharmacol.* **2013**, *72*, 879–887. [CrossRef] [PubMed]

43. Läppchen, T.; Dings, R.P.M.; Rossin, R.; Simon, J.F.; Visser, T.J.; Bakker, M.; Walhe, P.; van Mourik, T.; Donato, K.; van Beijnum, J.R.; et al. Novel analogs of antitumor agent calixarene 0118: Synthesis, cytotoxicity, click labeling with 2-[^{18}F]fluoroethylazide, and in vivo evaluation. *Eur. J. Med. Chem.* **2015**, *89*, 279–295.

44. Croci, D.O.; Salatino, M.; Rubinstein, N.; Cerliani, J.P.; Cavallin, L.E.; Leung, H.J.; Ouyang, J.; Ilarregui, J.M.; Toscano, M.A.; Domaica, C.I.; et al. Disrupting galectin-1 interactions with *N*-glycans suppresses hypoxia-driven angiogenesis and tumorigenesis in Kaposi's sarcoma. *J. Exp. Med.* **2012**, *209*, 1985–2000. [CrossRef] [PubMed]

45. Miller, M.C.; Klyosov, A.; Mayo, K.H. The α-galactomannan Davanat binds galectin-1 at a site different from the conventional galectin carbohydrate binding domain. *Glycobiology* **2009**, *19*, 1034–1045. [CrossRef] [PubMed]

46. Klyosov, A.; Zomer, E.; Platt, D. DAVANAT® (GM-CT-01) and Colon Cancer: Preclinical and Clinical (Phase I and II) Studies. In *Glycobiology and Drug Design*; ACS Symposium Series; American Chemical Society: Washington, DC, USA, 2012; Volume 1102, pp. 89–130.

47. Griffioen, A.W.; van der Schaft, D.W.; Barendsz-Janson, A.F.; Cox, A.; Struijker Boudier, H.A.; Hillen, H.F.; Mayo, K.H. Anginex, a designed peptide that inhibits angiogenesis. *Biochem. J.* **2001**, *354*, 233–242. [CrossRef] [PubMed]

48. Ju, W.; de, W.M.; En, L.; Dan, D. Peptides Advances and prospects of anginex as a promising anti-angiogenesis and anti-tumor agent. *Peptides* **2012**, *38*, 457–462.

49. Thijssen, V.L.J.L.; Postel, R.; Brandwijk, R.J.M.G.E.; Dings, R.P.M.; Nesmelova, I.; Satijn, S.; Verhofstad, N.; Nakabeppu, Y.; Baum, L.G.; Bakkers, J.; et al. Galectin-1 is essential in tumor angiogenesis and is a target for antiangiogenesis therapy. *Proc. Natl. Acad. Sci. USA* **2006**, *103*, 15975–15980. [CrossRef] [PubMed]

50. Thijssen, V.L.; Barkan, B.; Shoji, H.; Aries, I.M.; Mathieu, V.; Deltour, L.; Hackeng, T.M.; Kiss, R.; Kloog, Y.; Poirier, F.; et al. Tumor cells secrete galectin-1 to enhance endothelial cell activity. *Cancer Res.* **2010**, *70*, 6216–6224. [CrossRef] [PubMed]

51. Amano, M.; Suzuki, M.; Andoh, S.; Monzen, H.; Terai, K.; Williams, B.; Song, C.W.; Mayo, K.H.; Hasegawa, T.; Dings, R.P.M.; et al. Antiangiogenesis therapy using a novel angiogenesis inhibitor, anginex, following radiation causes tumor growth delay. *Int. J. Clin. Oncol.* **2007**, *12*, 42–47. [CrossRef] [PubMed]

52. Gorski, D.H.; Beckett, M.A.; Jaskowiak, N.T.; Calvin, D.P.; Mauceri, H.J.; Salloum, R.M.; Seetharam, S.; Koons, A.; Hari, D.M.; Kufe, D.W.; et al. Blockade of the vascular endothelial growth factor stress response increases the antitumor effects of ionizing radiation. *Cancer Res.* **1999**, *59*, 3374–3378. [PubMed]

53. Dings, R.P.M.; Williams, B.W.; Song, C.W.; Griffioen, A.W.; Mayo, K.H.; Griffin, R.J. Anginex synergizes with radiation therapy to inhibit tumor growth by radiosensitizing endothelial cells. *Int. J. Cancer* **2005**, *115*, 312–319. [CrossRef] [PubMed]

54. Dings, R.P.M.; Yokoyama, Y.; Ramakrishnan, S.; Griffioen, A.W.; Mayo, K.H. The designed angiostatic peptide anginex synergistically improves chemotherapy and antiangiogenesis therapy with angiostatin. *Cancer Res.* **2003**, *63*, 382–385. [PubMed]

55. Brandwijk, R.J.M.G.E.; Nesmelova, I.; Dings, R.P.M.; Mayo, K.H.; Thijssen, V.L.J.L.; Griffioen, A.W. Cloning an artificial gene encoding angiostatic anginex: From designed peptide to functional recombinant protein. *Biochem. Biophys. Res. Commun.* **2005**, *333*, 1261–1268. [CrossRef] [PubMed]

56. Brandwijk, R.J.M.G.E.; Mulder, W.J.M.; Nicolay, K.; Mayo, K.H.; Thijssen, V.L.J.L.; Griffioen, A.W. Anginex-Conjugated liposomes for targeting of angiogenic endothelial cells. *Bioconjug. Chem.* **2007**, *18*, 785–790. [CrossRef] [PubMed]

57. Dings, R.P.M.; van Laar, E.S.; Loren, M.; Webber, J.; Zhang, Y.; Waters, S.J.; Macdonald, J.R.; Mayo, K.H. Inhibiting tumor growth by targeting tumor vasculature with galectin-1 antagonist anginex conjugated to the cytotoxic acylfulvene, 6-hydroxylpropylacylfulvene. *Bioconjugate Chem.* **2010**, *21*, 20–27. [CrossRef] [PubMed]

58. Upreti, M.; Jyoti, A.; Johnson, S.E.; Swindell, E.P.; Sethi, P.; Chan, R.; Feddock, J.M.; Weiss, H.L.; O'Halloran, T.V.; Evers, B.M. Radiation-enhanced therapeutic targeting of galectin-1 enriched malignant stroma in triple negative breast cancer. *Oncotarget* **2016**, *7*, 41559–41574. [CrossRef] [PubMed]

59. Mayo, K.H.; Dings, R.P.M.; Flader, C.; Nesmelova, I.; Hargittai, B.; van der Schaft, D.W.J.; van Eijk, L.I.; Walek, D.; Haseman, J.; Hoye, T.R.; et al. Design of a partial peptide mimetic of anginex with antiangiogenic and anticancer activity. *J. Biol. Chem.* **2003**, *278*, 45746–45752. [CrossRef] [PubMed]

60. Astorgues-Xerri, L.; Riveiro, M.E.; Tijeras-Raballand, A.; Serova, M.; Rabinovich, G.A.; Bieche, I.; Vidaud, M.; de Gramont, A.; Martinet, M.; Cvitkovic, E.; et al. OTX008, a selective small-molecule inhibitor of galectin-1, downregulates cancer cell proliferation, invasion and tumour angiogenesis. *Eur. J. Cancer* **2014**, *50*, 2463–2477. [CrossRef] [PubMed]

61. Michael, J.V.; Wurtzel, J.G.T.; Goldfinger, L.E. Inhibition of galectin-1 sensitizes hras-driven tumor growth to rapamycin treatment. *Anticancer Res.* **2016**, *36*, 5053–5061. [CrossRef] [PubMed]

62. Dings, R.P.M.; Levine, J.I.; Brown, S.G.; Astorgues-Xerri, L.; MacDonald, J.R.; Hoye, T.R.; Raymond, E.; Mayo, K.H. Polycationic calixarene PTX013, a potent cytotoxic agent against tumors and drug resistant cancer. *Invest. New Drugs* **2013**, *31*, 1142–1150. [CrossRef] [PubMed]

63. Demotte, N.; Bigirimana, R.; Wieërs, G.; Stroobant, V.; Squifflet, J.L.; Carrasco, J.; Thielemans, K.; Baurain, J.F.; van der Smissen, P.; Courtoy, P.J.; et al. A short treatment with galactomannan GM-CT-01 corrects the functions of freshly isolated human tumor-infiltrating lymphocytes. *Clin. Cancer Res.* **2014**, *20*, 1823–1833. [CrossRef] [PubMed]

64. Cedeno-Laurent, F.; Opperman, M.J.; Barthel, S.R.; Hays, D.; Schatton, T.; Zhan, Q.; He, X.; Matta, K.L.; Supko, J.G.; Frank, M.H.; et al. Metabolic inhibition of galectin-1-binding carbohydrates accentuates antitumor immunity. *J. Invest. Dermatol.* **2012**, *132*, 410–420. [CrossRef] [PubMed]

65. Mathieu, V.; Le Mercier, M.; de Neve, N.; Sauvage, S.; Gras, T.; Roland, I.; Lefranc, F.; Kiss, R. Galectin-1 knockdown increases sensitivity to temozolomide in a B16F10 mouse metastatic melanoma model. *J. Invest. Dermatol.* **2007**, *127*, 2399–2410. [CrossRef] [PubMed]

66. Wu, X.; Li, J.; Connolly, E.M.; Liao, X.; Ouyang, J.; Giobbie-Hurder, A.; Lawrence, D.; McDermott, D.; Murphy, G.; Zhou, J.; et al. Combined Anti-VEGF and Anti–CTLA-4 therapy elicits humoral immunity to Galectin-1 which is associated with favorable clinical outcomes. *Cancer Immunol. Res.* **2017**, *5*, 446–454. [CrossRef] [PubMed]

67. Traber, P.G.; Zomer, E. Therapy of experimental NASH and fibrosis with galectin inhibitors. *PLoS ONE* **2013**, *8*, e83481. [CrossRef] [PubMed]

68. Harrison, S.A.; Marri, S.R.; Chalasani, N.; Kohli, R.; Aronstein, W.; Thompson, G.A.; Irish, W.; Miles, M.V.; Xanthakos, S.A.; Lawitz, E.; et al. Randomised clinical study: GR-MD-02, a galectin-3 inhibitor, vs. placebo in patients having non-alcoholic steatohepatitis with advanced fibrosis. *Aliment. Pharmacol. Ther.* **2016**, *44*, 1183–1198. [CrossRef] [PubMed]

69. Brown, E.R.; Doig, T.; Anderson, N.; Brenn, T.; Doherty, V.; Xu, Y.; Bartlett, J.M.S.; Smyth, J.F.; Melton, D.W. Association of galectin-3 expression with melanoma progression and prognosis. *Eur. J. Cancer* **2012**, *48*, 865–874. [CrossRef] [PubMed]

70. Vereecken, P.; Awada, A.; Suciu, S.; Castro, G.; Morandini, R.; Litynska, A.; Lienard, D.; Ezzedine, K.; Ghanem, G.; Heenen, M. Evaluation of the prognostic significance of serum galectin-3 in American Joint Committee on Cancer stage III and stage IV melanoma patients. *Melanoma Res.* **2009**, *19*, 316–320. [CrossRef] [PubMed]

71. Vereecken, P.; Zouaoui Boudjeltia, K.; Debray, C.; Awada, A.; Legssyer, I.; Sales, F.; Petein, M.; Vanhaeverbeek, M.; Ghanem, G.; Heenen, M. High serum galectin-3 in advanced melanoma: Preliminary results. *Clin. Exp. Dermatol.* **2006**, *31*, 105–109. [CrossRef] [PubMed]

72. Harazono, Y.; Nakajima, K.; Raz, A. Why anti-Bcl-2 clinical trials fail: A solution. *Cancer Metastasis Rev.* **2014**, *33*, 285–294. [CrossRef] [PubMed]

73. Leffler, H.; Carlsson, S.; Hedlund, M.; Qian, Y.; Poirier, F. Introduction to galectins. *Glycoconj. J.* **2002**, *19*, 433–440. [CrossRef] [PubMed]

74. Menon, R.P.; Hughes, R.C. Determinants in the N-terminal domains of galectin-3 for secretion by a novel pathway circumventing the endoplasmic reticulum–Golgi complex. *Eur. J. Biochem.* **1999**, *264*, 569–576. [CrossRef] [PubMed]

75. Tellez-Sanz, R.; Garcia-Fuentes, L.; Vargas-Berenguel, A. Human galectin-3 selective and high affinity inhibitors. Present state and future perspectives. *Curr. Med. Chem.* **2013**, *20*, 2979–2990. [CrossRef] [PubMed]

76. Newlaczyl, A.U.; Yu, L. Galectin-3—A jack-of-all-trades in cancer. *Cancer Lett.* **2011**, *313*, 123–128. [CrossRef] [PubMed]

77. Bartolazzi, A.; Gasbarri, A.; Papotti, M.; Bussolati, G.; Lucante, T.; Khan, A.; Inohara, H.; Marandino, F.; Orlandi, F.; Nardi, F.; et al. Application of an immunodiagnostic method for improving preoperative diagnosis of nodular thyroid lesions. *Lancet* **2001**, *357*, 1644–1650. [CrossRef]

78. Bartolazzi, A.; Orlandi, F.; Saggiorato, E.; Volante, M.; Arecco, F.; Rossetto, R.; Palestini, N.; Ghigo, E.; Papotti, M.; Bussolati, G.; et al. Galectin-3-expression analysis in the surgical selection of follicular thyroid nodules with indeterminate fine-needle aspiration cytology: A prospective multicentre study. *Lancet Oncol.* **2008**, *9*, 543–549. [CrossRef]

79. Bartolazzi, A.; Bellotti, C.; Sciacchitano, S. Methodology and technical requirements of the galectin-3 test for the preoperative characterization of thyroid nodules. *Appl. Immunohistochem. Mol. Morphol.* **2012**, *20*, 2–7. [CrossRef] [PubMed]

80. Trimboli, P.; Virili, C.; Romanelli, F.; Crescenzi, A.; Giovanella, L. Galectin-3 performance in histologic and cytologic assessment of thyroid nodules: A systematic review and Meta-analysis. *Int. J. Mol. Sci.* **2017**, *18*, 1756. [CrossRef] [PubMed]

81. Ruvolo, P.P. Galectin 3 as a guardian of the tumor microenvironment. *Biochim. Biophys. Acta Mol. Cell Res.* **2016**, *1863*, 427–437. [CrossRef] [PubMed]

82. Lin, H.-M.; Pestell, R.G.; Raz, A.; Kim, H.-R.C. Galectin-3 enhances cyclin D1 promoter activity through SP1 and a cAMP-responsive element in human breast epithelial cells. *Oncogene* **2002**, *21*, 8001–8010. [CrossRef] [PubMed]

83. Shimura, T.; Takenaka, Y.; Fukumori, T.; Tsutsumi, S.; Okada, K.; Hogan, V.; Kikuchi, A.; Kuwano, H.; Raz, A. Implication of galectin-3 in Wnt signaling. *Cancer Res.* **2005**, *65*, 3535–3537. [CrossRef] [PubMed]

84. Akahani, S.; Nangia-Makker, P.; Inohara, H.; Kim, H.-R.R.; Raz, A. Galectin-3: A novel antiapoptotic molecule with a functional BH1 (NWGR) domain of Bcl-2 family. *Cancer Res.* **1997**, *57*, 5272–5276. [PubMed]

85. Elad-Sfadia, G.; Haklai, R.; Balan, E.; Kloog, Y. Galectin-3 augments K-ras activation and triggers a ras signal that attenuates ERK but not phosphoinositide 3-kinase activity. *J. Biol. Chem.* **2004**, *279*, 34922–34930. [CrossRef] [PubMed]

86. Veschi, V.; Petroni, M.; Cardinali, B.; Dominici, C.; Screpanti, I.; Frati, L.; Bartolazzi, A.; Gulino, A.; Giannini, G. Galectin-3 impairment of MYCN-Dependent apoptosis-sensitive phenotype is antagonized by Nutlin-3 in Neuroblastoma cells. *PLoS ONE* **2012**, *7*, e49139. [CrossRef] [PubMed]

87. Xue, J.; Gao, X.; Fu, C.; Cong, Z.; Jiang, H.; Wang, W.; Chen, T.; Wei, Q.; Qin, C. Regulation of galectin-3-induced apoptosis of Jurkat cells by both *O*-glycans and *N*-glycans on CD45. *FEBS Lett.* **2013**, *587*, 3986–3994. [CrossRef] [PubMed]

88. Wang, W.; Guo, H.; Geng, J.; Zheng, X.; Wei, H.; Sun, R.; Tian, Z. Tumor-released galectin-3, a soluble inhibitory ligand of human NKp30, plays an important role in tumor escape from NK cell attack. *J. Biol. Chem.* **2014**, *289*, 33311–33319. [CrossRef] [PubMed]

89. Fukumori, T.; Oka, N.; Takenaka, Y.; Nangia-Makker, P.; Elsamman, E.; Kasai, T.; Shono, M.; Kanayama, H.; Ellerhorst, J.; Lotan, R.; et al. Galectin-3 regulates mitochondrial stability and antiapoptotic function in response to anticancer drug in prostate cancer. *Cancer Res.* **2006**, *66*, 3114–3119. [CrossRef] [PubMed]

90. Fukumori, T.; Kanayama, H.; Raz, A. The role of galectin-3 in cancer drug resistance. *Drug Resist. Updat.* **2007**, *10*, 101–108. [CrossRef] [PubMed]

91. Harazono, Y.; Kho, D.H.; Balan, V.; Nakajima, K.; Hogan, V.; Raz, A. Extracellular galectin-3 programs multidrug resistance through Na$^+$/K$^+$-ATPase and P-glycoprotein signaling. *Oncotarget* **2015**, *6*, 19592–19604. [CrossRef] [PubMed]

92. Chauhan, D.; Li, G.; Podar, K.; Hideshima, T.; Neri, P.; He, D.; Mitsiades, N.; Richardson, P.; Chang, Y.; Schindler, J.; et al. A novel carbohydrate-based therapeutic GCS-100 overcomes bortezomib resistance and enhances dexamethasone-induced apoptosis in multiple myeloma cells. *Cancer Res.* **2005**, *65*, 8350–8358. [CrossRef] [PubMed]

93. Newton-Northup, J.R.; Dickerson, M.T.; Ma, L.; Besch-Williford, C.L.; Deutscher, S.L. Inhibition of metastatic tumor formation in vivo by a bacteriophage display-derived galectin-3 targeting peptide. *Clin. Exp. Metastasis* **2013**, *30*, 119–132. [CrossRef] [PubMed]

94. Yang, Y.; Zhou, Z.; He, S.; Fan, T.; Jin, Y.; Zhu, X.; Chen, C.; Zhang, Z.; Huang, Y. Treatment of prostate carcinoma with (galectin-3)-targeted HPMA copolymer-(G3–C12)-5-Fluorouracil conjugates. *Biomaterials* **2012**, *33*, 2260–2271. [CrossRef] [PubMed]

95. Yang, Y.; Li, L.; Zhou, Z.; Yang, Q.; Liu, C.; Huang, Y. Targeting prostate carcinoma by G3–C12 peptide conjugated *N*-(2-hydroxypropyl)methacrylamide copolymers. *Mol. Pharm.* **2014**, *11*, 3251–3260. [CrossRef] [PubMed]

96. Yang, Q.; Yang, Y.; Li, L.; Sun, W.; Zhu, X.; Huang, Y. Polymeric nanomedicine for tumor-targeted combination therapy to elicit synergistic genotoxicity against prostate cancer. *ACS Appl. Mater. Interfaces* **2015**, *7*, 6661–6673. [CrossRef] [PubMed]

97. Sun, W.; Li, L.; Yang, Q.; Shan, W.; Zhang, Z.; Huang, Y. G3–C12 peptide reverses galectin-3 from foe to friend for active targeting cancer treatment. *Mol. Pharm.* **2015**, *12*, 4124–4136. [CrossRef] [PubMed]

98. Yan, J.; Katz, A. PectaSol-C Modified Citrus Pectin Induces Apoptosis and Inhibition of Proliferation in Human and Mouse Androgen-Dependent and -Independent Prostate Cancer Cells. *Integr. Cancer Ther.* **2010**, *9*, 197–203. [CrossRef] [PubMed]

99. Nangia-Makker, P.; Honjo, Y.; Sarvis, R.; Akahani, S.; Hogan, V.; Pienta, K.J.; Raz, A. Galectin-3 induces endothelial cell morphogenesis and angiogenesis. *Am. J. Pathol.* **2000**, *156*, 899–909. [CrossRef]

100. Nangia-makker, P.; Hogan, V.; Honjo, Y.; Baccarini, S.; Tait, L.; Bresalier, R.; Raz, A. Inhibition of human cancer cell growth and metastasis in nude mice by oral intake of modified citrus pectin. *J. Natl. Cancer Inst.* **2002**, *94*, 1854–1862. [CrossRef] [PubMed]

101. Azémar, M.; Hildenbrand, B.; Haering, B.; Heim, M.E.; Unger, C. Clinical Benefit in Patients with Advanced Solid Tumors Treated with Modified Citrus Pectin: A Prospective Pilot Study. *Clin. Med. Oncol.* **2007**, *1*, CMO.S285. [CrossRef]

102. Demotte, N.; Wieërs, G.; van der Smissen, P.; Moser, M.; Schmidt, C.; Thielemans, K.; Squifflet, J.-L.; Weynand, B.; Carrasco, J.; Lurquin, C.; et al. A galectin-3 ligand corrects the impaired function of human CD4 and CD8 tumor-infiltrating lymphocytes and favors tumor rejection in mice. *Cancer Res.* **2010**, *70*, 7476–7488. [CrossRef] [PubMed]

103. Tehranian, N.; Sepehri, H.; Mehdipour, P.; Biramijamal, F.; Hossein-Nezhad, A.; Sarrafnejad, A.; Hajizadeh, E. Combination effect of PectaSol and Doxorubicin on viability, cell cycle arrest and apoptosis in DU-145 and LNCaP prostate cancer cell lines. *Cell Biol. Int.* **2012**, *36*, 601–610. [CrossRef] [PubMed]

104. Hossein, G.; Keshavarz, M.; Ahmadi, S.; Naderi, N. Synergistic effects of PectaSol-C modified citrus pectin an inhibitor of galectin-3 and paclitaxel on apoptosis of human SKOV-3 ovarian cancer cells. *Asian Pac. J. Cancer Prev.* **2013**, *14*, 7561–7568. [CrossRef] [PubMed]

105. Streetly, M.J.; Maharaj, L.; Joel, S.; Schey, S.A.; Gribben, J.G.; Cotter, F.E. GCS-100, a novel galectin-3 antagonist, modulates MCL-1, NOXA, and cell cycle to induce myeloma cell death. *Blood* **2010**, *115*, 3939–3948. [CrossRef] [PubMed]

106. Wang, Y.; Balan, V.; Hogan, V.; Raz, A. Calpain activation through galectin-3 inhibition sensitizes prostate cancer cells to cisplatin treatment. *Cell Death Dis.* **2010**, *1*, e101–e110. [CrossRef] [PubMed]

107. Clark, M.C.; Pang, M.; Hsu, D.K.; Liu, F.; de Vos, S.; Gascoyne, R.D.; Said, J.; Baum, L.G. Galectin-3 binds to CD45 on diffuse large B-cell lymphoma cells to regulate susceptibility to cell death. *Blood* **2012**, *120*, 4635–4644. [CrossRef] [PubMed]

108. Ruvolo, P.P.; Ruvolo, V.R.; Benton, C.B.; Al-, A.; Burks, J.K.; Schober, W.; Rolke, J.; Tidmarsh, G.; Hail, N., Jr.; Davis, R.E.; et al. Combination of galectin inhibitor GCS-100 and BH3 mimetics eliminates both p53 wild type and p53 null AML cells. *BBA Mol. Cell Res.* **2015**, *1863*, 562–571. [CrossRef] [PubMed]

109. Cotter, F.; Smith, D.A.; Boyd, T.E.; Richards, D.A.; Alemany, C.; Loesch, D.; Salogub, G.; Tidmarsh, G.F.; Gammon, G.M.; Gribben, J. Single-agent activity of GCS-100, a first-in-class galectin-3 antagonist, in elderly patients with relapsed chronic lymphocytic leukemia. *J. Clin. Oncol.* **2009**, *27*, 7006. [CrossRef]

110. Straube, T.; von Mach, T.; Hönig, E.; Greb, C.; Schneider, D.; Jacob, R. pH-Dependent recycling of galectin-3 at the apical membrane of Epithelial cells. *Traffic* **2013**, *14*, 1014–1027. [CrossRef] [PubMed]

111. Salman, H.; Bergman, M.; Djaldetti, M.; Orlin, J.; Bessler, H. Citrus pectin affects cytokine production by human peripheral blood mononuclear cells. *Biomed. Pharmacother.* **2008**, *62*, 579–582. [CrossRef] [PubMed]

112. Zhang, W.; Xu, P.; Zhang, H. Pectin in cancer therapy: A review. *Trends Food Sci. Technol.* **2015**, *44*, 258–271. [CrossRef]

113. Huflejt, M.E.; Leffler, H. Galectin-4 in normal tissues and cancer. *Glycoconj. J.* **2003**, *20*, 247–255. [CrossRef] [PubMed]

114. Hayashi, T.; Saito, T.; Fujimura, T.; Hara, K.; Takamochi, K.; Mitani, K.; Mineki, R.; Kazuno, S.; Oh, S.; Ueno, T.; et al. Galectin-4, a Novel Predictor for Lymph Node Metastasis in Lung Adenocarcinoma. *PLoS ONE* **2013**, *8*, e81883. [CrossRef] [PubMed]

115. Barrow, H.; Rhodes, J.M.; Yu, L.-G. Simultaneous determination of serum galectin-3 and -4 levels detects metastases in colorectal cancer patients. *Cell. Oncol.* **2013**, *36*, 9–13. [CrossRef] [PubMed]

116. Kim, S.W.; Park, K.C.; Jeon, S.M.; Ohn, T.B.; Kim, T.I.; Kim, W.H.; Cheon, J.H. Abrogation of galectin-4 expression promotes tumorigenesis in colorectal cancer. *Cell. Oncol.* **2013**, *36*, 169–178. [CrossRef] [PubMed]

117. Satelli, A.; Rao, P.S.; Thirumala, S.; Rao, U.S. Galectin-4 functions as a tumor suppressor of human colorectal cancer. *Int. J. Cancer* **2011**, *129*, 799–809. [CrossRef] [PubMed]

118. Rustiguel, J.K.; Kumagai, P.S.; Dias-Baruffi, M.; Costa-Filho, A.J.; Nonato, M.C. Recombinant expression, purification and preliminary biophysical and structural studies of C-terminal carbohydrate recognition domain from human galectin-4. *Protein Expr. Purif.* **2016**, *118*, 39–48. [CrossRef] [PubMed]

119. Bum-Erdene, K.; Leffler, H.; Nilsson, U.J.; Blanchard, H. Structural characterisation of human galectin-4 N-terminal carbohydrate recognition domain in complex with glycerol, lactose, 3′-sulfo-lactose, and 2′-fucosyllactose. *Sci. Rep.* **2016**, *6*, 20289. [CrossRef] [PubMed]

120. Bum-Erdene, K.; Leffler, H.; Nilsson, U.J.; Blanchard, H. Structural characterization of human galectin-4 C-terminal domain: Elucidating the molecular basis for recognition of glycosphingolipids, sulfated saccharides and blood group antigens. *FEBS J.* **2015**, *282*, 3348–3367. [CrossRef] [PubMed]

121. Magnaldo, T.; Bernerd, F.; Darmon, M. Galectin-7, a human 14-kDa S-lectin, specifically expressed in keratinocytes and sensitive to retinoic acid. *Dev. Biol.* **1995**, *168*, 259–271. [CrossRef] [PubMed]

122. Demers, M.; Rose, A.A.N.; Grosset, A.-A.; Biron-Pain, K.; Gaboury, L.; Siegel, P.M.; St-Pierre, Y. Overexpression of galectin-7, a myoepithelial cell marker, enhances spontaneous metastasis of breast cancer cells. *Am. J. Pathol.* **2010**, *176*, 3023–3031. [CrossRef] [PubMed]

123. Rorive, S.; Eddafali, B.; Fernandez, S.; Decaestecker, C.; André, S.; Kaltner, H.; Kuwabara, I.; Liu, F.-T.; Gabius, H.-J.; Kiss, R.; et al. Changes in galectin-7 and cytokeratin-19 expression during the progression of malignancy in thyroid tumors: Diagnostic and biological implications. *Mod. Pathol.* **2002**, *15*, 1294–1301. [CrossRef] [PubMed]

124. Cada, Z.; Chovanec, M.; Smetana, K.; Betka, J.; Lacina, L.; Plzák, J.; Kodet, R.; Stork, J.; Lensch, M.; Kaltner, H.; et al. Galectin-7: Will the lectin's activity establish clinical correlations in head and neck squamous cell and basal cell carcinomas? *Histol. Histopathol.* **2009**, *24*, 41–48. [PubMed]

125. Demers, M.; Biron-Pain, K.; Hébert, J.; Lamarre, A.; Magnaldo, T.; St-Pierre, Y. Galectin-7 in lymphoma: Elevated expression in human lymphoid malignancies and decreased lymphoma dissemination by antisense strategies in experimental model. *Cancer Res.* **2007**, *67*, 2824–2829. [CrossRef] [PubMed]

126. Kim, S.; Hwang, J.; Ro, J.Y.; Lee, Y.; Chun, K. Galectin-7 is epigenetically-regulated tumor suppressor in gastric cancer. *Oncotarget* **2013**, *4*, 1461–1471. [CrossRef] [PubMed]

127. Ueda, S.; Kuwabara, I.; Liu, F. Suppression of tumor growth by galectin-7 gene transfer. *Cancer Res.* **2004**, *64*, 5672–5676. [CrossRef] [PubMed]

128. Zhu, H.; Wu, T.-C.; Chen, W.-Q.; Zhou, L.-J.; Wu, Y.; Zeng, L.; Pei, H.-P. Roles of galectin-7 and S100A9 in cervical squamous carcinoma: Clinicopathological and in vitro evidence. *Int. J. Cancer* **2013**, *132*, 1051–1059. [CrossRef] [PubMed]

129. Matsui, Y.; Ueda, S.; Watanabe, J.; Kuwabara, I.; Ogawa, O.; Nishiyama, H. Sensitizing effect of galectin-7 in urothelial cancer to cisplatin through the accumulation of intracellular reactive oxygen species. *Cancer Res.* **2007**, *67*, 1212–1220. [CrossRef] [PubMed]

130. Tsai, C.J.; Sulman, E.P.; Eifel, P.J.; Jhingran, A.; Allen, P.K.; Deavers, M.T.; Klopp, A.H. Galectin-7 levels predict radiation response in squamous cell carcinoma of the cervix. *Gynecol. Oncol.* **2013**, *131*, 645–649. [CrossRef] [PubMed]

131. Labrie, M.; Vladoiu, M.C.; Grosset, A.; Gaboury, L.; St-Pierre, Y. Expression and functions of galectin-7 in ovarian cancer. *Oncotarget* **2014**, *5*, 7705–7721. [CrossRef] [PubMed]

132. Masuyer, G.; Jabeen, T.; Öberg, C.T.; Leffler, H.; Nilsson, U.J.; Acharya, K.R. Inhibition mechanism of human galectin-7 by a novel galactose-benzylphosphate inhibitor. *FEBS J.* **2012**, *279*, 193–202. [CrossRef] [PubMed]

133. Vladoiu, M.C.; Labrie, M.; Létourneau, M.; Egesborg, P.; Gagné, D.; Billard, É.; Grosset, A.-A.; Doucet, N.; Chatenet, D.; St-Pierre, Y. Design of a peptidic inhibitor that targets the dimer interface of a prototypic galectin. *Oncotarget* **2015**, *6*, 40970–40980. [CrossRef] [PubMed]

134. Kaur, M.; Kaur, T.; Kamboj, S.S.; Singh, J. Roles of galectin-7 in cancer. *Asian Pac. J. Cancer Prev.* **2016**, *17*, 455–461. [CrossRef] [PubMed]

135. Si, Y.; Wang, Y.; Gao, J.; Song, C.; Feng, S.; Zhou, Y.; Tai, G.; Su, J. Crystallization of Galectin-8 Linker Reveals Intricate Relationship between the N-terminal Tail and the Linker. *Int. J. Mol. Sci.* **2016**, *17*, 2088. [CrossRef] [PubMed]

136. Bidon-Wagner, N.; le Pennec, J.-P. Human galectin-8 isoforms and cancer. *Glycoconj. J.* **2002**, *19*, 557–563. [CrossRef] [PubMed]

137. Troncoso, M.F.; Ferragut, F.; Bacigalupo, M.L.; Cardenas Delgado, V.M.; Nugnes, L.G.; Gentilini, L.; Laderach, D.; Wolfenstein-Todel, C.; Compagno, D.; Rabinovich, G.A.; et al. Galectin-8: A matricellular lectin with key roles in angiogenesis. *Glycobiology* **2014**, *24*, 907–914. [CrossRef] [PubMed]

138. Stowell, S.R.; Arthur, C.M.; Slanina, K.A.; Horton, J.R.; Smith, D.F.; Cummings, R.D. Dimeric galectin-8 induces phosphatidylserine exposure in leukocytes through polylactosamine recognition by the C-terminal domain. *J. Biol. Chem.* **2008**, *283*, 20547–20559. [CrossRef] [PubMed]

139. Liu, C.G.; Chien, C.H.; Lin, C.; Hsu, S.D. NMR assignments of the C-terminal domain of human galectin-8. *Biomol. NMR Assign.* **2015**, *9*, 427–430. [CrossRef] [PubMed]

140. Kumar, S.; Frank, M.; Schwartz-Albiez, R. Understanding the specificity of human galectin-8c domain interactions with its glycan ligands based on molecular dynamics simulations. *PLoS ONE* **2013**, *8*, e59761. [CrossRef] [PubMed]

141. Eshkar Sebban, L.; Ronen, D.; Levartovsky, D.; Elkayam, O.; Caspi, D.; Aamar, S.; Amital, H.; Rubinow, A.; Golan, I.; Naor, D.; et al. The involvement of CD44 and its novel ligand galectin-8 in apoptotic regulation of autoimmune inflammation. *J. Immunol.* **2007**, *179*, 1225–1235. [CrossRef] [PubMed]

142. Savin, S.; Cvejić, D.; Janković, M.; Isić, T.; Paunović, I.; Tatić, S. Evaluation of galectin-8 expression in thyroid tumors. *Med. Oncol.* **2009**, *26*, 314–318. [CrossRef] [PubMed]

143. Kramer, M.W.; Waalkes, S.; Serth, J.; Hennenlotter, J.; Tezval, H.; Stenzl, A.; Kuczyk, M.A.; Merseburger, A.S. Decreased galectin-8 is a strong marker for recurrence in urothelial carcinoma of the bladder. *Urol. Int.* **2011**, *87*, 143–150. [CrossRef] [PubMed]

144. Barrow, H.; Guo, X.; Wandall, H.H.; Pedersen, J.W.; Fu, B.; Zhao, Q.; Chen, C.; Rhodes, J.M.; Yu, L.-G. Serum galectin-2, -4, and -8 are greatly increased in colon and breast cancer patients and promote cancer cell adhesion to blood vascular endothelium. *Clin. Cancer Res.* **2011**, *17*, 7035–7046. [CrossRef] [PubMed]

145. Liu, Y.; Liu, Z.; Fu, Q.; Wang, Z.; Fu, H.; Liu, W.; Wang, Y.; Xu, J. Galectin-9 as a prognostic and predictive biomarker in bladder urothelial carcinoma. *Urol. Oncol. Semin. Orig. Investig.* **2017**, *35*, 349–355. [CrossRef] [PubMed]

146. Daniel Gentilini, L.; Martín Jaworski, F.; Tiraboschi, C.; González Pérez, I.; Lidia Kotler, M.; Chauchereau, A.; Jose Laderach, D.; Compagno, D. Stable and high expression of Galectin-8 tightly controls metastatic progression of prostate cancer. *Oncotarget* **2017**, *8*, 44654–44668. [CrossRef] [PubMed]

147. Danguy, A.; Rorive, S.; Decaestecker, C.; Bronckart, Y.; Kaltner, H.; Hadari, Y.R.; Goren, R.; Zich, Y.; Petein, M.; Salmon, I.; et al. Immunohistochemical profile of galectin-8 expression in benign and malignant tumors of epithelial, mesenchymatous and adipous origins, and of the nervous system. *Histol. Histopathol.* **2001**, *16*, 861–868. [PubMed]

148. Satelli, A.; Rao, P.S.; Gupta, P.K.; Lockman, P.R.; Srivenugopal, K.S.; Rao, U.S. Varied expression and localization of multiple galectins in different cancer cell lines. *Oncol. Rep.* **2008**, *19*, 587–594. [CrossRef] [PubMed]

149. Fernández, M.M.; Ferragut, F.; Cárdenas Delgado, V.M.; Bracalente, C.; Bravo, A.I.; Cagnoni, A.J.; Nuñez, M.; Morosi, L.G.; Quinta, H.R.; Espelt, M.V.; et al. Glycosylation-dependent binding of galectin-8 to activated leukocyte cell adhesion molecule (ALCAM/CD166) promotes its surface segregation on breast cancer cells. *Biochim. Biophys. Acta* **2016**, *1860*, 2255–2268. [CrossRef] [PubMed]

150. Türeci, Ö.; Schmitt, H.; Fadle, N.; Pfreundschuh, M.; Sahin, U. Molecular definition of a novel human galectin which is immunogenic in patients with Hodgkin's disease. *J. Biol. Chem.* **1997**, *272*, 6416–6422. [CrossRef] [PubMed]

151. Sato, M.; Nishi, N.; Shoji, H.; Seki, M.; Hashidate, T.; Hirabayashi, J.; Kasai, K.-I.; Hata, Y.; Suzuki, S.; Hirashima, M.; et al. Functional analysis of the carbohydrate recognition domains and a linker peptide of galectin-9 as to eosinophil chemoattractant activity. *Glycobiology* **2002**, *12*, 191–197. [CrossRef] [PubMed]

152. Zhu, C.; Anderson, A.C.; Schubart, A.; Xiong, H.; Imitola, J.; Khoury, S.J.; Zheng, X.X.; Strom, T.B.; Kuchroo, V.K. The Tim-3 ligand galectin-9 negatively regulates T helper type 1 immunity. *Nat. Immunol.* **2005**, *6*, 1245–1252. [CrossRef] [PubMed]

153. Katoh, S.; Ishii, N.; Nobumoto, A.; Takeshita, K.; Dai, S.; Shinonaga, R.; Niki, T.; Nishi, N.; Tominaga, A.; Yamauchi, A.; et al. Galectin-9 inhibits CD44–Hyaluronan interaction and suppresses a murine model of allergic asthma. *Am. J. Respir. Crit. Care Med.* **2007**, *176*, 27–35. [CrossRef] [PubMed]

154. Ohtsubo, K.; Takamatsu, S.; Minowa, M.T.; Yoshida, A.; Takeuchi, M.; Marth, J.D. Dietary and genetic control of glucose transporter 2 glycosylation promotes insulin secretion in suppressing diabetes. *Cell* **2005**, *123*, 1307–1321. [CrossRef] [PubMed]

155. Matsuura, A.; Tsukada, J.; Mizobe, T.; Higashi, T.; Mouri, F.; Tanikawa, R.; Yamauchi, A.; Hirashima, M.; Tanaka, Y. Intracellular galectin-9 activates inflammatory cytokines in monocytes. *Genes Cells* **2009**, *14*, 511–521. [CrossRef] [PubMed]

156. Kageshita, T.; Kashio, Y.; Yamauchi, A.; Seki, M.; Abedin, M.J.; Nishi, N.; Shoji, H.; Nakamura, T.; Ono, T.; Hirashima, M. Possible role of galectin-9 in cell aggregation and apoptosis of human melanoma cell lines and its clinical significance. *Int. J. Cancer* **2002**, *99*, 809–816. [CrossRef] [PubMed]

157. Irie, A.; Yamauchi, A.; Kontani, K.; Kihara, M.; Liu, D.; Shirato, Y.; Seki, M.; Nishi, N.; Nakamura, T.; Yokomise, H.; et al. Galectin-9 as a prognostic factor with antimetastatic potential in breast cancer. *Clin. Cancer Res.* **2005**, *11*, 2962–2968. [CrossRef] [PubMed]

158. Kashio, Y.; Nakamura, K.; Abedin, M.J.; Seki, M.; Nishi, N.; Yoshida, N.; Nakamura, T.; Hirashima, M. Galectin-9 induces apoptosis through the Calcium-Calpain-Caspase-1 pathway. *J. Immunol.* **2003**, *170*, 3631–3636. [CrossRef] [PubMed]

159. Saita, N.; Goto, E.; Yamamoto, T.; Cho, I.; Tsumori, K.; Kohrogi, H.; Maruo, K.; Ono, T.; Takeya, M.; Kashio, Y.; et al. Association of galectin-9 with eosinophil apoptosis. *Int. Arch. Allergy Immunol.* **2002**, *128*, 42–50. [CrossRef] [PubMed]

160. Seki, M.; Oomizu, S.; Sakata, K.; Sakata, A.; Arikawa, T.; Watanabe, K.; Ito, K.; Takeshita, K.; Niki, T.; Saita, N.; et al. Galectin-9 suppresses the generation of Th17, promotes the induction of regulatory T cells, and regulates experimental autoimmune arthritis. *Clin. Immunol.* **2008**, *127*, 78–88. [CrossRef] [PubMed]

161. Nagahara, K.; Arikawa, T.; Oomizu, S.; Kontani, K.; Nobumoto, A.; Tateno, H.; Watanabe, K.; Niki, T.; Katoh, S.; Miyake, M.; et al. Galectin-9 increases Tim-3+ dendritic cells and CD8+ T cells and enhances antitumor immunity via galectin-9-tim-3 interactions. *J. Immunol.* **2008**, *181*, 7660–7669. [CrossRef] [PubMed]

162. Zhang, Z.Y.; Dong, J.H.; Chen, Y.W.; Wang, X.Q.; Li, C.H.; Wang, J.; Wang, G.Q.; Li, H.L.; Wang, X.D. Galectin-9 acts as a prognostic factor with antimetastatic potential in hepatocellular carcinoma. *Asian Pac. J. Cancer Prev.* **2012**, *13*, 2503–2509. [CrossRef] [PubMed]

163. Lahm, H.; André, S.; Hoeflich, A.; Fischer, J.R.; Sordat, B.; Kaltner, H.; Wolf, E.; Gabius, H.J. Comprehensive galectin fingerprinting in a panel of 61 human tumor cell lines by RT-PCR and its implications for diagnostic and therapeutic procedures. *J. Cancer Res. Clin. Oncol.* **2001**, *127*, 375–386. [CrossRef] [PubMed]

164. Laderach, D.J.; Gentilini, L.D.; Giribaldi, L.; Delgado, V.C.; Nugnes, L.; Croci, D.O.; Al Nakouzi, N.; Sacca, P.; Casas, G.; Mazza, O.; et al. A unique galectin signature in human prostate cancer progression suggests galectin-1 as a key target for treatment of advanced disease. *Cancer Res.* **2013**, *73*, 86–96. [CrossRef] [PubMed]

165. Chan, S.W.; Kallarakkal, T.G.; Abraham, M.T. Changed expression of E-cadherin and galectin-9 in oral squamous cell carcinomas but lack of potential as prognostic markers. *Asian Pac. J. Cancer Prev.* **2014**, *15*, 2145–2152. [CrossRef] [PubMed]

166. Terris, B.; Blaveri, E.; Crnogorac-jurcevic, T.; Jones, M.; Missiaglia, E.; Ruszniewski, P.; Sauvanet, A.; Lemoine, N.R. Characterization of gene expression profiles in intraductal papillary-mucinous tumors of the pancreas. *Am. J. Pathol.* **2002**, *160*, 1745–1754. [CrossRef]

167. Fu, H.; Liu, Y.; Xu, L.; Liu, W.; Fu, Q.; Liu, H.; Zhang, W.; Xu, J. Galectin-9 predicts postoperative recurrence and survival of patients with clear-cell renal cell carcinoma. *Tumor Biol.* **2015**, *36*, 5791–5799. [CrossRef] [PubMed]

168. Kawashima, H.; Obayashi, A.; Kawamura, M.; Masaki, S.; Tamada, S.; Iguchi, T.; Uchida, J.; Kuratsukuri, K.; Tanaka, T.; Nakatani, T. Galectin 9 and PINCH, novel immunotherapy targets of renal cell carcinoma: A rationale to find potential tumour antigens and the resulting cytotoxic T lymphocytes induced by the derived peptides. *BJU Int.* **2014**, *113*, 320–332. [CrossRef] [PubMed]

169. Wiersma, V.R.; de Bruyn, M.; Wei, Y.; van Ginkel, R.J.; Hirashima, M.; Niki, T.; Nishi, N.; Zhou, J.; Pouwels, S.D.; Samplonius, D.F.; et al. The epithelial polarity regulator LGALS9/galectin-9 induces fatal frustrated autophagy in KRAS mutant colon carcinoma that depends on elevated basal autophagic flux. *Autophagy* **2015**, *11*, 1373–1388. [CrossRef] [PubMed]

170. Mishra, R.; Grzybek, M.; Niki, T.; Hirashima, M.; Simons, K. Galectin-9 trafficking regulates apical-basal polarity in Madin–Darby canine kidney epithelial cells. *Proc. Natl. Acad. Sci. USA* **2010**, *107*, 17633–17638. [CrossRef] [PubMed]

171. Kuroda, J.; Yamamoto, M.; Nagoshi, H.; Kobayashi, T.; Sasaki, N.; Shimura, Y.; Horiike, S.; Kimura, S.; Yamauchi, A.; Hirashima, M.; et al. Targeting activating transcription factor 3 by galectin-9 induces apoptosis and overcomes various types of treatment resistance in chronic myelogenous leukemia. *Mol. Cancer Res.* **2010**, *8*, 994–1001. [CrossRef] [PubMed]

172. Yang, R.; Hung, M.-C. The role of T-cell immunoglobulin mucin-3 and its ligand galectin-9 in antitumor immunity and cancer immunotherapy. *Sci. China Life Sci.* **2017**, *60*, 1058–1064. [CrossRef] [PubMed]

173. Sun, H.-W.; Li, C.-J.; Chen, H.-Q.; Lin, H.-L.; Lv, H.-X.; Zhang, Y.; Zhang, M. Involvement of integrins, MAPK, and NF-κB in regulation of the shear stress-induced MMP-9 expression in endothelial cells. *Biochem. Biophys. Res. Commun.* **2007**, *353*, 152–158. [CrossRef] [PubMed]

174. Lhuillier, C.; Barjon, C.; Niki, T.; Gelin, A.; Praz, F.; Morales, O.; Souquere, S.; Hirashima, M.; Wei, M.; Dellis, O.; et al. Impact of exogenous galectin-9 on human T cells. *J. Biol. Chem.* **2015**, *290*, 16797–16811. [CrossRef] [PubMed]

175. Peng, F.; Huang, Y.; Li, M.Y.; Li, G.Q.; Huang, H.C.; Guan, R.; Chen, Z.C.; Liang, S.P.; Chen, Y.H. Dissecting characteristics and dynamics of differentially expressed proteins during multistage carcinogenesis of human colorectal cancer. *World J. Gastroenterol.* **2016**, *22*, 4515–4528. [CrossRef] [PubMed]

176. Xue, H.; Yang, R.-Y.; Tai, G.; Liu, F.-T. Galectin-12 inhibits granulocytic differentiation of human NB4 promyelocytic leukemia cells while promoting lipogenesis. *J. Leukoc. Biol.* **2016**, *100*, 657–664. [CrossRef] [PubMed]

International Journal of
Molecular Sciences

MDPI

Review

Role of Galectins in Tumors and in Clinical Immunotherapy

Feng-Cheng Chou [1], Heng-Yi Chen [2], Chih-Chi Kuo [3] and Huey-Kang Sytwu [1,2,*]

[1]   Department and Graduate Institute of Microbiology and Immunology, National Defense Medical Center,
       Taipei 114, Taiwan; fengchengchou@gmail.com
[2]   Graduate Institute of Life Sciences, National Defense Medical Center, Taipei 114, Taiwan;
       vicky0128n@gmail.com
[3]   Teaching and Research Office, Tri-Service General Hospital Songshan Branch,
       National Defense Medical Center, Taipei 105, Taiwan; vampirelydia@mail.ndmctsgh.edu.tw
*   Correspondence: sytwu@mail.ndmctsgh.edu.tw; Tel.: +886-2-8792-3100 (ext. 18540)

Received: 30 December 2017; Accepted: 30 January 2018; Published: 1 February 2018

Abstract: Galectins are glycan-binding proteins that contain one or two carbohydrate domains and mediate multiple biological functions. By analyzing clinical tumor samples, the abnormal expression of galectins is known to be linked to the development, progression and metastasis of cancers. Galectins also have diverse functions on different immune cells that either promote inflammation or dampen T cell-mediated immune responses, depending on cognate receptors on target cells. Thus, tumor-derived galectins can have bifunctional effects on tumor and immune cells. This review focuses on the biological effects of galectin-1, galectin-3 and galectin-9 in various cancers and discusses anticancer therapies that target these molecules.

Keywords: galectin-1; galectin-3; galectin-9; immunotherapy; galectin inhibitors

1. Introduction

Galectins are a family of lectins composed of one or two carbohydrate-recognition domains (CRDs) that bind to beta-galactoside-containing glycans. To date, 15 galectins have been identified in mammals and 11 are found in humans, acting both intracellularly and extracellularly. Galectins can be classified into three groups based on their structure: (1) prototype galectins that contain one CRD that can form homodimers, including galectin-1, galectin-2, galectin-5, galectin-7, galectin-10, galectin-11 galectin-13, galectin-14 and galectin-15; (2) tandem repeat-type galectins that contain two CRDs and are connected by a flexible linker, including galectin-4, galectin-6, galectin-8, galectin-9 and galectin-12; and (3) chimeric-type galectin-3 that contains a CRD domain and an N-terminal extension that can form oligomers to increase their binding avidity (galectin-5, -11, -15 and -6 are not found in humans).

Galectins are soluble proteins that are widely expressed in various cell types and mediate their functions both intracellularly and extracellularly. Although galectins do not have any known signal sequence for their transport, it is likely to be a nonexocytotic pathway; for example, cells infected with the Epstein-Barr virus release galectin-9 via an exosome-mediated mechanism [1]. In general, the functions of galectins include the regulation of cell growth, apoptosis, pre-mRNA splicing, cell-cell and cell-matrix adhesion, cellular polarity, motility, differentiation, transformation, signal transduction and innate/adaptive immunity. Due to the diverse functions of galectins, such as in apoptosis, angiogenesis, cell migration and tumor-immune escape, altered levels have been implicated in cancer biology [2]. For example, in tumor cells, intracellular galectins can enhance oncogenic signals and reduce apoptosis that promote tumor transformation and proliferation (reviewed in [3]). By contrast, the extracellular galectins bind to cell surface glycoproteins and form galectin lattices, depending on the glycosylated sites. The galectin–receptor lattice can modulate the functions of the receptor and support either surface retention or transportation of the receptors [4].

The ligands of galectins are glycoproteins and glycolipids that contain different degrees of oligosaccharide modifications (*N*- and *O*-linked glycans). The selectivity and factors that influence how each galectin member binds to glycoproteins depend on the glycosylation sites (sequence-encoded information), on glycosylation levels in Golgi complex processing (depending on glycosyltransferase activity) and glycol-conjugate formations (e.g., LacNAc, 2-*O*-glycans, complex branched *N*-glycans or sialylated structures) [5,6]. Thus, given the preferred glycan branch of an individual galectin, each type can bind to a set of glycoconjugates on the cell surface that mediates specific functions. Interestingly, some galectins regulate innate and adaptive immune responses by binding to a panel of glycoproteins on immune cells. For example, galectin-1 binds to CD2, CD3, CD7, CD43 and CD45 on T cells, which downregulate immune responses by inducing apoptosis. By contrast, galectin-9 has a dual function, by binding to T cell immunoglobulin mucin 3 (TIM-3) expressed on T cells or dendritic cells, which induces apoptosis or inflammatory responses, respectively. In this review, we discuss the relationship between tumor-derived galectins and tumor prognosis. In addition, given the multiple immune regulatory functions of galectins on T cells, we also summarize the results of clinical trials that used galectin inhibitors combined with various forms of chemotherapy or immunotherapy.

2. Role of Galectins in Tumor Progression and Immune Surveillance

2.1. Human Galectins

Among the 11 galectins identified in humans, galectin-1, galectin-3 and galectin-9 have been the most extensively investigated in different fields including cell biology and immunology. Importantly, the roles of these three forms have been closely linked to cancer biology. The functions of these galectins in tumors include enhancing oncogenic signal pathways, regulating tumor cell growth or apoptosis, modulating cell migration and suppressing immune responses.

2.2. Correlation of Galectin-3 and/or Galectin-1 in Cancers

Galectin-3 and galectin-1 have been investigated extensively in various tumors [2]. Clinically, thyroid malignancies of epithelial origin display increased galectin-1 and galectin-3 expression compared with benign thyroid adenomas [7]. In human endometrial cancers, the expression of galectin-1 is upregulated in uterine adenocarcinomas compared with normal adjacent endometrium, whereas expression of galectin-3 is downregulated in endometrial cancer cells compared with normal mucosa. Interestingly, tumors with galectin-3 expressed in the cytoplasm were characterized by a deeper invasion of the myometrium compared with lesions where galectin-3 was found both in the nucleus and cytoplasm [8]. The localization of galectin-3 in normal cells and different stages of cancer cells were further investigated in cases of colorectal cancer progression. In line with previous findings, strong cytoplasmic expression of galectin-3 was associated with later phases of tumor progression and was inversely correlated with the survival of patients [9]. In cases of human bladder cancer, increased mRNA expression of galectin-1 in transitional-cell carcinomas was positively correlated with histological grade and clinical stage. However, the expression of galectin-3 only showed increases in carcinomas without being correlated with histological grade [10]. Then, the biological effects of galectins linked to tumors have been investigated. For example, increased levels of galectin-3 are involved in liver metastasis, venous invasion and lymph node metastasis of colorectal cancer [11]. Galectin-3 levels were also increased in the blood stream of cancer patients and this promoted cancer metastasis via binding to cell surface-associated mucin 1 (MUC1) on cancer cells, in turn leading to the exposure of smaller cell-surface adhesion molecules/ligands including CD44 and ligand(s) for E-selectin [12]. The interaction of galectin-3 with MUC1 has been reported to increase the association of MUC1 with the epidermal growth factor receptor (EGFR), which then promotes EGFR homo-/heterodimerization and subsequently increased and prolonged EGFR activation and signaling. These mechanisms might contribute to EGFR-associated tumorigenesis and cancer progression and could also influence the effectiveness of blocking the action of EGFR in patients undergoing cancer therapy [13]. Moreover, increased galectin-3 expression in gastric cancer cells contributes to cellular unresponsiveness to interferon gamma (IFN-γ) by facilitating the AKT/GSK-3β/SHP2 signaling cascade [14]. Importantly, several cytokines are also glycoproteins whose functions can be neutralized by extracellular galectins. One elegant study demonstrated that tumor-derived galectin-3 captured IFN-γ in tumor matrices and subsequently downregulated IFN-γ-induced chemokine gradients. Inhibition of galectin-3 in human tumor biopsies enhanced IFN-γ-induced CXCL-9 chemokine expression, suggesting that the blockade of galectin-3 in tumor cells may promote T cell tumor infiltration and activation [15].

2.3. Galectin-9 and Tumor Metastasis

Galectin-9 has attracted much attention because of its multiple biological functions and strong immunomodulatory effects. Expression of galectin-9 in solid tumors has been linked to tumor cell adhesion or metastasis. Thus, the galectin-9 expression level was correlated with cellular adhesion and aggregation in melanoma cells [16], oral squamous cell carcinomas [17], breast cancer [18] and hepatocellular carcinoma (HCC) [19]. These studies have demonstrated that the high expression level of galectin-9 in tumor cells promoted tumor cells aggregation in vitro and in mouse models, whereas downregulation of galectin-9 in these cells correlated with cell invasion. Additionally, the following studies further demonstrated that administration of galectin-9 in mice suppresses lung metastasis of melanoma cells via inhibition of the binding of adhesive molecules on tumor cells to ligands on vascular endothelium and extracellular matrix [20]. In clinical studies, an association of the expression level of galectin-9 with tumor metastasis has been established in breast cancer and in HCC. Patients with galectin-9 negative breast cancers showed high potential with distant metastases and correlated with higher histopathologic grades when compared with patients with galectin-9 positive breast cancer. In addition, the cumulative disease-free survival rate for galectin-9-positive patients was better than in a galectin-9-negative group [18]. Consistent with those findings, in patients with HCC, the decreased expression of galectin-9 was linked to lymph node metastasis, vascular invasion, intrahepatic metastasis and poor survival of patients [19]. In summary, the high expression level of galectin-9 in primary cancer lesions promoted cell-matrix interactions, and metastatic lesions displayed decreased galectin-9 expression, suggesting that galectin-9 might suppress tumor metastasis. Moreover, these results indicated that galectin-9 might serve as a prognostic factor with antimetastatic potential in patients with breast cancer and HCC; however, whether galectin-9 also has similar effects in other cancer types is still unclear.

A detailed summary of the differential expression levels of galectins in various tumors has been provided in [2]. Intriguingly, the relationship between the expression level of galectins and the malignancy of a given tumor is context dependent, suggesting that many factors should be taken into consideration, such as tumor type and stages, and the involvement of different galectins.

3. Galectins Are Involved in Immune Escape by Tumors

3.1. Galectin-1, -3 and -9 Regulate Immune Responses via Different Receptors

Despite the multiple functions of galectins in cancer cell growth, they also have strong immunomodulatory functions that regulate both innate and adaptive immunity by binding to surface glycoproteins. Tumor cells use various strategies to escape immune attack or even regulate immune responses. For example, galectin-1 and galectin-3 can induce T cell apoptosis by binding to CD45 and CD7 leukocyte proteins to induce apoptosis. Galectin-9, unlike other galectin members with many cellular receptors, binds specifically to the Tim-3 cell surface molecule on Th1 cells and induces apoptosis [6]. Ligands for galectin-1, 3 and 9 are listed in Table 1. Reports suggest that the expression of these galectins in tumor cells might help ablate the immune response to tumors.

Table 1. Galectin-1, -3, -9 and their binding partners in the regulation of immune responses and tumor biology.

	Ligand	Targeted Cells	Biological Function	Refs.
Galectin-1	CD2, CD3, CD7, CD43 and CD45	T cells	Apoptosis	[21,22]
	TCR	T cells	Signal transduction	[23]
	Neuropilin 1	Endothelial cells	Cell migration	[24]
	Pre-BCR	Pre-B cells	Signal transduction Cell maturation	[25,26]
	α4 integrins	Pre-B cells	Signal transduction Cell maturation	[25,26]
	H-Ras	Endometrial cancer cells	Membrane anchorage Cell transformation	[27]
	α5β1- integrins	Epithelial cancer cells	Epithelial integrity	[28]
Galectin-3	CD7, CD29, CD45, CD71	T cells	Apoptosis	[29,30]
	TCR	T cells	Signal transduction	[31]
	Alix	T cells	Signal transduction	[32]
	MUC1	Epithelial cancer cells	Signal transduction	[13]
	K-Ras	Breast carcinoma cells	Enhanced K-Ras stability	[33]
	TTF-1	Thyroid cancer cells	Tumor progression	[34]
	Mac-2BP	Melanoma cells	Cell-cell adhesion	[35]
Galectin-9	Dectin-1	Macrophages	Tolerogenic macrophage programming and adaptive immune suppression	[36]
	TIM-3	Dendritic cells, monocytes	Maturation and cytokine production	[37,38]
	TIM-3	T cells	Apoptosis	[39]
	4-1BB	T cells	Signal transduction	[40]
	CD40	T cells	Inducing cell death and suppressing proliferation	[41]

3.2. Galectin-1 Modulates the Antitumor T Cell Response

There have been reports that tumor microenvironments display enhanced expression of galectins linked to tolerogenic status. Galectin-1 exhibits strong immunoregulatory functions which have been demonstrated elegantly in a mouse melanoma model. Targeted inhibition of galectin-1 is correlated with enhanced T cell-mediated tumor clearance, demonstrating a strong immunosuppressive effect on T cells [42]. In clinical studies, increased galectin-1 levels have been reported in leukemic tumors, such as neoplastic Reed-Sternberg cells in patients with classic Hodgkin lymphomas [43] and in those with leukemic cutaneous T-cell lymphomas [44]. In those studies, increased galectin-1 promoted Th2 responses or expansion of Treg cells and inhibited the proliferation of other T cell types, suggesting that blockage of galectin-1 might enhance antitumor activity. In solid tumors, expression of galectin-1 was inversely correlated with the numbers of CD3+ T cells in tumor sections from patients with head and neck squamous cell carcinoma, and the expression of galectin-1 and CD3 also served as predictors for the prognosis of such patients [45].

3.3. Double-Edged Sword Role of Galectin-9 in Tumors

Unlike other galectins, galectin-9 both promotes and inhibits tumor activity, depending on its interactions with its ligands on T cells, antigen-presenting cells or tumor cells. Previous studies have demonstrated that Epstein-Barr virus-infected nasopharyngeal carcinoma cells (NPCs) release exosomes containing high amounts of galectin-9, which is able to induce TIM-3-expressing Th1 cell apoptosis and subsequently helps the tumor escape immune recognition [1]. The role of galectin-9 in the regulation of the immune response in the tumor microenvironment has been investigated further in patients with recurrent nasopharyngeal carcinoma. Such patients displayed increased galectin-9-positive tumor cells and FOXP3+ lymphocytes, whereas the TIM-3+ lymphocytes were decreased in the tumor microenvironment compared with primary NPCs, suggesting that the galectin-9–TIM-3 pathway mediates immune escape by NPCs [46]. This pathway also downregulates the antitumor response in hepatitis B virus-associated HCCs. Interestingly, IFN-γ produced by tumor-infiltrating lymphocytes induces galectin-9 production by Küpffer cells. Moreover, increased numbers of TIM-3+ T cells in tumors were inversely associated with patient survival, and blockade of the galectin-9–TIM-3 pathway promoted T cell proliferation and secretion of cytokines [47,48]. Besides, galectin-9 also interacts with dectin-1 expressed on macrophages and promotes tolerogenic macrophage programing in the microenvironment of pancreatic ductal adenocarcinoma [36]. These findings suggest that galectin-9 derived from the tumor microenvironment could attenuate antitumor effects and that blockage of this pathway could be a therapeutic target in further clinical applications. In contrast to the tumor-friendly role of galectin-9, results obtained from mouse models in vivo and in vitro indicated that galectin-9 can promote tumor cell apoptosis, including in chronic myelogenous leukemia cells [49], malignant melanomas [50,51], gallbladder carcinomas [52], HCCs [53], cholangiocarcinomas [54] and gastric cancer cells [55]. The mechanism of this galectin-9-mediated cancer cell apoptosis is likely linked to the carbohydrate-recognition function because administration of lactose blocked its proapoptotic effect. Besides, this effect might not act through the activation of the immune system, and the detailed mechanisms—for example, the involvement of endoplasmic reticulum stress-induced apoptosis and cellular receptors—need to be investigated further [56].

4. Targeting Galectins or their Ligands in Preclinical and Clinical Trials

As discussed above, galectins have multiple immunomodulatory effects and can be further applied to develop therapeutic strategies against autoimmunity, graft rejection and tumors. Besides, galectins also have other biological functions including regulating cell migration, adhesion and signal transduction. Given the complexity of the biological functions of galectins, increased levels of galectins in tumors have been reported to promote their growth or interfere with tumor therapies. These mechanisms include interfering with drug efficacy/delivery or dampening the antitumor effect of immune cells.

4.1. Galectins Interfere with Chemotherapy against Tumors

The multidrug resistance (MDR) phenotype is a major issue in the development of toxic chemotherapy for treating cancers. Cancer cells develop several mechanisms to combat anticancer drugs, including decreased drug uptake, increased efflux via ATP-binding cassette (ABC) transporters, and increased drug metabolism and/or resistance to drug-induced apoptosis [57]. Previous reports have demonstrated that the chemotherapy drugs adriamycin and imatinib upregulate galectin-1 expression in chronic myelogenous leukemia cells. Galectin-1 confers drug resistance via inducing the expression of MDR protein 1, which in turn helps tumor cells to pump out cytotoxic drugs [58]. Likewise, the expression level of galectin-3 was increased significantly in the sera of patients with various cancers and mediates MDR. Galectin-3 interacted with Na^+/K^+-ATPase and P-glycoprotein and enhanced ATPase activity, eventually leading to decreases in doxorubicin-induced cell death [59]. Moreover, in a preclinical mouse model of non-Hodgkin lymphoma using anti-CD20 target therapy, the expression of galectin-1 ablated antibody-dependent lymphoma phagocytosis in vitro and lymphoma cell sensitivity to CD20 immunotherapy in vivo. In addition, biopsies or blood samples from patients with Burkitt lymphoma, chronic lymphoid leukemia, diffuse large B-cell lymphoma, follicular lymphoma, hairy cell leukemia and mantle cell lymphoma displayed increased expression levels of galectin-1, suggesting that this confers resistance to anti-CD20 immunotherapy in humans [60].

Table 2. Targeting galectins and their ligands in clinical trials.

Status	Condition	Intervention	Phase	Sponsors and Collaborators
Completed	Colorectal, lung, breast, head and neck, and prostate cancers	GM-CT-01 (galectin-3 inhibitor) combined with 5-fluorouracil	Phase I NCT00054977	Galectin Therapeutics Inc.
Withdrawn	Cancers of the bile duct and gallbladder	GM-CT-01 (galectin-3 inhibitor) combined with 5-fluorouracil	Phase II NCT00386516	Galectin Therapeutics Inc.
Terminated	Colorectal cancer	GM-CT-01 (galectin-3 inhibitor) combined with 5-fluorouracil, leukovorin, bevacizumab	Phase II NCT00388700	Galectin Therapeutics Inc.
Unknown status	Metastatic melanoma	Tumor peptide vaccination combined with GM-CT-01 (galectin-3 inhibitor)	Phase I/II NCT01723813	Cliniques Universitaires Saint-Luc Université Catholique de Louvain
Unknown status	Solid tumors	OTX008 inhibitor of galectin-1 expression	Phase I NCT01724320	Oncoethix GmbH
Withdrawn	Diffuse large B-cell lymphoma	GCS-100 (galectin-3 inhibitor)	Phase I/II NCT00776802	La Jolla Pharmaceutical Co. investigators
Recruiting	Metastatic melanoma	GR-MD-02 (galectin-3 inhibitor) combined with ipilimumab (anti-CTLA-4)	Phase IB NCT02117362	Providence Health & Services, Providence Cancer Center, Earle A. Chiles Research Institute; Galectin Therapeutics Inc.
Recruiting	Melanoma, non-small cell lung cancer, and squamous cell head and neck cancers	GR-MD-02 combined with pembrolizumab (anti-PD-1; keytruda)	Phase IB NCT02575404	Galectin Therapeutics Inc.; Providence Health & Services
Recruiting	Advanced or metastatic solid tumors	TSR-022 (anti-TIM-3) combined with anti-PD-1 antibody	Phase I NCT02817633	Tesaro, Inc.
Recruiting	Solid tumors	LY3321367 (anti-TIM-3) combined with LY3300054 (anti-PD-L1)	Phase I NCT03099109	Eli Lilly and Co.
Recruiting	Advanced Malignancies	MBG453 (anti-TIM-3) combined with PDR001 (anti-PD-1)	Phase I/II NCT02608268	Novartis Pharmaceuticals
Recruiting	relapsed/refractory acute myeloid leukemia	MBG453 (anti-TIM-3) combined with PDR001 (anti-PD-1) and/or decitabine (5-aza-2'-deoxycytidine)	Phase I NCT03066648	Novartis Pharmaceuticals

In 11 clinical trials (updated to November 2017) registered by the United States National Institutes of Health (https://clinicaltrials.gov/, access on 30 November 2017), there are five using galectin inhibitors (e.g., GM-CT-01, and the Davanat® carbohydrate polymer) combined with chemotherapy drugs (5-fluorouracil) against solid tumors. In these trials, investigators have initiated multicenter phase I or II trials targeting various tumors including colorectal, lung, breast, head and neck, prostate, bile duct and gall bladder cancers, metastatic melanomas and diffuse large B-cell lymphomas. However, only one trial has been completed (Phase I trial, NCT00054977) and the others are neither withdrawn nor terminated, suggesting that not all galectin inhibitors are effective, and that efficacy might also depend on protocol design and galectin expression profiles in each individual (Table 2).

4.2. Immunotherapy Combined with Galectin Inhibition

Immunotherapy using monoclonal antibodies blocking immune checkpoint molecules has shown promising progress. However, to increase overall responsiveness, several investigators started to combine these with galectin inhibitors to enhance the therapeutic effect. Until November 2017, two clinical trials have been reported using the galectin-3 inhibitor DG-MD-02 along with ipilimumab (anti-CTLA-4) or pembrolizumab (anti-PD-1) to treat patients diagnosed with melanomas, non-small cell lung cancers, and squamous cell head and neck cancers (Table 2). Excitingly, one of these trials showed that the combination of pembrolizumab with the galectin-3 inhibitor GR-MD-02 gave promising early results in the treatment of patients with advanced melanomas in a phase Ib clinical trial. This mechanistic study showed that clinical responders to this combination might have reduced the numbers of myeloid-derived suppressor cells following treatment. Although this trial showed positive results, the detailed mechanisms are unclear and further clinical trials need to be conducted to demonstrate the efficacy of this approach and to evaluate any possible adverse effects. Despite the use of anti-PD-1 and anti-CTLA-4 antibodies that are effective clinically and the various combination therapies being conducted in clinical trials, several investigators are still working on using other immune checkpoint molecules as targets to enhance the antitumor functions of T cells. The galectin-9 binding partner, TIM-3, which is a negative regulator of T cells is now being used as a novel target in tumor immunotherapy [61,62]. Until November 2017, investigators had conducted three human clinical trials using an anti-TIM-3 monoclonal antibody combined with either anti-PD-1 or anti-PD-L1 antibodies in advanced solid tumors (Table 2). Although these clinical trials are still in their early stages, focusing on characterizing the safety, tolerability, pharmacokinetics, pharmacodynamics and antitumor activity, and blocking multiple negative regulators on T cells might unleash their antitumor effects and eventually help to control tumors.

5. Conclusions

Here we have summarized the roles of galectins in human cancer biology and focused on targeting the inhibition of these molecules in ongoing clinical trials. The functions of galectins have been investigated extensively in many fields including cancer cell biology, immunology and infectious diseases. In the earlier studies, researchers have put a lot of effort focus on the roles of galectins in tumor growth/metastasis and correlated the expression level with the prognosis of patients. However, the detailed mechanisms are still unclear, for example, galectin—tumor interaction or reciprocal interaction among galectins, tumor cells and immune cells have not been well examined. Galectin-1, -3 and -9 are three well-investigated galectins which can modulate tumor cells growth and regulate immune cells. To date, these galectins bind a panel of receptors on T cells which inhibit the functions of T cells (Table 1). Thus, tumor cell-derived galectins may not only regulate the growth of tumor cells but also dampen T cell responses. Moreover, galectins may bind to secreted cytokines or chemokines which interfere with immune cell communication [15]. Data obtained from in vitro cell culture or murine xenograft models are significant and informative; however, the results cannot yet be translated directly into clinical settings. Besides, the widely varying expression pattern of galectins and heterogeneity of tumors contribute to difficulties in these studies.

Among the clinical trials using galectin inhibitors, it seems that combined with immune intervention they showed positive results, suggesting that in the well-established tumor, the main function of the increased galectins is to interfere the anti-tumor effects of T cells. Therefore, it is reasonable to believe that using combination therapy (galectin inhibitors, chemotherapy and immune checkpoint blockades) may show great improvements in tumor therapy. However, the protocol (dosage, intervention strategy and side effects) for these clinical trials needs to be further investigated.

Given the complexity of the function of galectins, future works that aim to clarify the effects of galectin in tumors may focus on the following points: to analyze galectin profiles in different clinical samples systematically and to correlate these profiles with tumor stages; to characterize T cell phenotypes of patients with high galectin expression level in tumors, especially for the tumor-infiltrating lymphocytes; to explore the underline mechanism of the altered galectin expression in tumor growth and metastasis, for example, virus−infection−induced galectin overexpression in company with tumorigenesis. Altogether, these results might help in designing effective therapeutic approaches.

Acknowledgments: This work was supported by the Ministry of Science and Technology, ROC (MOST 104-2320-B-016-014-MY3, MOST 106-2320-B-016-009-MY3 and MOST 106-2321-B-016-003); Tri-Service General Hospital (TSGH-C106-004-006-008-S02, VTA105-T-1-1 and VTA106-T-1-1).

Author Contributions: Feng-Cheng Chou collected data and drafted the manuscript. Heng-Yi Chen collected data and prepared the section "Galectins Are Involved in Immune Escape by Tumors" and "Table 1". Chih-Chi Kuo collected data and prepared the section "Targeting Galectins or their Ligands in Preclinical and Clinical Trials" and "Table 2". Huey-Kang Sytwu critically reviewed and edited the manuscript.

Conflicts of Interest: The authors declare no conflicts of interest.

References

1. Klibi, J.; Niki, T.; Riedel, A.; Pioche-Durieu, C.; Souquere, S.; Rubinstein, E.; Moulec, S.L.; Guigay, J.; Hirashima, M.; Guemira, F.; et al. Blood diffusion and Th1-suppressive effects of galectin-9-containing exosomes released by Epstein-Barr virus-infected nasopharyngeal carcinoma cells. *Blood* **2009**, *113*, 1957–1966. [CrossRef] [PubMed]

2. Thijssen, V.L.; Heusschen, R.; Caers, J.; Griffioen, A.W. Galectin expression in cancer diagnosis and prognosis: A systematic review. *Biochim. Biophys. Acta* **2015**, *1855*, 235–247. [CrossRef] [PubMed]

3. Nakahara, S.; Raz, A. Biological modulation by lectins and their ligands in tumor progression and metastasis. *Anticancer Agents Med. Chem.* **2008**, *8*, 22–36. [PubMed]

4. Lau, K.S.; Partridge, E.A.; Grigorian, A.; Silvescu, C.I.; Reinhold, V.N.; Demetriou, M.; Dennis, J.W. Complex *N*-glycan number and degree of branching cooperate to regulate cell proliferation and differentiation. *Cell* **2007**, *129*, 123–134. [CrossRef] [PubMed]

5. Mendez-Huergo, S.P.; Blidner, A.G.; Rabinovich, G.A. Galectins: Emerging regulatory checkpoints linking tumor immunity and angiogenesis. *Curr. Opin. Immunol.* **2017**, *45*, 8–15. [CrossRef] [PubMed]

6. Rabinovich, G.A.; Toscano, M.A. Turning 'sweet' on immunity: Galectin-glycan interactions in immune tolerance and inflammation. *Nat. Rev. Immunol.* **2009**, *9*, 338–352. [CrossRef] [PubMed]

7. Xu, X.C.; el-Naggar, A.K.; Lotan, R. Differential expression of galectin-1 and galectin-3 in thyroid tumors. Potential diagnostic implications. *Am. J. Pathol.* **1995**, *147*, 815–822. [PubMed]

8. Van den Brule, F.A.; Buicu, C.; Berchuck, A.; Bast, R.C.; Deprez, M.; Liu, F.T.; Cooper, D.N.; Pieters, C.; Sobel, M.E.; Castronovo, V. Expression of the 67-kD laminin receptor, galectin-1, and galectin-3 in advanced human uterine adenocarcinoma. *Hum. Pathol.* **1996**, *27*, 1185–1191. [CrossRef]

9. Sanjuan, X.; Fernandez, P.L.; Castells, A.; Castronovo, V.; van den Brule, F.; Liu, F.T.; Cardesa, A.; Campo, E. Differential expression of galectin 3 and galectin 1 in colorectal cancer progression. *Gastroenterology* **1997**, *113*, 1906–1915. [CrossRef]

10. Cindolo, L.; Benvenuto, G.; Salvatore, P.; Pero, R.; Salvatore, G.; Mirone, V.; Prezioso, D.; Altieri, V.; Bruni, C.B.; Chiariotti, L. Galectin-1 and galectin-3 expression in human bladder transitional-cell carcinomas. *Int. J. Cancer* **1999**, *84*, 39–43. [CrossRef]

11. Nakamura, M.; Inufusa, H.; Adachi, T.; Aga, M.; Kurimoto, M.; Nakatani, Y.; Wakano, T.; Nakajima, A.; Hida, J.I.; Miyake, M.; et al. Involvement of galectin-3 expression in colorectal cancer progression and metastasis. *Int. J. Oncol.* **1999**, *15*, 143–148. [CrossRef] [PubMed]

12. Zhao, Q.; Guo, X.; Nash, G.B.; Stone, P.C.; Hilkens, J.; Rhodes, J.M.; Yu, L.G. Circulating galectin-3 promotes metastasis by modifying MUC1 localization on cancer cell surface. *Cancer Res.* **2009**, *69*, 6799–6806. [CrossRef] [PubMed]

13. Piyush, T.; Chacko, A.R.; Sindrewicz, P.; Hilkens, J.; Rhodes, J.M.; Yu, L.G. Interaction of galectin-3 with MUC1 on cell surface promotes EGFR dimerization and activation in human epithelial cancer cells. *Cell Death Differ.* **2017**, *24*, 1937–1947. [CrossRef] [PubMed]

14. Tseng, P.C.; Chen, C.L.; Shan, Y.S.; Lin, C.F. An increase in galectin-3 causes cellular unresponsiveness to IFN-gamma-induced signal transduction and growth inhibition in gastric cancer cells. *Oncotarget* **2016**, *7*, 15150–15160. [CrossRef] [PubMed]

15. Gordon-Alonso, M.; Hirsch, T.; Wildmann, C.; van der Bruggen, P. Galectin-3 captures interferon-gamma in the tumor matrix reducing chemokine gradient production and T-cell tumor infiltration. *Nat. Commun.* **2017**, *8*, 793. [CrossRef] [PubMed]

16. Kageshita, T.; Kashio, Y.; Yamauchi, A.; Seki, M.; Abedin, M.J.; Nishi, N.; Shoji, H.; Nakamura, T.; Ono, T.; Hirashima, M. Possible role of galectin-9 in cell aggregation and apoptosis of human melanoma cell lines and its clinical significance. *Int. J. Cancer* **2002**, *99*, 809–816. [CrossRef] [PubMed]

17. Kasamatsu, A.; Uzawa, K.; Nakashima, D.; Koike, H.; Shiiba, M.; Bukawa, H.; Yokoe, H.; Tanzawa, H. Galectin-9 as a regulator of cellular adhesion in human oral squamous cell carcinoma cell lines. *Int. J. Mol. Med.* **2005**, *16*, 269–273. [CrossRef] [PubMed]

18. Irie, A.; Yamauchi, A.; Kontani, K.; Kihara, M.; Liu, D.; Shirato, Y.; Seki, M.; Nishi, N.; Nakamura, T.; Yokomise, H.; et al. Galectin-9 as a prognostic factor with antimetastatic potential in breast cancer. *Clin. Cancer Res.* **2005**, *11*, 2962–2968. [CrossRef] [PubMed]

19. Zhang, Z.Y.; Dong, J.H.; Chen, Y.W.; Wang, X.Q.; Li, C.H.; Wang, J.; Wang, G.Q.; Li, H.L.; Wang, X.D. Galectin-9 Acts as a Prognostic Factor with Antimetastatic Potential in Hepatocellular Carcinoma. *Asian Pac. J. Cancer Prev.* **2012**, *13*, 2503–2509. [CrossRef] [PubMed]

20. Nobumoto, A.; Nagahara, K.; Oomizu, S.; Katoh, S.; Nishi, N.; Takeshita, K.; Niki, T.; Tominaga, A.; Yamauchi, A.; Hirashima, M. Galectin-9 suppresses tumor metastasis by blocking adhesion to endothelium and extracellular matrices. *Glycobiology* **2008**, *18*, 735–744. [CrossRef] [PubMed]

21. Walzel, H.; Fahmi, A.A.; Eldesouky, M.A.; Abou-Eladab, E.F.; Waitz, G.; Brock, J.; Tiedge, M. Effects of *N*-glycan processing inhibitors on signaling events and induction of apoptosis in galectin-1-stimulated Jurkat T lymphocytes. *Glycobiology* **2006**, *16*, 1262–1271. [CrossRef] [PubMed]

22. Pace, K.E.; Lee, C.; Stewart, P.L.; Baum, L.G. Restricted receptor segregation into membrane microdomains occurs on human T cells during apoptosis induced by galectin-1. *J. Immunol.* **1999**, *163*, 3801–3811. [PubMed]

23. Chung, C.D.; Patel, V.P.; Moran, M.; Lewis, L.A.; Miceli, M.C. Galectin-1 induces partial TCR zeta-chain phosphorylation and antagonizes processive TCR signal transduction. *J. Immunol.* **2000**, *165*, 3722–3729. [CrossRef] [PubMed]

24. Hsieh, S.H.; Ying, N.W.; Wu, M.H.; Chiang, W.F.; Hsu, C.L.; Wong, T.Y.; Jin, Y.T.; Hong, T.M.; Chen, Y.L. Galectin-1, a novel ligand of neuropilin-1, activates VEGFR-2 signaling and modulates the migration of vascular endothelial cells. *Oncogene* **2008**, *27*, 3746–3753. [CrossRef] [PubMed]

25. Gauthier, L.; Rossi, B.; Roux, F.; Termine, E.; Schiff, C. Galectin-1 is a stromal cell ligand of the pre-B cell receptor (BCR) implicated in synapse formation between pre-B and stromal cells and in pre-BCR triggering. *Proc. Natl. Acad. Sci. USA* **2002**, *99*, 13014–13019. [CrossRef] [PubMed]

26. Rossi, B.; Espeli, M.; Schiff, C.; Gauthier, L. Clustering of pre-B cell integrins induces galectin-1-dependent pre-B cell receptor relocalization and activation. *J. Immunol.* **2006**, *177*, 796–803. [CrossRef] [PubMed]

27. Paz, A.; Haklai, R.; Elad-Sfadia, G.; Ballan, E.; Kloog, Y. Galectin-1 binds oncogenic H-Ras to mediate Ras membrane anchorage and cell transformation. *Oncogene* **2001**, *20*, 7486–7493. [CrossRef] [PubMed]

28. Sanchez-Ruderisch, H.; Detjen, K.M.; Welzel, M.; Andre, S.; Fischer, C.; Gabius, H.J.; Rosewicz, S. Galectin-1 sensitizes carcinoma cells to anoikis via the fibronectin receptor alpha5beta1-integrin. *Cell Death Differ.* **2011**, *18*, 806–816. [CrossRef] [PubMed]

29. Fukumori, T.; Takenaka, Y.; Yoshii, T.; Kim, H.R.; Hogan, V.; Inohara, H.; Kagawa, S.; Raz, A. CD29 and CD7 mediate galectin-3-induced type II T-cell apoptosis. *Cancer Res.* **2003**, *63*, 8302–8311. [PubMed]

30. Stillman, B.N.; Hsu, D.K.; Pang, M.; Brewer, C.F.; Johnson, P.; Liu, F.T.; Baum, L.G. Galectin-3 and galectin-1 bind distinct cell surface glycoprotein receptors to induce T cell death. *J. Immunol.* **2006**, *176*, 778–789. [CrossRef] [PubMed]

31. Demetriou, M.; Granovsky, M.; Quaggin, S.; Dennis, J.W. Negative regulation of T-cell activation and autoimmunity by Mgat5 *N*-glycosylation. *Nature* **2001**, *409*, 733–739. [CrossRef] [PubMed]

32. Chen, H.Y.; Fermin, A.; Vardhana, S.; Weng, I.C.; Lo, K.F.; Chang, E.Y.; Maverakis, E.; Yang, R.Y.; Hsu, D.K.; Dustin, M.L.; et al. Galectin-3 negatively regulates TCR-mediated CD4$^+$ T-cell activation at the immunological synapse. *Proc. Natl. Acad. Sci. USA* **2009**, *106*, 14496–14501. [CrossRef] [PubMed]

33. Elad-Sfadia, G.; Haklai, R.; Balan, E.; Kloog, Y. Galectin-3 augments K-Ras activation and triggers a Ras signal that attenuates ERK but not phosphoinositide 3-kinase activity. *J. Biol. Chem.* **2004**, *279*, 34922–34930. [CrossRef] [PubMed]

34. Paron, I.; Scaloni, A.; Pines, A.; Bachi, A.; Liu, F.T.; Puppin, C.; Pandolfi, M.; Ledda, L.; Di Loreto, C.; Damante, G.; et al. Nuclear localization of Galectin-3 in transformed thyroid cells: A role in transcriptional regulation. *Biochem. Biophys. Res. Commun.* **2003**, *302*, 545–553. [CrossRef]

35. Inohara, H.; Akahani, S.; Koths, K.; Raz, A. Interactions between galectin-3 and Mac-2-binding protein mediate cell-cell adhesion. *Cancer Res.* **1996**, *56*, 4530–4534. [PubMed]

36. Daley, D.; Mani, V.R.; Mohan, N.; Akkad, N.; Ochi, A.; Heindel, D.W.; Lee, K.B.; Zambirinis, C.P.; Pandian, G.; Savadkar, S.; et al. Dectin 1 activation on macrophages by galectin 9 promotes pancreatic carcinoma and peritumoral immune tolerance. *Nat. Med.* **2017**, *23*, 556–567. [CrossRef] [PubMed]

37. Anderson, A.C.; Anderson, D.E.; Bregoli, L.; Hastings, W.D.; Kassam, N.; Lei, C.; Chandwaskar, R.; Karman, J.; Su, E.W.; Hirashima, M.; et al. Promotion of tissue inflammation by the immune receptor Tim-3 expressed on innate immune cells. *Science* **2007**, *318*, 1141–1143. [CrossRef] [PubMed]

38. Dai, S.Y.; Nakagawa, R.; Itoh, A.; Murakami, H.; Kashio, Y.; Abe, H.; Katoh, S.; Kontani, K.; Kihara, M.; Zhang, S.L.; et al. Galectin-9 induces maturation of human monocyte-derived dendritic cells. *J. Immunol.* **2005**, *175*, 2974–2981. [CrossRef] [PubMed]

39. Koguchi, K.; Anderson, D.E.; Yang, L.; O'Connor, K.C.; Kuchroo, V.K.; Hafler, D.A. Dysregulated T cell expression of TIM3 in multiple sclerosis. *J. Exp. Med.* **2006**, *203*, 1413–1418. [CrossRef] [PubMed]

40. Madireddi, S.; Eun, S.Y.; Lee, S.W.; Nemcovicova, I.; Mehta, A.K.; Zajonc, D.M.; Nishi, N.; Niki, T.; Hirashima, M.; Croft, M. Galectin-9 controls the therapeutic activity of 4-1BB-targeting antibodies. *J. Exp. Med.* **2014**, *211*, 1433–1448. [CrossRef] [PubMed]

41. Vaitaitis, G.M.; Wagner, D.H. Galectin-9 Controls CD40 Signaling through a Tim-3 Independent Mechanism and Redirects the Cytokine Profile of Pathogenic T Cells in Autoimmunity. *PLoS ONE* **2012**, *7*, e38708. [CrossRef] [PubMed]

42. Rubinstein, N.; Alvarez, M.; Zwirner, N.W.; Toscano, M.A.; Ilarregui, J.M.; Bravo, A.; Mordoh, J.; Fainboim, L.; Podhajcer, O.L.; Rabinovich, G.A. Targeted inhibition of galectin-1 gene expression in tumor cells results in heightened T cell-mediated rejection; A potential mechanism of tumor-immune privilege. *Cancer Cell* **2004**, *5*, 241–251. [CrossRef]

43. Juszczynski, P.; Ouyang, J.; Monti, S.; Rodig, S.J.; Takeyama, K.; Abramson, J.; Chen, W.; Kutok, J.L.; Rabinovich, G.A.; Shipp, M.A. The AP1-dependent secretion of galectin-1 by Reed Sternberg cells fosters immune privilege in classical Hodgkin lymphoma. *Proc. Natl. Acad. Sci. USA* **2007**, *104*, 13134–13139. [CrossRef] [PubMed]

44. Cedeno-Laurent, F.; Watanabe, R.; Teague, J.E.; Kupper, T.S.; Clark, R.A.; Dimitroff, C.J. Galectin-1 inhibits the viability, proliferation, and Th1 cytokine production of nonmalignant T cells in patients with leukemic cutaneous T-cell lymphoma. *Blood* **2012**, *119*, 3534–3538. [CrossRef] [PubMed]

45. Le, Q.T.; Shi, G.; Cao, H.; Nelson, D.W.; Wang, Y.; Chen, E.Y.; Zhao, S.; Kong, C.; Richardson, D.; O'Byrne, K.J.; et al. Galectin-1: A link between tumor hypoxia and tumor immune privilege. *J. Clin. Oncol.* **2005**, *23*, 8932–8941. [CrossRef] [PubMed]

46. Chen, T.C.; Chen, C.H.; Wang, C.P.; Lin, P.H.; Yang, T.L.; Lou, P.J.; Ko, J.Y.; Wu, C.T.; Chang, Y.L. The immunologic advantage of recurrent nasopharyngeal carcinoma from the viewpoint of Galectin-9/Tim-3-related changes in the tumour microenvironment. *Sci. Rep.* **2017**, *7*, 10349. [CrossRef] [PubMed]

47. Li, H.; Wu, K.; Tao, K.; Chen, L.; Zheng, Q.; Lu, X.; Liu, J.; Shi, L.; Liu, C.; Wang, G.; et al. Tim-3/galectin-9 signaling pathway mediates T-cell dysfunction and predicts poor prognosis in patients with hepatitis B virus-associated hepatocellular carcinoma. *Hepatology* **2012**, *56*, 1342–1351. [CrossRef] [PubMed]

48. Nebbia, G.; Peppa, D.; Schurich, A.; Khanna, P.; Singh, H.D.; Cheng, Y.; Rosenberg, W.; Dusheiko, G.; Gilson, R.; ChinAleong, J.; et al. Upregulation of the Tim-3/galectin-9 pathway of T cell exhaustion in chronic hepatitis B virus infection. *PLoS ONE* **2012**, *7*, e47648. [CrossRef] [PubMed]

49. Kuroda, J.; Yamamoto, M.; Nagoshi, H.; Kobayashi, T.; Sasaki, N.; Shimura, Y.; Horiike, S.; Kimura, S.; Yamauchi, A.; Hirashima, M.; et al. Targeting activating transcription factor 3 by Galectin-9 induces apoptosis and overcomes various types of treatment resistance in chronic myelogenous leukemia. *Mol. Cancer Res.* **2010**, *8*, 994–1001. [CrossRef] [PubMed]

50. Kobayashi, T.; Kuroda, J.; Ashihara, E.; Oomizu, S.; Terui, Y.; Taniyama, A.; Adachi, S.; Takagi, T.; Yamamoto, M.; Sasaki, N.; et al. Galectin-9 exhibits anti-myeloma activity through JNK and p38 MAP kinase pathways. *Leukemia* **2010**, *24*, 843–850. [CrossRef] [PubMed]

51. Wiersma, V.R.; de Bruyn, M.; van Ginkel, R.J.; Sigar, E.; Hirashima, M.; Niki, T.; Nishi, N.; Samplonius, D.F.; Helfrich, W.; Bremer, E. The glycan-binding protein galectin-9 has direct apoptotic activity toward melanoma cells. *J. Investig. Dermatol.* **2012**, *132*, 2302–2305. [CrossRef] [PubMed]

52. Tadokoro, T.; Morishita, A.; Fujihara, S.; Iwama, H.; Niki, T.; Fujita, K.; Akashi, E.; Mimura, S.; Oura, K.; Sakamoto, T.; et al. Galectin-9: An anticancer molecule for gallbladder carcinoma. *Int. J. Oncol.* **2016**, *48*, 1165–1174. [CrossRef] [PubMed]

53. Fujita, K.; Iwama, H.; Sakamoto, T.; Okura, R.; Kobayashi, K.; Takano, J.; Katsura, A.; Tatsuta, M.; Maeda, E.; Mimura, S.; et al. Galectin-9 suppresses the growth of hepatocellular carcinoma via apoptosis in vitro and in vivo. *Int. J. Oncol.* **2015**, *46*, 2419–2430. [CrossRef] [PubMed]

54. Kobayashi, K.; Morishita, A.; Iwama, H.; Fujita, K.; Okura, R.; Fujihara, S.; Yamashita, T.; Fujimori, T.; Kato, K.; Kamada, H.; et al. Galectin-9 suppresses cholangiocarcinoma cell proliferation by inducing apoptosis but not cell cycle arrest. *Oncol. Rep.* **2015**, *34*, 1761–1770. [CrossRef] [PubMed]

55. Takano, J.; Morishita, A.; Fujihara, S.; Iwama, H.; Kokado, F.; Fujikawa, K.; Fujita, K.; Chiyo, T.; Tadokoro, T.; Sakamoto, T.; et al. Galectin-9 suppresses the proliferation of gastric cancer cells in vitro. *Oncol. Rep.* **2016**, *35*, 851–860. [CrossRef] [PubMed]

56. Fujita, K.; Iwama, H.; Oura, K.; Tadokoro, T.; Samukawa, E.; Sakamoto, T.; Nomura, T.; Tani, J.; Yoneyama, H.; Morishita, A.; et al. Cancer Therapy Due to Apoptosis: Galectin-9. *Int. J. Mol. Sci.* **2017**, *18*, 74. [CrossRef] [PubMed]

57. Szakacs, G.; Paterson, J.K.; Ludwig, J.A.; Booth-Genthe, C.; Gottesman, M.M. Targeting multidrug resistance in cancer. *Nat. Rev. Drug Discov.* **2006**, *5*, 219–234. [CrossRef] [PubMed]

58. Luo, W.; Song, L.; Chen, X.L.; Zeng, X.F.; Wu, J.Z.; Zhu, C.R.; Huang, T.; Tan, X.P.; Lin, X.M.; Yang, Q.; et al. Identification of galectin-1 as a novel mediator for chemoresistance in chronic myeloid leukemia cells. *Oncotarget* **2016**, *7*, 26709–26723. [CrossRef] [PubMed]

59. Harazono, Y.; Kho, D.H.; Balan, V.; Nakajima, K.; Hogan, V.; Raz, A. Extracellular galectin-3 programs multidrug resistance through Na^+/K^+-ATPase and P-glycoprotein signaling. *Oncotarget* **2015**, *6*, 19592–19604. [CrossRef] [PubMed]

60. Lykken, J.M.; Horikawa, M.; Minard-Colin, V.; Kamata, M.; Miyagaki, T.; Poe, J.C.; Tedder, T.F. Galectin-1 drives lymphoma CD20 immunotherapy resistance: Validation of a preclinical system to identify resistance mechanisms. *Blood* **2016**, *127*, 1886–1895. [CrossRef] [PubMed]

61. Koyama, S.; Akbay, E.A.; Li, Y.Y.; Herter-Sprie, G.S.; Buczkowski, K.A.; Richards, W.G.; Gandhi, L.; Redig, A.J.; Rodig, S.J.; Asahina, H.; et al. Adaptive resistance to therapeutic PD-1 blockade is associated with upregulation of alternative immune checkpoints. *Nat. Commun.* **2016**, *7*, 10501. [CrossRef] [PubMed]

62. Romero, D. Immunotherapy: PD-1 says goodbye, TIM-3 says hello. *Nat. Rev. Clin. Oncol.* **2016**, *13*, 202–203. [CrossRef] [PubMed]

International Journal of
Molecular Sciences

MDPI

Review

Galectins as Molecular Targets for Therapeutic Intervention

Ruud P. M. Dings [1], Michelle C. Miller [2], Robert J. Griffin [1] and Kevin H. Mayo [2,*]

[1] Department of Radiation Oncology, University of Arkansas for Medical Sciences,
    Little Rock, AR 72205, USA; rpmdings@uams.edu (R.P.M.D.); rjgriffin@usma.edu (R.J.G.)
[2] Department of Biochemistry, Molecular Biology & Biophysics, University of Minnesota,
    Minneapolis, MN 55455, USA; mill0935@umn.edu
*   Correspondence: mayox001@umn.edu; Tel.: +1-612-625-9968; Fax: +1-612-624-5121

Received: 27 February 2018; Accepted: 15 March 2018; Published: 19 March 2018

Abstract: Galectins are a family of small, highly conserved, molecular effectors that mediate various biological processes, including chemotaxis and angiogenesis, and that function by interacting with various cell surface glycoconjugates, usually targeting β-galactoside epitopes. Because of their significant involvement in various biological functions and pathologies, galectins have become a focus of therapeutic discovery for clinical intervention against cancer, among other pathological disorders. In this review, we focus on understanding galectin structure-function relationships, their mechanisms of action on the molecular level, and targeting them for therapeutic intervention against cancer.

Keywords: galectins; carbohydrates; apoptosis; cell adhesion; protein structure; cancer; galectin-targeted therapeutics

---

1. Galectins from a Structural Perspective

Galectin-1 (Gal-1) has been the most studied and well-characterized galectin [1], since it was the first galectin discovered by its display of hemagglutinating activity [2]. Currently at least 14 mammalian galectins have been reported, and many more are found in different organisms, e.g., vertebrates, inter-vertebrates, and protists [3–5]. Early on following the discovery of galectins, it was proposed that they be divided into three groups [6,7]: prototype (galectin-1, -2,-5, -7, -10, -11, -13, -14), chimera (galectin-3), and tandem repeat (galectin-4, -6, -8, -9, -12). Prototype galectins consist of a single core domain, usually referred to as the carbohydrate recognition domain (CRD). Galectin-3 (Gal-3) is the only chimera galectin known, and it has a CRD and a collagen-like N-terminal tail with different properties. Tandem repeat galectins have two homologous, yet distinct, CRDs that are connected to each other via linker polypeptide chains.

All three types of galectins (prototype, chimera, and tandem repeat) have a well-defined CRD with a highly conserved amino acid sequence and β-sandwich structure [8]. The CRD β-sandwich structure is composed of eleven β-strands (β1 to β11) running in antiparallel fashion, with six of them (β1, β10, β3, β4, β5, β6) defining the sugar binding face (S-face) of the CRD and the remaining five (β11, β2, β7, β8, β9) defining the opposing F-face, as illustrated with the Gal-3 CRD in Figure 1A,B. High resolution structures of the CRDs of many galectins (usually bound to lactose or *N*-acetyl-lactosamine) have been reported: e.g., human galectin-1 (Gal-1, [9]), galectin-2 (Gal-2, [10]), CRD of Gal-3 [11,12], C-terminal CRD of galectin-4 (Gal-4, [13]), galectin-7 (Gal-7, [14]), galectin-10 (Gal-10, [15–17]), bovine Gal-1 [18,19], mouse galectin-9 (Gal-9, [20]), toad ovary galectin [21,22], chicken galectin-16 (Gal-16, [23], galectins from conger eel: congerin I [24] and congerin II [25,26], fungal galectins from *Coprinopsis cinerea* [27] and *Agrocybe cylindracea* [28].

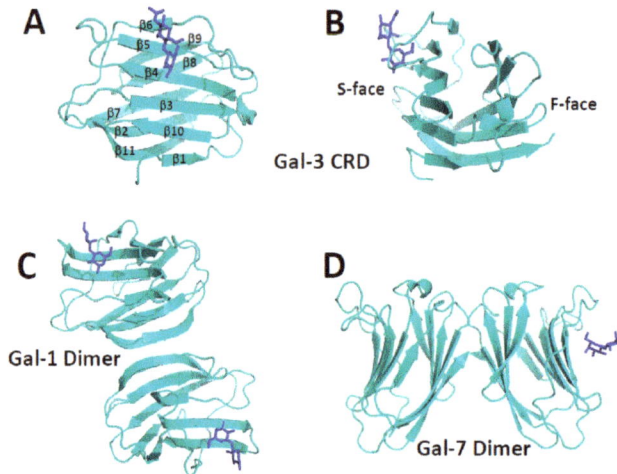

Figure 1. Galectin oligomer states. (**A,B**) Gal-3 carbohydrate recognition domain (CRD) (Protein Data Bank (PDB) access code 1A3K) as a monomer illustrating the β-sandwich fold common to all galectins. The 11 β-strands found within the CRD β-sandwich are labeled β1 to β11 in **A**, and the S- and F-faces of the CRD are identified in **B**. (**C,D**) Two prototype galectin dimers are shown. The CRD of the Gal-1 "terminal" dimer (PDB access code 1GZW) is shown in **C**, and the "symmetric sandwich" dimer of human Gal-7 (PDB access code 1BKZ) is shown in **D**. The carbohydrate binding sites in all structures are indicated by the lactose molecules shown in blue with a ball-and-stick structure.

The largest group of galectins (e.g., Gal-1, -2, -5, -7, -10, -13) belong to the prototype class and are generally known to self-associate, mostly as dimers [8]. Moreover, irrespective of their conserved monomer folds, galectins can form different types of dimers. The "terminal" dimer typified by Gal-1 is formed by hydrophobic interactions between N- and C-terminal residues of two subunits related by a 2-fold rotation axis perpendicular to the plane of the two β-sheets (Figure 1C). There are also "symmetric" and "non-symmetric" sandwich dimers. The former (e.g., Gal-7, Figure 1D) is stabilized by electrostatic interactions among charged residues on the F-faces of two monomers, and its inter-subunit contact surface is reduced compared to that in the non-symmetric dimer. The "non-symmetric" dimer (e.g., Gal-2) interface involving β-strands β1 and β6 from each subunit, is also formed primarily by electrostatic interactions at the inter-subunit interface of two monomers.

The Gal-1 dimer is the most thermodynamically stable of all galectins (dimer dissociation constant, K_d ~2–7 × 10^{-6} M) [29,30]. Dimers of other prototype galectins are generally less thermodynamically stable. For example, Gal-5 and Gal-7, even at intermediate concentrations, behave as monomers [31,32], even though Gal-5 can induce cell agglutination, suggesting the presence of self-association. In the crystal, Gal-7 appears to be a dimer [14], whereas in solution others have reported it to be either a monomer [4,14,33] or dimer [14,34,35]. In addition, Gal-10 can form Charcot-Leyden crystals in tissues and during secretion [16]. Whereas most galectins dimerize via non-covalent interactions, Gal-13 dimer subunits are covalently linked via disulfide bonds which when reduced abrogate cell agglutination function. In many of these instances, the solution environment can influence the degree of self-association.

Because CRD structures are highly conserved, formation and thermodynamic stability of a prototype galectin dimer result from the composition of amino acid residues at the inter-subunit interface [8,36]. When the free energy of interaction of one type of dimer is greater than that of another, the greater one will of course dominate in solution. Thus, the type of dimer formed is likely to be functionally important in terms of defining how different galectins bind to glyco-conjugates on the cell surface. Moreover, based on this same thermodynamic argument, different galectins have recently been reported to form heterodimers with potential biological consequences [37].

The only chimera Gal-3 (30 kDa) has a C-terminal CRD linked to a lengthy, collagen-like, dynamic and structurally aperiodic N-terminal tail (NT, 113 amino acid residues in human Gal-3) that is comprised of numerous proline and glycine residues (27 each in human Gal-3) usually found in "PGAY" tetrapeptide repeats [8]. Ippel et al. (2016) [38] found that the Gal-3 NT binds transiently to the F-face of the CRD with these tetrapeptide repeats being crucial to those interactions. Even though Gal-3 oligomerization has been proposed [39], its oligomeric state remain unclear. Gal-3 has been reported to be a monomer [40], a dimer [41,42], and a higher order oligomer state [43,44] that is possibly formed by chemically cross-linking [45] through the action of transglutaminase [46]. When bound to some synthetic carbohydrates, Gal-3 has been reported to precipitate from solution as a pentamer by interactions among its N-terminal non-lectin domain, presumably to enhance cross-linking of cell surface oligosaccharides [47]. Nevertheless, this model generally lacks experimental validation.

Tandem-repeat type Gal-4, -6, -8, -9 are comprised of two CRDs connected by a variable length linker peptide. Even though this class of galectins is usually reported to be monomeric, a few studies have reported that tandem-repeat Gal-9 self-associates as dimers (mouse Gal-9 [20]) or multimers (human Gal-9 [48]). Nevertheless, because tandem-type galectins have two CRDs, they effectively mimic the function of prototype galectin dimers in terms of cross-linking cell surface glycoconjugates. In any event, this suggests some level of biological control and/or evolutionary link, in that tandem-repeat type galectins cannot dissociate into single CRD monomers. The presence of two CRDs appears necessary to mediate full cross-linking function in terms of mediating cell adhesion and migration.

2. Carbohydrate Binding

At least extracellularly, galectins generally function by binding to the carbohydrate portion of glycoconjugates on the cell surface [8]. The galectin CRD carbohydrate binding site comprises highly conserved amino acids within the six-stranded β-sheet on the S-face (Figure 1A,B) [49]. Even though most studies have been performed with small β-galactosides such as lactose, glycoconjugates in situ on the cell surface to which galectins bind, are more complicated, which may also play a role in differentiating galectin function. Moreover, even though it appears that galectins generally recognize β-galactosides as the binding epitope, Gal-1 has been reported also to interact with some α-galactosides, albeit more weakly than to the β-galactoside lactose [50].

Lactose is the minimal carbohydrate ligand necessary for binding to galectins, and most structures of galectins are reported with lactose bound. Most CRD residues required for optimal interactions with carbohydrate ligands are conserved arginines and histidines, along with a single conserved tryptophan. Lactose is effectively "grabbed" by the peptide loop above the lactose molecule and the relatively large and flat tryptophan side chain at the bottom of the disaccharide. NMR structural studies indicate that this loop is relatively flexible when the disaccharide is absent and is more firmly positioned when the disaccharide is bound with other CRD residues becoming more mobile, thus contributing to conformational entropy and a more negative free energy of binding [36].

Lactose binding affinity to galectins usually falls in the micromolar to millimolar range (e.g., 64×10^{-6} M for Gal-1 [51] to 2×10^{-3} M for nematode galectin LEC-6 CRDs [52]. For Gal-3, the reported range is quite broad, e.g., 26×10^{-6} M [52], 1×10^{-3} M [53], and 0.6×10^{-3} M [45]. For Gal-2, this value was reported as 85×10^{-6} M [27], and for galectin-4, it is 0.9×10^{-3} M [54]. Overall, these K_d values indicate relatively weak binding of lactose, the minimal unit necessary for carbohydrate recognition by galectins. Binding affinity can be increased by modifying the disaccharide. *N*-acetyllactosamine binds e.g., about 5-fold better to Gal-3 (K_d of 0.2×10^{-3} M) and binds even more strongly to larger oligosaccharides [53,55], such as β(1,3)-linked polyNAc-lactosamino-glycan as found in the extracellular matrix and many cell surface glycoconjugates.

There are specific structural features in oligosaccharides that promote stronger binding to the CRD, and binding affinities can usually be explained by some carbohydrate recognition features [52]. The basic unit recognized by all galectins is Gal-β(1-4)-GlcNAc, although K_d values vary greatly with a particular galectin. Moreover, isomers of this disaccharide can modify binding affinities, e.g., Gal(1-3)GlcNAc. Structurally, configuration of the 3-OH group is essential for carbohydrate recognition, and substitution of 4-OH and 6-OH groups on the galactose ring usually attenuates binding. These three hydroxy groups in lactose (or *N*-acetyl-lactosamine) form hydrogen bonds with side chains of hydrophilic residues from galectins [25]. The galactose ring in particular forms several H-bonds between its oxygen atoms O4, O5 and O6, and H44, E71, and N61 of Gal-1. The O3 of the glucose ring forms H-bonds with residues R48 and E71. H52 and W68 make van der Waals interactions with both the glucose and galactose rings, respectively, in lactose. Because of this, galectins generally do not bind to terminal, non-reducing end mannoside or glucoside residues, and substitution at the 3(4)-OH of the penultimate saccharide can abolish binding.

The non-polar side of the galactose ring (i.e., H1, H3, and H5) interacts with the highly conserved tryptophan residue present with the carbohydrate binding sites in all galectins. In glucose, the C4-OH group is equatorial, which attenuates hydrophobic interactions with this tryptophan [56]. Due to these crucial interactions with the galactose ring, the remainder of the polysaccharide in longer carbohydrates is oriented away from the protein surface and out into solution [18,27]. Furthermore, changes at the 3′ position of the β-galactoside *N*-acetyllactosamine with sialic acid increases binding affinity compared to *N*-acetyl-lactosamine (LacNAc), and addition of the α(1-2)-fucoside increases it further. When immobilized, linear B2 trisaccharide and Galili pentasaccharide are some of the best ligands with K_d values ~1 × 10^{-6} M, along with more complex *N*-acetyllactosamine-based oligosaccharides (e.g., -3Galβ1-4GlcNAcβ1-)$_n$ sequences), complex-type biantennary *N*-glycans, and modified chitin-derived glycans that display similar K_d values ~2 to 4 × 10^{-6} M [57]. However, when free in solution, these glycans bind more weakly, suggesting that the binding epitope on surface-bound glycans is conformed for more favorable galectin interactions. In some galectins (e.g., Gal-1,-3 and -9), oligosaccharide branching can also enhance binding affinity [52], whereas in others (e.g., Gal-8) is can result in decreased affinity. Thus, branching is likely one other way in which galectins can modulate their activities.

Gal-1 and Gal-3 have also been reported to interact with relatively large polysaccharides. A 120 kDa rhamnogalactouronan was found to bind Gal-1 at the CRD S-face with the actual carbohydrate binding site being more extensive than for simple disaccharides, a finding that has implications for interactions between galectins and glycans on the cell surface [58]. Miller et al. [59] also reported that a ~60 kDa α-galactomannan binds Gal-1 at the F-face of the CRD, a site different from the S-face canonical carbohydrate binding site. Moreover, the binding epitope on this α-galactomannan most likely involves a disaccharide unit comprised of α-(1-6)-galactose-linked residues on the mannan backbone that is flanked by "naked" mannan regions [60]. More recently, Miller et al. [61] reported that this α-galactomannan interacts in a similar fashion with the Gal-3 CRD. Aside from this novel polysaccharide binding site on the CRD F-face of Gal-1 and -3, Miller et al. [62] found that another rhamnogalacturonan polysaccharide (RG-I-4, ~60 kDa) could bind relatively strongly to the N-terminal sequence of the Gal-3 NT, with strong binding occurring kinetically slowly that is most likely associated with proline cis trans isomerization.

3. Galectin Function

Galectin expression, which varies considerably from cell type to cell type, depends upon the activation state of a certain cell type. All cells appear to express at least one galectin, and each galectin tends to be expressed at high concentration in a few, but different cell types [44]. Galectins can be translocated to the nucleus or to other sub-cellular sites after being synthesized on cytosolic ribosomes. Galectins have several features in common with cytosolic proteins, such as being deficient in a secretion signal peptide or typical transmembrane segments, and they can have acetylated *N*-termini. This diversity in their occurrence is also reflected in the multimodal biological roles they exhibit in controlling cell-cell and cell-matrix interactions, adhesion, proliferation, apoptosis, pre-mRNA splicing, immunity, and inflammation [63], as illustrated in Figure 2. The underlying principle of all these functions is most often, but not always, carbohydrate recognition.

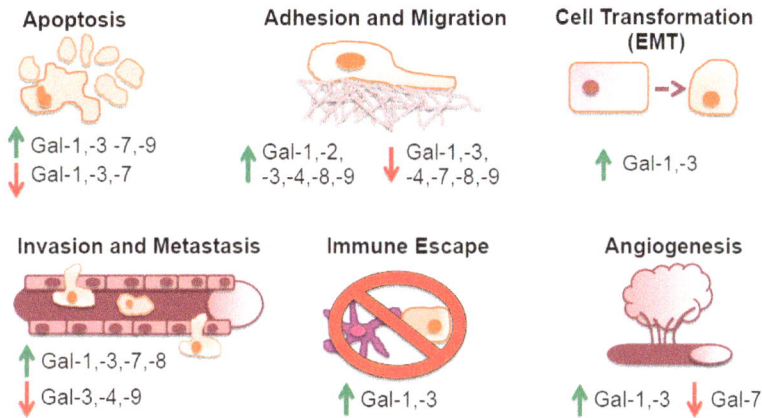

Figure 2. Galectins are involved in multiple processes of cancer initiation and development. A diversity of galectins is associated with key aspects of carcinogenesis, including apoptosis, adhesion and migration, cell transformation (EMT), invasion and metastasis, immune escape, and angiogenesis. Their tentative roles can be pro- and/or anti-tumorigenic, as indicated by green arrows up and red arrows down, respectively.

The extracellular mechanism of action of galectins generally starts with their binding to saccharides associated with cell surface glycol-conjugates [64]. However, the overall function of any given galectin can vary considerably. For example, Gal-1 can induce T-cell apoptosis, whereas Gal-3 can suppress apoptosis and increase T-cell proliferation [65,66]. Therefore, the activity of any galectin can be multi-faceted, and galectin self-association and interactions with cell surface glycans, as well as interactions with other biomolecules in situ, both extracellularly and intracellularly, can have significant impact on galectin function. For example, Gal-1-induced mitogenicity of human fibroblasts is attenuated as its concentration is increased (i.e., greater dimer population) [67], suggesting that monomers mediate mitogenic activity. On the other hand, the effect of Gal-1 on the growth of fibroblasts and human epithelial (HEP) 2 carcinoma cells is enhanced at high concentrations where the dimer population is greater [67].

Galectins bind numerous glycoconjugates on the surfaces of different cells, an event that impacts on their function. For example, Gal-1 interacts with various glycoconjugate ligands of the extracellular matrix (e.g., laminin, fibronectin, integrins, and ganglioside GM1), as well as those on endothelial cells (e.g., integrins, ROBO4, CD36, and CD13) [68] and on T lymphocytes (e.g., CD7, CD43, and CD45) where it promotes apoptosis [69]. Gal-2 can induce exposure of cell membrane surface phosphatidylserine in activated neutrophils, but not in activated T-cells [70], and has been associated with binding to lymphotoxin-α and myocardial infarction [71]. Gal-7 is associated with p53-induced apoptosis in keratinocytes [72], as well as in colon carcinoma [73]. Gal-8 is the most abundant galectin in tumor cells of different origin [74], and is closely related to prostate carcinoma tumor antigen-1 (PCTA-1) [75]. Gal-8 binds to gangolioside GM_3 (sialosyllactoseceramide) that associates with CD9 and CD82 to promote an anti-metastatic effect [76,77].

Gal-3 interacts with glycoconjugates in the extracellular matrix, such as laminin, fibronectin, vitronectin, elastin, neural cell adhesion molecule (N-CAM), lysosomal-associated membrane protein (LAMP) 1 and 2, and integrin $\alpha_3\beta_1$. Like Gal-1, Gal-3 can also bind to CD43 and CD45 on leukocytes, as well as to CD66, immunoglobulin E (IgE), IgE receptor, and Mac-2 binding protein. Besides its constitutive expression, Gal-3 can be induced by inflammatory mediators [78], such as chemokine CXCL8 [79]. Functionally, Gal-3 appears to be the most promiscuous galectin, exhibiting diverse biological activities from cell adhesion, apoptosis, immune regulation, to regulation of gene transcription.

The quaternary structure of prototype galectins can also lead to functional divergence [80]. For example, Nieminen et al. [81] reported that Gal-3 oligomerization mediates cell activation/repression and cell adhesion via three different modes of action: receptor clustering, lattice formation, and cell-cell interactions. In addition, for prototype galectins, formation of dimer type (terminal, symmetric, and non-symmetric) can be functionally important, since this may help differentiate how different galectins bind differently to complex glycans intra- or extracellularly on the surface of cells. These natural glycans are far more complex than simple disaccharides such as lactose that have been used to study galectin carbohydrate binding and function. Lactose binding to the Gal-1 and -7 dimers has also been shown to modify functional binding at one carbohydrate binding site can allosterically influence lactose binding to the other, either with positive or negative cooperativity, thus providing another angle for galectin functional divergence [36,82].

The functional importance of the Gal-1 oligomer state is also evidenced by a naturally-occurring form of Gal-1 (Gal-1β) that lacks the first 6 N-terminal residues [83] and remains essentially monomeric [76]. The Gal-1β monomer promotes axonal regeneration, but not Jurkat cell death, unlike dimer-forming Gal-1, which promotes both [76].

Galectins can also function intracellularly. e.g., Gal-3 can be found in the nucleus and cytoplasm as a multifunctional oncogenic protein that can associate with Ras [84], and other cytosolic moieties such as Bcl-2 to help regulate cell growth and apoptosis, an interaction that can be abrogated by carbohydrate binding [65]. Gal-1 can also interact with H-Ras to enhance its association with the intracellular membrane to modulate H-Ras-GTP loading [85], an activity of Ras that is dependent intracellular membrane anchorage via hydrophobic interactions, possibly with the farnesyl group covalently attached to Ras [84,86–88]. Gal-3 also interacts with thyroid-specific transcription factor TTF-1, suggesting a role for this lectin in controlling proliferation and tumor progression in thyroid cancer [89]. Gal-3 also interacts with synexin (annexin VII, a Ca^{2+} and phospholipid-binding protein) that mediates Gal-3 translocation/trafficking to the perinuclear mitochondrial membrane, where it regulates mitochondrial integrity and cytochrome *c* release critical for apoptosis regulation [90].

In regards specifically to cancer, multiple investigations have uncovered the various roles and mechanisms of action of galectins in tumor cell invasiveness and dissemination. The increased and decreased levels of some galectins in different cancers are illustrated in Figure 3. Based on clinical data the presence of e.g., Gal-1 has been correlated with increased rates of metastasis and poor patient survival outcome [91]. Gal-1 induces epithelial-mesenchymal transition (EMT) in multiple cancer types by various signaling pathways [91–97]. Tumor cells undergoing EMT differentiate into a mesenchymal state, often indicated by molecular markers such as vimentin, desmin and α-smooth muscle actin (α-SMA). Functionally, this mesenchymal state increases the cellular ability and probability for tumor cell migration and invasion via metastasis [91]. E.g., Bacigalupo et al. [92] noted that in hepatocellular carcinoma (HCC) cell line HepG2 Gal-1-associated EMT was mediated through β-catenin nuclear translocation, TCF4/LEF1 transcription activity and increased cyclin D1 and *c-Myc* gene expression, implying the involvement of the Wnt pathway. Zhang et al. elucidated that artificially inducing Gal-1 expression triggered EMT through the $α_v β_3$-integrin/FAK/PI3K/AKT signaling pathway [91].

Conversely in kidney cancer, elevated levels of Gal-1 induce nuclear factor (NF)-κB signaling thereby inducing chemokine CXCR4 expression [93]. This has also been demonstrated in glioblastoma in which EMT is triggered via the stromal cell-derived factor-1 (SDF-1)/CXCR4 axis [94]. In gastric cancer, however, overexpression of Gal-1 enhances the ability of gastric cancer cells to invade and metastasize via EMT through the non-canonical hedgehog pathway, increasing the transcription of glioma-associated oncogene-1 by a smoothened independent manner [95]. Aside from the effect of

Gal-1 on the gastric tumor cells, the stroma is also affected by Gal-1. In this regard, conditioned media from gastric cancer cells induces expression of Gal-1 and the EMT marker α-SMA in normal fibroblasts, thus causing normal fibroblast transformation into mesenchymal cancer-associated fibroblasts via a transforming growth factor-β (TGF-β) dependent mechanism and the progression of gastric tumors [96]. Thus, it appears that Gal-1 can induce EMT through multiple pathways on both the tumor parenchyma and stroma. In clinical samples of pancreatic ductal adenocarcinoma, immunohistochemical analysis has revealed a positive correlation of Gal-1 with the expression of EMT markers [97]. By means of knockdown and overexpression in pancreatic cancer cell line PANC-1 and co-cultures with activated pancreatic stellate cells, EMT was induced by the NF-κB pathway, stimulating malignant behavior of pancreatic ductal adenocarcinoma [97].

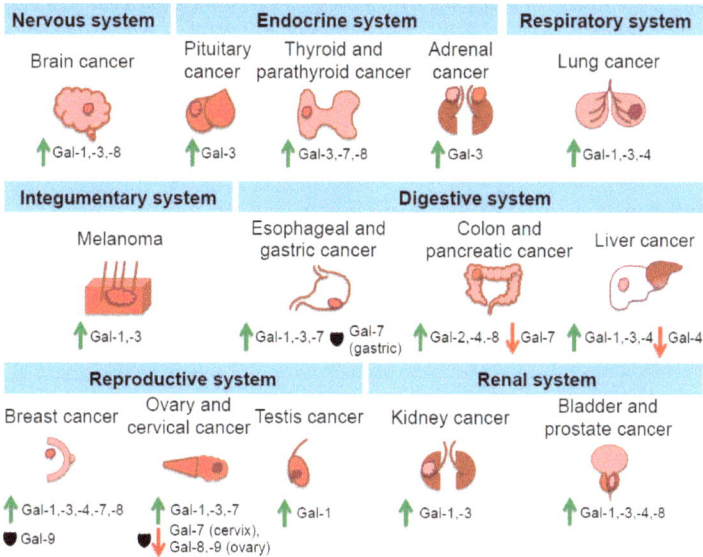

Figure 3. Involvement of galectins in organ-specific carcinogenesis within different physiological systems. Various elevated galectin expression is associated with tumor progression (green arrow up). In some instances, low galectin expression is correlated with the generation of neoplastic tissue (red arrow down). Additionally, galectins can have potential protective roles (black shield) during carcinogenesis.

Other galectins, such as Gal-3, have also been implicated in the EMT [98,99]. The Gal-3 EMT-associated phenotype was observed in patients with stage II colon cancer. Here, elevated tumoral expression of Gal-3 was positively correlated with vimentin expression and negatively correlated with E-cadherin expression when compared to the surrounding normal tissue [99]. Univariate analysis revealed that this EMT phenotype (i.e., elevated Gal-3 and vimentin expression) are predictors of tumor recurrence and survival [99]. Similarly, in patients with oral tongue squamous cell carcinoma (OTSCC), Gal-3 was found to be over-expressed in OTSCC compared to normal adjacent tissue, which was correlated strongly with the pathological stage, grade, and lymph node invasion in a Wnt/β-catenin dependent manner [100]. This has been similarly reported for Gal-1 in hepato carcinoma cells (HCC) [92]. At least in fibrosis, the EMT was counteracted by inhibition of Gal-3 by using modified citrus pectin (MCP), as shown by the reduction of mesenchymal molecules, fibronectin, smooth muscle actin and β-catenin, as well as hypertension and fibrosis [98]. Whether MCP can reverse Gal-3-induced EMT in cancer has yet to be demonstrated.

Gal-8 is the most abundant galectin in tumor cells of different origin [74], and is closely related to prostate carcinoma tumor antigen-1 (PCTA-1) [75]. Gal-3 is also reported to be a substrate for prostate-specific antigen (PSA) [101], and some of its glycoprotein ligands have been associated with prostasomes [102–104]. These findings suggest that Gal-8 and Gal-3 antagonists may be effective against reproductive and prostate cancer. On the other hand, low expression levels of Gal-3 have been associated with EMT and lymphovascular invasion and overall survival in lymph node positive breast cancer patients treated with doxorubicin [105]. In vitro assays using Gal-3 knockdown breast cancer stem cells have been shown to enhance tumorigenicity, which was confirmed in orthotopic mouse models [105]. Recently, Gal-8 has also been associated with EMT [106]. Non-tumurogenic Madin-Darby canine kidney (MDCK) cells acquired oncogenic potential after Gal-8 overexpression, displaying hallmarks of EMT alongside: downregulation of E-cadherin and upregulation of vimentin, fibronectin, β-catenin, and transcription factor Snail. Intriguingly, this EMT phenotype was considered partial and reversible because confluency was able to revert the EMT phenotype [106].

Overall, clinical data and experimental molecular studies have shown that galectins display various roles and mechanisms of actions within the EMT, ultimately causing increased cancer invasiveness and dissemination. Therefore, one or more of these galectins are potential molecular targets for therapeutic development.

4. Galectin Antagonists

For some time, many labs have sought to identify, discover, or design various galectin antagonists. Given its promiscuous nature, Gal-3 has been perhaps the most focused molecular target, and Gal-1 is a close second, because it is not usually involved in normal physiological processes such as wound healing [107] and yet is highly expressed in human tumors [108]. Nevertheless, antagonists against other galectins have also been identified.

Most reported galectin antagonists are based on the disaccharides lactose or *N*-acetyllactosamine, with the design aim of targeting a single galectin. Nevertheless, specificity and in vivo exposure remain as obstacles to developing highly effective galectin antagonists with therapeutic value in the anti-inflammatory [109] and anti-cancer arenas [110]. Some examples of galectin inhibitors are 3-(1,2,3-triazol-1-yl)-1-thio-galactosides (best $K_d \sim 107 \times 10^{-6}$ M) [111], *O*-galactosyl aldoximes (best $K_d \sim 180 \times 10^{-6}$ M) [112], and phenyl thio-β-D-galactopyranosides (best $K_d \sim 140 \times 10^{-6}$ M) [113]. Specificity with all of them was essentially absent. E.g., the phenyl thio-β-D-galacto-pyranosides interacted with all screened galectins (i.e., Gal-1, -3, -7, -8, and -9), with the best one binding most strongly to Gal-7. Thioureido *N*-acetyllactosamine derivatives were screened as inhibitors of Gal-7 and -9, with the best ones exhibiting K_d values of 23×10^{-6} M and 47×10^{-6} M, respectively [114].

Increased specificity for Gal-3 was observed upon addition of an aromatic (arene) group in lactose-based compounds having an aromatic 4-methoxy-2,3,5,6-tetrafluoro-benzamido moiety [12]. In the co-crystal structure with the Gal-3 CRD, the arene group improved affinity by stacking against Arg144 within the carbohydrate binding site of the CRD [12]. Further structure-based design of this compound class produced an analog with even greater affinity for Gal-3 ($K_d \sim 0.32 \times 10^{-6}$ M), and double arene thiodigalactoside bis-benzamido analogs improved affinity and specificity for Gal-3 further, because the two arene groups were observed in crystal structures to interact with two arginine residues (Arg144 and Arg186) [115]. In this case, the best analog had a K_d of 33×10^{-9} M [115]. The theme of adding a hydrophobic group to a lactose-based compound to improve Gal-3 inhibitor affinity and specificity was also used in the attachment of linear alkyl chains of varying length to the anomeric carbon of the glucose or *N*-acetyl glucose ring [116]. Although this class of compounds did increase affinity (best K_d values of 11×10^{-6} M to 73×10^{-6} M) over lactose [116], they were not as impressive as some compounds from the arene class [12,115].

Binding multivalency has also been exploited in designing galectin antagonists. By functionalizing unnatural amino acids (phenyl-bis-alanine and phenyl-tris-alanine) with 2-azidoethyl β-D-galactopyranosyl-(1-4)-β-D-glucopyranoside, a relatively effective compound targeting Gal-1 in particular has been found with a K_d of 3.2×10^{-6} M, about one order of magnitude higher affinity than for other galectins tested [117]. The multivalent design approach has been used in a number of other instances, namely with a trivalent "lactose" analog against Gal-4 (K_d of 22×10^{-6} M) [117], a bilactosylated steroid-based compound against Gal-1 [118], and lactulose amine compounds (i.e., polymethylene-spaced dilactoseamine derivatives) that show some selective effects linked to tumor cell apoptosis, cell aggregation, and endothelial cell morphogenesis [119].

The list of galectin antagonists is increased with the design of glycomimetics using high resolution structures of Gal-1 and -3 and computational approaches (e.g., Quantitative Structure-Activity Relationship (QSAR) models) to correlate molecular properties and binding affinities and to conclude that selective and potent inhibitors could be engineered by modifying the carbohydrate C-$3'$ and O-3 positions [120,121]. This approach led to the synthesis of aryl O- and S-galactosides and lactosides, as well as triazoles and isoxazoles, with the best compounds having K_d values in the 20 to 40×10^{-6} M range [117,121,122].

Many of these carbohydrate-based antagonists have provided evidence that subtle differences in saccharide structure can be used to fine tune binding affinity and specificity and may be potentially useful to block tumor growth. Nevertheless, galectin binding specificity and relatively low binding affinity remain problematic, in particular when developing a therapeutic with acceptable in vivo exposure. In this regard, clinical use has most often been limited to use in pathological disorders where topical administration can be done. For example, the Leffler lab has reported a small molecule and galactose-coumarin-based, Gal-3 inhibitor that reduces corneal neovascularization and pulmonary fibrosis in animal models [123,124]. Some other saccharide-based agents have potential to be developed as therapeutics for clinical use. Recently some reported thiodigalactoside, fluorine-amide and phenyl-arginine derivatives were reported with low nanomolar binding affinity and relatively high selectivity for Gal-3 [125,126].

A complex polysaccharide (a rhamnogalactouronan, termed GR-MD-02) has emerged as a relatively potent Gal-3 antagonist with considerable promise in preclinical animal models and clinical trials against non-alcoholic steatohepatitis (NASH) and toxic cirrhosis, demonstrating action at multiple pathophysiological processes [127–130]. GR-MD-02 also has recently been found to display good efficacy against cancer in pre-clinical cancer models when used in combination with immunotherapy agents. Moreover, an investigator-initiated phase 1b clinical trial with GR-MD-02 in combination with Merck's therapeutic drug KEYTRUDA (pembrolizumab) was shown to be effective against advanced melanoma with 5 of 8 responders (2 complete responders (CR) and 3 partial responders (PR)) in advanced melanoma (Galectin Therapeutics Inc., Atlanta, GA, USA). In addition, GR-MD-02 in combination with the anti-OX40 immunotherapy agonist was shown to improve survival and reduce lung metastases in the 4T1 breast cancer model, as well as improving survival in the mouse sarcoma cell (MCA-205) model in a CD8 T cell-dependent fashion. Another polysaccharide, an α-galactomannan, that targets and binds to the F-face of Gal-1 and -3 could be another potentially good cancer therapeutic [58,60].

Several galectin-targeting peptides have also been reported. For example, two peptides (G3-A9 and G3-C12, with amino acid sequences PQNSKIPGPTFLDPH and ANTPCGPYTHDCPVKR, respectively) were shown to bind relatively specifically to Gal-3 (K_d of 80×10^{-9} M) and recognize cell surface Gal-3 on carcinoma cells and monocytes, block the interaction between Gal-3 and TFAg (Thomsen-Friedenreich glycoantigen), and inhibit adhesion of human breast carcinoma cells to endothelial cells [131]. Another designed peptide anginex (33 amino acid residues) that targets Gal-1 (K_d ~90×10^{-9} M) [108] displays multimodal activities in terms of inhibiting tumor endothelial cell (EC) adhesion, migration and proliferation [132,133] and promoting leukocyte infiltration into tumors [134,135] leading to tumor growth inhibition in mouse models [136–138]. In this regard, anginex interferes with Gal-1 function by preventing tumor angiogenesis [108], abrogating tumor escape from immunity via blockade of Gal-1-induced apoptosis in activated T lymphocytes [139], and preventing metastasis via inhibition of Gal-1-facilitated tumor cell-EC interactions [140]. Anginex is unique in that rather than targeting the carbohydrate binding site on the CRD (S-face), it binds to a mostly hydrophobic patch on the CRD F-face [141]. A structure-based approach [142–144] was then used to design both a partial peptide mimetic [145] and a fully non-peptide, calixarene-based protein surface topomimetic of anginex called PTX008 [141,146]. Both agents are antagonists of Gal-1 function, with improved activity over anginex [144]. PTX008 has been used in the clinic in a Phase I trial against cancer and has shown some efficacy at inhibiting tumor growth.

5. Galectin Antagonists in Combination Therapy

Whereas galectin antagonists may take only a limited share of therapeutic space as stand-alone agents, their real forte is likely to be in combination therapy.

5.1. Improving Chemo- and Immunotherapy

Treatment modalities targeting tumor stroma have been shown to transiently normalize tumor vasculature, which can alleviate hypoxia, increase drug and anti-tumor immune cell delivery, and consequently improve clinical outcome [141,147–151]. Initially vessel normalization was thought to be achieved only by interfering with growth factor signaling, e.g., blocking vascular endothelial cell growth factor receptor 2 (VEGFR2) or basic growth factor signaling, causing destruction of "immature" vessels and thus improving the overall physiologic state of the tumor over a relatively short period of time [152–154]. Vessel normalization is defined by an increase in pericyte coverage of tumor vasculature and functionally by modifying tumor perfusion and interstitial pressure, resulting in increased tumor oxygenation. Anti-angiogenesis therapy can overcome endothelial cell anergy and promote leukocyte-endothelium interactions and infiltration in tumors [134].

In addition to growth factor interference, the inhibition of Gal-1 can be a potent approach to vessel normalization. The Gal-1 inhibitor anginex has been shown to normalize tumor vasculature and consequently elevates tumor oxygenation in multiple tumor models [137,155,156]. More recently, this finding was corroborated in mice bearing Kaposi's sarcomas by using a rabbit anti-Gal-1 IgG antibody (F8.G7), which was also able to transiently normalize tumor vasculature as evidenced by vasculature remodeling: increased pericyte coverage of vessels causing an improved tumor physiology indicated by reduced tumor hypoxia and improved T-cell infiltrate [157]. Anginex has also been conjugated to the cytotoxic acylfulvene, 6-hydroxylpropylacyl-fulvene to inhibit tumor growth more effectively [158]. In addition, the non-peptidic calixarene-based mimetic of anginex (PTX008), which specifically binds Gal-1 in an allosteric fashion, has been demonstrated to normalize tumor vasculature, thus promoting improved tumor oxygenation [141,146,159,160]. In addition, anginex and PTX008 in combination with the chemotherapeutic irofulven have been shown to lead to ovarian tumor growth regression [151]. The use of these anti-angiostatic agents has also been shown to enhance T-cell mediated anti-tumor response when used as an adjuvant to immunotherapy [161].

Thus, peptide-based, antibody-based, and small molecule-based Gal-1 inhibitors can transiently normalize tumor vasculature to increase the sensitivity of tumors to chemo- and immunotherapy and have the potential to significantly enhance clinical success.

5.2. Drug Resistance

One problem in the clinic with galectin-targeted therapeutics (as well as with any therapeutic agent) against cancer is the issue of drug resistance. This occurs for various reasons, e.g., tumor generated multiple isoforms via alternative splicing that may result in inhibitor-resistant galectins [162]. Overexpression of Gal-1 has been positively correlated with poor survival outcome, as well as drug and radiation resistance [91,163–165]. Additionally, multidrug resistant (MDR) breast cancer cell lines showed improved sensitivity to paclitaxel and adriamycin by knocking down Gal-1 expression [163]. This particular improvement of sensitivity appears to be regulated through P-glycoprotein expression via inhibiting the Raf-1/AP-1 pathway [163].

In triple-negative breast cancer (TNBC) cell lines, it was shown that Gal-1 was associated with doxorubicin sensitivity [164]. Doxorubicin is the first-line therapeutic in anti-breast cancer treatment, but toxicity and resistance remain an important concern clinically [166–168]. By silencing Gal-1, TNBC had decreased cell proliferation, migration, invasion, and doxorubicin resistance. This resistance was mediated through integrin β1/FAK/c-Src/ERK/STAT3/surviving pathway [164]. This implies that targeting Gal-1 in TNCB has great therapeutic potential and is likely to sensitize TNCB to doxorubin treatment at the same time. This is of importance as TNBC is particularly difficult to treat since it does not, or at low levels, express estrogen, progesterone, or Her2/Neu receptors—signaling pathways for many current USA Federal Drug Administration (FDA)-approved breast cancer therapeutics target. Overall this suggests that Gal-1 is a viable target to employ in MDR and TNBC, to be used for as a targeting strategy to deliver current conventional chemotherapeutics or via developing novel Gal-1 therapeutics.

In HCC, cells with genetically altered high expression of Gal-1 were more resistant to sorafenib as compared with isogenic low expressing Gal-1 cells, shown by knock-in as well as knock-out systems [91]. Sorafenib is a small molecule inhibitor of multiple tyrosine kinases, i.e., VEGFR, platelet derived growth factor receptor (PDGFR), C-and B-Raf FDA approved for the treatment of renal cell carcinoma and HCC. Survival curves for patients with low and high Gal-1 expression also correlated with sorafenib sensitivity and resistance, respectively.

Thus, these studies imply that targeting Gal-1 can be a powerful tool in combinatorial treatment strategies, overcoming single chemotherapy resistance, as well as in MDR cancers.

5.3. Radiation Resistance

Aside from Gal-1 overexpression in most cancers, this galectin is also upregulated by radiation [165,169,170] and hypoxia [171,172]. General hypoxia and local areas of hypoxia (<10 mmHg pO_2) are prevalent in solid tumors and are negatively associated with cancer therapy success, compromising radiotherapy, and driving malignant progression [173,174]. Radiation requires oxygen to induce DNA damage through the generation of hydroxyl radicals and H_2O_2 [175] and hypoxic conditions can require up to three times the amount of radiation to generate the same effects (oxygen enhancement ratio) [176]. Low oxygen tension leads to stabilization of the protein transcriptional regulating, hypoxia inducible factor (HIF-1α), inducing downstream signaling and protein generation [96,171].

Since Chen et al. reported that H-Ras and HIF-1 interact [177], Kuo et al. [165] suggested a possible HIF1/Gal-1-positive feedback loop where HIF1 signaling under hypoxia can enhance Gal-1, which acts through H-ras to further promote HIF1 transcriptional activity [177,178]. Thus, they hypothesized that tumors may utilize this positive feedback loop to maintain elevated Gal-1 expression and HIF1 signaling to drive radio-resistance and aggressive tumor phenotypes [165].

Another way in which Gal-1 promotes radio-resistance is through its effect on the immune system, particularly on T lymphocytes. Gal-1 induces apoptosis in activated T lymphocytes through a CD45-associated *N*-glycan dependent manner [69]. Conversely, radiation enhances the antitumor immune response by increasing the generation of tumor-specific peptide or antigen repertoire and recruitment of cytotoxic T lymphocytes into the tumor [179,180]. However, radiotherapy does not always result in complete protective immunity, as relapse or recurrence is still a major clinical concern. Gal-1 potently induces apoptosis in activated T-cell resulting in an overall Th2 cytokine profile (e.g., Interleukin (IL)-4, -5, and -13) over a tumoricidal Th1 cytokine profile (e.g., INFγ, IL-2 and TGF-β) blocking immune effector functions while promoting IL-10-producing regulatory T cells to create an immune privileged site at the tumor [139,181]. Thus, the radiation-induced increase of Gal-1 levels promotes tumor immune evasion [165], limiting therapeutic response arguing combinatory treatment strategies of gal-1 inhibitors and radiation.

Indeed, Gal-1-induced radiation resistance has been overcome by using the Gal-1-targeting, designed peptide inhibitor anginex in multiple human and syngeneic murine tumor models [108,136–138,169]. Anginex synergizes with radiation, e.g., in the aggressive and radiation-resistant murine mammary carcinoma cell (SCK) model [138,156], the head and neck squamous cell carcinoma (SCCVII) model [147], the B16F10 melanoma model [144,156], and in the human ovarian cancer xenograft MA148 [138,144,156]. Moreover, the synergy of Gal-1 inhibition was observed with multiple radiation modalities. Namely, fractionated relatively lower dose radiation, i.e., multiple doses of e.g., 2 Gy, as well as when combined in a hypo-fractionated approach, e.g., a single or few doses of >10 Gy [138]. Moreover, using a radiation microbeam approach, it was noted that anginex sensitized tumors preferentially when combined with wider beam spacing radiation. Beam geometries and doses capable of slowing tumor growth were also more effective when combined with anginex [148]. In addition, multiple myeloma growth and its relapse can be repressed by using the anti-angiogenic agent anginex in combination with radiotherapy [182].

Thus, the radiation resistance induced by Gal-1 can be overcome by combining radiation therapy with Gal-1 inhibitors and shows great promise also in the contemporary radiation approaches such as hypo-fractionated and microbeam radiation. In addition, the combination of chemotherapy with temozolomide and radiation therapy has been reported to induce the expression of both Gal-1 and -3 [183] (Bailey et al., 2015). Thus, targeting both Gal-3 and -1 could have therapeutic benefit.

6. Conclusions

Galectins are involved in many biological processes, generally functioning by interacting with various cell surface glycoconjugates, usually targeting β-galactoside epitopes. These small protein effector molecules mediate processes such as chemotaxis and angiogenesis, and thus have impact in various pathological disorders from cardiovascular disease to cancer. Whereas β-galactoside-directed glycan binding of galectins to various cell surface glycoconjugates is crucial to their extracellular biological functions, galectins are now also being recognized to mediate various intracellular functions via interactions with non-glycosylated nuclear and cytosolic biomolecules. Moreover, because their importance in biology has been growing rapidly in recent years, numerous efforts have been underway to identify effective antagonists of their function for use in the clinical setting. However only recently have galectins been fully accepted as valid therapeutic targets for clinical intervention. Even though this chapter has discussed a number of these drug discovery efforts, it is by no means exhaustive. It is likely that sometime soon, we will have one or more galectin antagonists available in the clinic to combat inflammatory diseases and cancer.

Note: Given the breadth of drug discovery and the galectin field of research with numerous labs involved in it, we apologize for the inadvertent omission of many excellent works.

Conflicts of Interest: The authors declare no conflict of interest.

References

1. Camby, I.; Le Mercier, M.; Lefranc, F.; Kiss, R. Galectin-1: A small protein with major functions. *Glycobiology* **2006**, *16*, 137–157. [CrossRef] [PubMed]
2. Teichberg, V.I.; Silman, I.; Beitsch, D.D.; Resheff, G. A beta-d-galactoside binding protein from electric organ tissue of Electrophorus electricus. *Proc. Natl. Acad. Sci. USA* **1975**, *72*, 1383–1387. [CrossRef] [PubMed]
3. Barondes, S.H.; Castronovo, V.; Cooper, D.N.; Cummings, R.D.; Drickamer, K.; Feizi, T.; Gitt, M.A.; Hirabayashi, J.; Hughes, C.; Kasai, K.; et al. Galectins: A family of animal beta-galactoside-binding lectins. *Cell* **1994**, *76*, 597–598. [CrossRef]
4. Cooper, D.N.; Barondes, S.H. God must love galectins; he made so many of them. *Glycobiology* **1999**, *9*, 979–984. [CrossRef] [PubMed]
5. Cooper, D.N. Galectinomics: Finding themes in complexity. *Biochim. Biophys. Acta* **2002**, *1572*, 209–231. [CrossRef]
6. Hirabayashi, J.; Kasai, K. The family of metazoan metal-independent beta-galactoside-binding lectins: Structure, function and molecular evolution. *Glycobiology* **1993**, *3*, 297–304. [CrossRef] [PubMed]
7. Kasai, K.; Hirabayashi, J. Galectins: A family of animal lectins that decipher glycocodes. *J. Biochem.* **1996**, *119*, 1–8. [CrossRef] [PubMed]

8.   Nesmelova, I.V.; Dings, R.P.M.; Mayo, K.H. Understanding galectin structure-function relationships to design effective antagonists. In *Galectins*; Klyosov, A., Ed.; Oxford University Press: New York, NY, USA, 2008; Chapter 2.

9.   Lopez-Lucendo, M.F.; Solis, D.; Andre, S.; Hirabayashi, J.; Kasai, K.; Kaltner, H.; Gabius, H.-J.; Romero, A. Growth-regulatory human galectin-1: Crystallographic characterisation of the structural changes induced by single-site mutations and their impact on the thermodynamics of ligand binding. *J. Mol. Biol.* **2004**, *343*, 957–970. [CrossRef] [PubMed]

10.   Lobsanov, Y.D.; Gitt, M.A.; Leffler, H.; Barondes, S.H.; Rini, J.M. X-ray crystal structure of the human dimeric S-Lac lectin, L-14-II, in complex with lactose at 2.9-A resolution. *J. Biol. Chem.* **1993**, *268*, 27034–27038. [PubMed]

11.   Seetharaman, J.; Kanigsberg, A.; Slaaby, R.; Leffler, H.; Barondes, S.H.; Rini, J.M. X-ray crystal structure of the human galectin-3 carbohydrate recognition domain at 2.1-A resolution. *J. Biol. Chem.* **1998**, *273*, 13047–13052. [CrossRef] [PubMed]

12.   Sorme, P.; Arnoux, P.; Kahl-Knutsson, B.; Leffler, H.; Rini, J.M.; Nilsson, U.J. Structural and thermodynamic studies on cation-Pi interactions in lectin-ligand complexes: High-affinity galectin-3 inhibitors through fine-tuning of an arginine-arene interaction. *J. Am. Chem. Soc.* **2005**, *127*, 1737–1743. [CrossRef] [PubMed]

13.   Tomizawa, T.; Kigawa, T.; Saito, K.; Koshiba, S.; Inoue, M.; Yokoyama, S. Solution structure of the C-terminal gal-bind lectin domain from human galectin-4. To be published. 2018.

14.   Leonidas, D.D.; Vatzaki, E.H.; Vorum, H.; Celis, J.E.; Madsen, P.; Acharya, K.R. Structural basis for the recognition of carbohydrates by human galectin-7. *Biochemistry* **1998**, *37*, 13930–13940. [CrossRef] [PubMed]

15.   Swaminathan, G.J.; Leonidas, D.D.; Savage, M.P.; Ackerman, S.J.; Acharya, K.R. Selective recognition of mannose by the human eosinophil Charcot-Leyden crystal protein (galectin-10): A crystallographic study at 1.8 A resolution. *Biochemistry* **1999**, *38*, 13837–13843. [CrossRef] [PubMed]

16.   Leonidas, D.D.; Elbert, B.L.; Zhou, Z.; Leffler, H.; Ackerman, S.J.; Acharya, K.R. Crystal structure of human Charcot-Leyden crystal protein, an eosinophil lysophospholipase, identifies it as a new member of the carbohydrate-binding family of galectins. *Structure* **1995**, *3*, 1379–1393. [CrossRef]

17.   Ackerman, S.J.; Liu, L.; Kwatia, M.A.; Savage, M.P.; Leonidas, D.D.; Swaminathan, G.J.; Acharya, K.R. Charcot-Leyden crystal protein (galectin-10) is not a dual function galectin with lysophospholipase activity but binds a lysophospholipase inhibitor in a novel structural fashion. *J. Biol. Chem.* **2002**, *277*, 14859–14868. [CrossRef] [PubMed]

18.   Bourne, Y.; Bolgiano, B.; Liao, D.I.; Strecker, G.; Cantau, P.; Herzberg, O.; Feizi, T.; Cambillau, C. Crosslinking of mammalian lectin (galectin-1) by complex biantennary saccharides. *Nat. Struct. Biol.* **1994**, *1*, 863–870. [CrossRef] [PubMed]

19.   Liao, D.I.; Kapadia, G.; Ahmed, H.; Vasta, G.R.; Herzberg, O. Structure of S-lectin, a developmentally regulated vertebrate beta-galactoside-binding protein. *Proc. Natl. Acad. Sci. USA* **1994**, *91*, 1428–1432. [CrossRef] [PubMed]

20.   Nagae, M.; Nishi, N.; Murata, T.; Usui, T.; Nakamura, T.; Wakatsuki, S.; Kato, R. Crystal structure of the galectin-9 N-terminal carbohydrate recognition domain from Mus musculus reveals the basic mechanism of carbohydrate recognition. *J. Biol. Chem.* **2006**, *281*, 35884–35893. [CrossRef] [PubMed]

21.   Bianchet, M.A.; Ahmed, H.; Vasta, G.R.; Amzel, L.M. Soluble beta-galactosyl-binding lectin (galectin) from toad ovary: Crystallographic studies of two protein-sugar complexes. *Proteins* **2000**, *40*, 378–388. [CrossRef]

22.   Ahmed, H.; Pohl, J.; Fink, N.E.; Strobel, F.; Vasta, G.R. The primary structure and carbohydrate specificity of a beta-galactosyl-binding lectin from toad (*Bufo arenarum* Hensel) ovary reveal closer similarities to the mammalian galectin-1 than to the galectin from the clawed frog *Xenopus laevis*. *J. Biol. Chem.* **1996**, *271*, 33083–33094. [CrossRef] [PubMed]

23.   Varela, P.F.; Solis, D.; Diaz-Maurino, T.; Kaltner, H.; Gabius, H.-J.; Romero, A. The 2.15 A crystal structure of CG-16, the developmentally regulated homodimeric chicken galectin. *J. Mol. Biol.* **1999**, *294*, 537–549. [CrossRef] [PubMed]

24.   Shirai, T.; Mitsuyama, C.; Niwa, Y.; Matsui, Y.; Hotta, H.; Yamane, T.; Kamiya, H.; Ishii, C.; Ogawa, T.; Muramoto, K. High-resolution structure of the conger eel galectin, congerin I, in lactose-liganded and ligand-free forms: Emergence of a new structure class by accelerated evolution. *Structure* **1999**, *7*, 1223–1233. [CrossRef]

25. Shirai, T.; Matsui, Y.; Shionyu-Mitsuyama, C.; Yamane, T.; Kamiya, H.; Ishii, C.; Ogawa, T.; Muramoto, K. Crystal structure of a conger eel galectin (congerin II) at 1.45 A resolution: Implication for the accelerated evolution of a new ligand-binding site following gene duplication. *J. Mol. Biol.* **2002**, *321*, 879–889. [CrossRef]

26. Shionyu-Mitsuyama, C.; Ito, Y.; Konno, A.; Miwa, Y.; Ogawa, T.; Muramoto, K.; Shirai, T. In vitro evolutionary thermostabilization of congerin II: A limited reproduction of natural protein evolution by artificial selection pressure. *J. Mol. Biol.* **2005**, *347*, 385–397. [CrossRef] [PubMed]

27. Walser, P.J.; Haebel, P.W.; Kunzler, M.; Sargent, D.; Kues, U.; Aebi, M.; Ban, N. Structure and functional analysis of the fungal galectin CGL2. *Structure* **2004**, *12*, 689–702. [CrossRef] [PubMed]

28. Ban, M.; Yoon, H.J.; Demirkan, E.; Utsumi, S.; Mikami, B.; Yagi, F. Structural basis of a fungal galectin from *Agrocybe cylindracea* for recognizing sialoconjugate. *J. Mol. Biol.* **2005**, *351*, 695–706. [CrossRef] [PubMed]

29. Cho, M.; Cummings, R.D. Galectin-1, a beta-galactoside-binding lectin in Chinese hamster ovary cells. I. Physical and chemical characterization. *J. Biol. Chem.* **1995**, *270*, 5198–5206. [CrossRef] [PubMed]

30. Giudicelli, V.; Lutomski, D.; Levi-Strauss, M.; Bladier, D.; Joubert-Caron, R.; Caron, M. Is human galectin-1 activity modulated by monomer/dimer equilibrium? *Glycobiology* **1997**, *7*, 8–10. [CrossRef]

31. Gitt, M.A.; Wiser, M.F.; Leffler, H.; Herrmann, J.; Xia, Y.R.; Massa, S.M.; Cooper, D.N.; Lusis, A.J.; Barondes, S.H. Sequence and mapping of galectin-5, a beta-galactoside-binding lectin, found in rat erythrocytes. *J. Biol. Chem.* **1995**, *270*, 5032–5038. [CrossRef] [PubMed]

32. Madsen, P.; Rasmussen, H.H.; Flint, T.; Gromov, P.; Kruse, T.A.; Honore, B.; Vorum, H.; Celis, J.E. Cloning, expression, and chromosome mapping of human galectin-7. *J. Biol. Chem.* **1995**, *270*, 5823–5829. [CrossRef] [PubMed]

33. Leffler, H. Introduction to galectins. *Trends Glycosci. Glycotechnol.* **1997**, *9*, 9–19. [CrossRef]

34. Morris, S.; Ahmad, N.; Andre, S.; Kaltner, H.; Gabius, H.-J.; Brenowitz, M.; Brewer, F. Quaternary solution structures of galectins-1, -3, and -7. *Glycobiology* **2004**, *14*, 293–300. [CrossRef] [PubMed]

35. Ahmad, N.; Gabius, H.J.; Kaltner, H.; André, S.; Kuwabara, I.; Liu, F.-T.; Oscarson, S.; Norberg, T.; Brewer, C.F. Thermodynamic binding studies of cell surface carbohydrate epitopes to galectins-1, -3, and -7: Evidence for differential binding specificities. *Can. J. Chem.* **2002**, *80*, 1096–1104. [CrossRef]

36. Nesmelova, I.V.; Ermakova, E.; Daragan, V.A.; Pang, M.; Menendez, M.; Lagartera, L.; Solis, D.; Baum, L.G.; Mayo, K.H. Lactose binding to galectin-1 modulates structural dynamics, increases conformational entropy, and occurs with apparent negative cooperativity. *J. Mol. Biol.* **2010**, *397*, 1209–1230. [CrossRef] [PubMed]

37. Miller, M.; Ludwig, A.K.; Wichapong, K.; Kaltner, H.; Kopitz, J.; Gabius, H.J.; Mayo, K. Adhesion/growth-regulatory galectins tested in combination: Evidence for formation of hybrids as heterodimers. *Biochem. J.* **2018**, *475*, 1003–1018. [CrossRef] [PubMed]

38. Ippel, H.; Miller, M.C.; Vértesy, S.; Zheng, Y.; Cañada, F.J.; Suylen, D.; Umemoto, K.; Romano, C.; Hackeng, T.; Tai, G.; et al. Intra- and intermolecular interactions of human galectin-3: Assessment by full-assignment-based NMR. *Glycobiology* **2016**, *26*, 888–903. [CrossRef] [PubMed]

39. Rini, J.M. Lectin structure. *Annu. Rev. Biophys. Biomol. Struct.* **1995**, *24*, 551–577. [CrossRef] [PubMed]

40. Massa, S.M.; Cooper, D.N.; Leffler, H.; Barondes, S.H. L-29, an endogenous lectin, binds to glycoconjugate ligands with positive cooperativity. *Biochemistry* **1993**, *32*, 260–267. [CrossRef] [PubMed]

41. Ochieng, J.; Platt, D.; Tait, L.; Hogan, V.; Raz, T.; Carmi, P.; Raz, A. Structure-function relationship of a recombinant human galactoside-binding protein. *Biochemistry* **1993**, *32*, 4455–4460. [CrossRef] [PubMed]

42. Kuklinski, S.; Probstmeier, R. Homophilic binding properties of galectin-3: Involvement of the carbohydrate recognition domain. *J. Neurochem.* **1998**, *70*, 814–823. [CrossRef] [PubMed]

43. Woo, H.J.; Lotz, M.M.; Jung, J.U.; Mercurio, A.M. Carbohydrate-binding protein 35 (Mac-2), a laminin-binding lectin, forms functional dimers using cysteine 186. *J. Biol. Chem.* **1991**, *266*, 18419–18422. [PubMed]

44. Leffler, H. Galectins structure and function—A synopsis. *Results Probl. Cell Differ.* **2001**, *33*, 57–83. [PubMed]

45. Hsu, D.K.; Zuberi, R.I.; Liu, F.T. Biochemical and biophysical characterization of human recombinant IgE-binding protein, an S-type animal lectin. *J. Biol. Chem.* **1992**, *267*, 14167–14174. [PubMed]

46. Mehul, B.; Bawumia, S.; Hughes, R.C. Cross-linking of galectin 3, a galactose-binding protein of mammalian cells, by tissue-type transglutaminase. *FEBS Lett.* **1995**, *360*, 160–164. [CrossRef]

47. Ahmad, N.; Gabius, H.J.; Andre, S.; Kaltner, H.; Sabesan, S.; Roy, R.; Liu, B.; Macaluso, F.; Brewer, C.F. Galectin-3 precipitates as a pentamer with synthetic multivalent carbohydrates and forms heterogeneous cross-linked complexes. *J. Biol. Chem.* **2004**, *279*, 10841–10847. [CrossRef] [PubMed]

48. Miyanishi, N.; Nishi, N.; Abe, H.; Kashio, Y.; Shinonaga, R.; Nakakita, S.; Sumiyoshi, W.; Yamauchi, A.; Nakamura, T.; Hirashima, M.; et al. Carbohydrate-recognition domains of galectin-9 are involved in intermolecular interaction with galectin-9 itself and other members of the galectin family. *Glycobiology* **2007**, *17*, 423–432. [CrossRef] [PubMed]

49. Kadoya, T.; Horie, H. Structural and functional studies of galectin-1: A novel axonal regeneration-promoting activity for oxidized galectin-1. *Curr. Drug Targets* **2005**, *6*, 375–383. [CrossRef] [PubMed]

50. Miller, M.C.; Ribeiro, J.P.; Roldós, V.; Martín-Santamaría, S.; Cañada, F.J.; Nesmelova, I.A.; André, S.; Pang, M.; Klyosov, A.A.; Baum, L.G.; et al. Structural aspects of binding of α-linked digalactosides to human galectin-1. *Glycobiology* **2011**, *21*, 1627–1641. [CrossRef] [PubMed]

51. Schwarz, F.P.; Ahmed, H.; Bianchet, M.A.; Amzel, L.M.; Vasta, G.R. Thermodynamics of bovine spleen galectin-1 binding to disaccharides: Correlation with structure and its effect on oligomerization at the denaturation temperature. *Biochemistry* **1998**, *37*, 5867–5877. [CrossRef] [PubMed]

52. Hirabayashi, J.; Hashidate, T.; Arata, Y.; Nishi, N.; Nakamura, T.; Hirashima, M.; Urashima, T.; Oka, T.; Futai, M.; Muller, W.E.; et al. Oligosaccharide specificity of galectins: A search by frontal affinity chromatography. *Biochim. Biophys. Acta* **2002**, *1572*, 232–254. [CrossRef]

53. Sparrow, C.P.; Leffler, H.; Barondes, S.H. Multiple soluble beta-galactoside-binding lectins from human lung. *J. Biol. Chem.* **1987**, *262*, 7383–7390. [PubMed]

54. Ideo, H.; Seko, A.; Ohkura, T.; Matta, K.L.; Yamashita, K. High-affinity binding of recombinant human galectin-4 to SO₃⁻→3Galβ1→3GalNAc pyranoside. *Glycobiology* **2002**, *12*, 199–208. [CrossRef] [PubMed]

55. Leffler, H.; Barondes, S.H. Specificity of binding of three soluble rat lung lectins to substituted and unsubstituted mammalian beta-galactosides. *J. Biol. Chem.* **1986**, *261*, 10119–10126. [PubMed]

56. Sujatha, M.S.; Sasidhar, Y.U.; Balaji, P.V. Insights into the role of the aromatic residue in galactose-binding sites: MP2/6-311G++** study on galactose- and glucose-aromatic residue analogue complexes. *Biochemistry* **2005**, *44*, 8554–8562. [CrossRef] [PubMed]

57. Leppanen, A.; Stowell, S.; Blixt, O.; Cummings, R.D. Dimeric galectin-1 binds with high affinity to α2,3-sialylated and non-sialylated terminal *N*-acetyllactosamine units on surface-bound extended glycans. *J. Biol. Chem.* **2005**, *280*, 5549–5562. [CrossRef] [PubMed]

58. Miller, M.C.; Klyosov, A.; Platt, D.; Mayo, K.H. Using pulse field gradient NMR diffusion measurements to define molecular size distributions in glycan preparations. *Carbohydr. Res.* **2009**, *344*, 1205–1212. [CrossRef] [PubMed]

59. Miller, M.; Nesmelova, I.V.; Klyosov, A.; Platt, D.; Mayo, K.H. The carbohydrate binding domain on galectin-1 is more extensive for a complex glycan than for simple saccharides: Implications for galectin-glycan interactions at the cell surface. *Biochem. J.* **2009**, *421*, 211–221. [CrossRef] [PubMed]

60. Miller, M.C.; Klyosov, A.; Mayo, K.H. Structural Features for α-galactomannan binding to galectin-1. *Glycobiology* **2012**, *22*, 543–551. [CrossRef] [PubMed]

61. Miller, M.C.; Ippel, H.; Suylen, D.; Klyosov, A.A.; Traber, P.G.; Hackeng, T.; Mayo, K.H. Binding of Polysaccharides to Human Galectin-3 at a Non-Canonical Site in its carbohydrate Recognition Domain. *Glycobiology* **2016**, *26*, 88–99. [CrossRef] [PubMed]

62. Miller, M.C.; Zheng, Y.; Yan, J.; Zhou, Y.; Tai, G.; Mayo, K.H. Novel polysaccharide binding to the *N*-terminal tail of galection-3 is likely modulated by proline isomerization. *Glycobiology* **2017**, *27*, 1038–1051. [CrossRef] [PubMed]

63. Liu, F.T.; Rabinovich, G.A. Galectins as modulators of tumour progression. *Nat. Rev. Cancer* **2005**, *5*, 29–41. [CrossRef] [PubMed]

64. Brewer, C.F.; Miceli, M.C.; Baum, L.G. Clusters, bundles, arrays and lattices: Novel mechanisms for lectin-saccharide-mediated cellular interactions. *Curr. Opin. Struct. Biol.* **2002**, *12*, 616–623. [CrossRef]

65. Yang, R.Y.; Hsu, D.K.; Liu, F.T. Expression of galectin-3 modulates T-cell growth and apoptosis. *Proc. Natl. Acad. Sci. USA* **1996**, *93*, 6737–6742. [CrossRef] [PubMed]

66. Akahani, S.; Nangia-Makker, P.; Inohara, H.; Kim, H.R.; Raz, A. Galectin-3: A novel antiapoptotic molecule with a functional BH1 (NWGR) domain of Bcl-2 family. *Cancer Res.* **1997**, *57*, 5272–5276. [PubMed]

67. Adams, L.; Scott, G.K.; Weinberg, C.S. Biphasic modulation of cell growth by recombinant human galectin-1. *Biochim. Biophys. Acta* **1996**, *1312*, 137–144. [CrossRef]

68. Neri, D.; Bicknell, R. Tumour vascular targeting. *Nat. Rev. Cancer* **2005**, *5*, 436–446. [CrossRef] [PubMed]

69. Perillo, N.L.; Pace, K.E.; Seilhamer, J.J.; Baum, L.G. Apoptosis of T cells mediated by galectin-1. *Nature* **1995**, *378*, 736–739. [CrossRef] [PubMed]

70. Stowell, S.R.; Karmakar, S.; Stowell, C.J.; Dias-Baruffi, M.; McEver, R.P.; Cummings, R.D. Human galectin-1, -2, and -4 induce surface exposure of phosphatidylserine in activated human neutrophils but not in activated T cells. *Blood* **2007**, *109*, 219–227. [CrossRef] [PubMed]

71. Ozaki, K.; Inoue, K.; Sato, H.; Iida, A.; Ohnishi, Y.; Sekine, A.; Sato, H.; Odashiro, K.; Nobuyoshi, M.; Hori, M.; et al. Functional variation in LGALS2 confers risk of myocardial infarction and regulates lymphotoxin-α secretion in vitro. *Nature* **2004**, *429*, 72–75. [CrossRef] [PubMed]

72. Bernerd, F.; Sarasin, A.; Magnaldo, T. Galectin-7 overexpression is associated with the apoptotic process in UVB-induced sunburn keratinocytes. *Proc. Natl. Acad. Sci. USA* **1999**, *96*, 11329–11334. [CrossRef] [PubMed]

73. Polyak, K.; Xia, Y.; Zweier, J.L.; Kinzler, K.W.; Vogelstein, B. A model for p53-induced apoptosis. *Nature* **1997**, *389*, 300–305. [CrossRef] [PubMed]

74. Lahm, H.; Andre, S.; Hoeflich, A.; Fischer, J.R.; Sordat, B.; Kaltner, H.; Wolf, E.; Gabius, H.J. Comprehensive galectin fingerprinting in a panel of 61 human tumor cell lines by RT-PCR and its implications for diagnostic and therapeutic procedures. *J. Cancer Res. Clin. Oncol.* **2001**, *127*, 375–386. [CrossRef] [PubMed]

75. Gopalkrishnan, R.V.; Roberts, T.; Tuli, S.; Kang, D.; Christiansen, K.A.; Fisher, P.B. Molecular characterization of prostate carcinoma tumor antigen-1, PCTA-1, a human galectin-8 related gene. *Oncogene* **2000**, *19*, 4405–4416. [CrossRef] [PubMed]

76. Miura, T.; Takahashi, M.; Horie, H.; Kurushima, H.; Tsuchimoto, D.; Sakumi, K.; Nakabeppu, Y. Galectin-1β, a natural monomeric form of galectin-1 lacking its six amino-terminal residues promotes axonal regeneration but not cell death. *Cell Death Differ.* **2004**, *11*, 1076–1083. [CrossRef] [PubMed]

77. Satoh, M.; Ito, A.; Nojiri, H.; Handa, K.; Numahata, K.; Ohyama, C.; Saito, S.; Hoshi, S.; Hakomori, S.I. Enhanced GM3 expression, associated with decreased invasiveness, is induced by brefeldin A in bladder cancer cells. *Int. J. Oncol.* **2001**, *19*, 723–731. [CrossRef] [PubMed]

78. Cherayil, B.J.; Weiner, S.J.; Pillai, S. The Mac-2 antigen is a galactose-specific lectin that binds IgE. *J. Exp. Med.* **1989**, *170*, 1959–1972. [CrossRef] [PubMed]

79. Gil, C.D.; La, M.; Perretti, M.; Oliani, S.M. Interaction of human neutrophils with endothelial cells regulates the expression of endogenous proteins annexin 1, galectin-1 and galectin-3. *Cell Biol. Int.* **2006**, *30*, 338–344. [CrossRef] [PubMed]

80. Andre, S.; Kaltner, H.; Lensch, M.; Russwurm, R.; Siebert, H.C.; Fallsehr, C.; Tajkhorshid, E.; Heck, A.J.; von Knebel Doeberitz, M.; Gabius, H.J.; et al. Determination of structural and functional overlap/divergence of five proto-type galectins by analysis of the growth-regulatory interaction with ganglioside GM1 in silico and in vitro on human neuroblastoma cells. *Int. J. Cancer* **2005**, *114*, 46–57. [CrossRef] [PubMed]

81. Nieminen, J.; Kuno, A.; Hirabayashi, J.; Sato, S. Visualization of galectin-3 oligomerization on the surface of neutrophils and endothelial cells using fluorescence resonance energy transfer. *J. Biol. Chem.* **2007**, *282*, 1374–1383. [CrossRef] [PubMed]

82. Ermakova, E.; Miller, M.C.; Nesmelova, I.V.; Lopez-Merino, L.; Berbís, M.A.; Nesmelov, Y.; Lagartera, L.; Daragan, V.A.; André, S.; Cañada, F.J.; et al. Lactose Binding to Human Galectin-7 (p53-induced gene 1) Induces Long-range Effects through the Protein Resulting in Increased Dimer Stability and Evidence for Positive Cooperativity. *Glycobiology* **2013**, *23*, 508–523. [CrossRef] [PubMed]

83. Nishioka, T.; Sakumi, K.; Miura, T.; Tahara, K.; Horie, H.; Kadoya, T.; Nakabeppu, Y. FosB gene products trigger cell proliferation and morphological alteration with an increased expression of a novel processed form of galectin-1 in the rat 3Y1 embryo cell line. *J. Biochem.* **2002**, *131*, 653–661. [CrossRef] [PubMed]

84. Ashery, U.; Yizhar, O.; Rotblat, B.; Elad-Sfadia, G.; Barkan, B.; Haklai, R.; Kloog, Y. Spatiotemporal organization of Ras signaling: Rasosomes and the galectin switch. *Cell. Mol. Neurobiol.* **2006**, *26*, 471–495. [CrossRef] [PubMed]

85. Paz, A.; Haklai, R.; Elad-Sfadia, G.; Ballan, E.; Kloog, Y. Galectin-1 binds oncogenic H-Ras to mediate Ras membrane anchorage and cell transformation. *Oncogene* **2001**, *20*, 7486–7493. [CrossRef] [PubMed]

86. Rotblat, B.; Niv, H.; Andre, S.; Kaltner, H.; Gabius, H.-J.; Kloog, Y. Galectin-1(L11A) predicted from a computed galectin-1 farnesyl-binding pocket selectively inhibits Ras-GTP. *Cancer Res.* **2004**, *64*, 3112–3118. [CrossRef] [PubMed]

87. Hoffman, G.R.; Nassar, N.; Cerione, R.A. Structure of the Rho family GTP-binding protein Cdc42 in complex with the multifunctional regulator RhoGDI. *Cell* **2000**, *100*, 345–356. [CrossRef]

88. Gorfe, A.A.; Hanzal-Bayer, M.; Abankwa, D.; Hancock, J.F.; McCammon, J.A. Structure and dynamics of the full-length lipid-modified H-Ras protein in a 1,2-dimyristoylglycero-3-phosphocholine bilayer. *J. Med. Chem.* **2007**, *50*, 674–684. [CrossRef] [PubMed]

89. Paron, I.; Scaloni, A.; Pines, A.; Bachi, A.; Liu, F.T.; Puppin, C.; Pandolfi, M.; Ledda, L.; Di Loreto, C.; Damante, G.; et al. Nuclear localization of Galectin-3 in transformed thyroid cells: A role in transcriptional regulation. *Biochem. Biophys. Res. Commun.* **2003**, *302*, 545–553. [CrossRef]

90. Yu, F.; Finley, R.L., Jr.; Raz, A.; Kim, H.R. Galectin-3 translocates to the perinuclear membranes and inhibits cytochrome c release from the mitochondria. A role for synexin in galectin-3 translocation. *J. Biol. Chem.* **2002**, *277*, 15819–15827. [CrossRef] [PubMed]

91. Zhang, P.F.; Li, K.S.; Shen, Y.H.; Gao, P.T.; Dong, Z.R.; Cai, J.B.; Zhang, C.; Huang, X.Y.; Tian, M.X.; Hu, Z.Q.; et al. Galectin-1 induces hepatocellular carcinoma EMT and sorafenib resistance by activating FAK/PI3K/AKT signaling. *Cell Death Dis.* **2016**, *7*, e2201. [CrossRef] [PubMed]

92. Bacigalupo, M.L.; Manzi, M.; Espelt, M.V.; Gentilini, L.D.; Compagno, D.; Laderach, D.J.; Wolfenstein-Todel, C.; Rabinovich, G.A.; Troncoso, M.F. Galectin-1 triggers epithelial-mesenchymal transition in human hepatocellular carcinoma cells. *J. Cell. Physiol.* **2015**, *230*, 1298–1309. [CrossRef] [PubMed]

93. Huang, C.S.; Tang, S.J.; Chung, L.Y.; Yu, C.P.; Ho, J.Y.; Cha, T.L.; Hsieh, C.C.; Wang, H.H.; Sun, G.H.; Sun, K.H. Galectin-1 upregulates CXCR4 to promote tumor progression and poor outcome in kidney cancer. *J. Am. Soc. Nephrol.* **2014**, *25*, 1486–1495. [CrossRef] [PubMed]

94. Lv, B.; Yang, X.; Lv, S.; Wang, L.; Fan, K.; Shi, R.; Wang, F.; Song, H.; Ma, X.; Tan, X.; et al. CXCR4 signaling induced epithelial-mesenchymal transition by PI3K/AKT and ERK pathways in glioblastoma. *Mol. Neurobiol.* **2015**, *52*, 1263–1268. [CrossRef] [PubMed]

95. Chong, Y.; Tang, D.; Gao, J.; Jiang, X.; Xu, C.; Xiong, Q.; Huang, Y.; Wang, J.; Zhou, H.; Shi, Y.; et al. Galectin-1 induces invasion and the epithelial-mesenchymal transition in human gastric cancer cells via non-canonical activation of the hedgehog signaling pathway. *Oncotarget* **2016**, *7*, 83611–83626. [CrossRef] [PubMed]

96. Zheng, L.; Xu, C.; Guan, Z.; Su, X.; Xu, Z.; Cao, J.; Teng, L. Galectin-1 mediates TGF-beta-induced transformation from normal fibroblasts into carcinoma-associated fibroblasts and promotes tumor progression in gastric cancer. *Am. J. Transl. Res.* **2016**, *8*, 1641–1658. [PubMed]

97. Tang, D.; Zhang, J.; Yuan, Z.; Zhang, H.; Chong, Y.; Huang, Y.; Wang, J.; Xiong, Q.; Wang, S.; Wu, Q.; et al. PSC-derived galectin-1 inducing epithelial-mesenchymal transition of pancreatic ductal adenocarcinoma cells by activating the NF-κB pathway. *Oncotarget* **2017**, *8*, 86488–86502. [CrossRef] [PubMed]

98. Martinez-Martinez, E.; Ibarrola, J.; Fernandez-Celis, A.; Calvier, L.; Leroy, C.; Cachofeiro, V.; Rossignol, P.; Lopez-Andres, N. Galectin-3 pharmacological inhibition attenuates early renal damage in spontaneously hypertensive rats. *J. Hypertens.* **2017**. [CrossRef] [PubMed]

99. Huang, Z.; Ai, Z.; Li, N.; Xi, H.; Gao, X.; Wang, F.; Tan, X.; Liu, H. Over expression of galectin-3 associates with short-term poor prognosis in stage II colon cancer. *Cancer Biomark.* **2016**, *17*, 445–455. [CrossRef] [PubMed]

100. Wang, L.P.; Chen, S.W.; Zhuang, S.M.; Li, H.; Song, M. Galectin-3 accelerates the progression of oral tongue squamous cell carcinoma via a Wnt/beta-catenin-dependent pathway. *Pathol. Oncol. Res.* **2013**, *19*, 461–474. [CrossRef] [PubMed]

101. Saraswati, S.; Block, A.S.; Davidson, M.K.; Rank, R.G.; Mahadevan, M.; Diekman, A.B. Galectin-3 is a substrate for prostate specific antigen (PSA) in human seminal plasma. *Prostate* **2011**, *71*, 197–208. [CrossRef] [PubMed]

102. Block, A.S.; Saraswati, S.; Lichti, C.F.; Mahadevan, M.; Diekman, A.B. Co-purification of Mac-2 binding protein with galectin-3 and association with prostasomes in human semen. *Prostate* **2011**, *71*, 711–721. [CrossRef] [PubMed]

103. Kovak, M.R.; Saraswati, S.; Goddard, S.D.; Diekman, A.B. Proteomic identification of galectin-3 binding ligands and characterization of galectin-3 proteolytic cleavage in human prostasomes. *Andrology* **2013**, *1*, 682–691. [CrossRef] [PubMed]

104. Kovak, M.R.; Saraswati, S.; Schoen, D.J.; Diekman, A.B. Investigation of galectin-3 function in the reproductive tract by identification of binding ligands in human seminal plasma. *Am. J. Reprod. Immunol.* **2014**, *72*, 403–412. [CrossRef] [PubMed]

105. Ilmer, M.; Mazurek, N.; Gilcrease, M.Z.; Byrd, J.C.; Woodward, W.A.; Buchholz, T.A.; Acklin, K.; Ramirez, K.; Hafley, M.; Alt, E.; et al. Low expression of galectin-3 is associated with poor survival in node-positive breast cancers and mesenchymal phenotype in breast cancer stem cells. *Breast Cancer Res.* **2016**, *18*, 97. [CrossRef] [PubMed]

106. Oyanadel, C.; Holmes, C.; Pardo, E.; Retamal, C.; Shaughnessy, R.; Smith, P.; Cortes, P.; Bravo-Zehnder, M.; Metz, C.; Feuerhake, T.; et al. Galectin-8 induces partial epithelial-mesenchymal transition with invasive tumorigenic capabilities involving a FAK/EGFR/proteasome pathway in MDCK cells. *Mol. Biol. Cell* **2018**. [CrossRef] [PubMed]

107. Cao, Z.; Said, N.; Amin, S.; Wu, H.K.; Bruce, A.; Garate, M.; Hsu, D.K.; Kuwabara, I.; Liu, F.T.; Panjwani, N. Galectins-3 and -7, but not galectin-1, play a role in re-epithelialization of wounds. *J. Biol. Chem.* **2002**, *277*, 42299–42305. [CrossRef] [PubMed]

108. Thijssen, V.L.; Postel, R.; Brandwijk, R.J.; Dings, R.P.; Nesmelova, I.; Satijn, S.; Verhofstad, N.; Nakabeppu, Y.; Baum, L.G.; Bakkers, J.; et al. Galectin-1 is essential in tumor angiogenesis and is a target for anti-angiogenesis therapy. *Proc. Natl. Acad. Sci. USA* **2006**, *103*, 15975–15980. [CrossRef] [PubMed]

109. Liu, F.T.; Patterson, R.J.; Wang, J.L. Intracellular functions of galectins. *Biochim. Biophys. Acta* **2002**, *1572*, 263–273. [CrossRef]

110. Ingrassia, L.; Camby, I.; Lefranc, F.; Mathieu, V.; Nshimyumukiza, P.; Darro, F.; Kiss, R. Anti-galectin compounds as potential anti-cancer drugs. *Curr. Med. Chem.* **2006**, *13*, 3513–3527. [CrossRef] [PubMed]

111. Salameh, B.A.; Leffler, H.; Nilsson, U.J. 3-(1,2,3-Triazol-1-yl)-1-thio-galactosides as small, efficient, and hydrolytically stable inhibitors of galectin-3. *Bioorg. Med. Chem. Lett.* **2005**, *15*, 3344–3346. [CrossRef] [PubMed]

112. Tejler, J.; Leffler, H.; Nilsson, U.J. Synthesis of *O*-galactosyl aldoximes as potent LacNAc-mimetic galectin-3 inhibitors. *Bioorg. Med. Chem. Lett.* **2005**, *15*, 2343–2345. [CrossRef] [PubMed]

113. Cumpstey, I.; Carlsson, S.; Leffler, H.; Nilsson, U.J. Synthesis of a phenyl thio-beta-d-galactopyranoside library from 1,5-difluoro-2,4-dinitrobenzene: Discovery of efficient and selective monosaccharide inhibitors of galectin-7. *Org. Biomol. Chem.* **2005**, *3*, 1922–1932. [CrossRef] [PubMed]

114. Salameh, B.A.; Sundin, A.; Leffler, H.; Nilsson, U.J. Thioureido *N*-acetyllactosamine derivatives as potent galectin-7 and 9N inhibitors. *Bioorg. Med. Chem.* **2006**, *14*, 1215–1220. [CrossRef] [PubMed]

115. Cumpstey, I.; Sundin, A.; Leffler, H.; Nilsson, U.J. C2-symmetrical thiodigalactoside bis-benzamido derivatives as high-affinity inhibitors of galectin-3: Efficient lectin inhibition through double arginine-arene interactions. *Angew. Chem.* **2005**, *44*, 5110–5112. [CrossRef] [PubMed]

116. Fort, S.; Kim, H.S.; Hindsgaul, O. Screening for galectin-3 inhibitors from synthetic lacto-*N*-biose libraries using microscale affinity chromatography coupled to mass spectrometry. *J. Org. Chem.* **2006**, *71*, 7146–7154. [CrossRef] [PubMed]

117. Tejler, J.; Tullberg, E.; Frejd, T.; Leffler, H.; Nilsson, U.J. Synthesis of multivalent lactose derivatives by 1,3-dipolar cycloadditions: Selective galectin-1 inhibition. *Carbohydr. Res.* **2006**, *341*, 1353–1362. [CrossRef] [PubMed]

118. Ingrassia, L.; Mathieu, V.; Mégalizzi, V.; Lefranc, F.; Darro, F.; Kiss, R. UNBS4209: A bilactosylated steroid with anti-galectin-1 activity. In Proceedings of the 98th Annual Meeting of American Association for Cancer Research, Los Angeles, CA, USA, 14–18 April 2007.

119. Rabinovich, G.A.; Cumashi, A.; Bianco, G.A.; Ciavardelli, D.; Iurisci, I.; D'Egidio, M.; Piccolo, E.; Tinari, N.; Nifantiev, N.; Iacobelli, S. Synthetic lactulose amines: Novel class of anticancer agents that induce tumor-cell apoptosis and inhibit galectin-mediated homotypic cell aggregation and endothelial cell morphogenesis. *Glycobiology* **2006**, *16*, 210–220. [CrossRef] [PubMed]

120. Sirois, S.; Giguere, D.; Roy, R. A first QSAR model for galectin-3 glycomimetic inhibitors based on 3D docked structures. *Med. Chem.* **2006**, *2*, 481–489. [CrossRef] [PubMed]

121. Giguere, D.; Patnam, R.; Bellefleur, M.A.; St-Pierre, C.; Sato, S.; Roy, R. Carbohydrate triazoles and isoxazoles as inhibitors of galectins-1 and -3. *Chem. Commun.* **2006**, *23*, 2379–2381. [CrossRef] [PubMed]

122. Giguere, D.; Sato, S.; St-Pierre, C.; Sirois, S.; Roy, R. Aryl *O*- and *S*-galactosides and lactosides as specific inhibitors of human galectins-1 and -3: Role of electrostatic potential at O-3. *Bioorg. Med. Chem. Lett.* **2006**, *16*, 1668–1672. [CrossRef] [PubMed]

123. Rajput, V.K.; MacKinnon, A.; Mandal, S.; Collins, P.; Blanchard, H.; Leffler, H.; Sethi, T.; Schambye, H.; Mukhopadhyay, B.; Nilsson, U.J. A Selective Galactose-Coumarin-Derived Galectin-3 Inhibitor Demonstrates Involvement of Galectin-3-glycan Interactions in a Pulmonary Fibrosis Model. *J. Med. Chem.* **2016**, *59*, 8141–8147. [CrossRef] [PubMed]

124. Chen, W.S.; Cao, Z.; Leffler, H.; Nilsson, U.J.; Panjwani, N. Galectin-3 inhibition by a small molecule inhibitor reduces both pathological corneal neovascularization and fibrosis. *Investig. Ophthalmol. Vis. Sci.* **2017**, *58*, 9–20. [CrossRef] [PubMed]

125. Peterson, K.; Kumar, R.; Stenström, O.; Verma, P.; Verma, P.R.; Håkansson, M.; Kahl-Knutsson, B.; Zetterberg, F.; Leffler, H.; Akke, M.; et al. Systematic Tuning of Fluoro-galectin-3 Interactions Provides Thiodigalactoside Derivatives with Single-Digit nM Affinity and High Selectivity. *J. Med. Chem.* **2018**, *61*, 1164–1175. [CrossRef] [PubMed]

126. Zetterberg, F.R.; Peterson, K.; Johnsson, R.E.; Brimert, T.; Håkansson, M.; Logan, D.T.; Leffler, H.; Nilsson, U.J. Monosaccharide Derivatives with Low-Nanomolar Lectin Affinity and High Selectivity Based on Combined Fluorine-Amide, Phenyl-Arginine, Sulfur-π, and Halogen Bond Interactions. *Chem. Med. Chem.* **2018**, *13*, 133–137. [CrossRef] [PubMed]

127. Henderson, N.C.; Mackinnon, A.C.; Farnworth, S.L.; Poirier, F.; Russo, F.P.; Iredale, J.P.; Haslett, C.; Simpson, K.J.; Sethi, T. Galectin-3 regulates myofibroblast activation and hepatic fibrosis. *Proc. Natl. Acad. Sci. USA* **2006**, *103*, 5060–5065. [CrossRef] [PubMed]

128. Iacobini, C.; Menini, S.; Ricci, C.; Blasetti Fantauzzi, C.; Scipioni, A.; Salvi, L.; Cordone, S.; Delucchi, F.; Serino, M.; Federici, M.; et al. Galectin-3 ablation protects mice from diet-induced NASH: A major scavenging role for galectin-3 in liver. *J. Hepatol.* **2011**, *54*, 975–983. [CrossRef] [PubMed]

129. Traber, P.G.; Chou, H.; Zomer, E.; Hong, F.; Klyosov, A.; Fiel, M.-I.; Friedman, S.L. Regression of fibrosis and reversal of cirrhosis in rats by galectin inhibitors in thioacetamide-induced liver disease. *PLoS ONE* **2013**, *8*, e75361. [CrossRef] [PubMed]

130. Traber, P.G.; Zomer, E. Therapy of experimental NASH and fibrosis with galectin inhibitors. *PLoS ONE* **2013**, *8*, e83481. [CrossRef] [PubMed]

131. Zou, J.; Glinsky, V.V.; Landon, L.A.; Matthews, L.; Deutscher, S.L. Peptides specific to the galectin-3 carbohydrate recognition domain inhibit metastasis-associated cancer cell adhesion. *Carcinogenesis* **2005**, *26*, 309–318. [CrossRef] [PubMed]

132. Griffioen, A.W.; van der Schaft, D.; Barandsz-Janson, A.; Cox, A.; Struijker-Boudier, H.A.; Hillen, H.F.P.; Mayo, K.H. Anginex, a designed Peptide that Inhibits Angiogenesis. *Biochem. J.* **2001**, *354*, 233–242. [CrossRef] [PubMed]

133. Rabinovich, G.A. Galectin-1 as a potential cancer target. *Br. J. Cancer* **2005**, *92*, 1188–1192. [CrossRef] [PubMed]

134. Dirkx, A.E.; Oude Egbrink, M.G.; Castermans, K.; Thijssen, V.L.; van der Schaft, D.W.; Kwee, L.; Mayo, K.H.; Wagstaff, J.; Bouma-ter Steege, J.C.; Dings, R.P.; et al. Anti-angiogenesis therapy can overcome endothelial cell anergy and promote leukocyte-endothelium interactions and infiltration in tumors. *FASEB J.* **2006**, *20*, 621–630. [CrossRef] [PubMed]

135. Hellebrekers, D.M.; Castermans, K.; Vire, E.; Dings, R.P.; Hoebers, N.T.; Mayo, K.H.; Oude Egbrink, M.G.; Molema, G.; Fuks, F.; van Engeland, M.; et al. Epigenetic regulation of tumor endothelial cell anergy: Silencing of intercellular adhesion molecule-1 by histone modifications. *Cancer Res.* **2006**, *66*, 10770–10777. [CrossRef] [PubMed]

136. Dings, R.P.; van der Schaft, D.W.; Hargittai, B.; Haseman, J.; Griffioen, A.W.; Mayo, K.H. Anti-tumor activity of the novel angiogenesis inhibitor anginex. *Cancer Lett.* **2003**, *194*, 55–66. [CrossRef]

137. Dings, R.P.; Yokoyama, Y.; Ramakrishnan, S.; Griffioen, A.W.; Mayo, K.H. The designed angiostatic peptide anginex synergistically improves chemotherapy and antiangiogenesis therapy with angiostatin. *Cancer Res.* **2003**, *63*, 382–385. [PubMed]

138. Dings, R.P.; Williams, B.W.; Song, C.W.; Griffioen, A.W.; Mayo, K.H.; Griffin, R.J. Anginex synergizes with radiation therapy to inhibit tumor growth by radio-sensitizing endothelial cells. *Int. J. Cancer* **2005**, *115*, 312–319. [CrossRef] [PubMed]

139. Rubinstein, N.; Alvarez, M.; Zwirner, N.W.; Toscano, M.A.; Ilarregui, J.M.; Bravo, A.; Mordoh, J.; Fainboim, L.; Podhajcer, O.L.; Rabinovich, G.A. Targeted inhibition of galectin-1 gene expression in tumor cells results in heightened T cell-mediated rejection; A potential mechanism of tumor-immune privilege. *Cancer Cell* **2004**, *5*, 241–251. [CrossRef]

140. Lotan, R.; Matsushita, Y.; Ohannesian, D.; Carralero, D.; Ota, D.M.; Cleary, K.R.; Nicolson, G.L.; Irimura, T. Lactose-binding lectin expression in human colorectal carcinomas. Relation to tumor progression. *Carbohydr. Res.* **1991**, *213*, 47–57. [CrossRef]

141. Dings, R.P.M.; Miller, M.C.; Nesmelova, I.; Astorgues-Xerri, L.; Kumar, N.; Serova, M.; Chen, X.; Raymond, E.; Hoye, T.R.; Mayo, K.H. Anti-tumor agent calixarene 0118 targets human galectin-1 as an allosteric inhibitor of carbohydrate binding. *J. Med. Chem.* **2012**, *55*, 5121–5129. [CrossRef] [PubMed]

142. Dings, R.P.; Arroyo, M.M.; Lockwood, N.A.; Van Eijk, L.I.; Haseman, J.R.; Griffioen, A.W.; Mayo, K.H. Beta-sheet is the bioactive conformation of the anti-angiogenic anginex peptide. *Biochem. J.* **2003**, *23*, 281–288. [CrossRef] [PubMed]

143. Dings, R.P.M.; Kumar, N.; Miller, M.C.; Loren, M.; Rangwala, H.; Hoye, T.R.; Mayo, K.H. Structure-Based Optimization of Angiostatic Agent 6DBF7, an Allosteric Inhibitor of Galectin-1. *J. Pharmacol. Exp. Ther.* **2013**, *344*, 589–599. [CrossRef] [PubMed]

144. Dings, R.P.M.; Mayo, K.H. A journey in structure-based discovery: From designed peptides to protein-surface topomimetics as antibiotic and antiangiogenic agents. *Acc. Chem. Res.* **2007**, *40*, 1057–1065. [CrossRef] [PubMed]

145. Mayo, K.H.; Dings, R.P.M.; Flader, C.; Nesmelova, I.; Hargittai, B.; van der Schaft, D.W.J.; van Eijk, L.I.; Walek, D.; Haseman, J.; Hoye, T.R.; et al. Design of a Partial-Peptide Mimetic of Anginex with Antiangiogenic and Anticancer Activity. *J. Biol. Chem.* **2003**, *278*, 45746–45752. [CrossRef] [PubMed]

146. Dings, R.P.M.; Chen, X.; Hellebrekers, D.M.E.I.; van Eijk, L.I.; Hoye, T.R.; Griffioen, A.W.; Mayo, K.H. Design of non-peptidic topomimetics of anti-angiogenic proteins with anti-tumor activities. *J. Natl. Can. Inst.* **2006**, *98*, 932–936. [CrossRef] [PubMed]

147. Amano, M.; Suzuki, M.; Andoh, S.; Monzen, H.; Terai, K.; Williams, B.; Song, C.W.; Mayo, K.H.; Hasegawa, T.; Dings, R.P.; et al. Antiangiogenesis therapy using a novel angiogenesis inhibitor, anginex, following radiation causes tumor growth delay. *Int. J. Clin. Oncol.* **2007**, *12*, 42–47. [CrossRef] [PubMed]

148. Griffin, R.J.; Koonce, N.A.; Dings, R.P.; Siegel, E.; Moros, E.G.; Brauer-Krisch, E.; Corry, P.M. Microbeam radiation therapy alters vascular architecture and tumor oxygenation and is enhanced by a galectin-1 targeted anti-angiogenic peptide. *Radiat. Res.* **2012**, *177*, 804–812. [CrossRef] [PubMed]

149. Willett, C.G.; Boucher, Y.; di Tomaso, E.; Duda, D.G.; Munn, L.L.; Tong, R.T.; Chung, D.C.; Sahani, D.V.; Kalva, S.P.; Kozin, S.V.; et al. Direct evidence that the VEGF-specific antibody bevacizumab has antivascular effects in human rectal cancer. *Nat. Med.* **2004**, *10*, 145–147. [CrossRef] [PubMed]

150. Dings, R.P.; Loren, M.L.; Zhang, Y.; Mikkelson, S.; Mayo, K.H.; Corry, P.; Griffin, R.J. Tumour thermotolerance, a physiological phenomenon involving vessel normalisation. *Int. J. Hyperth.* **2011**, *27*, 42–52. [CrossRef] [PubMed]

151. Dings, R.P.M.; van Laar, E.S.; Webber, J.; Zhang, Y.; Griffin, R.J.; Waters, S.J.; MacDonald, J.R.; Mayo, K.H. Ovarian tumor growth regression using a combination of vascular targeting agents anginex or 0118 and the chemotherapeutic irofulven. *Cancer Lett.* **2008**, *265*, 270–280. [CrossRef] [PubMed]

152. Jain, R.K. Normalizing tumor vasculature with anti-angiogenic therapy: A new paradigm for combination therapy. *Nat. Med.* **2001**, *7*, 987–989. [CrossRef] [PubMed]

153. Winkler, F.; Kozin, S.V.; Tong, R.T.; Chae, S.S.; Booth, M.F.; Garkavtsev, I.; Xu, L.; Hicklin, D.J.; Fukumura, D.; di Tomaso, E.; et al. Kinetics of vascular normalization by VEGFR2 blockade governs brain tumor response to radiation: Role of oxygenation, angiopoietin-1, and matrix metalloproteinases. *Cancer Cell* **2004**, *6*, 553–563. [CrossRef] [PubMed]

154. Ansiaux, R.; Baudelet, C.; Jordan, B.F.; Beghein, N.; Sonveaux, P.; De Wever, J.; Martinive, P.; Gregoire, V.; Feron, O.; Gallez, B. Thalidomide radiosensitizes tumors through early changes in the tumor microenvironment. *Clin. Cancer Res.* **2005**, *11*, 743–750. [PubMed]

155. Dings, R.P.; Klein, M.A.; Mayo, K.H. Cancer drug discovery and development. In *Sensitization of Cancer Cells for Chemo/Immuno/Radio-Therapy*; Teicher, B.A., Ed.; Humana Press: Totowa, NJ, USA, 2008; pp. 306–325.

156. Dings, R.P.M.; Loren, M.; Heun, H.; McNiel, E.; Griffioen, A.W.; Mayo, K.H.; Griffin, R.J. Scheduling of radiation with angiogenesis inhibitors Avastin and Anginex improves therapeutic outcome via vessel normalization. *Clin. Can. Res.* **2007**, *13*, 3395–3402. [CrossRef] [PubMed]

157. Croci, D.O.; Cerliani, J.P.; Dalotto-Moreno, T.; Mendez-Huergo, S.P.; Mascanfroni, I.D.; Dergan-Dylon, S.; Toscano, M.A.; Caramelo, J.J.; Garcia-Vallejo, J.J.; Ouyang, J.; et al. Glycosylation-dependent lectin-receptor interactions preserve angiogenesis in anti-VEGF refractory tumors. *Cell* **2014**, *156*, 744–758. [CrossRef] [PubMed]

158. Dings, R.P.M.; van Laar, E.S.; Loren, M.; Webber, J.; Zhang, Y.; Waters, S.J.; MacDonald, J.R.; Mayo, K.H. Inhibiting tumor growth by targeting tumor vasculature with galectin-1 antagonist anginex conjugated to the cytotoxic acylfulvene, 6-hydroxylpropylacylfulvene. *Bioconj. Chem.* **2010**, *21*, 20–27. [CrossRef] [PubMed]

159. Koonce, N.A.; Griffin, R.J.; Dings, R.P.M. Galectin-1 inhibitor OTX008 induces tumor vessel normalization and tumor growth inhibition in human head and neck squamous cell carcinoma models. *Int. J. Mol. Sci.* **2017**, *18*, 2671. [CrossRef] [PubMed]

160. Astorgues-Xerri, L.; Riveiro, M.E.; Tijeras-Raballand, A.; Serova, M.; Rabinovich, G.A.; Bieche, I.; Vidaud, M.; de Gramont, A.; Martinet, M.; Cvitkovic, E.; et al. Otx008, a selective small-molecule inhibitor of galectin-1, downregulates cancer cell proliferation, invasion and tumour angiogenesis. *Eur. J. Cancer* **2014**, *50*, 2463–2477. [CrossRef] [PubMed]

161. Dings, R.P.M.; Vang, K.B.; Castermans, K.; Popescu, F.E.; Zhang, Y.; Noesser, E.; Oude Egbrink, M.G.A.; Mescher, M.F.; Farrar, M.A.; Griffioen, A.W.; et al. Enhancement of T-cell mediated anti-tumor response: Angiostatic adjuvant to immunotherapy against cancer. *Clin. Cancer Res.* **2011**, *17*, 3134–3145. [CrossRef] [PubMed]

162. Thijssen, V.L.; Heusschen, R.; Caers, J.; Griffioen, A.W. Galectin expression in cancer diagnosis and prognosis: A systematic review. *Biochim. Biophys. Acta* **2015**, *1855*, 235–247. [CrossRef] [PubMed]

163. Wang, F.; Lv, P.; Gu, Y.; Li, L.; Ge, X.; Guo, G. Galectin-1 knockdown improves drug sensitivity of breast cancer by reducing P-glycoprotein expression through inhibiting the Raf-1/AP-1 signaling pathway. *Oncotarget* **2017**, *8*, 25097–25106. [CrossRef] [PubMed]

164. Nam, K.; Son, S.H.; Oh, S.; Jeon, D.; Kim, H.; Noh, D.Y.; Kim, S.; Shin, I. Binding of galectin-1 to integrin beta1 potentiates drug resistance by promoting survivin expression in breast cancer cells. *Oncotarget* **2017**, *8*, 35804–35823. [CrossRef] [PubMed]

165. Kuo, P.; Le, Q.T. Galectin-1 links tumor hypoxia and radiotherapy. *Glycobiology* **2014**, *24*, 921–925. [CrossRef] [PubMed]

166. Hortobagyi, G.N.; Frye, D.; Buzdar, A.U.; Ewer, M.S.; Fraschini, G.; Hug, V.; Ames, F.; Montague, E.; Carrasco, C.H.; Mackay, B.; et al. Decreased cardiac toxicity of doxorubicin administered by continuous intravenous infusion in combination chemotherapy for metastatic breast carcinoma. *Cancer* **1989**, *63*, 37–45. [CrossRef]

167. Doroshow, J.H. Doxorubicin-induced cardiac toxicity. *N. Engl. J. Med.* **1991**, *324*, 843–845. [CrossRef] [PubMed]

168. Jenkins, S.; Nima, Z.; Vang, K.B.; Nedosekin, D.; Zharov, V.P.; Griffin, R.J.; Biris, A.S.; Dings, R.P.M. Triple negative breast cancer targeting and killing by EpCAM-directed, plasmonically active nanodrug systems. *Precis. Oncol.* **2018**, *1*, 1–11. [CrossRef]

169. Upreti, M.; Jamshidi-Parsian, A.; Apana, S.; Berridge, M.; Fologea, D.A.; Koonce, N.A.; Henry, R.L.; Griffin, R.J. Radiation-induced galectin-1 by endothelial cells: A promising molecular target for preferential drug delivery to the tumor vasculature. *J. Mol. Med.* **2013**, *91*, 497–506. [CrossRef] [PubMed]

170. Strik, H.M.; Schmidt, K.; Lingor, P.; Tonges, L.; Kugler, W.; Nitsche, M.; Rabinovich, G.A.; Bahr, M. Galectin-1 expression in human glioma cells: Modulation by ionizing radiation and effects on tumor cell proliferation and migration. *Oncol. Rep.* **2007**, *18*, 483–488. [CrossRef] [PubMed]

171. Le, Q.T.; Shi, G.; Cao, H.; Nelson, D.W.; Wang, Y.; Chen, E.Y.; Zhao, S.; Kong, C.; Richardson, D.; O'Byrne, K.J.; et al. Galectin-1: A link between tumor hypoxia and tumor immune privilege. *J. Clin. Oncol.* **2005**, *23*, 8932–8941. [CrossRef] [PubMed]

172. Zhao, X.Y.; Chen, T.T.; Xia, L.; Guo, M.; Xu, Y.; Yue, F.; Jiang, Y.; Chen, G.Q.; Zhao, K.W. Hypoxia inducible factor-1 mediates expression of galectin-1: The potential role in migration/invasion of colorectal cancer cells. *Carcinogenesis* **2010**, *31*, 1367–1375. [CrossRef] [PubMed]

173. Koonce, N.A.; Levy, J.; Hardee, M.E.; Jamshidi-Parsian, A.; Vang, K.B.; Sharma, S.; Raleigh, J.A.; Dings, R.P.; Griffin, R.J. Targeting artificial tumor stromal targets for molecular imaging of tumor vascular hypoxia. *PLoS ONE* **2015**, *10*, e0135607. [CrossRef] [PubMed]

174. Nordsmark, M.; Bentzen, S.M.; Rudat, V.; Brizel, D.; Lartigau, E.; Stadler, P.; Becker, A.; Adam, M.; Molls, M.; Dunst, J.; et al. Prognostic value of tumor oxygenation in 397 head and neck tumors after primary radiation therapy. An international multi-center study. *Radiother. Oncol.* **2005**, *77*, 18–24. [CrossRef] [PubMed]

175. Prise, K.M.; Davies, S.; Michael, B.D. Cell killing and DNA damage in Chinese hamster V79 cells treated with hydrogen peroxide. *Int. J. Radiat. Biol.* **1989**, *55*, 583–592. [CrossRef] [PubMed]

176. Gray, L.H.; Chase, H.B.; Deschner, E.E.; Hunt, J.W.; Scott, O.C. The influence of oxygen and peroxides on the response of mammalian cells and tissues to ionizing radiations. *Prog. Nucl. Energy 6 Biol. Sci.* **1959**, *2*, 69–81. [PubMed]

177. Chen, C.; Pore, N.; Behrooz, A.; Ismail-Beigi, F.; Maity, A. Regulation of glut1 mRNA by hypoxia-inducible factor-1. Interaction between H-ras and hypoxia. *J. Biol. Chem.* **2001**, *276*, 9519–9525. [CrossRef] [PubMed]

178. Huang, E.Y.; Chen, Y.F.; Chen, Y.M.; Lin, I.H.; Wang, C.C.; Su, W.H.; Chuang, P.C.; Yang, K.D. A novel radioresistant mechanism of galectin-1 mediated by H-ras-dependent pathways in cervical cancer cells. *Cell Death Dis.* **2012**, *3*, e251. [CrossRef] [PubMed]

179. Lugade, A.A.; Moran, J.P.; Gerber, S.A.; Rose, R.C.; Frelinger, J.G.; Lord, E.M. Local radiation therapy of B16 melanoma tumors increases the generation of tumor antigen-specific effector cells that traffic to the tumor. *J. Immunol.* **2005**, *174*, 7516–7523. [CrossRef] [PubMed]

180. Liang, H.; Deng, L.; Chmura, S.; Burnette, B.; Liadis, N.; Darga, T.; Beckett, M.A.; Lingen, M.W.; Witt, M.; Weichselbaum, R.R.; et al. Radiation-induced equilibrium is a balance between tumor cell proliferation and T cell-mediated killing. *J. Immunol.* **2013**, *190*, 5874–5881. [CrossRef] [PubMed]

181. Toscano, M.A.; Commodaro, A.G.; Ilarregui, J.M.; Bianco, G.A.; Liberman, A.; Serra, H.M.; Hirabayashi, J.; Rizzo, L.V.; Rabinovich, G.A. Galectin-1 suppresses autoimmune retinal disease by promoting concomitant Th2- and T regulatory-mediated anti-inflammatory responses. *J. Immunol.* **2006**, *176*, 6323–6332. [CrossRef] [PubMed]

182. Jia, D.; Koonce, N.A.; Halakatti, R.; Li, X.; Yaccoby, S.; Swain, F.; Suva, L.; Hennings, L.; Berridge, M.S.; Apana, S.M.; et al. Repression of multiple myeloma growth and relapse with combined radiotherapy and anti-angiogenic agent. *Radiat. Res.* **2010**, *173*, 809–817. [CrossRef] [PubMed]

183. Bailey, L.A.; Jamshidi-Parsian, A.; Patel, T.; Koonce, N.A.; Diekman, A.B.; Cifarelli, C.P.; Marples, B.; Griffin, R.J. Combined temozolomide and ionizing radiation induces galectin-1 and galectin-3 expression in a model of human glioma. *Tumor Microenviron. Ther.* **2015**, *2*, 19–31. [CrossRef]

International Journal of
Molecular Sciences

MDPI

Communication

Galectin-1 Inhibitor OTX008 Induces Tumor Vessel Normalization and Tumor Growth Inhibition in Human Head and Neck Squamous Cell Carcinoma Models

Nathan A. Koonce [1,2], Robert J. Griffin [1,*] and Ruud P. M. Dings [1,*]

1   Department of Radiation Oncology, University of Arkansas for Medical Sciences, Little Rock, AR 72205, USA; Nathan.Koonce@fda.hhs.gov
2   National Center for Toxicological Research, Food and Drug Administration, Jefferson, AR 72079, USA
*   Correspondence: rjgriffin@uams.edu (R.J.G.); rpmdings@uams.edu (R.P.M.D.);
     Tel.: +1-501-526-7873 (R.J.G.); +1-501-526-7876 (R.P.M.D.); Fax: +1-501-526-5934 (R.J.G. & R.P.M.D.)

Received: 6 November 2017; Accepted: 5 December 2017; Published: 9 December 2017

Abstract: Galectin-1 is a hypoxia-regulated protein and a prognostic marker in head and neck squamous cell carcinomas (HNSCC). Here we assessed the ability of non-peptidic galectin-1 inhibitor OTX008 to improve tumor oxygenation levels via tumor vessel normalization as well as tumor growth inhibition in two human HNSCC tumor models, the human laryngeal squamous carcinoma SQ20B and the human epithelial type 2 HEp-2. Tumor-bearing mice were treated with OTX008, Anginex, or Avastin and oxygen levels were determined by fiber-optics and molecular marker pimonidazole binding. Immuno-fluorescence was used to determine vessel normalization status. Continued OTX008 treatment caused a transient reoxygenation in SQ20B tumors peaking on day 14, while a steady increase in tumor oxygenation was observed over 21 days in the HEp-2 model. A >50% decrease in immunohistochemical staining for tumor hypoxia verified the oxygenation data measured using a partial pressure of oxygen (pO_2) probe. Additionally, OTX008 induced tumor vessel normalization as tumor pericyte coverage increased by approximately 40% without inducing any toxicity. Moreover, OTX008 inhibited tumor growth as effectively as Anginex and Avastin, except in the HEp-2 model where Avastin was found to suspend tumor growth. Galectin-1 inhibitor OTX008 transiently increased overall tumor oxygenation via vessel normalization to various degrees in both HNSCC models. These findings suggest that targeting galectin-1—e.g., by OTX008—may be an effective approach to treat cancer patients as stand-alone therapy or in combination with other standards of care.

Keywords: galectin-1 inhibitor; OTX008; Avastin; Anginex; hypoxia; pimonidazole; vessel normalization

1. Introduction

Agents that target the tumor microenvironment and inhibit angiogenesis have shown promise as therapeutics in solid tumors. Although anti–vascular endothelial growth factor (VEGF) agents like bevacizumab (Avastin™, a humanized monoclonal antibody against VEGF; Genentech (South San Francisco, CA, USA)) are perhaps most discussed [1], many other compounds have been identified and are currently in various phases of clinical cancer trials [2,3]. Nevertheless, some angiogenesis inhibitors and vascular disruptors are ineffective or cause unwanted biological side effects [4], which underscores the need for more diverse compounds and/or treatment strategies. Although it is critical to recognize that anti-angiogenesis treatment can take the form of more than growth factor pathway inhibition alone, few other agents have made the clinical impact of Avastin.

Galectins are a family of carbohydrate-binding lectins that share a conserved carbohydrate recognition domain by binding to β-galactoside-containing glycoconjugates. Galectins promote cell-cell and cell-matrix interactions during cancer development and progression. Our recent results have validated (galectin-1) gal-1 as a viable cancer target with broad potential [5,6], as differential stromal elevation of gal-1 over the tumor parenchyma has been reported in several cancers, including those of head and neck, ovary, breast, brain, colon, skin, and prostate [7–11].

To date, most known galectin antagonists are glycomimetics and are derivatives or analogs of β-galactoside, targeting the canonical carbohydrate-binding site of galectins. These include aryl *O*- and *S*-galactosides and lactosides [12,13], carbohydrate-based triazoles and isoxazoles [14], *O*-galactosyl aldoximes [15], phenyl thio-β-D-galactopyranoside analogs [16], thioureido *N*-acetyllactosamine derivatives [17], talosides [18], and various multivalent sugar-based compounds [19]. We have previously reported on the design of Anginex, a 33mer amphipathic peptide that targets gal-1 [5,20]. While antibodies and peptides can be effective, small molecules tend to be advantageous in regards to bioavailability, immunogenic activation, degradation, and ability to scale up. We employed the calix[4]arene scaffold for the development of a peptidomimetic of Anginex because it approximated the molecular dimensions and surface topology of key spatially-related amino acid residues in Anginex [21]. We identified OTX008 to have more promising biological properties than Anginex [22] and discovered that OTX008 is an allosteric inhibitor of gal-1 as it binds at a different site than the canonical carbohydrate binding site [23].

Previous studies have identified gal-1 as a negative prognosticator in oral squamous cell carcinoma (OSCC) and human head and neck squamous cell carcinoma (HNSSC) [24,25]. Here, we investigated whether OTX008 can normalize tumor vasculature resulting in improved tumor oxygenation and subsequent tumor growth inhibition as compared to Anginex and Avastin in two HNSSC models, namely SQ20B and HEp-2. Whereas SQ20B has mutated *p53* and is resistant to radiation but sensitive to OTX008, HEp-2 is an HNSSC epithelial cell line derived from human epidermoid carcinoma of the larynx, which is resistant to both standard radiochemotherapy regiments as well as OTX008 [26].

2. Results and Discussion

2.1. OTX008 Increases Overall Tumor Oxygenation

Repeated daily treatment with OTX008 resulted in an increased overall tumor oxygenation (pO_2) up to day 14 of treatment (compared with vehicle-treated mice) in the SQ20B tumor model (Figure 1A). After day 14, tumor oxygenation decreased to levels in line with those measured in vehicle-treated mice. In contrast, in the HEp-2 tumor mouse model overall tumor oxygenation levels were steadily elevated throughout OTX008 treatment as compared to vehicle (control) treated mice (Figure 1B). In both models, this improvement in tumor oxygenation was comparable with the average increases induced by either Anginex or Avastin (Figure 1). Individual tumor oxygenation values for Anginex and Avastin in time for either model can be found in Figures S1 and S2. Overall, the results in the SQ20B and the HEp-2 models reiterated the theory that transient or sustained vessel normalization is not solely a VEGF-dependent phenomenon and moreover is seen in multiple solid tumor models including HNSSC. However, it is also interesting to note that the "tumor oxygenation window" did not occur in all tumor models, as the HEp-2 model displayed improved oxygenation throughout the study. Previously, other tumor mouse models and human clinical data have shown that this window occurs earlier, around day 3 through 5 after commencement of treatment [11,27]. The tumors in these previous studies were in general larger at the beginning of the treatment, which may have accounted for the more rapid change in oxygenation levels. The measurement of tumor oxygenation with various semi-invasive approaches may also influence the results obtained. Specifically, in the current study we sampled partial oxygenation pressure with an inserted fiber optic probe over several weeks at 1 to 3 points in each tumor were viable baseline oxygenation levels were found, thus intentionally avoiding necrotic or anoxic regions. Other invasive probes collect 50–100 point measurements and thus may be more influenced by the amount of necrotic volume being sampled [11,27]. It is also likely that this window varies due to origin, location, or growth rate of the particular cancer.

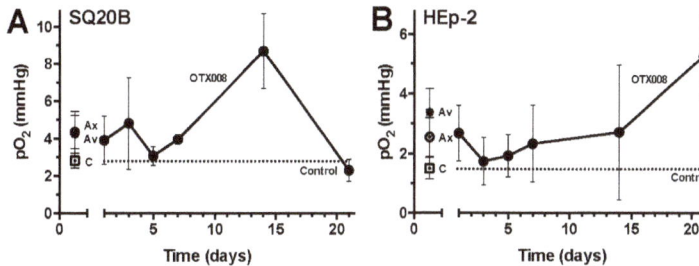

Figure 1. The effect of OTX008 on global tumor pO_2 over time. Global tumor pO_2 is transiently increased by OTX008 in SQ20B (**A**) and HEP2 (**B**) tumors as measured by fiber-optic oxygen sensor. Treatments: Control (C; -□-), OTX008 (-●-), Anginex (Ax; -○-), and Avastin (Av; -✱-). Points, average mean (± SEM) pO_2 value derived from the stabilized reading over a 60 s period at one location (four to seven mice per day per experimental group). The dotted line represents the average pO_2 mean of the control.

2.2. OTX008 Inhibits Tumor Growth without Apparent Toxicity

In a subsequent set of experiments, mice inoculated with SQ20B or HEp-2 tumor cells were randomized and treated with OTX008 (qd, 10 mg/kg, i.p. *n* = 7), Anginex (qd, 10 mg/kg, i.p. *n* = 7) or Avastin (once every 14 days, 10 mg/kg, i.v. *n* = 7) starting when tumors reached the size of ~ 100 mm³. In the SQ20B model OTX008 reduced tumor growth on average 25%, up to 35% on certain days (Figure 2). This inhibition was comparable to the effects of Avastin in this model and slightly better than Anginex (Figure 2). In the HEp-2 tumor model this was less pronounced, whereas Avastin caused tumor growth stabilization (Figure 3). As shown by Astorgues-Xerri et al., HEp-2 is inherently resistant to radiochemotherapy including OTX008 [26], which could explain the relatively limited HEp-2 tumor growth inhibition by OTX008. Additionally, in the current study OTX008 and Anginex was administered i.p. daily. As compared to previous reports, the data suggests that administering OTX008 or Anginex bis in die (BID) i.p. or subcutaneous (s.q.) continuously by osmotic mini-pump will result in greater tumor growth inhibition [20,22,28,29].

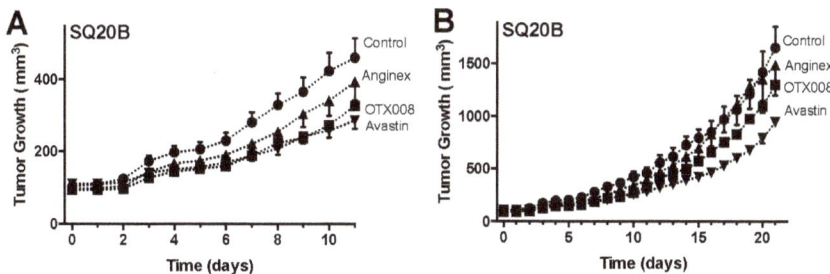

Figure 2. OTX008 inhibits SQ20B tumor growth in mice. (**A**) SQ20B tumor growth during the first 11 days and treatment with Control (vehicle), Anginex, OTX008 and Avastin; (**B**) SQ20B tumor growth during the whole treatment span with Control (vehicle), Anginex, OTX008 and Avastin. Treatments: Control (vehicle; -●-), Anginex (i.p. 10 mg/kg, q1d×21; *n* = 7; -▲-), OTX008 (i.p. 10 mg/kg, q1d×21; *n* = 7; -■-), and Avastin (i.v. 10 mg/kg, q14d×2; *n* = 7; -▼-). Data points represent means ± SEM (*n* = 7 each group).

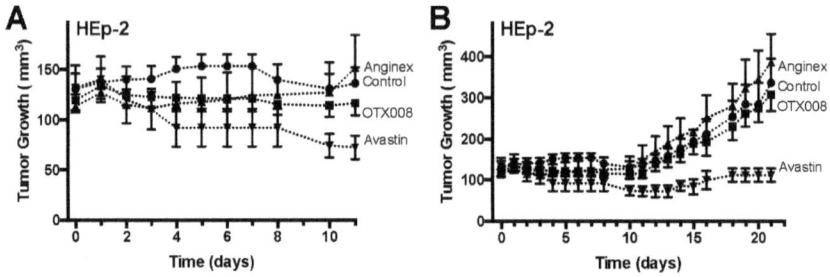

Figure 3. OTX008 inhibits HEp-2 tumor growth in mice. (**A**) HEp-2 tumor growth during the first 11 days and treatment with Control (vehicle), Anginex, OTX008 and Avastin; (**B**) HEp-2 tumor growth during the whole treatment span with Control (vehicle), Anginex, OTX008 and Avastin. Treatments: Control (vehicle; -●-), Anginex (i.p. 10 mg/kg, q1d×21; $n = 7$; -▲-), OTX008 (i.p. 10 mg/kg, q1d×21; $n = 7$; -■-), and Avastin (i.v. 10 mg/kg, q14d×2; $n = 7$; -▼-). Data points represent means ± SEM ($n = 7$ each group).

In both tumor models studied, prolonged treatment with OTX008, Anginex, or Avastin revealed no signs of toxicity as assessed by unaltered behavior and normal weight gain during experiments (Figure 4). Upon autopsy, macro- and microscopic morphologies of internal organs were also observed to be normal within all experimental groups of animals.

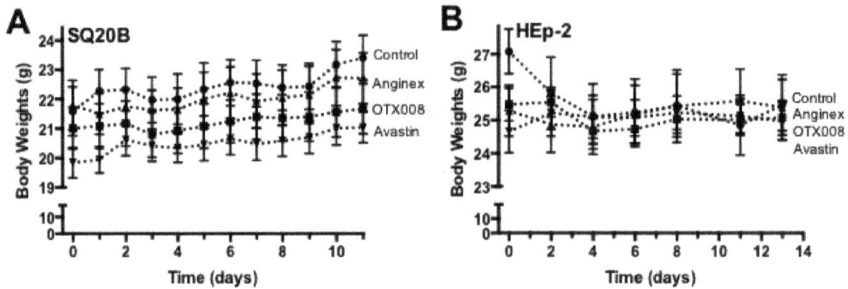

Figure 4. OTX008 does not show any sign of apparent toxicity. (**A**) In the SQ20B OTX008, Anginex and Avastin do not show any signs of apparent toxicity as measured by body weight fluctuation during treatment; (**B**) In the HEp-2 OTX008, Anginex and Avastin do not show any signs of apparent toxicity as measured by body weight fluctuation during treatment. Treatments: Control (vehicle; -●-), Anginex (i.p. 10 mg/kg, q1d×21; $n = 7$; -▲-), OTX008 (i.p. 10 mg/kg, q1d×21; $n = 7$; -■-), and Avastin (i.v. 10 mg/kg, q14d×2; $n = 7$; -▼-). Data points represent means ± SEM ($n = 7$ each group).

2.3. OTX008 Reduces Tumor Hypoxia and Improves Pericyte Coverage

OTX008 treatment resulted in a decrease of SQ20B tumor hypoxia, as measured by pimonidazole. Pimonidazole, a substituted 2-nitroimidazole, is preferentially reduced in hypoxic viable cells and forms irreversible protein adducts, which have been optimized for detection with immunohistochemistry. On average, OTX008 reduced hypoxia by ~70% in SQ20B tumors (Figure 5). These immunofluorescence results are in line with the improved oxygenation levels as determined by direct measurement of the partial pressure of oxygen (Figure 1). Besides the improvement of tumor oxygenation, vessel normalization was correlated with an improvement of pericyte coverage. As shown and quantified in Figure 5, OTX008 increased the amount of pericyte-covered vessels by ~40% in SQ20B tumors.

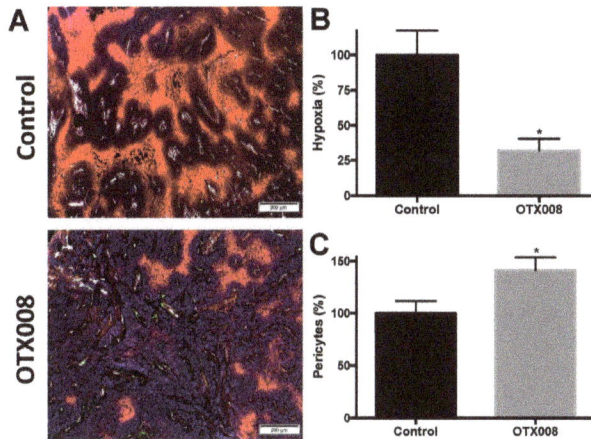

Figure 5. Hypoxia and pericyte content in SQ20B tumor tissue after OTX008 treatment. (**A**) Immunofluorescence analysis of SQ20B tumors cross-sections from OTX008 or vehicle treated animals at 14 days after the beginning of OTX008 therapy. Microvessel density is revealed by CD31 antibody staining (white), pericytes by alpha-smooth muscle actin (αSMA; green) and hypoxia by pimonidazole (red). Magnifications as indicated; (**B**) Quantification of hypoxia by pimonidazole positive regions; (**C**) Quantification of pericytes by α-SMA positive cells. Data is depicted as means $\pm$ SEM and normalized to the controls, set at 100%; * $p < 0.05$ (Student's *t*-test).

Overall, our results show that OTX008 can transiently increase overall tumor oxygenation via vessel normalization with varying kinetics and magnitude/duration of effects as tumors progress. Since Avastin, in combination with chemotherapy is already an Food and Drug Administration approved protocol for treatment of several tumor types [30], a gal-1 therapeutic such as Anginex or OTX008 could possibly avoid the noted systemic effects that Avastin usage promotes [2–4]. Moreover, gal-1 inhibitors may achieve a more permissive tumor microenvironment status for improved radiation or chemotherapeutic treatment.

Additionally, non-peptidic topomimetic OTX008 might be advantageous over a biologic such as Avastin based on the drug properties. Whereas biologics, antibodies included, are relatively large proteins (up to 150 kDa) they differ from small molecules in their cost, production, administration, and clinical efficacy. Larger proteins and antibodies are predominantly more difficult to produce and characterize as protein folding and tertiary structure are essential for their efficacy, resulting in increased costs. Due to these inherent challenges, specificity, scale up, purity, and batch consistency can also be more demanding [31]. Moreover biologics, Avastin included, are generally administered intravenously as they are likely to be digested when taken orally [32]. Small molecules or synthetics chemicals on the other hand are often orally available and absorbed into the bloodstream via the intestinal wall due to their small size and chemical composition. Additionally, larger proteins are assumed more likely to be immunogenic and less efficient in tissue penetration, tumor retention, and blood clearance than small molecules [32].

Subsequently, it would be pertinent to elucidate whether the increase in oxygenation caused by OTX008 enhances the effect of radiation therapy in vivo, especially if radiation at a large dose per fraction could be applied at the peak of tumor oxygenation observed. Furthermore, there may be opportunities to enhance the effects of various chemotherapeutic agents due to the improved physiological state of tumors during OTX008 treatment. By administering chemotherapy during the OTX008 oxygenation window, we would expect to maximize drug penetration/diffusion and overall coverage of the tumor. Recent preclinical and clinical reports on gal-1 playing a principal role in tumor progression, therapeutic resistance, and metastatic potential of multiple tumor types [33–35] highlight the translational potential of OTX008 and other gal-1 inhibitors in development.

3. Materials and Methods

3.1. Test Article and Control Reagents

OTX008 (also known as PTX008 or 0118) was reconstituted in H_2O at a stock concentration of 1 mg/ml and stored at −80 °C [22]. Anginex was manufactured and obtained from BioMedical Genomics Center at the University of Minnesota (Minneapolis, MN, USA) and formulated identical to OTX008 and stored at 2–8 °C, as before [36]. OTX008 and Anginex were given daily for 21 days (q1d×21) in alternating intraperitoneal (i.p.) locations at 10 mg/kg.

Avastin™ (bevacizumab, a humanized monoclonal antibody against VEGF) was obtained from the University of Arkansas for Medical Sciences (UAMS) pharmacy, manufactured by Genentech). The antibody is provided at a stock concentration of 25 mg/mL and stored at 2–8 °C. Avastin was given once every 14 days (q14d×2) intravenously (i.v.) at 10 mg/kg.

3.2. Tumor Models

Experiments were approved by the University of Arkansas for Medical Sciences Institutional Animal Care and Use Committee (Animal Usage Protocol #3157; 05 17 2013) and performed in accordance with relevant regulations and guidelines. For the xenograft models, female athymic nude mice (*Nu/Nu*, 5–6 weeks old) were inoculated into the right flank s.c. with tumor cells and when tumors reached a size of ~100 mm^3, animals were randomly assigned to treatment groups and administration of study drug was commenced. In total, 92 animals were inoculated and a total of 76 tumor-bearing animals were selected for randomization before the initiation of the intervention.

SQ20B—radio-resistant HNSCC SQ20B carries a *p53* mutation and has a constitutively active mutation in epidermal growth factor receptor (EGFR) resulting in robust Akt (protein kinase B) signaling [26,37–39]. It was shown previously that this cell line is sensitive to OTX008 in different in vitro and in vivo experiments [40].

HEp-2—an epithelial cell line derived from human epidermoid carcinoma of the larynx. This cell line is more resistant to current standard chemotherapeutic agents and it is more resistant in vitro to OTX008 as compared to HNSCC SQ20B [40].

3.3. Tumor Growth

Subcutaneous tumors were measured with a caliper prior to starting treatment (prior to randomization into treatment groups) and then once a day during treatment. Tumor volumes were calculated based on the equation $(a^2 \times b)/2$ where a is the shortest and b is the longest diameter of the tumor (width or length) [39].

3.4. Tumor Oxygenation Studies

Tumor partial pressure of oxygen (pO_2) was measured using a fiber-optic oxygen sensor (Oxylite™, Oxford Optronix, Oxford, UK) [41]. The probe was inserted via a needle track to approximately the center of the three dimensional tumor mass and readings were recorded for 3 min. Final values were based on the stabilized value over a 60 s period at one location. Once tumors grew to a larger size, the probe was moved to one or two more locations and readings recorded over 3 min at each position. The readings were based on four to seven mice per day per experimental group.

3.5. Tumor Vessel Density and Pericyte Staining

Extracted tumors were either fixed in 10% formalin or embedded in tissue freezing medium and snap frozen in liquid nitrogen. Formalin-fixed tumors embedded in paraffin or frozen tumors were cut into 5 µm sections. After rehydration and antigen retrieval, the slides were stained for either vessel density (CD31 from BD Pharmingen (San Jose, CA, USA); 2nd Ab αRat-647) or pericytes (anti-alpha-smooth muscle actin, αSMA-FITC from Sigma-Aldrich (St. Louis, MO, USA)) [22].

3.6. Tumor Hypoxia

An i.p. injection of 60 mg/kg pimonidazole was given to each mouse prior to euthanasia. Pimonidazole, a substituted 2-nitroimidazole, is preferentially reduced in hypoxic viable cells and forms irreversible protein adducts, which have been optimized for detection with immunohistochemistry [42]. At 1 h post-injection, the mice were euthanized, then the tumor dissected and immediately either fixed in 10% formalin or snap frozen. Following sectioning of tumor tissue, a FITC-conjugated monoclonal antibody against protein adducts of pimonidazole was added (hypoxyprobe-1 MAb; Chemicon International, Burlington, MA, USA); 2nd Ab αRabbit-594). Images of the section were acquired and subsequently digitally and differentially quantified by morphometric analysis [11].

3.7. Body Weight

As an indirect measure of general toxicity, body weights were measured prior to starting treatment and once daily during treatment. At the end of the study by autopsy, macro- and microscopic morphologies of internal organs were inspected and recorded in all experimental groups of animals [22].

Supplementary Materials: Supplementary materials can be found at www.mdpi.com/1422-0067/18/12/2671/s1.

Acknowledgments: This work was supported by a grant from the Arkansas Biosciences Institute and the Winthrop P. Rockefeller Cancer Institute (RPMD). The study was also supported in part by the Center for Microbial Pathogenesis and Host Inflammatory Responses grant P20GM103625 through the NIH National Institute of General Medical Sciences Centers of Biomedical Research Excellence (RPMD). The study was also supported in part by OncoEthix (Lausanne, Vaud, Switzerland). The content is solely the responsibility of the authors and does not necessarily represent the official views of the NIH or the Winthrop P. Rockefeller Cancer Institute.

Author Contributions: Nathan A. Koonce, Robert J. Griffin and Ruud P. M. Dings designed the work, analyzed and interpreted the data, drafted the article and critically revised the article; Nathan A. Koonce collected the data.

Conflicts of Interest: The authors declare no conflicts of interest.

References

1. Ferrara, N.; Davis-Smyth, T. The biology of vascular endothelial growth factor. *Endocr. Rev.* **1997**, *18*, 4–25.
2. Horsman, M.R.; Siemann, D.W. Pathophysiologic effects of vascular-targeting agents and the implications for combination with conventional therapies. *Cancer Res.* **2006**, *66*, 11520–11539. [CrossRef]
3. Potente, M.; Gerhardt, H.; Carmeliet, P. Basic and therapeutic aspects of angiogenesis. *Cell* **2011**, *146*, 873–887. [PubMed]
4. Lordick, F.; Geinitz, H.; Theisen, J.; Sendler, A.; Sarbia, M. Increased risk of ischemic bowel complications during treatment with bevacizumab after pelvic irradiation: Report of three cases. *Int. J. Radiat. Oncol. Biol. Phys.* **2006**, *64*, 1295–1298. [CrossRef]
5. Thijssen, V.L.; Postel, R.; Brandwijk, R.J.; Dings, R.P.; Nesmelova, I.; Satijn, S.; Verhofstad, N.; Nakabeppu, Y.; Baum, L.G.; Bakkers, J.; et al. Galectin-1 is essential in tumor angiogenesis and is a target for antiangiogenesis therapy. *Proc. Natl. Acad. Sci. USA* **2006**, *103*, 15975–15980. [CrossRef] [PubMed]
6. Thijssen, V.L.; Barkan, B.; Shoji, H.; Aries, I.M.; Mathieu, V.; Deltour, L.; Hackeng, T.M.; Kiss, R.; Kloog, Y.; Poirier, F.; et al. Tumor cells secrete galectin-1 to enhance endothelial cell activity. *Cancer Res.* **2010**, *70*, 6216–6224. [PubMed]
7. Van den Brule, F.; Califice, S.; Castronovo, V. Expression of galectins in cancer: A critical review. *Glycoconj. J.* **2004**, *19*, 537–542.
8. Camby, I.; Le Mercier, M.; Lefranc, F.; Kiss, R. Galectin-1: A small protein with major functions. *Glycobiology* **2006**, *16*, 137R–157R. [PubMed]
9. Liu, F.T.; Rabinovich, G.A. Galectins as modulators of tumour progression. *Nat. Rev. Cancer* **2005**, *5*, 29–41. [CrossRef] [PubMed]

10. Lefranc, F.; Mathieu, V.; Kiss, R. Galectin-1-mediated biochemical controls of melanoma and glioma aggressive behavior. *World J. Biol. Chem.* **2011**, *2*, 193–201. [PubMed]

11. Dings, R.P.; Loren, M.; Heun, H.; McNiel, E.; Griffioen, A.W.; Mayo, K.H.; Griffin, R.J. Scheduling of radiation with angiogenesis inhibitors anginex and avastin improves therapeutic outcome via vessel normalization. *Clin. Cancer Res.* **2007**, *13*, 3395–3402. [CrossRef] [PubMed]

12. Sirois, S.; Giguere, D.; Roy, R. A first QSAR model for galectin-3 glycomimetic inhibitors based on 3D docked structures. *Med. Chem.* **2006**, *2*, 481–489. [CrossRef] [PubMed]

13. St-Pierre, C.; Ouellet, M.; Giguere, D.; Ohtake, R.; Roy, R.; Sato, S.; Tremblay, M.J. Galectin-1-specific inhibitors as a new class of compounds to treat hiv-1 infection. *Antimicrob. Agents Chemother.* **2012**, *56*, 154–162. [CrossRef] [PubMed]

14. Salameh, B.A.; Leffler, H.; Nilsson, U.J. 3-(1,2,3-triazol-1-yl)-1-thio-galactosides as small, efficient, and hydrolytically stable inhibitors of galectin-3. *Bioorg. Med. Chem. Lett.* **2005**, *15*, 3344–3346. [CrossRef] [PubMed]

15. Tejler, J.; Leffler, H.; Nilsson, U.J. Synthesis of O-galactosyl aldoximes as potent LacNAc-mimetic galectin-3 inhibitors. *Bioorg. Med. Chem. Lett.* **2005**, *15*, 2343–2345. [CrossRef] [PubMed]

16. Cumpstey, I.; Sundin, A.; Leffler, H.; Nilsson, U.J. C2-symmetrical thiodigalactoside bis-benzamido derivatives as high-affinity inhibitors of galectin-3: Efficient lectin inhibition through double arginine-arene interactions. *Angew. Chem. Int. Ed. Engl.* **2005**, *44*, 5110–5112. [CrossRef] [PubMed]

17. Salameh, B.A.; Sundin, A.; Leffler, H.; Nilsson, U.J. Thioureido N-acetyllactosamine derivatives as potent galectin-7 and 9N inhibitors. *Bioorg. Med. Chem.* **2006**, *14*, 1215–1220. [CrossRef] [PubMed]

18. Collins, P.M.; Oberg, C.T.; Leffler, H.; Nilsson, U.J.; Blanchard, H. Taloside inhibitors of galectin-1 and galectin-3. *Chem. Biol. Drug Des.* **2012**, *79*, 339–346. [PubMed]

19. Rabinovich, G.A.; Cumashi, A.; Bianco, G.A.; Ciavardelli, D.; Iurisci, I.; D'Egidio, M.; Piccolo, E.; Tinari, N.; Nifantiev, N.; Iacobelli, S. Synthetic lactulose amines: Novel class of anticancer agents that induce tumor-cell apoptosis and inhibit galectin-mediated homotypic cell aggregation and endothelial cell morphogenesis. *Glycobiology* **2006**, *16*, 210–220. [CrossRef] [PubMed]

20. Dings, R.P.; van der Schaft, D.W.; Hargittai, B.; Haseman, J.; Griffioen, A.W.; Mayo, K.H. Anti-tumor activity of the novel angiogenesis inhibitor anginex. *Cancer Lett.* **2003**, *194*, 55–66. [CrossRef]

21. Dings, R.P.; Arroyo, M.M.; Lockwood, N.A.; Van Eijk, L.I.; Haseman, J.R.; Griffioen, A.W.; Mayo, K.H. Beta-sheet is the bioactive conformation of the anti-angiogenic anginex peptide. *Biochem. J.* **2003**, *23*, 281–288. [CrossRef] [PubMed]

22. Dings, R.P.; Chen, X.; Hellebrekers, D.M.; van Eijk, L.I.; Zhang, Y.; Hoye, T.R.; Griffioen, A.W.; Mayo, K.H. Design of nonpeptidic topomimetics of antiangiogenic proteins with antitumor activities. *J. Natl. Cancer Inst.* **2006**, *98*, 932–936. [CrossRef] [PubMed]

23. Dings, R.P.; Miller, M.C.; Nesmelova, I.; Astorgues-Xerri, L.; Kumar, N.; Serova, M.; Chen, X.; Raymond, E.; Hoye, T.R.; Mayo, K.H. Antitumor agent calixarene 0118 targets human galectin-1 as an allosteric inhibitor of carbohydrate binding. *J. Med. Chem.* **2012**, *55*, 5121–5129. [PubMed]

24. Valach, J.; Fik, Z.; Strnad, H.; Chovanec, M.; Plzak, J.; Cada, Z.; Szabo, P.; Sachova, J.; Hroudova, M.; Urbanova, M.; et al. Smooth muscle actin-expressing stromal fibroblasts in head and neck squamous cell carcinoma: Increased expression of galectin-1 and induction of poor prognosis factors. *Int. J. Cancer* **2012**, *131*, 2499–2508. [PubMed]

25. Noda, Y.; Kondo, Y.; Sakai, M.; Sato, S.; Kishino, M. Galectin-1 is a useful marker for detecting neoplastic squamous cells in oral cytology smears. *Hum. Pathol.* **2016**, *52*, 101–109. [CrossRef] [PubMed]

26. Astorgues-Xerri, L.; Serova, M.; Bieche, I.; Vivier, S.; Cvitkovic, E.; Noel, K.; Raymond, E.; Faivre, S. Effects of PTX-008, a non-peptidic topomimetic targeting galectin-1 in human epithelial cancer cells. *Mol. Cancer Ther.* **2009**, *8*. [CrossRef]

27. Willett, C.G.; Boucher, Y.; di Tomaso, E.; Duda, D.G.; Munn, L.L.; Tong, R.T.; Chung, D.C.; Sahani, D.V.; Kalva, S.P.; Kozin, S.V.; et al. Direct evidence that the VEGF-specific antibody bevacizumab has antivascular effects in human rectal cancer. *Nat. Med.* **2004**, *10*, 145–147. [PubMed]

28. Dings, R.P.; Vang, K.B.; Castermans, K.; Popescu, F.; Zhang, Y.; Oude Egbrink, M.G.; Mescher, M.F.; Farrar, M.A.; Griffioen, A.W.; Mayo, K.H. Enhancement of T-cell-mediated antitumor response: Angiostatic adjuvant to immunotherapy against cancer. *Clin. Cancer Res.* **2011**, *17*, 3134–3145. [PubMed]

29. Dings, R.P.; Mayo, K.H. A journey in structure-based drug discovery: From designed peptides to protein surface topomimetics as antibiotic and antiangiogenic agents. *Acc. Chem. Res.* **2007**, *40*, 1057–1065. [CrossRef] [PubMed]

30. Hudis, C.A. Clinical implications of antiangiogenic therapies. *Oncology* **2005**, *19*, 26–31. [PubMed]

31. Bordeaux, J.; Welsh, A.; Agarwal, S.; Killiam, E.; Baquero, M.; Hanna, J.; Anagnostou, V.; Rimm, D. Antibody validation. *BioTechniques* **2010**, *48*, 197–209. [CrossRef] [PubMed]

32. Imai, K.; Takaoka, A. Comparing antibody and small-molecule therapies for cancer. *Nat. Rev. Cancer* **2006**, *6*, 714–727. [PubMed]

33. Zhang, P.F.; Li, K.S.; Shen, Y.H.; Gao, P.T.; Dong, Z.R.; Cai, J.B.; Zhang, C.; Huang, X.Y.; Tian, M.X.; Hu, Z.Q.; et al. Galectin-1 induces hepatocellular carcinoma emt and sorafenib resistance by activating FAK/PI3K/AKT signaling. *Cell Death Dis.* **2016**, *7*, e2201. [CrossRef] [PubMed]

34. Shih, T.-C.; Liu, R.; Fung, G.; Bhardwaj, G.; Ghosh, P.M.; Lam, K.S. A novel galectin-1 inhibitor discovered through one-bead two-compound library potentiates the antitumor effects of paclitaxel in vivo. *Mol. Cancer Ther.* **2017**, *16*, 1212–1223. [CrossRef] [PubMed]

35. Martínez-Bosch, N.; Fernández-Barrena, M.G.; Moreno, M.; Ortiz-Zapater, E.; Munné-Collado, J.; Iglesias, M.; André, S.; Gabius, H.-J.; Hwang, R.F.; Poirier, F.; et al. Galectin-1 drives pancreatic carcinogenesis through stroma remodeling and hedgehog signaling activation. *Cancer Res.* **2014**, *74*, 3512–3524. [PubMed]

36. Dings, R.P.; Yokoyama, Y.; Ramakrishnan, S.; Griffioen, A.W.; Mayo, K.H. The designed angiostatic peptide anginex synergistically improves chemotherapy and antiangiogenesis therapy with angiostatin. *Cancer Res.* **2003**, *63*, 382–385. [PubMed]

37. Astorgues-Xerri, L.; De Groot, D.; Serova, M.; Huet, E.; Menashi, S.; Riveiro, M.E.; Cvitkovic, E.; Noel, K.; Raymond, E.; Faivre, S. Targeting galectin-1 with PTX-008 inhibits invasion, tumor growth and metastasis in human carcinoma models. *Cancer Res.* **2011**, *71*. [CrossRef]

38. Astorgues-Xerri, L.; Tijeras-Raballand, A.; Riveiro, M.E.; Serova, M.; Martinet, M.; Huet, E.; Menashi, S.; Herait, P.; Cvitkovic, E.; Noel, K.; et al. Antitumor and antiangiogenic effects of OTX-008, a novel calyx[4]arene, are mediated by galectin-1 inhibition in human tumor xenograft models. *Cancer Res.* **2012**, *72*. [CrossRef]

39. Dings, R.P.; Levine, J.I.; Brown, S.G.; Astorgues-Xerri, L.; MacDonald, J.R.; Hoye, T.R.; Raymond, E.; Mayo, K.H. Polycationic calixarene PTX013, a potent cytotoxic agent against tumors and drug resistant cancer. *Investig. New Drugs* **2013**, *31*, 1142–1150. [CrossRef] [PubMed]

40. Astorgues-Xerri, L.; Riveiro, M.E.; Tijeras-Raballand, A.; Serova, M.; Rabinovich, G.A.; Bieche, I.; Vidaud, M.; de Gramont, A.; Martinet, M.; Cvitkovic, E.; et al. OTX008, a selective small-molecule inhibitor of galectin-1, downregulates cancer cell proliferation, invasion and tumour angiogenesis. *Eur. J. Cancer* **2014**, *50*, 2463–2477. [PubMed]

41. Pagan, J.; Przybyla, B.; Jamshidi-Parsian, A.; Gupta, K.; Griffin, R.J. Blood outgrowth endothelial cells increase tumor growth rates and modify tumor physiology: Relevance for therapeutic targeting. *Cancers* **2013**, *5*, 205–217. [CrossRef] [PubMed]

42. Koonce, N.A.; Levy, J.; Hardee, M.E.; Jamshidi-Parsian, A.; Vang, K.B.; Sharma, S.; Raleigh, J.A.; Dings, R.P.; Griffin, R.J. Targeting artificial tumor stromal targets for molecular imaging of tumor vascular hypoxia. *PLoS ONE* **2015**, *10*, e0135607.

International Journal of
Molecular Sciences

MDPI

Review

Galectin-3: One Molecule for an Alphabet of Diseases, from A to Z

Salvatore Sciacchitano [1,2,*], Luca Lavra [2], Alessandra Morgante [2], Alessandra Ulivieri [2], Fiorenza Magi [2], Gian Paolo De Francesco [3], Carlo Bellotti [4], Leila B. Salehi [2,5] and Alberto Ricci [1]

[1] Department of Clinical and Molecular Medicine, Sapienza University, Policlinico Umberto I, Viale Regina Elena 324, 00161 Rome, Italy; alberto.ricci@uniroma1.it
[2] Laboratory of Biomedical Research, Niccolò Cusano University Foundation, Via Don Carlo Gnocchi 3, 00166 Rome, Italy; luca.lavra@fondazioneniccolocusano.it (L.L.); alessandra.morgante@fondazioneniccolocusano.it (A.M.); alessandra.ulivieri@fondazioneniccolocusano.it (A.U.); fiorenza.magi@fondazioneniccolocusano.it (F.M.); leila.b.salehi@fondazioneniccolocusano.it (L.B.S.)
[3] Department of Oncological Science, Breast Unit, St Andrea University Hospital, Via di Grottarossa, 1035/39, 00189 Rome, Italy; Gianpaolo.defrancesco@gmail.com
[4] Operative Unit Surgery of Thyroid and Parathyroid, Sapienza University of Rome, S. Andrea Hospital, Via di Grottarossa, 1035/39, 00189 Rome, Italy; carlo_bellotti@hotmail.com
[5] Department of Biopathology and Diagnostic Imaging, Tor Vergata University, Via Montpellier 1, 00133 Rome, Italy
[*] Correspondence: salvatore.sciacchitano@fondazioneniccolocusano.it; Tel.: +39-06-4567-8373

Received: 31 December 2017; Accepted: 22 January 2018; Published: 26 January 2018

Abstract: Galectin-3 (Gal-3) regulates basic cellular functions such as cell–cell and cell–matrix interactions, growth, proliferation, differentiation, and inflammation. It is not surprising, therefore, that this protein is involved in the pathogenesis of many relevant human diseases, including cancer, fibrosis, chronic inflammation and scarring affecting many different tissues. The papers published in the literature have progressively increased in number during the last decades, testifying the great interest given to this protein by numerous researchers involved in many different clinical contexts. Considering the crucial role exerted by Gal-3 in many different clinical conditions, Gal-3 is emerging as a new diagnostic, prognostic biomarker and as a new promising therapeutic target. The current review aims to extensively examine the studies published so far on the role of Gal-3 in all the clinical conditions and diseases, listed in alphabetical order, where it was analyzed.

Keywords: Galectin-3; Gal-3

---

1. Introduction

Galectins were initially isolated as β-galactoside-binding proteins. They represent a family of widely expressed lectins in metazoans. In mammalians a total of 15 galectins have been identified so far, containing either one or two ~130 amino-acid-long conserved carbohydrate recognition domains (CRD)s [1–3]. Galectins are synthesized in the cytoplasm and interact with cell surface glycans following their secretion by a non-classical exocytic pathway (i.e., not via the ER/Golgi secretory route), that is likely to be an exosome-mediated secretory route [4,5]. Their ability to crosslink glycosylated ligands allows them to form a dynamic lattice. The galectin lattice regulates many different functions such as the diffusion, compartmentalization and endocytosis of plasma membrane glycoproteins and glycolipids, the selection, activation and arrest of T cells, receptor kinase signaling and the functionality of membrane receptors, (including the glucagon receptor, glucose and amino acid transporters, cadherins and integrins) [6].

Although the galectins are cytosolic proteins, there is abundant evidence for their secretion from the cytosol via nonclassical pathways or translocation to the nucleus or the other cellular compartments [4,7]. The galectin family can be subdivided into three different groups:

1. Prototypic single-CRD galectins that can form non-covalent homodimers. The following galectins belong to this group: Gal-1, Gal-2, Gal-5, Gal-7, Gal-10, Gal-11, Gal-13, Gal-14 and Gal-15.
2. Tandem-repeats of two CRD motifs, similar but not identical, including Gal-4, Gal-6, Gal-8, Gal-9 and Gal-12.
3. The chimera-type. The only member of this group is represented by Gal-3, with a single CRD and an intrinsically disordered sequence at the N-terminal domain that promotes oligomerization. In addition, Gal-3 is the only one to be able to pentamerize. It exhibits specific pleiotropic biological function, playing a key role in many physiological and pathological processes (Figure 1).

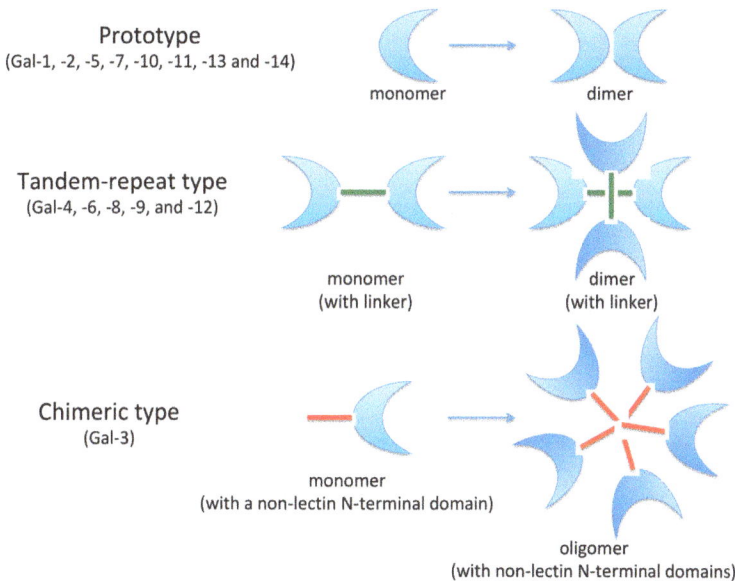

Figure 1. The structure of the galectin family members. The galectin family members are divided into three types: The prototype with one carbohydrate recognition domain (CRD), the tandem-repeat type with two CRDs connected by a non-conserved linker, and the chimeric type with one CRD and a non-lectin N-terminal domain (ND). Some galectins can self-associate into dimers or oligomers.

1.1. Gal-3 Identification

Gal-3 is a member of a growing family of β-galactoside binding lectins [8], with an approximate molecular weight of 30,000. It was originally identified in murine peritoneal macrophages and named MAC-2 [9]. Later, it was also described as CBP-35, a 35-kDa carbohydrate-binding protein in mouse 3T3 fibroblasts [10]. Since then it was detected in numerous type of tumoral tissues [11], basophilic leukemia cells [12], human and murine tumor cells [13,14], human lung [15] rat intestine [16] and human HeLa cells [17].

Gal-3 is encoded by the *LGALS3* Gene, which has previously been known under different names: IgE-Binding Protein [18], MAC-2 [9], Lectin L-29 [19], L-34 [20] CBP-30 [21], Galectin 3 (Gal-3), Advanced Glycation End-Product Receptor 3, Lectin, Galactoside-Binding, Soluble, 3, Carbohydrate-Binding Protein 35 (CBP 35) [22], Galactose-Specific Lectin 3, Laminin-Binding Protein [23], 35 KDa Lectin, Galactoside-Binding Protein, GALBP [12].

1.2. Gal-3 Tissue Distribution

In adults, Gal-3 is ubiquitously expressed, while experiments performed in mice demonstrated during embryogenesis its expression is tissue- and time-dependent. Although its expression is mainly related to the epithelial cells and myeloid/amoeboid cells, Gal-3 expression was detected in many different types of cells, including: Small intestinal epithelial cells, colonic epithelia, corneal and conjuctival epithelia, olfactory epithelium, epithelial cells of kidney, lung, thymus, breast, and prostate. It was also detected in ductal cells of salivary glands, pancreas, kidney, eye, in intrahepatic bile ducts, in fibroblasts, chondrocytes and osteoblasts, osteoclasts, keratinocytes, Schwann cells and gastric mucosa, as well as in the endothelial cells from various tissues and organs [24]. In addition, there are numerous data on Gal-3 expression in the cells involved in immune response, such as neutrophils, eosinophils, basophils and mast cells, Langerhans cells, dendritic cells, as well as monocytes and macrophages from different tissues. In some other cell types, such as lymphocytes, Gal-3 is not normally expressed, yet it expression can be induced by various stimuli [4,7]. Moreover, Gal-3 displays pathological expression in many tumors, such as those affecting the pancreas, the liver, the colonic mucosa, the breast, the lung, the prostate, the head and neck, the nervous system and the thyroid [24–26].

1.3. Gal-3 Protein/Gene Structure and Carbohydrate Binding

Gal-3 is the most studied member of the galectin family. It is the sole member of chimera-type family of galectins [27]. Gal-3 (m.w. 31 kDa) is found in solution as a monomer with two functional domains [22,28–30]. Gal-3 is so far unique in the family in having an extra long and flexible N-terminal domain consisting of 100–150 amino acid residues, according to species of origin, made up of repetitive sequence of nine amino acid residues rich in proline, glycine, tyrosine and glutamine and lacking charged or large side-chain hydrophobic residues [4,7,29,30]. The N-terminal domain contains sites for phosphorylation (Ser 6, Ser 12) [31,32] and other determinants important for the secretion of the lectin by a novel, nonclassical mechanism [33]. The C-terminus is the CRD, consisting of about 135 amino acid residues; this is what defines the molecule as a galectin. The structure of the *LGALS3* gene is consistent with the multi-domain organization of the protein. The gene for Gal-3 is composed of six exons and five introns (human locus 14q21-22). Exon I encodes the major part of the 5′ untranslated sequence mRNA. Exon II contains the remaining part of the 5′ untranslated sequence, the protein translation initiation site and the first six amino acids including the initial methionine. The repetitive sequence in the N-terminal half of the gene product is encoded within exon III. Exons IV, V and VI code the C-terminal half of the protein [7,34]. Gal-3, like most members of the galectin family, acts as a receptor for ligands containing poly-N-acetyllactosamine sequences which consist of many disaccharide units: Gal β1,4 GlcNAc bond to each other by β1,3 linkage. However, Gal-3 appears to have an increased affinity for the more complex oligosaccharides [4,34].

1.4. Gal-3 Subcellular Localization

Although Gal-3 is predominantly located in the cytoplasm, it has also been detected in the nucleus, on the cell surface and in the extracellular environment, suggesting a multifunctionality of this molecule.

Extracellular Gal-3 modulates important interactions between epithelial cells and extracellular matrix, and plays a role in the embryonic development of collecting ducts [35]. In contrast, intracellular Gal-3 is important for cell survival due to its ability to block the intrinsic apoptotic pathway [36,37]. In this regard, it has been demonstrated that cytoplasmic Gal-3 in cancer cells is usually associated with an aggressive phenotype, in opposition to nuclear Gal-3 [38].

Nuclear localization of Gal-3 was first described in in vitro proliferating fibroblasts [39–41], but not in fibroblasts with replicative deficiencies [42]. Nuclear Gal-3 is localized mainly in interchromatin spaces, at the border of condensed chromatin, on the dense fibrillar component, and at the periphery of the fibrillar centers of nucleoli [43].

1.5. Gal-3 Secretion

Gal-3 is synthesized on free ribosomes in the cytoplasm and lacks any signal sequence for translocation into the endoplasmic reticulum (ER) [4]. Although it does not traverse the endoplasmic reticulum/Golgi network, there is abundant evidence for Gal-3 also having an extracellular location. This protein has been shown to be secreted from cells by a novel, incompletely understood mechanism called ectocytosis, which is independent of the classical secretory pathway through the ER and Golgi system [4,44,45]. A short segment of the Gal-3 N-terminal sequence comprising residues 89–96 (Tyr-Pro-Ser-Ala-Pro-Gly-Ala-Tyr) has been found to play a critical role in Gal-3 secretion. However, this sequence is not sufficient on its own to cause the direct secretion of the CAT fusion protein, indicating that it is operative in the context of a large stretch of the N-terminal sequence of Gal-3 [33]. Immunohistochemical studies have indicated that the first step in Gal-3 secretion is its accumulation at the cytoplasmic side of the plasma membrane [44–46].

1.6. Gal-3 Ligands

There are several known ligands for Gal-3 [47]. They can be subdivided, according to their localization, into extracellular, intracellular and nuclear ligands.

1.6.1. Gal-3 Extracellular Ligands

Gal-3 binds to glycosylated extracellular matrix components, including laminin, fibronectin, tenascin, chondroitin sulfate and Mac-2 binding protein [21,48–51]. Some cell-surface adhesion molecules, such as integrins, are also ligands for Gal-3 and galectins, through binding to the extracellular domains of one or both subunits of an integrin, may positively or negatively modulate integrin activation, and affect binding with extracellular ligands [52].

A major ligand for Gal-3 on mouse macrophage is the α-subunit of the integrin $\alpha M\beta2$, otherwise known as CD11b/18 [53,54]. Moreover, Gal-3 also interacts with integrin $\alpha1\beta1$ via its CRD domain in a lactose dependent manner [55]. Gal-3 also seems to be an endogenous cross-linker of the CD98 antigen, leading to the activation of integrin-mediated adhesion [53].

1.6.2. Gal-3 Intracellular Ligands

Gal-3 is also known to have an intracellular location and to interact with several proteins inside the cell, such as: Cytokeratins [56], CBP70 [57], Chrp [58], Gemin4 [59], Alix/AIP-1 [60], and Bcl-2 [61]. It is worth noting that almost all of the mentioned intracellular ligands interact with Gal-3 via protein–protein rather than lectin-glycoconjugate interactions. The only exception is the cytokeratins. In fact, it has been shown that cytokeratins of MCF7, as well as other human cells, carry a novel posttranslational modification, a glycan with a terminal α linked GalNAc and that these residues are recognized in vitro by mammalian Gal-3 [56].

1.6.3. Gal-3 Nuclear Ligands

Nuclear Gal-3 can interact with Gemin-4, a component of a macromolecular complex containing approximately 15 polypeptides, among them SMN (survival of motor neuron) protein, Gemin-2, Gemin-3, some of the Sm core proteins of snRNPs and other as yet unidentified proteins. These complexes are implicated in processes directly or indirectly related to pre-mRNA splicing. The identification of Gemin-4 as an interacting partner of Gal-3 provides strong evidence that this galectin can play a role in splicesome assembly in vivo [59,60]. The screening of a Jurkat cell cDNA library, by the yeast two-hybrid method using Gal-3 as bait allowed the identification of another Gal-3 binding protein. It is a human homologue of ALG-2 linked protein x (Alix) or ALG-2 interacting protein-1 (AIP-1) [60]. This protein interacts with ALG-2, a calcium binding protein necessary for cell death induced by different stimuli. AIP-1 cooperates with ALG-2 in executing the calcium dependent requirements along the cell death pathway [62,63]. Alix/AIP-1 contains a proline, glycine, alanine and tyrosine rich sequence in the C-terminal region, which is highly homologous to the tandem repeat sequence in the N-terminal part of Gal-3 [60].

2. Role of Gal-3 in Different Clinical Conditions and Diseases (Listed in Alphabetical Order)

Gal-3 is a multifunctional protein engaged in different biological events, in different tissues, playing a relevant role in many different clinical conditions and diseases. Since its discovery as an IgE-binding protein in 1971 and subsequently its recognition as Mac-2, a 32-kDa cell surface antigen expressed on murine peritoneal macrophages in 1982, more than 8200 papers have been published in the literature so far (Figure 2).

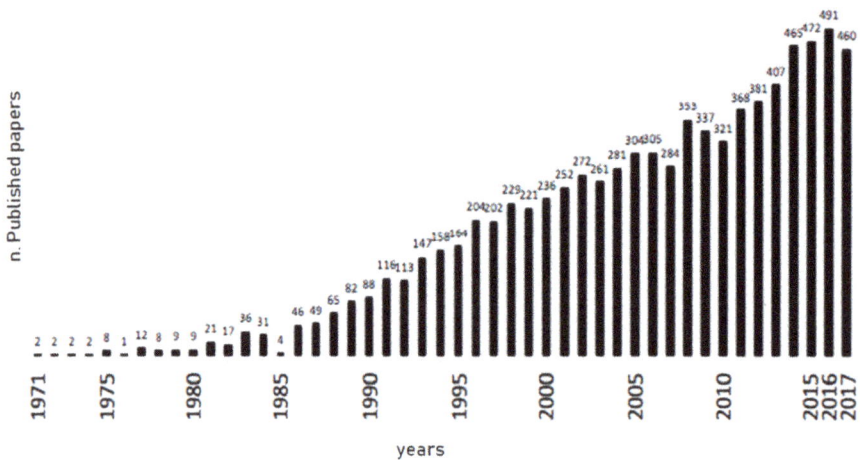

Figure 2. PubMed (U.S. National Library of Medicine, Medical Institutes of Health) records on academic publications on Gal-3, total 8289 items by December 2017. Search Strategy MESH terms: Galectin 3 or Gal-3 or Advanced Glycation End-Product Receptor 3 or Lectin, Galactoside-Binding, Soluble, 3 or Carbohydrate-Binding Protein 35 (CBP 35) or Galactose-Specific Lectin 3 or Laminin-Binding Protein or IgE-Binding Protein or Lectin L-29.

The constant increase in the number of papers published in the most recent years and concerning the role of this protein in many clinical conditions and diseases testify the great clinical relevance of Gal-3 in human pathophysiology.

Numerous previous studies have indicated that Gal-3 may be used as a diagnostic or prognostic biomarker for many types of diseases and conditions. Recently, the potential clinical uses of Gal-3 as a diagnostic biomarker and as a therapeutic target were described in a comprehensive review [64].

The current review represents an attempt to examine the many clinical implications of Gal-3 protein, based on a classification of clinical diseases, listed in alphabetical order (Figure 3).

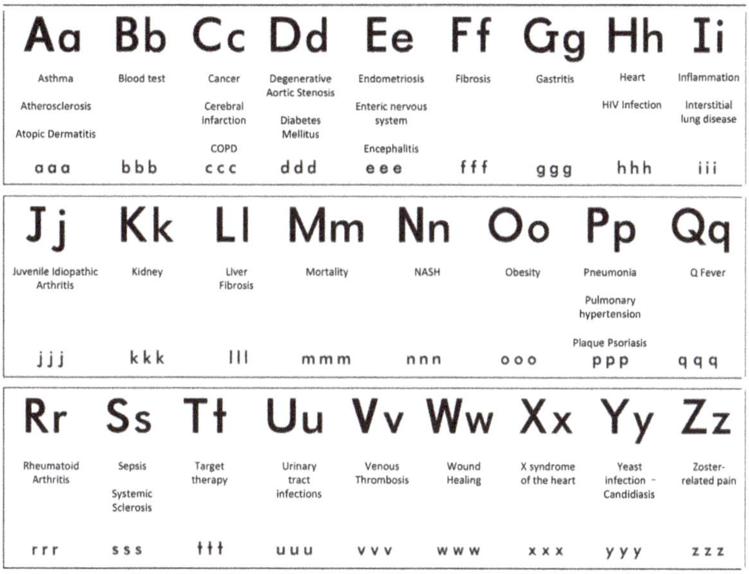

Aa	Bb	Cc	Dd	Ee	Ff	Gg	Hh	Ii
Asthma	Blood test	Cancer	Degenerative Aortic Stenosis	Endometriosis	Fibrosis	Gastritis	Heart	Inflammation
Atherosclerosis		Cerebral Infarction	Diabetes Mellitus	Enteric nervous system			HIV Infection	Interstitial lung disease
Atopic Dermatitis								
		COPD		Encephalitis				
a a a	b b b	c c c	d d d	e e e	f f f	g g g	h h h	i i i

Jj	Kk	Ll	Mm	Nn	Oo	Pp	Qq
Juvenile Idiopathic Arthritis	Kidney	Liver Fibrosis	Mortality	NASH	Obesity	Pneumonia	Q Fever
						Pulmonary hypertension	
						Plaque Psoriasis	
j j j	k k k	l l l	m m m	n n n	o o o	p p p	q q q

Rr	Ss	Tt	Uu	Vv	Ww	Xx	Yy	Zz
Rheumatoid Arthritis	Sepsis	Target therapy	Urinary tract infections	Venous Thrombosis	Wound Healing	X syndrome of the heart	Yeast infection – Candidiasis	Zoster-related pain
	Systemic Sclerosis							
r r r	s s s	t t t	u u u	v v v	w w w	x x x	y y y	z z z

Figure 3. Conditions and diseases in which a role for Gal-3 has been postulated, listed in alphabetical order.

3. A

3.1. Asthma

Asthma is a heterogeneous disease characterized by chronic airway inflammation, variable and reversible airway obstruction and airway hyperresponsiveness (AHR) [65]. Since their recognition, almost fifty years ago, the class of human antibody officially named IgE was unequivocally considered as the factor capable of transferring sensitivity to allergens and plays a central player in the allergic response [66]. It is now clear that IgE acts as part of a complex protein network. The interaction of genes belonging to such network with environmental factors is able to determine the risk of allergic sensitization [67]. In addition to the two major IgE receptors, namely the FcεRI (high affinity) and the FcεRII/CD23 (low affinity), Gal-3 has received increasing interest as a major player in this network and is generally thought to be potentially relevant in the pathogenesis of inflammation in asthma and its phenotypes. In this regard it is relevant to note that Gal-3 was initially identified as an IgE-binding protein [68]. Several experimental studies, using murine animal models, indicate that Gal-3 has a role in the pathophysiological mechanisms in immune response and, in particular, in asthma [69–75]. The specific role of Gal-3 in asthma is related to its function in the recruitment, activation and removal of neutrophils [73]. Experimental data from Gal-3 null mice showed reduced neutrophil recruitment during infection [74] and Gal-3 increases uptake of apoptotic neutrophils. Therefore, any Gal-3 dysfunction may influence the ability to remove apoptotic neutrophils from the site of inflammation [75]. This has several consequences: Reduction in the removal of apoptotic cells can result in the release of damaging enzymes and oxidants, which can promote persistence of inflammation. In humans, a significant reduced level of Gal-3 was detected in the sputum of neutrophilic asthma compared to eosinophilic and paucigranulocytic asthma [76]. However, the most interesting finding was the recent observation that Gal-3 may be considered as a new biomarker predictive of long-term responsiveness of pulmonary function in patients with severe asthma treated with the monoclonal antibody, anti-immunoglobulin E (IgE), named omalizumab [77]. According to this study, the analysis of bronchial biopsy specimens, obtained from patients with severe asthma, may help in predicting response to treatment. In particular, patients that were classified as Gal-3 positive before treatment initiation were responsive to the treatment with omalizumab while Gal-3 negative patients were non-responders to the same treatment with omalizumab. This study needs to be confirmed by others but it indicates that Gal-3 may be considered as new asthma-related biomarker, useful for patient stratification for therapy, in addition to the other biomarkers already known, such as the fractional exhaled nitric oxide (FeNO) levels, the blood/sputum eosinophil counts, the serum periostin levels, the smooth muscle proteins and the keratins.

Inflammation driven by type-2 helper T lymphocytes (Th2) is relevant in asthma and understanding the mechanisms that underline Th2 cytokine production is crucial for the establishment of novel therapeutic options. At present, there are limited data on the role of Gal-3 in allergic inflammation. As a multifunctional protein expressed by different types of inflammatory cells, Gal-3 has been considered as an IgE-binding protein and a regulator of infiltrating cell inflammation, activation and clearance [78]. Gal-3 activity may be operative in no-eosinophilic asthma [78]. Moreover, low levels of Gal-3 were seen in neutrophilic asthma compared with eosinophilic asthma [76]. Furthermore, galectins in relationship with their location (extracellular or intracellular, endogenous or exogenously delivered) may differentially regulate eosinophil behavior exerting a pro- or anti-inflammatory outcome. In severe asthma Gal-3 may be considered as a biomarker of lung remodeling and its levels are related to response to omalizumab treatment and improve lung function [77]. Moreover, Gal-3 deficient mice developed less airway hyperresponsiveness and eosinophils infiltration, a reduction of Th2 cytokines in the airways, during acute allergen-challenge if compared to WT mice [69–71], suggesting a role of Gal-3 in the chronic asthma. In Brown-Norway rats the intra-tracheal deposition of a vector with the gene encoding for Gal-3 causes the impairment of some Th2 effects, eosinophil influx and the down-regulation of IL-5 gene expression, IL5 that is one of the main regulatory cytokines that modulate eosinophilic behavior [79].

3.2. Atherosclerosis

Atherosclerosis is considered an inflammatory-proliferative response of the intima to injury [80,81]. The participation of monocytes in atherogenesis is well known [82,83] and their presence in the artery plaques and in the lipid core was recognized in 1993 [84]. More recently, macrophages were recognized as the principal inflammatory cell mediator in the atheromatous plaque that may participate in intima remodeling, immune response, and scavenger functions through the synthesis and secretion of a variety of enzymes and factors [80,85]. The role of Gal-3 in atherosclerosis was analyzed in specimens obtained from carotid endarterectomies, lower limb amputations, and thoracic aortas from autopsies of young adult trauma victims [86]. According to this study, a positive staining for Gal-3 can be detected close to a lipid core or near areas with fibrosis, neovascularization, calcification, hemorrhage, or thrombosis. In the carotid and lower limb arteries, foam cells were the cell type that most frequently stained positive for Gal-3. Mononuclear round cells and foam cells containing Gal-3 were also observed in the sub-endothelium, sometimes at sites of thrombosis. Endothelial cells and cells lining neo-vessels within plaques were not stained for Gal-3. Spindle cells from the tunica media of atherosclerotic arteries were not stained for Gal-3, but isolated Gal-3-positive cells were seen outside the lamina elastica externa. Gal-3 can be detected by immunoperoxidase staining in advanced atherosclerotic lesions of carotid and lower limb arteries and in early aortic atherosclerotic changes. In atherosclerotic lesions, Gal-3 is often localized in macrophages and foam cells and rarely in smooth muscle cells. Gal-3 is not detectable by immunohistochemistry in the tunica media of umbilical cord, carotid, limb, and aorta arteries. Finally, Gal-3 mRNA can be detected in advanced atherosclerotic lesions from carotid and lower limb arteries and a higher level of Gal-3 was demonstrated by Western blot analysis in these arteries in comparison with umbilical cord arteries.

The presence and the role of Gal-3 in most foam cells into the atherosclerotic lesions needs to be elucidated. It could be related to an association with lipid loading or alternatively the accumulation of Gal-3 together with advanced glycation endproducts (AGEs), may suggest that Gal-3 could act as a receptor for modified lipoproteins, thus participating in the formation of foam cells. More studies are needed to unravel any pathogenic or diagnostic role of Gal-3 in atherosclerosis.

3.3. Atopic Dermatitis

Atopic dermatitis (AD) is a chronic inflammatory skin condition that involves an inherited epidermal barrier defect, which may allow for greater penetration by allergens and irritants.

Atopic dermatitis, commonly called atopic eczema, usually arises in early childhood, often in infancy. While it usually resolves by early teenage years, approximately 5–10% of patients have the disease extend into adulthood. Classic symptoms are itching and burning of the skin, resulting in thickening of the skin in response to the scratching. In some adults, it can be severe with debilitating itching, inability to sleep, and social stigmatization due to skin damage and thickening, often on the face. IgE-associated atopic eczema/dermatitis syndrome (AEDS) contains abnormal B-cells that are surface IgE-positive (sIgE+) and produce IgE spontaneously in culture. Gal-3 is highly expressed in epithelial cells including keratinocytes, hair follicles, sebaceous, and sweat glands [87]. It has been reported that the IgE production was amplified by PMNs in patients AEDS, but not in healthy controls, and this effect was mediated by Gal-3 release by PMNs. In this regard, it has been demonstrated that IgE release was enhanced by the addition of Gal-3 to the cultures of AEDS B cells admixed with PMNs from healthy controls. These results suggest that Gal-3 can enhance PMN-induced IgE synthesis by B cells that are already committed to IgE production. These observations indicate that Gal-3 may be involved in the hyperactive immune response in AEDS [88]. Although initial evidence, based on the results of immunohistochemical analysis of skin biopsies from AD patients, argued against a critical role of Gal-3 in the binding of monoclonal or serum Ig-E [89], many experimental results obtained using mouse models of AD suggest that Gal-3 plays a critical role in the development of the allergic inflammatory response in AD [90]. The possible involvement of Gal-3 in the development of an allergic inflammatory response was, in fact, analyzed in an experimental mouse model in which AD was induced by repeated epicutaneous sensitization with ovalbumin (OVA) [90]. In this mouse model an upregulation of Gal-3 expression in the epidermis was observed after such sensitization. These results suggest that Gal-3 is a pro-inflammatory mediator of skin inflammation in atopic skin disease and that is required for the development of the Th2 inflammatory response to epicutaneously-introduced antigens through its effects on both DCs and T cells [90]. Based on these experimental results a company named "Galectin Therapeutics Inc." (NASDAQ:GALT, Norcross, GA, USA) has developed a new Gal-3 inhibitor, capable of binding to Gal-3 and disrupting its function. A small, open label, investigator-initiated protocol was recently started and three adult patients with severe atopic dermatitis, classified as refractory, meaning that they have not responded to medical treatment beyond topical steroids, were successfully treated without any significant adverse reaction [91]. A phase 2 controlled trial, using various doses and regimens to further explore the drug's potential, is now under way.

4. B

Blood Test

Clinical parameters, such as advanced age, higher New York Heart Association (NYHA) functional class, reduced left ventricular ejection fraction (LVEF), lower body mass index, renal dysfunction, and anemia, have all been associated with poor outcomes in heart failure (HF), but are not significant predictors of mortality. Many efforts have been made recently to find biomarkers that might help in the risk stratification, and prognostication of acute and chronic heart failure beside the well-established markers, such as the brain natriuretic peptide (BNP) and its N-terminal part (NT-proBNP), the soluble suppression of tumorigenicity 2 (ST2), the growth differentiation factor-15, and the highly sensitive troponins.

In a pivotal paper published in 2004 by Sharma and colleagues [92] Gal-3 was identified as the most upregulated protein in an animal model of left ventricular hypertrophy (LVH) and HF suggesting that Gal-3 is a critical player in the pathogenesis and progression of HF. Since then an acceleration of research was observed, especially in the field of cardiology with many studies published concerning the role of Gal-3 in the evaluation and management of patients with heart failure. Gal-3 levels in plasma were, in fact, found to be positively associated with known cardiovascular risk factors, as age, sex, diabetes, hypertension, hypercholesterolemia, body mass index, renal function and smoking and was proposed as an additional prognostic factors, useful to identify high-risk patients in whom additional care management efforts and advanced therapies could be warranted.

Gal-3 was proposed in many studies as a new biomarker of heart fibrosis that could predict outcome of heart failure (HF) [93] and, in particular, occurrence of mortality [94,95]. In several cohorts of acute HF [96,97], as well as chronic HF [94], Gal-3 was shown to be a powerful predictor of mortality. Clinical effectiveness and prognostic value of this test have been demonstrated in several heart-failure trials, including the "Coordinating Study Evaluating Outcomes of Advising and Counseling in Heart Failure (COACH)" [97], the "Pro-BNP Investigation of Dyspnea in the Emergency Department (PRIDE)" [98], the University of Maryland Pro-BNP for Diagnosis and Prognosis in Patients Presenting with Dyspnea (UMD)" [99] and the "Heart Failure: A Controlled Trial Investigating Outcomes of Exercise Training (HF-ACTION)" [100]. Gal-3 was also associated with mortality when evaluated in the general population, as reported in the "Prevention of Renal and Vascular End-Stage Disease (PREVEND) study [101], in the "Framingham Offspring Study" [102] and in the "Rancho Bernardo Study" [103].

Finally, Gal-3 was associated with increased risk of developing AF in the Framingham study, but this association was not found to be significant after adjustment for traditional clinical AF risk factors [104].

Gal-3 is currently used in clinical practice and it is measured using many different blood assays by manual or automated assays. The enzyme linked immunosorbent assay manual assay (ELISA) is the most frequently used method in the published studies. Recently, FDA has approved for market new automated blood assays, with faster delivery of the results, the BGM Gal-3 assay, manufactured by BG Medicine, Inc. (Waltham, MA, USA) and the ARCHITECT Gal-3 assay, by Abbott Diagnostics (Abbott Park, IL, USA) and more are on their ways. According to this tests plasma levels lower than 17.8 ng/mL indicate low risk, levels greater than 17.8 ng/mL but lower than 25.9 ng/mL are indicative of moderate risk and levels above 25.9 ng/mL are associated with high risk of adverse outcomes and need more frequent follow-up and intensified as well as advanced strategies of cure. Plasma Gal-3 levels greater than 25.9 ng/mL, independent of symptoms, clinical findings, and other laboratory measures, predict a patient who is likely to have rapid progression of heart failure, eventually resulting in hospitalization and death [105].

5. C

5.1. Cancer

A large body of reported data about galectin expression in cancer is available in the literature and the chapter on Gal-3 action on cancer development and progression represents one of the most relevant ones. Gal-3 has been shown to influence many significant biological processes linked to cancer development and progression, including cell adhesion, proliferation, differentiation, mRNA splicing, cell-cycle progression, immune system evasion, inflammation, angiogenesis, apoptosis and metastasis [25,38,106–110]. In addition, Gal-3 exerts a role as a pro-tumor factor by acting within the tumor microenvironment, the so-called "tumor niche", to suppress immune surveillance. Its ability to promote tumor cell proliferation and survival of the cancer cell both directly and indirectly, by acting on cell adhesion and cell contact with mesenchymal stromal cells, has lead to consider Gal-3 as a guardian of the tumor microenvironment [111].

Gal-3 influences oncogenesis, cancer progression and metastasis through a variety of pathways both inside and outside the cell [112–118]. The studies on the role of Gal-3 in different types of cancers indicate that such effects may be specific for the different types of cancer. In fact, Gal-3 may have both positive and negative effects on cancer cell survival depending of the type of tumor considered and its subcellular localization likely influences the role of Gal-3 in different cells [24–26].

5.1.1. Gal-3 Subcellular Localization in Tumor Tissues

In tumor tissues Gal-3 can be detected in the nucleus [119,120], in the cytoplasm [121], and eventually can be secreted into the extracellular spaces [4] or in the circulation [122]. The localization of Gal-3 is essential to its function. Gal-3's function, in fact, will vary depending on whether it is in the nucleus, cytoplasm, or extracellular spaces [123]. In addition, as observed for Gal-1, the effects of Gal-3 seem to be cell type specific [24,119,124–126]. All these factors make extremely difficult to simplify the role of Gal-3 in cancer and may explain the variable and sometimes opposite effects reported in different cells and in different cellular compartments.

5.1.2. Gal-3 and Apoptosis in Tumor Tissues

The most intensively studied mechanism through which Gal-3 influence cancer cell is represented by its anti-apoptotic function [24,61,110,127]. The elucidation of the mechanisms through which Gal-3 exerts its effect on apoptosis has led to identify a new apoptotic pathway in which different types of DNA damage triggers the HIPK2 activation and induces p53 mediated repression of Gal-3 expression [128]. Dysregulation of this pathway due to the occurrence of p53 mutations or due to HIPK2 loss may alter such pathway thus leading to Gal-3 overexpression and [129]. Gal-3 influences also many other signal transduction cascades and pro-survival processes that include major oncogenes such as *RAS*, *BCL2*, and *MYC* [130–137].

The ability of Gal-3 to form lattices with glycoproteins and glycolipids has been implicated in regulating cell adhesion, metastasis, endocytosis, and other biological processes [138–143]. Aberrant glycosylation could influence transport and trafficking of many different proteins involved in cancer development and progression [143]. Interaction of Gal-3 with glycosphingolipid (GSL) resulted in invaginations in the membrane that occurred during the endocytotic process and has been referred to as "membrane bending" [140–143]. "Membrane bending" requires both Gal-3 and GSL. As of now, integrin β1 endocytosis is mediated by GSL and Gal-3 and it is likely that other glycoproteins will also be transported by this mechanism [143].

5.1.3. Gal-3 Immune Surveillance and Angiogenesis in Tumor Tissues

Gal-3 is able to modulate a variety of immune cell processes by co-opting selected inhibitory receptors, disrupting co-stimulatory pathways and/or controlling activation, differentiation, and survival of immune cells. On the other hand, Gal-3 can regulate vascular signaling programs through binding to integrin avb3 or by sustaining the pro-angiogenic capacity of tumor-associated macrophages. Therefore, Gal-3 serves as a bifunctional messenger that has emerged as a novel regulatory checkpoint able to couple immunosuppression and angiogenesis, two events that occur simultaneously during tumor growth and are often mediated by the same cytokines and the same cell types [144].

5.1.4. Gal-3 in Various Types of Human Cancers

The large body of studies, regarding the expression of Gal-3 in various types of cancers that have been published in the literature were obtained using different types of cells/tissues as well as various animal models [38,106,145]. The specific techniques used, the inclusion of appropriate controls and the heterogeneity of the tissues examined may explain the discrepancies observed in the different studies. It is relevant to note that the percentage of normal epithelial, stromal, vascular and inflammatory cells may constitute a strong bias in the evaluation of the expression of Gal-3 in a given cancerous tissue specimens. This is especially true for Gal-3, a protein that is highly expressed in inflammatory cells. In general, histological techniques such as immunohistochemistry or in situ hybridization that allow a cell type-specific analysis of Gal-3 expression are most reliable, compared to Western blot and RT-PCR techniques that may be give results that are potentially contaminated or diluted by the presence of noncancerous cells. Gal-3 was analyzed as oncologic marker both by immunohistochemistry and by measuring its blood levels.

An increased expression of Gal-3 was observed at immunohistochemistry or at RT-PCR, in many different types of cancers [25,146,147] and most of the authors agree that detection of Gal-3 could help to better define diagnosis and/or prognosis of cancer. Increased expression of Gal-3 was reported breast carcinoma, in colon carcinomas, in gastric cancers, in hepatocellular carcinoma, in well differentiated thyroid carcinomas in anaplastic large-cell lymphoma, in head and neck squamous cell carcinomas, in tongue carcinomas and in non-small cell lung cancer. Other studies demonstrated an opposite picture, with decreased Gal-3 expression in breast, ovarian cancer prostate tumors, advanced uterine adenocarcinoma, basal cell carcinoma of the skin, epithelial skin cancer and malignant salivary gland neoplasms, compared to the corresponding normal tissue. Several studies have investigated the expression of Gal-3 in in human colon carcinomas, but conflicting results have been reported [148–151]. It seems that Gal-3 expression is down-regulated in the initial stages of colon neoplastic progression [152,153], whereas a dissociated cytoplasmic expression increases in later phases of tumor progression [154].

Among the cancers that express high level of Gal-3 there are those affecting the thyroid gland [155–161], the central nervous system in adults [162] and in children [163,164], the head and neck squamous cell [165], the pancreas [166], the stomach [112], the bladder [167], the kidney [168,169], the liver [170], in the parathyroid [171,172] and in the salivary glands [173]. In addition, elevated Gal-3 expression levels were also observed in lymphoma [174] and in melanoma [175]. Finally, Gal-3 was found to play a relevant prognostic role in neuroblastoma [164].

The analysis of the expression levels of Gal-3 was extensively studied in the thyroid tissue. In this tissue, Gal-3 expression analysis was suggested as a presurgical marker of cancer [156] and the detection of Gal-3 expression, specifically located in the cytoplasm of the thyroid cells was used a diagnostic hallmark of thyroid malignancy. On such basis a ThyroTests was setup [176–178], clinically validated [160,179] and used in many countries [180]. Its good accuracy and its low cost has led to propose its inclusion in the diagnostic algorithm of thyroid cancer as a screening test especially useful for suspicious thyroid nodules with a cytologically indeterminate pattern [181,182].

In addition to the analysis of the expression levels in neoplastic tissues, Gal-3 was also investigated as a potential new oncologic scintigraphic as well as blood marker in many different types of tumors, including the thyroid [183–185], the prostate [186] and the ovarian cancer [187].

5.2. Cerebral Infarction

It is often difficult to determine the prognosis and clarify the pathophysiology of patients referred to emergency department for suspected ischemic stroke. Computerized cranial tomography (CCT) and diffusion-weighted magnetic resonance imaging (MRI) are crucial in the diagnosis of cerebral infarction. However, a lot of efforts have been made to search for new biomarkers and, in particular, for serum markers specifically associated with brain damage, that may allow discrimination of ischemic stroke and intracerebral hemorrhage. The levels of Gal-3 have been reported to increase in ischemic brain damage [188,189]. In a recent study, Gal-3 serum levels were found to be increased in clinically-suspected ischemic stroke patients with normal CCT, seen in the emergency department. On such basis, it has been suggested that its measurement may contribute to the diagnosis of cerebral infarction [190]. Such promising studies need to be confirmed in large numbers of patients.

5.3. Chronic Obstructive Pulmonary Disease

At present there are limited data available on the possible role of Gal-3 in chronic obstructive pulmonary disease (COPD). COPD is a chronic, progressive, irreversible disease associated with high morbidity and mortality. It is characterized by bronchial epithelial changes and neutrophil infiltration. Gal-3 has been demonstrated to be higher in patients with COPD with elevated systolic pulmonary artery pressure compared to healthy controls [191]. In this subset of patients, Gal-3 levels seem to be a predictive marker of right ventricular dysfunction [191]. In small airways of severe COPD patients, epithelial Gal-3 immunostaining was increased when compared with healthy nonsmokers or smokers [192]. In addition, in patients with severe COPD, Gal-3 expression and neutrophil accumulation in the small airway epithelium correlate with epithelial proliferation and airway obstruction [192]. Finally, in the airway of current or ex-smoker COPD patients, Gal-3 and its receptor CD98 were decreased. Furthermore, Gal-3 levels measured in broncho alveolar lavage (BAL) were reduced in healthy smokers suggesting a potential role of smoke in regulating Gal-3 expression although in COPD ex-smokers the defect in Gal-3 expression persists despite cessation of smoking [193]. Administration of exogenous Gal-3 to alveolar macrophages from BAL, cultivated in vitro and obtained from patients affected by COPD, caused an increase in efferocytosis, the process by which dying/dead cells (e.g., apoptotic or necrotic) are removed by phagocytic cells, thus indicating a possible role of Gal-3 as a novel macrophage-targeted therapies for COPD/Emphysema [194]. Very recently, the level of serum Gal-3 was significantly increased in acute exacerbation of COPD compared with the level in COPD convalescence phase, suggesting that Gal-3 might be a valuable diagnostic biomarker for this condition [195]. In such patients, the serum levels of Gal-3 positively correlated with systemic inflammation and smoking [195].

6. D

6.1. Degenerative Aortic Stenosis

The first clinical report concerning the relationship between myocardial Gal-3 expression and severe aortic stenosis (AS) was published in 2004 [92]. In this study, Gal-3 expression was analyzed in cardiac biopsies from patients undergoing aortic valve replacement for aortic stenosis. Gal-3 myocardial expression was increased in aortic stenosis patients with cardiac hypertrophy and relatively depressed ejection fraction compared to aortic stenosis subjects with LV hypertrophy and normal or elevated ejection fraction.

Recently, another study was conducted to test the potential prognostic value of Gal-3 in patients with degenerative AS [196]. In patients treated with balloon aortic valvuloplasty (BAV) Gal-3 was associated with worse survival and this was unrelated to estimated glomerular filtration rate (eGFR). Measurement of circulating levels of Gal-3 may therefore be useful in risk stratification in elderly patients affected by degenerative AS with coexistent diseases to choose the optimal therapy.

6.2. Diabetes Mellitus

Whereas most of the experiments revealed a protective role of Gal-3 in acute inflammation, its function in chronic inflammatory diseases, like diabetes mellitus (DM) and in particular in type 2 diabetes mellitus (T2DM) is less clear [197,198]. The role of Gal-3 in the development of Diabetes of its complication is still an area of debate. Several studies indicate that Gal-3 is involved in the regulation of glucose homeostasis by acting at the level of adipose tissue and pancreatic islets, thus participating in the pathogenesis of obesity and T2DM, but its effects may be influenced by differences in genetic background, diet, age, sex, as well as experimental end points. In fact, in the literature controversial results have been reported regarding to the roles of Gal-3 in the pathogenesis of T2DM. In some studies Gal-3 was considered as a protective factor, while in others it was found to be associated with an increase in the severity of T2DM.

6.2.1. Gal-3 Increases Severity of DM

In humans, systemic Gal-3 is elevated in obesity and negatively correlates with glycated hemoglobin in T2DM patients [199]. In a population-based cross-sectional survey in Zhejiang, China involving large number of patients, high Gal-3 serum levels were associated with increased odds of developing heart failure, nephropathy, and peripheral arterial disease in patients with T2DM [200]. In other clinical studies, circulating Gal-3 was higher in T2DM patients [201–203] and thought to be a risk factor for vascular complications, such as heart failure, nephropathy, peripheral artery disease, and other vascular complications. Patients with Gal-3 level > 25 ng/mL exhibited a 11.4-fold higher risk of microvascular complications (retinopathy and/or nephropathy) and a 8.5-fold increased risk of macrovascular complications (myocardial infraction, angina pectoris, cerebrovascular event, and peripheral artery disease) compared to patients with Gal-3 level < 10 ng/mL [200]. The order of circulating level of glectin-3 was as follows: Diabetes > prediabetes > normal control. Therefore, it has since been considered as a marker for prediabetes [201]. In diabetic patients, Gal-3 concentrations were significantly elevated in subjects with coronary artery disease and associated with the formation of diseased vessels and plaques [203]. In agreement with human studies, also in mice models in which hyperglycemia was induced after high-fat diet feeding with a corresponding threefold increase in homeostasis model assessment for insulin resistance (HOMA-IR) index and a 1.5-fold decrease in Akt activation, indicate that Gal-3 expression was increased in endothelial cells as well as in blood, suggesting that Gal-3 plays a role in the vascular response in DM [204]. Another report found increased Gal-3 expression in mice fed with a high-fat diet and recombinant human Gal-3 promoted preadipocyte proliferation [205]. Moreover, galecitin-3 knockout mice exhibited less weight gain when fed with a high-fat diet. The expression levels of PPAR-γ, CCAAT-enhancer-binding protein alpha (C/EBPα), CCAAT-enhancer-binding protein beta (C/EBPβ), and fatty acid binding protein 4 (Fabp4) were also reduced in the adipose tissue of Gal-3 knockout mice fed with a high-fat diet [206]. Gal-3 upregulation was also found in hepatocytes treated with 100 mM d-glucose and in the sera of patients with T2DM [207]. Consistently, Gal-3 ablation protected mice from the development of streptozotocin-induced type 1 DM (T1DM) [208].

6.2.2. Gal-3 Decreases Severity of DM

Although most of the studies agree in considering Gal-3 as a marker of inflammation and fibrosis, many studies suggest that the increased expression of Gal-3 may be part of an adaptive response to tissue injury, favoring resolution of inflammation and opposing to chronification of the inflammatory process [209]. Both clinical studies in humans and experimental studies in mice, indicate that Gal-3 may indeed play a protective role in some metabolic diseases, such as T2DM. In a study of 20 patients, Ohkura et al. measured the glucose disposal rate, fasting insulin, HOMA-IR, insulin sensitivity index, Gal-3, and adiponectin. Notably, Gal-3 levels negatively correlated with fasting insulin and HOMA-IR, but positively correlated with glucose disposal rate, insulin sensitivity index, and serum adiponectin level [210]. Moreover, body weight, amount of total visceral adipose tissue, fasting blood glucose, and insulin levels increased in high-fat diet-fed Gal-3 knockout mice. These mice also displayed enhanced Th1 T-cells and natural killer T (NKT) cells, as well as pro-inflammatory CD11c⁺CD11b⁺ macrophages, whereas the number of anti- inflammatory CD4⁺CD25⁺FoxP3⁺ regulatory T-cells and M2 macrophages were reduced. Infiltrating pancreatic and peritoneal macrophages also exhibited significant increases in NLR family, pyrin domain containing 3 (NLRP3) inflammasome, nuclear factor-κB activity, and IL-1β production [211]. Gal-3 deficiency in mice led to the accumulation of fat as well as increased inflammation in adipose tissue, which is important in promoting insulin resistance. There were higher numbers of infiltrated macrophages and inflammatory cytokines, IL-6 and TNF-α, in the adipose tissue. High-fat diet also attenuated the expression of adiponectin and PPAR-γ in the adipose tissue [212]. Furthermore, galecitin-3 deficiency was also found to suppress endothelial glucose transporter, type 4 (GLUT4) expression and promote insulin resistance in high-fat diet-fed mice, as suggested by comparing Gal-3 knockout mice to wild-type controls [213]. Experimental studies, performed in mice, demonstrated a protective role of Gal-3 toward obesity and T2DM, via the modulation of the responsiveness of innate and adaptive immunity to overnutrition [212,214]. In accordance with such hypothesis, Gal-3 KO in mice induces a pro-inflammatory phenotype characterized by an increased systemic, pancreatic and VAT inflammatory response to metabolic stimuli and an exacerbated vascular and renal tissue damage induced by diabetes and related disorders [215,216]. Moreover, Gal-3 deficient mice exhibit greater hyperglycemia and impaired glucose tolerance compared to control wild type mice, indicating that Gal-3 deficiency contributes to the pathogenesis of diabetes [213].

The conflicting data reported so far regarding the role Gal-3 in obesity and T2DM suggest that we need of further attention and more clinical studies in order to clarify its role as a potential player and therapeutic target in these conditions.

7. E

7.1. Endometriosis

Endometriosis is an inflammatory disease of reproductive-aged women, and it is strongly related to consequent infertility [217]. Endometriosis is caused by a dysregulation of inflammatory and vascular signaling [218]. Gal-3 plays a relevant role in both these processes and it is overexpressed in endometriotic tissue [219]. In addition, higher levels of Gal-3 can be detected in the peritoneal fluids from women with endometriosis [220]. Another aspect in which Gal-3 may be involved in the pathogenesis of endometriosis and its complication in the development of pain associated with inflammation. Gal-3 was found to be involved in myelin phagocytosis, in Wallerian degeneration of neurons, and is able to trigger neuronal apoptosis after nerve injury [221]. It has been hypothesized that Gal-3, overexpressed in endometriotic foci, could be responsible for the induction of nerve degeneration and pain [222]. In this regard it is interesting to note that neurotrophin, a nerve growth factor strongly expressed in endometriosis, upregulates galcetin-3 expression [219]. All of these studies suggest that Gal-3, and other Galectins as well, play important roles in immune responses in inflammatory disease and in the pathogenesis of endometriosis, but there is still need of high-quality clinical studies to confirm these observations.

7.2. Enteric Nervous System

The enteric nervous system (ENS) governs the function of the gastrointestinal (GI) tract. It is embedded in the lining of the gastrointestinal system, beginning in the esophagus and extending down to the anus [223]. It is derived from neural crest cells and is often named as the second brain [224,225]. Dysfunction of ENS, due to ischemic cerebral stroke, may be responsible for the occurrence of neurological manifestation that complicate stroke, including dysphagia, dysmotility and colorectal dysfunction.

It is well known that Gal-3 in the GI tract stabilizes cell-cell junctions and enables polarization of intestinal epithelial cells [226,227]. Furthermore, it has been associated with GI cancers [228], as well as with Crohn's disease and ulcerative colitis [229].

Induction of stroke in experimental murine models by permanent middle cerebral artery occlusion (pMCAO) triggers central and peripheral Gal-3 release by microglia. Increased levels of Gal-3 are, in turn, responsible for enteric neuronal loss [230]. Gal-3, therefore, appears to be a crucial mediator of the GI damage observed after brain injuries, directly at the level of enteric neuronal cells and it may represent a potential new target for the treatment of post-stroke gastrointestinal complications. In this regard, it has been recently reported that elevated serum levels of Gal-3 correlates with poor outcome after acute heart failure and high Gal-3 values are observed in heart failure patients with a history of stroke [231].

7.3. Encephalitis

Gal-3 is expressed in a variety of glial cells of the central nervous system (CNS) tissues [232] and contributes to the migration of neuroblasts during brain development [233] and to the differentiation of oligodendrocytes [232]. Since Gal-3 is involved in inflammatory response and in the interaction with extracellular matrix, a potential role has been postulated in various types of CNS inflammation and in particular in viral encephalitis. Its expression was analyzed in some neuroinflammation animal models. It has been reported that Gal-3 is upregulated after long-term virus inoculation in a *Junin virus*-induced encephalitis model using newborn mice [234]. Gal-3 is rapidly upregulated in activated microglia, in the *encephalomyocarditis virus*-induced adult mouse encephalitis model of viral infection [235]. Gal-3 was found to be expressed in macrophages in remote spinal cord dorsal horn lesions caused by peripheral inoculation of *herpes virus* [236].

8. F

Fibrosis

Tissue fibrosis is characterized by the occurrence of tissue overgrowth, hardening, and/or scarring [237]. Tissue fibrosis may progressively evolve toward a severely debilitating disease, characterized by superabundant accumulation of extracellular matrix (ECM) including collagen. Tissue fibrosis represents the end result of chronic inflammatory reactions induced by a variety of stimuli including persistent infections, autoimmune reactions, allergic responses, chemical insults, radiation, and tissue injury and ultimately may lead to excessive tissue scarring, organ injury, function decline, and even failure [238–240]. Fibrosis is generally considered a major cause of morbidity and mortality worldwide because therapeutic options for the treatment of tissue fibrosis are rather limited. The only effective treatment in end-stage fibrotic disease is often represented by the organ transplantation. It is well-known that the mechanisms driving fibrogenesis are distinct from those regulating acute inflammatory reactions but few clear pathogenic and therapeutic targets have been identified so far. Recently, Gal-3 has emerged as a key regulator of chronic inflammation and tissue fibrogenesis and as an attractive new therapeutic target in the search for effective anti-fibrotic therapies [241].

Many studies indicate that Gal-3 is increased in patients with fibrosis affecting different tissues. In particular, Gal-3 is associated with fibrosis affecting the liver [170,242], the kidney [243,244], the lung [245,246], the heart [247,248] as well as the nervous system [249]. In addition, Gal-3 was found to be associated with systemic sclerosis (SSc), and in particular with skin fibrosis and proliferative vasculopathy observed in such condition [250]. Recently, another study indicated that the serum Gal-3 levels were higher in the SSc patients group compared with the controls group and they were higher in the active SSc group than in the inactive SSc group [251]. These data suggest that Gal-3 may be considered as a prominent biomarker of disease activity in SSc. In these studies examining the pathogenic role of Gal-3 in fibrosis, Gal-3 is identified as the link between macrophages, fibroblasts and the pro-fibrotic phenotype. It is known, in fact, that Gal-3 mediates IL-4 induced alternative macrophage activation and that IL-4/IL-13 activated macrophages upregulate profibrotic genes, stimulating matrix production and enhancing fibrosis [252–254]. These observations have led to the concept of a Gal-3/macrophage/fibroblast "axis".

On the basis of such observations regarding the profibrotic role of Gal-3, the measurement of its serum levels was proposed as a new prognostic factor [101]. In the Prevention of REnal and Vascular ENd-stage Disease (PREVEND), serum Gal-3 levels were measured in 7968 subjects that were followed-up for of approximately 10 years. Gal-3 serum levels were correlated with established risk factors for CV disease and mortality in subjects from the general population and gradually increased with age. After correction for such classical CV risk factors (smoking, blood pressure, cholesterol and diabetes) the fibrosis marker Gal-3 independently predicted all-cause mortality. The same observation was recently reported for patients with systemic sclerosis where Gal-3 was found to be the best predictor of the all-cause mortality in a group of 152 patients with systemic sclerosis that were followed-up for up to 10 years [255]. All these results indicate that Gal-3 may be considered as a useful new simple biomarker in the prediction of future CV events and mortality and its inhibition may represent a promising therapeutic strategy against tissue fibrosis.

9. G

Gastritis

Numerous infectious agents, such as *Trypanosma cruzi* [256], *human immunodeficiency virus-1* [257] and the *human T-lymphotropic virus-1* [258], have been reported to be able to induce Gal-3 upregulation. The first report regarding the association between Gal-3 and gastritis due to *Helicobacter pylori* infection was published in 2006 [259]. This study demonstrated that the expression of Gal-3 was upregulated by gastric epithelial cells following adhesion of *Helicobacter pylori*, suggesting that in addition to colonization this protein also plays a role in the host response to infection. Murine experimental models have recently confirmed the key role played by Gal-3 in the innate immunity against infection and in the colonization of gastric mucosa by *Helicobacter pylori* [260]. Administration of extracellular recombinant Gal-3 (rGal-3) was shown to inhibit the adhesion of H. pylori to the gastric epithelial cells resulting in a decrease in apoptosis. Gal-3 acts as a chemoattractant in the recruitment of THP-1 monocytes, working as a negative regulator of *Helicobacter pylori* infection and its effects on gastric mucosa [261]. Since there is a strong association between *Helicobacter pylori* infection and the development of gastric adenocarcinoma and mucosa-associated gastric lymphoma [262,263], it has been suggested that the expression and upregulation of Gal-3 may act a critical endogenous event in *Helicobacter pylori* infection that interferes with various intracellular events, causing prolonged cell survival, which is characteristic in carcinogenesis [264].

10. H

10.1. Heart

Although Gal-3 is expressed at a low constitutive expression level in the heart, under pathophysiologic conditions, the level of expression of Gal-3 may change substantially and, therefore, may play a prominent function.

The first observation regarding the role of Gal-3 in heart failure was published in 2004 [92]. In this study, performed on a rat model of severe hypertension and overt heart failure, the Gal-3 was the strongest regulated gene, being overexpressed in decompensated hearts. A more than five-fold upregulation was observed in rat with decompensated hearts compared to those with compensated hearts. These results were confirmed in human heart tissues, in the same study. Gal-3 expression analysis, performed in ventricular biopsies from patients with aortic stenosis with preserved versus depressed ejection fraction demonstrated an upregulation in the myocardial biopsies from patients with depressed ejection fraction.

The demonstration that Gal-3 is responsible for the induction of left ventricular remodeling and development of heart failure came from experimental studies, using rats in which Gal-3 was infused in the pericardial sac [265]. The structural and functional modifications of the rat myocardial tissues were prevented by the co-administration of N-acetyl-seryl-aspartyl-lysyl-proline (Ac-SDKP), a naturally occurring tetrapeptide, acting as a Gal-3 inhibitor. The mechanism of action of Gal-3 at the myocardium level is not entirely clarified yet and it may be a general event in left ventricular dysfunction caused by different etiologies. Once Gal-3 is overexpressed it is responsible for activation of fibroblasts and macrophages and consequent induction of fibrosis, scar production and, ultimately, of cardiac remodeling. Gal-3 appears to be involved in the control of interstitial fibrosis, especially in the overloaded hearts [266]. The demonstration of the role of myocardial Gal-3 expression in heart failure has prompted some authors to evaluate the possibility that Gal-3 serum levels may be useful in the diagnosis of acute heart failure in patients presenting with dyspnea [96]. According to what suggested by Kramer [267], all the clinical trials evaluating Gal-3 in cardiovascular disease patients can be categorized into four groups of studies, according to their aim:

- assessing the utility of the serum/plasma concentration of the lectin for diagnosis,
- helping in stratifying patients for therapy,

- monitoring therapy response, and
- predicting short- and long-term morbidity and mortality.

The clinical utility of Gal-3 in the diagnosis of heart failure was not clearly demonstrated and, in particular, Gal-3 did not show correlation with the NYHA functional classification of patients and, in addition, it showed lower specificity and sensitivity in identifying heart failure, as compare to the NT-proBNP [96,268]. Gal-3 proved to be more useful in as a marker for patient stratification for therapy. Low levels of serum Gal-3 (<19 ng/mL) were associated with better survival rates and lower cardiovascular event rates in patients treated with rosuvastatin therapy [269]. Other evidence regarding the clinical utility of Gal-3 in stratification for therapy came from the Multicenter Automatic Defibrillator Implantation Trial with Cardiac Resynchronization Therapy (MADIT-CRT) study [270]. In this study, patients with high baseline Gal-3 levels (in the top quartile of the distribution) benefited more (65% reduction in risk of the primary end point) from cardiac resynchronization therapy compared to those with lower baseline Gal-3 values (25% decrease in risk). The clinical utility of Gal-3 measurement in the serum as a marker for monitoring therapy was evaluated in patients that underwent to left-ventricular assist device (LVAD) implantation [271,272]. In such patients, however, unloading of the heart did not influence plasma Gal-3 levels within the first 30 days after device implantation. Gal-3, therefore, did not provide sufficient discrimination for prediction of outcome.

Much more promising are the studies regarding the clinical utility of Gal-3 measurement in the serum in predicting short- and long-term morbidity and mortality [94,101,105,273,274]. Many different clinical trials have been conducted so far to test the prognostic value of Gal-3 in patients with heart failure as well as in the general population. Gal-3 proved to be associated with mortality in 232 patients with chronic heart failure, followed for 6.5 years. Gal-3, therefore, represents a significant predictor of mortality risk in patients with heart failure, after adjustment for age and sex, and severity of HF and renal dysfunction, as assessed by NT-proBNP and estimated glomerular filtration rate, respectively [94]. On the basis of such evidence the Gal-3 blood test was proposed in the evaluation of patients with heart failure [105] and in the management of acute heart failure patients presenting to the emergency department [275]. Much interest came form the Prevention of REnal and Vascular END stage disease (PREVEND) clinical trial, the largest study with the longest observation period, aimed to investigate Gal-3 levels to date [101]. In such study a total of 7968 Caucasian individuals from the general population were examined and followed up for 10 years. It is interesting to note that Gal-3 serum levels increase with age and that females showed higher values than males. This study demonstrated that there is a clear association of Gal-3 serum levels and increased all-cause mortality, suggesting that measurement of Gal-3 in the serum may be useful not only to predict outcome in patients with heart failure but also in the general population.

At the present time, current guidelines indicate that the measurement of biomarkers of fibrosis, such as soluble ST2 and Gal-3, may be considered for additive risk stratification (class of recommendation IIa, level of evidence A) [276–278].

10.2. HIV

Human immunodeficiency virus (HIV) triggers decreases in both the number and function of CD4+ T lymphocytes. Its replication cycle consists of three stages: Attachment and entry, gene and protein expression and assembly and budding. Gal-3 was found to play a role in HIV-1 viral budding [279]. After infection HIV-1 can survive for a long period without gene expression in host cells or can cause continuous viral expression without cytopathic effects on the cells. HIV-1 principally targets CD4+ T cells and cells of the monocyte/macrophage lineage [280].

A specific role of Gal-3 expression in HIV infection was hypothesized many years ago. Gal-3 was found to be associated with early stages of HIV infection, in particular during transport and/or splicing of HIV mRNA. Gal-3 expression greatly increases after infection of Molt-3 cells with HIV type 1, concomitantly with the onset of expression of Tat protein [281]. The increase in Gal-3 expression was mediated at the transcriptional level. Gal-3 promoter, in fact, was significantly upregulated by expression vectors encoding the 40-kDa Tax protein, a potent transactivator in HTLV-1 [258]. Subsequent studies demonstrated that Tat protein expression is able to induce an upregulation of Gal-3 in several human cell lines, by means of Tat transactivation of Sp-1-rich regulatory sequences upstream of the Gal-3 gene [257]. The role of Gal-3 in HIV infection however, is not entirely elucidated yet. The effect on HIV infection, however, seems to be different, and perhaps opposite, from those exerted by Gal-1 [282]. Recently, it has been suggested that Gal-3 may be involved in the host response to HIV infection. According to this study, Gal-3 is induced in HIV-1-infected macrophages and may promote death of HIV-1/SIV-infected macrophages, thus contributing to eradicate HIV infection [283].

11. I

11.1. Inflammation

Besides its known expression in epithelial and endothelial cells, Gal-3 is also expressed in many different cells involved in mediating inflammatory response, such as macrophages [284], monocytes [285], dendritic cells [286], eosinophils [287], mast cells [288], natural killer cells [289], as well as T lymphocytes [290] and activated B lymphocytic cells [291]. In addition, Gal-3 is overexpressed in phagocytic macrophages and is, therefore, considered as a macrophage activation marker [292]. The role of Gal-3 in inflammation has been extensively reviewed in many studies [197,291,293–295].

Gal-3 is involved in many processes during the acute inflammatory response including neutrophil activation and adhesion [296], chemoattraction of monocytes/macrophages [285], opsonization of apoptotic neutrophils [75], and activation of mast cells [297]. Gal-3 is abundantly expressed and secreted by macrophages [298]. Furthermore, expression of Gal-3 is increased when monocytes differentiate into macrophages [298] and decreased when immature dendritic cells differentiate into mature dendritic cells [299]. Gal-3 also augments monocyte–monocyte interactions that lead to polykaryon (multinucleated giant cell) formation, a phenotype associated with alternative macrophage activation [300].

Since the role of neutrophils is considered crucial with respect to the immune response to microbial agents, several studies have investigated also the relationship between Gal-3 and neutrophils activation and migration. By using a murine pneumonia model system, Gal-3 was found to be accumulated into the alveolar space, following induction of pneumonia with *Streptococcus pneumoniae* and that this event correlated with the onset of neutrophil extravasation [301]. Interestingly, the accumulation of Gal-3 in lung was not observed during neutrophil emigration into alveoli induced by *Escherichia coli* infection. Thus, this effect appears to be specific for the alveolar infection caused by *Streptococcus pneumoniae*.

In addition, Gal-3 was found to display antimicrobial activity against the fungus *Candida albicans* [302].

Gal-3 effects on monocyte–monocyte interactions are also involved in chronic inflammatory and fibrotic diseases [303]. However, its role is complex and both pro-fibrotic and protective roles for Gal-3 have been reported in models of chronic inflammation and fibrogenesis. To further complicate the story, the effects of Gal-3 appear to be dependent on both the organ and type of injury model analyzed.

Chronic inflammation with resultant fibrosis, loss of tissue architecture, scarring and subsequent organ failure characterizes the pathogenesis of many chronic human diseases and is a major cause of morbidity and mortality. Tissue fibrosis is mediated by the fibroblasts and myofibroblasts, which represent key cells in the initiation and perpetuation of tissue fibrogenesis [304] and by the macrophages themselves, which are involved, in the same time, in both the evolution and resolution of tissue fibrosis [305]. Gal-3 and Gal-1 are potent mitogens for fibroblasts in vitro. Gal-3 is able to stimulate DNA synthesis and cell proliferation in quiescent cultures of normal human lung fibroblast [306] as well as in fibroblast from different tissues [307].

11.2. Interstitial Lung Disease

Interstitial lung diseases are a wide group of lung diseases in which the pathologic process involves the interstitium. Among these diseases, idiopathic pulmonary fibrosis (IPF) represents a separate nosographic entity, characterized by rapid progression toward lung failure and dead. IPF is characterized by restriction in lung capacity, due to scarring of lung tissue. IPF is an orphan condition that affects between 200,000 and 300,000 subjects in the Western world. Current therapies have little effect on the natural course of the disease and only partial treatment options are available. There are few studies in experimental mouse models and in humans on the role of Gal-3 addressing this issue. Gal-3 was increased in a mouse model of bleomycin induced lung fibrosis [308]. In humans, Gal-3 was increased in the serum of patients affected by IPF, and higher Gal-3 concentrations were associated with decreased lung volumes and altered gas exchange [309]. Furthermore, the increase of Gal-3 has been documented in the broncho-alveolar lavage fluid (BALF) from patients affected with IPF and this increase appear to be specific for IPF and interstitial pneumonia associated with collagen vascular disease (CVD-IP) because patients with other interstitial lung disease such as the cryptogenic organizing pneumonia/bronchiolitis obliterance organizing pneumonia (COP/BOOP), the acute hypersensitive pneumonia and the *Pneumocystis jiroveci* infection did not show any significant increase of Gal-3 in the BAL fluid [246]. The increase in Gal-3, measured in the BAL fluid was associated with an increase of Gal-3 protein expression inside the alveolar macrophages, detected by immunofluorescence. Gal-3 is produced and secreted by alveolar macrophages and its induction may be determined by pro-inflammatory cytokines. It is well known that TNF alpha, a pro-fibrogenetic cytokine, is able to induce in vitro Gal-3 production by macrophage/monocyte cell line model [310,311]. A positive feedback between TNF alpha and Gal-3 was hypothesized based on the ability of Gal-3 to induce TNF alpha production sustaining inflammation [246]. It was also reported that Gal-3 might participate in vascular remodeling, known to be present in fibrotic lung [312]. In this regard, Gal-3 was able to induce capillary tube formation of endothelial cells in vitro and to stimulate angiogenesis in vivo [313]. Gal-3 action on neovascularization observed in IPF may be due to both a direct effect on endothelial cells and an indirect effect through IL 8 production [246]. Based on the assumption that Gal-3 is responsible for a promoting effect on IPF and considering the lack of effective treatment for this pulmonary condition, it has been postulated that Gal-3 may represent a possible new target for treatment. According to this hypothesis, the inhibition of Gal-3 expression and activity would result in an amelioration of the clinical conditions of patients affected. Taking advantage of a new compound named TD139, a highly potent, specific inhibitor of the galactoside binding pocket of Gal-3 a novel target therapy for IPF was proposed and a clinical trial process was started in 2014. TD139 was formulated for inhalation, which enables direct targeting the fibrotic tissue in the lungs, while minimizing systemic exposure. A randomized, double-blind, multi-centre, placebo-controlled, multiple dose expansion cohort study, was designed to assess the safety, tolerability, PK and PD of TD139 (ClinicalTrials.gov Identifier: NCT02257177). Galecto Biotech Company recently announced the successful completion of phase Ib/IIa clinical trial of inhaled TD139, administered to human volunteers and IPF patients. The possibility to treat IPF by targeting Gal-3 molecule offer the chance to better treatment of such devastating disease and open a new era on the clinical use of Gal-3 antagonists.

12. J

Juvenile Idiopathic Arthritis

Juvenile idiopathic arthritis (JIA) is characterized by hyperplasia of synovial cells and accumulation of mononuclear inflammatory infiltrates. Such accumulation is generally thought to be due to the alteration of the local balance between cell proliferation and apoptosis. In agreement with this hypothesis, patients with polyarticular juvenile arthritis showed low apoptotic indexes, accompanied by low Bcl-2 expression, low proliferation rates and upregulation of Gal-3 expression in their synovial tissue [314]. Gal-3 upregulation in the synovial tissues was suggested to be the cause of the defective mononuclear apoptosis observed in synovial inflammatory infiltrates from these patients. Other evidence derives from the studies that have measured the serum levels in patients affected by JIA. Gal-3 serum levels were increased in a group of 50 active JIA children, compared to control subjects [315]. In addition, a positive correlation was demonstrated between serum Gal-3 levels and disease activity, i.e., the higher the level during exacerbations the higher it remained after remission. Finally, in such juvenile patients, the synovial fluid Gal-3 levels were found to be increased (approximately three times higher) compared to the serum level in paired samples. All these observations indicate that Gal-3 plays a crucial role in the pathophysiology of Juvenile idiopathic arthritis. Further studies including larger number of patients are needed, but it can be postulated that Gal-3 would be used in the future as a new biomarker for JIA disease, indicating activity, severity and progression of disease.

13. K

Kidney

Gal-3 is expressed during embryogenesis and regulates branching morphogenesis and metanephrogenesis [316]. Gal-3 plays a role also in the adult human kidney [317,318]. In normal adult kidneys, its expression is restricted to the collecting tubules and primary cilium. However, it has been reported in rats that its expression can be temporarily re-induced during regeneration after tissue injury [319]. Experimental studies, performed in mice and rats, demonstrated that Gal-3 is markedly upregulated in acute tubular injury and the subsequent regeneration [319] as well as in progressive renal fibrosis [243]. Gal-3, indeed, appears to be a potent activator of fibroblasts in the kidneys [320]. However, its absence is protective against renal myofibroblast accumulation and activation [243]. Many studies, performed on mouse models of diabetic nephropathy and acute renal failure, concordantly indicate that Gal-3 is upregulated in such conditions [215,243,319]. Interesting results were obtained in studies performed also in humans with renal failure and especially with those affected by diabetes. Gal3, in fact, was found to be upregulated in the glomeruli of diabetic patients and the number of Gal-3 positive cells in the glomeruli correlated with the observed urinary protein excretion [321], suggesting a potential prognostic role of Gal-3 in predicting a poor outcome of such patients. In this regard, one of the most interesting observation regarding the correlation between Gal-3 and kidney function comes from the LUdwigshafen RIsk and Cardiovascular health (LURIC) study, based on data from 2578 patients [322] and from 4D studies based on data obtained from 1168 patients [323]. The analysis of Gal-3 plasma levels, measured in patients from these two large prospective studies, demonstrated a significant association of Gal-3 levels with cardiovascular end points, infections, and all-cause death in patients with impaired renal function [324]. The correlation was observed during a period of follow-up extended to 10 years (LURIC study) or four years (4D study) and the increase of Gal-3 levels was observed in parallel with the decrease in kidney function, with a marked elevation in dialysis patients affected by T2DM. Based on such evidence, Gal-3 was proposed a novel treatment opportunity for renal disease. Kolatsi-Joannou et al. investigated changes in the expression of Gal-3 in mice with acute kidney injury induced by folic acid, using modified citrus pectin (MCP), a pectin derivative that can bind to the Gal-3 carbohydrate recognition domain [325]. At two weeks of the recovery phase, MCP-treated mice demonstrated reduced Gal-3, with decreased renal fibrosis, macrophages, pro-inflammatory cytokine expression, and apoptosis. Their findings reveal that pharmacologic inhibition of Gal-3 with MCP was protective in experimental nephropathy by modulating early proliferation and later, Gal-3 expression, apoptosis, and fibrosis. Thus, MCP may be considered and proposed as a novel target to reduce long-term renal injuries, attenuate renal fibrosis and preserve kidney function, possibly via the effect of Gal-3 on carbohydrate binding-related functions [325]. A recent phase II study comparing treatment with another Gal-3 inhibitor, namely the GCS-100, and a placebo in patients with CKD stage 3b showed that GCS-100 significantly improved the eGFR [326]. Taken together, Gal-3 may be used to assess the prognosis, guide therapy, and potentially suggest specific anti-Gal-3 therapy. However, prospective studies are needed to validate these hypotheses.

14. L

Liver Fibrosis

Liver fibrosis represents the final pathway of chronic damage to the liver, in conjunction with the accumulation of extracellular matrix proteins [327]. Chronic inflammation with the formation of scar tissue, loss of tissue architecture, and organ failure is a characteristic feature of this disease. An increased in Gal-3 expression has been previously reported in fibrosis in different types of tissues [170,308,328]. The main causes of liver fibrosis include chronic HCV infection, alcohol abuse, and nonalcoholic steatohepatitis (NASH). Hepatic stellate cells (HSCs) are considered the main collagen-producing cells in the liver. Gal-3 was found to play a critical role in the regulation of HSC activation, both in vitro and in vivo [329]. In particular, the HSC activation to a myofibroblast phenotype, a critical event in extracellular matrix deposition and in the development of cirrhosis, is dependent on the upregulation of Gal-3 expression within liver cells, in the areas of fibrotic tissue injury [327]. Based on such experimental and clinical evidence, a Phase 2b NASH-CX trial is ongoing (ClinicalTrials.gov Identifier: NCT02462967) to treat NASH cirrhosis patients without esophageal varices [330]. It is a randomized, double-blind, placebo-controlled clinical trial in patients with NASH cirrhosis without varices and is aimed to target Gal-3 expression by means of an antagonist developed by Galectin Therapeutics Inc. (Norcross, GA, USA) and named GR-MD-02. Preliminary results, released by the company on 2 December 2017, indicate a positive trend in treated patients versus placebo control patients, with an improvement in portal hypertension or liver biopsy. The results of this trial could lead to a better treatment of such condition and to effective prevention of its progression toward cirrhosis.

15. M

Mortality

Many different studies have reported that Gal-3 concentrations are associated with incident HF and mortality in the general population and measurement of serum Gal-3 levels may be useful in predicting mortality for all-cause and cardiovascular disease (CVD) [101–103,331]. Gal-3, in fact, was a significant predictor of CVD mortality and of all-cause mortality in a large population study, based on 1393 participants without CVD with a mean age of 70 years, who were followed up for a mean of 11 years for the occurrence of coronary heart disease, CVD mortality, and all-cause mortality [103]. The same results were reported by the Framingham Heart Study, based on 3353 participants in the Framingham Offspring Cohort (mean age 59 years, 53% women), in which Gal-3 was associated with risk of all-cause mortality [102] and by the Prevention of Renal and Vascular ENd-stage Disease (PREVEND), based on 7968 subjects where Gal-3 levels independently predicted all-cause mortality [101].

16. N

Non-Alcoholic Steatohepatitis (NASH)

Non-alcoholic steatohepatitis (NASH) is a condition in which fat builds up in the liver along with inflammation, composed predominantly of lymphocytes and Kupffer cells, and liver damage, in the absence of a history of alcoholism. It is more frequent in patients with obesity, metabolic syndrome and T2DM and is the most common chronic liver disease, with a rapidly increasing prevalence worldwide. The hallmarks of the disease are the presence, at histology, of steatosis, lobular inflammation, hepatocellular ballooning and fibrosis, which typically begin in zone 3 and show a characteristic perisinusoidal/pericellular ("chicken wire") pattern [332]. A link between Gal-3 and hepatic fibrosis was established when mice that lack the Gal-3 gene were found to be resistant to liver fibrosis induced by toxin administration [242]. Such observations are in agreement with studies in other organs and systems regarding the role of Gal-3 in activating a variety of profibrotic factors, promoting fibroblast proliferation and transformation, and mediating collagen production [333]. The effects of Gal-3 knockout on NASH are rather controversial. It has been demonstrated that such mice are resistant to fat accumulation, inflammation and fibrosis when fed a high-fat diet [334]. However, other groups have demonstrated that Gal-3 null mice show an increased propensity to develop NASH [335]. An attempt to explain such discrepancy was recently given [336]. According to this study, Gal-3 null animals are more prone to develop greater liver steatosis but there is decreased inflammation, liver cell injury and fibrosis. Since Gal-3 appears to be a critical regulator of liver fibrosis attempts have been made to inhibit its activity by means of a specific inhibitor, named GR-MB-02, in mice [337], in rats [338] and in humans [339]. In particular, the human study was a sequential dose-ranging, placebo controlled, double-blinded study and was undertaken to evaluate the safety, pharmacokinetics profiles and to evaluate changes in potential serum biomarkers and liver stiffness as assessed by FibroScan in subjects having NASH with bridging fibrosis. GR-MD-02 proved to be safe and well tolerated with evidence of a pharmacodynamic effect. Such results constitute the basis for a Phase 2 development program of treatment with GR-MD-02 in advanced fibrosis due to NASH. At the present time, the smaller four-month long phase 2a trial, involving a small set of 30 NASH patients with advanced fibrosis (NASH-FX) at a single site, indicate that the results missed the primary goal and a few secondary endpoints [340], but additional positive results are expected from the larger in-progress phase 2b clinical trial (NASH-FX) [341].

17. O

Obesity

Obesity is a common risk factor for metabolic syndrome, a group of several diseases that include insulin resistance, hypertension, glucose intolerance and toxicity, hepatic steatosis, atherogenic dyslipidemia, and T2DM [342]. Obesity correlates with chronic, low-grade inflammation, which appears to be necessary for obesity-associated insulin resistance and the subsequent development of T2DM [343]. During the development of obesity, macrophages permeate white adipose tissue and, together with adipocytes, secrete various pro-inflammatory cytokines and chemokines [344,345]. In obese individuals the visceral adipose tissue is infiltrated by macrophages that cooperate in generating and sustaining the inflammatory responses [346]. Gal-3 is expressed in all types of macrophages that permeate white adipose tissue and it is an important regulator of both polarization and activity, thus mediating both pro-inflammatory and anti-inflammatory effects in such cells [347,348]. Experimental evidence, using Gal-3$^{-/-}$ mice indicate that in vivo Gal-3 administration causes glucose intolerance and insulin resistance whereas in vitro treatment with selective small molecule Gal-3 inhibitors can directly induce decreased insulin sensitivity in myocytes, hepatocytes, and adipocytes [349]. Such experiments suggest a possible interesting link between inflammation and decreased insulin sensitivity.

18. P

18.1. Pneumonia

The Gram-positive bacterium *Streptococcus pneumoniae* is the most common cause of community-acquired pneumonia, still responsible of a high mortality rate, especially in children and in developing Countries [350]. In the lung, resident alveolar macrophages are the first line of cellular defense and play a phagocytic role during the early stages of infection. Gal-3 is abundantly expressed and secreted by macrophages [298] and clearance of apoptotic neutrophils by macrophages is a key step in the resolution of inflammation. Using an in vivo pneumonia mouse model in which Gal-3 was knocked out, it has been demonstrated that Gal-3 is involved in the recruitment of neutrophils during lung infections with *Streptococcus pneumoniae* [301]. In addition, it was demonstrated that the absence of Gal-3 is responsible for an inadequate host response to pneumococcal infection, resulting in high mortality of the Gal-3-null mice [74,351,352]. In agreement with this observation, the exogenous administration of Gal-3 was able to enhance phagocytosis of *Streptococcus pneumoniae* by both WT and Gal-3$^{-/-}$ neutrophils [74,351]. In the same studies it has been demonstrated that administration of exogenous Gal-3 had a direct antimicrobial, bacteriostatic effect on *Streptococcus pneumoniae*. Gal-3, in fact, is able to inhibit *in vitro* growth of *Streptococcus pneumoniae* and to reduce the severity of bacteremia and of the pneumonia in vivo. These very interesting and promising effects of Gal-3 may have relevant clinical implications in the treatment of *Pneumococcal pneumonia* in humans but needs to be further investigated.

18.2. Pulmonary Hypertension

Pulmonary arterial hypertension (PAH) is a heterogeneous group of diseases. It includes the idiopathic pulmonary hypertension characterized by a persistent increase in pulmonary arterial resistance often related to small pulmonary arterial vessel remodeling and endothelial metabolic alteration with consequent deterioration of lung and heart functions. During PAH increase of pulmonary pressure is determined by microvascular inflammation, endothelial dysfunction, activation of vascular smooth muscle cells and vascular thrombosis [353]. In this context, Gal-3 is able to modulate fibroblast and endothelial behavior acting on fibroblast and vascular endothelial growth factors [354]. Currently, there are few and contrasting studies addressing the precise role of Gal-3 in PAH development and progression. This confusion may be due to the different PAH patient characteristics like the presence of comorbidities or diseases secondary complicated by PAH like connective tissue diseases, respiratory or hearth diseases. Song et al. reported in 2016 that Gal-3 plasma levels are decreased in patients with PAH in comparison with healthy controls [355]. Moreover, Gal-3 mRNA was down-regulated in the pulmonary vessels in pulmonary hypertensive rats [355]. Contrarily, elevated level of Gal-3 has been reported in PAH independently to etiology and associated with severity of PAH [356,357]. Gal-3 plasma levels are correlated with right ventricle dysfunction [358] and with extracellular matrix markers [356,357]. Gal-3 seems to play a key role in acute and chronic inflammatory responses to endothelial injury, vascular remodeling and fibrosis in PAH [359]. The serial measurement of Gal-3 serum levels was suggested as a useful marker to improve prediction and progression of PAH [360]. A preclinical study demonstrated that Gal-3 might be also useful as a target for the treatment of this condition. In fact, the treatment with Gal-3 inhibitors was able to counteract lung and heart remodeling and fibrosis in a rat model of PAH [361].

18.3. Plaque Psoriasis

Gal-3 is abundantly expressed in the cytoplasm of normal keratinocytes, in hair follicles, sebaceous, and sweat glands [87]. Moreover, Gal-3 is associated with keratinocyte differentiation and maturation [362].

Plaque psoriasis is one of the most prevalent autoimmune skin diseases. It is characterized by the presence of well-demarcated red silvery scales on the skin surface. Gal-3 was expressed in a high proportion of Langerhans cells from skin punch biopsies, obtained from lesional plaque-type psoriatic skin [363]. In addition, a strong expression of Gal-3/Gal-3-reactive glycoligands was detected in capillaries of psoriatic dermis [364].

However, the most relevant indication regarding the role of Gal-3 in psoriasis was obtained unexpectedly during the exploratory phase 2 clinical trials on the use of GR-MD-02 in nonalcoholic steatohepatitis ("NASH") (ClinicalTrials.gov Identifier: NCT01899859). A side effect of the treatment was, in fact, the evidence of long-term remission or significant improvements in disease activity of psoriasis in patients treated with GR-MD-02. This finding suggested that blocking Gal-3 could be effective in treating moderate to severe plaque psoriasis patients. That is why a new single-center, open-label phase 2a clinical trial was initiated to evaluate safety and efficacy of GR-MD-02 for treatment of psoriasis (ClinicalTrials.gov Identifier: NCT02407041).

19. Q

Q Fever

The etiologic agent of Q-fever is a highly infectious human pathogen named *Coxiella burnetii* [365]. *Coxiella burnetii* is an obligate intracellular bacterium that inhabits mainly monocytes/macrophages. It is known that galectins are recruited to injured membranes by their interactions with N-glycans that are exposed to the cytoplasm when the membrane of a vesicle is pierced and, therefore, are considered useful molecules to detect damage of intracellular membranous compartments. In this regards, there is evidence derived from infection by two bacterial pathogens, namely the *Shigella flexneri*, a rod-shaped, Gram-negative bacterium responsible for bacillary dysentary and the *Listeria monocytogenes*, a rod-shaped Garm-positive bacterium that causes severe meningitis and bacteriemia in immunocompromised hosts, that demonstrated that Gal-3 accumulates in structures in close proximity to phagosomes containing bacterial pathogens and are useful to detect membrane damage in bacteria's vacuoles [366–368]. These bacteria share similar lifestyles in the sense that by producing surface proteins they both lyse their phagosome and induce their own phagocytosis by host cells [369]. In such bacteria it has been postulated that Gal-3 may function as a danger receptor for membrane damages [366,367]. Similar to the behavior of these two pathogens also *Coxiella burnetii* was found to transits through the endocytic pathway, which delivers the bacteria to the low pH environment of a lysosome. Several members of the galectin family interact with the *Coxiella burnetii* containing vacuole, at different times of infection. The role of the expression of Galectins, and in particular of Gal-3 in the lysosomes of the infected cells appears to be crucial. It is believed that when lysosomes are damaged they are marked with Gal-3 and, as a consequence, can be targeted to autophagy [370]. This study points to Gal-3 as a sensor of damage at the *Coxiella burnetii* limiting membrane and indicates a possible role of Gal-3 in the mechanisms of autophagy during membrane repair.

20. R

20.1. Rheumatoid Arthritis

In contrast to Gal-1 that act as a negative regulator of autoimmunity in rheumatoid arthritis (RA), Gal-3 promotes inflammation in RA. The role of Gal-3 as a pro-inflammatory mediator in RA is based on several lines of evidence, both in animals and in humans [371].

20.1.1. Experimental Studies in Animals

Studies with collagen-induced arthritic (CIA) rats found increased Gal-3 secretion into the plasma over time, which correlated with the disease progression, implicating that Gal-3 promotes the development of experimental arthritis [372]. Recent studies with Gal-3-deficient mice further confirmed the stimulating role of Gal-3 in arthritis [373]. The joint inflammation and bone erosion of antigen-induced arthritis were markedly suppressed in Gal-3-deficient mice as compared with the wild type mice. The reduced arthritis in Gal-3-deficient mice was accompanied by a decreased level of antigen-specific IgG and pro-inflammatory cytokines, including TNFα, IL-6, and IL-17. Furthermore, an exogenous supply of recombinant Gal-3 restored the reduced arthritis and cytokine production in Gal-3-deficient mice [373]. This study provided the direct evidence that Gal-3 plays a crucial role in the development of arthritis in animal models.

20.1.2. Clinical Studies in Humans

The role of Gal-3 in RA was also demonstrated in humans where Gal-3 was found to be increased in the serum and synovial fluid of RA patients with long-standing disease, compared with osteoarthritis

(OA) as well as in JIA patients [374–376]. The serum levels of Gal-3 were elevated in patients with RA, JIA, Behçet's disease, or systemic sclerosis [250,315,374,377]. Although the increased Gal-3 is not specific for RA, the serum levels of Gal-3 were significantly associated with the C-reactive protein (CRP) levels and the disease activity scores in patients with JIA, suggesting that Gal-3 may be utilized as a biomarker for the disease progression of JIA [315]. In addition, the *LGALS3* gene allele (*LGALS3 + 292C*) is more prevalent in RA patients than in healthy controls, indicating that genetic polymorphisms of Gal-3 may influence the susceptibility to RA [378]. In addition to immune cells, fibroblast-like synoviocytes (FLS) in the synovium of RA patients also express Gal-3 at high levels [379–381]. While floating FLS only express low levels of Gal-3 [379], adhesion of FLS to cartilage components through CD51/CD61 induces Gal-3 expression [375]. In RA patients, about 39% of FLS are cartilage-adhering cells, which is four times more than in osteoarthritis (OA) patients. The increased numbers of adhering FLS contribute to the elevated Gal-3 levels in the RA synovium [375]. Moreover, Gal-3 can induce rheumatoid FLS to secret a set of pro-inflammatory cytokines and chemokines including IL-6, granulocyte-macrophage colony-stimulating factor (GMCSF), TNF, CXCL8, CCL2, CCL3, and CCL5 [381]. The induction of cytokines and chemokines by Gal-3 appears to involve different signaling pathways. The MAPK-ERK pathway was necessary for cyotokine IL-6 production, while phosphatidylinositol 3-kinase (PI3K) was required for chemokine CCL5 induction [381]. These studies using human materials further suggest a promotional role of Gal-3 in the pathogenesis of RA. In addition, Gal-3 mRNA and Gal-3 binding protein were highly expressed in RA joints, particularly at the sites of cartilage and bone destruction [377]. Gal-3 was reported to contribute to the activation of RA synovial fibroblasts [377] and to promote the production of pro-inflammatory cytokines, such as interleukin-6 and tumor necrosis factor-α, in synovial fibroblasts [381,382]. For all these reasons the Gal-3 serum level was proposed as a predictive marker of future RA [383]. According to this study, the increase of serum levels of Gal-3 may help in discriminating between patients with undifferentiated arthritis who subsequently developed RA and those with undifferentiated arthritis who do not develop RA and in this sense it works better than other serological markers of cartilage regeneration (such as the type IIA collagen N-terminal propeptide, PIIANP) or synovial swellings (such as the hyaluronan, HYA). The therapeutic administration of lentiviral vectors encoding Gal-3 small hairpin RNA (shRNA), directly injected intra-articularly, into the ankle joints of rats with collagen-induced arthritis, significantly ameliorates the disease activity [384]. This treatment, in fact, reduced articular index scores, radiographic scores and histological scores and was accompanied with decreased T-cell infiltrates and reduced microvessel density in the ankle joints. All these experimental findings indicate that Gal-3 emerges not only as a key player in the pathogenesis of RA [385], but also as a novel potential therapeutic target and implicate that down-regulation of Gal-3 may represent novel a therapeutic strategy for RA.

21. S

21.1. Sepsis

Sepsis is a complex immune disorder with a high mortality rate of 20–50% that currently has no therapeutic interventions [386]. There are few studies that suggested a possible role of Gal-3 in inflammatory conditions, including endotoxemia and airway inflammation [71,387,388].

The relationship between Gal-3 and sepsis was recently studied using a murine inhalation model, infected with *Francisella novicida*, a Gram negative bacterial pathogen, responsible for the development of severe sepsis characterized by hyperinflammation, T cell depletion, and extensive cell death in systemic organs. The measurement of both mRNA and protein levels of Gal-3 in lungs of mice undergoing lethal pulmonary infection with such pathogen, indicated a significant increase in Gal-3 expression [386]. This increase was evident 3 days after lethal injection and was associated with the appearance of other sepsis features, such as extensive cell death, hyperinflammatory response and increased vascular injury. When the same treatment was given to Gal-3$^{-/-}$ mice, a significant reduction in levels of several vascular injury markers and of several inflammatory cytokines was observed. In addition, Gal-3$^{-/-}$ mice showed significantly improved survival as compared to the infected wild-type mice. These results indicate a strong immune-stimulatory role of Gal-3 during pulmonary lethal infection and indicate that Gal-3 may represent a potential target for treatment of sepsis, at least for that induced by *Francisella novicida*.

However, the situation may be different when other pathogens are involved. In fact, lipopolysaccharide expressed on *Escherichia coli* has been shown to down-regulate Gal-3 expression [298,389].

Results similar to those observed with *Francisella novicida* were obtained also in the case of infection with *Streptococcus pneumonia*. In this regard, it has been demonstrated that, after pneumococcal infection of the lungs, Gal-3 accumulates in the alveolar space and this accumulation correlates with the onset of neutrophil activation [301]. Infection of mice with the Gram-positive *Streptococcus pneumoniae* was more severe when it was induced in Gal-3$^{-/-}$ mice, compared to wild-type, and Gal-3 released from resident alveolar macrophages enhance their bacterial killing and phagocytic capability of neutrophils, aiding pathogen clearance [351]. These data suggest that, conversely to what reported in sepsis induced by *Francisella novicida*, Gal-3 may play a protective role against sepsis due to pulmonary infection by *Streptococcus pneumoniae*. Recently, the hypothesis that Gal-3 could be useful as a marker to predict mortality was tested in a group of 273 patients affected by sepsis and analyzed retrospectively [390]. Gal-3 was measured in association with other biomarkers, such as procalcitonin, presepsin, and soluble ST2. Gal-3 showed the strongest prognostic value in predicting mortality in septic patients. Further studies are required to elucidate the real diagnostic and prognostic usefulness of Gal-3 in septic patients as well as its potential therapeutic use.

21.2. Systemic Sclerosis

Systemic sclerosis (SSc), also known as scleroderma, is a connective tissue disease characterized by vasculopathy and progressive fibrosis of the skin and certain internal organs, due to activation of fibroblasts and progressive accumulation of extracellular matrix molecules, eventually leading to internal organ dysfunction [391]. Considering the known relevant role of Gal-3 in fibrosis [333] it is obvious that it was analyzed in a disease such as SSc in which progressive fibrosis of the skin and of visceral organs represents a major component of the disease. Serum Gal-3 levels were, in fact, analyzed in patients affected by SSc in some studies. They were found to be significantly decreased in early diffuse cutaneous SSc, compared with the control subjects, but showed increase during the course of the disease [250]. In addition, in diffuse cutaneous SSc, a significant correlation was reported between serum Gal-3 levels and the developmental process of skin sclerosis and of digital ulcers [250]. In another study, serum Gal-3 levels were found to be higher in SSc patients than in healthy controls [251]. In order to evaluate the possible associations between Gal-3 levels and patient characteristics and to investigate its long-term prognostic value, a large population of 152 patients that fulfilled the American College of Rheumatology criteria for SSc [392] was recently analyzed [255]. In multivariate Cox regression analyses and in a long-term follow-up (7.2 ± 2.3 years), Gal-3 was an independent predictor both of the all-cause and of cardiovascular-related mortality, even after adjustment for age, gender, BSA, creatinine and NT-proBNP levels. In particular, patients with elevated serum Gal-3 levels (>10.25 ng/mL) showed a clear association with the signs of the advanced organ sclerosis and with the laboratory parameters of inflammation. Gal-3 appears to be a reliable biomarker to predict all-cause mortality in patients affected by SSc.

22. T

Target Therapy

Gal-3 is involved in the regulations of a myriad of cell activities. As a multifunctional protein widely expressed by many types of human cells, it interacts with a range of different binding partners at different locations. In addition, its overexpression in many different conditions and changes of its sub-

and inter-cellular localization are commonly seen in various types of human diseases. Gal-3 research is one of the fastest growing fields within medicine today and Gal-3 is becoming the most informative new biomarker for our most serious health epidemics for which early detection is critical to successful clinical prognosis. A growing body of evidence demonstrates, in fact, how Gal-3 is involved in numerous degenerative processes within the body, most notably cancer proliferation/metastasis, heart failure, atherosclerosis, diabetes, chronic inflammation, fibrosis and related organ failure. In addition, Gal-3 was found to be a critical prognostic factor even in all-cause mortality. Therefore, it is not surprising that Gal-3 represents an attractive molecule to be targeted for the development of new strategies in the treatment of many different diseases [64]. The studies concerning the possibility to inhibit Gal-3 function were driven by the experiments using Gal-3 null mice. Basically, two different approaches were taken in developing drugs that target the Gal-3 molecule. One approach is based on the use of modified disaccharides. One example of such compound is the TD139 (Galecto Biotech). Another approach is to use naturally occurred large polysaccharides that contain galactose. Example of such naturally occurred large polysaccharides are the modified citrus pectins (MCP), and in particular: The GR-MD-02, belonging to the galacto-rhamnogalacturonate (GR) class and the GM-CT-01, belonging to the galactomannan (GM) class, by Galectin therapeutics Inc. (Norcross, GA, USA) [393]. Recently, another MCP compound, named PectaSol-C, was introduced and commercialized by EcoNugenics (Santa Rosa, CA, USA). It was demonstrated to be able to reduce Gal-3 serum levels in vivo in both mice [394] and humans [395].

The Gal-3 antagonist named GCS-100 is a MCP, capable of binding to and antagonizing Gal-3, developed (La Jolla Pharmaceutical Company, San Diego, CA, USA) and utilized in a clinical trial (ClinicalTrials.gov Identifier: NCT00776802) for the treatment of refractory diffuse large B-cell lymphoma and refractory solid tumors [396]. The trial was discontinued in 2015 because of lack of funding. The same drug was used in another interventional trial (ClinicalTrials.gov Identifier: NCT01843790) for the treatment of patients with chronic kidney disease. Pectins are available as various types of dietary fibers (e.g., fruit, vegetables, sugar beets). They constitute the fibers found in the pith (the bitter, white stuff) in the peels of fruit, particularly rich in oranges, grapefruits, and lemons. Pectins are commonly used as a thickening agent in jams and baked goods. Pectins have been identified as Gal-3 inhibitors. Both pectins and MCP bind to the CRD of Gal-3 and neutralize its activity [397]. They were reported to be able to inhibit cancer growth and to be an effective anti-fibrotic agent in mouse models [337,338,398–400].

A large numbers of clinical trials are now ongoing for many different clinical conditions. Currently, there are 9 interventional clinical trials ongoing regarding the therapeutic use of Gal-3 inhibitors, registered at the web site *ClinicalTrials.gov*, maintained by the National Library of Medicine (NLM) at the National Institutes of Health (NIH) [401]. They include chronic inflammation and fibrosis affecting the lung (ClinicalTrials.gov Identifier: NCT02257177) and the kidney (ClinicalTrials.gov Identifier: NCT01717248 and NCT01843790), cirrhosis and portal hypertension (ClinicalTrials.gov Identifier: NCT02462967), high blood pressure (ClinicalTrials.gov Identifier: NCT01960946), osteoarthritis (ClinicalTrials.gov Identifier: NCT02800629), and combination immunotherapy for cancer, including melanoma (ClinicalTrials.gov Identifier: NCT02117362 and NCT01723813), large B-cell lymphoma (NCT00776802, withdrawn prior to enrollment for lack of funding). In addition, the occurrence of serendipity has led to expand the potential clinical use of Gal-3 inhibitors also to skin diseases. In fact, administration of GR-MD-02 to a patient enrolled in a Phase I NASH trial induced a remarkable remission of psoriasis, indicating the potential importance of Gal-3 in such disease. Following this observation, an open-label, phase 2a study was started in 2015 to evaluate safety and efficacy of GR-MD-02 for the treatment of psoriasis (ClinicalTrials.gov Identifier: NCT02407041). One could argue that a new science is born, called "Pectin Science" or "Pectinology" [400]. The analysis of health promoting capacities of Pectins is progressing fast and will provide interesting results in the near future.

23. U

Urinary Tract Infections

Various galectins have been shown to bind a wide range of pathogens, such as Gram-positive bacteria (e.g., *Streptococcus pneumoniae*), Gram-negative bacteria (e.g., *Klebsiella pneumoniae, Neisseria meningitidis, Neisseria gonorrhoeae, Haemophilus influenzae, Pseudomonas aeruginosa, Porphyromonas gingivalis, Helicobacter pylori* and *Escherichia coli*), enveloped viruses (*Nipah* and *Hendra* paramyxoviruses, *human immunodeficiency virus-1, HIV-1,* and *influenza virus A*), retrovirus (*human T-lymphotropic virus, HTLV-1*), hantavirus, fungi (*Candida albicans*) and parasites (*Toxoplasma gondii, Leishmania major, Schistosoma mansoni, Trypanosoma cruzi,* and *Trichomonas vaginalis*) [259,282,402–410]. The specific role of Gal-3 in urinary tract infections in humans was analyzed in case of viral and bacterial infections. Gal-3 serum levels were increased in patients with chronic viral infectious disease such as hepatitis C [411] and HIV [412]. In particular, an elevated Gal-3 serum level was associated with unresponsiveness to alpha-interferon treatment and a strong association was observed between viral load and Gal-3 expression in HIV-infected patients. Gal-3 was also investigated in case of urinary tract infection due to bacterial infection. Gal-3 was found to play a crucial role in bacterial infections, and in particular in infections due to *Proteus mirabilis*, a bacterium frequently detected in the urinary tract. Gal-3 influences adhesion of such bacterium to the extracellular surface of the plasma membrane of different kidney epithelial cell lines and its crucial role in *Proteus mirabilis* infection was demonstrated using specific monoclonal antibodies directed against Gal-3 [413]. These studies indicate that Gal-3 plays a key role of in renal pathology due to infections [414].

24. V

Venous Thrombosis

Gal-3 is associated with thrombogenesis and in particular, it is characterized by prothrombotic and pro-inflammatory properties in the context of experimental venous thrombosis. Serum levels of Gal-3 and of Gal-3 binding protein are elevated in the course of venous thrombosis in both mice and humans [415]. Circulating Gal-3 is particularly elevated during the early phases of venous thrombosis and its level correlates with the size of the thrombus. Gal-3 appears to be increased both locally, in the thrombus and in the vein wall, and systemically, in blood-circulating elements, such as red blood cells, with blood cells, platelets and microparticles. Therefore, Gal-3 may represent an ideal target for therapies aimed to prevent venous thrombosis/pulmonary embolism [415].

25. W

25.1. Wound Healing

Many studies, performed in epithelium form different tissues, indicate that one of the best-known functions of Gal-3 is to promote cell migration and wound re-epithelialization [416–419]. Impaired wound healing and healing defects may affect the epithelium in many different organs systems, including the skin, the gastrointestinal tract and the corneal epithelium [420–424]. In addition, the damage of delayed re-epithelialization and the occurrence of persistent epithelial defects are also observed in chronic wounds of the elderly, decubitus ulcers as well as in venous statis ulcers of the skin. The major defect responsible for the failure to re-epithelialize in generally thought to be the reduced ability of the epithelial cells to migrate across the wound healing bed more than a defect in cell proliferation [425] and Gal-3 is able to modulate cell migration through its effects on cell adhesion and its interactions with cell-matrix. For these reasons Gal-3 is considered a key player in the re-epithelialization of wounds in many different tissues [426].

25.1.1. Wound Healing in the Cornea

The effects of Gal-3 in the process of corneal re-epithelialization of wounds have been analyzed using a Gal-3 null mouse model [416]. In such mouse model, the absence of Gal-3 impairs keratinocyte migration and skin wound re-epithelialization. Using a mouse model in which corneas were injured using both laser or alkali treatment it has been demonstrated that corneal epithelial wound closure rate was significantly slower in Gal3$^{-/-}$ mice compared with wild-type mice [416]. Since the rate of corneal epithelial cell proliferation is not altered in Gal3$^{-/-}$ mice, it is likely that the effect is due to impairment in the cell migration process. To confirm this observation the mouse corneas were incubated in in serum-free media, in the presence of different amounts of recombinant Gal-3 protein. Exogenous Gal-3 was able to stimulate the rate of wound closure in a concentration-dependent manner [416]. The same effect was demonstrated the corneal epithelium of rats [427] and monkeys [428].

25.1.2. Wound Healing in the Skin

It is known that the EGF-stimulated EGFR-ERK signaling is essential for keratinocyte migration, a crucial step for skin wound re-epithelialization [429–431]. The role of Gal-3 in skin wound re-epithelialization was analyzed in a Gal-3$^{-/-}$ mouse model [432]. Keratinocyte migration is impaired in Gal-3$^{-/-}$ mice. Gal-3 exerts this effect by regulating the intracellular trafficking of EGFR.

25.1.3. Wound Healing in the Intestinal Tract

Using the in vitro model of T84 human colonic adenocarcinoma cell line, it was demonstrated that the Gal-3 promotes wound healing and that the cleavage of Gal-3 by Matrilysin-1 (MMP7) results in the overall abrogation of such wound healing capacity [433]. Based on this observation it has been hypothesized that cleavage of Gal-3 is one mechanism by which MMP7 inhibits wound healing. The relevance of such finding and its possible clinical role in conditions such as intestinal ulcers and chronic inflammatory bowel diseases still needs to be elucidated.

26. X

X Syndrome of the Heart

The X syndrome of the heart, also worded as cardiac syndrome X (CSX) and differing from the other X syndrome of the metabolism, also known as metabolic syndrome X (MSX), refers to the triad of typical effort angina, positive stress test and normal epicardial coronary angiogram [434]. It can be diagnosed in about 3–11% of patients undergoing coronary angiography because of typical chest pain [435]. Follow-up studies of patients with CSX generally report good prognosis [436,437]. A systematic review of the literature shows that the CSX prognosis in terms of the overall cardiac event rate is excellent, with a risk of myocardial infarction or cardiovascular death of 1.5% per five years [438]. However, even if such patients have been considered to be a low-risk population the occurrence of cardiovascular events in such patients is not infrequent, especially in women [439]. Many attempts have been made to find useful markers for identifying the CSX patients at high risk of cardiovascular disease, and CRP, interleukin-6 and uric acid as well as the occurrence of perivascular fibrosis, with higher carotid mean intima-media thickness, and apoptosis of endothelial cells in the small blood vessels were found to be associated with such syndrome [440–442]. Serum levels of Gal-3 were measured in a cohort of 115 patients affected by CSX demonstrated. They were significantly higher in the CSX patients in comparison with the healthy controls [443]. This preliminary study needs to be validated in larger number of patients. However, it suggests that Gal-3 may play a key role in the CSX and may represent a useful marker for identifying those CSX patients that are at high risk of developing cardiovascular disease.

27. Y

Yeast Infection-Candidiasis

Galectin family of lectins participates in the innate immune defense against pathogens [404]. Gal-3 was shown to have antimicrobial activity toward the pathogenic fungus *Candida albicans* [302]. Candida species are the fourth leading cause of nosocomial bloodstream infections. They are responsible for life threatening disease in immunocompromised individuals. Invasive candidiasis affects more than 250,000 people each year and leads to more than 50,000 deaths worldwide. In the United States mortality due to systemic, invasive infections by *Candida albicans* can be as high as 40% even after receiving antifungal therapy [444–446]. Vulvovaginal candidiasis affects 75% of women of childbearing age [447]. It is interesting to note that Gal-3 is normally expressed by epithelia lining the body surfaces that are exposed to *Candida* colonization, such as vulvar and oral mucosa, corneal, conjunctival and intestinal epithelia [448–451]. Gal-3 appears to be present in the fungal granulomata, it binds to *Candida albicans* in a carbohydrate-specific manner and it displays a direct fungicidal activity that may be restricted to specific Candida species. This is the first observation regarding the direct antifungicidal activity for a member of the galectin family of mammalian lectins. The ability to induce Candida death is a unique function of Gal-3, because galectin-1 is not able to bind to any Candida species examined. More recently Gal-3, secreted from neutrophils, was also demonstrated to be also active against other *non-albicans* species, such as the *Candida parapsilosis* and the *Candida albicans hyphae* [452]. Gal-3, in fact, acts as a pro-inflammatory autocrine/paracrine signal in neutrophil phagocytosis. This observation suggests that Gal-3 has a unique role in neutrophil response to *Candida parapsilosis* yeast and *Candida albicans hyphae* distinct from *Candida albicans* yeast. In addition, Gal-3 plays an important role in host recognition and response to candida infection, as demonstrated using a murine model of disseminated candidiasis [453]. In such murine model, knockout of the Gal-3 gene (Gal-3$^{-/-}$ mice) conferred susceptibility to *Candida* infection compared to wild type ones (Gal-3 WT-mice) [453]. When infected with the less virulent specie, namely the *Candida parapsilosis*, Gal-3$^{-/-}$ mice had significantly higher renal fungal burdens and abscess formation compared to Gal-3 WT-mice. This study suggests a potential mechanism of neonatal susceptibility to these infections.

Additional informations will be obtained from studies aimed to analyze the gene polymorphisms of candidate genes associated with yeast colonization [454,455]. In this regard, a new clinical trial is now ongoing to analyze the innate immunity gene polymorphisms of the so-called "Candigene". In particular, aim of the study will be the genotyping of the following lectins: Mannose-binding lectin (MBL), the Dectin-1 and the Gal-3 (ClinicalTrials.gov Identifier: NCT02888860).

The mechanism of Gal-3 secretion by the neutrophils and how it influences host defense was very recently investigated [456]. According to this study Gal-3 deficiency ameliorates systemic candidiasis by reducing fungal burden, renal pathology, and mortality. These studies indicate that blockade of Gal-3 in neutrophils may be a promising therapeutic strategy for systemic *Candida* infection. The inhibition of Gal-3, obtained by knocking down the gene, by silencing its expression or by administration of inhibitors such as the newly identified and TD139 molecule [457], patented by Galecto Biotech [458], is able to inhibit neutrophil reactive oxygen species (ROS) production and enhance the ability of neutrophils to kill when encountering "hard-to-kill" Candida.

28. Z

Zoster-Related Pain (Allodynia)

Herpes zoster (HZ) is caused by the reactivation of *varicella zoster virus* (VZV) after being latent in the sensory ganglia [459,460] and is characterized by blistering skin eruption and neuropathic pain in the affected dermatome [461]. It has been calculated that approximately 30% of population will suffer of HZ during their lifetime [462]. Persistence of pain that lasts after the rash and blisters have healed, a condition known as postherpetic neuralgia (PHN) or zoster-related pain, is considered as the most common and severe complication of HZ [463]. The sensory changes within the affected dermatome, associated with both HZ and PHN, may induce reduced sensitivity of the skin as well as allodynia, that is occurrence of pain due to stimulus which does not normally provoke pain [464]. Gal-3 is highly expressed in peripheral nervous tissues of adult rodents, where is localized to cell bodies of sensory neurons, axons and Schwann cells [465] and in the autonomic preganglionic neurons in the spinal cord [466]. There is growing evidence suggesting that Gal-3 is involved in the process of Wallerian degeneration distal to the injury and subsequent axonal regrowth and functional re-innervation [221]. To identify the genes involved in acute herpetic pain and analyze the role of Gal-3, a mouse model (C57BL/6J) was subjected to transdermal *Herpes simplex virus-1* (HSV-1) inoculation in the hind paw, to produce *Herpes zoster*-like skin lesions and mechanical allodynia. In such mouse model Gal-3 expression levels as well as the global gene expression profiles of the dorsal cord, at the L4 to L6 levels, were analyzed by GeneChip oligonucleotide expression arrays [236]. Gal-3 mRNA levels were slightly induced on day 5 after inoculation, rapidly increased from day six to day eight, and then gradually decreased. HSV-1 induced mechanical allodynia (day six after inoculation) caused a strong induction of Gal-3 protein expression, analyzed by immunohistochemical methods, in the superficial part and sparsely in the deep part of the lumbar dorsal horn on the HSV-1-inoculated side. To test whether Gal-3 contributes to herpetic allodynia, the effects of deficiency in the Gal-3 gene were analyzed using Gal-3$^{-/-}$ mice. In such mice, the onset (day five) and peak (around day seven) of HSV-1 induced allodynia were similar to those of wild-type mice, but the intensity of the pain was significantly reduced. *Zoster*-like skin lesions were not affected by the Gal-3 gene deficiency. Finally, the intrathecal injection of rat and goat anti-Gal-3 antibodies significantly ameliorated HSV-1-induced allodynia. Considering all these observations is not surprising that Gal-3 was considered as a key molecule in the pathogenesis of neuropathic pain following peripheral nerve injury and tested for its possible therapeutic use. In a rat model (Sprague–Dawley) in which neuroinflammation and neuropathic pain were induced by L5 spinal nerve ligation (SNL), an increase in the mRNA and protein expression levels of Gal-3 was observed in dorsal root ganglions [467]. Intrathecal administration of modified citrus pectin (MCP), a Gal-3 inhibitor, was able to inhibit such effect, causing a reduction of Gal-3 expression in dorsal root ganglions. MCP treatment also inhibits SNL-induced Gal-3 expression in primary rat microglia and decreases LPS-induced expression of pro-inflammatory mediators, such as IL-1β, TNF-α and IL-6. Gal-3 inhibition resulted in a significant attenuation of neurophathic pain and in a decreased mechanical and cold hypersensitivity. Gal-3 emerged as a new target for the treatment and clinical management of peripheral nerve injury-induced neuropathic pain. No clinical studies regarding the potential role of Gal-3 in postherpetic neuralgia in humans have been published yet.

29. Conclusions

This review represents an attempt to describe all the known clinical conditions in which a role of Gal-3 was postulated or demonstrated. All the various clinical diseases and conditions, reported in this review in alphabetical order, in which Gal-3 was analyzed and for which a specific clinical pathogenic, diagnostic or prognostic role of Gal-3 was proposed are summarized in Table 1. However, this list should not be considered completed yet. It is likely that many other diseases are still waiting to be analyzed and added to it. It is impressive to realize how relevant is the clinical impact of all such studies on Gal-3 and how fast was the introduction of clinical tests based on Gal-3 into clinical practice to be used as diagnostic or prognostic biomarkers in many different clinical contexts. Even much impressive is the prompt use of all the information regarding the pathogenic role of Gal-3 in severe diseases where few or no therapeutic options are available to design new target treatments focused on Gal-3. Many different clinical trials targeting Gal-3 molecule are still ongoing to treat severe neoplastic, fibrotic, metabolic and degenerative disease and many pharmaceutical companies are investing money and efforts to develop and test drugs targeting the Gal-3 molecule and acting as antagonist or agonist. The astonishing clinical history of this extraordinary lectin and the new science called "Pectin Science" or "Pectinology" are just at the very earl stages and the dawning of a new era with new exciting evolutions are expected soon.

Table 1. Conditions and diseases, listed in alphabetical order, in which a possible role of Gal-3 was postulated in pathogenesis, diagnosis, prognosis, stratification for therapy and in therapy.

Letter Names	Disease/Condition	Possible Role in Pathogenesis	Possible Role in Diagnosis/Prognosis	Possible Role in Patients Stratification for Therapy	Possible Role in Therapy
A	Asthma			Gal-3 positive patients are more responsive to omalizumab	
	Atherosclerosis	Gal-3 is expressed in advanced atherosclerotic lesions			
	Atopic Dermatitis	Gal-3 released by PMNs amplifies IgE production			Gal-3 inhibitors used in a phase 2 controlled trial for treatment of severe atopic dermatitis
B	Blood test		Blood assays (ELISA) to predict outcome of acute and chronic heart failure		
C	Cancer	Gal-3 controls oncogenesis, cancer progression and metastasis	Diagnostic role in thyroid, prostate, and ovarian cancers		Clinical trials using Gal-3 inhibitors for melanoma and large B-cell lymphoma
	Cerebral infarction	Serum levels of Gal-3 are increased in ischemic brain damage	Measurement of serum Gal-3 levels may help in diagnosis of cerebral infarction		
	COPD	Gal-3 levels measured in BAL are reduced in healthy smokers	In patients with COPD serum Gal-3 levels may predict right ventricular dysfunction and acute exacerbation of COPD		
D	Degenerative Aortic Stenosis			Serum Gal-3 levels may be useful in stratification for therapy in elderly patients affected by degenerative aortic stenosis	
	Diabetes Mellitus	Conflicting results: Gal-3 plays a rotective role; Gal-3 is a risk factor for vascular/renal/cardiac complications			
E	Endometriosis	Gal-3 is overexpressed in endometriotic tissue			
	Enteric nervous system	Gal-3 is crucial mediator of the GI damage observed after brain injuries, at the level of enteric neuronal cells			
	Encephalitis	Gal-3 is involved in various types of CNS inflammation and in particular in viral encephalitis			
F	Fibrosis	Gal-3 is key regulator of chronic inflammation and tissue fibrogenesis	Serum Gal-3 levels are increased in fibrosis affecting the liver, the kidney, the lung, the heart, the nervous system and in the systemic sclerosis. In such conditions Gal-3 has been proposed as a new prognostic factor		

Table 1. *Cont.*

Letter Names	Disease/Condition	Possible Role in Pathogenesis	Possible Role in Diagnosis/Prognosis	Possible Role in Patients Stratification for Therapy	Possible Role in Therapy
G	Gastritis	Gal-3 is upregulated in gastric epithelial cells following adhesion of *Helicobacter pylori*			Experimental evidence indicates that administration of extracellular recombinant Gal-3 (rGal-3) is able to inhibit the adhesion of *Helicobacter pylori* to the gastric epithelial cells
H	Heart		Serum Gal-3 levels are associated with mortality in patients with chronic heart failure	Gal-3 is useful as a marker for stratification for therapy (rosuvastatin and cardiac resynchronization) in patient with heart failure	
	HIV infection	Gal-3 is associated with early stages of HIV infection and is involved in the host response to HIV infection.			
I	Inflammation	Gal-3 is involved in many processes during both acute and chronic inflammatory response			
	Interstitial lung disease	Gal-3 is increased in the serum of patients affected by IPF, and is associated with decreased lung volumes and altered gas exchange			A novel target therapy for IPF was proposed using a Gal-3 inhibitor (TD139) formulated for inhalation and a clinical trial process was started
J	Juvenile Idiopathic Arthritis	Gal-3 serum levels are increased in children affected by JIA and correlate with disease activity			
K	Kidney	Gal-3 serum levels are upregulated in diabetic nephropathy and acute renal failure			Experimental evidence indicates that pharmacologic inhibition of Gal-3 with MCP is protective toward nephropathy. Another Gal-3 inhibitor, the GCS-100, significantly improved the eGFR in patients with CKD stage 3b.
L	Liver fibrosis	Gal-3 plays a critical role in hepatic stellate cells activation to a myofibroblast phenotype, a critical event in extracellular matrix deposition and in the development of cirrhosis			A Gal-3 antagonist, named GR-MD-02, is current on a phase 2 Clinical Trial to prevent progression toward cirrhosis in patients with NASH
M	Mortality		Gal-3 is associated with risk of coronary heart disease, CVD mortality, and all-cause mortality		
N	Non-alcoholic steatohepatitis (NASH)	Experimental evidence indicates that mice lacking the Gal-3 gene are resistant to liver fibrosis			The specific Gal-3 inhibitor, named GR-MB-02, is on Phase 2 Clinical Trial for the treatment of advanced fibrosis due to NASH
O	Obesity	Gal-3 is expressed in all types of macrophages that permeate white adipose tissue and it is an important regulator of their polarization and activity. Gal3 can bind directly to the insulin receptor (IR) and inhibit downstream IR signaling.			Experimental evidence in mice indicates that inhibition of Gal-3 improves insulin sensitivity

Table 1. *Cont.*

Letter Names	Disease/Condition	Possible Role in Pathogenesis	Possible Role in Diagnosis/Prognosis	Possible Role in Patients Stratification for Therapy	Possible Role in Therapy
P	Pneumonia	Gal-3 is involved in the recruitment of neutrophils during lung infections with *Streptococcus pneumoniae*			Experimental evidence indicates that exogenous administration of Gal-3 to mice has a direct antimicrobial, bacteriostatic effect on *Streptococcus pneumonia*
	Pulmonary hypertension	Conflicting results:Gal-3 plasma levels are decreased in patients with PAH;Gal-3 plasma levels are increased in PAH independently to etiology and are are associated with severity of PAH			
	Plaque Psoriasis	Gal-3 is expressed in a high proportion of Langerhans cells from skin punch biopsies, obtained from lesional plaque-type psoriatic skin			An open-label phase 2a clinical trial is ongoing using the Gal-3 antagonist GR-MD-02 for the treatment of psoriasis
Q	Q Fever	Glactin-3 is a sensor of damage caused by the *Coxiella burnetii*			
R	Rheumatoid Arthritis	Gal-3 plays a key role as pro-inflammatory mediator in RA	Gal-3 serum level has been proposed as a predictive marker of future RA		Experimental evidence indicates that therapeutic intra-articularly administration of lentiviral vectors encoding Gal-3 small hairpin RNA (shRNA), in rats with collagen-induced arthritis, significantly ameliorates disease activity
S	Sepsis		Serum Gal-3 levels may represent a useful marker to predict mortality in patients affected by sepsis		Experimental evidence indicates that Gal-3 may represent a potential target for treatment of sepsis, at least for that induced by *Francisella novicida*
	Systemic Sclerosis	Serum Gal-3 levels correlate with the developmental process of skin sclerosis and of digital ulcers in diffuse cutaneous SSc	Gal-3 appears to be a reliable biomarker to predict all-cause mortality in patients affected by SSc		
T	Target therapy	Gal-3 is involved in numerous degenerative processes within the body, including cancer proliferation/metastasis, heart failure, atherosclerosis, diabetes, chronic inflammation, fibrosis and related organ failure.			Numerous compounds that are able to inhibit Gal-3 function have been proposed and are currently in Clinical Trials for many different diseases/conditions
U	Urinary tract infections	Gal-3 plays a role in urinary tract viral and bacterial infections in humans			
V	Venous Thrombosis	Serum levels of Gal-3 and of Gal-3 binding protein are elevated in the course of venous thrombosis			
W	Wound Healing	Gal-3 plays a key player in the re-epithelialization of wounds in many different tissues, including the cornea, the skin and the GI tract			

Table 1. *Cont.*

Letter Names	Disease/Condition	Possible Role in Pathogenesis	Possible Role in Diagnosis/Prognosis	Possible Role in Patients Stratification for Therapy	Possible Role in Therapy
X	X syndrome of the heart	Serum Gal-3 levels are significantly higher in the CSX patients in comparison with the healthy controls			
Y	Yeast infection-Candidiasis	Gal-3 is present in the fungal granulomata, it binds to *Candida albicans* in a carbohydrate-specific manner and it displays a direct fungicidal activity			Experimental evidence indicates that the inhibition of Gal-3, obtained by knocking down the gene, by silencing its expression or by administration of the specific inhibitor TD139 is able to inhibit neutrophil reactive oxygen species (ROS) production and enhance the ability of neutrophils to kill *Candida albicans*
Z	Zoster-related pain (allodynia)	Gal-3 is involved in the process of Wallerian degeneration distal to the injury and subsequent axonal regrowth and functional re-innervation			Experimental evidence indicates that in mice the intrathecal administration of the Gal-3 inhibitor modified citrus pectin (MCP) is able to significantly attenuate neurophathic pain

Acknowledgments: This work was entirely supported by Niccolò Cusano University Foundation, Via Don Carlo Gnocchi 3, 00166 Rome, Italy.

Author Contributions: Salvatore Sciacchitano and Alberto Ricci conceived and designed the study; Fiorenza Magi, Alessandra Morgante and Gian Paolo De Francesco performed the literature search; Alessandra Ulivieri, Luca Lavra, Leila B. Salehi and Carlo Bellotti reviewed and analyzed the studies; Salvatore Sciacchitano and Alberto Ricci wrote the paper.

Conflicts of Interest: The authors declare no conflict of interest. The founding sponsors had no role in the design of the study; in the collection, analyses, or interpretation of data; in the writing of the manuscript, and in the decision to publish the results.

References

1. Cooper, D.N.W. Galectinomics: Finding themes in complexity. *Biochim. Biophys. Acta* **2002**, *1572*, 209–231. [CrossRef]

2. Drickamer, K.; Fadden, A.J. Genomic analysis of C-type lectins. *Biochem. Soc. Symp.* **2002**, *69*, 59–72. [CrossRef]

3. Seetharaman, J.; Kanigsberg, A.; Slaaby, R.; Leffler, H.; Barondes, S.H.; Rini, J.M. X-ray crystal structure of the human galectin-3 carbohydrate recognition domain at 2.1-A resolution. *J. Biol. Chem.* **1998**, *273*, 13047–13052. [CrossRef] [PubMed]

4. Hughes, R.C. Secretion of the galectin family of mammalian carbohydrate-binding proteins. *Biochim. Biophys. Acta* **1999**, *1473*, 172–185. [CrossRef]

5. Jones, J.L.; Saraswati, S.; Block, A.S.; Lichti, C.F.; Mahadevan, M.; Diekman, A.B. Galectin-3 is associated with prostasomes in human semen. *Glycoconj. J.* **2010**, *27*, 227–236. [CrossRef] [PubMed]

6. Nabi, I.R.; Shankar, J.; Dennis, J.W. The galectin lattice at a glance. *J. Cell Sci.* **2015**, *128*, 2213–2219. [CrossRef] [PubMed]

7. Hughes, R.C. The galectin family of mammalian carbohydrate-binding molecules. *Biochem. Soc. Trans.* **1997**, *25*, 1194–1198. [CrossRef] [PubMed]

8. Barondes, S.H.; Castronovo, V.; Cooper, D.N.; Cummings, R.D.; Drickamer, K.; Feizi, T.; Gitt, M.A.; Hirabayashi, J.; Hughes, C.; Kasai, K.; et al. Galectins: A family of animal beta-galactoside-binding lectins. *Cell* **1994**, *76*, 597–598. [CrossRef]

9. Ho, M.K.; Springer, T.A. Mac-2, a novel 32,000 Mr mouse macrophage subpopulation-specific antigen defined by monoclonal antibodies. *J. Immunol.* **1982**, *128*, 1221–1228. [PubMed]

10. Roff, C.F.; Wang, J.L. Endogenous lectins from cultured cells: Isolation and characterization of carbohydrate-binding proteins from 3T3 fibroblasts. *J. Biol. Chem.* **1983**, *258*, 10657–10663. [PubMed]

11. Da, Y.; Leffler, H.; Sakakura, Y.; Kasai, K.; Barondes, S.H. Human breast carcinoma cDNA encoding a galactoside-binding lectin homologous to mouse Mac-2 antigen. *Gene* **1991**, *99*, 279–283.

12. Cherayil, B.J.; Weiner, S.J.; Pillai, S. The Mac-2 antigen is a galactose-specific lectin that binds IgE. *J. Exp. Med.* **1989**, *170*, 1959–1972. [CrossRef] [PubMed]

13. Raz, A.; Pazerini, G.; Carmi, P. Identification of the metastasis-associated, galactoside-binding lectin as a chimeric gene product with homology to an IgE-binding protein. *Cancer Res.* **1989**, *49*, 3489–3493. [PubMed]

14. Raz, A.; Carmi, P.; Raz, T.; Hogan, V.; Mohamed, A.; Wolman, S.R. Molecular cloning and chromosomal mapping of a human galactoside-binding protein. *Cancer Res.* **1991**, *51*, 2173–2178. [PubMed]

15. Sparrow, C.P.; Leffler, H.; Barondes, S.H. Multiple soluble beta-galactoside-binding lectins from human lung. *J. Biol. Chem.* **1987**, *262*, 7383–7390. [PubMed]

16. Leffler, H.; Masiarz, F.R.; Barondes, S.H. Soluble lactose-binding vertebrate lectins: A growing family. *Biochemistry* **1989**, *28*, 9222–9229. [CrossRef] [PubMed]

17. Robertson, M.W.; Albrandt, K.; Keller, D.; Liu, F.T. Human IgE-binding protein: A soluble lectin exhibiting a highly conserved interspecies sequence and differential recognition of IgE glycoforms. *Biochemistry* **1990**, *29*, 8093–8100. [CrossRef] [PubMed]

18. Albrandt, K.; Orida, N.K.; Liu, F.T. An IgE-binding protein with a distinctive repetitive sequence and homology with an IgG receptor. *Proc. Natl. Acad. Sci. USA* **1987**, *84*, 6859–6863. [CrossRef] [PubMed]

19. Cerra, R.F.; Gitt, M.A.; Barondes, S.H. Three soluble rat beta-galactoside-binding lectins. *J. Biol. Chem.* **1985**, *260*, 10474–10477. [PubMed]

20. Lotan, R.; Carralero, D.; Lotan, D.; Raz, A. Biochemical and immunological characterization of K-1735P melanoma galactoside-binding lectins and their modulation by differentiation inducers. *Cancer Res.* **1989**, *49*, 1261–1268. [PubMed]

21. Sato, S.; Hughes, R.C. Binding specificity of a baby hamster kidney lectin for H type I and II chains, polylactosamine glycans, and appropriately glycosylated forms of laminin and fibronectin. *J. Biol. Chem.* **1992**, *267*, 6983–6990. [PubMed]

22. Jia, S.; Wang, J.L. Carbohydrate binding protein 35. Complementary DNA sequence reveals homology with proteins of the heterogeneous nuclear RNP. *J. Biol. Chem.* **1988**, *263*, 6009–6011. [PubMed]

23. Woo, H.J.; Shaw, L.M.; Messier, J.M.; Mercurio, A.M. The major nonintegrin laminin binding protein of macrophages is identical to carbohydrate binding protein 35 (Mac-2). *J. Biol. Chem.* **1990**, *265*, 7097–7099. [PubMed]

24. Dumic, J.; Dabelic, S.; Flögel, M. Galectin-3: An open-ended story. *Biochim. Biophys. Acta* **2006**, *1760*, 616–635. [CrossRef] [PubMed]

25. Van den Brûle, F.; Califice, S.; Castronovo, V. Expression of galectins in cancer: A critical review. *Glycoconj. J.* **2004**, *19*, 537–542. [CrossRef] [PubMed]

26. Song, L.; Tang, J.W.; Owusu, L.; Sun, M.Z.; Wu, J.; Zhang, J. Galectin-3 in cancer. *Clin. Chim. Acta* **2014**, *431*, 185–191. [CrossRef] [PubMed]

27. Hirabayashi, J.; Kasai, K. The family of metazoan metal-independent beta-galactoside-binding lectins: Structure, function and molecular evolution. *Glycobiology* **1993**, *4*, 297–304. [CrossRef]

28. Wang, J.L.; Laing, J.G.; Anderson, R.L. Lectins in the cell nucleus. *Glycobiology* **1991**, *3*, 243–252. [CrossRef]

29. Hughes, R.C. Mac-2: A versatile galactose-binding protein of mammalian tissues. *Glycobiology* **1994**, *4*, 5–12. [CrossRef] [PubMed]

30. Birdsall, B.; Feeney, J.; Burdett, I.D.J.; Bawumia, S.; Barboni, E.A.M.; Hughes, R.C. NMR solution studies of hamster Gal-3 and electron microscopic visualization of surface-adsorbed complexes: Evidence for interactions between the N- and C-terminal domains. *Biochemistry* **2001**, *40*, 4859–4866. [CrossRef] [PubMed]

31. Huflejt, M.E.; Turck, C.W.; Lindstedt, R.; Barondes, S.H.; Leffler, H. L-29, a soluble lactose-binding lectin, is phosphorylated on serine 6 and serine 12 in vivo and by casein kinase I. *J. Biol. Chem.* **1993**, *268*, 26712–26718. [PubMed]

32. Mazurek, N.; Conklin, J.; Byrd, J.C.; Raz, A.; Bresalier, R.S. Phosphorylation of the β-galactoside-binding protein Gal-3 modulates binding to its ligands. *J. Biol. Chem.* **2000**, *275*, 36311–36315. [CrossRef] [PubMed]

33. Menon, R.P.; Hughes, R.C. Determinants in the N-terminal domains of Gal-3 for secretion by a novel pathway circumventing the endoplasmic reticulum-Golgi complex. *Eur. J. Biochem.* **1999**, *264*, 569–576. [CrossRef] [PubMed]

34. Hirabayashi, J.; Hashidate, T.; Arata, Y.; Nishi, N.; Nakamura, T.; Hirashima, M.; Urashima, T.; Oka, T.; Futai, M.; Muller, W.E.G.; et al. Oligosaccharide specificity of galectins: A search by frontal affinity chromatography. *Biochim. Biophys. Acta* **2002**, *1572*, 232–254. [CrossRef]

35. Ochieng, J.; Furtak, V.; Lukyanov, P. Extracellular functions of Gal-3. *Glycoconj. J.* **2004**, *19*, 527–535. [CrossRef] [PubMed]

36. Davidson, P.J.; Davis, M.J.; Patterson, R.J.; Ripoche, M.A.; Poirier, F.; Wang, J.L. Shuttling of Gal-3 between the nucleus and cytoplasm. *Glycobiology* **2002**, *12*, 329–337. [CrossRef] [PubMed]

37. Hsu, D.K.; Liu, F.T. Regulation of cellular homeostasis by galectins. *Glycoconj. J.* **2004**, *19*, 507–515. [CrossRef] [PubMed]

38. Califice, S.; Castronovo, V.; Van Den Brûle, F. Galectin-3 and cancer (Review). *Int. J. Oncol.* **2004**, *25*, 983–992. [PubMed]

39. Moutsatsos, I.K.; Davis, J.M.; Wang, J.L. Endogenous lectins from cultured cells: Subcellular localization of carbohydrate binding protein 35 in 3T3 fibroblasts. *J. Cell Biol.* **1986**, *102*, 477–483. [CrossRef] [PubMed]

40. Moutsatsos, I.K.; Wade, M.; Schindler, M.; Wang, J.L. Endogenous lectins from cultured cells: Nuclear localization of carbohydrate binding protein 35 in proliferating 3T3 fibroblasts. *Proc. Natl. Acad. Sci. USA* **1987**, *84*, 6452–6466. [CrossRef] [PubMed]

41. Agrwal, N.; Wang, J.L.; Voss, P.G. Carbohydrate-binding protein 35. Levels of transcription and mRNA accumulation in quiescent and proliferating cells. *J. Biol. Chem.* **1989**, *264*, 17236–17242. [PubMed]

42. Cowles, E.A.; Moutsatsos, I.K.; Wang, J.L.; Anderson, R.L. Expression of carbohydrate binding protein 35 in human fibroblasts: Comparisons between cells with different proliferative capacities. *Exp. Gerontol.* **1989**, *24*, 577–585. [CrossRef]

43. Hubert, M.; Wang, S.Y.; Wang, J.L.; Seve, A.P.; Hubert, J. Intranuclear distribution of galectin-3 in mouse 3T3 fibroblasts: Comparative analyses by immunofluorescence and immunoelectron microscopy. *Exp. Cell Res.* **1995**, *220*, 397–406. [CrossRef] [PubMed]

44. Sato, S.; Burdett, I.; Hughes, R.C. Secretion of the baby hamster kidney 30-kDa galactose-binding lectin from polarized and nonpolarized cells: A pathway independent of the endoplasmic reticulum-Golgi complex. *Exp. Cell Res.* **1993**, *207*, 8–18. [CrossRef] [PubMed]

45. Nickel, W. The mystery of nonclassical protein secretion. A current view on cargo proteins and potential export routes. *Eur. J. Biochem.* **2003**, *270*, 2109–2119. [CrossRef] [PubMed]

46. Mehul, B.; Hughes, R.C. Plasma membrane targeting, vesicular budding and release of galectin 3 from the cytoplasm of mammalian cells during secretion. *J. Cell Sci.* **1997**, *110*, 1169–1178. [PubMed]

47. Gupta, A. Ligands for Galectin-3: Binding interactions. In *Animal Lectins: Form, Function and Clinical Applications*; Gupta, G.S., Ed.; Springer Verlag Wien: Vienna, Austria; Springer Science & Business Media: Berlin, Germany, 2012; Chapter 12.2, Volume 1, pp. 268–271.

48. Talaga, M.L.; Fan, N.; Fueri, A.L.; Brown, R.K.; Bandyopadhyay, P.; Dam, T.K. Multitasking Human Lectin Galectin-3 Interacts with Sulfated Glycosaminoglycans and Chondroitin Sulfate Proteoglycans. *Biochemistry* **2016**, *55*, 4541–4551. [CrossRef] [PubMed]

49. Rosenberg, I.; Cherayil, B.J.; Isselbacher, K.J.; Pillai, S. Mac-2-binding glycoproteins. Putative ligands for a cytosolic β-galactoside lectin. *J. Biol. Chem.* **1991**, *266*, 18731–18736. [PubMed]

50. Koths, K.; Taylor, E.; Halenbeck, R.; Casipit, C.; Wang, A. Cloning and characterization of a human Mac-2-binding protein, a new member of the superfamily defined by the macrophage scavenger receptor cysteine-rich domain. *J. Biol. Chem.* **1993**, *268*, 14245–14249. [PubMed]

51. Probstmeier, R.; Montag, D.; Schachner, M. Gal-3, a β- galactoside-binding animal lectin, binds to neural recognition molecules. *J. Neurochem.* **1995**, *64*, 2465–2472. [CrossRef] [PubMed]

52. Ochieng, J.; Warfield, P. Gal-3 binding potentials of mouse tumor RHS and human placental laminins. *Biochem. Biophys. Res. Commun.* **1995**, *217*, 402–406. [CrossRef] [PubMed]

53. Hughes, R.C. Galectins as modulators of cell adhesion. *Biochimie* **2001**, *83*, 667–676. [CrossRef]

54. Dong, S.; Hughes, R.C. Macrophage surface glycoproteins binding to Gal-3 (Mac-2-antigen). *Glycoconj. J.* **1997**, *14*, 267–274. [CrossRef] [PubMed]

55. Ochieng, J.; Leite-Browning, M.L.; Warfield, P. Regulation of cellular adhesion to extracellular matrix proteins by Gal-3. *Biochem. Biophys. Res. Commun.* **1998**, *246*, 788–791. [CrossRef] [PubMed]

56. Goletz, S.; Hanisch, F.G.; Karsten, U. Novel αGalNAc containing glycans on cytokeratins are recognized in vitro by galectins with type II carbohydrate recognition domains. *J. Cell Sci.* **1997**, *110*, 1585–1596. [PubMed]

57. Sève, A.-P.; Felin, M.; Doyennette-Moyne, M.A.; Sahraoui, T.; Aubery, M.; Hubert, J. Evidence for a lactose-mediated association between two nuclear carbohydrate-binding proteins. *Glycobiology* **1993**, *3*, 23–30. [CrossRef] [PubMed]

58. Menon, R.P.; Strom, M.; Hughes, R.C. Interaction of a novel cysteine and histidine-rich cytoplasmic protein with Gal-3 in a carbohydrate independent manner. *FEBS Lett.* **2000**, *470*, 227–231. [CrossRef]

59. Park, J.W.; Voss, P.G.; Grabski, S.; Wang, J.L.; Patterson, R.J. Association of galectin-1 and Gal-3 with Gemin4 in complexes containing the SMN protein. *Nucleic Acids Res.* **2001**, *27*, 3595–3602. [CrossRef]

60. Liu, F.T.; Patterson, R.J.; Wang, J.L. Intracellular functions of galectins. *Biochim. Biophys. Acta* **2002**, *1572*, 263–273. [CrossRef]

61. Akahani, S.; Nangia-Makker, P.; Inohara, H.; Kim, H.R.C.; Raz, A. Gal-3: A novel antiapoptotic molecule with a functional BH1 (NWGR) domain of Bcl-2 family. *Cancer Res.* **1997**, *57*, 5272–5276. [PubMed]

62. Missotten, M.; Nichols, A.; Rieger, K.; Sadoul, R. Alix, a novel mouse protein undergoing calcium-dependent interaction with the apoptosis-linked-gene 2 (ALG-2) protein. *Cell Death Differ.* **1999**, *6*, 124–129. [CrossRef] [PubMed]

63. Vito, P.; Pellegrini, L.; Guiet, C.; D'Adamio, L. Cloning of AIP1, a novel protein that associates with the apoptosis-linked gene ALG-2 in a Ca2+-dependent reaction. *J. Biol. Chem.* **1999**, *274*, 1533–1540. [CrossRef] [PubMed]

64. Dong, R.; Zhang, M.; Hu, Q.; Zheng, S.; Soh, A.; Zheng, Y.; Yuan, H. Galectin-3 as a novel biomarker for disease diagnosis and a target for therapy (Review). *Int. J. Mol. Med.* **2018**, *41*, 599–614. [CrossRef] [PubMed]
65. Busse, W.W.; Lemanske, R.F., Jr. Asthma. *N. Engl. J. Med.* **2001**, *344*, 350–362. [CrossRef] [PubMed]
66. Stanworth, D.R. The discovery of IgE. *Allergy* **1993**, *48*, 67–71. [CrossRef] [PubMed]
67. Finkelman, F.D.; Vercelli, D. Advances in asthma, allergy mechanisms, and genetics in 2006. *J. Allergy Clin. Immunol.* **2007**, *120*, 544–550. [CrossRef] [PubMed]
68. Sullivan, A.L.; Grimley, P.M.; Metzger, H. Electron Microscopic Localization of Immunoglobulin E on the Surface Membrane of Human Basophils. *J. Exp. Med.* **1971**, *134*, 1403–1416. [CrossRef] [PubMed]
69. Zuberi, R.I.; Hsu, D.K.; Kalayci, O.; Chen, H.Y.; Sheldon, H.K.; Yu, L.; Apgar, J.R.; Kawakami, T.; Lilly, C.M.; Liu, F.T. Critical role for Gal-3 in airway inflammation and bronchial hyperresponsiveness in a murine model of asthma. *Am. J. Pathol.* **2004**, *165*, 2045–2053. [CrossRef]
70. Ge, X.N.; Bahaie, N.S.; Kang, B.N.; Hosseinkhani, M.R.; Ha, S.G.; Frenzel, E.M.; Liu, F.T.; Rao, S.P.; Sriramarao, P. Allergen-induced airway remodeling is impaired in Gal-3-deficient mice. *J. Immunol.* **2010**, *185*, 1205–1214. [CrossRef] [PubMed]
71. Lopez, E.; del Pozo, V.; Miguel, T.; Sastre, B.; Seoane, C.; Civantos, E.; Llanes, E.; Baeza, M.L.; Palomino, P.; Cárdaba, B.; et al. Inhibition of Chronic Airway Inflammation and Remodeling by Gal-3 Gene Therapy in a Murine Model. *J. Immunol.* **2006**, *176*, 1943–1950. [CrossRef] [PubMed]
72. López, E.; Zafra, M.P.; Sastre, B.; Gámez, C.; Lahoz, C.; del Pozo, V. Gene Expression Profiling in Lungs of Chronic Asthmatic Mice Treated with Gal-3: Downregulation of Inflammatory and Regulatory Genes. *Mediat. Inflamm.* **2011**, *2011*, 823279. [CrossRef] [PubMed]
73. Nieminen, J.; St-Pierre, C.; Sato, S. Gal-3 interacts with naive and primed neutrophils, inducing innate immune responses. *J. Leukoc. Biol.* **2005**, *78*, 1127–1135. [CrossRef] [PubMed]
74. Nieminen, J.; St-Pierre, C.; Bhaumik, P.; Poirier, F.; Sato, S. Role of Gal-3 in leukocyte recruitment in a murine model of lung infection by Streptococcus pneumoniae. *J. Immunol.* **2008**, *180*, 2466–2473. [CrossRef] [PubMed]
75. Karlsson, A.; Christenson, K.; Matlak, M.; Bjorstad, A.; Brown, K.L.; Telemo, E.; Salomonsson, E.; Leffler, H.; Bylund, J. Gal-3 functions as an opsonin and enhances the macrophage clearance of apoptotic neutrophils. *Glycobiology* **2009**, *19*, 16–20. [CrossRef] [PubMed]
76. Gao, P.; Gibson, P.G.; Baines, K.J.; Yang, I.A.; Upham, J.W.; Reynolds, P.N.; Hodge, S.; James, A.L.; Jenkins, C.; Peters, M.J.; et al. Anti-inflammatory deficiencies in neutrophilic asthma: Reduced Gal-3 and IL-1RA/IL-1β. *Respir. Res.* **2015**, *16*, 5. [CrossRef] [PubMed]
77. Riccio, A.M.; Mauri, P.; De Ferrari, L.; Rossi, R.; Di Silvestre, D.; Benazzi, L.; Chiappori, A.; Dal Negro, R.W.; Micheletto, C.; Canonica, G.W. Galectin-3: An early predictive biomarker of modulation of airway remodeling in patients with severe asthma treated with omalizumab for 36 months. *Clin. Transl. Allergy* **2017**, *7*, 6. [CrossRef] [PubMed]
78. Gao, P.; Simpson, J.L.; Zhang, J.; Gibson, P.G. Galectin-3: Its role in asthma and potential as an anti-inflammatory target. *Respir. Res.* **2013**, *14*, 136. [CrossRef] [PubMed]
79. Del Pozo, V.; Rojo, M.; Rubio, M.L.; Cortegano, I.; Cárdaba, B.; Gallardo, S.; Ortega, M.; Civantos, E.; López, E.; Martín-Mosquer, C.; et al. Gene therapy with Galectin-3 inhibits bronchial obstruction and Inflammation in Antigen-challenged rats through Interleukin-5 Gene downregulation. *Am. J. Respir. Crit. Care Med.* **2002**, *166*, 732–737. [CrossRef] [PubMed]
80. Ross, R. The pathogenesis of atherosclerosis: A perspective for the 1990s. *Nature* **1993**, *362*, 801–809. [CrossRef] [PubMed]
81. Munro, J.M.; Cotran, R.S. The pathogenesis of atherosclerosis: Atherogenesis and inflammation. *Lab. Investig.* **1988**, *58*, 249–261. [PubMed]
82. Geer, J.C.; McGill, H.C.; Strong, J.P. The fine structure of human atherosclerotic lesions. *Am. J. Pathol.* **1961**, *38*, 263–287. [PubMed]
83. Gerrity, R.G. The role of the monocyte in atherogenesis. I. Transition of blood-borne monocytes into foam cells in fatty lesions. *Am. J. Pathol.* **1981**, *103*, 181–190. [PubMed]
84. Faruqi, R.M.; Di Corleto, P.E. Mechanisms of monocyte recruitment and accumulation. *Br. Heart J.* **1993**, *69*, S19–S29. [CrossRef] [PubMed]

85. Stary, H.C.; Blankenhorn, D.H.; Chandler, A.B.; Glagov, S.; Insull, W., Jr.; Richardson, M.; Rosenfeld, M.E.; Shaffer, S.A.; Schwartz, C.J.; Wagner, W.D.; et al. A definition of the intima of human arteries and of its atherosclerosis-prone regions: A report from the Committee on Vascular Lesions of the Council on Arteriosclerosis, American Heart Association. *Circulation* **1992**, *85*, 391–404. [CrossRef] [PubMed]

86. Nachtigal, M.; Al-Assaad, Z.; Mayer, E.P.; Kim, K.; Monsigny, M. Gal-3 Expression in Human Atherosclerotic Lesions. *Am. J. Patol.* **1998**, *152*, 1199–1208.

87. Konstantinov, K.N.; Shames, B.; Izuno, G.; Liu, F.T. Expression of epsilon BP, a beta-galactoside-binding soluble lectin, in normal and neoplastic epidermis. *Exp. Dermatol.* **1994**, *3*, 9–16. [CrossRef] [PubMed]

88. Kimata, H. Enhancement of IgE production in B cells by neutrophils via Gal-3 in IgE-associated atopic eczema/dermatitis syndrome. *Int. Arch. Allergy Immunol.* **2002**, *128*, 168–170. [CrossRef] [PubMed]

89. Klubal, R.; Osterhoff, B.; Wang, B.; Kinet, J.P.; Maurer, D.; Stingl, G. The high-affinity receptor for Ige is the predominant IgE-binding structure in lesional skin of atopic dermatitis patients. *J. Investig. Dermatol.* **1997**, *108*, 336–342. [CrossRef] [PubMed]

90. Saegusa, J.; Hsu, D.K.; Chen, H.Y.; Yu, L.; Fermin, A.; Fung, M.A.; Liu, F.T. Gal-3 Is Critical for the Development of the Allergic Inflammatory Response in a Mouse Model of Atopic Dermatitis. *Am. J. Pathol.* **2009**, *174*, 922–931. [CrossRef] [PubMed]

91. GR-MD-02 Demonstrates Clinically Significant Effect in Patients with Severe and Refractory Atopic Dermatitis (Eczema). Available online: http://globenewswire.com/news-release/2017/03/14/936280/0/en/GR-MD-02Demonstrates-Clinically-Significant-Effect-in-Patients-with-Severe-and-Refractory-Atopic-Dermatitis-Eczema.html (accessed on 22 December 2017).

92. Sharma, U.C.; Pokharel, S.; van Brakel, T.J.; van Berlo, J.H.; Cleutjens, J.P.; Schroen, B.; André, S.; Crijns, H.J.; Gabius, H.J.; Maessen, J.; et al. Gal-3 marks activated macrophages in failure-prone hypertrophied hearts and contributes to cardiac dysfunction. *Circulation* **2004**, *110*, 3121–3128. [CrossRef] [PubMed]

93. De Boer, R.A.; Voors, A.A.; Muntendam, P.; van Gilst, W.H.; van Veldhuisen, D.J. Gal-3: A novel mediator of heart failure development and progression. *Eur. J. Heart Fail.* **2009**, *11*, 811–817. [CrossRef] [PubMed]

94. Lok, D.J.; Van Der Meer, P.; de la Porte, P.W.; Lipsic, E.; Van Wijngaarden, J.; Hillege, H.L.; van Veldhuisen, D.J. Prognostic value of galectin-3, a novel marker of fibrosis, in patients with chronic heart failure: Data from the DEAL-HF study. *Clin. Res. Cardiol.* **2010**, *99*, 323–328. [CrossRef] [PubMed]

95. De Boer, R.A.; Lok, D.J.; Jaarsma, T.; van der Meer, P.; Voors, A.A.; Hillege, H.L.; van Veldhuisen, D.J. Predictive value of plasma Gal-3 levels in heart failure with reduced and preserved ejection fraction. *Ann. Med.* **2011**, *43*, 60–68. [CrossRef] [PubMed]

96. Van Kimmenade, R.R.; Januzzi, J.L., Jr.; Ellinor, P.T.; Sharma, U.C.; Bakker, J.A.; Low, A.F.; Martinez, A.; Crijns, H.J.; MacRae, C.A.; Menheere, P.P.; et al. Utility of amino-terminal pro-brain natriuretic peptide, Gal-3, and apelin for the evaluation of patients with acute heart failure. *J. Am. Coll. Cardiol.* **2006**, *48*, 1217–1224. [CrossRef] [PubMed]

97. Jaarsma, T.; van der Wal, M.H.; Lesman-Leegte, I.; Luttik, M.L.; Hogenhuis, J.; Veeger, N.J.; Sanderman, R.; Hoes, A.W.; van Gilst, W.H.; Lok, D.J.; et al. Effect of moderate or intensive disease management program on outcome in patients with heart failure: Coordinating Study Evaluating Outcomes of Advising and Counseling in Heart Failure (COACH). *Arch. Intern. Med.* **2008**, *168*, 316–324. [CrossRef] [PubMed]

98. Januzzi, J.L., Jr.; Camargo, C.A.; Anwaruddin, S.; Baggish, A.L.; Chen, A.A.; Krauser, D.G.; Tung, R.; Cameron, R.; Nagurney, J.T.; Chae, C.U.; et al. The N-terminal pro-BNP investigation of dyspnea in the emergency department (PRIDE) study. *Am. J. Cardiol.* **2005**, *95*, 948–954. [CrossRef] [PubMed]

99. Meijers, W.C.; Januzzi, J.L.; de Filippi, C.; Adourian, A.S.; Shah, S.J.; van Veldhuisen, D.J.; de Boer, R.A. Elevated plasma Gal-3 is associated with near-term rehospitalization in heart failure: A pooled analysis of 3 clinical trials. *Am. Heart J.* **2014**, *167*, 853–860. [CrossRef] [PubMed]

100. Felker, G.M.; Fiuzat, M.; Shaw, L.K.; Clare, R.; Whellan, D.J.; Bettari, L.; Shirolkar, S.C.; Donahue, M.; Kitzman, D.W.; Zannad, F.; et al. Galectin-3 in ambulatory patients with heart failure: Results from the HF-ACTION study. *Cir. Heart Fail.* **2012**, *5*, 72–78. [CrossRef] [PubMed]

101. De Boer, R.A.; van Veldhuisen, D.J.; Gansevoort, R.T.; Muller Kobold, A.C.; van Gilst, W.H.; Hillege, H.L.; Bakker, S.J.; van der Harst, P. The fibrosis marker Gal-3 and outcome in the general population. *J. Intern. Med.* **2012**, *272*, 55–64. [CrossRef] [PubMed]

102. Ho, J.E.; Liu, C.; Lyass, A.; Courchesne, P.; Pencina, M.J.; Vasan, R.S.; Larson, M.G.; Levy, D. Gal-3, a marker of cardiac fibrosis, predicts incident heart failure in the community. *J. Am. Coll. Cardiol.* **2012**, *60*, 1249–1256. [CrossRef] [PubMed]

103. Daniels, L.B.; Clopton, P.; Laughlin, G.A.; Maisel, A.S.; Barrett-Connor, E. Gal-3 is independently associated with cardiovascular mortality in community-dwelling older adults without known cardiovascular disease: The Rancho Bernardo Study. *Am. Heart J.* **2014**, *167*, 674–682. [CrossRef] [PubMed]

104. Ho, J.E.; Yin, X.; Levy, D.; Vasan, R.S.; Magnani, J.W.; Ellinor, P.T.; McManus, D.D.; Lubitz, S.A.; Larson, M.G.; Benjamin, E.J. Galectin 3 and incident atrial fibrillation in the community. *Am. Heart J.* **2014**, *167*, 729–734. [CrossRef] [PubMed]

105. McCullough, P.A.; Olobatoke, A. Gal-3: A novel blood test for the evaluation and management of patients with heart failure. *Rev. Cardiovasc. Med.* **2011**, *12*, 200–210. [PubMed]

106. Danguy, A.; Camby, I.; Kiss, R. Galectins and cancer. *Biochim. Biophys. Acta* **2002**, *1572*, 285–293. [CrossRef]

107. Takenaka, Y.; Fukumori, T.; Raz, A. Gal-3 and metastasis. *Glycoconj. J.* **2004**, *19*, 543–549. [CrossRef] [PubMed]

108. Liu, F.T.; Rabinovich, G.A. Galectins as modulators of tumour progression. *Nat. Rev. Cancer* **2005**, *5*, 29–41. [CrossRef] [PubMed]

109. Dagher, S.F.; Wang, J.L.; Patterson, R.J. Identification of galectin-3 as a factor in pre-mRNA splicing. *Proc. Natl. Acad. Sci. USA* **1995**, *92*, 1213–1217. [CrossRef] [PubMed]

110. Nakahara, S.; Oka, N.; Raz, A. On the role of Gal-3 in cancer apoptosis. *Apoptosis* **2005**, *10*, 267–275. [CrossRef] [PubMed]

111. Ruvolo, P.R. Galectin 3 as a guardian of the tumor microenvironment. *Biochim. Biophys. Acta* **2016**, *1863*, 427–437. [CrossRef] [PubMed]

112. Baldus, S.E.; Zirbes, T.K.; Weingarten, M.; Fromm, S.; Glossmann, J.; Hanisch, F.G.; Mönig, S.P.; Schröder, W.; Flucke, U.; Thiele, J.; et al. Increased Gal-3 expression in gastric cancer: Correlations with histopathological subtypes, galactosylated antigens and tumor cell proliferation. *Tumour Biol.* **2000**, *21*, 258–266. [CrossRef] [PubMed]

113. Radosavljevic, G.; Jovanovic, I.; Majstorovic, I.; Mitrovic, M.; Lisnic, V.J.; Arsenijevic, N.; Jonjic, S.; Lukic, M.L. Deletion of Gal-3 in the host attenuates metastasis of murine melanoma by modulating tumor adhesion and NK cell activity. *Clin. Exp. Metastasis* **2011**, *28*, 451–462. [CrossRef] [PubMed]

114. Prieto, V.G.; Mourad-Zeidan, A.A.; Melnikova, V.; Johnson, M.M.; Lopez, A.; Diwan, A.H.; Lazar, A.J.; Shen, S.S.; Zhang, P.S.; Reed, J.A.; et al. Gal-3 expression is associated with tumor progression and pattern of sun exposure in melanoma. *Clin. Cancer Res.* **2006**, *12*, 6709–6715. [CrossRef] [PubMed]

115. Shibata, T.; Noguchi, T.; Takeno, S.; Takahashi, Y.; Fumoto, S.; Kawahara, K. Impact of nuclear Gal-3 expression on histological differentiation and vascular invasion in patients with esophageal squamous cell carcinoma. *Oncol. Rep.* **2005**, *139*, 235–239.

116. Califice, S.; Castronovo, V.; Bracke, M.; van den Brûle, F. Dual activities of Gal-3 in human prostate cancer: Tumor suppression of nuclear Gal-3 vs tumor promotion of cytoplasmic Gal-3. *Oncogene* **2004**, *23*, 7527–7536. [CrossRef] [PubMed]

117. Van den Brûle, F.A.; Waltregny, D.; Liu, F.T.; Castronovo, V. Alteration of the cytoplasmic/nuclear expression pattern of Gal-3 correlates with prostate carcinoma progression. *Int. J. Cancer* **2000**, *89*, 361–367. [CrossRef]

118. Puglisi, F.; Minisini, A.M.; Barbone, F.; Intersimone, D.; Aprile, G.; Puppin, C.; Damante, G.; Paron, I.; Tell, G.; Piga, A.; et al. Gal-3 expression in non-small cell lung carcinoma. *Cancer Lett.* **2004**, *212*, 233–239. [CrossRef] [PubMed]

119. Nakahara, S.; Hogan, V.; Inohara, H.; Raz, A. Importin-mediated nuclear translocation of galectin-3. *J. Biol. Chem.* **2006**, *281*, 39649–39659. [CrossRef] [PubMed]

120. Funasaka, T.; Raz, A.; Nangia-Makker, P. Nuclear transport of galectin-3 and its ther- apeutic implications. *Semin. Cancer Biol.* **2014**, *27*, 30–38. [CrossRef] [PubMed]

121. Haudek, K.C.; Spronk, K.J.; Voss, P.G.; Patterson, R.J.; Wang, J.L.; Arnoys, E.J. Dynamics of Galectin-3 in the Nucleus and Cytoplasm. *Biochim. Biophys. Acta* **2010**, *1800*, 181–189. [CrossRef] [PubMed]

122. Kapucuoglu, N.; Basak, P.Y.; Bircan, S.; Sert, S.; Akkaya, V.B. Immunohistochemical galectin-3 expression in non-melanoma skin cancers. *Pathol. Res. Pract.* **2009**, *205*, 97–103. [CrossRef] [PubMed]

123. Thijssen, V.L.; Heuschen, R.; Caers, J.; Griffioen, A.W. Galectin expression in cancer diagnosis and prognosis: A systematic review. *Biochim. Biophys. Acta* **2015**, *1855*, 235–247. [CrossRef] [PubMed]

124. Fortuna-Costa, A.; Gomes, A.M.; Kozlowski, E.O.; Stelling, M.P.; Pavão, M.S. Extracellular galectin-3 in tumor progression and metastasis. *Front. Oncol.* **2014**, *4*, 138. [CrossRef] [PubMed]

125. Xin, M.; Dong, X.W.; Guo, X.L. Role of the interaction between galectin-3 and cell adhesion molecules in cancer metastasis. *Biomed. Pharmacother.* **2015**, *69*, 179–185. [CrossRef] [PubMed]

126. Smetana, K., Jr.; André, S.; Kaltner, H.; Kopitz, J.; Gabius, H.J. Context-dependent multifunctionality of galectin-1: A challenge for defining the lectin as therapeutic target. *Expert Opin. Ther. Targets* **2013**, *17*, 379–392. [CrossRef] [PubMed]

127. Yoshii, T.; Fukumori, T.; Honjo, Y.; Inohara, H.; Kim, H.R.; Raz, A. Galectin-3 phosphorylation is required for its anti-apoptotic function and cell cycle arrest. *J. Biol. Chem.* **2002**, *277*, 6852–6857. [CrossRef] [PubMed]

128. Cecchinelli, B.; Lavra, L.; Rinaldo, C.; Iacovelli, S.; Gurtner, A.; Gasbarri, A.; Ulivieri, A.; Del Prete, F.; Trovato, M.; Piaggio, G.; et al. Repression of the antiapoptotic molecule Gal-3 by homeodomain-interacting protein kinase 2-activated p53 is required for p53-induced apoptosis. *Mol. Cell. Biol.* **2006**, *26*, 4746–4757. [CrossRef] [PubMed]

129. Lavra, L.; Ulivieri, A.; Rinaldo, C.; Dominici, R.; Volante, M.; Luciani, E.; Bartolazzi, A.; Frasca, F.; Soddu, S.; Sciacchitano, S. Gal-3 is stimulated by gain-of-function p53 mutations and modulates chemoresistance in anaplastic thyroid carcinomas. *J. Pathol.* **2009**, *218*, 66–75. [CrossRef] [PubMed]

130. Fukumori, T.; Kanayama, H.O.; Raz, A. The role of galectin-3 in cancer drug resistance. *Drug Resist. Updates* **2007**, *10*, 101–108. [CrossRef] [PubMed]

131. Harazono, Y.; Nakajima, K.; Raz, A. Why anti-Bcl-2 clinical trials fail: A solution. *Cancer Metastasis Rev.* **2014**, *33*, 285–294. [CrossRef] [PubMed]

132. Hoyer, K.K.; Pang, M.; Gui, D.; Shintaku, I.P.; Kuwabara, I.; Liu, F.T.; Said, J.W.; Baum, L.G.; Teitell, M.A. An Anti-Apoptotic Role for Galectin-3 in Diffuse Large B-Cell Lymphoma. *Am. J. Pathol.* **2004**, *164*, 893–902. [CrossRef]

133. Harazono, Y.; Kho, D.H.; Balan, V.; Nakajima, K.; Zhang, T.; Hogan, V.; Raz, A. Galectin-3 leads to attenuation of apoptosis through Bax heterodimerization in human thyroid carcinoma cells. *Oncotarget* **2014**, *5*, 9992–10001. [CrossRef] [PubMed]

134. Song, S.; Ji, B.; Ramachandran, V.; Wang, H.; Hafley, M.; Logsdon, C.; Bresalier, R.S. Overexpressed galectin-3 in pancreatic cancer induces cell proliferation and invasion by binding Ras and activating Ras signaling. *PLoS ONE* **2012**, *7*, e42699. [CrossRef] [PubMed]

135. Elad-Sfadia, G.; Haklai, R.; Balan, E.; Kloog, Y. Galectin-3 augments K-Ras activation and triggers a Ras signal that attenuates ERK but not phosphoinositide 3-kinase activity. *J. Biol. Chem.* **2004**, *279*, 34922–34930. [CrossRef] [PubMed]

136. Levy, R.; Grafi-Cohen, M.; Kraiem, Z.; Kloog, Y. Galectin-3 promotes chronic activation of K-Ras and differentiation block in malignant thyroid carcinomas. *Mol. Cancer Ther.* **2010**, *9*, 2208–2219. [CrossRef] [PubMed]

137. Veschi, V.; Petroni, M.; Cardinali, B.; Dominici, C.; Screpanti, I.; Frati, L.; Bartolazzi, A.; Gulino, A.; Giannini, G. Galectin-3 impairment of MYCN-dependent apoptosis-sensitive phenotype is antagonized by nutlin-3 in neuroblastoma cells. *PLoS ONE* **2012**, *7*, e49139. [CrossRef] [PubMed]

138. Collins, P.M.; Bum-Erdene, K.; Yu, X.; Blanchard, H. Galectin-3 interactions with glycosphingolipids. *J. Mol. Biol.* **2014**, *426*, 1439–1451. [CrossRef] [PubMed]

139. Hauselmann, I.; Borsig, L. Altered tumor-cell glycosylation promotes metastasis. *Front. Oncol.* **2014**, *4*, 28. [CrossRef] [PubMed]

140. Du Toit, A. Endocytosis: Bend it like galectin 3. *Nat. Rev. Mol. Cell Biol.* **2014**, *15*, 430–431. [CrossRef] [PubMed]

141. Stanley, P. Galectins CLIC cargo inside. *Nat. Cell Biol.* **2014**, *16*, 506–507. [CrossRef] [PubMed]

142. Carlsson, M.C.; Bengtson, P.; Cucak, H.; Leffler, H. Galectin-3 guides intracellular trafficking of some human serotransferrin glycoforms. *J. Biol. Chem.* **2013**, *288*, 28398–28408. [CrossRef] [PubMed]

143. Lakshminarayan, R.; Wunder, C.; Becken, U.; Howes, M.T.; Benzing, C.; Arumugam, S.; Sales, S.; Ariotti, N.; Chambon, V.; Lamaze, C.; et al. Galectin-3 drives glycosphingolipid-dependent biogenesis of clathrinindependent carriers. *Nat. Cell Biol.* **2014**, *16*, 595–606. [CrossRef] [PubMed]

144. Mendez-Huergo, S.P.; Blidner, A.G.; Rabinovich, G.A. Galectins: Emerging regulatory checkpoints linking tumor immunity and angiogenesis. *Curr. Opin. Immunol.* **2017**, *45*, 8–15. [CrossRef] [PubMed]

145. Newlaczyl, A.U.; Yu, L.G. Galectin-3—A jack-of-all-trades in cancer. *Cancer Lett.* **2011**, *313*, 123–128. [CrossRef] [PubMed]

146. Cay, T. Immunohistochemical expression of Gal-3 in cancer: A review of the literature. *Turk Patoloji Derg.* **2012**, *28*, 1–10. [PubMed]

147. Ebrahim, A.H.; Alalawi, Z.; Mirandola, L.; Rakhshanda, R.; Dahlbeck, S.; Nguyen, D.; Jenkins, M.; Grizzi, F.; Cobos, E.; Figueroa, J.A.; et al. Galectins in cancer: Carcinogenesis, diagnosis and therapy. *Ann. Transl. Med.* **2014**, *2*, 88. [CrossRef] [PubMed]

148. Irimura, T.; Matsushita, Y.; Sutton, R.C.; Carralero, D.; Ohannesian, D.W.; Cleary, K.R.; Ota, D.M.; Nicolson, G.L.; Lotan, R. Increased content of an endogenous lactose-binding lectin in human colorectal carcinoma progressed to metastatic stages. *Cancer Res.* **1991**, *51*, 387–393. [PubMed]

149. Lotan, R.; Matsushita, Y.; Ohannesian, D.; Carralero, D.; Ota, D.M.; Cleary, K.R.; Nicolson, G.L.; Irimura, T. Lactose-binding lectin expression in human colorectal carcinomas. Relation to tumor progression. *Carbohydr. Res.* **1991**, *213*, 47–57. [CrossRef]

150. Nakamura, M.; Inufusa, H.; Adachi, T.; Aga, M.; Kurimoto, M.; Nakatani, Y.; Wakano, T.; Nakajima, A.; Hida, J.I.; Miyaké, M.; et al. Involvement of galectin-3 expression in colorectal cancer progression and metastasis. *Int. J. Oncol.* **1999**, *15*, 143–148. [CrossRef] [PubMed]

151. Schoeppner, H.L.; Raz, A.; Ho, S.B.; Bresalier, R.S. Expression of an endogenous galactose-binding lectin correlates with neoplastic progression in the colon. *Cancer* **1995**, *75*, 2818–2826. [CrossRef]

152. Castronovo, V.; Campo, E.; van den Brule, F.A.; Claysmith, A.P.; Cioce, V.; Liu, F.T.; Fernandez, P.L.; Sobel, M.E. Inverse modulation of steady-state messenger RNA levels of two nonintegrin laminin-binding proteins in human colon carcinoma. *J. Natl. Cancer Inst.* **1992**, *84*, 1161–1169. [CrossRef] [PubMed]

153. Lotz, M.M.; Andrews, C.W., Jr.; Korzelius, C.A.; Lee, E.C.; Steele, G.D., Jr.; Clarke, A.; Mercurio, A.M. Decreased expression of Mac-2 (carbohydrate binding protein 35) and loss of its nuclear localization are associated with the neoplastic progression of colon carcinoma. *Proc. Natl. Acad. Sci. USA* **1993**, *90*, 3466–3470. [CrossRef] [PubMed]

154. Sanjuan, X.; Fernandez, P.L.; Castells, A.; Castronovo, V.; van den Brule, F.; Liu, F.T.; Cardesa, A.; Campo, E. Differential expression of galectin 3 and galectin 1 in colorectal cancer progression. *Gastroenterology* **1997**, *113*, 1906–1915. [CrossRef]

155. Xu, X.C.; el-Naggar, A.K.; Lotan, R. Differential expression of galectin-1 and Gal-3 in thyroid tumors. Potential diagnostic implications. *Am. J. Pathol.* **1995**, *147*, 815–822. [PubMed]

156. Orlandi, F.; Saggiorato, E.; Pivano, G.; Puligheddu, B.; Termine, A.; Cappia, S.; De Giuli, P.; Angeli, A. Gal-3 is a presurgical marker of human thyroid carcinoma. *Cancer Res.* **1998**, *58*, 3015–3020. [PubMed]

157. Gasbarri, A.; Martegani, M.P.; Del Prete, F.; Lucante, T.; Natali, P.G.; Bartolazzi, A. Gal-3 and CD44v6 isoforms in the preoperative evaluation of thyroid nodules. *J. Clin. Oncol.* **1999**, *17*, 3494–3502. [CrossRef] [PubMed]

158. Inohara, H.; Honjo, Y.; Yoshii, T.; Akahani, S.; Yoshida, J.; Hattori, K.; Okamoto, S.; Sawada, T.; Raz, A.; Kubo, T. Expression of Gal-3 in fine-needle aspirates as a diagnostic marker differentiating benign from malignant thyroid neoplasms. *Cancer* **1999**, *85*, 2475–2484. [CrossRef]

159. Cvejic, D.; Savin, S.; Golubovic, S.; Paunovic, I.; Tatic, S.; Havelka, M. Gal-3 and carcinoembryonic antigen expression in medullary thyroid carcinoma: Possible relation to tumour progression. *Histopathology* **2000**, *37*, 530–535. [CrossRef] [PubMed]

160. Bartolazzi, A.; Gasbarri, A.; Papotti, M.; Bussolati, G.; Lucante, T.; Khan, A.; Inohara, H.; Marandino, F.; Orlandi, F.; Nardi, F.; et al. Application of an immunodiagnostic method for improving preoperative diagnosis of nodular thyroid lesions. *Lancet* **2001**, *357*, 1644–1650. [CrossRef]

161. Saggiorato, E.; Cappia, S.; De Giuli, P.; Mussa, A.; Pancani, G.; Caraci, P.; Angeli, A.; Orlandi, F. Gal-3 as a presurgical immunocytodiagnostic marker of minimally invasive follicular thyroid carcinoma. *J. Clin. Endocrinol. Metab.* **2001**, *86*, 5152–5158. [CrossRef] [PubMed]

162. Bresalier, R.S.; Yan, P.S.; Byrd, J.C.; Lotan, R.; Raz, A. Expression of the endogenous galactose-binding protein Gal-3 correlates with the malignant potential of tumors in the central nervous system. *Cancer* **1997**, *80*, 776–787. [CrossRef]

163. Borges, C.B.; Bernardes, E.S.; Latorraca, E.F.; Becker, A.P.; Neder, L.; Chammas, R.; Roque-Barreira, M.C.; Machado, H.R.; de Oliveira, R.S. Gal-3 expression: A useful tool in the differential diagnosis of posterior fossa tumors in children. *Childs Nerv. Syst.* **2011**, *27*, 253–257. [CrossRef] [PubMed]

164. Veschi, V.; Petroni, M.; Bartolazzi, A.; Altavista, P.; Dominici, C.; Capalbo, C.; Boldrini, R.; Castellano, A.; McDowell, H.P.; Pizer, B.; et al. Galectin-3 is a marker of favorable prognosis and a biologically relevant molecule in neuroblastic tumors. *Cell. Death Dis.* **2014**, *5*, e1100. [CrossRef] [PubMed]

165. Gillenwater, A.; Xu, X.C.; el-Naggar, A.K.; Clayman, G.L.; Lotan, R. Expression of galectins in head and neck squamous cell carcinoma. *Head Neck* **1996**, *18*, 422–432. [CrossRef]

166. Schaffert, C.; Pour, P.M.; Chaney, W.G. Localization of Gal-3 in normal and diseased pancreatic tissue. *Int. J. Pancreatol.* **1998**, *23*, 1–9. [PubMed]

167. Cindolo, L.; Benvenuto, G.; Salvatore, P.; Pero, R.; Salvatore, G.; Mirone, V.; Prezioso, D.; Altieri, V.; Bruni, C.B.; Chiariotti, L. Galectin-1 and Gal-3 expression in human bladder transitional-cell carcinomas. *Int. J. Cancer* **1999**, *84*, 39–43. [CrossRef]

168. Dancer, J.Y.; Truong, L.D.; Zhai, Q.; Shen, S.S. Expression of Gal-3 in renal neoplasms: A diagnostic, possible prognostic marker. *Arch. Pathol. Lab. Med.* **2010**, *134*, 90–94. [PubMed]

169. Young, A.N.; Amin, M.B.; Moreno, C.S.; Lim, S.D.; Cohen, C.; Petros, J.A.; Marshall, F.F.; Neish, A.S. Expression profiling of renal epithelial neoplasms: A method for tumor classification and discovery of diagnostic molecular markers. *Am. J. Pathol.* **2001**, *158*, 1639–1651. [CrossRef]

170. Hsu, D.K.; Dowling, C.A.; Jeng, K.C.G.; Chen, J.T.; Yang, R.Y.; Liu, F.T. Gal-3 expression is induced in cirrhotic liver and hepatocellular carcinoma. *Int. J. Cancer* **1999**, *81*, 519–526. [CrossRef]

171. Karaarslan, S.; Yurum, F.N.; Kumbaraci, B.S.; Pala, E.E.; Sivrikoz, O.N.; Akyildiz, M.; Bugdayci, M.H. The role of parafibromin, Gal-3, HBME-1, and Ki-67 in the differential diagnosis of parathyroid tumors. *Oman Med. J.* **2015**, *30*, 421–427. [CrossRef] [PubMed]

172. Saggiorato, E.; Bergero, N.; Volante, M.; Bacillo, E.; Rosas, R.; Gasparri, G.; Orlandi, F.; Papotti, M. Gal-3 and Ki-67 expression in multiglandular parathyroid lesions. *Am. J. Clin. Pathol.* **2006**, *126*, 59–66. [CrossRef] [PubMed]

173. Remmelink, M.; de Leval, L.; Decaestecker, C.; Duray, A.; Crompot, E.; Sirtaine, N.; André, S.; Kaltner, H.; Leroy, X.; Gabius, H.J.; et al. Quantitative immunohistochemical fingerprinting of adhesion/growth-regulatory galectins in salivary gland tumours: Divergent profiles with diagnostic potential. *Histopathology* **2011**, *58*, 543–556. [CrossRef] [PubMed]

174. Kim, S.J.; Lee, S.J.; Sung, H.J.; Choi, I.K.; Choi, C.W.; Kim, B.S.; Kim, J.S.; Yu, W.; Hwang, H.S.; Kim, I.S. Increased serum 90 K and Gal-3 expression are associated with advanced stage and a worse prognosis in diffuse large B-cell lymphomas. *Acta Haematol.* **2008**, *120*, 211–216. [CrossRef] [PubMed]

175. Abdou, A.G.; Hammam, M.A.; Farargy, S.E.; Farag, A.G.; El Shafey, E.N.; Farouk, S.; Elnaidany, N.F. Diagnostic and Prognostic Role of Galectin 3 Expression in Cutaneous Melanoma. *Am. J. Dermatopathol.* **2010**, *32*, 809–814. [CrossRef] [PubMed]

176. Bartolazzi, A.; Bellotti, C.; Sciacchitano, S. Methodology and technical requirements of the Gal-3 test for the preoperative characterization of thyroid nodules. *Appl. Immunohistochem. Mol. Morphol.* **2012**, *20*, 2–7. [CrossRef] [PubMed]

177. Carpi, A.; Naccarato, A.G.; Iervasi, G.; Nicolini, A.; Bevilacqua, G.; Viacava, P.; Collecchi, P.; Lavra, L.; Marchetti, C.; Sciacchitano, S.; et al. Large needle aspiration biopsy and Gal-3 determination in selected thyroid nodules with indeterminate FNA-cytology. *Br. J. Cancer* **2006**, *95*, 204–209. [CrossRef] [PubMed]

178. Carpi, A.; Di Coscio, G.; Iervasi, G.; Antonelli, A.; Mechanick, J.; Sciacchitano, S.; Nicolini, A. Thyroid fine needle aspiration: How to improve clinicians' confidence and performance with the technique. *Cancer Lett.* **2008**, *264*, 163–171. [CrossRef] [PubMed]

179. Bartolazzi, A.; Orlandi, F.; Saggiorato, E.; Volante, M.; Arecco, F.; Rossetto, R.; Palestini, N.; Ghigo, E.; Papotti, M.; Bussolati, G.; et al. Italian Thyroid Cancer Study Group (ITCSG). Gal-3-expression analysis in the surgical selection of follicular thyroid nodules with indeterminate fine-needle aspiration cytology: A prospective multicentre study. *Lancet Oncol.* **2008**, *9*, 543–549. [CrossRef]

180. De Matos, L.L.; Del Giglio, A.B.; Matsubayashi, C.O.; de Lima Farah, M.; Del Giglio, A.; da Silva Pinhal, M.A. Expression of ck-19, Gal-3 and hbme-1 in the differentiation of thyroid lesions: Systematic review and diagnostic meta-analysis. *Diagn. Pathol.* **2012**, *7*, 97. [CrossRef] [PubMed]

181. Sciacchitano, S.; Lavra, L.; Ulivieri, A.; Magi, F.; De Francesco, G.P.; Bellotti, C.; Salehi, L.B.; Trovato, M.; Drago, C.; Bartolazzi, A. Comparative analysis of diagnostic performance, feasibility and cost of different test-methods for thyroid nodules with indeterminate cytology. *Oncotarget* **2017**, *8*, 49421–49442. [CrossRef] [PubMed]

182. Sciacchitano, S.; Lavra, L.; Ulivieri, A.; Magi, F.; Porcelli, T.; Amendola, S.; De Francesco, G.P.; Bellotti, C.; Trovato, M.C.; Salehi, L.B.; et al. Combined clinical and ultrasound follow-up assists in malignancy detection in Gal-3 negative Thy-3 thyroid nodules. *Endocrine* **2016**, *54*, 139–147. [CrossRef] [PubMed]

183. Bartolazzi, A.; D'Alessandria, C.; Parisella, M.G.; Signore, A.; Del Prete, F.; Lavra, L.; Braesch-Andersen, S.; Massari, R.; Trotta, C.; Soluri, A.; et al. Thyroid cancer imaging in vivo by targeting the anti-apoptotic molecule Gal-3. *PLoS ONE* **2008**, *3*, e3768. [CrossRef] [PubMed]

184. Iurisci, I.; Tinari, N.; Natoli, C.; Angelucci, D.; Cianchetti, E.; Iacobelli, S. Concentrations of Gal-3 in the sera of normal controls and cancer patients. *Clin. Cancer Res.* **2000**, *6*, 1389–1393. [PubMed]

185. Chiu, C.G.; Strugnell, S.S.; Griffith, O.L.; Jones, S.J.; Gown, A.M.; Walker, B.; Nabi, I.R.; Wiseman, S.M. Diagnostic utility of Gal-3 in thyroid cancer. *Am. J. Pathol.* **2010**, *176*, 2067–2081. [CrossRef] [PubMed]

186. Balan, V.; Wang, Y.; Nangia-Makker, P.; Kho, D.; Bajaj, M.; Smith, D.; Heilbrun, L.; Raz, A.; Heath, E. Gal-3: A possible complementary marker to the PSA blood test. *Oncotarget* **2013**, *4*, 542–549. [CrossRef] [PubMed]

187. Eliaz, I. The role of Gal-3 as a marker of cancer and inflammation in a stage IV ovarian cancer patient with underlying pro-inflammatory comorbidities. *Case Rep. Oncol.* **2013**, *6*, 343–349. [CrossRef] [PubMed]

188. Lalancette-Hébert, M.; Swarup, V.; Beaulieu, J.M.; Bohacek, I.; Abdelhamid, E.; Weng, Y.C.; Sato, S.; Kriz, J. Gal-3 is required for resident microglia activation and proliferation in response to ischemic injury. *J. Neurosci.* **2012**, *32*, 10383–10395. [CrossRef] [PubMed]

189. Doverhag, C.; Hedtjärn, M.; Poirier, F.; Mallard, C.; Hagberg, H.; Karlsson, A.; Sävman, K. Gal-3 contributes to neonatal hypoxic-ischemic brain injury. *Neurobiol. Dis.* **2010**, *38*, 36–46. [CrossRef] [PubMed]

190. Ekingen, E.; Yilmaz, M.; Yildiz, M.; Atescelik, M.; Goktekin, M.C.; Gurger, M.; Alatas, O.D.; Basturk, M.; Ilhan, N. Utilization of glial fibrillary acidic protein and Gal-3 in the dagnosis of cerebral infarction patients with normal cranial tomography. *Niger. J. Clin. Pract.* **2017**, *20*, 433–437. [PubMed]

191. Agoston-Coldea, L.; Lupu, S.; Petrovai, D.; Mocan, T.; Mousseaux, E. Correlations between echocardiographic parameters of right ventricular dysfunction and Galectin-3 in patients with chronic obstructive pulmonary disease and pulmonary hypertension. *Med. Ultrason.* **2015**, *17*, 487–495. [PubMed]

192. Pilette, C.; Colinet, B.; Kiss, R.; André, S.; Kaltner, H.; Gabius, H.J.; Delos, M.; Vaerman, J.P.; Decramer, M.; Sibille, Y. Increased galectin-3 expression and intra-epithelial neutrophils in small airways in severe COPD. *Eur. Respir. J.* **2007**, *29*, 914–922. [CrossRef] [PubMed]

193. Mukaro, V.R.; Bylund, J.; Hodge, G.; Holmes, M.; Jersmann, H.; Reynolds, P.N.; Hodge, S. Lectins offer new perspectives in the development of macrophage-targeted therapies for COPD/emphysema. *PLoS ONE* **2013**, *8*, e56147. [CrossRef] [PubMed]

194. Hodge, S.; Dean, M.; Eisen, D.P. Lectins as Potential Adjunct Therapeutics for COPD/Emphysema? *J. Pulm. Respir. Med.* **2013**, *3*, 5. [CrossRef]

195. Feng, W.; Wu, X.; Li, S.; Zhai, C.; Wang, J.; Shi, W.; Li, M. Association of serum Galectin-3 with the acute exacerbation of chronic obstructive pulmonary disease. *Med. Sci. Monit.* **2017**, *23*, 4612–4618. [CrossRef] [PubMed]

196. Bobrowska, B.; Wieczorek-Surdacka, E.; Kruszelnicka, O.; Chyrchel, B.; Surdacki, A.; Dudek, D. Clinical correlates and prognostic value of plasma Gal-3 levels in degenerative aortic stenosis: A single-center prospective study of patients referred for invasive treatment. *Int. J. Mol. Sci.* **2017**, *18*, 947. [CrossRef] [PubMed]

197. Henderson, N.C.; Sethi, T. The regulation of inflammation by Gal-3. *Immunol. Rev.* **2009**, *230*, 160–171. [CrossRef] [PubMed]

198. Wan, L.; Liu, F.T. Gal-3 and inflammation. *Glycobiol. Insights* **2016**, *6*, 1–9.

199. Weigert, J.; Neumeier, M.; Wanninger, J.; Bauer, S.; Farkas, S.; Scherer, M.N.; Schnitzbauer, A.; Schäffler, A.; Aslanidis, C.; Schölmerich, J.; et al. Serum Gal-3 is elevated in obesity and negatively correlates with glycosylated hemoglobin in type 2 diabetes. *J. Clin. Endocrinol. Metab.* **2010**, *95*, 1404–1411. [CrossRef] [PubMed]

200. Jin, Q.H.; Lou, Y.F.; Li, T.L.; Chen, H.H.; Liu, Q.; He, X.J. Serum Gal-3: A risk factor for vascular complications in type 2 diabetes mellitus. *Chin. Med. J.* **2013**, *126*, 2109–2115. [PubMed]

201. Yilmaz, H.; Cakmak, M.; Inan, O.; Darcin, T.; Akcay, A. Increased levels of Gal-3 were associated with prediabetes and diabetes: New risk factor? *J. Endocrinol. Investig.* **2014**, *38*, 527–533. [CrossRef] [PubMed]

202. Seferovic, J.P.; Lalic, N.M.; Floridi, F.; Tesic, M.; Seferovic, P.M.; Giga, V.; Lalic, K.; Jotic, A.; Jovicic, S.; Colak, E.; et al. Structural myocardial alterations in diabetes and hypertension: The role of Gal-3. *Clin. Chem. Lab. Med.* **2014**, *52*, 1499–1505. [CrossRef] [PubMed]

203. Ozturk, D.; Celik, O.; Satilmis, S.; Aslan, S.; Erturk, M.; Cakmak, H.A.; Kalkan, A.K.; Ozyilmaz, S.; Diker, V.; Gul, M. Association between serum Gal-3 levels and coronary atherosclerosis and plaque burden/structure in patients with type 2 diabetes mellitus. *Coron. Artery Dis.* **2015**, *26*, 396–401. [CrossRef] [PubMed]

204. Darrow, A.L.; Shohet, R.V.; Maresh, J.G. Transcriptional analysis of the endothelial response to diabetes reveals a role for Gal-3. *Physiol. Genom.* **2011**, *43*, 1144–1152. [CrossRef] [PubMed]

205. Kiwaki, K.; Novak, C.M.; Hsu, D.K.; Liu, F.T.; Levine, J.A. Gal-3 stimulates preadipocyte proliferation and is up-regulated in growing adipose tissue. *Obesity* **2007**, *15*, 32–39. [CrossRef] [PubMed]

206. Baek, J.H.; Kim, S.J.; Kang, H.G.; Lee, H.W.; Kim, J.H.; Hwang, K.A.; Song, J.; Chun, K.H. Gal-3 activates PPARgamma and supports white adipose tissue formation and high-fat diet-induced obesity. *Endocrinology* **2015**, *156*, 147–156. [CrossRef] [PubMed]

207. Chen, J.Y.; Chou, H.C.; Chen, Y.H.; Chan, H.L. High glucose-induced proteome alterations in hepatocytes and its possible relevance to diabetic liver disease. *J. Nutr. Biochem.* **2013**, *24*, 1889–1910. [CrossRef] [PubMed]

208. Mensah-Brown, E.P.; Al Rabesi, Z.; Shahin, A.; Al Shamsi, M.; Arsenijevic, N.; Hsu, D.K.; Liu, F.T.; Lukic, M.L. Targeted disruption of the Gal-3 gene results in decreased susceptibility to multiple low dose streptozotocin-induced diabetes in mice. *Clin. Immunol.* **2009**, *130*, 83–88. [CrossRef] [PubMed]

209. Pugliese, G.; Iacobini, C.; Ricci, C.; Blasetti Fantauzzi, C.; Menini, S. Gal-3 in diabetic patients. *Clin. Chem. Lab. Med.* **2014**, *52*, 1413–1423. [CrossRef] [PubMed]

210. Ohkura, T.; Fujioka, Y.; Nakanishi, R.; Shiochi, H.; Sumi, K.; Yamamoto, N.; Matsuzawa, K.; Izawa, S.; Ohkura, H.; Ueta, E.; et al. Low serum Gal-3 concentrations are associated with insulin resistance in patients with type 2 diabetes mellitus. *Diabetol. Metab. Syndr.* **2014**, *6*, 106. [CrossRef] [PubMed]

211. Pejnovic, N.N.; Pantic, J.M.; Jovanovic, I.P.; Radosavljevic, G.D.; Milovanovic, M.Z.; Nikolic, I.G.; Zdravkovic, N.S.; Djukic, A.L.; Arsenijevic, N.N.; Lukic, M.L. Gal-3 deficiency accelerates high-fat diet-induced obesity and amplifies inflammation in adipose tissue and pancreatic islets. *Diabetes* **2013**, *62*, 1932–1944. [CrossRef] [PubMed]

212. Pang, J.; Rhodes, D.H.; Pini, M.; Akasheh, R.T.; Castellanos, K.J.; Cabay, R.J.; Cooper, D.; Perretti, M.; Fantuzzi, G. Increased adiposity, dysregulated glucose metabolism and systemic inflammation in Gal-3 KO mice. *PLoS ONE* **2013**, *8*, e57915. [CrossRef] [PubMed]

213. Darrow, A.L.; Shohet, R.V. Gal-3 deficiency exacerbates hyperglycemia and the endothelial response to diabetes. *Cardiovasc. Diabetol.* **2015**, *14*, 73. [CrossRef] [PubMed]

214. Rhodes, D.H.; Pini, M.; Castellanos, K.J.; Montero-Melendez, T.; Cooper, D.; Perretti, M.; Fantuzzi, G. Adipose Tissue-Specific Modulation of Galectin Expression in Lean and Obese Mice: Evidence for Regulatory Function. *Obesity* **2013**, *21*, 310–319. [CrossRef] [PubMed]

215. Pugliese, G.; Pricci, F.; Iacobini, C.; Leto, G.; Amadio, L.; Barsotti, P.; Frigeri, L.; Hsu, D.K.; Vlassara, H.; Liu, F.T.; et al. Accelerated diabetic glomerulopathy in Gal-3/AGE receptor-3 knockout mice. *FASEB J.* **2001**, *15*, 2471–2479. [CrossRef] [PubMed]

216. Iacobini, C.; Menini, S.; Oddi, G.; Ricci, C.; Amadio, L.; Pricci, F.; Olivieri, A.; Sorcini, M.; Di Mario, U.; Pesce, C.; et al. Gal-3/AGE-receptor 3 knockout mice show accelerated AGE-induced glomerular injury. Evidence for a protective role of Gal-3 as an AGE-receptor. *FASEB J.* **2004**, *18*, 1773–1775. [CrossRef] [PubMed]

217. Lebovic, D.I.; Mueller, M.D.; Taylor, R.N. Immunobiology of endometriosis. *Fertil. Steril.* **2001**, *75*, 1–10. [CrossRef]

218. Omwandho, C.O.; Konrad, L.; Halis, G.; Oehmke, F.; Tinneberg, H.R. Role of TGF-betas in normal human endometrium and endometriosis. *Hum. Reprod.* **2010**, *25*, 101–109. [CrossRef] [PubMed]

219. Noël, J.C.; Chapron, C.; Borghese, B.; Fayt, I.; Anaf, V. Gal-3 is over- expressed in various forms of endometriosis. *Appl. Immunohistochem. Mol. Morphol.* **2011**, *19*, 253–257. [PubMed]

220. Caserta, D.; Di Benedetto, L.; Bordi, G.; D'Ambrosio, A.; Moscarini, M. Levels of Gal-3 and stimulation expressed gene 2 in the peritoneal fluid of women with endometriosis: A pilot study. *Gynecol. Endocrinol.* **2014**, *30*, 877–880. [CrossRef] [PubMed]

221. Chen, H.L.; Liao, F.; Lin, T.N.; Liu, F.T. Galectins and neuroinflammation. *Adv. Neurobiol.* **2014**, *9*, 517–542. [PubMed]

222. Borghese, B.; Vaiman, D.; Mondon, F.; Mbaye, M.; Anaf, V.; Noël, J.C.; de Ziegler, D.; Chapron, C. Neurotrophins and pain in endometriosis. *Gynecol. Obstet. Fertil.* **2010**, *38*, 442–446. [CrossRef] [PubMed]

223. Furness, J.B. *The Enteric Nervous System*; John Wiley & Sons: Hoboken, NJ, USA, 2008; pp. 35–38.

224. Young, E. Gut instincts: The secrets of your second brain. *New Sci.* **2012**, *216*, 38–42. [CrossRef]

225. Gershon, M. *The Second Brain: The Scientific Basis of Gut Instinct and a Groundbreaking New Understanding of Nervous Disorders of the Stomach and Intestines*; Harper Collins: New York, NY, USA, 1998.

226. Jiang, K.; Rankin, C.R.; Nava, P.; Sumagin, R.; Kamekura, R.; Stowell, S.R.; Feng, M.; Parkos, C.A.; Nusrat, A. Gal-3 regulates desmoglein-2 and intestinal epithelial intercellular adhesion. *J. Biol. Chem.* **2014**, *289*, 10510–10517. [CrossRef] [PubMed]

227. Delacour, D.; Koch, A.; Ackermann, W.; Eude-Le Parco, I.; Elsässer, H.P.; Poirier, F.; Jacob, R. Loss of Gal-3 impairs membrane polarisation of mouse enterocytes in vivo. *J. Cell Sci.* **2008**, *121*, 458–465. [CrossRef] [PubMed]

228. Dudas, S.P.; Yunker, C.K.; Sternberg, L.R.; Byrd, J.C.; Bresalier, R.S. Expression of human intestinal mucin is modulated by the beta-galactoside binding protein Gal-3 in colon cancer. *Gastroenterology* **2002**, *123*, 817–826. [CrossRef] [PubMed]

229. Lippert, E.; Gunckel, M.; Brenmoehl, J.; Bataille, F.; Falk, W.; Scholmerich, J.; Obermeier, F.; Rogler, G. Regulation of Gal-3 function in mucosal fibroblasts: Potential role in mucosal inflammation. *Clin. Exp. Immunol.* **2008**, *152*, 285–297. [CrossRef] [PubMed]

230. Cheng, X.; Boza-Serrano, A.; Turesson, M.F.; Deierborg, T.; Ekblad, E.; Voss, U. Gal-3 causes enteric neuronal loss in mice after left sided permanent middle cerebral artery occlusion, a model of stroke. *Sci. Rep.* **2016**, *6*, 32893. [CrossRef] [PubMed]

231. He, X.-W.; Li, W.L.; Li, C.; Liu, P.; Shen, Y.G.; Zhu, M.; Jin, X.P. Serum levels of galectin-1, galectin-3, and galectin-9 are associated with large artery atherosclerotic stroke. *Sci Rep.* **2017**, *7*, 40994. [CrossRef] [PubMed]

232. Pasquini, L.A.; Millet, V.; Hoyos, H.C.; Giannoni, J.P.; Croci, D.O.; Marder, M.; Liu, F.T.; Rabinovich, G.A.; Pasquini, J.M. Galectin-3 drives oligodendrocyte differentiation to control myelin integrity and function. *Cell Death Differ.* **2011**, *18*, 1746–1756. [CrossRef] [PubMed]

233. Comte, I.; Kim, Y.; Young, C.C.; van der Harg, J.M.; Hockberger, P.; Bolam, P.J.; Poirier, F.; Szele, F.G. Galectin-3 maintains cell motility from the subventricular zone to the olfactory bulb. *J. Cell Sci.* **2011**, *124*, 2438–2447. [CrossRef] [PubMed]

234. Jaquenod De Giusti, C.; Alberdi, L.; Frik, J.; Ferrer, M.F.; Scharrig, E.; Schattner, M.; Gomez, R.M. Gal-3 is upregulated in activated glia during Junin virus-induced murine encephalitis. *Neurosci. Lett.* **2011**, *501*, 163–166. [CrossRef] [PubMed]

235. Kobayashi, K.; Niwa, M.; Hoshi, M.; Saito, K.; Hisamatsu, K.; Hatano, Y.; Tomita, H.; Miyazaki, T.; Hara, A. Early microlesion of viral encephalitis confirmed by Gal-3 expression after a virus inoculation. *Neurosci. Lett.* **2015**, *592*, 107–112. [CrossRef] [PubMed]

236. Takasaki, I.; Taniguchi, K.; Komatsu, F.; Sasaki, A.; Andoh, T.; Nojima, H.; Shiraki, K.; Hsu, D.K.; Liu, F.T.; Kato, I.; et al. Contribution of spinal galectin-3 to acute herpetic allodynia in mice. *Pain* **2012**, *153*, 585–592. [CrossRef] [PubMed]

237. Wynn, T.A. Cellular and molecular mechanisms of fibrosis. *J. Pathol.* **2008**, *214*, 199–210. [CrossRef] [PubMed]

238. Insel, P.A.; Murray, F.; Yokoyama, U.; Romano, S.; Yun, H.; Brown, L.; Snead, A.; Lu, D.; Aroonsakool, N. cAMP and Epac in the regulation of tissue fibrosis. *Br. J. Pharmacol.* **2012**, *166*, 447–456. [CrossRef] [PubMed]

239. Speca, S.; Giusti, I.; Rieder, F.; Latella, G. Cellular and molecular mechanisms of intestinal fibrosis. *World J. Gastroenterol.* **2012**, *18*, 3635–3661. [CrossRef] [PubMed]

240. Friedman, S.L.; Sheppard, D.; Duffield, J.S.; Violette, S. Therapy for fibrotic diseases: Nearing the starting line. *Sci. Transl. Med.* **2013**, *5*, 167sr1. [CrossRef] [PubMed]

241. Henderson, N.C.; Mackinnon, A.C.; Rooney, C.; Sethi, T. Gal-3: A Central Regulator of Chronic Inflammation and Tissue Fibrosis. In *Galectins and Disease Implications for Targeted Therapeutics*; ACS Symposium Series; American Chemical Society: Washington, DC, USA, 2012; Chapter 22, Volume 1115, pp. 377–390.

242. Henderson, N.C.; Mackinnon, A.C.; Farnworth, S.L.; Poirier, F.; Russo, F.P.; Iredale, J.P.; Haslett, C.; Simpson, K.J.; Sethi, T. Gal-3 regulates myofibroblast activation and hepatic fibrosis. *Proc. Natl. Acad. Sci. USA* **2006**, *103*, 5060–5065. [CrossRef] [PubMed]

243. Henderson, N.C.; Mackinnon, A.C.; Farnworth, S.L.; Kipari, T.; Haslett, C.; Iredale, J.P.; Liu, F.T.; Hughes, J.; Sethi, T. Gal-3 expression and secretion links macrophages to the promotion of renal fibrosis. *Am. J. Pathol.* **2008**, *172*, 288–298. [CrossRef] [PubMed]

244. Dang, Z.; MacKinnon, A.; Marson, L.P.; Sethi, T. Tubular atrophy and interstitial fibrosis after renal transplantation is dependent on Gal-3. *Transplantation* **2012**, *93*, 477–484. [CrossRef] [PubMed]

245. Mackinnon, A.C.; Gibbons, M.A.; Farnworth, S.L.; Leffler, H.; Nilsson, U.J.; Delaine, T.; Simpson, A.J.; Forbes, S.J.; Hirani, N.; Gauldie, J.; et al. Regulation of transforming growth factor-beta1-driven lung fibrosis by Gal-3. *Am. J. Respir. Crit. Care Med.* **2012**, *185*, 537–546. [CrossRef] [PubMed]

246. Nishi, Y.; Sano, H.; Kawashima, T.; Okada, T.; Kuroda, T.; Kikkawa, K.; Kawashima, S.; Tanabe, M.; Goto, T.; Matsuzawa, Y.; et al. Role of Gal-3 in human pulmonary fibrosis. *Allergol. Int.* **2007**, *56*, 57–65. [CrossRef] [PubMed]

247. Thandavarayan, R.A.; Watanabe, K.; Ma, M.; Veeraveedu, P.T.; Gurusamy, N.; Palaniyandi, S.S.; Zhang, S.; Muslin, A.J.; Kodama, M.; Aizawa, Y. 14-3-3 Protein regulates Ask1 signaling and protects against diabetic cardiomyopathy. *Biochem. Pharmacol.* **2008**, *75*, 1797–1806. [CrossRef] [PubMed]

248. Sharma, U.; Rhaleb, N.E.; Pokharel, S.; Harding, P.; Rasoul, S.; Peng, H.; Carretero, O.A. Novel anti-inflammatory mechanisms of Nacetyl-Ser-Asp-Lys-Pro in hypertension-induced target organ damage. *Am. J. Physiol.* **2008**, *294*, H1226–H1232.

249. Burguillos, M.A.; Svensson, M.; Schulte, T.; Boza-Serrano, A.; Garcia-Quintanilla, A.; Kavanagh, E.; Santiago, M.; Viceconte, N.; Oliva-Martin, M.J.; Osman, A.M.; et al. Microglia-Secreted Gal-3 Acts as a Toll-like Receptor 4 Ligand and Contributes to Microglial Activation. *Cell Rep.* **2015**, *10*, 1626–1638. [CrossRef] [PubMed]

250. Taniguchi, T.; Asano, Y.; Akamata, K.; Noda, S.; Masui, Y.; Yamada, D.; Takahashi, T.; Ichimura, Y.; Toyama, T.; Tamaki, Z.; et al. Serum Levels of Gal-3: Possible Association with Fibrosis, Aberrant Angiogenesis, and Immune Activation in Patients with Systemic Sclerosis. *J. Rheumatol.* **2012**, *39*, 539–544. [CrossRef] [PubMed]

251. Koca, S.S.; Akbas, F.; Ozgen, M.; Yolbas, S.; Ilhan, N.; Gundogdu, B.; Isik, A. Serum Gal-3 level in systemic sclerosis. *Clin. Rheumatol.* **2014**, *33*, 215–220. [CrossRef] [PubMed]

252. Fichtner-Feigl, S.; Strober, W.; Kawakami, K.; Puri, R.K.; Kitani, A. IL-13 signaling through the IL-13alpha2 receptor is involved in induction of TGF-beta1 production and fibrosis. *Nat. Med.* **2006**, *12*, 99–106. [CrossRef] [PubMed]

253. Liu, T.; Jin, H.; Ullenbruch, M.; Hu, B.; Hashimoto, N.; Moore, B.; McKenzie, A.; Lukacs, N.W.; Phan, S.H. Regulation of found in inflammatory zone 1 expression in bleomycin-induced lung fibrosis: Role of IL-4/IL-13 and mediation via STAT-6. *J. Immunol.* **2004**, *173*, 3425–3431. [CrossRef] [PubMed]

254. Wynn, T.A. Fibrotic disease and the T(H)1/T(H)2 paradigm. *Nat. Rev. Immunol.* **2004**, *4*, 583–594. [CrossRef] [PubMed]

255. Faludi, R.; Nagy, G.; Tőkés-Füzesi, M.; Kovács, K.; Czirják, L.; Komócsi, A. Gal-3 is an independent predictor of survival in systemic sclerosis. *Int. J. Cardiol.* **2017**, *233*, 118–124. [CrossRef] [PubMed]

256. Vray, B.; Camby, I.; Vercruysse, V.; Mijatovic, T.; Bovin, N.V.; Ricciardi-Castagnoli, P.; Kaltner, H.; Salmon, I.; Gabius, H.J.; Kiss, R. Up-regulation of Gal-3 and its ligands by Trypanosoma cruzi infection with modulation of adhesion and migration of murine dendritic cells. *Glycobiology* **2004**, *14*, 647–657. [CrossRef] [PubMed]

257. Fogel, S.; Guittaut, M.; Legrand, A.; Monsigny, M.; Hebert, E. The tat protein of HIV-1 induces galectin-3 expression. *Glycobiology* **1999**, *9*, 383–387. [CrossRef] [PubMed]

258. Hsu, D.; Hammes, S.R.; Kuwabara, I.; Greene, W.C.; Liu, F.T. Human T lymphotropic virus-I infection of human T lymphocytes induces expression of the beta- galactoside-binding lectin, Gal-3. *Am. J. Pathol.* **1996**, *148*, 1661–1670. [PubMed]

259. Fowler, M.; Thomas, R.J.; Atherton, J.; Roberts, I.S.; High, N.J. Gal-3 binds to Helicobacter pylori O-antigen: It is upregulated and rapidly secreted by gastric epithelial cells in response to H. pylori adhesion. *Cell. Microbiol.* **2006**, *8*, 44–54. [CrossRef] [PubMed]

260. Park, A.M.; Hagiwara, S.; Hsu, D.K.; Liu, F.T.; Yoshie, O. Gal-3 Plays an important role in innate immunity to gastric infection by helicobacter pylori. *Infect. Immun.* **2016**, *84*, 1184–1193. [CrossRef] [PubMed]

261. Subhash, V.V.; Ling, S.S.; Ho, B. Extracellular galectin-3 counteracts adhesion and exhibits chemoattraction in Helicobacter pylori infected gastric cancer cells. *Microbiology* **2016**, *162*, 1360–1366. [CrossRef] [PubMed]

262. Parsonnet, J.; Friedman, G.D.; Vandersteen, D.P.; Chang, Y.; Vogelman, J.H.; Orentreich, N.; Sibley, R.K. Helicobacter pylori infection and the risk of gastric carcinoma. *N. Engl. J. Med.* **1991**, *325*, 1127–1131. [CrossRef] [PubMed]

263. Uemura, N.; Okamoto, S.; Yamamoto, S.; Matsumura, N.; Yamaguchi, S.; Yamakido, M.; Taniyama, K.; Sasaki, N.; Schlemper, R.J. Helicobacter pylori infection and the development of gastric cancer. *N. Engl. J. Med.* **2001**, *345*, 784–789. [CrossRef] [PubMed]

264. Subhash, V.V.; Ho, B. Galectin 3 acts as an enhancer of survival responses in H. pylori-infected gastric cancer cells. *Cell Biol. Toxicol.* **2016**, *32*, 23–25. [CrossRef] [PubMed]

265. Liu, Y.H.; D'Ambrosio, M.; Liao, T.D.; Peng, H.; Rhaleb, N.E.; Sharma, U.; André, S.; Gabius, H.J.; Carretero, O.A. N-acetyl-seryl-aspartyllysyl-proline prevents cardiac remodeling and dysfunction induced by Gal-3, a mammalian adhesion/growth-regulatory lectin. *Am. J. Physiol. Heart Circ. Physiol.* **2009**, *296*, H404–H412. [CrossRef] [PubMed]

266. Creemers, E.E.; Pinto, Y.M. Molecular mechanisms that control interstitial fibrosis in the pressure overloaded heart. *Cardiovasc. Res.* **2011**, *89*, 265–272. [CrossRef] [PubMed]

267. Kramer, F. Galectin-3: Clinical utility and prognostic value in patients with heart failure. *Res. Rep. Clin. Cardiol.* **2013**, *4*, 13–22. [CrossRef]

268. Shah, R.V.; Chen-Tournoux, A.A.; Picard, M.H.; van Kimmenade, R.R.; Januzzi, J.L. Galectin-3, cardiac structure and function, and long-term mortality in patients with acutely decompensated heart failure. *Eur. J. Heart Fail.* **2010**, *12*, 826–832. [CrossRef] [PubMed]

269. Gullestad, L.; Ueland, T.; Kjekshus, J.; Nymo, S.H.; Hulthe, J.; Muntendam, P.; Adourian, A.; Böhm, M.; van Veldhuisen, D.J.; Komajda, M.; et al. Galectin-3 predicts response to statin therapy in the Controlled Rosuvastatin Multinational Trial in Heart Failure (CORONA). *Eur. Heart J.* **2012**, *33*, 2290–2296. [CrossRef] [PubMed]

270. Stolen, C.M.; Adourian, A.; Meyer, T.E.; Stein, K.M.; Solomon, S.D. Plasma Galectin-3 and Heart Failure Outcomes in MADIT-CRT (Multicenter Automatic Defibrillator Implantation Trial with Cardiac Resynchronization Therapy). *J. Card. Fail.* **2014**, *20*, 793–799. [CrossRef] [PubMed]

271. Milting, H.; Ellinghaus, P.; Seewald, M.; Cakar, H.; Bohms, B.; Kassner, A.; Körfer, R.; Klein, M.; Krahn, T.; Kruska, L.; et al. Plasma biomarkers of myocardial fibrosis and remodeling in terminal heart failure patients supported by mechanical circulatory support devices. *J. Heart Lung Transplant.* **2008**, *27*, 589–596. [CrossRef] [PubMed]

272. Erkilet, G.; Özpeker, C.; Böthig, D.; Kramer, F.; Röfe, D.; Bohms, B.; Morshuis, M.; Gummert, J.; Milting, H. The biomarker plasma galectin-3 in advanced heart failure and survival with mechanical circulatory support devices. *J. Heart Lung Transplant.* **2013**, *32*, 221–230. [CrossRef] [PubMed]

273. Grandin, E.W.; Jarolim, P.; Murphy, S.A.; Ritterova, L.; Cannon, C.P.; Braunwald, E.; Morrow, D.A. Galectin-3 and the development of heart failure after acute coronary syndrome: Pilot experience from PROVE IT-TIMI 22. *Clin. Chem.* **2012**, *58*, 267–273. [CrossRef] [PubMed]

274. Lok, D.J.; Lok, S.I.; Bruggink-André de la Porte, P.W.; Badings, E.; Lipsic, E.; van Wijngaarden, J.; de Boer, R.A.; van Veldhuisen, D.J.; van der Meer, P. Galectin-3 is an independent marker for ventricular remodeling and mortality in patients with chronic heart failure. *Clin. Res. Cardiol.* **2013**, *102*, 103–110. [CrossRef] [PubMed]

275. Peacock, W.F.; Di Somma, S. Emergency department use of Galectin-3. *Crit. Pathw. Cardiol.* **2014**, *13*, 73–77. [CrossRef] [PubMed]

276. Teichman, S.L.; Maisel, A.S.; Storrow, A.B. Challenges in acute heart failure clinical management: Optimizing care despite incomplete evidence and imperfect drugs. *Crit. Pathw. Cardiol.* **2015**, *14*, 12–24. [CrossRef] [PubMed]

277. Yancy, C.W.; Jessup, M.; Bozkurt, B.; Butler, J.; Casey, D.E., Jr.; Drazner, M.H.; Fonarow, G.C.; Geraci, S.A.; Horwich, T.; Januzzi, J.L.; et al. 2013 ACCF/AHA guideline for the management of heart failure: A report of the American College of Cardiology Foundation/American Heart Association Task Force on Practice Guidelines. *Circulation* **2013**, *128*, e240–e327. [CrossRef] [PubMed]

278. Yancy, C.W.; Jessup, M.; Bozkurt, B.; Butler, J.; Casey, D.E., Jr.; Colvin, M.M.; Drazner, M.H.; Filippatos, G.S.; Fonarow, G.C.; Givertz, M.M.; et al. 2017 ACC/AHA/HFSA focused update of the 2013 ACCF/AHA guideline for the management of heart failure: A report of the American College of Cardiology/American Heart Association Task Force on Clinical Practice Guidelines and the Heart Failure Society of America. *J. Am. Coll. Cardiol.* **2017**, *70*, 776–803. [PubMed]

279. Wang, S.F.; Tsao, C.H.; Lin, Y.T.; Hsu, D.K.; Chiang, M.L.; Lo, C.H.; Chien, F.C.; Chen, P.; Arthur Chen, Y.M.; Chen, H.Y.; et al. Galectin-3 promotes HIV-1 budding via association with Alix and Gag p6. *Glycobiology* **2014**, *24*, 1022–1035. [CrossRef] [PubMed]

280. Abbas, W.; Herbein, G. Plasma membrane signaling in HIV-1 infection. *Biochim. Biophys. Acta* **2014**, *1838*, 1132–1142. [CrossRef] [PubMed]

281. Schröder, H.C.; Ushijima, H.; Theis, C.; Sève, A.P.; Hubert, J.; Müller, W.E.G. Expression of nuclear lectin carbohydrate-binding protein 35 in human immunodeficiency virus type 1-infected molt-3 cells. *J. Acquir. Immune Defic. Syndr. Hum. Retrovirol.* **1995**, *9*, 340–348. [PubMed]

282. Ouellet, M.; Mercier, S.; Pelletier, I.; Bounou, S.; Roy, J.; Hirabayashi, J.; Sato, S.; Tremblay, M.J. Galectin-1 Acts as a Soluble Host Factor That Promotes HIV-1 Infectivity through Stabilization of Virus Attachment to Host Cells. *J. Immunol.* **2005**, *174*, 4120–4126. [CrossRef] [PubMed]

283. Xue, J.; Fu, C.; Cong, Z.; Peng, L.; Peng, Z.; Chen, T.; Wang, W.; Jiang, H.; Wei, Q.; Qin, C. Galectin-3 promotes caspase-independent cell death of HIV-1-infected macrophages. *FEBS J.* **2017**, *284*, 97–113. [CrossRef] [PubMed]

284. Dabelic, S.; Supraha, S.; Dumic, J. Gal-3 in macrophage-like cells exposed to immunomodulatory drugs. *Biochim. Biophys. Acta* **2006**, *1760*, 701–709. [CrossRef] [PubMed]

285. Sano, H.; Hsu, D.K.; Yu, L.; Apgar, J.R.; Kuwabara, I.; Yamanaka, T.; Hirashima, M.; Liu, F.T. Human Gal-3 is a novel chemoattractant for monocytes and macrophages. *J. Immunol.* **2000**, *165*, 2156–2164. [CrossRef] [PubMed]

286. Swarte, V.V.; Mebius, R.E.; Joziasse, D.H.; Van den Eijnden, D.H.; Kraal, G. Lymphocyte triggering via L-selectin leads to enhanced Gal-3-mediated binding to dendritic cells. *Eur. J. Immunol.* **1998**, *28*, 2864–2871. [CrossRef]

287. Truong, M.J.; Gruart, V.; Liu, F.T.; Prin, L.; Capron, A.; Capron, M. IgE-binding molecules (Mac-2/epsilon BP) expressed by human eosinophils. Implication in IgE-dependent eosinophil cytotoxicity. *Eur. J. Immunol.* **1993**, *23*, 3230–3235. [CrossRef] [PubMed]

288. Frigeri, L.G.; Liu, F.T. Surface expression of function- al IgE binding protein, an endogenous lectin, on mast cells and macrophages. *J. Immunol.* **1992**, *148*, 861–867. [PubMed]

289. Tsuboi, S.; Sutoh, M.; Hatakeyama, S.; Hiraoka, N.; Habuchi, T.; Horikawa, Y.; Hashimoto, Y.; Yoneyama, T.; Mori, K.; Koie, T.; et al. A novel strategy for evasion of NK cell immunity by tumours expressing core2 O-glycans. *EMBO J.* **2011**, *30*, 3173–3185. [CrossRef] [PubMed]

290. Peng, W.; Wang, H.Y.; Miyahara, Y.; Peng, G.; Wang, R.F. Tumor-associated Gal-3 modulates the function of tumor-reactive T cells. *Cancer Res.* **2008**, *68*, 7228–7236. [CrossRef] [PubMed]

291. Acosta-Rodriguez, E.V.; Montes, C.L.; Motran, C.C.; Zuniga, E.I.; Liu, F.T.; Rabinovich, G.A.; Gruppi, A. Gal-3 mediates IL-4-induced survival and differentiation of B cells: Functional cross-talk and implications during Trypanosoma cruzi infection. *J. Immunol.* **2004**, *172*, 493–502. [CrossRef] [PubMed]

292. Elliott, M.J.; Strasser, A.; Metcalf, D. Selective up-regulation of macrophage function in granulocyte-macrophage colony-stimulating factor transgenic mice. *J. Immunol.* **1991**, *147*, 2957–2963. [PubMed]

293. Rabinovich, G.A.; Toscano, M.A. Turning 'sweet' on immunity: Galectin–glycan interactions in immune tolerance and inflammation. *Nat. Rev. Immunol.* **2009**, *9*, 338–352. [CrossRef] [PubMed]

294. Thiemann, S.; Baum, L.G. Galectins and immune responses-just how do they do those things they do? *Annu. Rev. Immunol.* **2016**, *34*, 243–264. [CrossRef] [PubMed]

295. Chen, H.Y.; Liu, F.T.; Yang, R.Y. Roles of Gal-3 in immune responses. *Arch. Immunol. Ther. Exp.* **2005**, *53*, 497–504.

296. Kuwabara, I.; Liu, F.T. Gal-3 promotes adhesion of human neutrophils to laminin. *J. Immunol.* **1996**, *156*, 3939–3944. [PubMed]

297. Frigeri, L.G.; Zuberi, R.I.; Liu, F.T. Epsilon BP, a beta-galactoside-binding animal lectin, recognizes IgE receptor (Fc epsilon RI) and activates mast cells. *Biochemistry* **1993**, *32*, 7644–7649. [CrossRef] [PubMed]

298. Liu, F.T.; Hsu, D.K.; Zuberi, R.I.; Kuwabara, I.; Chi, E.Y.; Henderson, W.R., Jr. Expression and function of Gal-3, a beta-galactoside binding lectin, in human monocytes and macrophages. *Am. J. Pathol.* **1995**, *147*, 1016–1028. [PubMed]

299. Dietz, A.B.; Bulur, P.A.; Knutson, G.J.; Matasic, R.; Vuk-Pavlovic, S. Maturation of human monocyte-derived dendritic cells studied by microarray hybridization. *Biochem. Biophys. Res. Commun.* **2000**, *275*, 731–738. [CrossRef] [PubMed]

300. Helming, L.; Gordon, S. Macrophage fusion induced by IL-4 alternative activation is a multistage process involving multiple target molecules. *Eur. J. Immunol.* **2007**, *37*, 33–42. [CrossRef] [PubMed]

301. Sato, S.; Ouellet, N.; Pelletier, I.; Simard, M.; Rancourt, A.; Bergeron, M.G. Role of Gal-3 as an adhesion molecule for neutrophil extravasation during streptococcal pneumonia. *J. Immunol.* **2002**, *168*, 1813–1822. [CrossRef] [PubMed]

302. Kohatsu, L.; Hsu, D.K.; Jegalian, A.G.; Liu, F.T.; Baum, L.G. Gal-3 induces death of Candida species expressing specific beta-1,2-linked mannans. *J. Immunol.* **2006**, *177*, 4718–4726. [CrossRef] [PubMed]

303. Okamoto, H.; Mizuno, K.; Horio, T. Monocyte derived multinucleated giant cells and sarcoidosis. *J. Dermatol. Sci.* **2003**, *31*, 119–128. [CrossRef]

304. Friedman, S.L. Molecular regulation of hepatic fibrosis, an integrated cellular response to tissue injury. *J. Biol. Chem.* **2000**, *275*, 2247–2250. [CrossRef] [PubMed]

305. Duffield, J.S.; Forbes, S.J.; Constandinou, C.M.; Clay, S.; Partolina, M.; Vuthoori, S.; Wu, S.; Lang, R.; Iredale, J.P. Selective depletion of macrophages reveals distinct, opposing roles during liver injury and repair. *J. Clin. Investig.* **2005**, *115*, 56–65. [CrossRef] [PubMed]

306. Inohara, H.; Akahani, S.; Raz, A. Gal-3 stimulates cell proliferation. *Exp. Cell Res.* **1998**, *245*, 294–302. [CrossRef] [PubMed]

307. Kristensen, D.B.; Kawada, N.; Imamura, K.; Miyamoto, Y.; Tateno, C.; Seki, S.; Kuroki, T.; Yoshizato, K. Proteome analysis of rat hepatic stellate cells. *Hepatology* **2000**, *32*, 268–277. [CrossRef] [PubMed]

308. Kasper, M.; Hughes, R.C. Immunocytochemical evidence for a modulation of galectin 3 (Mac-2), a carbohydrate binding protein, in pulmonary fibrosis. *J. Pathol.* **1996**, *179*, 309–316. [CrossRef]

309. Ho, J.E.; Gao, W.; Levy, D.; Santhanakrishnan, R.; Araki, T.; Rosas, I.O.; Hatabu, H.; Latourelle, J.C.; Nishino, M.; Dupuis, J.; et al. Galectin-3 Is Associated with Restrictive Lung Disease and Interstitial Lung Abnormalities. *Am. J. Respir. Crit. Care Med.* **2016**, *194*, 77–83. [CrossRef] [PubMed]

310. Piguet, P.F.; Collart, M.A.; Grau, G.E.; Kapanci, Y.; Vassalli, P. Tumor necrosis factor/cachectin plays a key role in bleomycin-induced pneumopathy and fibrosis. *J. Exp. Med.* **1989**, *170*, 655–663. [CrossRef] [PubMed]

311. Piguet, P.F.; Ribaux, C.; Karpuz, V.; Grau, G.E.; Kapanci, Y. Expression and localization of tumor necrosis factor-alpha and its mRNA in idiopathic pulmonary fibrosis. *Am. J. Pathol.* **1993**, *143*, 651–655. [PubMed]

312. Turner-warwick, M. Precapillary systemic-pulmonary anastomoses. *Thorax* **1963**, *18*, 225–237. [CrossRef] [PubMed]

313. Nangia-Makker, P.; Honjo, Y.; Sarvis, R.; Akahani, S.; Hogan, V.; Pienta, K.J.; Raz, A. Galectin-3 induces endothelial cell morphogenesis and angiogenesis. *Am. J. Pathol.* **2000**, *156*, 899–909. [CrossRef]

314. Harjacek, M.; Diaz-Cano, S.; De Miguel, M.; Wolfe, H.; Maldonado, C.A.; Rabinovich, G.A. Expression of galectins-1 and -3 correlates with defective mononuclear cell apoptosis in patients with juvenile idiopathic arthritis. *J. Rheumatol.* **2001**, *28*, 1914–1922. [PubMed]

315. Ezzat, M.H.; El-Gammasy, T.M.; Shaheen, K.Y.; Osman, A.O. Elevated production of Gal-3 is correlated with juvenile idiopathic arthritis disease activity, severity, and progression. *Int. J. Rheum. Dis.* **2011**, *14*, 345–352. [CrossRef] [PubMed]

316. Bullock, S.L.; Johnson, T.M.; Bao, Q.; Hughes, R.C.; Winyard, P.J.; Woolf, A.S. Gal-3 modulates ureteric bud branching in organ culture of the developing mouse kidney. *J. Am. Soc. Nephrol.* **2001**, *12*, 515–523. [PubMed]

317. Bichara, M.; Attmane-Elakeb, A.; Brown, D.; Essig, M.; Karim, Z.; Muffat-Joly, M.; Micheli, L.; Eude-Le Parco, I.; Cluzeaud, F.; Peuchmaur, M.; et al. Exploring the role of galectin 3 in kidney function: A genetic approach. *Glycobiology* **2006**, *16*, 36–45. [CrossRef] [PubMed]

318. Chen, S.Z.; Kuo, P.L. The Role of Gal-3 in the Kidneys. *Int. J. Mol. Sci.* **2016**, *17*, 565. [CrossRef] [PubMed]

319. Nishiyama, J.; Kobayashi, S.; Ishida, A.; Nakabayashi, I.; Tajima, O.; Miura, S.; Katayama, M.; Nogami, H. Up-regulation of Gal-3 in acute renal failure of the rat. *Am. J. Pathol.* **2000**, *157*, 815–823. [CrossRef]

320. Sasaki, S.; Bao, Q.; Hughes, R.C. Gal-3 modulates rat mesangial cell proliferation and matrix synthesis during experimental glomerulonephritis induced by anti-Thy1.1 antibodies. *J. Pathol.* **1999**, *187*, 481–489. [CrossRef]

321. Kikuchi, Y.; Kobayashi, S.; Hemmi, N.; Ikee, R.; Hyodo, N.; Saigusa, T.; Namikoshi, T.; Yamada, M.; Suzuki, S.; Miura, S. Gal-3-positive cell infiltration in human diabetic nephropathy. *Nephrol. Dial. Transplant.* **2004**, *19*, 602–607. [CrossRef] [PubMed]

322. Winkelmann, B.R.; März, W.; Boehm, B.O.; Zotz, R.; Hager, J.; Hellstern, P.; Senges, J.; LURIC Study Group (LUdwigshafen RIsk and Cardiovascular Health). Rationale and design of the LURIC study-a resource for functional genomics, pharmacogenomics and long-term prognosis of cardiovascular disease. *Pharmacogenomics* **2001**, *2* (Suppl. 1), S1–S73. [CrossRef] [PubMed]

323. Wanner, C.; Krane, V.; März, W.; Olschewski, M.; Asmus, H.G.; Krämer, W.; Kühn, K.W.; Kütemeyer, H.; Mann, J.F.; Ruf, G.; et al. Randomized controlled trial on the efficacy and safety of atorvastatin in patients with type 2 diabetes on hemodialysis (4D study): Demographic and baseline characteristics. *Kidney Blood Press. Res.* **2004**, *27*, 259–266. [CrossRef] [PubMed]

324. Drechsler, C.; Delgado, G.; Wanner, C.; Blouin, K.; Pilz, S.; Tomaschitz, A.; Kleber, M.E.; Dressel, A.; Willmes, C.; Krane, V.; et al. Gal-3, Renal Function, and Clinical Outcomes: Results from the LURIC and 4D Studies. *J. Am. Soc. Nephrol.* **2015**, *26*, 2213–2221. [CrossRef] [PubMed]

325. Kolatsi-Joannou, M.; Price, K.L.; Winyard, P.J.; Long, D.A. Modified citrus pectin reduces Gal-3 expression and disease severity in experimental acute kidney injury. *PLoS ONE* **2011**, *6*, e18683. [CrossRef] [PubMed]

326. La Jolla Begins Phase IIb Trial of GCS-100 to Treat Advanced CKD. Available online: http://www.drugdevelopment-technology.com/news/newsla-jolla-begins-phase-iib-trial-of-gcs-100-totreat-advanced-ckd-4535972 (accessed on 22 December 2017).

327. Bataller, R.; Brenner, D.A. Liver fibrosis. *J. Clin. Investig.* **2005**, *115*, 209–218. [CrossRef] [PubMed]

328. Wang, L.; Friess, H.; Zhu, Z.; Frigeri, L.; Zimmermann, A.; Korc, M.; Berberat, P.O.; Buchler, M.W. Galectin-1 and galectin-3 in chronic pancreatitis. *Lab. Investig.* **2000**, *80*, 1233–1241. [CrossRef] [PubMed]

329. Friedman, S.L.; Roll, F.J.; Boyles, J.; Bissell, D.M. Hepatic lipocytes: The principal collagen-producing cells of normal rat liver. *Proc. Natl. Acad. Sci. USA* **1985**, *82*, 8681–8685. [CrossRef] [PubMed]

330. Galectin Therapeutics Announces Results from Phase 2b NASH-CX Trial. Available online: http://investor.galectintherapeutics.com/news-releases/news-release-details/galectin-therapeutics-announces-results-phase-2b-nash-cx-trial (accessed on 22 December 2017).

331. Djoussé, L.; Matsumoto, C.; Petrone, A.; Weir, N.L.; Tsai, M.Y.; Gaziano, J.M. Plasma galectin 3 and heart failure risk in the Physicians' Health Study. *Eur. J. Heart Fail.* **2014**, *16*, 350–354. [CrossRef] [PubMed]

332. Takahashi, Y.; Fukusato, T. Histopathology of nonalcoholic fatty liver disease/nonalcoholic steatohepatitis. *World J. Gastroenterol.* **2014**, *20*, 15539–15548. [CrossRef] [PubMed]

333. Li, L.C.; Li, J.; Gao, J. Functions of Galectin-3 and Its Role in Fibrotic Diseases. *J. Pharmacol. Exp. Ther.* **2014**, *351*, 336–343. [CrossRef] [PubMed]

334. Iacobini, C.; Menini, S.; Ricci, C.; Blasetti Fantauzzi, C.; Scipioni, A.; Salvi, L.; Cordone, S.; Delucchi, F.; Serino, M.; Federici, M.; et al. Gal-3 ablation protects mice from diet-induced NASH: A major scavenging role for Gal-3 in liver. *J. Hepatol.* **2011**, *54*, 975–983. [CrossRef] [PubMed]

335. Nomoto, K.; Tsuneyama, K.; Abdel Aziz, H.O.; Takahashi, H.; Murai, Y.; Cui, Z.G.; Fujimoto, M.; Kato, I.; Hiraga, K.; Hsu, D.K.; et al. Disrupted Gal-3 causes non-alcoholic fatty liver disease in male mice. *J. Pathol.* **2006**, *210*, 469–477. [CrossRef] [PubMed]

336. Jeftic, I.; Jovicic, N.; Pantic, J.; Arsenijevic, N.; Lukic, M.L.; Pejnovic, N. Gal-3 ablation enhances liver steatosis, but attenuates Inflammation and IL-33-dependent fibrosis in obesogenic mouse model of nonalcoholic steatohepatitis. *Mol. Med.* **2015**, *21*, 453–465. [CrossRef] [PubMed]

337. Traber, P.G.; Zomer, E. Therapy of experimental NASH and fibrosis with galectin inhibitors. *PLoS ONE* **2013**, *8*, e83481. [CrossRef] [PubMed]

338. Traber, P.G.; Chou, H.; Zomer, E.; Hong, F.; Klyosov, A.; Fiel, M.I.; Friedman, S.L. Regression of fibrosis and reversal of cirrhosis in rats by galectin inhibitors in thioacetamide-induced liver disease. *PLoS ONE* **2013**, *8*, e75361. [CrossRef] [PubMed]

339. Harrison, S.A.; Marri, S.R.; Chalasani, N.; Kohli, R.; Aronstein, W.; Thompson, G.A.; Irish, W.; Miles, M.V.; Xanthakos, S.A.; Lawitz, E.; et al. Randomised clinical study: GR-MD-02, a Gal-3 inhibitor, vs. placebo in patients having non-alcoholic steatohepatitis with advanced fibrosis. *Aliment. Pharmacol. Ther.* **2016**, *44*, 1183–1198. [CrossRef] [PubMed]

340. Galectin Tanks on Failure of NASH Treatment (GALT). Available online: https://www.investopedia.com/news/galectin-tanks-failure-nash-treatment-galt/ (accessed on 22 December 2017).

341. Results from the NASH-FX Study Underscore the Importance of Completing the NASH-CX Clinical Trial for Patients with NASH Cirrhosis. Available online: http://perspectives.galectintherapeutics.com/importance-completing-nash-cx-trial-for-patients-with-nash-cirrhosis/ (accessed on 22 December 2017).

342. Shoelson, S.E.; Lee, J.; Goldfine, A.B. Inflammation and insulin resistance. *J. Clin. Investig.* **2006**, *116*, 1793–1801. [CrossRef] [PubMed]

343. Van Greevenbroek, M.M.; Schalkwijk, C.G.; Stehouwer, C.D. Obesity-associated low-grade inflammation in type 2 diabetes mellitus: Causes and consequences. *Neth. J. Med.* **2013**, *71*, 174–187. [PubMed]

344. Xu, L.; Kitade, H.; Ni, Y.; Ota, T. Roles of chemokines and chemokine receptors in obesity-associated insulin resistance and nonalcoholic fatty liver disease. *Biomolecules* **2015**, *5*, 1563–1579. [CrossRef] [PubMed]

345. Herrero, L.; Shapiro, H.; Nayer, A.; Lee, J.; Shoelson, S.E. Inflammation and adipose tissue macrophages in lipodystrophic mice. *Proc. Natl. Acad. Sci. USA* **2010**, *107*, 240–245. [CrossRef] [PubMed]

346. Xu, H.; Barnes, G.T.; Yang, Q.; Tan, G.; Yang, D.; Chou, C.J.; Sole, J.; Nichols, A.; Ross, J.S.; Tartaglia, L.A.; et al. Chronic inflammation in fat plays a crucial role in the development of obesity-related insulin resistance. *J. Clin. Investig.* **2003**, *112*, 1821–1830. [CrossRef] [PubMed]

347. Huang, J.Y.; Chiang, M.T.; Yet, S.F.; Chau, L.Y. Myeloid heme oxygenase-1 haploin- sufficiency reduces high fat diet-induced insulin resistance by affecting adipose macrophage infiltration in mice. *PLoS ONE* **2012**, *7*, e38626. [CrossRef]

348. Lumeng, C.N.; Del Proposto, J.B.; Westcott, D.J.; Saltiel, A.R. Phenotypic switching of adipose tissue macrophages with obesity is generated by spatiotemporal differences in macrophage subtypes. *Diabetes* **2008**, *57*, 3239–3246. [CrossRef] [PubMed]

349. Li, P.; Liu, S.; Lu, M.; Bandyopadhyay, G.; Oh, D.; Imamura, T.; Johnson, A.M.F.; Sears, D.; Shen, Z.; Cui, B.; et al. Hematopoietic-derived Galectin-3 Causes Cellular and Systemic Insulin Resistance. *Cell* **2016**, *167*, 973–984. [CrossRef] [PubMed]

350. Rudan, I.; Tomaskovic, L.; Boschi-Pinto, C.H.C. Global estimate of the incidence of clinical pneumonia among children under five years of age. *Bull. World Health Org.* **2004**, *82*, 895–903. [PubMed]

351. Farnworth, S.L.; Henderson, N.C.; Mackinnon, A.C.; Atkinson, K.M.; Wilkinson, T.; Dhaliwal, K.; Hayashi, K.; Simpson, A.J.; Rossi, A.G.; Haslett, C.; et al. Galectin-3 Reduces the Severity of Pneumococcal Pneumonia by Augmenting Neutrophil Function. *Am. J. Pathol.* **2008**, *172*, 395–405. [CrossRef] [PubMed]

352. Wright, R.D.; Souza, P.R.; Flak, M.B.; Thedchanamoorthy, P.; Norling, L.V.; Cooper, D. Galectin-3–null mice display defective neutrophil clearance during acute inflammation. *J. Leukoc. Biol.* **2017**, *101*, 717–726. [CrossRef] [PubMed]

353. Morrell, N.W.; Adnot, S.; Archer, S.L.; Dupuis, J.; Jones, P.L.; MacLean, M.R.; McMurtry, I.F.; Stenmark, K.R.; Thistlethwaite, P.A.; Weissmann, N.; et al. Cellular and Molecular Basis of Pulmonary Arterial Hypertension. *J. Am. Coll. Cardiol.* **2009**, *54* (Suppl. 1), S20–S31. [CrossRef] [PubMed]

354. Markowska, A.I.; Liu, F.T.; Panjwani, N. Galectin-3 is an important mediator of VEGF- and bFGF-mediated angiogenic response. *J. Exp. Med.* **2010**, *207*, 1981–1993. [CrossRef] [PubMed]

355. Song, J.; Li, X.; Liu, B.; Li, T.; Yu, Z. Galectin-3: A potential biomarker in pulmonary arterial hypertension. *Cardiol. Plus* **2016**, *1*, 14–20.

356. Calvier, L.; Legchenko, E.; Grimm, L.; Sallmon, H.; Hatch, A.; Plouffe, B.D.; Schroeder, C.; Bauersachs, J.; Murthy, S.K.; Hansmann, G. Galectin-3 and aldosterone as potential tandem biomarkers in pulmonary arterial hypertension. *Heart* **2016**, *102*, 390–396. [CrossRef] [PubMed]

357. He, J.; Li, X.; Luo, H.; Li, T.; Zhao, L.; Qi, Q.; Liu, Y.; Yu, Z. Galectin-3 mediates the pulmonary arterial hypertension-induced right ventricular remodeling through interacting with NADPH oxidase 4. *J. Am. Soc. Hypertens.* **2017**, *11*, 275–289. [CrossRef] [PubMed]

358. Fenster, B.E.; Lasalvia, L.; Schroeder, J.D.; Smyser, J.; Silveira, L.J.; Buckner, J.K.; Brown, K.K. Galectin-3 levels are associated with right ventricular functional and morphologic changes in pulmonary arterial hypertension. *Heart Vessels* **2016**, *31*, 939–946. [CrossRef] [PubMed]

359. Berezin, A.E. Elevated Galectin-3 Level Predicts Pulmonary Artery Hypertension. *Biol. Mark. Guid. Ther.* **2016**, *3*, 89–97. [CrossRef]

360. Berezin, A.E. Is elevated circulating galectin-3 level a predictor of pulmonary artery hypertension development and progression? *Clin. Med. Biochem.* **2016**, *2*, 114. [CrossRef]

361. Barman, S.A.; Chen, F.; Su, Y.; Stepp, D.; Zhou, J.; Meadows, L.; Wang, Y.; Li, X.; Haigh, S.; Bordan, Z.; et al. Gal-3 mediates vascular remodeling in pulmonary arterial hypertension. *Am. J. Respir. Crit. Care Med.* **2016**, *193*, A7918.

362. Larsen, L.; Chen, H.-Y.; Saegusa, J.; Liu, F.T. Gal-3 and the skin. *J. Dermatolol. Sci.* **2011**, *64*, 85–91. [CrossRef] [PubMed]

363. De la Fuente, H.; Perez-Gala, S.; Bonay, P.; Cruz-Adalia, A.; Cibrian, D.; Sanchez-Cuellar, S.; Dauden, E.; Fresno, M.; García-Diez, A.; Sanchez-Madrid, F. Psoriasis in humans is associated with down-regulation of galectins in dendritic cells. *J. Pathol.* **2012**, *228*, 193–203. [CrossRef] [PubMed]

364. Lacina, L.; Plzáková, Z.; Smetana, K., Jr.; Štork, J.; Kaltner, H.; André, S. Glycophenotype of Psoriatic Skin. *Folia Biol.* **2006**, *52*, 10–15.

365. Voth, D.E.; Heinzen, R.A. Lounging in a lysosome: The intracellular lifestyle of Coxiella burnetii. *Cell. Microbiol.* **2007**, *9*, 829–840. [CrossRef] [PubMed]

366. Dupont, N.; Lafont, F. How autophagy regulates the host cell signaling associated with the postpartum bacteria cocoon experienced as a danger signal. *Autophagy* **2009**, *5*, 1222–1223. [CrossRef] [PubMed]

367. Paz, I.; Sachse, M.; Dupont, N.; Mounier, J.; Cederfur, C.; Enninga, J.; Leffler, H.; Poirier, F.; Prevost, M.C.; Lafont, F.; et al. Gal-3, a marker for vacuole lysis by invasive pathogens. *Cell. Microbiol.* **2010**, *12*, 530–544. [CrossRef] [PubMed]

368. Thurston, T.L.M.; Wandel, M.P.; vonMuhlinen, N.; Foeglein, A.; Randow, F. Galectin 8 targets damaged vesicles for autophagy to defend cells against bacterial invasion. *Nature* **2012**, *482*, 414–418. [CrossRef] [PubMed]

369. Southwick, F.S.; Purich, D.L. Listeria and Shigella actin-based motility in host cells. *Trans. Am. Clin. Climatol. Assoc.* **1998**, *109*, 160–173. [PubMed]

370. Maejima, I.; Takahashi, A.; Omori, H.; Kimura, T.; Takabatake, Y.; Saitoh, T.; Yamamoto, A.; Hamasaki, M.; Noda, T.; Isaka, Y.; et al. Autophagy sequesters damaged lysosomes to control lysosomal biogenesis and kidney injury. *EMBO J.* **2013**, *32*, 2336–2347. [CrossRef] [PubMed]

371. Li, S.; Yu, Y.; Koehn, C.D.; Zhang, Z.; Su, K. Galectins in the pathogenesis of rheumatoid arthritis. *J. Clin. Cell. Immunol.* **2013**, *4*, 1000164. [CrossRef] [PubMed]

372. Shou, J.; Bull, C.M.; Li, L.; Qian, H.R.; Wei, T.; Luo, S.; Perkins, D.; Solenberg, P.J.; Tan, S.L.; Chen, X.Y.; et al. Identification of blood biomarkers of rheumatoid arthritis by transcript profiling of peripheral blood mononuclear cells from the rat collagen induced arthritis model. *Arthritis Res. Ther.* **2006**, *8*, R28. [CrossRef] [PubMed]

373. Forsman, H.; Islander, U.; Andreasson, E.; Andersson, A.; Onnheim, K.; Karlstrom, A.; Savman, K.; Magnusson, M.; Brown, K.L.; Karlsson, A. Galectin 3 aggravates joint inflammation and destruction in antigen-induced arthritis. *Arthritis Rheum.* **2011**, *63*, 445–454. [CrossRef] [PubMed]

374. Ohshima, S.; Kuchen, S.; Seemayer, C.A.; Kyburz, D.; Hirt, A.; Klinzing, S.; Michel, B.A.; Gay, R.E.; Liu, F.T.; Gay, S.; et al. Galectin 3 and its binding protein in rheumatoid arthritis. *Arthritis Rheum.* **2003**, *48*, 2788–2795. [CrossRef] [PubMed]

375. Neidhart, M.; Zaucke, F.; von Knoch, R.; Jungel, A.; Michel, B.A.; Gay, R.E.; Gay, S. Gal-3 is induced in rheumatoid arthritis synovial fibroblasts after adhesion to cartilage oligomeric. *Ann. Rheum. Dis.* **2005**, *64*, 419–424. [CrossRef] [PubMed]

376. Neidhart, M.; Rethage, J.; Kuchen, S.; Künzler, P.; Crowl, R.M.; Billingham, M.E.; Gay, R.E.; Gay, S. Retrotransposable L1 elements expressed in rheumatoid arthritis synovial tissue: Association with genomic DNA hypomethylation and influence on gene expression. *Arthritis Rheum.* **2000**, *43*, 2634–2647. [CrossRef]

377. Lee, Y.J.; Kang, S.W.; Song, J.K.; Park, J.J.; Bae, Y.D.; Lee, E.Y.; Lee, E.B.; Song, Y.W. Serum Gal-3 and Gal-3 binding protein levels in Behçet's disease and their association with disease activity. *Clin. Exp. Rheumatol.* **2007**, *25*, S41–S45. [PubMed]

378. Hu, C.Y.; Chang, S.K.; Wu, C.S.; Tsai, W.I.; Hsu, P.N. Gal-3 gene (LGALS3) +292C allele is a genetic predisposition factor for rheumatoid arthritis in Taiwan. *Clin. Rheumatol.* **2011**, *30*, 1227–1233. [CrossRef] [PubMed]

379. Neidhart, M.; Seemayer, C.A.; Hummel, K.M.; Michel, B.A.; Gay, R.E.; Gay, S. Functional characterization of adherent synovial fluid cells in rheumatoid arthritis: Destructive potential in vitro and in vivo. *Arthritis Rheum.* **2003**, *4*, 1873–1880. [CrossRef] [PubMed]

380. Dasuri, K.; Antonovici, M.; Chen, K.; Wong, K.; Standing, K.; Ens, W.; El-Gabalawy, H.; Wilkins, J.A. The synovial proteome: Analysis of fibroblast-like synoviocytes. *Arthritis Res. Ther.* **2004**, *6*, R161–R168. [CrossRef] [PubMed]

381. Filer, A.; Bik, M.; Parsonage, G.N.; Fitton, J.; Trebilcock, E.; Howlett, K.; Cook, M.; Raza, K.; Simmons, D.L.; Thomas, A.M.; et al. Galectin 3 induces a distinctive pattern of cytokine and chemokine production in rheumatoid synovial fibroblasts via selective signaling pathways. *Arthritis Rheum.* **2009**, *60*, 1604–1614. [CrossRef] [PubMed]

382. Stowell, S.R.; Qian, Y.; Karmakar, S.; Koyama, N.S.; Dias-Baruffi, M.; Leffler, H.; McEver, R.P.; Cummings, R.D. Differential roles of galectin-1 and Gal-3 in regulating leukocyte viability and cytokine secretion. *J. Immunol.* **2008**, *180*, 3091–4102. [CrossRef] [PubMed]

383. Issa, S.F.; Duer, A.; Østergaard, M.; Hørslev-Petersen, K.; Hetland, M.L.; Hansen, M.S.; Junker, K.; Lindegaard, H.M.; Møller, J.M.; Junker, P. Increased galectin-3 may serve as a serologic signature of pre-rheumatoid arthritis while markers of synovitis and cartilage do not differ between early undifferentiated arthritis subsets. *Arthritis Res. Ther.* **2017**, *19*, 80. [CrossRef] [PubMed]

384. Wang, C.R.; Shiau, A.L.; Chen, S.Y.; Cheng, Z.S.; Li, Y.T.; Lee, C.H.; Yo, Y.T.; Lo, C.W.; Lin, Y.S.; Juan, H.Y.; et al. Intra-articular lentivirus-mediated delivery of Gal-3 shRNA and galectin-1 gene ameliorates collagen-induced arthritis. *Gene Ther.* **2010**, *17*, 1225–1233. [CrossRef] [PubMed]

385. Hu, Y.; Yéléhé-Okouma, M.; Ea, H.K.; Jouzeau, J.Y.; Reboul, P. Gal-3: A key player in arthritis. *Joint Bone Spine* **2017**, *84*, 15–20. [CrossRef] [PubMed]

386. Mishra, B.B.; Li, Q.; Steichen, A.L.; Binstock, B.J.; Metzger, D.W.; Teale, J.M.; Sharma, J. Galectin-3 Functions as an Alarmin: Pathogenic role for sepsis development in murine respiratory tularemia. *PLoS ONE* **2013**, *8*, e59616. [CrossRef] [PubMed]

387. Li, Y.; Komai-Koma, M.; Gilchrist, D.S.; Hsu, D.K.; Liu, F.T.; Springall, T.; Xu, D. Gal-3 is a negative regulator of lipopolysaccharide-mediated inflammation. *J. Immunol.* **2008**, *181*, 2781–2789. [CrossRef] [PubMed]

388. Rabinovich, G.A.; Baum, L.G.; Tinari, N.; Paganelli, R.; Natoli, C.; Liu, F.T.; Iacobelli, S. Galectins and their ligands: Amplifiers, silencers or tuners of the inflammatory response? *Trends Immunol.* **2002**, *23*, 313–320. [CrossRef]

389. Sato, S.; Hughes, R.C. Regulation of secretion and surface expression of Mac-2, a galactoside-binding protein of macrophages. *J. Biol. Chem.* **1994**, *269*, 4424–4430. [PubMed]

390. Kim, H.; Hur, M.; Moon, H.W.; Yun, Y.M.; Di Somma, S.; On behalf of GREAT Network. Multi-marker approach using procalcitonin, presepsin, Gal-3, and soluble suppression of tumorigenicity 2 for the prediction of mortality in sepsis. *Ann. Intensive Care* **2017**, *7*, 27. [CrossRef] [PubMed]

391. Gabrielli, A.; Avvedimento, E.V.; Krieg, T. Scleroderma. *N. Engl. J. Med.* **2009**, *360*, 1989–2003. [CrossRef] [PubMed]

392. Van den Hoogen, F.; Khanna, D.; Fransen, J.; Johnson, S.R.; Baron, M.; Tyndall, A.; Matucci-Cerinic, M.; Naden, R.P.; Medsger, T.A., Jr.; Carreira, P.E.; et al. 2013 classification criteria for systemic sclerosis: An American college of rheumatology/European league against rheumatism collaborative initiative. *Ann. Rheum. Dis.* **2013**, *72*, 1747–1755. [CrossRef] [PubMed]

393. Glinsky, V.V.; Raz, A. Modified citrus pectin anti-metastatic properties: One bullet, multiple targets. *Carbohydr. Res.* **2009**, *344*, 1788–1791. [CrossRef] [PubMed]

394. Abu-Elsaad, N.M.; Elkashef, W.F. Modified citrus pectin stops progression of liver fibrosis by inhibiting Gal-3 and inducing apoptosis of stellate cells. *Can. J. Physiol. Pharmacol.* **2016**, *94*, 554–562. [CrossRef] [PubMed]

395. Delphi, L.; Sepehri, H.; Khorramizadeh, M.R.; Mansoori, F. Pectic-oligoshaccharides from apples induce apoptosis and cell cycle arrest in MDA-MB-231 cells, a model of human breast cancer. *Asian Pac. J. Cancer Prev.* **2015**, *16*, 5265–5271. [CrossRef] [PubMed]

396. Grous, J.J.; Redfern, C.H.; Mahadevan, D.; Schindler, J. GCS-100, a galectin-3 antagonist, in refractory solid tumors: A phase I study. *J. Clin. Oncol.* **2006**, *24* (Suppl. 18), 13023. [CrossRef]

397. Ahmad, N.; Gabius, H.J.; Andre, S.; Kaltner, H.; Sabesan, S.; Roy, R.; Liu, B.; Macaluso, F.; Brewer, C.F. Gal-3 precipitates as a pentamer with synthetic multivalent carbohydrates and forms heterogeneous cross-linked complexes. *J. Biol. Chem.* **2004**, *279*, 10841–10847. [CrossRef] [PubMed]

398. Streetly, M.J.; Maharaj, L.; Joel, S.; Schey, S.A.; Gribben, J.G.; Cotter, F.E. GCS-100, a novel galectin-3 antagonist, modulates MCL-1, NOXA, and cell cycle to induce myeloma cell death. *Blood* **2010**, *115*, 3939–3948. [CrossRef] [PubMed]

399. Chauhan, D.; Li, G.; Podar, K.; Hideshima, T.; Neri, P.; He, D.; Mitsiades, N.; Richardson, P.; Chang, Y.; Schindler, J.; et al. A novel carbohydrate-based therapeutic GCS-100 overcomes bortezomib resistance and enhances dexamethasone-induced apoptosis in multiple myeloma cells. *Cancer Res.* **2005**, *65*, 8350–8358. [CrossRef] [PubMed]

400. De Boer, R.A.; van der Velde, A.R.; Mueller, C.; van Veldhuisen, D.J.; Anker, S.D.; Peacock, W.F.; Adams, K.F.; Maisel, A.S. Gal-3: A Modifiable Risk Factor in Heart Failure. *Cardiovasc. Drugs Ther.* **2014**, *28*, 237–246. [CrossRef] [PubMed]

401. National Institutes of Health (NIH). Available online: https://www.clinicaltrials.gov/ct2/search (accessed on 22 December 2017).

402. Sato, S.; St-Pierre, C.; Bhaumik, P.; Nieminen, J. Galectins in innate immunity: Dual functions of host soluble beta-galactoside-binding lectins as damage-associated molecular patterns (DAMPs) and as receptors for pathogen-associated molecular patterns (PAMPs). *Immunol. Rev.* **2009**, *230*, 172–187. [CrossRef] [PubMed]

403. John, C.M.; Jarvis, G.A.; Swanson, K.V.; Leffler, H.; Cooper, M.D.; Huflejt, M.E.; Griffiss, J.M. Gal-3 binds lactosaminylated lipooligosaccharides from Neisseria gonorrhoeae and is selectively expressed by mucosal epithelial cells that are infected. *Cell. Microbiol.* **2002**, *4*, 649–662. [CrossRef] [PubMed]

404. Rabinovich, G.A.; Gruppi, A. Galectins as immunoregulators during infectious processes: From microbial invasion to the resolution of the disease. *Parasite Immunol.* **2005**, *27*, 103–114. [CrossRef] [PubMed]

405. Kasamatsu, A.; Uzawa, K.; Shimada, K.; Shiiba, M.; Otsuka, Y.; Seki, N.; Abiko, Y.; Tanzawa, H. Elevation of galectin-9 as an ammatory response in the periodontal ligament cells exposed to Porphylomonas gingivalis lipopolysaccharide in vitro and in vivo. *Int. J. Biochem. Cell Biol.* **2005**, *37*, 397–408. [CrossRef] [PubMed]

406. Sato, S.; Ouellet, M.; St-Pierre, C.; Tremblay, M.J. Glycans, galectins, and HIV-1 infection. *Ann. N.Y. Acad. Sci.* **2012**, *1253*, 133–148. [CrossRef] [PubMed]

407. Gauthier, S.; Pelletier, I.; Ouellet, M.; Vargas, A.; Tremblay, M.J.; Sato, S.; Barbeau, B. Induction of galectin-1 expression by HTLV-I Tax and its impact on HTLV-I infectivity. *Retrovirology* **2008**, *5*, 105. [CrossRef] [PubMed]

408. Okumura, C.Y.; Baum, L.G.; Johnson, P.J. Galectin-1 on cervical epithelial cells is a receptor for the sexually transmitted human parasite Trichomonas vaginalis. *Cell. Microbiol.* **2008**, *10*, 2078–2090. [CrossRef] [PubMed]

409. Fichorova, R.N. Impact of *T. vaginalis* infection on innate immune responses and reproductive outcome. *J. Reprod. Immunol.* **2009**, *83*, 185–189. [CrossRef] [PubMed]

410. Hepojoki, J.; Strandin, T.; Hetzel, U.; Sironen, T.; Klingström, J.; Sane, J.; Mäkelä, S.; Mustonen, J.; Meri, S.; Lundkvist, A.; et al. Acute hantavirus infection induces Gal-3-binding protein. *J. Gen. Virol.* **2014**, *95*, 2356–2364. [CrossRef] [PubMed]

411. Artini, M.; Natoli, C.; Tinari, N.; Costanzo, A.; Marinelli, R.; Balsano, C.; Porcari, P.; Angelucci, D.; D'Egidio, M.; Levrero, M.; et al. Elevated serum levels of 90K/MAC-2 BP predict unresponsiveness to alpha-interferon therapy in chronic HCV hepatitis patients. *J. Hepatol.* **1996**, *25*, 212–217. [CrossRef]

412. Longo, G.; Natoli, C.; Rafanelli, D.; Tinari, N.; Morfini, M.; Rossi-Ferrini, P.; D'Ostilio, N.; Iacobelli, S. Prognostic value of a novel circulating serum 90K antigen in HIV-infected haemophilia patients. *Br. J. Haematol.* **1993**, *85*, 207–209. [CrossRef] [PubMed]

413. Altman, E.; Harrison, B.A.; Latta, R.K.; Lee, K.K.; Kelly, J.F.; Thibault, P. Gal-3-mediated adherence of Proteus mirabilis to Madin-Darby canine kidney cells. *Biochem. Cell Biol.* **2001**, *79*, 783–788. [CrossRef] [PubMed]

414. Desmedt, V.; Desmedt, S.; Delanghe, J.R.; Speeckaert, R.; Speeckaert, M.M. Gal-3 in renal pathology: More than just an innocent bystander. *Am. J. Nephrol.* **2016**, *43*, 305–317. [CrossRef] [PubMed]

415. DeRoo, E.P.; Wrobleski, S.K.; Shea, E.M.; Al-Khalil, R.K.; Hawley, A.E.; Henke, P.K.; Myers, D.D., Jr.; Wakefield, T.W.; Diaz, J.A. The role of Gal-3 and Gal-3–binding protein in venous thrombosis. *Blood* **2015**, *125*, 1813–1821. [CrossRef] [PubMed]

416. Cao, Z.; Said, N.; Amin, S.; Wu, H.K.; Bruce, A.; Garate, M.; Hsu, D.K.; Kuwabara, I.; Liu, F.T.; Panjwani, N. Galectins-3 and -7, but not galectin-1, play a role in re- epithelialization of wounds. *J. Biol. Chem.* **2002**, *277*, 42299–42305. [CrossRef] [PubMed]

417. Ochieng, J.; Green, B.; Evans, S.; James, O.; Warfield, P. Modulation of the biological functions of Gal-3 by matrix metalloproteinases. *Biochim. Biophys. Acta* **1998**, *1379*, 97–106. [CrossRef]

418. Saravanan, C.; Cao, Z.; Head, S.R.; Panjwani, N. Detection of differentially expressed wound-healing-related glycogenes in Gal-3-deficient mice. *Investig. Ophthalmol. Vis. Sci.* **2009**, *50*, 5690–5696. [CrossRef] [PubMed]

419. Paret, C.; Bourouba, M.; Beer, A.; Miyazaki, K.; Schnölzer, M.; Fiedler, S.; Zöller, M. Ly6 family member C4.4A binds laminins 1 and 5, associates with Gal-3 and supports cell migration. *Int. J. Cancer* **2005**, *115*, 724–733.

420. Singer, A.J.; Clark, R.A. Cutaneous wound healing. *N. Engl. J. Med.* **1999**, *341*, 738–746. [CrossRef] [PubMed]

421. Sivamani, R.K.; Garcia, M.S.; Isseroff, R.R. Wound re-epithelialization: Modulating keratinocyte migration in wound healing. *Front. Biosci.* **2007**, *12*, 2849–2868.

422. Fonder, M.A.; Lazarus, G.S.; Cowan, D.A.; Aronson-Cook, B.; Kohli, A.R.; Mamelak, A.J. Treating the chronic wound: A practical approach to the care of nonhealing wounds and wound care dressings. *J. Am. Acad. Dermatol.* **2008**, *58*, 185–206. [CrossRef] [PubMed]

423. Eaglstein, W.H.; Falanga, V. Chronic wounds. *Surg. Clin. N. Am.* **1997**, *77*, 689–700. [CrossRef]

424. Ljubimov, A.V.; Saghizadeh, M. Progress in corneal wound healing. *Prog. Retin. Eye Res.* **2015**, *49*, 17–45. [CrossRef] [PubMed]

425. Seiler, W.O.; Stahelin, H.B.; Zolliker, R.; Kallenberger, A.; Lüscher, N.J. Impaired migration of epidermal cells from decubitus ulcers in cell cultures. A cause of protracted wound healing? *Am. J. Clin. Pathol.* **1989**, *92*, 430–434. [CrossRef] [PubMed]

426. Panjwani, N. Role of galectins in re-epithelialization of wounds. *Ann. Transl. Med.* **2014**, *2*, 89. [CrossRef] [PubMed]

427. Yabuta, C.; Yano, F.; Fujii, A.; Shearer, T.R.; Azuma, M. Gal-3 enhances epithelial cell adhesion and wound healing in rat cornea. *Ophthalmic Res.* **2014**, *51*, 96–103. [CrossRef] [PubMed]

428. Fujii, A.; Shearer, T.R.; Azuma, M. Gal-3 enhances extracellular matrix associations and wound healing in monkey corneal epithelium. *Exp. Eye Res.* **2015**, *137*, 71–78. [CrossRef] [PubMed]

429. Santoro, M.M.; Gaudino, G. Cellular and molecular facets of keratinocyte reepithelization during wound healing. *Exp. Cell Res.* **2005**, *304*, 274–286. [CrossRef] [PubMed]

430. McCawley, L.J.; O'Brien, P.; Hudson, L.G. Epidermal growth factor (EGF)- and scatter factor/hepatocyte growth factor (SF/HGF)-mediated keratinocyte migration is coincident with induction of matrix metalloproteinase (MMP)-9. *J. Cell. Physiol.* **1998**, *176*, 255–265. [CrossRef]

431. Shirakata, Y.; Kimura, R.; Nanba, D.; Iwamoto, R.; Tokumaru, S.; Morimoto, C.; Yokota, K.; Nakamura, M.; Sayama, K.; Mekada, E.; et al. Heparin-binding EGF-like growth factor accelerates keratinocyte migration and skin wound healing. *J. Cell Sci.* **2005**, *118*, 2363–2370. [CrossRef] [PubMed]

432. Liu, W.; Hsu, D.K.; Chen, H.Y.; Yang, R.Y.; Carraway, K.L., 3rd.; Isseroff, R.R.; Liu, F.T. Galectin-3 regulates intracellular trafficking of egfr through alix and promotes keratinocyte migration. *J. Investig. Dermatol.* **2012**, *132*, 2828–2837. [CrossRef] [PubMed]

433. Puthenedam, M.; Wu, F.; Shetye, A.; Michaels, A.; Rhee, K.J.; Kwon, J.H. Matrilysin-1 (MMP7) cleaves Gal-3 and inhibits wound healing in intestinal epithelial cells. *Inflamm. Bowel Dis.* **2011**, *17*, 260–267. [CrossRef] [PubMed]

434. Kemp, H.G. Left ventricular function in patients with anginal syndrome and normal coronary arteriograms. *Am. J. Cardiol.* **1973**, *32*, 375–376. [CrossRef]

435. Vermeltfoort, I.A.; Raijmakers, P.G.; Riphagen, I.I.; Odekerken, D.A.M.; Kuijper, A.F.M.; Zwijnenburg, A.; Teule, G.J.J. Definitions and incidence of cardiac syndrome X: Review and analysis of clinical data. *Clin. Res. Cardiol.* **2010**, *99*, 475–481. [CrossRef] [PubMed]

436. Kaski, J.C.; Rosano, G.M.; Collins, P.; Nihoyannopoulos, P.; Maseri, A.; Poole-Wilson, P.A. Cardiac syndrome X: Clinical characteristics and left ventricular function. Long-term follow-up study. *J. Am. Coll. Cardiol.* **1995**, *25*, 807–814. [CrossRef]

437. Lamendola, P.; Lanza, G.A.; Spinelli, A.; Sgueglia, G.A.; Di Monaco, A.; Barone, L.; Sestito, A.; Crea, F. Long-term prognosis of patients with cardiac syndrome X. *Int. J. Cardiol.* **2010**, *140*, 197–199. [CrossRef] [PubMed]

438. Vermeltfoort, I.A.C.; Teule, G.J.J.; van Dijk, A.B.; Muntinga, H.J.; Raijmakers, P.G.H.M. Long-term prognosis of patients with cardiac syndrome X: A review. *Neth. Heart J.* **2012**, *20*, 365–371. [CrossRef] [PubMed]

439. Gulati, M.; Cooper-DeHoff, R.M.; McClure, C.; Johnson, B.D.; Shaw, L.J.; Handberg, E.M.; Zineh, I.; Kelsey, S.F.; Arnsdorf, M.F.; Black, H.R.; et al. Adverse cardiovascular outcomes in women with nonobstructive coronary artery disease: A report from the Women's Ischemia Syndrome Evaluation Study and the St James Women Take Heart Project. *Arch. Intern. Med.* **2009**, *169*, 843–850. [CrossRef] [PubMed]

440. Li, J.J.; Zhu, C.G.; Nan, J.L.; Li, J.; Li, Z.C.; Zeng, H.S.; Gao, Z.; Qin, X.W.; Zhang, C.Y. Elevated circulating inflammatory markers in female patients with cardiac syndrome X. *Cytokine* **2007**, *40*, 172–176. [CrossRef] [PubMed]

441. Arroyo-Espliguero, R.; Mollichelli, N.; Avanzas, P.; Zouridakis, E.; Newey, V.R.; Nassiri, D.K.; Kaski, J.C. Chronic inflammation and increased arterial stiffness in patients with cardiac syndrome X. *Eur. Heart J.* **2003**, *24*, 2006–2011. [CrossRef] [PubMed]

442. Eroglu, S.; Sade, L.E.; Yildirir, A.; Demir, O.; Bozbas, H.; Muderrisoglu, H. Serum levels of C-reactive protein and uric acid in patients with cardiac syndrome X. *Acta Cardiol.* **2009**, *64*, 207–211. [CrossRef] [PubMed]

443. Bozcali, E.; Polat, V.; Aciksari, G.; Opan, S.; Bayrak, I.H.; Paker, N.; Karakaya, O. Serum concentrations of galectin-3 in patients with cardiac syndrome X. *Atherosclerosis* **2014**, *237*, 259–263. [CrossRef] [PubMed]

444. Wenzel, R.P.; John, E. Bennett Forum on Deep Mycoses Study Design 2004: Candidiasis and salvage therapy for aspergillosis. *Clin. Infect. Dis.* **2005**, *41* (Suppl. 6), S369–S370. [CrossRef] [PubMed]

445. Bassetti, M.; Mikulska, M.; Viscoli, C. Bench-to-bedside review: Therapeutic management of invasive candidiasis in the intensive care unit. *Crit. Care* **2010**, *14*, 244. [CrossRef] [PubMed]

446. Benjamin, D.K., Jr.; Stoll, B.J.; Gantz, M.G.; Walsh, M.C.; Sanchez, P.J.; Das, A.; Shankaran, S.; Higgins, R.D.; Auten, K.J.; Miller, N.A.; et al. Eunice Kennedy shriver national institute of child health and human development neonatal research network. Neonatal candidiasis: Epidemiology, risk factors, and clinical judgment. *Pediatrics* **2010**, *126*, e865–e873. [CrossRef] [PubMed]

447. Sobel, J.D. Pathogenesis and treatment of recurrent vulvovaginal candidiasis. *Clin. Infect. Dis.* **1992**, *14* (Suppl. 1), S148–S153. [CrossRef] [PubMed]

448. Holikova, Z.; Hrdlickova-Cela, E.; Plzak, J.; Smetana, K., Jr.; Betka, J.; Dvorankova, B.; Esner, M.; Wasano, K.; Andre, S.; Kaltner, H.; et al. Defining the glycophenotype of squamous epithelia using plant and mammalian lectins: Differentiation-dependent expression of 2,6- and 2,3-linked N-acetylneura-minic acid in squamous epithelia and carcinomas, and its differential effect on binding of the endogenous lectins galectins-1 and -3. *APMIS* **2002**, *110*, 845–856. [PubMed]

449. Hrdlickova-Cela, E.; Plzak, J.; Smetana, K.; Melkova, Z.; Kaltner, H.; Filipec, M.; Liu, F.; Gabius, H. Detection of Gal-3 in tear fluid at disease states and immunohistochemical and lectin histochemical analysis in human corneal and conjunctival epithelium. *Br. J. Ophthalmol.* **2001**, *85*, 1336–1340. [CrossRef] [PubMed]

450. Honjo, Y.; Inohara, H.; Akahani, S.; Yoshii, T.; Takenaka, Y.; Yoshida, J.; Hattori, K.; Tomiyama, Y.; Raz, A.; Kubo, T. Expression of cytoplasmic Gal-3 as a prognostic marker in tongue carcinoma. *Clin. Cancer Res.* **2000**, *6*, 4635–4640. [PubMed]

451. Jensen-Jarolim, E.; Gscheidlinger, R.; Oberhuber, G.; Neuchrist, C.; Lucas, T.; Bises, G.; Radauer, C.; Willheim, M.; Scheiner, O.; Liu, F.T.; et al. The constitutive expression of Gal-3 is downregulated in the intestinal epithelia of Crohn's disease patients, and tumour necrosis factor decreases the level of Gal-3-specific mRNA in HCT-8 cells. *Eur. J. Gastroenterol. Hepatol.* **2002**, *14*, 145–152. [CrossRef] [PubMed]

452. Linden, J.R.; Kunkel, D.; Laforce-Nesbitt, S.S.; Bliss, J.M. The Role of Gal-3 in Phagocytosis of Candida albicans and Candida parapsilosis by Human Neutrophils. *Cell. Microbiol.* **2013**, *15*, 1127–1142. [CrossRef] [PubMed]

453. Linden, J.R.; De Paepe, M.E.; Laforce-Nesbitt, S.S.; Bliss, J.M. Galectin-3 plays an important role in protection against disseminated candidiasis. *Med. Mycol.* **2013**, *51*, 641–651. [CrossRef] [PubMed]

454. Plantinga, T.S.; Johnson, M.D.; Scott, W.K.; Joosten, L.A.; van der Meer, J.W.; Perfect, J.R.; Kullberg, B.J.; Netea, M.G. Human genetic susceptibility to Candida infections. *Med. Mycol.* **2012**, *50*, 785–794. [CrossRef] [PubMed]

455. Smeekens, S.P.; van de Veerdonk, F.L.; Kullberg, B.J.; Netea, M.G. Genetic susceptibility to Candida infections. *EMBO Mol. Med.* **2013**, *5*, 805–813. [CrossRef] [PubMed]

456. Wu, S.Y.; Huang, J.H.; Chen, W.Y.; Chan, Y.C.; Lin, C.H.; Chen, Y.C.; Liu, F.T.; Wu-Hsieh, B.A. Cell Intrinsic Gal-3 Attenuates Neutrophil ROS-Dependent Killing of Candida by Modulating CR3 Downstream Syk Activation. *Front. Immunol.* **2017**, *8*, 1–15. [CrossRef] [PubMed]

457. Hsieh, T.J.; Lin, H.Y.; Tu, Z.; Lin, T.C.; Wu, S.C.; Tseng, Y.Y.; Liu, F.T.; Hsu, S.T.; Lin, C.H. Dual thio-digalactoside-binding modes of human galectins as the structural basis for the design of potent and selective inhibitors. *Sci. Rep.* **2016**, *6*, 29457. [CrossRef] [PubMed]

458. Nilsson, U.; Leffler, H.; Mukhopadhyay, B.; Rajput, V. Inventor; Galecto Biotech Ab, Assignee. Galectoside Inhibitors of Galectins. U.S. Patent 9,353,141, 2016. Available online: http://www.google.com/patents/US9353141 (accessed on 22 December 2017).

459. Sampathkumar, P.; Drage, L.A.; Martin, D.P. Herpes zoster (shingles) and postherpetic neuralgia. *Mayo Clin. Proc.* **2009**, *84*, 274–280. [CrossRef] [PubMed]

460. Gershon, A.A.; Gershon, M.D.; Breuer, J.; Levin, M.J.; Oaklander, A.L.; Griffiths, P.D. Advances in the understanding of the pathogenesis and epidemiology of herpes zoster. *J. Clin. Virol.* **2010**, *48*, S2–S7. [CrossRef]

461. Park, J.; Jang, W.S.; Park, K.Y.; Li, K.; Seo, S.J.; Hong, C.K.; Lee, J.B. Thermography as a predictor of postherpetic neuralgia in acute herpes zoster patients: A preliminary study. *Skin Res. Technol.* **2012**, *18*, 88–93. [CrossRef] [PubMed]

462. Yawn, B.P.; Gilden, D. The global epidemiology of herpes zoster. *Neurology* **2013**, *81*, 928–930. [CrossRef] [PubMed]

463. Kawai, K.; Gebremeskel, B.G.; Acosta, C.J. Systematic review of incidence and complications of herpes zoster: Towards a global perspective. *BMJ Open* **2014**, *4*, e004833. [CrossRef] [PubMed]

464. Rowbotham, M.C.; Fields, H.L. The relationship of pain, allodynia and thermal sensation in post-herpetic neuralgia. *Brain* **1996**, *119*, 347–354. [CrossRef] [PubMed]

465. Regan, L.J.; Dodd, J.; Barondes, S.H.; Jessell, T.M. Selective expression of endogenous lactose-binding lectins and lactoseries glycoconjugates in subsets of rat sensory neurons. *Proc. Natl. Acad. Sci. USA* **1986**, *83*, 2248–2252. [CrossRef] [PubMed]

466. Park, M.J.; Chung, K. Endogenous lectin (RL-29) expression of the autonomic preganglionic neurons in the rat spinal cord. *Anat. Rec.* **1999**, *254*, 53–60. [CrossRef]

467. Ma, Z.; Han, Q.; Wang, X.; Ai, Z.; Zheng, Y. Galectin-3 Inhibition Is Associated with Neuropathic Pain Attenuation after Peripheral Nerve Injury. *PLoS ONE* **2016**, *11*, e0148792.

International Journal of
Molecular Sciences

MDPI

Review

Galectin-3 in Atrial Fibrillation: Mechanisms and Therapeutic Implications

Nicolas Clementy [1,*], Eric Piver [2], Arnaud Bisson [1], Clémentine Andre [1], Anne Bernard [1], Bertrand Pierre [1], Laurent Fauchier [1] and Dominique Babuty [1]

[1] Cardiology Department, Trousseau Hospital, François Rabelais University, 37000 Tours, France; arnaud.bisson37@gmail.com (A.B.); clementine2andre@gmail.com (C.A.); anne.bernard@univ-tours.fr (A.B.); b.pierre@chu-tours.fr (B.P.); laurent.fauchier@univ-tours.fr (L.F.); d.babuty@chu-tours.fr (D.B.)
[2] Biochemistry Department, Trousseau Hospital, François Rabelais University, 37000 Tours, France; piver_e@univ-tours.fr
* Correspondence: nclementy@yahoo.fr; Tel.: +33-0247-474-687; Fax: +33-0247-475-919

Received: 21 February 2018; Accepted: 23 March 2018; Published: 25 March 2018

Abstract: Maintenance of atrial fibrillation is a complex mechanism, including extensive electrical and structural remodeling of the atria which involves progressive fibrogenesis. Galectin-3 is a biomarker of fibrosis, and, thus, may be involved in atrial remodeling in atrial fibrillation patients. We review the role of galectin-3 in AF mechanisms and its potential therapeutic implications.

Keywords: galectin-3; atrial fibrillation; ablation

1. Introduction

Atrial fibrillation (AF), the most frequent of all arrhythmia is associated with structural, electrical, genomic, hormonal and autonomic atrial remodeling. Structural atrial remodeling includes hibernation phenomena, degenerative processes (aptotosis and fibrosis), and gap junction expression alteration [1]. More recently, the concept of fibrotic atrial cardiomyopathy has materialized, defined by a progressive invasion of atrial myocardium by fibrosis, which favors initiation and maintenance of AF [2]. Galectin-3 (Gal-3), a biomarker of fibrosis, has been shown to play a role in promoting fibrosis in patients with AF, and has emerged as a prognostic marker in this population. We review here the histologic and molecular mechanisms involved, and the subsequent issues in clinical practice.

2. Clinical Atrial Fibrillation

Atrial fibrillation, characterized by irregular, rapid and anarchical atrial electrical activity, either silent or responsible for symptoms (especially palpitations and/or dyspnea), is associated with a significant morbidity, an increased risk of thrombo-embolic events, especially strokes, hospitalizations for heart failure, and thus, an increased all-cause mortality [3]. Management of this disease has become a major issue in our healthcare systems.

Clinically, AF typically begins with paroxysmal episodes, which by arbitrary definition last less than seven days and may self-terminate spontaneously, or may need cardioversion, either with medications or an electrical shock [4]. It then may evolve towards so-called "persistent" AF, defined by more prolonged episodes, typically lasting more than 7 days. Finally, long-persistent AF is defined by a sustained arrhythmia lasting one year or more. The term of "permanent" AF is restrained to patients in whom normal sinus rhythm will not to be restored. Patients do evolve very differently. Some patients will present persistent AF as a first manifestation of the disease, while some will pass through several years of paroxysmal episodes, suggesting that the clinical history of the disease and its underlying pathophysiological mechanisms are not correlated. Although the mechanisms involved in the evolution of the disease remain unclear, AF has been shown to lead to electrical remodeling and progressive fibrosis of the atria, which in turn favor AF perpetuation and chronicization: "AF begets AF" [5,6].

3. Atrial Fibrillation Mechanisms

Different triggers can induce AF: abnormal automaticity, generally originated from pulmonary veins, induced by stretch or left atrium dilatation; functional localized reentry, usually induced by fibers anisotropy [7,8]. Perpetuation of AF may be related to rotors (localized reentries), multiple circuits of reentry, favored by atrial remodeling (multiple wavelet hypothesis), and electrophysiological remodeling secondary to autonomic nervous system, endocrinous and/or hemodynamic modifications [9]. Tissue heterogeneities, i.e., structural remodeling, are associated to slow conduction velocities and short refractory periods, i.e., electrical remodeling, the electrophysiological substrate for maintenance of arrhythmia [10–13].

3.1. Electrical Remodeling

At the cellular level, AF mechanisms are still controversial. They involve electrical remodeling of ion channels, gap junctions and abnormal Ca^{2+} handling from the sarcoplasmic reticulum. There is also evidence supporting the role of autonomic nervous system and inflammation [14]. Martins et al. have evaluated the rate of electrical and structural remodeling in an animal model of persistent AF. The increase of fibrillatory frequency varied between animals but its slope predicted the time at which AF becomes persistent, reflecting an abbreviation of action potential duration, changes in densities of ion currents, and a progressive increase in atrial dilatation, myocyte hypertrophy and atrial fibrosis [15].

Electrical remodeling involves major changes in action potentials: shortening of action potentials duration, acceleration of repolarization, but also a mismatching in the repolarization time course during rate changes [16]. Shortening of repolarization and refractory periods lead to a decrease of the wavelength, allowing formation of small circuits of reentry (micro-reentries). These changes alone may not explain the trigger of the arrhythmia. Gap junction cell coupling remodeling might be involved in the creation of the substrate needed to sustain reentry in single or multiple small circuits [17].

Autonomic remodeling plays a major role in electrical remodeling. The atria are innervated by parasympathetic and sympathetic systems: the extrinsic ganglia, the intrinsic ganglionated plexi network, and the baroreceptors. AF was found to lead to a heterogeneous increase in atrial sympathetic innervation [18]. Vagal activation is known to contribute to shortening of atrial refractoriness and subsequent AF [19]. Chen and colleagues have shown that the mechanisms by which autonomic changes promote AF include an enhanced automaticity, early/delayed after-depolarizations, and heterogenous shortening of refractory periods [20].

In an animal model study, high-resolution mapping demonstrated reduced left atrial AF electrical complexity along with reduced sympathetic activity after renal denervation [21]. Neuromodulation of autonomic tone is then to be investigated for treating AF.

3.2. Structural Remodeling

It is well known for several decades that AF is associated atrial cavity dilatation, atrial myocardium hypertrophy, and variable rarefaction of atrial cardiomyocytes, either focal or diffuse, replaced with fibrotic tissue. Whether structural remodeling precedes or follows the development of the arrhythmia is unclear. Interstitial fibrosis forces depolarization wave-front to slow down. The electrical impulse propagates through alternative pathways, in which cardiomyocytes have recovered excitability, inducing multiple reentrant circuits [1]. Atria are more prone to fibrogenesis than the ventricles. The relative abundance of collagen creates heterogeneous areas with different electrophysiological properties. Matrix metalloproteinases and pro-fibrotic signals stimulate the proliferation of fibroblasts and subsequent collagen synthesis and degradation. Mechanisms also involve altered expression of matrix proteins other than collagen (fibronectin 1, fibrillin 1), and deposition of proteoglycans and other extracellular matrix components.

At the stage of AF perpetuation (clinical persistent AF), left atrium is usually characterized by the presence of large areas of fibrosis. Magnetic resonance imaging (MRI), with the help of a specific dedicated software, remains the gold standard to identify atrial fibrotic tissue extend, but this technique lacks adapted availability in regular practice. The lack of feasibility due the rather scarce thickness of the atrial wall may also limit the analyses [22].

Another quantitative method to assess atrial fibrosis is the invasive measurement of low voltage areas by electroanatomic mapping. Electroanatomic bipolar voltage mapping has been used as a surrogate to define scar [23–25]. A mapping diagnostic catheter is moved inside the atrial cavity, and dedicated software translates local mean signal amplitude into a color-coded image, i.e., a voltage map. Low voltage areas are then considered invaded by fibrosis (Figure 1). Extent of fibrosis is highly variable, and does not correlate with the clinical presentation of AF [26–28]. Haissaguerre et al. have shown that the fibrosis burden, particularly left atrial fibrosis, correlates independently with the number of electrophysiologically reentrant regions and the great majority of atrial fibrillation electrical drivers were found at the borders of fibrosis areas [29]. They have also shown a positive correlation with left atrial volume, and with uninterrupted AF duration. Sites with reentrant drivers showed a significantly higher disorganization index (i.e., a greater entropy), and a greater density of local fibrosis, showing that not all fibrosis carries the same proarrhythmic effect. Distinct tissue characteristics may form distinct electrical remodeling, favoring or not the creation of reentrant drivers.

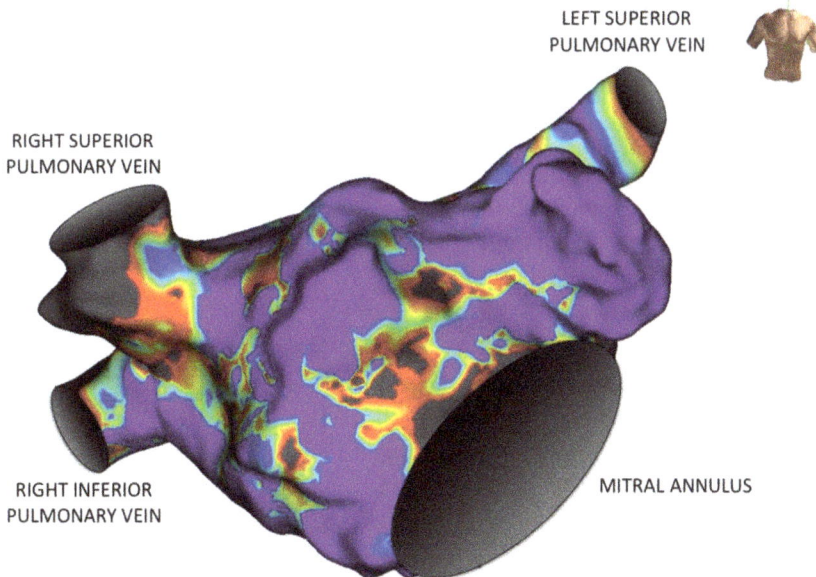

Figure 1. Three-dimensional voltage map of the left atrium obtained during invasive electrophysiological mapping (anterior view). The amplitude of local electrical activity is color-coded: from grey for low voltage areas to mauve for high voltage areas. Local amplitude correlates to the underlying atrial myocardial thickness, so that fibrotic areas translate into low voltage heterogeneous ("patchy") anisotropic areas, here localized on the inter-atrial septum and the anterior wall.

Fibrosis tridimensional geometry may have an impact in AF perpetuation. Diffuse fibrosis has little effect on electrical wave fronts propagation. Conversely, "patchy" fibrosis can produce severe disruptions to propagating wave-fronts and rotor dynamics [30]. Propagation through highly fibrotic tissue is an example of percolation, the movement of a substance through a medium with random structure [31]. As demonstrated by Haissaguerre, "a wave-front passing through a medium with random fibrosis may then produce rate-dependent disruption of propagation and various reentries within a series of loosely coupled islands connected by strands" [29]. This phenomenon contributes to anisotropy. The cellular architecture of the myocardium is indeed anisotropic: myocardial cells are markedly elongated and form layers of tissue with sharply demarcated fiber orientation. Such fiber orientation induces a clear discrepancy between transverse and longitudinal electrical conduction velocities, translating into anisotropic properties in the propagation of the action potential and cells repolarization [32,33].

Cardiac magnetic resonance imaging lacking of availability and spatial resolution, and electro-anatomic mapping requiring an invasive procedure, simpler tools are needed to evaluate atrial structural remodeling in AF patients. Biomarkers seem to be promising in that setting.

4. Atrial Remodeling and Biomarkers

Atrial tissue is known to be more reactive to fibrosis than the ventricular one. Multiple interdependent signaling pathways lead to cardiac fibrosis: oxidative stress; inflammation via chemokines and cytokines; the renin-angiotensin-aldosterone system; transforming growth factor β (TGF-β) pathway; mechanical stretch via the expression of matrix metalloproteinases and their inhibitors [7]. Another signal pathway has been recently identified: epicardial adipocytes induce atrial fibrosis via adipo-fibrokines secretion [34].

In the biochemistry domain, fibrosis can be identified by evaluation of different biomarkers involved in extracellular matrix formation, such as collagen, fibronectin, laminin, fibrillin, or matrix metalloproteinases [35]. Among these markers, MMP-9, GDF-15, PIIINP and galectin-3 (Gal-3) appear as most promising [36].

Matrix metalloproteinase-9 (MMP-9) is an endopeptidase, synthesized and secreted in a monomeric form as zymogen. It can degrade the extracellular matrix components [37]. MMP-9 plasma levels are higher in AF patients and show a gradual increase from paroxysmal AF through persistent AF to permanent AF [38].

AF patients have also been shown to display higher values of GDF-15, a protein member of the transforming growth factor beta superfamily, than non-AF patients, GDF-15 being independently associated with paroxysmal AF [39].

Levels of PIIINP, the amino-terminal peptide of type III procollagen released into the blood during both synthesis and degradation of collagen type III, showed a nonlinear relationship with the risk of incident AF, both before and after risk adjustment [40].

Finally, studies on Gal-3 seem very encouraging regarding a clinical application in daily practice.

5. Galectin-3 Profibrotic Signaling

Atrial fibrillation induces tissue injuries, which lead to increased Gal-3 synthesis and secretion. Gal-3 belongs to the galectins family, which are ß-galactoside-binding proteins, specifically binding to glycoproteins with N-acetyl-D-lactosamine disaccharide. In mammalians, 15 galectins were identified and named in accordance with their sequential discovery. They are classified in three groups according to the organization of their carbohydrate recognition domain (CRD). The first group of galectins (including Gal-1, -2, -5, -7, -10, -11, -13, -14, and -15) consists of homodimers through their single CRD, while in the second group galectins (Gal-4, -6, -8, -9, and -12) are structured in tandem-repeat with two distinct CRDs at their N- and C-termini. Gal-3, with a chimera structure, constitutes the third group. Gal-3 gene, named LGALS3, belongs to the chromosome 14 (locus q21-22). The *LGALS3* gene is composed of six exons and five introns spanning about 17 kilobases. Gal-3 molecule is a soluble protein of 29–35 kDa, and has a single CRD with a unique N-terminal domain (NTD) (Figure 2). The NTD consists of a relatively flexible structure of 7–14 repeating sequences and totals 110–130 amino acids (AA). It contains a N-terminal region (NTR) of 12 AA. In this NTR, the serine 6 can be phosphorylated by casein kinases 1 and 2, this regulation contributing to nuclear translocation and a reduction of affinity to its ligands [41]. Between the CRD and the NTD, a collagen-like sequence (CLS) consists of about 100 AA. This long tail contains the collagenase cleavable H-domain, in which histidine 64 is the site of action of matrix metalloproteinases (MMPs) such as MMP9 and MMP2 [42,43]. The CRD, consisting of 130 AA, forms a globular structure containing the binding site to carbohydrates, similar to other galectins. It also contains an Asp-Trp-Gly-Arg (NWGR) motif, similar to those described in anti-apoptotic BCL-2 proteins [44]. This sequence is also involved in the aggregation of Gal-3 molecules in the absence of ligand. Gal-3 affinity to its ligands is proportional to the number of lactosamine repeating units from the oligosaccharide structure.

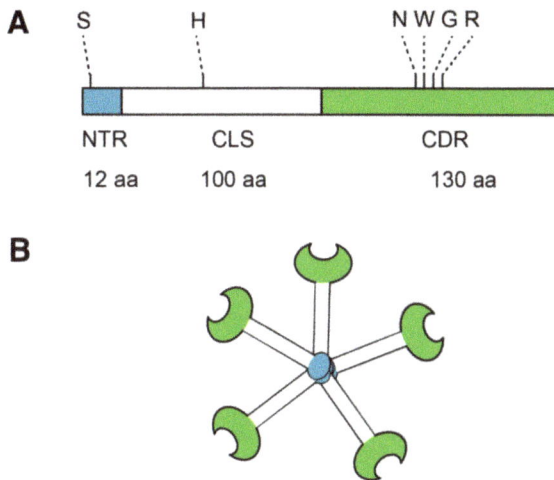

Figure 2. Structure of Galectin-3. (**A**) Galectin-3 protein structure consists of N terminal Domain (NTD), which has a N terminal Region of 12 amino acids (aa) and contains serine 6 (S) phosphorylation site. The carbohydrate recognition domain (CRD) 130 aa comprise the C-terminal and contains the NWGR motif; (**B**) Pentameric structure of Galectin-3.

Gal-3 was initially described as Mac-2 antigen in 1982 for its capacity to identify a macrophage sub-population. Actually, its identification in different pathway leads has given to Gal-3 a variety of names such as IgE binding protein, L-29, CBP30 or CBP35 [45]. It was eventually cloned in 1991 and subsequently recognized as a β-galactoside-binding lectin. Gal-3 has a pleiotropic distribution and could be found in the cytoplasm, in the nucleus and at the cell membrane, or as a pentameric circulating form [46]. The Gal-3 NTD drives self-oligomerization into a pentamer and may interact with extracellular and intracellular components. In the extracellular space, Gal-3 has the capacity to bind to different cells surface and extracellular matrix glycans in order to induce cell adhesion, migration, and growth regulation, mainly pro-apoptotic effects. In the intracellular space, Gal-3 regulates the cell cycle, inducing proliferation and anti-apoptotic effects. These combined actions, specific to Gal-3 in the galectin family, participate to a variety of pathophysiological processes involved in atrial fibrosis genesis: apoptosis, angiogenesis and inflammation [47]. Gal-3 is mainly produced by activated macrophages, mast cells, neutrophils and eosinophils [48]. In the heart, it is mainly expressed in fibroblasts.

The respective mechanisms by which Gal-3 exerts fibrogenic activity (fibroblast proliferation and collagen deposition) are not completely depicted. Extracellular pentameric Gal-3 interaction with profibrotic effectors such as TGF-β/SMAD could be a part of the pathway that initiates fibrogenesis. One of the hypotheses for activation of this pathway is the capacity of Gal-3 to form lectin-saccharide lattices on cell surfaces. TGF-β receptor entrapment within the lattice may amplify profibrotic signaling [49]. The TGF-β/SMAD pathway is well known to induce the recruitment, the activation and the transition of macrophages and mast cells, and the production of extracellular matrix by tubular epithelial cells, endothelial cells, mesengial cells, podocytes, fibroblasts and myofibroblasts [50] (Figure 3).

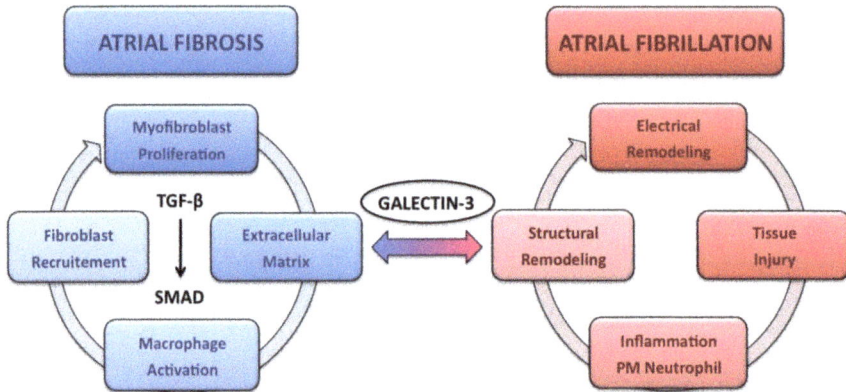

Figure 3. "Atrial fibrillation begets atrial fibrillation". Galectin-3 production, promoted by atrial remodeling in atrial fibrillation patients, induces extracellular matrix production, mainly through the TGF-β/SMAD signaling pathway. Atrial fibrosis in return is associated with atrial dilatation, tissue anisotropy with heterogeneous electrical properties, which in turn favor atrial fibrillation initiation and maintenance.

TGF-β is consistently activated in cardiac fibrosis [51]. TGF-β, a pleiotropic cytokine, exerts its effects by binding to two distinct receptors (TβRI and TβRII) with intrinsic serine/threonine kinase activity. It can be found in three isoforms (TGF-β1, 2 and 3) encoded by three distinct genes [52]. These three isoforms signal through the same cell surface receptors and share common cellular targets. However, TGF-β1 remains the predominant form in the heart, so that the role of TGF-β in cardiac fibrosis is probably limited to the β1 isoform. TGF-β1 is present in the normal heart as a latent complex, unable to associate with its receptors. Following cardiac injury, a relatively small amount of latent TGF-β is activated, sufficient to induce a maximal cellular response [53]. Proteases, such as plasmin, MMP-2 and MMP-9, and the matricellular protein TSP-1, can activate TGF-β and thus play a major role in atrial structural remodeling. Reactive oxygen species can also trigger TGF-β activation.

TGF-β stimulation induces myofibroblast trans-differentiation and improves extracellular matrix synthesis [54]. TGF-β has also matrix-preserving effects. It can induce the expression of protease inhibitors: plasminogen activator inhibitor (PAI-1), tissue inhibitors of metalloproteinases (TIMPs) [52]. Activated TGF-β binds to the active kinase domain of TβRII. It induced subsequent activation of TβRI by transphosphorylation of serine and threonine residues in the GS box of its cytoplasmic domain. This phosphorylation induces binding to R-SMADs (receptor-regulated). Receptor-associated R-SMADs then heterodimerize with common mediator SMAD. They convey signals from membrane to nucleus, where they bind to SMAD binding element (SBE) sequences located in different promoter regions of various genes regulating cell growth. After ligand stimulation, inhibitory SMADs (I-SMADs) translocate from the nucleus to the cytoplasm and bind to TβRI to deactivate R-SMADs, causing inhibition of TGF-β signaling [55].

The role of SMAD-independent pathways in fibrotic cardiac remodeling remains unclear. Protein kinase B (Akt) consists of three closely related isoforms (Akt1, Akt2, and Akt3). It modulates cell growth, proliferation, and survival. Stimulation of PI3-K (p85 catalytic and p110 regulatory subunits) activates Akt, which phosphorylates proteins involved in the control of cell cycle, stimulating cell growth. The activation of Akt by TGF-β can inhibit the glycogen synthase kinase GSK-3β by increasing its phosphorylation. Increased phosphorylation of GSK-3β is associated with increased α-SMA (alpha smooth muscle actin, ACTA2), an increased transformation of fibroblasts into myofibroblasts, and subsequent fibrosis [56]. TGF-β1 can also increase protein kinase C α (PKCα) activity. TGF-β1-induced increased expression of α-SMA has been shown to be PKCα activity dependent [57]. In addition, the increased fibroblast—myofibroblast transdifferentiation via the TGF-β1–PKCα pathway also leads to an increased expression of phosphodiesterase 1A (PDE1A). Pharmacological inhibition of PDE1A can reduce the TGF-β1-induced increased expression of α-SMA, whereas phorbol-12-myristate-13-acetate (a PKCα activator) increases it. Upregulation of PKCα expression by TGF-β1 is also inhibited by PDE1A inhibition. TGF-β1 can then increase α-SMA expression and subsequent myofibroblast formation via a PDE1A-PKCα-dependent mechanism [58]. Negative regulation of TGF-β signaling may be a major therapeutic target in order to limit the development of atrial fibrosis [59].

Gal-3 may regulate the effects of TGF-β1, and other cytokines, by promoting the retention of their receptors on the atrial myofibroblast surface membrane, leading to an increased transcription of profibrotic molecules via stimulation of phosphorylation and nuclear translocation of the SMAD complex. Gal-3 may also act as a promoter of nuclear translocation of β-catenin via an inhibition of GSK-3β phosphorylating activity [47].

Gal-3 is then a pivotal actor of cardiac fibrosis, highly expressed in fibrotic tissues, and upregulated in chronic inflammatory and fibrotic conditions in human [60].

Mammalian models of cardiac fibrosis demonstrate high level of Gal-3, and its inhibition may prevent cardiac fibrosis [61]. Commercially available Gal-3 inhibitors may be needed. GM-CT-01 is a galactomannan that binds to Gal-3 carbohydrate-binding domain [62,63]. Takemoto et al. have shown that GM-CT reduced both electrical and structural remodeling in an AF sheep model [64]. The authors suggested a role of TGF-β1/SMAD pathway inhibition, reducing activation of myofibroblasts and subsequent collagen deposition [65].

Gal-3 levels correlate with risk factors of cardiovascular disease associated to chronic inflammatory conditions such as hypertension, diabetes or obesity, and were shown elevated in heart failure [66]. Gal-3 appears now with the recommended characteristics of a biomarker that could be useful in the cardiovascular domain [67–70].

6. Disease Progression and Therapeutic Implications

Galectin-3 has been well studied in AF and associated with the disease progression. Ho et al. have shown in a large cohort that higher Gal-3 levels were associated with increased risk of developing AF in the general population, although not after accounting for other risk factors [71]. In that study, the mean value of Gal-3 in patients developing AF was 15 ng/mL. We found in a population of AF patients that Gal-3 levels ≥15 ng/mL were present in older patients, female, and those with hypertension or diabetes [72]. Moreover, Gal-3 levels were higher in patients with a persistent form of AF (non self-terminating AF), suggesting a role in the maintenance of this arrhythmia. Similar results were found following a myocardial infarction [73].

In the supplementary data of the DECAAF study, left atrial volume was similar whatever the extent of atrial fibrosis [22]. In our initial study, no correlation between LA volume measured by echocardiography was established with the level of Gal-3 [72]. However, Gal-3 levels were shown to independently correlate with the extent of left atrial fibrosis detected with MRI [74].

Gal-3 may then help guiding the medical therapy. Renin-angiotensin-aldosterone system inhibitors, known to limit AF progression, may be used in patients with higher Gal-3 serum levels [75]. Gal-3 may even participate in the future to AF therapy as a therapeutic target. Takemoto and colleagues have indeed shown that Gal-3 inhibition decreased AF inducibility, but also increased the probability of spontaneous conversion to sinus rhythm during persistent AF [64]. Gal-3 inhibition may then act on structural remodeling by decreasing atrial dilatation and atrial fibrogenesis, but also on electrical remodeling (action potential duration shortening and atrial dominant frequency increase).

Therapy in AF also includes anticoagulation, and disease progression may be associated with an increased risk of stroke. It was shown that increased plasma levels of galectin-3 were associated with occurrence of postoperative strokes among female patients who undergo carotid endarterectomy [76]. Higher Gal-3 levels were also found in patients with acute ischemic stroke in a case-control study, and independently associated with a more severe disease and a poorer outcome [77,78]. Specifically in AF patients, an ARISTOTLE sub-study showed that patients in higher quartiles of Gal-3 had a higher risk of ischemic stroke, although the association was not independent [79]. Higher Gal-3 levels are actually associated with diabetes, hypertension, or heart failure, and thus higher CHADSVASC scores and subsequent risk of stroke. Whether patients with higher Gal-3 levels would benefit more from anticoagulation remains speculative.

Finally, Gal-3 may predict the substrate that needs to be treated during radiofrequency ablation. Patients with paroxysmal AF but higher Gal-3 levels might require more extensive intervention in addition to the simple isolation of pulmonary veins [80].

7. Galectin-3 and Atrial Fibrillation Ablation

Ablation of atrial fibrillation (AF) can be recommended in symptomatic patients, after failure of initial medical therapy [4]. Ablation consists mainly in an electrical isolation of pulmonary veins, using either radiofrequency energy or cryotherapy. This procedure can be associated with a defragmentation, i.e., an ablation of the proarrhythmic substrate with homogenization of low voltage patchy areas and of slow conduction. Depending on the clinical presentation of the disease, and the procedure method and endpoint, outcomes differ, with a much higher rate of recurrence after ablation in patients with a persistent form of AF [81]. Wu et al. found in a small cohort of 50 patients with persistent AF that Gal-3 concentration was an independent predictor of AF recurrence after ablation [82]. We also identified Gal-3 and left atrial diameter as independent predictive factors of recurrence, both parameters being related to structural atrial remodeling [83]. Left atrial diameter was known to be a strong predictor of recurrences after AF ablation, probably due to a parallel progression of fibrosis and LA enlargement [84]. But we found a distinct role of these 2 parameters, confirming the results of the DECAAF study in which the extent of fibrosis was not correlated to left atrial size [22,85].

The DECAAF study emphasized the role of atrial fibrosis quantified by MRI in AF ablation patients. However, MRI appears cost ineffective when compared to a simple dosage of Gal-3 serum level. There is also no clear evidence that the presence of late gadolinium enhancement actually identifies histological fibrosis within left atrium. Gal-3 level is directly related to fibrotic tissue activity [86]. Finally, MRI is not feasible in all centers, as it necessitates specific dedicated software, and not feasible in all patients, especially patients with implanted cardiac devices such as pacemakers and defibrillators.

The clinical history of AF appears less important than the nature and the amount of the atrial arrhythmogenic substrate, which determine the results after AF ablation. Higher Gal-3 levels in AF patients are thought to be related to obesity, hypertension or diabetes (metabolic syndrome), conditions also associated with an increased incidence of AF [87]. These metabolic conditions actually favor fibrogenesis. Gal-3 level may reflect the importance of arrhythmogenic substrate, no matter what initially prompted it [88].

An extensive ablation may not always be required, or may even be deleterious, in the absence of a significant atrial remodeling [89]. Gal-3 level might help predefining the ablation strategy, either more conservative or more extensive, in advance of the potential intervention.

8. Limitations

Galectin-3, as a ubiquitous protein, is not a specific of cardiac fibrosis. It is elevated in several conditions such as liver cirrhosis, lung fibrosis or chronic inflammatory diseases. Moreover, it is not specific of atrial myocardium and is elevated in heart failure and cardiomyopathies with an underlying ventricular structural disease. This is particularly important as AF and heart failure may be inter-dependant, one condition leading to the other one and vice versa. Analyses of Gal-3 levels in AF patients must then always be cautious of any of these other conditions potentially associated with extra-atrial fibrotic invasion.

9. Conclusions

Galectin-3 plays a major role in the progression of atrial fibrosis, and has thus become an essential prognostic biomarker in AF patients. Elevated Gal-3 levels correlate with a more advanced form of the disease, associated severe comorbidities, less efficacy of treatment, and worse outcomes. Galectin-3 inhibition may be a therapeutic target for AF patients in the future.

Author Contributions: Nicolas Clementy drafted the manuscript; Eric Piver, Arnaud Bisson, Clémentine Andre, Anne Bernard, Bertrand Pierre, Laurent Fauchier, and Dominique Babuty contributed substantially to the manuscript revision.

Conflicts of Interest: The authors declare no conflict of interest.

Abbreviations

AF	Atrial fibrillation
Gal-3	Galectin-3
MRI	Magnetic resonance imaging

References

1. Allessie, M.; Ausma, J.; Schotten, U. Electrical, contractile and structural remodeling during atrial fibrillation. *Cardiovasc. Res.* **2002**, *54*, 230–246. [CrossRef]
2. Goldberger, J.J.; Arora, R.; Green, D.; Greenland, P.; Lee, D.; Lloyd-Jones, D.M.; Marl, M.; Ng, J.; Shah, S.J. Evaluating the atrial myopathy underlying atrial fibrillation: Identifying the arrhythmogenic and thrombogenic substrate. *Circulation* **2015**, *132*, 278–291. [CrossRef] [PubMed]
3. Melgaard, L.; Rasmussen, L.H.; Skjøth, F.; Lip, G.Y.; Larsen, T.B. Age Dependence of RiskFactors for Stroke and Death in Young Patients With Atrial Fibrillation: A NationwideStudy. *Stroke* **2014**, *45*, 1331–1337. [CrossRef] [PubMed]
4. Kirchhof, P.; Benussi, S.; Kotecha, D.; Ahlsson, A.; Atar, D.; Casadei, B.; Castella, M.; Diener, H.C.; Heidbuchel, H.; Hendriks, J.; et al. 2016 ESC Guidelines for the management of atrial fibrillation developed in collaboration with EACTS. *Eur. Heart J.* **2016**, *37*, 2893–2962. [CrossRef] [PubMed]
5. Jalife, J. Mechanisms of persistent atrial fibrillation. *Curr. Opin. Cardiol.* **2014**, *29*, 20–27. [CrossRef] [PubMed]
6. Nattel, S.; Harada, M. Atrial Remodeling and Atrial Fibrillation: Recent Advances and Translational Perspectives. *J. Am. Coll. Cardiol.* **2014**, *63*, 2335–2345. [CrossRef] [PubMed]
7. Schotten, U.; Verheule, S.; Kirchhof, P.; Goette, A. Pathophysiological mechanisms of atrial fibrillation. A translational appraisal. *Physiol. Rev.* **2011**, *91*, 265–325. [CrossRef] [PubMed]
8. Haïssaguerre, M.; Jaïs, P.; Shah, D.C.; Takahashi, A.; Hocini, M.; Quiniou, G.; Garrigue, S.; Le Mouroux, A.; Le Métayer, P.; Clémenty, J. Spontaneous initiation of atrial fibrillation by ectopic beats originating in the pulmonary veins. *N. Engl. J. Med.* **1998**, *339*, 659–566. [CrossRef] [PubMed]
9. Nguyen, B.L.; Fisbein, M.C.; Chen, L.S.; Chen, P.S.; Masroor, S. Histopathological substrate for chronic atrial fibrillation in humans. *Heart Rhythm.* **2009**, *6*, 454–460. [CrossRef] [PubMed]
10. Sánchez-Quintana, D.; López-Mínguez, J.R.; Pizarro, G.; Murillo, M.; Cabrera, J.A. Triggers and anatomical substrates in the genesis and perpetuation of atrial fibrillation. *Curr. Cardiol. Rev.* **2012**, *8*, 310–326. [CrossRef] [PubMed]
11. Stiles, M.K.; John, B.; Wong, C.X.; Kuklik, P.; Brooks, A.G.; Lau, D.H.; Dimitri, H.; Roberts-Thomson, K.C.; Wilson, L.; De Sciscio, P.; et al. Paroxysmal lone atrial fibrillation is associated with an abnormal atrial substrate: Characterizing the "second factor". *J. Am. Coll. Cardiol.* **2009**, *53*, 1182–1191. [CrossRef] [PubMed]
12. Teh, A.W.; Kistler, P.M.; Lee, G.; Medi, C.; Heck, P.M.; Spence, S.J.; Sparks, P.B.; Morton, J.B.; Kalman, J.M. Electroanatomic remodeling of the left atrium in paroxysmal and persistent atrial fibrillation patients without structural heart disease. *J. Cardiovasc. Electrophysiol.* **2012**, *23*, 232–238. [CrossRef] [PubMed]

13. Xu, J.; Cui, G.; Esmailian, F.; Plunkett, M.; Marelli, D.; Ardehali, A.; Odim, J.; Laks, H.; Sen, L. Atrial extracellular matrix remodeling and the maintenance of atrial fibrillation. *Circulation* **2004**, *109*, 363–368. [CrossRef] [PubMed]

14. Hu, Y.F.; Chen, Y.J.; Lin, Y.J.; Chen, S.A. Inflammation and the pathogenesis of atrial fibrillation. *Nat. Rev. Cardiol.* **2015**, *12*, 230–243. [CrossRef] [PubMed]

15. Martins, R.P.; Kaur, K.; Hwang, E.; Ramirez, R.J.; Willis, B.C.; Filgueiras-Rama, D.; Ennis, S.R.; Takemoto, Y.; Ponce-Balbuena, D.; Zarzoso, M.; et al. Dominant frequency increase rate predicts transition from paroxysmal to long-term persistent atrial fibrillation. *Circulation* **2014**, *129*, 1472–1482. [CrossRef] [PubMed]

16. Nattel, S.; Maguy, A.; Le Bouter, S.; Yeh, Y.H. Arrhythmogenic ion-channel remodeling in the heart: Heart failure, myocardial infarction, and atrial fibrillation. *Physiol. Rev.* **2007**, *87*, 425–456. [CrossRef] [PubMed]

17. Duffy, H.S.; Wit, A.L. Is there a role for remodeled connexins in AF? No simple answers. *J. Mol. Cell Cardiol.* **2008**, *44*, 4–13. [CrossRef] [PubMed]

18. Jayachandran, J.V.; Sih, H.J.; Winkle, W.; Zipes, D.P.; Hutchins, G.D.; Olgin, J.E. Atrial fibrillation produced by prolonged rapid atrial pacing is associated with heterogeneous changes in atrial sympathetic innervation. *Circulation* **2000**, *101*, 1185–1191. [CrossRef] [PubMed]

19. Linz, D.; Schotten, U.; Neuberger, H.R.; Bohm, M.; Wirth, K. Negative tracheal pressure during obstructive respiratory events promotes atrial fibrillation by vagal activation. *Heart Rhythm.* **2011**, *8*, 1436–1443. [CrossRef] [PubMed]

20. Chen, P.S.; Chen, L.S.; Fishbein, M.C.; Lin, S.F.; Nattel, S. Role of the autonomic nervous system in atrial fibrillation: Pathophysiology and therapy. *Circ. Res.* **2014**, *114*, 1500–1515. [CrossRef] [PubMed]

21. Linz, D.; Van Hunnik, A.; Hohl, M.; Mahfoud, F.; Wolf, M.; Neuberger, H.R.; Casadei, B.; Reilly, S.N.; Verheule, S.; Bohm, M.; et al. Catheter-based renal denervation reduces atrial nerve sprouting and complexity of atrial fibrillation in goats. *Circ. Arrhythm. Electrophysiol.* **2015**, *8*, 466–474. [CrossRef] [PubMed]

22. Marrouche, N.F.; Wilber, D.; Hindricks, G.; Jais, P.; Akoum, N.; Marchlinski, F.; Kholmovski, E.; Burgon, N.; Hu, N.; Mont, L.; et al. Association of atrial tissue fibrosisidentified by delayedenhancement MRI and atrial fibrillation catheter ablation: The DECAAF study. *JAMA* **2014**, *311*, 498–506. [CrossRef] [PubMed]

23. Jadidi, A.S.; Cochet, H.; Shah, A.J.; Kim, S.J.; Duncan, E.; Miyazaki, S.; Sermesant, M.; Lehrmann, H.; Lederlin, M.; Linton, N.; et al. Inverse relationship between fractionated electrograms and atrial fibrosis in persistentatrial fibrillation: Combined magnetic resonance imaging and high density mapping. *J. Am. Coll. Cardiol.* **2013**, *62*, 802–812. [CrossRef] [PubMed]

24. Malcolme-Lawes, L.C.; Juli, C.; Karim, R.; Bai, W.; Quest, R.; Lim, P.B.; Jamil-Copley, S.; Kojodjojo, P.; Ariff, B.; Davies, D.W.; et al. Automated analysis of atrial late gadolinium enhancement imaging that correlates with endocardial voltage and clinical outcomes: A 2-center study. *Heart Rhythm.* **2013**, *10*, 1184–1191. [CrossRef] [PubMed]

25. Kapa, S.; Desjardins, B.; Callans, D.J.; Marchlinski, F.E.; Dixit, S. Contact electroanatomic mapping derived voltage criteria for characterizing left atrial scar in patients undergoing ablation for atrial fibrillation. *J. Cardiovasc. Electrophysiol.* **2014**, *25*, 1044–1052. [CrossRef] [PubMed]

26. Kottkamp, H. Human atrial fibrillation substrate: Towards a specific fibrotic atrial cardiomyopathy. *Eur. Heart J.* **2013**, *34*, 2731–2738. [CrossRef] [PubMed]

27. Rolf, S.; Kircher, S.; Arya, A.; Eitel, C.; Sommer, P.; Richter, S.; Gaspar, T.; Bollmann, A.; Altmann, D.; Piedra, C.; Hindricks, G.; Piorkowski, C. Tailored atrial substrate modification based on low-voltage areas in catheter ablation of atrial fibrillation. *Circ. Arrhythm. Electrophysiol.* **2014**, *7*, 825–833. [CrossRef] [PubMed]

28. Arentz, T.; Lehrmann, H.; Sorrel, J.; Markstein, V.; Park, C.; Allgeier, J.; Weber, R.; Jadidi, A. Selective substrate based ablation compared to pulmonary vein isolation alone in patients with persistent atrial fibrillation. *Heart Rhythm.* **2015**, *12*, S347.

29. Haissaguerre, M.; Shah, A.J.; Cochet, H.; Hocini, M.; Dubois, R.; Efimov, I.; Vigmond, E.; Bernus, O.; Trayanova, N. Intermittent drivers anchoring to structural heterogeneities as a major pathophysiological mechanism of human persistent atrial fibrillation. *J. Physiol.* **2016**, *594*, 2387–2398. [CrossRef] [PubMed]

30. Comtois, P.; Nattel, S. Interactions between cardiac fibrosis spatial pattern and ionic remodeling on electrical wave propagation. In Proceedings of the 2011 Annual International Conference of the IEEE Engineering in Medicine and Biology Society, Boston, MA, USA, 30 August–3 September 2011; pp. 4669–4672.

31. Alonso, S.; Bär, M. Reentry near the percolation threshold in a heterogeneous discrete model for cardiac tissue. *Phys. Rev. Lett.* **2013**, *110*, 158101. [CrossRef] [PubMed]

32. Sano, T.; Takayama, N.; Shimamoto, T. Directional difference of conduction velocity in the cardiac ventricular syncytium studied by microelectrodes. *Circ. Res.* **1959**, *7*, 262–267. [CrossRef] [PubMed]

33. Valderrábano, M. Influence of anisotropic conduction properties in the propagation of the cardiac action potential. *Prog. Biophys. Mol. Biol.* **2007**, *94*, 144–168. [CrossRef] [PubMed]

34. Venteclef, N.; Guglielmi, V.; Balse, E.; Gaborit, B.; Cotillard, A.; Atassi, F.; Amour, J.; Leprince, P.; Dutour, A.; Clément, K.; et al. Human epicardial adipose tissue induces fibrosis of the atrial myocardium through the secretion of adipo-fibrokines. *Eur. Heart J.* **2015**, *36*, 795–805. [CrossRef] [PubMed]

35. Pellman, J.; Lyon, R.C.; Sheikh, F. Extracellular matrix remodeling in atrial fibrosis: Mechanisms and implications in atrial fibrillation. *J. Mol. Cell. Cardiol.* **2010**, *48*, 461–467. [CrossRef] [PubMed]

36. Szegedi, I.; Szapáry, L.; Csécsei, P.; Csanádi, Z.; Csiba, L. Potential Biological Markers of Atrial Fibrillation: A Chance to Prevent Cryptogenic Stroke. *BioMed Res. Int.* **2017**, *2017*, 8153024. [CrossRef] [PubMed]

37. Van Den Steen, P.E.; Dubois, B.; Nelissen, I.; Rudd, P.M.; Dwek, R.A.; Opdenakker, G. Biochemistry and molecular biology of gelatinase B or matrix metalloproteinase-9 (MMP-9). *Crit. Rev. Biochem. Mol. Biol.* **2002**, *37*, 375–536. [CrossRef] [PubMed]

38. Li, M.; Yang, G.; Xie, B.; Babu, K.; Huang, C. Changes in matrix metalloproteinase-9 levels during progression of atrial fibrillation. *J. Int. Med. Res.* **2014**, *42*, 224–230. [CrossRef] [PubMed]

39. Shao, Q.; Liu, H.; Ng, C.Y.; Xu, G.; Liu, E.; Li, G.; Liu, T. Circulating serum levels of growth differentiation factor-15 and neuregulin-1 in patients with paroxysmal non-valvular atrial fibrillation. *Int. J. Cardiol.* **2014**, *172*, e311–e313. [CrossRef] [PubMed]

40. Rosenberg, M.A.; Maziarz, M.; Tan, A.Y.; Glazer, N.L.; Zieman, S.J.; Kizer, J.R.; Ix, J.H.; Djousse, L.; Siscovick, D.S.; Heckbert, S.R.; et al. Circulating fibrosis biomarkers and risk of atrial fibrillation: The Cardiovascular Health Study (CHS). *Am. Heart J.* **2014**, *167*, 723–728. [CrossRef] [PubMed]

41. Funasaka, T.; Raz, A.; Nangia-Makker, P. Nuclear transport of galectin-3 and its therapeutic implications. *Semin. Cancer Biol.* **2014**, *27*, 30–38. [CrossRef] [PubMed]

42. Hughes, R.C. Mac-2: A versatile galactose-binding protein of mammalian tissues. *Glycobiology* **1994**, *4*, 5–12. [CrossRef] [PubMed]

43. Barondes, S.H.; Castronovo, V.; Cooper, D.N.; Cummings, R.D.; Drickamer, K.; Feizi, T.; Gitt, M.A.; Hirabayashi, J.; Hughes, C.; Kasai, K.; et al. Galectins: A family of animal beta-galactoside-binding lectins. *Cell* **1994**, *76*, 597–598. [CrossRef]

44. Akahani, S.; Nangia-Makker, P.; Inohara, H.; Kim, H.R.; Raz, A. Galectin-3: A novel antiapoptotic molecule with a functional BH1 (NWGR) domain of Bcl-2 family. *Cancer Res.* **1997**, *57*, 5272–5276. [PubMed]

45. Suthahar, N.; Meijers, W.C.; Silljé, H.H.W.; Ho, J.E.; Liu, F.T.; de Boer, R.A. Galectin-3 activation and inhibition in heart failure and cardiovascular disease: An update. *Theranostics* **2018**, *8*, 593–609. [CrossRef] [PubMed]

46. Li, L.C.; Li, J.; Gao, J. Functions of galectin-3 and its role in fibrotic diseases. *J. Pharmacol. Exp. Ther.* **2014**, *351*, 336–343. [CrossRef] [PubMed]

47. Mackinnon, A.C.; Gibbons, M.A.; Farnworth, S.L.; Leffler, H.; Nilsson, U.J.; Delaine, T.; Simpson, A.J.; Forbes, S.J.; Hirani, N.; Gauldie, J.; et al. Regulation of transforming growth factor-β1-driven lung fibrosis by galectin-3. *Am. J. Respir. Crit. Care Med.* **2012**, *185*, 537–546. [CrossRef] [PubMed]

48. Sharma, U.C.; Pokharel, S.; van Brakel, T.J.; van Berlo, J.H.; Cleutjens, J.P.; Schroen, B.; André, S.; Crijns, H.J.; Gabius, H.J.; Maessen, J.; et al. Galectin-3 marks activated macrophages in failure-prone hypertrophied hearts and contributes to cardiac dysfunction. *Circulation* **2004**, *110*, 3121–3128. [CrossRef] [PubMed]

49. Nabi, I.R.; Shankar, J.; Dennis, J.W. The galectin lattice at a glance. *J. Cell Sci.* **2015**, *128*, 2213–2219. [CrossRef] [PubMed]

50. Meng, X.M.; Tang, P.M.; Li, J.; Lan, H.Y. TGF-β/Smad signaling in renal fibrosis. *Front. Physiol.* **2015**, *6*, 82. [CrossRef] [PubMed]

51. Pauschinger, M.; Knopf, D.; Petschauer, S.; Doerner, A.; Poller, W.; Schwimmbeck, P.L.; Kuhl, U.; Schultheiss, H.P. Dilated cardiomyopathy is associated with significant changes in collagen type I/III ratio. *Circulation* **1999**, *99*, 2750–2756. [CrossRef] [PubMed]

52. Schiller, M.; Javelaud, D.; Mauviel, A. TGF-beta-induced SMAD signaling and gene regulation: Consequences for extracellular matrix remodeling and wound healing. *J. Dermatol. Sci.* **2004**, *35*, 83–92. [CrossRef] [PubMed]

53. Annes, J.P.; Munger, J.S.; Rifkin, D.B. Making sense of latent TGFbeta activation. *J. Cell Sci.* **2003**, *116*, 217–224. [CrossRef] [PubMed]

54. Desmouliere, A.; Geinoz, A.; Gabbiani, F.; Gabbiani, G. Transforming growth factor-beta 1 induces alpha-smooth muscle actin expression in granulation tissue myofibroblasts and in quiescent and growing cultured fibroblasts. *J. Cell Biol.* **1993**, *122*, 103–111. [CrossRef] [PubMed]

55. Bujak, M.; Ren, G.; Kweon, H.J.; Dobaczewski, M.; Reddy, A.; Taffet, G.; Wang, X.F.; Frangogiannis, N.G. Essential Role of Smad3 in Infarct Healing and in the Pathogenesis of Cardiac Remodeling. *Circulation* **2007**, *116*, 2127–2138. [CrossRef] [PubMed]

56. Lal, H.; Ahmad, F.; Zhou, J.; Yu, J.E.; Vagnozzi, R.J.; Guo, Y.; Yu, D.; Tsai, E.J.; Woodgett, J.; Gao, E.; et al. Cardiac fibroblast glycogen synthase kinase-3β regulates ventricular remodeling and dysfunction in ischemic heart. *Circulation* **2014**, *130*, 419–430. [CrossRef] [PubMed]

57. Gao, P.J.; Li, Y.; Sun, A.J.; Liu, J.J.; Ji, K.D.; Zhang, Y.Z.; Sun, W.L.; Marche, P.; Zhu, D.L. Differentiation of vascular myofibroblasts induced by transforming growth factor-beta1 requires the involvement of protein kinase Calpha. *J. Mol. Cell. Cardiol.* **2003**, *35*, 1105–1112. [CrossRef]

58. Zhou, H.Y.; Chen, W.D.; Zhu, D.L.; Wu, L.Y.; Zhang, J.; Han, W.Q.; Li, J.D.; Yan, C.; Gao, P.J. The PDE1A-PKCalpha signaling pathway is involved in the upregulation of alpha-smooth muscle actin by TGF-beta1 in adventitial fibroblasts. *J. Vasc. Res.* **2010**, *47*, 9–15. [CrossRef] [PubMed]

59. Villar, A.V.; Garcia, R.; Llano, M.; Cobo, M.; Merino, D.; Lantero, A.; Tramullas, M.; Hurle, J.M.; Hurle, M.A.; Nistal, J.F. BAMBI (BMP and activin membrane-bound inhibitor) protects the murine heart from pressure-overload biomechanical stress by restraining TGF-beta signaling. *Biochim. Biophys. Acta* **2013**, *1832*, 323–335. [CrossRef] [PubMed]

60. Calvier, L.; Miana, M.; Reboul, P.; Cachofeiro, V.; Martinez-Martinez, E.; de Boer, R.A.; Poirier, F.; Lacolley, P.; Zannad, F.; Rossignol, P.; et al. Galectin-3 mediates aldosterone-induced vascular fibrosis. *Arterioscler. Thromb. Vasc. Biol.* **2013**, *33*, 67–75. [CrossRef] [PubMed]

61. Yu, L.; Ruifrok, W.P.; Meissner, M.; Bos, E.M.; van Goor, H.; Sanjabi, B.; van der Harst, P.; Pitt, B.; Goldstein, I.J.; Koerts, J.A.; et al. Genetic and pharmacological inhibition of galectin-3 prevents cardiac remodeling by interfering with myocardial fibrogenesis. *Circ. Heart Fail.* **2013**, *6*, 107–117. [CrossRef] [PubMed]

62. Traber, P.G.; Zomer, E. Therapy of experimental NASH and fibrosis with galectin inhibitors. *PLoS ONE* **2013**, *8*, e83481. [CrossRef] [PubMed]

63. Ahmad, N.; Gabius, H.J.; André, S.; Kaltner, H.; Sabesan, S.; Roy, R.; Liu, B.; Macaluso, F.; Brewer, C.F. Galectin-3 precipitates as a pentamer with synthetic multivalent carbohydrates and forms heterogeneous cross-linked complexes. *J. Biol. Chem.* **2004**, *279*, 10841–10847. [CrossRef] [PubMed]

64. Takemoto, Y.; Ramirez, R.J.; Yokokawa, M.; Kaur, K.; Ponce-Balbuena, D.; Sinno, M.C.; Willis, B.C.; Ghanbari, H.; Ennis, S.R.; Guerrero-Serna, G.; et al. Galectin-3 Regulates Atrial Fibrillation Remodeling and Predicts Catheter Ablation Outcomes. *JACC Basic. Transl. Sci.* **2016**, *1*, 143–154. [CrossRef] [PubMed]

65. Petrov, V.V.; Fagard, R.H.; Lijnen, P.J. Stimulation of collagen production by transforming growth factor-beta1 during differentiation of cardiac fibroblasts to myofibroblasts. *Hypertension* **2002**, *39*, 258–263. [CrossRef] [PubMed]

66. De Boer, R.A.; van Veldhuisen, D.J.; Gansevoort, R.T.; Muller Kobold, A.C.; van Gilst, W.H.; Hillege, H.L.; Bakker, S.J.; van der Harst, P. The fibrosis marker galectin-3 and outcome in the general population. *J. Intern. Med.* **2012**, *272*, 55–64. [CrossRef] [PubMed]

67. De Couto, G.; Ouzounian, M.; Liu, P.P. Early detection of myocardial dysfunction and heart failure. *Nat. Rev. Cardiol.* **2010**, *7*, 334–344. [CrossRef] [PubMed]

68. Chow, S.L.; Maisel, A.S.; Anand, I.; Bozkurt, B.; de Boer, R.A.; Felker, G.M.; Fonarow, G.C.; Greenberg, B.; Januzzi, J.L., Jr.; Kiernan, M.S.; et al. Role of Biomarkers for the Prevention, Assessment, and Management of Heart Failure: A Scientific Statement From the American Heart Association. *Circulation* **2017**, *135*, e1054–e1091. [CrossRef] [PubMed]

69. Dong, R.; Zhang, M.; Hu, Q.; Zheng, S.; Soh, A.; Zheng, Y.; Yuan, H. Galectin-3 as a novel biomarker for disease diagnosis and a target for therapy (Review). *Int. J. Mol. Med.* **2018**, *41*, 599–614. [CrossRef] [PubMed]

70. Sciacchitano, S.; Lavra, L.; Morgante, A.; Ulivieri, A.; Magi, F.; De Francesco, G.P.; Bellotti, C.; Salehi, L.B.; Ricci, A. Galectin-3: One Molecule for an Alphabet of Diseases, from A to Z. *Int. J. Mol. Sci.* **2018**, *19*, 379. [CrossRef] [PubMed]

71. Ho, J.E.; Yin, X.; Levy, D.; Vasan, R.S.; Magnani, J.W.; Ellinor, P.T.; McManus, D.D.; Lubitz, S.A.; Larson, M.G.; Benjamin, E.J. Galectin 3 and incident atrial fibrillation in the community. *Am. Heart J.* **2014**, *167*, 729–734. [CrossRef] [PubMed]

72. Clementy, N.; Piver, E.; Benhenda, N.; Bernard, A.; Pierre, B.; Simeon, E.; Fauchier, L.; Pages, J.C.; Babuty, D. Galectin-3 in patients undergoing ablation of atrial fibrillation. *IJC Metabolic. Endocrine.* **2014**, *5*, 56–60. [CrossRef]

73. Szadkowska, I.; Wlazeł, R.N.; Migała, M.; Szadkowski, K.; Zielińska, M.; Paradowski, M.; Pawlicki, L. The association between galectin-3 and clinical parameters in patients with first acute myocardial infarction treated with primary percutaneous coronary angioplasty. *Cardiol. J.* **2013**, *20*, 577–582. [CrossRef] [PubMed]

74. Yalcin, M.U.; Gurses, K.M.; Kocyigit, D.; Canpinar, H.; Canpolat, U.; Evranos, B.; Yorgun, H.; Sahiner, M.L.; Kaya, E.B.; Hazirolan, T.; et al. The Association of Serum Galectin-3 Levels with Atrial Electrical and Structural Remodeling. *J. Cardiovasc. Electrophysiol.* **2015**, *26*, 635–640. [CrossRef] [PubMed]

75. Wachtell, K.; Lehto, M.; Gerdts, E.; Olsen, M.H.; Hornestam, B.; Dahlöf, B.; Ibsen, H.; Julius, S.; Kjeldsen, S.E.; Lindholm, L.H.; et al. Angiotensin II receptor blockade reduces new-onset atrial fibrillation and subsequent stroke compared to atenolol: The Losartan Intervention For End Point Reduction in Hypertension (LIFE) study. *J. Am. Coll. Cardiol.* **2005**, *45*, 712–719. [CrossRef] [PubMed]

76. Edsfeldt, A.; Bengtsson, E.; Asciutto, G.; Dunér, P.; Björkbacka, H.; Fredrikson, G.N.; Nilsson, J.; Goncalves, I. High plasma levels of Galectin-3 are associated with increased risk for stroke after carotid endarterectomy. *Cerebrovasc. Dis.* **2016**, *41*, 199–203. [CrossRef] [PubMed]

77. Dong, H.; Wang, Z.H.; Zhang, N.; Liu, S.D.; Zhao, J.J.; Liu, S.Y. Serum Galectin-3 level, not Galectin-1, is associated with the clinical feature and outcome in patients with acute ischemic stroke. *Oncotarget* **2017**, *8*, 109752–109761. [CrossRef] [PubMed]

78. Wang, A.; Zhong, C.; Zhu, Z.; Xu, T.; Peng, Y.; Xu, T.; Peng, H.; Chen, C.S.; Wang, J.; Ju, Z.; et al. Serum Galectin-3 and Poor Outcomes Among Patients With Acute Ischemic Stroke. *Stroke* **2018**, *49*, 211–214. [CrossRef] [PubMed]

79. Asberg, S.; Hijazi, Z.; Siegbahn, A.; Andersson, U.; Granger, C.B.; Hanna, M.; Horowitz, J.D.; Lopes, R.D.; Lindahl, B.; Wallentin, L. Galectin-3 is associated with worse clinical outcome in patients with atrial fibrillation: A substudy from the ARISTOTLE trial. *Eur. Heart J.* **2014**, *35*, 426.

80. Verma, A.; Mantovan, R.; Macle, L.; De Martino, G.; Chen, J.; Morillo, C.A.; Novak, P.; Calzolari, V.; Guerra, P.G.; Nair, G.; et al. Substrate and Trigger Ablation for Reduction of Atrial Fibrillation (STAR AF): A randomized, multicentre, international trial. *Eur. Heart J.* **2010**, *31*, 1344–1356. [CrossRef] [PubMed]

81. Kornej, J.; Hindricks, G.; Shoemaker, M.B.; Husser, D.; Arya, A.; Sommer, P.; Rolf, S.; Saavedra, P.; Kanagasundram, A.; Patrick Whalen, S.; et al. The APPLE score: A novel and simple score for the prediction of rhythm outcomes after catheter ablation of atrial fibrillation. *Clin. Res. Cardiol.* **2015**, *104*, 871–876. [CrossRef] [PubMed]

82. Wu, X.Y.; Li, S.N.; Wen, S.N.; Nie, J.G.; Deng, W.N.; Bai, R.; Liu, N.; Tang, R.B.; Zhang, T.; Du, X.; et al. Plasma galectin-3 predicts clinical outcomes after catheter ablation in persistent atrial fibrillation patients without structural heart disease. *Europace* **2015**, *17*, 1541–1547. [CrossRef] [PubMed]

83. Clementy, N.; Benhenda, N.; Piver, E.; Pierre, B.; Bernard, A.; Fauchier, L.; Pages, J.C.; Babuty, D. Serum Galectin-3 Levels Predict Recurrences after Ablation of Atrial Fibrillation. *Sci. Rep.* **2016**, *6*, 34357. [CrossRef] [PubMed]

84. Zhuang, J.; Wang, Y.; Tang, K.; Li, X.; Peng, W.; Liang, C.; Xu, Y. Association between left atrial size and atrial fibrillation recurrence after single circumferential pulmonary vein isolation: A systematic review and meta-analysis of observational studies. *Europace* **2012**, *14*, 638–645. [CrossRef] [PubMed]

85. Sramko, M.; Peichl, P.; Wichterle, D.; Tintera, J.; Weichet, J.; Maxian, R.; Pasnisinova, S.; Kockova, R.; Kautzner, J. Clinical value of assessment of left atrial late gadolinium enhancement in patients undergoing ablation of atrial fibrillation. *Int. J. Cardiol.* **2015**, *179*, 351–357. [CrossRef] [PubMed]

86. Kornej, J.; Schmidl, J.; Bollmann, A. Galectin-3 in atrial fibrillation: A novel marker of atrial remodeling or just bystander? *Am. J. Cardiol.* **2015**, *116*, 163. [CrossRef] [PubMed]

87. Mohanty, S.; Mohanty, P.; Di Biase, L.; Bai, R.; Pump, A.; Santangeli, P.; Burkhardt, D.; Gallinghouse, J.G.; Horton, R.; Sanchez, J.E.; et al. Impact of metabolic syndrome on procedural outcomes in patients with atrial fibrillation undergoing catheter ablation. *J. Am. Coll. Cardiol.* **2012**, *59*, 1295–1301. [CrossRef] [PubMed]

88. Lippi, G.; Cervellin, G.; Sanchis-Gomar, F. Galectin-3 in atrial fibrillation: Simple bystander, player or both? *Clin. Biochem.* **2015**, *48*, 818–822. [CrossRef] [PubMed]

89. Verma, A.; Jiang, C.Y.; Betts, T.R.; Chen, J.; Deisenhofer, I.; Mantovan, R.; Macle, L.; Morillo, C.A.; Haverkamp, W.; Weerasooriya, R.; et al. STAR AF II Investigators. Approaches to catheter ablation for persistent atrial fibrillation. *N. Engl. J. Med.* **2015**, *372*, 1812–1822. [CrossRef] [PubMed]

International Journal of
Molecular Sciences

MDPI

Review

Translational Implication of Galectin-9 in the Pathogenesis and Treatment of Viral Infection

Jenn-Haung Lai [1,2,*], Shue-Fen Luo [1], Mei-Yi Wang [1] and Ling-Jun Ho [3,*]

[1] Division of Allergy, Immunology, and Rheumatology, Department of Internal Medicine, Chang Gung Memorial Hospital, Chang Gung University, Tao-Yuan 33305, Taiwan; lsf00076@adm.cgmh.org.tw (S.-F.L.); meiyiwang800106@gmail.com (M.-Y.W.)
[2] Graduate Institute of Medical Science, National Defense Medical Center, Taipei 11490, Taiwan
[3] Institute of Cellular and System Medicine, National Health Research Institute, Zhunan 35053, Taiwan
[*] Correspondence: laiandho@gmail.com (J.-H.L.); lingjunho@nhri.org.tw (L.-J.H.);
Tel.: +886-2-8792-7135 (J.-H.L.); +886-2-8791-8382 (L.-J.H.);
Fax: +886-2-8792-7136 (J.-H.L.); +886-2-8791-8382 (L.-J.H.)

Received: 11 September 2017; Accepted: 6 October 2017; Published: 8 October 2017

Abstract: The interaction between galectin-9 and its receptor, Tim-3, triggers a series of signaling events that regulate immune responses. The expression of galectin-9 has been shown to be increased in a variety of target cells of many different viruses, such as hepatitis C virus (HCV), hepatitis B virus (HBV), herpes simplex virus (HSV), influenza virus, dengue virus (DENV), and human immunodeficiency virus (HIV). This enhanced expression of galectin-9 following viral infection promotes significant changes in the behaviors of the virus-infected cells, and the resulting events tightly correlate with the immunopathogenesis of the viral disease. Because the human immune response to different viral infections can vary, and the lack of appropriate treatment can have potentially fatal consequences, understanding the implications of galectin-9 is crucial for developing better methods for monitoring and treating viral infections. This review seeks to address how we can apply the current understanding of galectin-9 function to better understand the pathogenesis of viral infection and better treat viral diseases.

Keywords: translational medicine; galectin-9; pathogenesis; treatment; viral infection

---

1. Introduction of Galectin-9

Galectins, as part of a group of pattern recognition receptors, are highly conserved lectins and are widely expressed in different tissues cells and immune cells [1]. The characteristic structure of galectins includes carbohydrate recognition domains of approximately 130 amino acids in length that bind β-galactosides [1]. To date, 15 galectins have been identified and were found to contain various numbers of carbohydrate recognition domains and associated structures [2]. Galectin-9 was first cloned and characterized as a T cell-derived eosinophil-specific chemoattractant, and was found to be involved in various cellular processes such as differentiation, aggregation, adhesion, and death, as well as various phases of immune responses [3–5]. Galectin-9 contains two distinct carbohydrate recognition domains connected by a linker peptide and is a natural ligand for the T cell immunoglobulin domain and mucin domain protein 3 (Tim-3) [6]. Previous studies revealed the increased expression of galectin-9 in different target cells in response to various stimuli, such as mitogens, agonists of toll-like receptors, and pro-inflammatory cytokines like interferon-γ (IFN-γ) and interleukin-1β (IL-1β) [5,7]. Interestingly, stimulation with interferon-γ, but not with IL-1β or lipopolysaccharide, effectively induces galectin-9 expression in macrophages derived from macrophage colony-stimulating factor-treated CD14[+] primary monocytes [8]. In contrast, the intracellular signals that regulate galectin-9 expression are largely unknown. A previous study showed that stimulating phorbol 12-myristate 13-acetate results in the upregulation of galectin-9 expression and that this effect can be partially blocked by treatment with inhibitors of either protein kinase C inhibitor or matrix metalloproteinase [9]. Aside from endogenous and exogenous regulatory factors, the differentiation process from monocyte to macrophage can also induce the production of galectin-9 [7]. However, galectin-9 may play a role in inducing the maturation of monocyte-derived dendritic cells [10]. Harwood et al. demonstrated that maximal galectin-9 induction in monocytes requires direct contact with, or proximity to, the interacting cells [7].

Many signals can induce the expression and production of galectin-9. As such, the induced expression of galectin-9 has been widely demonstrated in infections caused by several different viruses, including hepatitis C virus (HCV) [8,11,12], hepatitis B virus (HBV) [13], herpes simplex virus (HSV) [14,15], influenza virus [16], Epstein-Barr virus [17], dengue virus (DENV) [18,19], and human immunodeficiency virus (HIV) [20]. Although the level varies, the increased expression or production of galectin-9 can be observed following the infection of a variety of tissue cells and immune effector cells. Li et al. reported the variable expression levels of galectin-9 on different subsets of antigen-presenting cells, such as Kupffer cells, myeloid dendritic cells, and plasmacytoid dendritic cells in patients with HBV-associated hepatocellular carcinoma (HCC), with the highest expression observed in Kupffer cells [13]. In addition, $CD14^{low/high}CD16^+$ monocytes produce higher amounts of galectin-9 when compared to classic $CD14^+CD16^-$ monocytes, which are associated with the severity of liver injury and fibrosis in patients with chronic HBV infection [7,21]. The enhanced expression of galectin-9 promotes significant changes in behaviors of the virus-infected cells, and the resulting events may tightly correlate with immunopathogenic processes of the viral disease. As anticipated, the variable expression of galectin-9 in different populations of immune effector cells may lead to different disease pathogenesis. The galectin-9-mediated effects can be even more complex given that galectin-9 exists and functions both intracellularly and extracellularly to regulate inflammatory responses [6].

Tim-3 is the most well-known binding receptor for galectin-9, and its expression is upregulated on the surface of immune effector cells such as $CD4^+$ and $CD8^+$ T cells in patients with HIV infection [22,23]. A mutual regulation of galectin-9 and Tim-3 has been reported: the knockdown of galectin-9 downregulates Tim-3 levels, and the knockdown of Tim-3 inhibits galectin-9 secretion in U-937 cells [24]. However, previous studies suggest the presence of Tim-3-independent galectin-9-mediated mechanisms [25,26], and thus, Tim-3 may not be the only receptor for galectin-9. Intriguingly, given that the secretion of galectin-9 is dependent on Tim-3 in human acute myeloid leukemia cells, this Tim-3-independent mechanism does not exist in lymphoid cells [24]. In contrast, the galectin-9 ligation-induced changes of many transcriptional and functional genes are independent of Tim-3 in natural killer (NK) cells [12]. Clayton et al. reported that galectin-9 binds to multiple immune accessory molecules, such as the receptor phosphatases CD45 and CD148; moreover, by binding to these two molecules, galectin-9 enhances their interaction with Tim-3 within the CD3 signaling complexes [27]. Galectin-9 binds to CD44 and blocks the interaction between CD44 and hyaluronan, which consequently inhibits airway inflammation and airway hyperresponsiveness [28]. Program death-1 (PD-1), rather than Tim-3, is more crucial for galectin-9 to regulate T cell function and migration in HIV infection [29]. Furthermore, treatment with galectin-9 induces apoptosis in plasma cells purified from MRL/lpr lupus-prone mice, and the effect cannot be prevented with anti-Tim-3 monoclonal antibodies (mAb); this suggests the involvement of a non-Tim-3 receptor for galectin-9-mediated apoptosis in plasma cells [30]. Lhuillier et al. also observed that galectin-9 induces apoptosis of Jurkat T cells in a Tim-3-independent manner [31]. Biochemical and biophysical studies on Tim-3 revealed a unique structural feature that is conserved in TIM family members that may mediate a galectin-9-independent binding process [32].

In addition to galectin-9, other molecules may serve as Tim-3 ligands. According to Nakayama et al., Tim-3–immunoglobulin fusion protein can weakly but substantially bind to phosphatidylserine, but not to phosphatidylethanolamine, phosphatidylinositol, or phosphatidylcholine on apoptotic cells through an FG loop in the immunoglobulin-like variable (IgV) domain [33]. Tim-3 binds to carcinoembryonic antigen-related cell adhesion molecule 1 to induce T cell tolerance and exhaustion [34]. Given that

Tim-3 suppresses innate responses to nucleic acids, the mechanisms appear to involve the interaction between Tim-3 and alarmin HMGB1, which inhibits the recruitment of nucleic acids into the endosomes of dendritic cells [35].

TIM family proteins play roles in enhancing the entry of a wide range of enveloped viruses through the interaction between the IgV domain of TIM proteins and the virion-associated phosphatidylserine [36]. However, not all TIM family members effectively promote viral entry; Tim-3 contains an IgV domain that binds phosphatidylserine, but it cannot effectively enhance viral entry. Overexpression of human Tim-3 causes HeLa cells to become partially susceptible to hepatitis A viral (HAV) infection [37]. In addition, in sharp contrast to the overexpression of Tim-1 or Tim-4, overexpression of Tim-3 in 293T cells only mildly increases the percentage of DENV-infected cells [38]. Many factors may determine the binding avidity between lectins and glycans, such as the structure and the density of glycan epitopes on glycoproteins and their cell surface density [39]. Accordingly, the interaction between different TIM proteins and viruses provides a good example to explain this phenomenon.

2. Galectin-9/Tim-3 Interaction Regulates the Immune Response

Homeostasis of the immune system requires both positive and negative machineries to tightly control various phases of cellular interaction or cell-environment communication. The binding between galectins and glycans leads to sequential signaling events and plays important roles in immune tolerance and inflammation [40]. Depending on the specific model or target examined, galectin-9 may either induce or suppress inflammation [41,42]. Expression of a proteolysis-resistant galectin-9 construct induces apoptosis in hematological, dermatological, and gastrointestinal malignant cells [43] (Figure 1). When compared to control mice, galectin-9-deficient mice are more susceptible to developing collagen-induced arthritis, which is reflected by an increase in the numbers of CD4$^+$ Tim-3$^+$ T cells and a decrease in the numbers of Foxp3$^+$ regulatory T cells (Tregs) in these mice. In vitro treatment of naïve T cells with galectin-9 promotes differentiation into Tregs and inhibits differentiation into T helper (Th)17 cells [44]. In addition, the interaction between galectin-9 and Tim-3 inhibits T cell proliferation and triggers T cell apoptosis [22]. Moreover, treatment with recombinant human galectin-9 inhibits the release of Th1/Th2/Th17 cytokines in co-cultures of autologous monocyte-derived dendritic cells and peripheral blood lymphocytes from healthy controls and from patients with autoimmune thyroid disease [45]. The negative correlation between galectin-9 mRNA levels in peripheral blood mononuclear cells of patients with rheumatoid arthritis and the disease activity provides additional evidence that galectin-9 is anti-inflammatory [46]. A similar conclusion was also reached in a xenograft-interaction model. A co-culture of human galectin-9-expressing porcine kidney epithelial cells and M1-differentiated THP-1 cells suppresses the production of pro-inflammatory cytokines and reduces the cytotoxic effects of the cells [47]. Together, these results from different studies suggest that galectin-9-mediated signaling events may lead to the inhibition of inflammatory responses [48,49].

In contrast, some results suggest that galectin-9 may mediate pro-inflammatory responses. Galectin-9 was originally found to be produced by T cells to serve as an eosinophil chemoattractant that mediates pro-inflammatory responses [3–5]. The concentrations of galectin-9 produced by activated astrocytes in cerebrospinal fluid positively correlate with the number of lesions on T1 weighted images of MRI, but not with gadolinium enhancing lesions, in patients with secondary progressive multiple sclerosis [50]. Treatment with recombinant human galectin-9 induces the production of IFN-γ in a Tim-3-overexpressing NK cell line, and in Tim-3$^+$ primary NK cells induced with low-dose IL-12 and IL-18 [51]. However, galectin-9 ligation has also been shown to reduce the proportion of IFN-γ-producing NK cells stimulated with IL-12/IL-15, downregulate NK cell-mediated cytotoxicity, and inhibit lymphokine-activated killing. Interestingly, these observed effects appear to be independent of Tim-3 [12], and together, these studies suggest that the effects of galectin-9 may differ depending on the interacting molecules. Furthermore, galectin-9 induces T helper cells to produce pro-inflammatory cytokines in a dose-dependent Tim-3-independent manner [25]. By enhancing

IFN-γ-induced pro-inflammatory effects, galectin-9 potently mediates neuroinflammation in the mouse hippocampus [52].

Structural biology studies have investigated the differential roles of galectin-9-mediated immune responses. Crystal or X-ray structural analysis of both the N- and C-terminal carbohydrate recognition domains of galectin-9 has been conducted [53,54]. The studies indicate that the N-terminal and the C-terminal carbohydrate recognition domains have different binding specificities in recognizing branched and α 2-3-sialylated oligosaccharides [54]. These differences in binding specificity were further investigated by generating various mutant, wild-type, or homodimer constructs containing N- or C-terminal regions of the carbohydrate recognition domains of galectin-9 [55]. The results showed that all of the examined galectin-9 constructs differentially regulate dendritic cell activation and T cell death: the C-terminal carbohydrate recognition domain is more potent in inducing T cell death, and the N-terminal carbohydrate recognition domain is more effective in activating dendritic cells to induce the production of pro-inflammatory cytokines. According to computer analysis, these constructs also preserve different patterns and affinities in binding to Tim-3 [55].

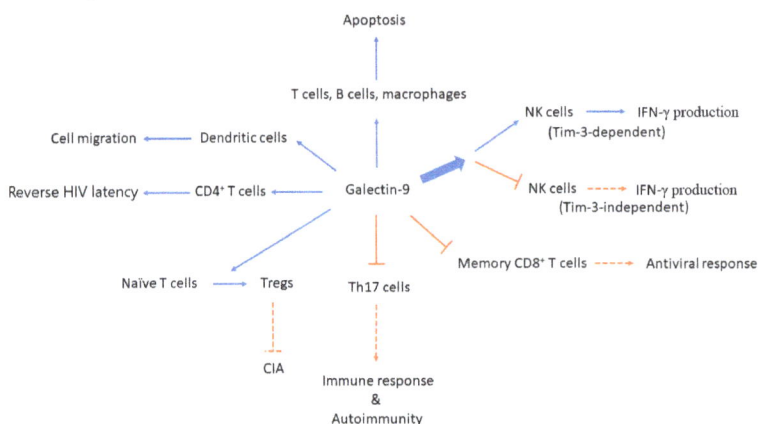

Figure 1. A simplified diagram illustrates the effects of galectin-9 on immune cells and the potential consequences. Expression of a proteolysis-resistant galectin-9 construct induces apoptosis in immune cells as well as in malignant cells. Galectin-9 causes the differentiation of naive T cells to Tregs and inhibits differentiation towards T helper (Th)17 cells. Both effects lead to the inhibition of the development of immune response and autoimmune arthritis in a CIA animal model. While treatment with recombinant human galectin-9 induces IFN-γ production in a Tim-3-overexpressing NK cell line and in Tim-3⁺ primary NK cells induced with low-dose IL-12 and IL-18, galectin-9 ligation has also been shown to reduce the proportion of IFN-γ-producing NK cells stimulated with IL-12/IL-15 in a Tim-3-independent manner. Galectin-9 treatment extensively regulates gene expression of molecules involved in the processes of human immunodeficiency virus (HIV) latency. Galectin-9 may also regulate dendritic cell migration in the example of dengue virus (DENV) infection. NK, natural killer; CIA, collagen-induced arthritis; IFN-γ, interferon-γ; and, Tregs, regulatory T cells. Blue arrow: induce; bold blue arrow: simply means there are two diverging effects out of this direction; orange T-bar arrow/orange dotted T-bar arrow: inhibit; orange arrow with dotted line: reduce.

3. Galectin-9 in Viral Pathogenesis

The effects of galectin-9 greatly affect the immune defense mechanisms against viral infection. Some of the galectin-9-mediated effects have been investigated in several examples of viral infection. Depending on the target cells and the mechanisms of disease pathogenesis in individual viral infections examined, the roles of galectin-9 in viral pathogenesis may moderately differ. Below, we will discuss some reported features of galectin-9/Tim-3 in the immunopathogenesis of different viral infections.

A previous study found that the reduction in the numbers of CD4$^+$ T helper cells and the impaired virus-specific CD8$^+$ T cell response are both major factors responsible for persistent infection in HCV-infected patients [56]. To identify factors contributing to the impaired virus-specific CD8$^+$ T cell response, the authors examined the expression of two negative regulatory receptors, namely, Tim-3 and PD-1, on cytotoxic T cells (CTLs) in HCV-infected patients. The results showed that the population of PD-1$^-$Tim-3$^-$ HCV-specific CTLs significantly outnumbers the population of PD-1$^+$Tim-3$^+$ CTLs in patients with acute resolving infection [11]. Moreover, the population of PD-1$^+$Tim-3$^+$ T cells are enriched within the central memory T cell subset and within the liver [11]. While inhibiting either PD-1 or Tim-3 similarly results in the increased proliferation of HCV-specific CTLs, only the suppression of Tim-3 increases cytotoxicity against a hepatocyte cell line expressing cognate HCV epitopes [11]. Kared et al. also demonstrated that, during acute HCV infection, the progression to persistent infection is associated with increased plasma levels of galectin-9 and the expansion of galectin-9-expressing Tregs [57].

A co-culture of purified healthy CD4$^+$ T cells and HCV-infected hepatocytes, which express higher levels of galectin-9 and transforming growth factor-β than that of control cells, promotes CD25$^+$Foxp3$^+$ Tregs differentiation [58] in a Tim-3-dependent manner. However, galectin-9 induces apoptosis of HCV-specific CTLs [8]. In addition, co-incubation with HCV-infected hepatocytes also induces apoptosis of CD4$^+$CD25$^+$Foxp3$^-$ effector T cells [58]. Moreover, the interaction between Tim-3 on the surface of virus-specific CD8$^+$ T cells and galectin-9 on Kupffer cells may induce apoptosis of T cells infiltrating the HCV-infected liver [59]. The work from these studies suggests that galectin-9 negatively impacts the immune response, thus causing liver damage during HCV infection. Nevertheless, according to Mengshol et al., treating hepatic and peripheral blood mononuclear cells with galectin-9 induces the production of pro-inflammatory cytokines, including IL-1β, TNF-α, and IFN-γ; however, galectin-9 also induces anti-inflammatory cytokines, such as IL-4, IL-10, and IL-13, in peripheral mononuclear cells [8].

In response to HSV infection, when compared to the wild-type animals, galectin-9-deficient mice have increased the frequencies and numbers of IFN-γ^+TNF-α^+ T cells suggesting stronger virus-specific CD8$^+$ T cell responses [14]. In addition, the galectin-9 knockout mice but not the control mice developed sustained virus-specific CD8$^+$ memory T cell responses [14]. Administering α-lactose to HSV-infected animals to block galectin-9 not only reproduces the effects observed in galectin-9 knockout, but also diminishes Tregs responses in animals [14]. In contrast, intraperitoneal injection of galectin-9 reduces the HSV-specific CD8$^+$ T cell response and delays viral clearance [14]. In a different animal model, HSV-infected galectin-9 knockout mice have reduced numbers of Tregs in trigeminal ganglion as compared to the control animals [15]. In contrast, the galectin-9 knockout mice have stronger CD8$^+$ T cell responses in trigeminal ganglion [15]. Although spontaneous reactivation occurs equally well in ex vivo trigeminal ganglion cultures of both wild-type and galectin-9 knockout mice, the reactivation was delayed in galectin-9 knockout mice when compared to wild-type mice [15].

A previous study revealed that, compared to cells in pre-reactivation status, there is a significant increase in galectin-9 transcription in the peripheral blood mononuclear cells of transplant recipients with post-reactivation of human cytomegalovirus infection [60]. Additionally, this study did not find any significant change in the expression of galectin-9 in transplant recipients without reactivation of viral infection [60], but did find that IFN-β is likely to play roles in mediating this effect [60]. Cytomegalovirus infection also induces higher production of IFN-γ from hepatic NK cells in galectin-9 knockout mice than in control mice [12]. Interestingly, the mechanisms involved in cytomegalovirus infection are reminiscent of the influenza A virus infection, such that galectin-9 knockout mice have a stronger virus-specific CD8$^+$ T cell response that develops during the acute phase than do control mice [16]. In addition, the plasma levels of virus-specific immunoglobulins, including IgM, IgG, and IgA, increase in combination with enhanced Ab-secreting B220^{int} CD138$^+$ cells [16]. Furthermore, better viral clearance is observed in galectin-9 knockout mice than in wild-type mice [16].

Of note, in addition to regulating immunocompetency and apoptosis of T cells, there have been recent publications that have comprehensively examined galectin-9 function. In one study delineating the mechanisms of establishing HIV latency, recombinant galectin-9 treatment extensively regulates the gene expression of molecules involved in key transcription initiation, chromatin remodeling, and promoter-proximal pausing processes, which are critical in the processes of HIV latency [61]. In addition, the interaction between galectin-9/Tim-3 signaling and follicular CD4[+] T helper cells also plays a role mediating the persistence of chronic HCV infection [62]. Galectin-9 may have a role in regulating dendritic cell migration. A recent study found that galectin-9 knockdown impairs DENV-induced dendritic cell migration towards the chemoattractants CCL19 and CCL21, and that the receptor of these two chemokines CCR7 was not affected [63]. This suggests that defective migration of galectin-9-deficient cells may be caused by the inhibition of CCR7-mediated downstream signaling [63].

4. Galectin-9/Tim-3 Signaling-Related Markers Predict the Severity of Viral Infection and Prognosis

A population of HBV-infected patients may progress to develop HCC. Blocking the galectin-9/Tim-3 signaling pathway increases the proliferation and cytokine production from tumor-infiltrating Tim-3[+] T cells [13]. Importantly, the numbers of Tim-3[+] tumor-infiltrating cells are negatively associated with survival in patients with HBV-associated HCC [13]. The study raises an interesting observation that the galectin-9/Tim-3 expression may serve as a useful prognostic marker in patients with HBV-associated HCC [13].

The increased expression of galectin-9 in peripheral blood mononuclear cells positively correlates with viremia and negatively correlates with CD4[+] T cell counts in patients with HIV infection [20]. However, although the loss of Tim-3[+] NK cells over time is more pronounced in HIV-infected patients with low CD4[+] T cell counts, there is no difference in plasma galectin-9 levels between HAART (highly active antiretroviral therapy)-treated and untreated subjects with progressive or controlled HIV infection [20]. In contrast, Tandon et al. found that the elevation of plasma galectin-9 is not only detected in the acute stage of HIV infection, but is also detected in chronic HIV-infected patients during suppressive antiretroviral therapy and in a select number of control patients [64]. Statistical analysis revealed that plasma galectin-9 levels not only correlate with viral RNA loads in CD4[+] T cells in HIV-infected individuals, but also associate with the quantity and binding avidity of circulating anti-HIV antibodies [61]. In addition, the plasma levels of Tim-3 increase in patients with early and chronic untreated HIV infection [65]. Furthermore, plasma Tim-3 levels and the frequency of Tim-3-expressing plasmacytoid dendritic cells in patients with HIV infection were found to positively correlate with viral load and to negatively correlate with CD4[+] T cell counts [65,66]. Tim-3-positive plasmacytoid dendritic cells that produce IFN-α or TNF-α in response to TLR7 or TLR9 agonist stimulation were identified infrequently [66]. Thus, Tim-3 can serve as a useful biomarker of dysfunctional plasmacytoid dendritic cells in HIV-infected patients.

The serum levels of galectin-9 but not galectin-1 or galectin-3 increase in DENV-infected patients [18]. However, there is no significant correlation between serum levels of galectin-9 and the white blood cell or platelet count [18]. Studies from another group of researchers demonstrated that serum galectin-9 levels are much higher in patients with dengue hemorrhagic fever than in patients with dengue fever only [67]. In these studies, stepwise discriminative analysis on serum levels of multiple cytokines and chemokines was then undertaken to determine the possible differences between dengue hemorrhagic fever patients and dengue fever patients. The results showed that the combined analysis of several markers, including eotaxin, galectin-9, IFN-α2, and monocyte chemoattractant protein-1, can help to detect 92% of dengue hemorrhagic fever and 79.3% of dengue fever patients. Importantly, serum levels of galectin-9 were found to significantly correlate with the levels of several pro-inflammatory mediators, such as IL-1α, IL-8, IFN-γ-inducible protein 10, and vascular endothelial growth factor, during the critical phase of infection. Even during the recovery phase, there is a significant association between galectin-9 and several other DENV infection-induced mediators such as epidermal growth factor, IL-10, IL-8, and vascular endothelial growth factor [67]. Moreover, the knockdown of galectin-9 in dendritic cells results in the suppression of IL-12p40 production [63].

5. Targeting Galectin-9/Tim-3 in Treating Viral Infection

Given the broad-spectrum roles of Tim-3 and galectin-9 in regulating immune responses, the collected data indicate that targeting Tim-3 or galectin-9 may have a potential therapeutic use against viral infection [68]. The administration of Tim-3 Fc fusion protein significantly rescues a proportion of CD8$^+$ T cells to produce IFN-γ and TNF-α when challenged with HBV peptides [59]. In addition, blocking Tim-3 rescues HBV-specific cytotoxic CD8$^+$ T cell and IL-2-producing CD4$^+$ T cell responses better than blocking PD ligand-1/2 [59]. Furthermore, blocking the galectin-9/Tim-3 interaction provides complementary effects to PD-1 inhibition in inducing HBV-specific T cells [59]. In response to influenza A virus challenge, blocking galectin-9 signaling by Tim-3 fusion protein treatment effectively induces and amplifies the intensity and quality of virus-specific CD8$^+$ T cell responses in mice [16]. All together, these results suggest that disrupting the galectin-9/Tim-3 interaction can improve the immune response against viruses and may be a useful approach in developing therapeutics for HBV infection [69].

The treatment of HSV1-infected mice with MAbT25, a monoclonal antibody that induces the expansion of cells expressing the TNF receptor superfamily member 25 and the Tregs population. Recombinant galectin-9 significantly reduces the severity of stromal keratitis [70] by reducing the number of Th1 cells, reducing the total number of CD4$^+$ T cells, upregulating anti-inflammatory cytokines, such as transforming growth factor-β and IL-10, and downregulating pro-inflammatory cytokines, such as IFN-γ and IL-6 [70]. The administration of galectin-9 to mice with HSV-induced Behçet's disease inhibits pro-inflammatory cytokine production, attenuates inflammation, and reduces disease severity; however, galectin-9 administration also increases the number of Tregs [71]. Using a gene delivery approach, the introduction of a recombinant adenovirus 9-galectin-9 adenoviral plasmid intra-nasally into respiratory syncytial virus-infected mice results in the significant reduction of viral load, mucus secretion, and lung pathology [72]. Investigating the underlying mechanisms reveals several possible contributing factors such as the inhibition of Th17 cell production, induction of Tregs expansion, and regulation of CD8$^+$ T cell apoptosis [72].

The interaction between soluble galectin-9 and Tim-3 expressed on the surface of activated CD4$^+$ T cells results in the reduction of HIV-1 coreceptor expression and renders these cells less susceptible to HIV-1 infection and replication [73]. Because Tim-3 is consistently expressed on dysfunctional T cells during chronic HIV infection, recombinant galectin-9 administration provides another therapeutic option for HIV infection. Treatment with either anti-human Tim-3 or anti-human galectin-9 mAb suppresses bacterial colony forming units in co-cultures of monocyte-derived macrophages infected by mycobacterium tuberculosis and T cells from HIV-infected patients [23]. Given the high risk of mycobacterium tuberculosis infection in HIV-infected patients, especially in endemic areas, the results suggest the beneficial effects of inhibiting galectin-9/Tim-3 interaction in controlling mycobacterium tuberculosis growth in HIV-infected patients with co-infection by mycobacterium tuberculosis. Furthermore, recombinant galectin-9 treatment reverses HIV latency in vitro in the J-Lat HIV latency model [61]. In ex vivo experiments, the same group of researchers also demonstrated that recombinant galectin-9 reverses HIV latency in primary CD4$^+$ T cells from HIV-infected and antiretroviral therapy-suppressed individuals with a potency stronger than vorinostat, a compound that inhibits histone deacetylase. Moreover, combining recombinant galectin-9 and a bromodomain inhibitor JQ1 was reported to have synergistic antiviral activity [61]. The recombinant galectin-9 administration also induces the expression of several anti-HIV host restriction factors, reduces the infectivity of progeny virus, and importantly, minimizes the possibility of recovery of the HIV reservoir once latency is reversed [61].

6. Perspectives

Administering galectin-9 to induce cellular apoptosis is a well-accepted approach in cancer therapy [43]; however, the opportunity to apply this concept for antiviral therapy remains to be investigated. The great challenge of this idea is that, in contrast to inducing apoptosis in cancer cells, causing cell death in virus-infected cells may result in severe tissue damage and thus is not favorable. According to structural analysis, different regions of galectin-9 may mediate different functions in terms of promoting Tregs or effector T cells activity. By introducing the appropriate galectin-9 component, we may be able to enhance the effector activity to fight against viruses or to reduce the generation of Tregs activity to increase immunity [55]. Along with the growing knowledge about this molecule, targeting galectin-9 may provide a direction of applying glycobiology knowledge for the treatment of viral infection. Furthermore, targeting galectin-9 may highlight a potential benefit for certain diseases, like DENV infection, for which the development of therapeutics has been challenging [74]. An additional advantage of applying galectin-9 for a therapeutic purpose is gained by the relatively clear pharmacokinetics of galectin-9 when introduced by the subcutaneous or intraperitoneal route in mice [44].

Overall, much remains to be understood about the galectin-9/Tim-3 signaling pathway. A previous study revealed that the galectin-9/Tim-3 interaction results in phosphorylation of the residues Y256 and Y263 in the Tim-3 tail and leads to the release of Bat3, thus initiating the Tim-3-mediated inhibitory signal [75]. The study also demonstrated that Bat3 can promote IFN-γ production and rendering Th1 cells resistant to galectin-9-mediated cell death [75]. Furthermore, Bat3 suppresses the development of T cell exhaustion, a phenomenon observed in chronic viral infection, and promotes the differentiation of effector T cells, which subsequently proliferate and produce IL-2 [75]. Because galectin-9 is a secreted and a resident intracellular molecule, additional studies investigating galectin-9-dependent intracellular signaling events may be helpful in delineating the different consequences generated in galectin-9-mediated effects. Given the complex consequences on immune responses following galectin-9/Tim-3 interaction, appropriate strategies targeting galectin-9 require more clear knowledge about the molecular mechanisms of galectin-9/Tim-3-mediated immune responses.

Acknowledgments: This work was supported in part by grants from the Ministry of Science and Technology (MOST 105-2314-B-182A-136-MY2) and Chang Gung Memorial Hospital (CMRPG3F0571 and CMRPG3E2161), Taiwan.

Author Contributions: Jenn-Haung Lai and Ling-Jun Ho wrote the manuscript; Shue-Fen Luo and Mei-Yi Wang helped in discussing and organizing the report.

Conflicts of Interest: The authors declare no conflict of interest.

References

1. Vasta, G.R. Galectins as pattern recognition receptors: Structure, function, and evolution. *Adv. Exp. Med. Biol.* **2012**, *946*, 21–36. [PubMed]
2. Nio-Kobayashi, J. Tissue- and cell-specific localization of galectins, β-galactose-binding animal lectins, and their potential functions in health and disease. *Anat. Sci. Int.* **2017**, *92*, 25–36. [CrossRef] [PubMed]
3. Wada, J.; Kanwar, Y.S. Identification and characterization of galectin-9, a novel β-galactoside-binding mammalian lectin. *J. Biol. Chem.* **1997**, *272*, 6078–6086. [CrossRef] [PubMed]
4. Hirashima, M. Ecalectin/galectin-9, a novel eosinophil chemoattractant: Its function and production. *Int. Arch. Allergy Immunol.* **2000**, *122* (Suppl. 1), 6–9. [CrossRef] [PubMed]
5. Hirashima, M.; Kashio, Y.; Nishi, N.; Yamauchi, A.; Imaizumi, T.A.; Kageshita, T.; Saita, N.; Nakamura, T. Galectin-9 in physiological and pathological conditions. *Glycoconj. J.* **2004**, *19*, 593–600. [CrossRef] [PubMed]
6. Rabinovich, G.A.; Croci, D.O. Regulatory circuits mediated by lectin-glycan interactions in autoimmunity and cancer. *Immunity* **2012**, *36*, 322–335. [CrossRef] [PubMed]

7. Harwood, N.M.; Golden-Mason, L.; Cheng, L.; Rosen, H.R.; Mengshol, J.A. HCV-infected cells and differentiation increase monocyte immunoregulatory galectin-9 production. *J. Leukoc. Biol.* **2016**, *99*, 495–503. [CrossRef] [PubMed]

8. Mengshol, J.A.; Golden-Mason, L.; Arikawa, T.; Smith, M.; Niki, T.; McWilliams, R.; Randall, J.A.; McMahan, R.; Zimmerman, M.A.; Rangachari, M.; et al. A crucial role for kupffer cell-derived galectin-9 in regulation of T cell immunity in hepatitis c infection. *PLoS ONE* **2010**, *5*, e9504. [CrossRef]

9. Chabot, S.; Kashio, Y.; Seki, M.; Shirato, Y.; Nakamura, K.; Nishi, N.; Nakamura, T.; Matsumoto, R.; Hirashima, M. Regulation of galectin-9 expression and release in jurkat T cell line cells. *Glycobiology* **2002**, *12*, 111–118. [CrossRef] [PubMed]

10. Dai, S.Y.; Nakagawa, R.; Itoh, A.; Murakami, H.; Kashio, Y.; Abe, H.; Katoh, S.; Kontani, K.; Kihara, M.; Zhang, S.L.; et al. Galectin-9 induces maturation of human monocyte-derived dendritic cells. *J. Immunol.* **2005**, *175*, 2974–2981. [CrossRef] [PubMed]

11. McMahan, R.H.; Golden-Mason, L.; Nishimura, M.I.; McMahon, B.J.; Kemper, M.; Allen, T.M.; Gretch, D.R.; Rosen, H.R. Tim-3 expression on PD-1+ HCV-specific human CTLs is associated with viral persistence, and its blockade restores hepatocyte-directed in vitro cytotoxicity. *J. Clin. Invest.* **2010**, *120*, 4546–4557. [CrossRef] [PubMed]

12. Golden-Mason, L.; McMahan, R.H.; Strong, M.; Reisdorph, R.; Mahaffey, S.; Palmer, B.E.; Cheng, L.; Kulesza, C.; Hirashima, M.; Niki, T.; et al. Galectin-9 functionally impairs natural killer cells in humans and mice. *J. Virol.* **2013**, *87*, 4835–4845. [CrossRef] [PubMed]

13. Li, H.; Wu, K.; Tao, K.; Chen, L.; Zheng, Q.; Lu, X.; Liu, J.; Shi, L.; Liu, C.; Wang, G.; et al. Tim-3/galectin-9 signaling pathway mediates T-cell dysfunction and predicts poor prognosis in patients with hepatitis b virus-associated hepatocellular carcinoma. *Hepatology* **2012**, *56*, 1342–1351. [CrossRef] [PubMed]

14. Sehrawat, S.; Reddy, P.B.; Rajasagi, N.; Suryawanshi, A.; Hirashima, M.; Rouse, B.T. Galectin-9/Tim-3 interaction regulates virus-specific primary and memory CD8 T cell response. *PLoS Pathog.* **2010**, *6*, e1000882. [CrossRef] [PubMed]

15. Reddy, P.B.; Sehrawat, S.; Suryawanshi, A.; Rajasagi, N.K.; Mulik, S.; Hirashima, M.; Rouse, B.T. Influence of galectin-9/Tim-3 interaction on herpes simplex virus-1 latency. *J. Immunol.* **2011**, *187*, 5745–5755. [CrossRef] [PubMed]

16. Sharma, S.; Sundararajan, A.; Suryawanshi, A.; Kumar, N.; Veiga-Parga, T.; Kuchroo, V.K.; Thomas, P.G.; Sangster, M.Y.; Rouse, B.T. T cell immunoglobulin and mucin protein-3 (Tim-3)/galectin-9 interaction regulates influenza a virus-specific humoral and CD8 T-cell responses. *Proc. Natl. Acad. Sci. USA* **2011**, *108*, 19001–19006. [CrossRef] [PubMed]

17. Klibi, J.; Niki, T.; Riedel, A.; Pioche-Durieu, C.; Souquere, S.; Rubinstein, E.; Le Moulec, S.; Guigay, J.; Hirashima, M.; Guemira, F.; et al. Blood diffusion and TH1-suppressive effects of galectin-9-containing exosomes released by epstein-barr virus-infected nasopharyngeal carcinoma cells. *Blood* **2009**, *113*, 1957–1966. [CrossRef] [PubMed]

18. Liu, K.T.; Liu, Y.H.; Chen, Y.H.; Lin, C.Y.; Huang, C.H.; Yen, M.C.; Kuo, P.L. Serum galectin-9 and galectin-3-binding protein in acute dengue virus infection. *Int. J. Mol. Sci.* **2016**, *17*, 832. [CrossRef] [PubMed]

19. Dapat, I.C.; Pascapurnama, D.N.; Iwasaki, H.; Labayo, H.K.; Chagan-Yasutan, H.; Egawa, S.; Hattori, T. Secretion of galectin-9 as a damp during dengue virus infection in thp-1 cells. *Int. J. Mol. Sci.* **2017**, *18*, 1644. [CrossRef] [PubMed]

20. Jost, S.; Moreno-Nieves, U.Y.; Garcia-Beltran, W.F.; Rands, K.; Reardon, J.; Toth, I.; Piechocka-Trocha, A.; Altfeld, M.; Addo, M.M. Dysregulated Tim-3 expression on natural killer cells is associated with increased galectin-9 levels in HIV-1 infection. *Retrovirology* **2013**, *10*, 74. [CrossRef] [PubMed]

21. Zhang, J.Y.; Zou, Z.S.; Huang, A.; Zhang, Z.; Fu, J.L.; Xu, X.S.; Chen, L.M.; Li, B.S.; Wang, F.S. Hyper-activated pro-inflammatory CD16+ monocytes correlate with the severity of liver injury and fibrosis in patients with chronic hepatitis B. *PLoS ONE* **2011**, *6*, e17484. [CrossRef] [PubMed]

22. Zhu, C.; Anderson, A.C.; Schubart, A.; Xiong, H.; Imitola, J.; Khoury, S.J.; Zheng, X.X.; Strom, T.B.; Kuchroo, V.K. The Tim-3 ligand galectin-9 negatively regulates T helper type 1 immunity. *Nat. Immunol.* **2005**, *6*, 1245–1252. [CrossRef] [PubMed]

23. Sada-Ovalle, I.; Ocana-Guzman, R.; Perez-Patrigeon, S.; Chavez-Galan, L.; Sierra-Madero, J.; Torre-Bouscoulet, L.; Addo, M.M. Tim-3 blocking rescue macrophage and T cell function against mycobacterium tuberculosis infection in HIV+ patients. *J. Int. AIDS Soc.* **2015**, *18*, 20078. [CrossRef] [PubMed]

24. Goncalves Silva, I.; Ruegg, L.; Gibbs, B.F.; Bardelli, M.; Fruehwirth, A.; Varani, L.; Berger, S.M.; Fasler-Kan, E.; Sumbayev, V.V. The immune receptor Tim-3 acts as a trafficker in a Tim-3/galectin-9 autocrine loop in human myeloid leukemia cells. *Oncoimmunology* **2016**, *5*, e1195535. [CrossRef] [PubMed]

25. Su, E.W.; Bi, S.; Kane, L.P. Galectin-9 regulates T helper cell function independently of Tim-3. *Glycobiology* **2011**, *21*, 1258–1265. [CrossRef] [PubMed]

26. Leitner, J.; Rieger, A.; Pickl, W.F.; Zlabinger, G.; Grabmeier-Pfistershammer, K.; Steinberger, P. Tim-3 does not act as a receptor for galectin-9. *PLoS Pathog.* **2013**, *9*, e1003253. [CrossRef] [PubMed]

27. Clayton, K.L.; Haaland, M.S.; Douglas-Vail, M.B.; Mujib, S.; Chew, G.M.; Ndhlovu, L.C.; Ostrowski, M.A. T cell ig and mucin domain-containing protein 3 is recruited to the immune synapse, disrupts stable synapse formation, and associates with receptor phosphatases. *J. Immunol.* **2014**, *192*, 782–791. [CrossRef] [PubMed]

28. Katoh, S.; Ishii, N.; Nobumoto, A.; Takeshita, K.; Dai, S.Y.; Shinonaga, R.; Niki, T.; Nishi, N.; Tominaga, A.; Yamauchi, A.; et al. Galectin-9 inhibits CD44-hyaluronan interaction and suppresses a murine model of allergic asthma. *Am. J. Respir. Crit. Care Med.* **2007**, *176*, 27–35. [CrossRef] [PubMed]

29. Bi, S.; Hong, P.W.; Lee, B.; Baum, L.G. Galectin-9 binding to cell surface protein disulfide isomerase regulates the redox environment to enhance T-cell migration and HIV entry. *Proc. Natl. Acad. Sci. USA* **2011**, *108*, 10650–10655. [CrossRef] [PubMed]

30. Moritoki, M.; Kadowaki, T.; Niki, T.; Nakano, D.; Soma, G.; Mori, H.; Kobara, H.; Masaki, T.; Kohno, M.; Hirashima, M. Galectin-9 ameliorates clinical severity of MRL/LPR lupus-prone mice by inducing plasma cell apoptosis independently of Tim-3. *PLoS ONE* **2013**, *8*, e60807. [CrossRef] [PubMed]

31. Lhuillier, C.; Barjon, C.; Niki, T.; Gelin, A.; Praz, F.; Morales, O.; Souquere, S.; Hirashima, M.; Wei, M.; Dellis, O.; et al. Impact of exogenous galectin-9 on human T cells: Contribution of the T cell receptor complex to antigen-independent activation but not to apoptosis induction. *J. Biol. Chem.* **2015**, *290*, 16797–16811. [CrossRef] [PubMed]

32. Cao, E.; Zang, X.; Ramagopal, U.A.; Mukhopadhaya, A.; Fedorov, A.; Fedorov, E.; Zencheck, W.D.; Lary, J.W.; Cole, J.L.; Deng, H.; et al. T cell immunoglobulin mucin-3 crystal structure reveals a galectin-9-independent ligand-binding surface. *Immunity* **2007**, *26*, 311–321. [CrossRef] [PubMed]

33. Nakayama, M.; Akiba, H.; Takeda, K.; Kojima, Y.; Hashiguchi, M.; Azuma, M.; Yagita, H.; Okumura, K. Tim-3 mediates phagocytosis of apoptotic cells and cross-presentation. *Blood* **2009**, *113*, 3821–3830. [CrossRef] [PubMed]

34. Huang, Y.H.; Zhu, C.; Kondo, Y.; Anderson, A.C.; Gandhi, A.; Russell, A.; Dougan, S.K.; Petersen, B.S.; Melum, E.; Pertel, T.; et al. Ceacam1 regulates Tim-3-mediated tolerance and exhaustion. *Nature* **2015**, *517*, 386–390. [CrossRef] [PubMed]

35. Chiba, S.; Baghdadi, M.; Akiba, H.; Yoshiyama, H.; Kinoshita, I.; Dosaka-Akita, H.; Fujioka, Y.; Ohba, Y.; Gorman, J.V.; Colgan, J.D.; et al. Tumor-infiltrating DCs suppress nucleic acid-mediated innate immune responses through interactions between the receptor TIM-3 and the alarmin HMGB1. *Nat. Immunol.* **2012**, *13*, 832–842. [CrossRef] [PubMed]

36. Jemielity, S.; Wang, J.J.; Chan, Y.K.; Ahmed, A.A.; Li, W.; Monahan, S.; Bu, X.; Farzan, M.; Freeman, G.J.; Umetsu, D.T.; et al. Tim-family proteins promote infection of multiple enveloped viruses through virion-associated phosphatidylserine. *PLoS Pathog.* **2013**, *9*, e1003232. [CrossRef] [PubMed]

37. Sui, L.; Zhang, W.; Chen, Y.; Zheng, Y.; Wan, T.; Zhang, W.; Yang, Y.; Fang, G.; Mao, J.; Cao, X. Human membrane protein Tim-3 facilitates hepatitis a virus entry into target cells. *Int. J. Mol. Med.* **2006**, *17*, 1093–1099. [CrossRef] [PubMed]

38. Meertens, L.; Carnec, X.; Lecoin, M.P.; Ramdasi, R.; Guivel-Benhassine, F.; Lew, E.; Lemke, G.; Schwartz, O.; Amara, A. The TIM and TAM families of phosphatidylserine receptors mediate dengue virus entry. *Cell. Host Microbe* **2012**, *12*, 544–557. [CrossRef] [PubMed]

39. Dam, T.K.; Brewer, C.F. Maintenance of cell surface glycan density by lectin-glycan interactions: A homeostatic and innate immune regulatory mechanism. *Glycobiology* **2010**, *20*, 1061–1064. [CrossRef] [PubMed]

40. Rabinovich, G.A.; Toscano, M.A. Turning 'sweet' on immunity: Galectin-glycan interactions in immune tolerance and inflammation. *Nat. Rev. Immunol.* **2009**, *9*, 338–352. [CrossRef] [PubMed]

41. Anderson, A.C.; Anderson, D.E.; Bregoli, L.; Hastings, W.D.; Kassam, N.; Lei, C.; Chandwaskar, R.; Karman, J.; Su, E.W.; Hirashima, M.; et al. Promotion of tissue inflammation by the immune receptor Tim-3 expressed on innate immune cells. *Science* **2007**, *318*, 1141–1143. [CrossRef] [PubMed]

42. Anderson, A.C.; Anderson, D.E. Tim-3 in autoimmunity. *Curr. Opin. Immunol.* **2006**, *18*, 665–669. [CrossRef] [PubMed]

43. Fujita, K.; Iwama, H.; Oura, K.; Tadokoro, T.; Samukawa, E.; Sakamoto, T.; Nomura, T.; Tani, J.; Yoneyama, H.; Morishita, A.; et al. Cancer therapy due to apoptosis: Galectin-9. *Int. J. Mol. Sci.* **2017**, *18*, 74. [CrossRef] [PubMed]

44. Seki, M.; Oomizu, S.; Sakata, K.M.; Sakata, A.; Arikawa, T.; Watanabe, K.; Ito, K.; Takeshita, K.; Niki, T.; Saita, N.; et al. Galectin-9 suppresses the generation of TH17, promotes the induction of regulatory T cells, and regulates experimental autoimmune arthritis. *Clin. Immunol.* **2008**, *127*, 78–88. [CrossRef] [PubMed]

45. Leskela, S.; Serrano, A.; de la Fuente, H.; Rodriguez-Munoz, A.; Ramos-Levi, A.; Sampedro-Nunez, M.; Sanchez-Madrid, F.; Gonzalez-Amaro, R.; Marazuela, M. Graves' disease is associated with a defective expression of the immune regulatory molecule galectin-9 in antigen-presenting dendritic cells. *PLoS ONE* **2015**, *10*, e0123938. [CrossRef] [PubMed]

46. Lee, J.; Oh, J.M.; Hwang, J.W.; Ahn, J.K.; Bae, E.K.; Won, J.; Koh, E.M.; Cha, H.S. Expression of human Tim-3 and its correlation with disease activity in rheumatoid arthritis. *Scand. J. Rheumatol.* **2011**, *40*, 334–340. [CrossRef] [PubMed]

47. Jung, S.H.; Hwang, J.H.; Kim, S.E.; Kim, Y.K.; Park, H.C.; Lee, H.T. Human galectin-9 on the porcine cells affects the cytotoxic activity of M1-differentiated THP-1 cells through inducing a shift in M2-differentiated THP-1 cells. *Xenotransplantation* **2017**. [CrossRef] [PubMed]

48. Sanchez-Fueyo, A.; Tian, J.; Picarella, D.; Domenig, C.; Zheng, X.X.; Sabatos, C.A.; Manlongat, N.; Bender, O.; Kamradt, T.; Kuchroo, V.K.; et al. Tim-3 inhibits T helper type 1-mediated auto- and alloimmune responses and promotes immunological tolerance. *Nat. Immunol.* **2003**, *4*, 1093–1101. [CrossRef] [PubMed]

49. Johnson, J.L.; Jones, M.B.; Ryan, S.O.; Cobb, B.A. The regulatory power of glycans and their binding partners in immunity. *Trends Immunol.* **2013**, *34*, 290–298. [CrossRef] [PubMed]

50. Burman, J.; Svenningsson, A. Cerebrospinal fluid concentration of galectin-9 is increased in secondary progressive multiple sclerosis. *J. Neuroimmunol.* **2016**, *292*, 40–44. [CrossRef] [PubMed]

51. Gleason, M.K.; Lenvik, T.R.; McCullar, V.; Felices, M.; O'Brien, M.S.; Cooley, S.A.; Verneris, M.R.; Cichocki, F.; Holman, C.J.; Panoskaltsis-Mortari, A.; et al. Tim-3 is an inducible human natural killer cell receptor that enhances interferon gamma production in response to galectin-9. *Blood* **2012**, *119*, 3064–3072. [CrossRef] [PubMed]

52. Brooks, A.K.; Lawson, M.A.; Rytych, J.L.; Yu, K.C.; Janda, T.M.; Steelman, A.J.; McCusker, R.H. Immunomodulatory factors galectin-9 and interferon-gamma synergize to induce expression of rate-limiting enzymes of the kynurenine pathway in the mouse hippocampus. *Front. Immunol.* **2016**, *7*, 422. [CrossRef] [PubMed]

53. Nagae, M.; Nishi, N.; Murata, T.; Usui, T.; Nakamura, T.; Wakatsuki, S.; Kato, R. Crystal structure of the galectin-9 n-terminal carbohydrate recognition domain from mus musculus reveals the basic mechanism of carbohydrate recognition. *J. Biol. Chem.* **2006**, *281*, 35884–35893. [CrossRef] [PubMed]

54. Yoshida, H.; Teraoka, M.; Nishi, N.; Nakakita, S.; Nakamura, T.; Hirashima, M.; Kamitori, S. X-ray structures of human galectin-9 C-terminal domain in complexes with a biantennary oligosaccharide and sialyllactose. *J. Biol. Chem.* **2010**, *285*, 36969–36976. [CrossRef] [PubMed]

55. Li, Y.; Feng, J.; Geng, S.; Geng, S.; Wei, H.; Chen, G.; Li, X.; Wang, L.; Wang, R.; Peng, H.; et al. The N- and C-terminal carbohydrate recognition domains of galectin-9 contribute differently to its multiple functions in innate immunity and adaptive immunity. *Mol. Immunol.* **2011**, *48*, 670–677. [CrossRef] [PubMed]

56. Bowen, D.G.; Walker, C.M. Adaptive immune responses in acute and chronic hepatitis C virus infection. *Nature* **2005**, *436*, 946–952. [CrossRef] [PubMed]

57. Kared, H.; Fabre, T.; Bedard, N.; Bruneau, J.; Shoukry, N.H. Galectin-9 and IL-21 mediate cross-regulation between TH17 and Treg cells during acute hepatitis C. *PLoS Pathog.* **2013**, *9*, e1003422. [CrossRef] [PubMed]

58. Ji, X.J.; Ma, C.J.; Wang, J.M.; Wu, X.Y.; Niki, T.; Hirashima, M.; Moorman, J.P.; Yao, Z.Q. HCV-infected hepatocytes drive CD4$^+$ CD25$^+$ foxp3$^+$ regulatory T-cell development through the Tim-3/Gal-9 pathway. *Eur. J. Immunol.* **2013**, *43*, 458–467. [CrossRef] [PubMed]

59. Nebbia, G.; Peppa, D.; Schurich, A.; Khanna, P.; Singh, H.D.; Cheng, Y.; Rosenberg, W.; Dusheiko, G.; Gilson, R.; ChinAleong, J.; et al. Upregulation of the Tim-3/galectin-9 pathway of T cell exhaustion in chronic hepatitis B virus infection. *PLoS ONE* **2012**, *7*, e47648. [CrossRef] [PubMed]

60. McSharry, B.P.; Forbes, S.K.; Cao, J.Z.; Avdic, S.; Machala, E.A.; Gottlieb, D.J.; Abendroth, A.; Slobedman, B. Human cytomegalovirus upregulates expression of the lectin galectin 9 via induction of β interferon. *J. Virol.* **2014**, *88*, 10990–10994. [CrossRef] [PubMed]

61. Abdel-Mohsen, M.; Chavez, L.; Tandon, R.; Chew, G.M.; Deng, X.; Danesh, A.; Keating, S.; Lanteri, M.; Samuels, M.L.; Hoh, R.; et al. Human galectin-9 is a potent mediator of HIV transcription and reactivation. *PLoS Pathog.* **2016**, *12*, e1005677. [CrossRef] [PubMed]

62. Zhuo, Y.; Zhang, Y.F.; Wu, H.J.; Qin, L.; Wang, Y.P.; Liu, A.M.; Wang, X.H. Interaction between galectin-9/Tim-3 pathway and follicular helper CD4$^+$ T cells contributes to viral persistence in chronic hepatitis C. *Biomed. Pharmacother.* **2017**, *94*, 386–393. [CrossRef] [PubMed]

63. Hsu, Y.L.; Wang, M.Y.; Ho, L.J.; Huang, C.Y.; Lai, J.H. Up-regulation of galectin-9 induces cell migration in human dendritic cells infected with dengue virus. *J. Cell. Mol. Med.* **2015**, *19*, 1065–1076. [CrossRef] [PubMed]

64. Tandon, R.; Chew, G.M.; Byron, M.M.; Borrow, P.; Niki, T.; Hirashima, M.; Barbour, J.D.; Norris, P.J.; Lanteri, M.C.; Martin, J.N.; et al. Galectin-9 is rapidly released during acute HIV-1 infection and remains sustained at high levels despite viral suppression even in elite controllers. *AIDS Res. Hum. Retroviruses.* **2014**, *30*, 654–664. [CrossRef] [PubMed]

65. Clayton, K.L.; Douglas-Vail, M.B.; Nur-ur Rahman, A.K.; Medcalf, K.E.; Xie, I.Y.; Chew, G.M.; Tandon, R.; Lanteri, M.C.; Norris, P.J.; Deeks, S.G.; et al. Soluble T cell immunoglobulin mucin domain 3 is shed from CD8$^+$ T cells by the sheddase adam10, is increased in plasma during untreated HIV infection, and correlates with HIV disease progression. *J. Virol.* **2015**, *89*, 3723–3736. [CrossRef] [PubMed]

66. Schwartz, J.A.; Clayton, K.L.; Mujib, S.; Zhang, H.; Rahman, A.K.; Liu, J.; Yue, F.Y.; Benko, E.; Kovacs, C.; Ostrowski, M.A. Tim-3 is a marker of plasmacytoid dendritic cell dysfunction during HIV infection and is associated with the recruitment of IRF7 and p85 into lysosomes and with the submembrane displacement of TLR9. *J. Immunol.* **2017**, *198*, 3181–3194. [CrossRef] [PubMed]

67. Chagan-Yasutan, H.; Ndhlovu, L.C.; Lacuesta, T.L.; Kubo, T.; Leano, P.S.; Niki, T.; Oguma, S.; Morita, K.; Chew, G.M.; Barbour, J.D.; et al. Galectin-9 plasma levels reflect adverse hematological and immunological features in acute dengue virus infection. *J. Clin. Virol.* **2013**, *58*, 635–640. [CrossRef] [PubMed]

68. Golden-Mason, L.; Rosen, H.R. Galectin-9: Diverse roles in hepatic immune homeostasis and inflammation. *Hepatology* **2017**, *66*, 271–279. [CrossRef] [PubMed]

69. Merani, S.; Chen, W.; Elahi, S. The bitter side of sweet: The role of galectin-9 in immunopathogenesis of viral infections. *Rev. Med. Virol.* **2015**, *25*, 175–186. [CrossRef] [PubMed]

70. PB, J.R.; Schreiber, T.H.; Rajasagi, N.K.; Suryawanshi, A.; Mulik, S.; Veiga-Parga, T.; Niki, T.; Hirashima, M.; Podack, E.R.; Rouse, B.T. TNFRSF25 agonistic antibody and galectin-9 combination therapy controls herpes simplex virus-induced immunoinflammatory lesions. *J. Virol.* **2012**, *86*, 10606–10620.

71. Shim, J.A.; Park, S.; Lee, E.S.; Niki, T.; Hirashima, M.; Sohn, S. Galectin-9 ameliorates herpes simplex virus-induced inflammation through apoptosis. *Immunobiology* **2012**, *217*, 657–666. [CrossRef] [PubMed]

72. Lu, X.; McCoy, K.S.; Xu, J.; Hu, W.; Chen, H.; Jiang, K.; Han, F.; Chen, P.; Wang, Y. Galectin-9 ameliorates respiratory syncytial virus-induced pulmonary immunopathology through regulating the balance between TH17 and regulatory T cells. *Virus Res.* **2015**, *195*, 162–171. [CrossRef] [PubMed]

73. Elahi, S.; Niki, T.; Hirashima, M.; Horton, H. Galectin-9 binding to Tim-3 renders activated human CD4$^+$ T cells less susceptible to HIV-1 infection. *Blood* **2012**, *119*, 4192–4204. [CrossRef] [PubMed]

74. Lai, J.H.; Lin, Y.L.; Hsieh, S.L. Pharmacological intervention for dengue virus infection. *Biochem. Pharmacol.* **2017**, *129*, 14–25. [CrossRef] [PubMed]

75. Rangachari, M.; Zhu, C.; Sakuishi, K.; Xiao, S.; Karman, J.; Chen, A.; Angin, M.; Wakeham, A.; Greenfield, E.A.; Sobel, R.A.; et al. Bat3 promotes T cell responses and autoimmunity by repressing Tim-3-mediated cell death and exhaustion. *Nat. Med.* **2012**, *18*, 1394–1400. [CrossRef] [PubMed]

International Journal of
Molecular Sciences

MDPI

Review

Role of Galectin-3 in Bone Cell Differentiation, Bone Pathophysiology and Vascular Osteogenesis

Carla Iacobini, Claudia Blasetti Fantauzzi, Giuseppe Pugliese * and Stefano Menini

Department of Clinical and Molecular Medicine, La Sapienza University, 00185 Rome, Italy;
carla.iacobini@uniroma1.it (C.I.); claudia.blasettifantauzzi@uniroma1.it (C.B.F.);
stefano.menini@uniroma1.it (S.M.)
* Correspondence: giuseppe.pugliese@uniroma1.it; Tel.: +39-063-377-5440; Fax: +39-063-377-6327

Received: 31 October 2017; Accepted: 19 November 2017; Published: 21 November 2017

Abstract: Galectin-3 is expressed in various tissues, including the bone, where it is considered a marker of chondrogenic and osteogenic cell lineages. Galectin-3 protein was found to be increased in the differentiated chondrocytes of the metaphyseal plate cartilage, where it favors chondrocyte survival and cartilage matrix mineralization. It was also shown to be highly expressed in differentiating osteoblasts and osteoclasts, in concomitance with expression of osteogenic markers and Runt-related transcription factor 2 and with the appearance of a mature phenotype. Galectin-3 is expressed also by osteocytes, though its function in these cells has not been fully elucidated. The effects of galectin-3 on bone cells were also investigated in galectin-3 null mice, further supporting its role in all stages of bone biology, from development to remodeling. Galectin-3 was also shown to act as a receptor for advanced glycation endproducts, which have been implicated in age-dependent and diabetes-associated bone fragility. Moreover, its regulatory role in inflammatory bone and joint disorders entitles galectin-3 as a possible therapeutic target. Finally, galectin-3 capacity to commit mesenchymal stem cells to the osteoblastic lineage and to favor transdifferentiation of vascular smooth muscle cells into an osteoblast-like phenotype open a new area of interest in bone and vascular pathologies.

Keywords: galectin-3; osteoblasts; osteoclasts; bone remodeling; vascular osteogenesis

---

1. Introduction

Galectin-3 is a 29- to 35-kDa protein belonging to the family of the β-galactoside binding animal lectins and is constitutively expressed in various tissues, including the bone [1]. It is the only chimera-type galectin in vertebrates and comprises one conserved carbohydrate recognition domain linked to a non-lectin domain through a collagen-like linker region [2]. This lectin has been recognized over the past two decades as being involved in many physiological and pathological processes [2,3]. In quiescent cells, galectin-3 shows prominent cytoplasmic localization, whereas it is found predominantly in the nucleus of replicating cells [3]. Galectin-3 is also secreted into the extracellular space through a non-classical secretory pathway [4]. Here, it interacts with the β-galactoside residues of several glycoproteins, thus forming higher order supramolecular structures resulting in galectin-ligand lattices on the cell surface [5]. These lattices have been shown to play a major role in the regulation of receptor clustering, endocytosis and signaling, thereby controlling important cell functions such as cell transdifferentiation, migration, and fibrogenesis [5,6]. As a component of the cell surface lattice, galectin-3 also regulates the biogenesis of a subpopulation of clathrin-independent carriers involved in the endocytosis of specific cargo proteins, which could represent one of the main mechanisms behind its functions [7].

Intracellular galectin-3 has been implicated in several basic cellular processes related to control of cell differentiation, growth, and apoptosis, as well as in specific cell biosynthetic activities. Galectin-3 has been found in the spliceosome, where it is a required factor in the splicing of nuclear pre-mRNA [8]. This lectin also regulates cell cycle by modulating the activity of cyclins and their inhibitors, as well as the phosphorylation status of retinoblastoma protein [9]. Of great interest for bone biology, intracellular galectin-3 is a key regulator of the Wnt/β-catenin signaling pathway, both through its interaction with β-catenin and because of its structural similarities with it [10]. Intracellular galectin-3 has also been shown to promote cell proliferation [11,12] and favor survival by its anti-apoptotic activity, which is related to its sequence homology and association with bcl-2 [11,13]. However, depending on the cell type, galectin-3 can also promote apoptosis, as demonstrated by its involvement in T-cell and neutrophil death [14].

Extracellular galectin-3 also participates in the control of cell cycle and division. Growth factors immobilization into the galectin-3 lattice is, in fact, an additional mechanism for the regulation of cell growth and differentiation [15]. In addition, cell surface galectin-3 has been shown to regulate cell adhesion in opposite fashions, by both promoting homo- and heterotypic cell-to-cell interactions [16,17] and down-regulating cell adhesion to the extracellular matrix component laminin, thus producing an anti-adhesive effect [18,19].

Another important function of galectin-3 is the uptake and removal of advanced glycation endproducts (AGEs) [20]. AGEs are a heterogeneous class of nonenzymatically glycated proteins, lipids and nucleic acids, which accumulate in tissues during aging and, at a faster rate, in metabolic disorders such as diabetes and obesity [20]. AGEs are toxic molecules inducing tissue injury by direct and indirect mechanisms. In fact, they can exert detrimental direct physicochemical effects by interacting with several molecules, thus inducing changes in enzymatic activity, ligand half-life, binding, and immunogenicity [21]. Moreover, AGEs display indirect deleterious effects by binding to several cell surface receptors, of which the most studied is the receptor for AGEs (RAGE), a 35-kDa member of the immunoglobulin superfamily of receptors [22]. RAGE ligation is associated with cellular oxidative stress [23] and activation of proinflammatory signaling pathways, eventually culminating in tissue inflammation and fibrosis as well as in cell damage and death [20,24]. Noteworthy, galectin-3 and RAGE appear to exert opposite actions as AGE-receptors, with RAGE mediating the injurious effects of AGEs and galectin-3 playing a protective role by favoring removal and degradation of these toxic by-products [2,25].

Finally, galectin-3 plays an important role in the modulation of the immune/inflammatory response, as evidenced by the numerous scientific reports showing a regulatory activity on both innate and adaptive immunity [26]. Galectin-3 has been shown to exert both pro-inflammatory actions, prevailing in an acute setting, and anti-inflammatory effects, particularly in chronic inflammatory conditions, as recently reviewed [20]. In detail, well-established effects of galectin-3 on immune cells include the stimulation of T-cell apoptosis [27], inhibition of T-cell growth and T-helper 1 differentiation [28], down-regulation of T-cell receptor-mediated T-cell activation [29], and induction of alternative (M2) macrophage activation [30].

As much as concerning the bone, galectin-3 has been shown to be a marker of both chondrogenic and osteogenic cell lineages. Aubin et al. were the first to report that chondroblasts and osteoblasts share the expression of several molecules, including markers such as galectin-3 [31]. Later on, other groups confirmed and extended this initial observation showing that galectin-3 is also found in osteoblasts and osteocytes of the woven trabecular and cortical bone as well as in osteoclasts at the front of ossification and mononuclear cells within bone marrow cavities [1]. In developing and mature bone, galectin-3 expression seems to be restricted to the cytoplasm of chondrocytes and bone cells, although it is occasionally detected in the nuclei of dense non-hypertrophic chondrocytes in the zone of calcification and of young osteoblasts; conversely, no extracellular galectin-3 is detected in developing or mature bone tissue [1]. Noteworthy, Stock et al. demonstrated that the skeletal expression pattern of galectin-3 overlaps at many sites with that of Runt-related transcription factor 2 (RUNX2), which is a key regulator of osteoblast differentiation and chondrocyte maturation [32]. In summary, these early studies showed that galectin-3 is widely distributed both in the developing and mature bone and that its expression is under control of the master regulator of bone growth RUNX2. These findings suggest that galectin-3 may be a key player in all stages of bone biology, thus deserving to be thoroughly investigated as a potential target in bone disorders. In this extensive review, we will present the results of studies that investigated the role of galectin-3 in bone cell differentiation and function, in bone development and remodeling, and in the pathogenesis of inflammatory bone diseases. We will also outline problems and aspects that have not been addressed yet, or addressed only partially, which may represent relevant future research directions. Finally, the role of galectin-3 in vascular osteogenesis, which was recently reported by us and other investigators [33–35], will also be discussed. Figure 1 and Table 1 recapitulate the main findings discussed in this review, as well as topics deserving further investigation in the future.

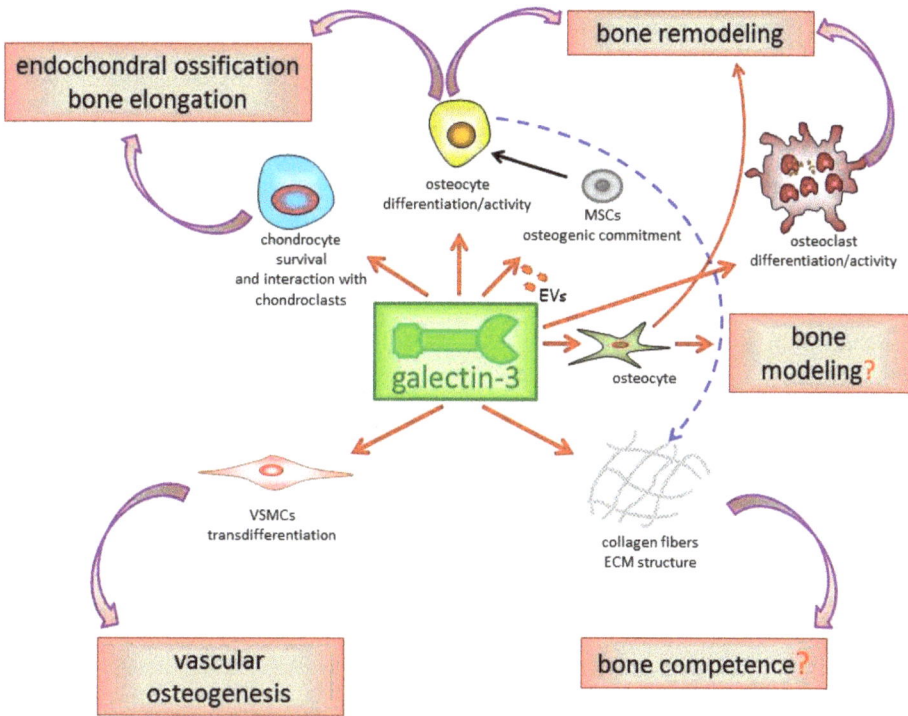

Figure 1. Galectin-3 osteogenic activities in bone and vascular tissue. Galectin-3 is considered a marker of chondrogenic and osteogenic cell lineages and is up-regulated in conjunction with other osteogenic markers during differentiation of bone cells. Consistent with a role in bone physiology, galectin-3 deficiency affects endochondral ossification and bone remodeling. Galectin-3 is also critical for maintaining proper extracellular matrix (ECM) structure and, possibly, bone competence. Being also expressed by osteocytes, it is possible to hypothesize its participation in the regulation of mechanosensory function of these cells and promotion of bone modeling and remodeling also through this mechanism. Finally, galectin-3 activity is central for vascular smooth muscle cells (VSMCs) transdifferentiation into an osteoblast-like phenotype and vascular osteogenesis. EVs = extracellular vesicles; MSCs = mesenchymal stem cells.

Table 1. Proven roles of galectin-3 in bone biology and supposed functions, to be investigated, in the pathophysiology of bone metabolism.

Cell	Proven	To Be Investigated
Chondrocyte	cell marker and pro-survival factor [1,36] regulator of cell-to-cell interaction with chondroclasts [37] target and regulator of MMPs activity [38–40]	mechanisms underlying the regulation of chondrocyte activity and survival
Osteoblast	RUNX2 target gene [32] differentiation marker [41] proper bone ECM structure by regulating collagen fibers synthesis [42] protection against AGE mediated toxic effects by its AGE-receptor scavenger activity [43–45]	mechanisms underlying the regulation of osteoblast differentiationprotection from age- and diabetes-related bone fragility
Osteocyte	cell marker [1,41]	mechanosensory function and promotion of bone modelling and remodeling
MSC	increased osteoblastogenic differentiation capacity [46] positive regulation of the master transcription factor RUNX2 [46] stabilization and increase of β-catenin levels [46]	promotion of bone repair and homeostasis through modulation of β-catenin
Osteoclast	differentiation marker [47] mediator of cell matrix adhesion [48] regulation of differentiation from progenitors and pro-survival factor [40] downstream regulator of MMP-9 activity [40] and ECM degradation [37] negative regulation of osteoclastogenesis [49,50]	opposite effects depending on intra- and extracellular localization

MMPs = matrix metalloproteinases; RUNX2 = Runt-related transcription factor 2; ECM = extracellular matrix; AGEs = advanced glycation endproducts.

2. Galectin-3 in Chondrocyte Differentiation and Endochondral Bone Formation

Endochondral ossification is an essential process by which most of the bones, particularly long bones, of the mammalian skeletal system are formed during embryonic development. At variance with intramembranous ossification, endochondral ossification involves formation of cartilage elements (chondrocytes) from mesenchymal condensation, secretion of extracellular matrix by mature chondrocytes and blood vessels invasion. During bone growth, cartilage anlage is progressively replaced by bone tissue in a highly coordinate process, which is achieved also through a tight balance between the rates of chondrocyte proliferation and apoptosis at the growth plate cartilage. This is a specialized developmental tissue located at the metaphyseal level of long bones in which chondrocytes are organized in three characteristic zones: proliferative, mature, and hypertrophic. Chondrocytes are responsible for the synthesis of the major constituents of the matrix and the enzymes that degrade cartilage matrix. Hence, their function is essential to regulate cartilage synthesis and degradation, thereby coordinating the timing of vascular invasion and subsequent bone deposition [51]. The elongation process takes place through a series of coordinated and complex events, which include apoptosis of the chondrocytes in the terminal hypertrophic zone, partial degradation of the calcified cartilage matrix, and bone matrix deposition by osteoblasts on the remaining cartilage septa, which serve as a scaffold [52]. Failure in any of these events, including perturbations in chondrocyte hypertrophy, would severely affect endochondral ossification and, consequently, normal bone development [53].

Galectin-3 has been initially involved in the process of endochondral bone formation as a regulator of chondrocyte survival. In fact, high levels of galectin-3 protein were demonstrated in the differentiated chondrocytes of the metaphyseal plate cartilage of long bones of both fetal and neonatal mice. The highest concentrations of galectin-3 were found in the cytoplasm of mature and early hypertrophic chondrocytes [1]. At variance, very little expression was detected in the late hypertrophic chondrocytes undergoing terminal maturation and cell death in the zone of calcification, both at protein and mRNA levels. These findings are consistent with the anti-apoptotic function of galectin-3 [11,13,54] and with the observation that galectin-3 null ($Lgals3^{-/-}$) mice show accelerated terminal differentiation and death of chondrocytes [1,36]. In addition to chondrocytes, Colnot et al. reported high concentrations of galectin-3 also in the cytoplasm of osteoblasts, osteocytes, and osteoclasts of trabecular and cortical bone, and in monocytes in the bone marrow cavity, thus suggesting a role for this lectin in the differentiation and/or activity of all the cell types present in mature bone [1].

Subsequent comprehensive histological and ultrastructural analyses performed by the same group identified a number of abnormalities in epiphyseal femurs and tibias of fetal $Lgals3^{-/-}$ mice, affecting both chondrocytes of the proliferative, mature, and hypertrophic zones and the extracellular matrix of the hypertrophic zone. The greatest abnormalities were detected at the chondrovascular junction, where premature and increased cell death was associated with, and might account for, the increased empty lacunae observed in mutant mice. Consistently, more numerous condensed chondrocytes exhibiting characteristic signs of apoptosis were found in the late hypertrophic zone, indicating that the rate of chondrocyte death was increased in $Lgals3^{-/-}$ mice. Taken together, these findings suggest a role for galectin 3 as a regulator of chondrocyte survival and indicate that galectin 3 may also participate in the coordination between chondrocyte death and metaphyseal vascularization. At histological examination, the process of cartilage matrix mineralization appeared to be disturbed in $Lgals3^{-/-}$ mice, both in terms of quantity and quality. However, despite these numerous abnormalities of the metaphyseal growth plate cartilage, X-ray analysis conducted in adult $Lgals3^{-/-}$ mice did not reveal significant macroscopic anatomical differences (i.e., final size) of long bones, indicating that the initial defects are transient or, alternatively, have no impact on the process of bone elongation [42].

Another mechanism by which galectin-3 may regulate the process of endochondral ossification emerged from studies investigating the cartilaginous bar of the mandibular arch (Meckel's cartilage), the middle portion of which is known to degrade via hypertrophy and death of chondrocytes and cartilage resorption by chondroclasts/osteoclasts [37]. Sakakura et al. showed that galectin-3 levels were especially higher in hypertrophic chondrocytes adjacent to chondroclasts. This observation suggests that galectin-3 may also coordinate the resorption of calcified cartilage through cell-to-cell interaction with chondroclasts, which are responsible for the synthesis and secretion of matrix metalloproteinases (MMPs) involved in extracellular matrix degradation and in growth plate angiogenesis [37]. Therefore, in addition to controlling chondrocyte cell cycling and survival, evidence suggested that galectin-3 might participate in endochondral bone formation also through regulation of MMPs activity, or vice versa [38,39]. According to this interpretation, deficiency of MMP-9 was shown to induce accumulation of galectin-3 in the expanded hypertrophic cartilage of developing bone, which was associated with a decreased rate of chondrocyte apoptosis and accumulation of late hypertrophic chondrocytes in the same zone [40]. This finding is consistent with the observation that galectin-3 is an endogenous substrate of MMP-9, as indicated by in vitro experiments showing that treatment of wild-type embryonic metatarsals with full-length galectin-3, but not MMP-9-cleaved galectin-3, was able to reproduce the embryonic phenotype of $Mmp-9^{-/-}$ mice, characterized by an expanded hypertrophic zone [40]. Altogether, these early studies indicated that galectin-3 and MMP-9 represent two members of the same pathway implicated in the regulation of hypertrophic chondrocyte clearance, degradation of extracellular matrix, and vascular invasion.

3. Galectin-3 in Bone Cell Differentiation and Function and Bone Homeostasis

3.1. Galectin-3 in Osteoblast Biology and Pathology

Osteoblasts are specialized, not terminally differentiated, mononuclear cells involved in secretion and mineralization of the bone matrix, thus being responsible for new bone formation [55]. Osteoblasts also control bone resorption through regulation of osteoclast activity, thus having a primary role in modeling and preserving skeletal architecture [56]. Osteoblasts derive from the same mesenchymal stem cell (MSC) precursors from which originate chondrocytes, adipocytes, and myoblasts [57], depending on which transcription factor is expressed [55,57,58]. Among the most important transcription factors recognized as specific and necessary to commit MSCs to the osteoblastic lineage are RUNX2, also called core-binding factor subunit alpha-1, and osterix. Both these factors are essential for osteoblastogenesis and, in their absence, no osteoblasts can be formed. Other important factors are activating transcription factor 4 and bone morphogenic proteins (BMPs), which provide important signals that are essential for complete osteoblastogenic differentiation [57,58].

In culture, osteoblasts resemble the features of fibroblasts, except for the expression of RUNX2 and the bone-building protein osteocalcin (OCN). At the morphological level, osteoblasts differ from fibroblasts for the unique, osteoblast-specific feature represented by the production of a mineralized extracellular matrix [59]. In the bone microenvironment, as osteoblasts differentiate, they start to secrete the organic portion of the bone matrix (osteoid) and to produce hydroxyapatite crystals that are deposited into the organic matrix to form the bone's mineralized structure. Ultimately, the osteoblast becomes enclosed in the mineralized matrix and differentiate into mature osteocyte, a star-shaped cell representing the most abundant cell type of mature bone [59]. This terminally differentiated bone cell is involved in many aspects of bone biology, from the regulation of bone metabolism through the expression of the negative regulator of bone mass sclerostin, to detection and transduction of mechanical load into biological responses, thus adapting bone shape to daily mechanical forces [60]. However, the exact role of osteocytes in bone homeostasis is not completely defined, as reviewed by Prideaux et al. [61].

The first report of a possible role for galectin-3 in osteoblast differentiation came from Aubin's laboratory in 1995. Using immunohistochemical and molecular approaches, this research group demonstrated that osteoblasts and hypertrophic chondrocytes share the expression of galectin-3, in addition to other markers such as alkaline phosphatase (ALP), OCN, bone sialoprotein (BSP), osteopontin (OPN), and collagen type I. In addition, they observed that, in cultures of fetal rat calvaria, differentiation of osteoprogenitor cells into mature osteoblasts forming bone nodules is preceded by a phase of intense cell division, followed by loss of proliferative capacity and progressively increased expression of the osteoblast markers mentioned above, including galectin-3 [31].

Subsequently, the same group confirmed and extended the initial observations by investigating galectin-3 (both protein and mRNA) expression and regulation in several osteoblastic model systems, including an in vitro model of osteogenesis and bone tissue in vivo [41]. The most striking evidence in favor of a key role of galectin-3 in osteogenic differentiation was the observation that while galectin-3 mRNA levels increased during long-term culture in rat calvaria cells, the mRNA expression of the lectin fell down with time in culture of rat skin fibroblasts. Moreover, the amount of galectin-3 mRNA and protein increased with time concomitant with the increase of osteoblast differentiation markers and bone nodules formation. Finally, a role for galectin-3 in osteoblast differentiation was also supported by the observations that (a) galectin-3 expression in differentiating osteoblasts was modulated by two hormones known to affect osteogenesis, such as glucocorticoids and 1,25-dihydroxyvitamin D; and (b) mature osteoblasts and osteocytes of calvaria bone showed strong galectin-3 positivity, while faint staining was detected in immature osteogenic cells residing in periosteum [41]. Although these early studies clearly demonstrated that galectin-3 is expressed by osteoblast and that its levels increase during osteogenesis in parallel with the increased expression of established osteogenic markers, the mechanisms by which galectin-3 may regulate differentiation and activity of osteoblast or bone metabolism remains to be determined. Surprisingly, no further studies have been conducted to pursue this aim. The only additional information on the role of galectin-3 in osteoblast biology has come from a study investigating the molecular mechanisms through which the mechanical load is converted into molecular signals in bone. Briefly, galectin-3 expression was found to be up-regulated in the human osteoblastic HOBIT cell line exposed to extracellular nucleotides concomitant to RUNX2 DNA-binding activity, thus suggesting that ATP and/or UTP released by osteocytes in response to mechanical stimuli may account for the mechanosensory function of these cells [62]. These data also confirmed indirectly the finding by Stock et al. that galectin-3 is a RUNX2 target gene in bone [32].

Though these studies suggest that galectin-3 is critical for osteoblast differentiation and function, a recent report showed that exogenous recombinant galectin-3 inhibited terminal differentiation of a human pre-osteoblast cell line [63]. This finding may suggest different, or even opposite effects of galectin-3 on osteoblastogenesis, depending on its intracellular or extracellular localization, as already reported for other important regulatory functions in different cell types [20].

Regarding the possible involvement of galectin-3 in osteoblast and bone pathology, a couple of studies investigated the AGE-receptor role of this lectin in the deleterious effects exerted by AGEs on osteoblasts, as the possible molecular mechanism implicated in the pathogenesis of bone remodeling disorders associated with aging and diabetes. These in vitro studies confirmed that galectin-3, also known as AGE-receptor 3, and the best-characterized AGE receptor RAGE, are both expressed in osteoblastic cell lines, including the MC3T3E1 mouse calvaria-derived osteoblasts, and showed that AGEs are able to induce mRNA and protein levels of both AGE receptors [43,44]. In a previous study, the same investigators showed that exposure to AGEs elicits a biphasic response in osteoblasts, with an early increase in cellular proliferation and expression of differentiation markers, followed by induction of apoptosis and reduced differentiation in longer exposure to AGEs [45]. Interestingly, the effect of AGE exposure on galectin-3 and RAGE up-regulation was also time-dependent, with a prompt induction of galectin-3 expression and a delayed increase of RAGE levels, which coincided with the opposite biological effects exerted by AGEs in osteoblasts [44]. Therefore, galectin-3 could be involved in the early effects of AGEs on osteoblasts, i.e., proliferation and differentiation, while RAGE could be responsible for the long-term harmful effects of these by-products, such as radical oxygen species (ROS)-induced apoptosis, activation of the Extracellular Signal-regulated Kinase signal transduction pathway and inhibition of differentiation. These results are also consistent with the opposite effects exerted by galectin-3 and RAGE in AGE-mediated disease conditions, with the former playing a protective role through clearance of these toxic by-products via endocytosis, and the second implicated in tissue injury [20,64]. Altogether, these findings point to a possible role of the AGE/RAGE axis in the disorders of bone remodeling associated with aging and diabetes, with galectin-3 as a putative protecting factor through scavenging of AGEs and promotion of osteoblast differentiation and function.

3.2. Galectin-3 in Osteogenic Differentiation Capacity of Mesenchymal Stem Cells

MSCs are important for tissue and organ regeneration, including the bone. Consistently, it has been reported that MSC commitment towards the osteogenic phenotype decreases with age, an observation that might partly explain the impaired osteoblastogenesis observed in age-associated osteoporosis [65,66]. However, though the number of mature osteoblasts is progressively decreasing with advancing age, it was reported that the number of MSCs in mature bone remains constant throughout life [67]. This suggests that the age-dependent gradual reduction of ostoblastogenesis might be the consequence of a loss of function of MSCs or, alternatively, a perturbation of the environmental signals involved in the osteogenic commitment of MSCs [46], such as growth hormone [68] and estrogens [69].

Interestingly, it has been recently demonstrated that exposure to the systemic environment of young, but not elderly individuals favors MSC functionality and bone repair through modulation of β-catenin [70]. Therefore, attention has recently focused on the search for secreted circulating factors favoring stem and progenitor cell function. In this endeavor, special consideration has been paid to extracellular vesicles (EVs), which are small vesicles carrying proteins, mRNAs, and other molecules through the circulation [71]. EVs are released by most of the body's cells [72] and their cargo is delivered to specific recipient cells [73]. Recently, Weilner et al. demonstrated that EVs isolated from young donors were more effective in inducing osteoblastogenesis in vitro compared to vesicles derived from elderly individuals. Interestingly, this osteoblastogenic donor age-dependent effect mediated by EVs was directly associated with galectin-3 content, which was higher in the EVs from young individuals [46]. These data indicated that galectin-3 not only plays an important role in the late stage of osteoblast differentiation and maturation, as previously shown [32,41], but also enhances the osteogenic differentiation capacity of MSCs favoring β-catenin activity. Mechanistically, galectin-3's serine-96 phosphorylation site was found to compete for the glycogen synthase kinase-3β phosphorylation site on β-catenin, which represents a crucial step in the degradation of this transcriptional coactivator [46].

Finally, Weilner et al. also showed that even a moderate increase of galectin-3 expression induced ALP and RUNX2 expression in MSCs, thus suggesting that galectin-3 is not only a target gene, but it is also a positive regulator of the master osteogenic regulator RUNX2 [46].

In conclusion, these data confirm and expand previous reports on the importance of galectin-3 in osteoblast differentiation and propose an additional role for this lectin in bone remodeling, demonstrating that MSCs osteogenic differentiation is dependent on galectin-3 levels in circulating EVs. Therefore, it is tempting to speculate that the age-dependent reduction of galectin-3 EVs content might contribute to the reduction of bone formation, development of osteoporosis, and increased facture risk in the elderly [46].

3.3. Galectin-3 in Osteoclast Biology

Osteoclasts are large multinucleated cells responsible for bone resorption by degrading mineralized matrix. This cell type is critical for normal bone remodeling and is responsible for bone loss in pathologic conditions by increasing resorptive activity. Osteoclasts are terminally differentiated cells from immature hematopoietic monocyte/macrophage progenitors, a process regulated by multiple factors, including receptor activator of NF-κB ligand (RANKL), its decoy receptor osteoprotogerin (OPG), and macrophage colony-stimulating factor released by osteoblasts [74]. Once recruited to the bone, osteoclast precursors crawl to the resorption site, fuse with other precursors, and attach to the bone surface. Attachment to the bone matrix is a critical event in osteoclast activation, leading to osteoclast actin cytoskeletal reorganization and formation of sealing zones, which are composed of specialized integrin-mediated adhesive structures called podosomes [75]. It is well-known that protein-bound carbohydrates on the outer cell surface contribute to specific cell-cell and cell-matrix recognition [47]. Therefore, receptors for carbohydrates on osteoclasts and their precursors likely have a pivotal role in mediating cell-matrix interactions and, possibly, in regulating other related activities, such as migration, differentiation, and bone resorption. Since galectin-3 has been recognized as an abundant laminin-binding protein of macrophages [76], which share with osteoclasts a common lineage, its possible function in osteoclast activity has attracted some attention.

Galectin-3 expression by osteoclasts was originally described by Niida et al. in 1994. They reported that mouse tartrate-resistant acid phosphatase (TRAP)-positive mononuclear precursors (pre-osteoclasts) and mature osteoclasts are positive for galectin-3, but negative for the specific murine macrophage marker F4/80. In more detail, galectin-3 was detected in the cytoplasm and nucleus of pre-osteoclasts, as well as on their plasma membrane, suggesting a potential role of this lectin in cell-matrix adhesion during differentiation [48]. These initial findings were confirmed by Colnot et al. who reported that mature resorbing cells present an immunostaining for galectin-3 in their cytoplasm, including the multinucleated osteoclasts actively involved in resorbing the calcified cartilage core in the metaphysis and those present in the trabecular bone of neonatal mice [1]. They also noted positive staining for galectin-3 in mononuclear cells inside the bone marrow cavity [1]. Therefore, from these initial reports, it appeared that galectin-3 is expressed by bone osteoclasts and by their monocyte progenitors. Since galectin-3 is a well-known macrophage marker and is essential for phagocytosis [77], this lectin can be considered as a marker of the entire monocyte-macrophage-osteoclast cell lineage and it can be assumed that it is critical for some important osteoclast function. A few years later, Gorski et al. found two new alternatively spliced sequences of galectin-3 expressed by chicken osteoclasts [47]. Interestingly, galectin-3 expression was found to rise dramatically in bone marrow cells after 5 days of in vitro culture in an osteoclast differentiation media, suggesting a role for this lectin in the acquisition of the main phenotypic characteristics of osteoclasts, such as displaying TRAP activity, multinuclearity, and owning the capacity to resorb bone [47].

The role of galectin-3 in regulating osteoclast differentiation and activity was also investigated by Ortega et al. These Authors found that, besides playing a role in hypertrophic chondrocyte differentiation, galectin-3 also regulates osteoclast recruitment to the primary ossification center during endochondral bone formation, as well as differentiation of mononuclear osteoclast progenitors and osteoclasts survival at the chondro-osseous junction [40]. Consistent with a key role for endogenous galectin-3 in the biology of cells of the osteoclast lineage, Ortega et al. reported that mice null for MMP-9, the prominent MMP for cleavage of the N-terminal domain of galectin-3, displayed an increased number of galectin-3 positive cells, in parallel with an increase of TRAP immunostaining at the front of ossification. They also confirmed these data in an embryonic metatarsal culture system, observing that an excess of exogenous recombinant galectin-3, besides from increasing the number of TRAP-positive cells at the front of ossification, also interfered with recruitment of osteoclast precursors. Altogether, these findings suggest that aberrant persistence of higher levels of extracellular galectin-3, both exogenous or as a consequence of reduced cleavage by MMP-9, may induce abnormal osteoclast survival and, eventually, excessive bone remodeling [40].

3.4. Galectin-3 in Bone Remodeling

Bone remodeling is a lifelong process by which bone is renewed to maintain bone mass. Bone remodeling involves finely orchestrated cellular and molecular events and is commonly viewed as a two-step process carried on by bone cells: it involves old bone resorption by osteoclasts followed by new bone formation by osteoblasts [78]. These two arms of bone remodeling must occur in a balanced and coordinated manner in order to maintain bone mass and shape largely unchanged throughout adult life. Therefore, both increased bone resorption by osteoclasts or reduced bone formation by osteoblasts can lead to bone loss [79,80]. Indeed, bone metabolism is more complex than a two-step process, and there is still a lot to be learned about cellular and molecular mechanisms involved in remodeling. Particularly, unbalance between bone resorption and bone formation leading to reduced bone mass might be the result of multifaceted events affecting simultaneously osteoblasts and osteoclasts. Unfortunately, our knowledge of the molecular regulators that modulate differentiation and activity of osteoclasts and osteoblasts is still insufficient [81], thus hampering the identification of new therapeutic strategies to reduce the health burden and costs related to osteoporotic fractures, which are expected to increase in the future [82].

Although galectin-3 has been found to play a critical role in the differentiation and/or function of chondrocytes, osteoblasts osteocytes and osteoclasts [1,31,32,36–42,44,45,47,48,62,63], and even in favoring MSC osteogenic differentiation [46], so far, only one study has addressed the role of this lectin in bone homeostasis [49]. Simon et al. demonstrated that $Lgals3^{-/-}$ mice exhibited altered bone homeostasis, as attested by histomorphometric analysis and micro-computer tomography measurements revealing a decreased trabecular bone volume in 12-week-old male $Lgals3^{-/-}$ mice compared to wild-type littermates. Surprisingly, however, the role of galectin-3 in osteoblast differentiation was not investigated in this study, nor it was evaluated the effect of galectin-3 deficiency on the ability of osteoblast to form calcified nodules in vitro and new bone (bone formation rate) in vivo. However, at this age, the number of osteoblasts in bones of $Lgals3^{-/-}$ mice was not different compared with wild-type mice [49]. At variance, bones of $Lgals3^{-/-}$ mice exhibited an increased number of osteoclasts in the absence of changed RANKL/OPG ratio, a finding suggesting a direct inhibitory effect of galectin-3 on osteoclastogenesis. Consistent with this hypothesis, but at variance with previous studies indicating a positive effect of galectin-3 on osteoclast differentiation and maturation [40,47,48], bone marrow cells from $Lgals3^{-/-}$ mice displayed an increased osteoclastogenic ability in ex vivo differentiation assays combined with a higher resorption activity. From a mechanistic perspective, molecular analysis revealed higher mRNA levels of the tumor necrosis factor receptor associated factor 6, a critical factor in RANKL signaling and terminal differentiation of osteoclast progenitors. Also in contrast to previous studies [40,47,48], exogenous galectin-3 added to wild-type murine or human osteoclasts was able to inhibit osteoclast differentiation. Moreover, co-culture assays of wild-type osteoclasts with galectin-3 deficient osteoblasts favored osteoclast maturation (size) and number [49]. Based on these findings, Simon et al. concluded that galectin-3 may be part of a molecular mechanisms through which osteoblasts control osteoclastogenesis at sites of mineralization. However, further research is needed to characterize the bone phenotype of $Lgals3^{-/-}$ mice during growth and aging, to clarify the role of endogenous galectin-3 in osteoclast differentiation and function and, importantly, to elucidate the role of this lectin in regulating osteoblast activity in vivo. Regarding this last topic, previous reports described irregular structure of the extracellular matrix in bones from $Lgals3^{-/-}$ mice, which displayed loss and reduced length of collagen fibers compared with wild-type animals [42]. Interestingly, these structural abnormalities of the bone matrix may account for the empirical observation of bone femur fragility of these mice [83]. Altogether, these findings suggest a critical role for galectin-3 also in the molecular architecture of the mineralized matrix, in maintaining bone quality, and in the optimization of bone strength. However, these assumptions were not confirmed by Simon et al. who reported unchanged mechanical stiffness of femurs in $Lgals3^{-/-}$ mice compared with wild-type mice, as assessed by four-point bending test [49].

4. Galectin-3 in Inflammatory Bone and Joint Disorders

Arthritis is an inflammatory disorder of the joints. The two most common form of arthritis are rheumatoid arthritis (RA) and osteoarthritis (OA). They share common symptoms, such as pain, swelling, and stiffness of the joints, leading to disability. However, these two inflammatory joint disorders differ in etiological factors and pathogenesis [84]. In fact, while RA is an autoimmune disease characterized by autoantibody production, no specific causes have been identified for OA, which is generally considered a degenerative disease, being the result of "wear and tear" of the joints. Moreover, though chronic inflammation is a feature of both joint diseases, synovitis is a primary event in RA [85], whereas inflammation of the synovial membrane is a late event in OA, secondary to cartilage destruction and erosion [86]. In the clinical setting, synovial levels of galectin-3 are elevated in both RA and OA [87–89], though to a greater extent in the former [87,88]. However, it is unclear how galectin-3 is involved in the pathogenesis and progression of RA and OA, also because of the contradictory results obtained so far from diverse studies on different experimental models of these disorders. For the purpose of the review, we will only discuss in detail the most important experimental studies on the role of galectin-3 in OA and RA and the reader is referred to a recent review on the subject [84].

The role of galectin-3 in articular tissues and OA has been recently investigated using human OA chondrocytes and mice with OA induced by injection of mono-iodoacetate (MIA). Besides confirming the critical role of galectin-3 in chondrocyte survival [1,32,36,37,40,42], this study demonstrated that intracellular galectin-3 preserves cartilage structures, as evidenced by the finding of an accelerated age-dependent cartilage erosion in $Lgals3^{-/-}$ mice. In fact, cartilage score of 4-month-old $Lgals3^{-/-}$ mice was the same as in coeval wild-type mice injected with MIA, which also displayed cartilage lesions similar to those found in $Lgals3^{-/-}$ mice. Moreover, the decrease of subchondral bone surface induced by MIA in wild-type mice was further accentuated in $Lgals3^{-/-}$ mice, suggesting that galectin-3 may be essential for remodeling of the subchondral bone. Thus, this study showed a protective role of intracellular galectin-3 in OA and suggested that it is essential for cartilage homeostasis during ageing as its absence induced OA-like lesions. In addition, lack of galectin-3 during the OA process accelerated tissue injury in the joint [90].

In RA, the chronic inflammation of the synovial lining is accompanied by severe destruction of articular cartilage and bone erosion [85], which is mainly mediated by osteoclasts [50,91]. The prominent role of osteoclasts was also demonstrated using mice deficient in RANKL, an essential factor for osteoclast differentiation, showing that they were protected from articular bone loss in a serum transfer model of arthritis [92]. However, to make things more complicated, osteoblast function and new bone formation may also be negatively affected at sites of inflammation and erosion, thereby contributing to the net loss of bone [93]. Li et al. investigated the involvement of galectin-3 in bone destruction around the ankle joints in rats with adjuvant-induced arthritis (AA), which is considered an experimental model of human RA [94]. In this study, galectin-3 was shown to be abundantly expressed by macrophages and granulocytes accumulated in the areas of severe bone destruction in rats with AA, but only weakly by osteoclasts. This observation indicated an association between galectin-3 expression and incidence of arthritis. Moreover, galectin-3 injection in the affected joints was able to suppress bone destruction and reduce osteoclast recruitment, suggesting a protective role of the lectin. In vitro experiments confirmed the in vivo findings, demonstrating that exogenous recombinant galectin-3 was able to inhibit both formation of osteoclasts and dentine resorption by mature osteoclasts. Therefore, this study demonstrated a protective role of galectin-3 in AA-induced bone destruction and suggested that the protective effect of galectin-3 may be mediated by its negative regulation of osteoclastogenesis [94]. However, this conclusion is in contrast with previous data suggesting a critical role for this lectin in osteoclast differentiation and function [40,47]. Moreover, from these data, it cannot be ruled-out a protective role mediated by the anti-inflammatory, pro-resolution effect of galectin-3 [20,27–30,95]. Likewise, it cannot be excluded the possibility that the pro-osteogenic properties of galectin-3 may have contributed to protection from net bone loss. It must also be noticed that a recent work has put into question the results obtained by Li et al., proving that *Lgals3*$^{-/-}$ mice are protected from joint injury associated with AA [96].

To our knowledge, the only study on the role of galectin-3 in a specific, non-tumoral, bone pathology was performed by Wehrhan et al. who investigated the molecular mechanisms involved in the pathogenesis of bisphosphonate-associated osteonecrosis of the jaw (BRONJ). BRONJ is an oral disorder characterized by an impairment in hard- and soft tissue repair. It is characterized by bisphosphonate-related impairment of oral tissues remodeling, involving bone cells, fibroblasts, and keratinocytes [97]. In addition, Reid et al. reported that BRONJ was associated to increased osteogenic differentiation of mucoperiosteal progenitors, anergy of the affected tissues, and increased galectin-3 expression in the periosteum and fibrous tissue stroma cells [97], thereby differing from osteoradionecrosis-related mucoperiosteal tissue, which was characterized by local inflammation and significantly less galectin-3 staining [98]. On the basis of the evidences indicating that galectin-3 is involved in the differentiation of osteoblasts and chondroblasts [1,31,32,36,37,40–42,44,45] and mediates inhibition of intraoral inflammation [97,99], the Authors concluded that the higher levels of galectin-3 might be involved in both the increased osteogenic differentiation of the mucoperiosteal precursors and the inflammatory anergy observed in BRONJ-affected soft tissues.

5. Galectin-3 in Vascular Osteogenesis

A large body of evidence indicates that vascular calcification is an unfavorable event in the natural history of the atherosclerotic disease. Consistently, coronary calcium score independently predicts cardiovascular morbidity and mortality, thus directly linking the amount of calcium in the vessel wall with unfavorable outcomes [100,101]. However, recent findings suggest that vascular calcification may either promote plaque progression and vulnerability or favor plaque stability, depending on the elemental composition [102] and pattern [33] of calcium deposition within the vessel wall. For instance, intravascular ultrasound studies indicate that spotty, granular and less dense calcification (commonly referred to as "microcalcification") within a fibro-atheromatous plaque is the main pattern of calcium deposition in patients with acute myocardial infarction, whereas extensive dense calcification (commonly referred to as "macrocalcification") prevails in subjects with stable angina pectoris [103–105]. However, despite the increasing body of evidence supporting the dual role of calcification in plaque stability and its clinical implications, little is known about the biological mechanisms that favor one type of vascular calcification over the other.

Though previously considered as a passive degenerative phenomenon [106], vascular calcification is now proven to be an active, tightly regulated process, similar to bone mineralization, activating an osteogenic gene regulatory program and involving various bone-related proteins. Many mechanisms have been proposed to explain the phenomenon of ectopic calcification within the vessel wall, including the osteogenic transdifferentiation of vascular smooth muscle cells (VSMCs) [107]. These specialized contracting cells are an important structural component of the artery wall, playing a key role in maintaining vascular tone and participating to various important functions, such as the regulation of blood pressure. VSMCs also play a key role in different pathological processes affecting the vasculature, including the pathobiology of vascular calcification. The process of VSMCs transdifferentiation towards osteochondrogenic precursors is a process characterized by progressive expression of the osteochondrogenic markers OCN, ALP, and OPN, associated with the expression of the osteochondrogenic transcription factor RUNX2. This process is also accompanied by phenotypic transition of VSMCs, attested by progressive reduction of expression of α smooth muscle actin, the main smooth muscle lineage marker, and loss of the contractile phenotype [108,109]. Several pro-inflammatory and pro-osteogenic stimuli have been shown to promote vascular calcification in the setting of atherosclerosis, including ROS [110], AGEs [111], advanced lipoxidation end-products [112], calgranulins [113], BMPs [114], and OPG [115]. However, most of the in vitro and in vivo studies performed to investigate the role of these stimuli in atherosclerotic vascular calcification have only assessed the quantitative aspect of the phenomenon, irrespective of the type of calcification process in which they are involved.

Some insight into the mechanism regulating the pattern of calcification in the process of atherogenesis has been recently provided [33]. Interestingly, this mechanism involves galectin-3 and RAGE, which were previously shown to modulate in opposite ways osteoblast differentiation and function in the presence of AGEs. In fact, while galectin-3 was shown to favor osteoblast differentiation, RAGE was found to be responsible for the detrimental effects of AGEs on osteoblasts, such as apoptosis and reduced differentiation [44]. Consistent with a pro-osteogenic effect of galectin-3 also in VSMC transdifferentiation, the increased expression of osteogenic markers induced by incubation in osteogenic medium was accompanied by a parallel increase in galectin-3 expression, in accordance with what was observed in differentiating osteoblasts [31]. In addition, galectin-3 deficiency was associated with defective osteogenic transdifferentiation of VSMCs, as attested by reduced expression of the osteogenic transcription factor RUNX2 and the bone cell marker ALP, enhanced apoptosis, reduced proliferation, and disorganized mineralization [33]. $Lgals3^{-/-}$ VSMCs cells also showed decreased activation of Wnt/β-catenin signaling, which is in agreement with galectin-3's ability to stabilize and increase β-catenin levels in MSCs differentiating towards the osteogenic lineage [46]. Noteworthy, the reduced ability of $Lgals3^{-/-}$ VSMCs to acquire an osteogenic phenotype was also associated with RAGE up-regulation, and was exacerbated by treatment with AGEs, once again in keeping with previous data on osteoblast differentiation [44]. In vivo analysis of carotid plaques from patients undergoing carotid endarterectomy supported the hypothesis that galectin-3 and RAGE regulate in opposite ways the pattern of calcification, with RAGE expression observed in plaque regions with diffuse microcalcification and galectin-3 being mainly expressed by VSMCs in regions adjacent to extended and dense macrocalcification, where it colocalized with the osteogenic marker ALP [33].

Altogether, these data demonstrated that galectin-3 is essential for full transdifferentiation of VSMCs into an osteoblast-like phenotype via modulation of the Wnt/β-catenin pathway. They also provided evidence for a role of galectin-3 and RAGE in determining the pattern of calcification in atherosclerotic lesions. Noteworthy, the roles played by galectin-3 and RAGE in regulating VSMC transdifferentiation and vascular calcification recapitulate the effects they have at the skeletal level. In fact, galectin-3 was shown to regulate chondroblast/osteoblast differentiation/function and bone development [1,31,41]. Conversely, RAGE was shown to be essential for osteoclast maturation and function in skeletal tissue both in vivo and in vitro [116].

Finally, an ancillary observation is that a large body of literature indicates a role of galectin-3 as a prognostic marker in cardiovascular disease. In particular, elevated levels of circulating galectin-3 have been found to be associated with higher risk of death in heart failure patients [117]. Moreover, some experimental evidence suggests an active role of this lectin in cardiac remodeling and atherosclerosis, mainly through its recognized fibrogenic properties [20,117]. However, given the important role played by vascular calcification in atherosclerosis and cardiovascular disease, and that of galectin-3 in VSMCs osteogenic transdifferentiation, it cannot be rule out the possibility that circulating galectin-3 might affect cardiovascular disease also through its osteogenic properties.

6. Conclusions

Most of the study conducted to investigate the role of galectin-3 in the biology of bone and its cells date back to the late 90s of last century and the beginning of 2000s. Although they produced convincing evidence about the involvement of this lectin in multiple biologic processes of the bone, such as differentiation and activity of all major bone cells, they were not followed by an in-depth analysis of the role of galectin-3 in the bone physiology and pathology. Only recently, the skeleton of *Lgals3*$^{-/-}$ mice has been analyzed and results demonstrated that, in fact, adult (3-month-old) mice have a bone phenotype, characterized by reduced trabecular bone compared with wild-type mice [49]. However, the analysis of the mechanisms underlying the structural abnormalities only investigated the resorption arm of the bone remodeling process, leaving completely unexplored the bone deposition arm and the cells responsible for bone formation. Since, galectin-3 has been shown to have a critical role for chondroblast and osteoblast differentiation and activity [1,31,32,36,37,40–42,44,45], an in-depth analysis of these cells and their function in *Lgals3*$^{-/-}$ mice is necessary to understand the net effect of this lectin on bone homeostasis. It would also be desirable to extend the analysis to the skeleton of younger and older *Lgals3*$^{-/-}$ to get information on the role of this lectin in growth and age-dependent changes of the bony structure. These analyses could provide valuable information possibly helping in identifying galectin-3 as potential therapeutic target in some bone pathologies, primarily senile osteoporosis. This concept is strengthened by the evidence that, in addition to its role in osteogenesis, (a) galectin-3 is able to stimulate osteogenic differentiation of MSCs [46]; (b) in osteoblast lineage cells [43–45], it acts as a receptor capable of degrading AGEs, which have been implicated in age-dependent [118] and diabetes-associated [119] bone fragility; and (c) it may favor resolution of chronic inflammation [20], also in the bone tissue [20,97–99].

Finally, the osteogenic properties of galectin-3 deserve to also be investigated in the process of ectopic calcification, primarily vascular osteogenesis. The possibility of modifying the amount and especially the type of calcification in the atherosclerotic plaque by modulating the expression of galectin-3 could be very useful in stabilizing the atherosclerotic lesions, thus reducing the risk of fatal events.

Acknowledgments: This work was supported by a research grant of the Research Foundation of the Italian Diabetes Society (Diabete Ricerca 2013) and Sapienza University of Rome—Progetti di Ateneo 2013 to Stefano Menini.

Author Contributions: Carla Iacobini, Blasetti Fantauzzi, and Giuseppe Pugliese critically revised the article for important intellectual content; Stefano Menini drafted the article.

Conflicts of Interest: The authors declare no conflict of interest.

Abbreviations

AGEs	Advanced Glycation
RAGE	Receptor for AGEs
RUNX2	Runt-Related Transcription Factor 2
Lgals3$^{-/-}$	Galectin-3 Null
MMPs	Metalloproteinases
MSC	Mesenchymal Stem Cell
BMPs	Bone Morphogenic Proteins
OCN	Osteocalcin
ALP	Alkaline Phosphatase
BSP	Bone Sialoprotein
OPN	Osteopontin
EVs	Extracellular Vesicles
RANKL	Receptor Activator of NF-κB Ligand
OPG	Osteoprotogerin
TRAP	Tartrate-Resistant Acid Phosphatase
RA	Rheumatoid Arthritis
OA	Osteoarthritis
MIA	Mono-Iodoacetate
AA	Adjuvant-Induced Arthritis
BRONJ	Bisphosphonate-Associated Osteonecrosis of the Jaw

References

1. Colnot, C.; Sidhu, S.S.; Poirier, F.; Balmain, N. Cellular and subcellular distribution of galectin-3 in the epiphyseal cartilage and bone of fetal and neonatal mice. *Cell. Mol. Biol.* **1999**, *45*, 1191–1202. [PubMed]
2. Iacobini, C.; Amadio, L.; Oddi, G.; Ricci, C.; Barsotti, P.; Missori, S.; Sorcini, M.; Di Mario, U.; Pricci, F.; Pugliese, G. Role of galectin-3 in diabetic nephropathy. *J. Am. Soc. Nephrol.* **2003**, *14*, S264–S270. [CrossRef] [PubMed]
3. Liu, F.T.; Patterson, R.J.; Wang, J.L. Intracellular functions of galectins. *BBA Gen. Subj.* **2002**, *1572*, 263–273. [CrossRef]
4. Menon, R.P.; Hughes, R.C. Determinants in the N-terminal domains of galectin-3 for secretion by a novel pathway circumventing the endoplasmic reticulum-Golgi complex. *Eur. J. Biochem.* **1999**, *264*, 569–576. [CrossRef] [PubMed]
5. Dennis, J.W.; Nabi, I.R.; Demetriou, M. Metabolism, cell surface organization, and disease. *Cell* **2009**, *139*, 1229–1241. [CrossRef] [PubMed]
6. Partridge, E.A.; Le Roy, C.; Di Guglielmo, G.M.; Pawling, J.; Cheung, P.; Granovsky, M.; Nabi, I.R.; Wrana, J.L.; Dennis, J.W. Regulation of cytokine receptors by golgi *N*-glycan processing and endocytosis. *Science* **2004**, *306*, 120–124. [CrossRef] [PubMed]
7. Lakshminarayan, R.; Wunder, C.; Becken, U.; Howes, M.T.; Benzing, C.; Arumugam, S.; Sales, S.; Ariotti, N.; Chambon, V.; Lamaze, C.; et al. Galectin-3 drives glycosphingolipid-dependent biogenesis of clathrinin-dependent carriers. *Nat. Cell Biol.* **2014**, *16*, 595–606. [CrossRef] [PubMed]
8. Haudek, K.C.; Patterson, R.J.; Wang, J.L. SR proteins and galectins: What's in a name? *Glycobiology* **2010**, *20*, 1199–1207. [CrossRef] [PubMed]
9. Kim, H.R.; Lin, H.M.; Biliran, H.; Raz, A. Cell cycle arrest and inhibition of anoikis by galectin-3 in human breast epithelial cells. *Cancer Res.* **1999**, *59*, 4148–4154. [PubMed]
10. Shimura, T.; Takenaka, Y.; Tsutsumi, S.; Hogan, V.; Kikuchi, A.; Raz, A. Galectin-3, a novel binding partner of beta-catenin. *Cancer Res.* **2004**, *64*, 6363–6367. [CrossRef] [PubMed]
11. Yang, R.Y.; Hsu, D.K.; Liu, F.-T. Expression of galectin-3 modulates T-cell growth and apoptosis. *Proc. Natl. Acad. Sci. USA* **1996**, *93*, 6737–6742. [CrossRef] [PubMed]
12. Inohara, H.; Akahani, S.; Raz, A. Galectin-3 stimulates cell proliferation. *Exp. Cell Res.* **1998**, *245*, 294–302. [CrossRef] [PubMed]
13. Akahani, S.; Nangia-Makker, P.; Inohara, H.; Kim, H.R.; Raz, A. Galectin-3: A novel antiapoptotic molecule with a functional BH1 [NWGR] domain of Bcl-2 family. *Cancer Res.* **1997**, *57*, 5272–5276. [PubMed]
14. Yang, R.; Rabinovich, G.; Liu, F. Galectins: Structure, function and therapeutic potential. *Expert Rev. Mol. Med.* **2008**, *10*, 1–24. [CrossRef] [PubMed]
15. Lau, K.S.; Partridge, E.A.; Grigorian, A.; Silvescu, C.I.; Reinhold, V.N.; Demetriou, M.; Dennis, J.W. Complex *N*-glycan number and degree of branching cooperate to regulate cell proliferation and differentiation. *Cell* **2007**, *129*, 123–134. [CrossRef] [PubMed]
16. Inohara, H.; Raz, A. Functional evidence that cell surface galectin-3 mediates homotypic cell adhesion. *Cancer Res.* **1995**, *55*, 3267–3271. [PubMed]
17. Friedrichs, J.; Torkko, J.M.; Helenius, J.; Teräväinen, T.P.; Füllekrug, J.; Muller, D.J.; Simons, K.; Manninen, A. Contributions of galectin-3 and -9 to epithelial cell adhesion analyzed by single cell force spectroscopy. *J. Biol. Chem.* **2007**, *282*, 29375–29383. [CrossRef] [PubMed]
18. Ochieng, J.; Leite-Browning, M.L.; Warfield, P. Regulation of cellular adhesion to extracellular matrix proteins by galectin-3. *Biochem. Biophys. Res. Commun.* **1998**, *246*, 788–791. [CrossRef] [PubMed]
19. Friedrichs, J.; Manninen, A.; Muller, D.J.; Helenius, J. Galectin-3 regulates integrin alpha2beta1-mediated adhesion to collagen-I and -IV. *J. Biol. Chem.* **2008**, *283*, 32264–32272. [CrossRef] [PubMed]
20. Pugliese, G.; Iacobini, C.; Pesce, C.M.; Menini, S. Galectin-3: An emerging all-out player in metabolic disorders and their complications. *Glycobiology* **2015**, *25*, 136–150. [CrossRef] [PubMed]
21. Brownlee, M. Advanced protein glycosylation in diabetes and aging. *Annu. Rev. Med.* **1995**, *46*, 223–234. [CrossRef] [PubMed]
22. Schmidt, A.M.; Yan, S.D.; Yan, S.F.; Stern, D.M. The multiligand receptor RAGE as a progression factor amplifying immune and inflammatory responses. *J. Clin. Investig.* **2001**, *108*, 949–955. [CrossRef] [PubMed]

23. Basta, G.; Lazzerini, G.; Del Turco, S.; Ratto, G.M.; Schmidt, A.M.; De Caterina, R. At least 2 distinct pathways generating reactive oxygen species mediate vascular cell adhesion molecule-1 induction by advanced glycation end products. *Arterioscler. Thromb. Vasc. Biol.* **2005**, *25*, 1401–1407. [CrossRef] [PubMed]

24. Bierhaus, A.; Humpert, P.M.; Morcos, M.; Wendt, T.; Chavakis, T.; Arnold, B.; Stern, D.M.; Nawroth, P.P. Understanding RAGE, the receptor for advanced glycation end products. *J. Mol. Med.* **2005**, *83*, 876–886. [CrossRef] [PubMed]

25. Pricci, F.; Leto, G.; Amadio, L.; Iacobini, C.; Romeo, G.; Cordone, S.; Gradini, R.; Barsotti, P.; Liu, F.T.; Di Mario, U.; et al. Role of galectin-3 as a receptor for advanced glycosylation end products. *Kidney Int.* **2000**, *77*, S31–S39. [CrossRef]

26. Norling, L.V.; Perretti, M.; Cooper, D. Endogenous galectins and the control of the host inflammatory response. *J. Endocrinol.* **2009**, *201*, 169–184. [CrossRef] [PubMed]

27. Fukumori, T.; Takenaka, Y.; Yoshii, T.; Kim, H.R.; Hogan, V.; Inohara, H.; Kagawa, S.; Raz, A. CD29 and CD7 mediate galectin-3-induced type II T-cell apoptosis. *Cancer Res.* **2003**, *63*, 8302–8311. [PubMed]

28. Morgan, R.; Gao, G.; Pawling, J.; Dennis, J.W.; Demetriou, M.; Li, B. *N*-acetylglucosaminyltransferase V [Mgat5]-mediated *N*-glycosylation negatively regulates Th1 cytokine production by T cells. *J. Immunol.* **2004**, *173*, 7200–7208. [CrossRef] [PubMed]

29. Demetriou, M.; Granovsky, M.; Quaggin, S.; Dennis, J.W. Negative regulation of T-cell activation and autoimmunity by Mgat5 *N*-glycosylation. *Nature* **2001**, *409*, 733–739. [CrossRef] [PubMed]

30. MacKinnon, A.C.; Farnworth, S.L.; Hodkinson, P.S.; Henderson, N.C.; Atkinson, K.M.; Leffler, H.; Nilsson, U.J.; Haslett, C.; Forbes, S.J.; Sethi, T. Regulation of alternative macrophage activation by galectin-3. *J. Immunol.* **2008**, *180*, 2650–2658. [CrossRef] [PubMed]

31. Aubin, J.E.; Liu, F.; Malaval, L.; Gupta, A.K. Osteoblast and chondroblast differentiation. *Bone* **1995**, *17*, 77S–83S. [CrossRef]

32. Stock, M.; Schäfer, H.; Stricker, S.; Gross, G.; Mundlos, S.; Otto, F. Expression of galectin-3 in skeletal tissues is controlled by Runx2. *J. Biol. Chem.* **2003**, *278*, 17360–17367. [CrossRef] [PubMed]

33. Menini, S.; Iacobini, C.; Ricci, C.; Blasetti Fantauzzi, C.; Salvi, L.; Pesce, C.M.; Relucenti, M.; Familiari, G.; Taurino, M.; Pugliese, G. The galectin-3/RAGE dyad modulates vascular osteogenesis in atherosclerosis. *Cardiovasc. Res.* **2013**, *100*, 472–480. [CrossRef] [PubMed]

34. Pugliese, G.; Iacobini, C.; Blasetti Fantauzzi, C.; Menini, S. The dark and bright side of atherosclerotic calcification. *Atherosclerosis* **2015**, *238*, 220–230. [CrossRef] [PubMed]

35. Sádaba, J.R.; Martínez-Martínez, E.; Arrieta, V.; Álvarez, V.; Fernández-Celis, A.; Ibarrola, J.; Melero, A.; Rossignol, P.; Cachofeiro, V.; López-Andrés, N. Role for galectin-3 in calcific aortic valve stenosis. *J. Am. Heart Assoc.* **2016**, *5*, e004360. [CrossRef] [PubMed]

36. Roach, H.I. New aspects of endochondral ossification in the chick: Chondrocyte apoptosis, bone formation by former chondrocytes, and acid phosphatase activity in the endochondral bone matrix. *J. Bone Miner. Res.* **1997**, *12*, 795–805. [CrossRef] [PubMed]

37. Sakakura, Y.; Hosokawa, Y.; Tsuruga, E.; Irie, K.; Nakamura, M.; Yajima, T. Contributions of matrix metalloproteinases toward Meckel's cartilage resorption in mice: Immunohistochemical studies, including comparisons with developing endochondral bones. *Cell Tissue Res.* **2007**, *328*, 137–151. [CrossRef] [PubMed]

38. Dange, M.C.; Agarwal, A.K.; Kalraiya, R.D. Extracellular galectin-3 induces MMP9 expression by activating p38 MAPK pathway via lysosome-associated membrane protein-1 [LAMP1]. *Mol. Cell. Biochem.* **2015**, *404*, 79–86. [CrossRef] [PubMed]

39. Mauris, J.; Woodward, A.M.; Cao, Z.; Panjwani, N.; Argüeso, P. Molecular basis for MMP9 induction and disruption of epithelial cell-cell contacts by galectin-3. *J. Cell Sci.* **2014**, *127*, 3141–3148. [CrossRef] [PubMed]

40. Ortega, N.; Behonick, D.J.; Colnot, C.; Cooper, D.N.; Werb, Z. Galectin-3 is a downstream regulator of matrix metalloproteinase-9 function during endochondral bone formation. *Mol. Biol. Cell* **2005**, *16*, 3028–3039. [CrossRef] [PubMed]

41. Aubin, J.E.; Gupta, A.K.; Bhargava, U.; Turksen, K. Expression and regulation of galectin-3 in rat osteoblastic cells. *J. Cell Physiol.* **1996**, *169*, 468–480. [CrossRef]

42. Colnot, C.; Sidhu, S.S.; Balmain, N.; Poirier, F. Uncoupling of chondrocyte death and vascular invasion in mouse galectin 3 null mutant bones. *Dev. Biol.* **2001**, *229*, 203–214. [CrossRef] [PubMed]

43. Mercer, N.; Ahmed, H.; McCarthy, A.D.; Etcheverry, S.B.; Vasta, G.R.; Cortizo, A.M. AGE-R3/galectin-3 expression in osteoblast-like cells: Regulation by AGEs. *Mol. Cell. Biochem.* **2004**, *266*, 17–24. [CrossRef] [PubMed]

44. Mercer, N.; Ahmed, H.; Etcheverry, S.B.; Vasta, G.R.; Cortizo, A.M. Regulation of advanced glycation end product [AGE] receptors and apoptosis by AGEs in osteoblast-like cells. *Mol. Cell. Biochem.* **2007**, *306*, 87–94. [CrossRef] [PubMed]

45. McCarthy, A.D.; Etcheverry, S.B.; Bruzzone, L.; Cortizo, A.M. Effects of advanced glycation end-products on the proliferation and differentiation of osteoblast-like cells. *Mol. Cell. Biochem.* **1997**, *170*, 43–51. [CrossRef]

46. Weilner, S.; Keider, V.; Winter, M.; Harreither, E.; Salzer, B.; Weiss, F.; Schraml, E.; Messner, P.; Pietschmann, P.; Hildner, F.; et al. Vesicular Galectin-3 levels decrease with donor age and contribute to the reduced osteo-inductive potential of human plasma derived extracellular vesicles. *Aging* **2016**, *8*, 16–33. [CrossRef] [PubMed]

47. Gorski, J.P.; Liu, F.T.; Artigues, A.; Castagna, L.F.; Osdoby, P. New alternatively spliced form of galectin-3, a member of the beta-galactoside-binding animal lectin family, contains a predicted transmembrane-spanning domain and a leucine zipper motif. *J. Biol. Chem.* **2002**, *277*, 18840–18848. [CrossRef] [PubMed]

48. Niida, S.; Amizuka, N.; Hara, F.; Ozawa, H.; Kodama, H. Expression of Mac-2 antigen in the preosteoclast and osteoclast identified in the op/op mouse injected with macrophage colony-stimulating factor. *J. Bone Miner. Res.* **1994**, *9*, 873–881. [CrossRef] [PubMed]

49. Simon, D.; Derer, A.; Andes, F.T.; Lezuo, P.; Bozec, A.; Schett, G.; Herrmann, M.; Harre, U. Galectin-3 as a novel regulator of osteoblast-osteoclast interaction and bone homeostasis. *Bone* **2017**, *105*, 35–41. [CrossRef] [PubMed]

50. Gough, A.; Sambrook, P.; Devlin, J.; Huissoon, A.; Njeh, C.; Robbins, S.; Nguyen, T.; Emery, P. Osteoclastic activation is the principal mechanism leading to secondary osteoporosis in rheumatoid arthritis. *J. Rheumatol.* **1998**, *25*, 1282–1289. [PubMed]

51. Caplan, A.I.; Pechak, D.G. The cellular and molecular embryology of bone formation. In *Bone and Mineral Research*; Peck, W.A., Ed.; Elsevier: New York, NY, USA, 1997; Volume 5, pp. 117–183.

52. Poole, A.R. *The Growth Plate: Cellular Physiology, Cartilage Assembly and Mineralization*; CRC Press: Boca Raton, FL, USA, 1991.

53. Volk, S.W.; Leboy, P.S. Regulating the regulators of chondrocyte hypertrophy. *J. Bone Miner. Res.* **1999**, *14*, 483–486. [CrossRef] [PubMed]

54. Yu, F.; Finley, R.L., Jr.; Raz, A.; Kim, H.R. Galectin-3 translocates to the perinuclear membranes and inhibits cytochrome c release from the mitochondria. A role for synexin in galectin-3 translocation. *J. Biol. Chem.* **2002**, *277*, 15819–15827. [CrossRef] [PubMed]

55. Harada, S.; Rodan, G.A. Control of osteoblast function and regulation of bone mass. *Nature* **2003**, *423*, 349–355. [CrossRef] [PubMed]

56. Mackie, E.J. Osteoblasts: Novel roles in orchestration of skeletal architecture. *Int. J. Biochem. Cell Biol.* **2003**, *35*, 1301–1305. [CrossRef]

57. Karsenty, G. The complexities of skeletal biology. *Nature* **2003**, *423*, 316–318. [CrossRef] [PubMed]

58. Krane, S.M. Identifying genes that regulate bone remodeling as potential therapeutic targets. *J. Exp. Med.* **2005**, *201*, 841–843. [CrossRef] [PubMed]

59. Ducy, P.; Schinke, T.; Karsenty, G. The osteoblast: A sophisticated fibroblast under central surveillance. *Science* **2000**, *289*, 1501–1504. [CrossRef] [PubMed]

60. Rochefort, G.Y.; Pallu, S.; Benhamou, C.L. Osteocyte: The unrecognized side of bone tissue. *Osteoporos. Int.* **2010**, *21*, 1457–1469. [CrossRef] [PubMed]

61. Prideaux, M.; Findlay, D.M.; Atkins, G.J. Osteocytes: The master cells in bone remodelling. *Curr. Opin. Pharmacol.* **2016**, *28*, 24–30. [CrossRef] [PubMed]

62. Costessi, A.; Pines, A.; D'Andrea, P.; Romanello, M.; Damante, G.; Cesaratto, L.; Quadrifoglio, F.; Moro, L.; Tell, G. Extracellular nucleotides activate Runx2 in the osteoblast-like HOBIT cell line: A possible molecular link between mechanical stress and osteoblasts' response. *Bone* **2005**, *36*, 418–432. [CrossRef] [PubMed]

63. Nakajima, K.; Kho, D.H.; Yanagawa, T.; Harazono, Y.; Gao, X.; Hogan, V.; Raz, A. Galectin-3 inhibits osteoblast differentiation through notch signaling. *Neoplasia* **2014**, *16*, 939–949. [CrossRef] [PubMed]

64. Menini, S.; Iacobini, C.; Blasetti Fantauzzi, C.; Pesce, C.M.; Pugliese, G. Role of galectin-3 in obesity and impaired glucose homeostasis. *Oxid. Med. Cell. Longev.* **2016**, *2016*, 9618092. [CrossRef] [PubMed]

65. Roholl, P.J.; Blauw, E.; Zurcher, C.; Dormans, J.A.; Theuns, H.M. Evidence for a diminished maturation of preosteoblasts into osteoblasts during aging in rats: An ultrastructural analysis. *J. Bone Miner. Res.* **1994**, *9*, 355–366. [CrossRef] [PubMed]

66. Föger-Samwald, U.; Patsch, J.M.; Schamall, D.; Alaghebandan, A.; Deutschmann, J.; Salem, S.; Mousavi, M.; Pietschmann, P. Molecular evidence of osteoblast dysfunction in elderly men with osteoporotic hip fractures. *Exp. Gentol.* **2014**, *57C*, 114–121. [CrossRef] [PubMed]

67. Almeida, M.; Han, L.; Martin-Millan, M.; Plotkin, L.I.; Stewart, S.A.; Roberson, P.K.; Kousteni, S.; O'Brien, C.A.; Bellido, T.; Parfitt, A.M.; et al. Skeletal involution by age associated oxidative stress and its acceleration by loss of sex steroids. *J. Biol. Chem.* **2007**, *282*, 27285–27297. [CrossRef] [PubMed]

68. Corpas, E.; Harman, S.M.; Blackman, M.R. Human growth hormone and human aging. *Endocr. Rev.* **1993**, *14*, 20–39. [CrossRef] [PubMed]

69. Wilson, S.; Thorp, B.H. Estrogen and cancellous bone loss in the fowl. *Calcif. Tissue Int.* **1998**, *62*, 506–511. [CrossRef] [PubMed]

70. Baht, G.S.; Silkstone, D.; Vi, L.; Nadesan, P.; Amani, Y.; Whetstone, H.; Wei, Q.; Alman, B.A. Exposure to a youthful circulation rejuvenates bone repair through modulation of β-catenin. *Nat. Commun.* **2015**, *6*, 7131. [CrossRef] [PubMed]

71. Nolte-'t Hoen, E.N.; Buermans, H.P.; Waasdorp, M.; Stoorvogel, W.; Wauben, M.H.M.; 'T Hoen, P.A. Deep sequencing of RNA from immune cell-derived vesicles uncovers the selective incorporation of small non-coding RNA biotypes with potential regulatory functions. *Nucleic Acids Res.* **2012**, *40*, 9272–9285. [CrossRef] [PubMed]

72. Fais, S.; Logozzi, M.; Lugini, L.; Federici, C.; Azzarito, T.; Zarovni, N.; Chiesi, A. Exosomes: The ideal nanovectors for biodelivery. *Biol. Chem.* **2013**, *394*, 1–15. [CrossRef] [PubMed]

73. Fevrier, B.; Raposo, G. Exosomes: Endosomal-derived vesicles shipping extracellular messages. *Curr. Opin. Cell Biol.* **2004**, *16*, 415–421. [CrossRef] [PubMed]

74. Boyle, W.J.; Simonet, W.S.; Lacey, D.L. Osteoclast differentiation and activation. *Nature* **2003**, *423*, 337–342. [CrossRef] [PubMed]

75. Teitelbaum, S.L. Osteoclasts, integrins, and osteoporosis. *J. Bone Miner. Metab.* **2000**, *18*, 344–349. [CrossRef] [PubMed]

76. Woo, H.J.; Shaw, L.M.; Messier, J.M.; Mercurio, A.M. The major non-integrin laminin binding protein of macrophages is identical to carbohydrate binding protein 35 [Mac-2]. *J. Biol. Chem.* **1990**, *265*, 7097–7099. [PubMed]

77. Sano, H.; Hsu, D.K.; Apgar, J.R.; Yu, L.; Sharma, B.B.; Kuwabara, I.; Izui, S.; Liu, F.T. Critical role of galectin-3 in phagocytosis by macrophages. *J. Clin. Investig.* **2003**, *112*, 389–397. [CrossRef] [PubMed]

78. Rachner, T.D.; Khosla, S.; Hofbauer, L.C. Osteoporosis: Now and the future. *Lancet* **2011**, *377*, 1276–1287. [CrossRef]

79. Ducy, P.; Amling, M.; Takeda, S.; Priemel, M.; Schilling, A.F.; Beil, F.T.; Shen, J.; Vinson, C.; Rueger, J.M.; Karsenty, G.; et al. Leptin inhibits bone formation through a hypothalamic relay: A central control of bone mass. *Cell* **2000**, *100*, 197–207. [CrossRef]

80. Delaisse, J.M. The reversal phase of the bone-remodeling cycle: Cellular prerequisites for coupling resorption and formation. *BoneKEY Rep.* **2014**, *3*, 561. [CrossRef] [PubMed]

81. Lewiecki, E.M.; Binkley, N. What we do not know about osteoporosis. *J. Endocrinol. Investig.* **2016**, *39*, 491–493. [CrossRef] [PubMed]

82. Blume, S.W.; Curtis, J.R. Medical costs of osteoporosis in the elderly Medicare population. *Osteoporos. Int.* **2011**, *22*, 1835–1844. [CrossRef] [PubMed]

83. Brand, C.; Oliveira, F.L.; Ricon, L.; Fermino, M.L.; Boldrini, L.C.; Hsu, D.K.; Liu, F.T.; Chammas, R.; Borojevic, R.; Farina, M.; et al. The bone marrow compartment is modified in the absence of galectin-3. *Cell Tissue Res.* **2011**, *346*, 427–437. [CrossRef] [PubMed]

84. Hu, Y.; Yéléhé-Okouma, M.; Ea, H.K.; Jouzeau, J.Y.; Reboul, P. Galectin-3: A key player in arthritis. *Jt. Bone Spine* **2017**, *84*, 15–20. [CrossRef] [PubMed]

85. Feldmann, M.; Brennan, F.M.; Maini, R.N. Rheumatoid arthritis. *Cell* **1996**, *85*, 307–310. [CrossRef]

86. Sellam, J.; Berenbaum, F. The role of synovitis in pathophysiology and clinical symptoms of osteoarthritis. *Nat. Rev. Rheumatol.* **2010**, *6*, 625–635. [CrossRef] [PubMed]

87. Ohshima, S.; Kuchen, S.; Seemayer, C.A.; Kyburz, D.; Hirt, A.; Klinzing, S.; Michel, B.A.; Gay, R.E.; Liu, F.T.; Gay, S.; et al. Galectin-3 and its binding protein in rheumatoid arthritis. *Arthritis Rheum.* **2003**, *48*, 2788–2795. [CrossRef] [PubMed]

88. Alturfan, A.A.; Eralp, L.; Emekli, N. Investigation of inflammatory and hemostatic parameters in female patients undergoing total knee arthroplasty surgery. *Inflammation* **2008**, *31*, 414–421. [CrossRef] [PubMed]

89. Neidhart, M.; Zaucke, F.; von Knoch, R.; Jüngel, A.; Michel, B.A.; Gay, R.E.; Gay, S. Galectin-3 is induced in rheumatoid arthritis synovial fibroblasts after adhesion to cartilage oligomeric matrix protein. *Ann. Rheum. Dis.* **2005**, *64*, 419–424. [CrossRef] [PubMed]

90. Boileau, C.; Poirier, F.; Pelletier, J.P.; Guévremont, M.; Duval, N.; Martel-Pelletier, J.; Reboul, P. Intracellular localisation of galectin-3 has a protective role in chondrocyte survival. *Ann. Rheum. Dis.* **2008**, *67*, 175–181. [CrossRef] [PubMed]

91. Gravallese, E.M.; Harada, Y.; Wang, J.T.; Gorn, A.H.; Thornhill, T.S.; Goldring, S.R. Identification of cell types responsible for bone resorption in rheumatoid arthritis and juvenile rheumatoid arthritis. *Am. J. Pathol.* **1998**, *152*, 943–951. [PubMed]

92. Pettit, A.R.; Ji, H.; von Stechow, D.; Müller, R.; Goldring, S.R.; Choi, Y.; Benoist, C.; Gravallese, E.M. TRANCE/RANKL knockout mice are protected from bone erosion in a serum transfer model of arthritis. *Am. J. Pathol.* **2001**, *159*, 1689–1699. [CrossRef]

93. Karmakar, S.; Kay, J.; Gravallese, E.M. Bone damage in rheumatoid arthritis: Mechanistic insights and approaches to prevention. *Rheum. Dis. Clin. N. Am.* **2010**, *36*, 385–404. [CrossRef] [PubMed]

94. Li, Y.J.; Kukita, A.; Teramachi, J.; Nagata, K.; Wu, Z.; Akamine, A.; Kukita, T. A possible suppressive role of galectin-3 in upregulated osteoclastogenesis accompanying adjuvant-induced arthritis in rats. *Lab Investig.* **2009**, *89*, 26–37. [CrossRef] [PubMed]

95. Wright, R.D.; Souza, P.R.; Flak, M.B.; Thedchanamoorthy, P.; Norling, L.V.; Cooper, D. Galectin-3-null mice display defective neutrophil clearance during acute inflammation. *J. Leukoc. Biol.* **2017**, *101*, 717–726. [CrossRef] [PubMed]

96. Forsman, H.; Islander, U.; Andréasson, E.; Andersson, A.; Onnheim, K.; Karlström, A.; Sävman, K.; Magnusson, M.; Brown, K.L.; Karlsson, A. Galectin 3 aggravates joint inflammation and destruction in antigen-induced arthritis. *Arthritis Rheum.* **2011**, *63*, 445–454. [CrossRef] [PubMed]

97. Reid, I.R. Osteonecrosis of the jaw: Who gets it, and why? *Bone* **2009**, *44*, 4–10. [CrossRef] [PubMed]

98. Wehrhan, F.; Hyckel, P.; Guentsch, A.; Nkenke, E.; Stockmann, P.; Schlegel, K.A.; Neukam, F.W.; Amann, K. Bisphosphonate-associated osteonecrosis of the jaw is linked to suppressed TGFβ1-signaling and increased Galectin-3 expression: A histological study on biopsies. *J. Transl. Med.* **2011**, *9*, 102. [CrossRef] [PubMed]

99. Kato, T.; Uzawa, A.; Ishihara, K. Inhibitory effect of galectin-3 on the cytokine-inducing activity of periodontopathic Aggregatibacter actinomycetemcomitans endotoxin in splenocytes derived from mice. *FEMS Immunol. Med. Microbiol.* **2009**, *57*, 40–45. [CrossRef] [PubMed]

100. Detrano, R.; Guerci, A.D.; Carr, J.J.; Bild, D.E.; Burke, G.; Folsom, A.R.; Liu, K.; Shea, S.; Szklo, M.; Bluemke, D.A.; et al. Coronary calcium as a predictor of coronary events in four racial or ethnic groups. *N. Engl. J. Med.* **2008**, *358*, 1336–1345. [CrossRef] [PubMed]

101. Taylor, A.J.; Bindeman, J.; Feuerstein, I.; Cao, F.; Brazaitis, M.; O'Malley, P.G. Coronary calcium independently predicts incident premature coronary heart disease over measured cardiovascular risk factors: Mean three-year outcomes in the Prospective Army Coronary Calcium [PACC] project. *J. Am. Coll. Cardiol.* **2005**, *46*, 807–814. [CrossRef] [PubMed]

102. Bischetti, S.; Scimeca, M.; Bonanno, E.; Federici, M.; Anemona, L.; Menghini, R.; Casella, S.; Cardellini, M.; Ippoliti, A.; Mauriello, A. Carotid plaque instability is not related to quantity but to elemental composition of calcification. *Nutr. Metab. Cardiovasc. Dis.* **2017**, *27*, 768–774. [CrossRef] [PubMed]

103. Ehara, S.; Kobayashi, Y.; Yoshiyama, M.; Shimada, K.; Shimada, Y.; Fukuda, D.; Nakamura, Y.; Yamashita, H.; Yamagishi, H.; Takeuchi, K.; et al. Spotty calcification typifies the culprit plaque in patients with acute myocardial infarction: An intravascular ultrasound study. *Circulation* **2004**, *110*, 3424–3429. [CrossRef] [PubMed]

104. Kataoka, Y.; Wolski, K.; Uno, K.; Puri, R.; Tuzcu, E.M.; Nissen, S.E.; Nicholls, S.J. Spotty calcification as a marker of accelerated progression of coronary atherosclerosis: Insights from serial intravascular ultrasound. *J. Am. Coll. Cardiol.* **2012**, *59*, 1592–1597. [CrossRef] [PubMed]

105. Vazquez-Figueroa, J.G.; Rinehart, S.; Qian, Z.; Joshi, P.H.; Sharma, A.; Lee, J.; Anderson, H.; Murrieta, L.; Wilmer, C.; Carlson, H.; et al. Prospective validation that vulnerable plaque associated with major adverse outcomes have larger plaque volume, less dense calcium, and more non-calcified plaque by quantitative, three-dimensional measurements using intravascular ultrasound with radiofrequency backscatter analysis: Results from the ATLANTA I Study. *J. Cardiovasc. Transl. Res.* **2013**, *6*, 762–771. [PubMed]

106. Virchow, R. Cellular pathology. As based upon physiological and pathological histology. Lecture XV Id—Atheromatous affection of arteries. 1858. *Nutr. Rev.* **1989**, *47*, 23–25. [CrossRef] [PubMed]

107. Giachelli, C.M. Vascular calcification mechanisms. *J. Am. Soc. Nephrol.* **2004**, *15*, 2959–2964. [CrossRef] [PubMed]

108. Speer, M.Y.; Yang, H.Y.; Brabb, T.; Leaf, E.; Look, A.; Lin, W.L.; Frutkin, A.; Dichek, D.; Giachelli, C.M. Smooth muscle cells give rise to osteochondrogenic precursors and chondrocytes in calcifying arteries. *Circ. Res.* **2009**, *104*, 733–741. [CrossRef] [PubMed]

109. Steitz, S.A.; Speer, M.Y.; Curinga, G.; Yang, H.Y.; Haynes, P.; Aebersold, R.; Schinke, T.; Karsenty, G.; Giachelli, C.M. Smooth muscle cell phenotypic transition associated with calcification: Upregulation of Cbfa1 and downregulation of smooth muscle lineage markers. *Circ. Res.* **2001**, *89*, 1147–1154. [CrossRef] [PubMed]

110. Liberman, M.; Bassi, E.; Martinatti, M.K.; Lario, F.C.; Wosniak, J., Jr.; Pomerantzeff, P.M.; Laurindo, F.R. Oxidant generation predominates around calcifying foci and enhances progression of aortic valve calcification. *Arterioscler. Thromb. Vasc. Biol.* **2008**, *28*, 463–470. [CrossRef] [PubMed]

111. Wang, Z.; Jiang, Y.; Liu, N.; Ren, L.; Zhu, Y.; An, Y.; Chen, D. Advanced glycation end-product Nε-carboxymethyl-Lysine accelerates progression of atherosclerotic calcification in diabetes. *Atherosclerosis* **2012**, *221*, 387–396. [CrossRef] [PubMed]

112. Demer, L.; Tintut, Y. The roles of lipid oxidation products and receptor activator of nuclear factor-κB signaling in atherosclerotic calcification. *Circ. Res.* **2011**, *108*, 1482–1493. [CrossRef] [PubMed]

113. Hofmann Bowman, M.A.; Gawdzik, J.; Bukhari, U.; Husain, A.N.; Toth, P.T.; Kim, G.; Earley, J.; McNally, E.M. S100A12 in vascular smooth muscle accelerates vascular calcification in apolipoprotein E-null mice by activating an osteogenic gene regulatory program. *Arterioscler. Thromb. Vasc. Biol.* **2011**, *31*, 337–344. [CrossRef] [PubMed]

114. Derwall, M.; Malhotra, R.; Lai, C.S.; Beppu, Y.; Aikawa, E.; Seehra, J.S.; Zapol, W.M.; Bloch, K.D.; Yu, P.B. Inhibition of bone morphogenetic protein signaling reduces vascular calcification and atherosclerosis. *Arterioscler. Thromb. Vasc. Biol.* **2012**, *32*, 613–622. [CrossRef] [PubMed]

115. Bennett, B.J.; Scatena, M.; Kirk, E.A.; Rattazzi, M.; Varon, R.M.; Averill, M.; Schwartz, S.M.; Giachelli, C.M.; Rosenfeld, M.E. Osteoprotegerin inactivation accelerates advanced atherosclerotic lesion progression and calcification in older ApoE$^{-/-}$ mice. *Arterioscler. Thromb. Vasc. Biol.* **2006**, *26*, 2117–2124. [CrossRef] [PubMed]

116. Zhou, Z.; Immel, D.; Xi, C.X.; Bierhaus, A.; Feng, X.; Mei, L.; Nawroth, P.; Stern, D.M.; Xiong, W.C. Regulation of osteoclast function and bone mass by RAGE. *J. Exp. Med.* **2006**, *203*, 1067–1080. [CrossRef] [PubMed]

117. Suarez, G.; Meyerrose, G. Heart failure and galectin 3. *Ann. Transl. Med.* **2014**, *2*, 86. [PubMed]

118. Ural, A.; Janeiro, C.; Karim, L.; Diab, T.; Vashishth, D. Association between non-enzymatic glycation, resorption, and microdamage in human tibial cortices. *Osteoporos. Int.* **2015**, *26*, 865–873. [CrossRef] [PubMed]

119. Napoli, N.; Chandran, M.; Pierroz, D.D.; Abrahamsen, B.; Schwartz, A.V.; Ferrari, S.L. IOF Bone and Diabetes Working Group. Mechanisms of diabetes mellitus-induced bone fragility. *Nat. Rev. Endocrinol.* **2017**, *13*, 208–219. [CrossRef] [PubMed]

International Journal of
Molecular Sciences

MDPI

Review

Galectin-12 in Cellular Differentiation, Apoptosis and Polarization

Lei Wan [1,2,3,4], Ri-Yao Yang [5] and Fu-Tong Liu [6,7,*]

1   School of Chinese Medicine, China Medical University, Taichung 40402, Taiwan; lei.joseph@gmail.com
2   Department of Biotechnology, Asia University, Taichung 41354, Taiwan
3   Department of Gynecology, China Medical University Hospital, Taichung 40402, Taiwan
4   Research Center for Chinese Medicine & Acupuncture, China Medical University, Taichung 40402, Taiwan
5   Department of Molecular and Cellular Oncology, University of Texas MD Anderson Cancer Center, Houston, TX 77030, USA; ryang@mdanderson.org
6   Institute of Biomedical Sciences, Academia Sinica, Taipei 11529, Taiwan
7   Department of Dermatology, University of California-Davis, School of Medicine, Sacramento, CA 95816, USA
*   Correspondence: fliu@ucdavis.edu or ftliu@ibms.sinica.edu.tw; Tel.: +886-02-2789-9402

Received: 13 November 2017; Accepted: 3 January 2018; Published: 7 January 2018

Abstract: Galectin-12 is a member of a family of mammalian lectins characterized by their affinity for β-galactosides and consensus amino acid sequences. The protein structure consists of a single polypeptide chain containing two carbohydrate-recognition domains joined by a linker region. Galectin-12 is predominantly expressed in adipose tissue, but is also detected in macrophages and other leukocytes. Downregulation of galectin-12 in mouse 3T3-L1 cells impairs their differentiation into adipocytes. Conversely, overexpression of galectin-12 in vitro induces cell cycle arrest in G1 and apoptosis. Upregulation of galectin-12 and initiation of G1 cell cycle arrest are associated with driving pre-adipocytes toward terminal differentiation. Galectin-12 deficiency increases insulin sensitivity and glucose tolerance in obese animals. Galectin-12 inhibits macrophage polarization to the M2 population, enhancing inflammation and decreasing insulin sensitivity in adipocytes. Galectin-12 also affects myeloid differentiation, which is associated with chemotherapy resistance. In addition to highlighting the above-mentioned aspects, this review also discusses the potential clinical applications of modulating the function of galectin-12.

Keywords: galectin-12; differentiation; adipogenesis; cellular plasticity

---

1. The Family of Galectins

Galectins are known for their β-galactoside-binding activities and, to date, 17 galectins have been identified. Galectins share consensus sequences in the carbohydrate-recognition domains (CRDs). Galectin-1, -2, -5, -7, -10, -11, -13, -14, -15, -16, and -17 are known as one-CRD-type galectins. Galectin-4, -6, -8, -9, and -12 are two-CRD-type galectins, containing two homologous CRDs in a single polypeptide chain. Galectin-3, the only chimeric galectin, contains one CRD and a non-lectin region comprising proline- and glycine-rich short tandem repeats. Some galectins are widely distributed among different tissues, whereas others are tissue-specific. One-CRD galectins often exist as homodimers, whereas two-CRD galectins have two carbohydrate-binding sites. In addition, galectin-3 can form oligomers, which exhibit multivalent carbohydrate-binding activity. Moreover, the carbohydrate-binding specificities and affinities differ between galectin family members [1,2].

Galectins do not contain a classical leader sequence and mainly exist as intracellular proteins. However, they can also be detected on the cell surface and in the extracellular space [1,2], although how these proteins are transported to these areas remains unknown. Galectins do not have specific cell-surface receptors, but bind to various glycoproteins via their carbohydrate moieties. Mostly through the use of recombinant proteins and cultured single cell types, galectins have been shown to exhibit various in vitro activities by interacting with cell-surface glycoproteins or extracellular matrix proteins in a carbohydrate-dependent manner. However, these may not represent the natural biological functions of endogenous galectins. Whether these activities can occur in multi-cellular situations in vivo is uncertain. Moreover, the concentrations of recombinant proteins used to demonstrate extracellular functions might not be reachable under physiological or pathological conditions. Additionally, many studies have revealed intracellular biological activities of galectins, some of which are independent of carbohydrate binding. Indeed, galectins bind to a number of intracellular signaling molecules to regulate signal transduction [1,2]. Galectins have been shown to participate in cell adhesion, migration, and growth, and have been implicated in the pathogenesis of a variety of diseases, including cancer initiation, progression, and metastasis [3,4]. Importantly, galectins also influence the activation and regulation of both innate and adaptive immune responses.

2. Biology of Galectin-12

Galectin-12 was first discovered in 2001 by two independent groups [5,6]. This protein contains two CRDs joined by a linker region in a single polypeptide chain. Northern blot analysis revealed that galectin-12 is predominantly expressed in adipose tissue. It is also detected at low levels in the heart, pancreas, spleen, thymus, and peripheral blood leukocytes [6]. The amino acid sequence of galectin-12 shows considerable variation when compared with the consensus sequences of galectin family proteins. Comparison of the N-terminal CRD protein sequence with other CRDs from galectins 1–11 revealed 32–43% (an average of 38.8%) identity; however, the sequence identity of the C-terminal CRD with these other CRDs is 28–41% (an average of 33.7%). In addition, the C-terminal CRD of galectin-12 exhibits significant differences from other members of the galectin family in the conserved amino acid residues important for binding to β-galactosides. The amino acids matching Asn-160, Arg-162, Asn-174, and Glu-184, which are important residues for carbohydrate binding in galectin-3, are changed to Pro-251, Thr-253, Arg-261, and Lys-272 in the C-terminal CRD of galectin-12. The binding of galectin-12 to lactosyl-agarose is much weaker than that of galectin-8 [5]. Although the possibility of galectin-12 secretion and binding to certain physiologically or pathologically relevant glycan targets under some specific conditions cannot be totally ruled out, the data above are consistent with its being a galectin that functions intracellularly, independent of glycan binding. The biological functions of galectin-12 have only started to emerge; in this review, the bioactivities of galectin-12 in apoptosis and differentiation are discussed (Figure 1).

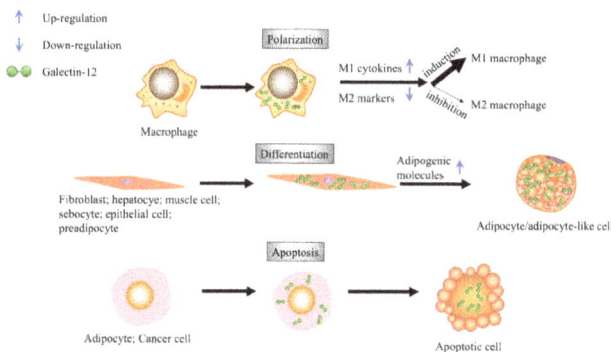

Figure 1. Biological functions of galectin-12. Galectin-12 is expressed in macrophages and promotes the M1 polarization of macrophages. Galectin-12 expression promotes apoptosis in adipocytes, as well as cervical cancer cells. Galectin-12 expression promotes cell differentiation into adipocytes or adipocyte-like cells.

3. Galectin-12 Regulates Cell Growth and Apoptosis

The expression of galectin-12 is tightly regulated, as evidenced by the fact that (1) the start codon of galectin-12 is known to be a weak initiation site for translation according to the Kozak rule, and (2) the 3' untranslated region of galectin-12 mRNA consists of an AU (AUUUA)-rich motif, suggesting mRNA instability. The AU-rich motif has been detected in many proto-oncogenes, transcription factors, and cytokines, which play critical roles in cell growth and differentiation [6]. Consistently, galectin-12 has been shown to play a role in cell cycle regulation. Jurkat T cells synchronized at the G1/S boundary or G1 phase showed increased levels of galectin-12; moreover, overexpression of galectin-12 in the human cervical cell line HeLa caused cell cycle arrest at G1 phase [6]. Furthermore, ectopic expression of galectin-12 induced apoptosis in COS-1 cells. The number of apoptotic cells was also elevated in the adipose tissue of Zucker rats treated with troglitazone, which induces the expression of galectin-12. A positive correlation between galectin-12 mRNA level and the number of apoptotic cells in adipose tissue has been observed [5]. Overexpression of galectin-12 in vitro induces cell cycle arrest in G1 and apoptosis. Moreover, upregulation of galectin-12 and initiation of G1 cell cycle arrest are associated with driving pre-adipocytes towards terminal differentiation. In acute myeloid leukemia, the expression level of galectin-12 is downregulated (88.7% downregulation compared to normal controls) and patients with lower *galectin-12* levels have a lower survival rate compared to those with higher galectin-12 levels [7]. These results suggest anti-proliferative and pro-apoptotic function of galectin-12, although further mechanistic studies are required to establish which cell death and proliferation pathways are involved, and under what physiological/pathological circumstances such function of galectin-12 is evoked.

4. Galectin-12 in Cellular Differentiation

The level of galectin-12 in 3T3-L1 cells is low in the subconfluent state but increased in a state of growth arrest when cells reach confluence. Galectin-12 level was further increased when confluent 3T3-L1 cells were treated with adipogenic hormones to promote differentiation into adipocytes. Galectin-12 is required for adipocyte differentiation, as downregulation of galectin-12 by siRNA inhibited the adipocyte differentiation of 3T3-L1 cells initiated by adipogenic hormones [8]. Expression levels of the adipogenic transcription factors Ccaat-enhancer-binding protein α (C/EBPα), C/EBPβ, and peroxisome proliferator-activated receptor γ (PPARγ), as well as the early adipocyte differentiation marker adipocyte protein 2 (aP2), and late adipocyte differentiation marker adipsin, were downregulated in 3T3-L1 cells treated with galectin-12 siRNA (Figure 2). The activation of protein kinase B (Akt), extracellular receptor signal-related kinase (ERK), and cyclic AMP-responsive element-binding protein 1 (CREB1), which are important signaling molecules in the adipogenic signaling pathway, was also inhibited (Figure 2) [8].

Galectin-12 has been shown to bind to VPS13C, a member of the VPS13 family of proteins, which are associated with vacuolar protein sorting (VPS), and are highly conserved throughout eukaryotic evolution. The protein has been shown to be required for galectin-12 stability. In VPS13C knockdown cells, galectin-12 is degraded through the lysosomal pathway and adipocyte differentiation is suppressed [9]. Whether this protein otherwise mediates the function of galectin-12 is unknown. It is also not known whether galectin-12 interacts with VPS13C in other cell types. Loss of function of VPS13C is known to cause an early-onset form of Parkinson disease with a distinct phenotype of severe and rapid progression and a cellular manifestation of mitochondrial dysfunction and increased mitophagy [10]. Expression of galectin-12 in normal or diseased brain has not been carefully characterized, and whether it is involved in this disease remains to be clarified.

Galectin-12 is predominantly located on large lipid droplets. Galectin-12-deficient mice exhibited a ~40% reduction in whole body lipid content with smaller adipocyte sizes compared with wild-type mice. Low adiposity in galectin-12-deficient mice are due to increased lipolysis, which is initiated by defective phosphodiesterase activity to increase cyclic AMP (cAMP) levels. Increased cAMP subsequently increases protein kinase A (PKA) activity and the activity of hormone-sensitive lipase, which functions with adipocyte triglyceride lipase to promote the hydrolysis of lipids stored in adipocytes. The metabolic rate of galectin-12-deficient mice is high, as indicated by a higher oxygen consumption in galectin-12-deficient adipocytes than in wild-type cells [11,12].

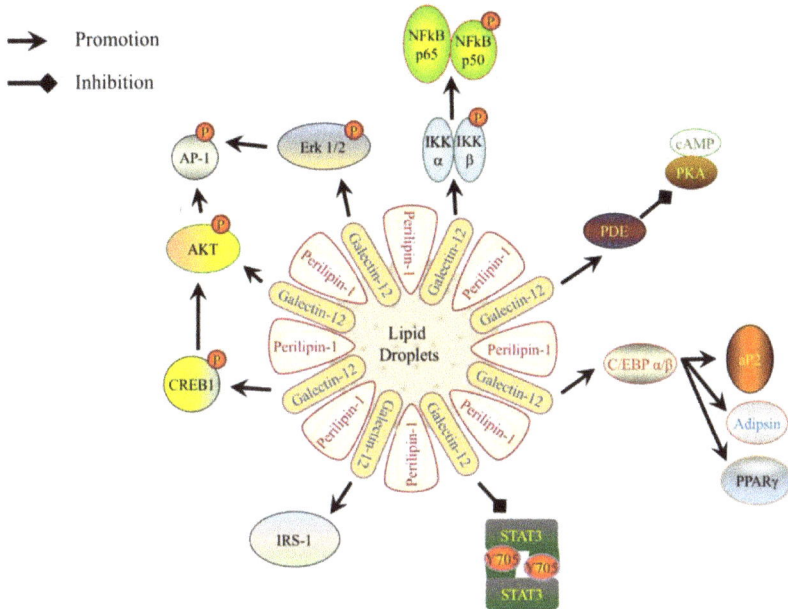

Figure 2. A schematic representation of how galectin-12 modulates the signaling pathways to elicit its biological activities. Galectin-12 is localized on lipid droplets and is co-localized with Perilipin-1. Galectin-12 promotes the expression of the adipogenic transcription factors Ccaat-enhancer-binding protein α (C/EBPα), C/EBPβ, and peroxisome proliferator-activated receptor γ (PPARγ), as well as the early adipocyte differentiation marker adipocyte protein 2 (aP2), and late adipocyte differentiation marker adipsin. Galectin-12 also enhances the activation of protein kinase B (Akt), extracellular receptor signal-related kinase (ERK) and cyclic AMP-responsive element-binding protein 1 (CREB1), which in turn promote activator protein 1 (AP-1). The decreased size of adipocytes in galectin-12-deficient mice is due to increased lipolysis. This is because galectin-12 negatively regulates phosphodiesterase (PDE) activity and cyclic AMP (cAMP) levels, which is upstream of protein kinase A (PKA) activity. Galectin-12 also promotes expression of insulin receptor substrate 1 (IRS-1), as well as activation of IκB kinases (IKK)α/β; the latter in turn enhances the activation of nuclear factor-κB (NF-κB). Finally, galectin-12 suppresses the activation of signal transducer and activator of transcription 3 (STAT3).

In addition, galectin-12 has also been suggested as an important molecule regulating glucose homeostasis, as galectin-12 downregulation lowered the expression of insulin receptor, as well as insulin receptor substrate 1 (IRS-1) (Figure 2). Galectin-12 deficiency increases insulin sensitivity and glucose tolerance in obese animals [11]. Collectively, these data show that galectin-12 is important in adipose tissue fatty acid metabolism and whole-body energy metabolism, and may be a promising therapeutic target in combating metabolic disorders [13].

Galectin-12 was also detected in sebaceous glands of human scalp skin and subsequently confirmed to be expressed in sebocytes, which are sebum-secreting cells. C/EBPα and PPARγ, two molecules regulated by galectin-12, were shown to have important roles in sebocyte differentiation. This suggests that galectin-12 is important in sebocyte differentiation and in the biological function of sebocytes [14]. The differentiation of acute promyelocytic leukemia cells was investigated by

Xue et al. [15], who found that galectin-12 is involved in all-trans retinoic acid-induced granulocytic differentiation. Downregulation of galectin-12 promoted all-trans retinoic acid-induced neutrophil differentiation, but impeded the formation of lipid droplets in terminally differentiated neutrophils, suggesting that lipogenesis and other aspects of myeloid differentiation can be independently regulated by galectin-12. Consistently, downregulation of galectin-12 interferes with the lipogenic pathway by lowering the levels of lipogenic transcription factors C/EBPα, C/EBPβ, PPARγ, as well as the lipogenic signaling pathway by inhibiting the activation of CREB1 and ERK (Figure 2).

5. Galectin-12 as a Modulator of Cell Plasticity?

Differentiation is not a permanent event, but rather is associated with a great deal of plasticity [16]. In our previous study, we found that galectin-12 was expressed in macrophages. Ablation of galectin-12 did not affect the differentiation of bone marrow cells into macrophages, but decreased phagocytic activity against *Escherichia coli* and reduced the secretion of nitric oxide. Ablation of galectin-12 also resulted in polarization of macrophages to the M2 subset, as indicated by increased expression levels of M2 markers, namely resistin-like molecule beta and chitinase 3-like protein 3, as well as reduced expression levels of numerous M1 pro-inflammatory cytokines. The lowered expression of pro-inflammatory cytokines in macrophages by galectin-12 deletion was due to lowered activation of IKKα/β, Akt, and extracellular receptor signal-related kinase, which in turn reduced the activation of nuclear factor-κB and activator protein 1. STAT3 activation was significantly higher in Gal12$^{-/-}$ macrophages induced by lipopolysaccharide, which corresponded to higher levels of interleukin-10 (Figure 2). Adipocytes showed higher insulin sensitivity when treated with Gal12$^{-/-}$ macrophage-conditioned media than those treated with Gal12$^{+/+}$ macrophage-conditioned media. Galectin-12 negatively regulates macrophage polarization into the M2 population, causing enhanced inflammation and decreased insulin sensitivity in adipocytes [17]. The reduced inflammation in Gal12$^{-/-}$ macrophages may also result from an increase in cAMP level which is mediated by PKA-A kinase-anchoring protein (AKAP) 95-p105 pathway [18]. Galectin-12 has been shown to affect inflammation-related signaling pathways, which suggests that galectin-12 may serve as an adaptor protein that brings signaling molecules in close proximity to promote signal transduction.

Plasticity allows cells to switch morphology and function in response to internal or external stimulus, which promotes differentiation from one phenotype to another [16]. The functions of galectin-12 suggest that this molecule serves as a cell plasticity modulator to promote cellular reprogramming to change the biological function. Lipid droplets are universal organelles composed of neutral lipids as an energy source or for membrane synthesis [19]. Fatty liver is characterized by the accumulation of fat in hepatocytes. Large lipid droplets may eventually displace the nucleus of hepatocytes, which is a characteristic of adipocytes with large lipid droplets. Excessive lipids in hepatocytes may confer hepatocytes with adipocyte-like activities. The lipogenesis genes coding for C/EBPα, C/EBPβ, and PPARγ are elevated in the liver with excess lipid accumulation [20,21]. The levels of inflammatory cytokines tumor necrosis factor α [22], interleukin-6 [23], and monocyte chemoattractant protein-1 [24] are also increased in fatty liver. The expression of galectin-12 promotes the secretion of inflammatory cytokines [11]. In our previous study, we found a very low level of galectin-12 in normal hepatocytes, but an increase in galectin-12 in the liver of C57BL/6J mice with steatosis induced by high-fat diet. We hypothesized that galectin-12 is not expressed in hepatocytes under normal conditions (with small lipid droplet). However, under conditions of energy excess, lipid biosynthesis increases and lipid droplets are enlarged because of excessive fat, which promotes hepatocytes to differentiate into an adipocyte-like phenotype and induces the expression of galectin-12. It remains unknown whether the expression of galectin-12 is induced by excessive fat in hepatocytes or increased fatty acids in the blood stream, which confers hepatocytes with an adipocyte-like phenotype to store more energy. This condition may apply to several different cell types in obesity, in which many cells store large lipid droplets.

6. Conclusions

Galectin-12 has been shown to play important roles in regulating growth, differentiation, and lipolysis, as well as in modulating inflammation. However, many questions remain unanswered. For example, whether galectin-12 is involved in cellular plasticity is unclear. In many cases, endogenous galectin-12 is upregulated in a way that correlates with lipid accumulation, suggesting that a hydrophobic environment may be required for its proper localization and function. VPS13C has been identified as a galectin-12-binding protein required for galectin-12 protein stability by preventing its degradation through the lysosomal pathway [9]. While the function of galectin-12 in adipocytes is reasonably well studied, much remains to be explored about its roles in other cell types. Further studies on the regulation of galectin-12 expression and downstream signaling pathways will yield a more complete picture on the functions of this important protein.

Acknowledgments: This study was supported in part by the Ministry of Science and Technology, Taiwan, R.O.C. (MOST103-2314-B-039-035-MY3 and MOST105-2628-B-039-008-MY3), China Medical University Hospital, Taichung, Taiwan (DMR-106-084), and China Medical University, Taichung, Taiwan (CMU106-ASIA-24). This work was also supported by China Medical University under the Aim for Top University Plan of the Ministry of Education, Taiwan. The sponsor or funding organization had no role in the design or conduct of this research.

Author Contributions: All three authors contributed to preparation of the paper.

Conflicts of Interest: The authors declare no conflict of interest.

References

1. Liu, F.T.; Rabinovich, G.A. Galectins: Regulators of acute and chronic inflammation. *Ann. N. Y. Acad. Sci.* **2010**, *1183*, 158–182. [CrossRef] [PubMed]
2. Yang, R.Y.; Liu, F.T. Galectins in cell growth and apoptosis. *Cell. Mol. Life Sci. CMLS* **2003**, *60*, 267–276. [CrossRef] [PubMed]
3. Kaltner, H.; Toegel, S.; Caballero, G.G.; Manning, J.C.; Ledeen, R.W.; Gabius, H.J. Galectins: Their network and roles in immunity/tumor growth control. *Histochem. Cell Biol.* **2017**, *147*, 239–256. [CrossRef] [PubMed]
4. Arthur, C.M.; Baruffi, M.D.; Cummings, R.D.; Stowell, S.R. Evolving mechanistic insights into galectin functions. *Methods Mol. Biol.* **2015**, *1207*, 1–35. [PubMed]
5. Hotta, K.; Funahashi, T.; Matsukawa, Y.; Takahashi, M.; Nishizawa, H.; Kishida, K.; Matsuda, M.; Kuriyama, H.; Kihara, S.; Nakamura, T.; et al. Galectin-12, an Adipose-expressed Galectin-like Molecule Possessing Apoptosis-inducing Activity. *J. Biol. Chem.* **2001**, *276*, 34089–34097. [CrossRef] [PubMed]
6. Yang, R.Y.; Hsu, D.K.; Yu, L.; Ni, J.; Liu, F.T. Cell cycle regulation by galectin-12, a new member of the galectin superfamily. *J. Biol. Chem.* **2001**, *276*, 20252–20260. [CrossRef] [PubMed]
7. El Leithy, A.A.; Helwa, R.; Assem, M.M.; Hassan, N.H. Expression profiling of cancer-related galectins in acute myeloid leukemia. *Tumour Biol.* **2015**, *36*, 7929–7939. [CrossRef] [PubMed]
8. Yang, R.Y.; Hsu, D.K.; Yu, L.; Chen, H.Y.; Liu, F.T. Galectin-12 is required for adipogenic signaling and adipocyte differentiation. *J. Biol. Chem.* **2004**, *279*, 29761–29766. [CrossRef] [PubMed]
9. Yang, R.Y.; Xue, H.; Yu, L.; Velayos-Baeza, A.; Monaco, A.P.; Liu, F.T. Identification of VPS13C as a Galectin-12-Binding Protein That Regulates Galectin-12 Protein Stability and Adipogenesis. *PLoS ONE* **2016**, *11*, e0153534. [CrossRef] [PubMed]
10. Lesage, S.; Drouet, V.; Majounie, E.; Deramecourt, V.; Jacoupy, M.; Nicolas, A.; Cormier-Dequaire, F.; Hassoun, S.M.; Pujol, C.; Ciura, S.; et al. Loss of VPS13C Function in Autosomal-Recessive Parkinsonism Causes Mitochondrial Dysfunction and Increases PINK1/Parkin-Dependent Mitophagy. *Am. J. Hum. Genet.* **2016**, *98*, 500–513. [CrossRef] [PubMed]
11. Yang, R.Y.; Yu, L.; Graham, J.L.; Hsu, D.K.; Kent Lloyd, K.C.; Havel, P.J.; Liu, F.T. Ablation of a galectin preferentially expressed in adipocytes increases lipolysis, reduces adiposity, and improves insulin sensitivity in mice. *Proc. Natl. Acad. Sci. USA* **2011**, *108*, 18696–18701. [CrossRef] [PubMed]
12. Yang, R.Y.; Havel, P.J.; Liu, F.T. Galectin-12: A protein associated with lipid droplets that regulates lipid metabolism and energy balance. *Adipocyte* **2012**, *1*, 96–100. [CrossRef] [PubMed]
13. Baum, L.G. Burn control, an adipocyte-specific function for galectin-12. *Proc. Natl. Acad. Sci. USA* **2011**, *108*, 18575–18576. [CrossRef] [PubMed]

14. Harrison, W.J.; Bull, J.J.; Seltmann, H.; Zouboulis, C.C.; Philpott, M.P. Expression of lipogenic factors galectin-12, resistin, SREBP-1, and SCD in human sebaceous glands and cultured sebocytes. *J. Investig. Dermatol.* **2007**, *127*, 1309–1317. [CrossRef] [PubMed]

15. Xue, H.; Yang, R.Y.; Tai, G.; Liu, F.T. Galectin-12 inhibits granulocytic differentiation of human NB4 promyelocytic leukemia cells while promoting lipogenesis. *J. Leukoc. Biol.* **2016**, *100*, 657–664. [CrossRef] [PubMed]

16. Jessen, K.R.; Mirsky, R.; Arthur-Farraj, P. The Role of Cell Plasticity in Tissue Repair: Adaptive Cellular Reprogramming. *Dev. Cell* **2015**, *34*, 613–620. [CrossRef] [PubMed]

17. Wan, L.; Lin, H.J.; Huang, C.C.; Chen, Y.C.; Hsu, Y.A.; Lin, C.H.; Lin, H.C.; Chang, C.Y.; Huang, S.H.; Lin, J.M.; et al. Galectin-12 enhances inflammation by promoting M1 polarization of macrophages and reduces insulin sensitivity in adipocytes. *Glycobiology* **2016**, *26*, 732–744. [CrossRef] [PubMed]

18. Wall, E.A.; Zavzavadjian, J.R.; Chang, M.S.; Randhawa, B.; Zhu, X.; Hsueh, R.C.; Liu, J.; Driver, A.; Bao, X.R.; Sternweis, P.C.; et al. Suppression of LPS-Induced TNF-α Production in Macrophages by cAMP Is Mediated by PKA-AKAP95-p105. *Sci. Signal.* **2009**, *2*, ra28. [CrossRef] [PubMed]

19. Walther, T.C.; Chung, J.; Farese, R.V., Jr. Lipid Droplet Biogenesis. *Annu. Rev. Cell Dev. Biol.* **2017**, *33*, 491–510. [CrossRef] [PubMed]

20. Chen, Y.H.; Yang, C.M.; Chang, S.P.; Hu, M.L. C/EBP beta and C/EBP delta expression is elevated in the early phase of ethanol-induced hepatosteatosis in mice. *Acta Pharmacol. Sin.* **2009**, *30*, 1138–1143. [CrossRef] [PubMed]

21. Samuel, V.T.; Shulman, G.I. Nonalcoholic Fatty Liver Disease as a Nexus of Metabolic and Hepatic Diseases. *Cell Metab.* **2017**. [CrossRef] [PubMed]

22. Kakino, S.; Ohki, T.; Nakayama, H.; Yuan, X.; Otabe, S.; Hashinaga, T.; Wada, N.; Kurita, Y.; Tanaka, K.; Hara, K.; et al. Pivotal Role of TNF-alpha in the Development and Progression of Nonalcoholic Fatty Liver Disease in a Murine Model. *Horm. Metab. Res.* **2017**. [CrossRef]

23. Gerst, F.; Wagner, R.; Kaiser, G.; Panse, M.; Heni, M.; Machann, J.; Bongers, M.N.; Sartorius, T.; Sipos, B.; Fend, F.; et al. Metabolic crosstalk between fatty pancreas and fatty liver: Effects on local inflammation and insulin secretion. *Diabetologia* **2017**, *60*, 2240–2251. [CrossRef] [PubMed]

24. Mridha, A.R.; Wree, A.; Robertson, A.A.B.; Yeh, M.M.; Johnson, C.D.; Van Rooyen, D.M.; Haczeyni, F.; Teoh, N.C.; Savard, C.; Ioannou, G.N.; et al. NLRP3 inflammasome blockade reduces liver inflammation and fibrosis in experimental NASH in mice. *J. Hepatol.* **2017**, *66*, 1037–1046. [CrossRef] [PubMed]

MDPI

St. Alban-Anlage 66

4052 Basel

Switzerland

Tel. +41 61 683 77 34

Fax +41 61 302 89 18

www.mdpi.com

International Journal of Molecular Sciences Editorial Office

E-mail: ijms@mdpi.com

www.mdpi.com/journal/ijms

www.ingramcontent.com/pod-product-compliance
Lightning Source LLC
Chambersburg PA
CBHW051711210326
41597CB00032B/5444